T0312195

This book provides a thorough description of classical electromagnetic radiation, starting from Maxwell's equations and moving on to show how fundamental concepts are applied in a wide variety of examples from areas such as classical optics, antenna analysis, electromagnetic scattering, and particle accelerators. Theoretical and experimental results are interwoven throughout to help give insight into the physical and historical foundations of the subject.

Following introductory chapters covering the basic theory of classical electromagnetism and the properties of plane waves, the concept of a plane-wave spectrum is developed and applied to the radiation from apertures. Radiation from a moving point charge is described in depth, as is that from a variety of thin-wire antennas. A key feature of the book is that pulsed and time-harmonic fields are presented on an equal footing.

Mathematical and physical explanations are enhanced by a wealth of illustrations (over 300), and the book includes more than 140 problems. It will be of great interest to advanced undergraduate and graduate students of electrical engineering and physics, as well as to scientists and engineers working in applied electromagnetics.

H.C. OERSTED

A.M. AMPÈRE

J.C. MAXWELL

M. FARADAY

H.R. HERTZ

Picture Credits, Frontispiece:

A.M. Ampère, AIP Emilio Segrè Visual Archives
M. Faraday, AIP Emilio Segrè Visual Archives, E. Scott Barr Collection
H.R. Hertz, Deutsches Museum, Courtesy AIP, Emilio Segrè Visual Archives
J.C. Maxwell, AIP Emilio Segrè Visual Archives
H.C. Oersted, The Burndy Library, Dibner Institute for the History of Science and Technology

An introduction to classical electromagnetic radiation

An introduction to classical electromagnetic radiation

GLENN S. SMITH

Regents' Professor of Electrical Engineering
Georgia Institute of Technology
Atlanta, GA

CAMBRIDGE
UNIVERSITY PRESS

32 Avenue of the Americas, New York NY 10013-2473, USA

Cambridge University Press is part of the University of Cambridge.

It furthers the University's mission by disseminating knowledge in the pursuit of education, learning and research at the highest international levels of excellence.

www.cambridge.org
Information on this title: www.cambridge.org/9780521586986

First published 1997

A catalogue record for this publication is available from the British Library

Library of Congress Cataloguing in Publication data

Smith, Glenn S. (Glenn Stanley), 1945–
An introduction to classical electromagnetic radiation / Glenn S. Smith.
p. cm.
Includes bibliographical references.
ISBN 0 521 58093 5 hardback
ISBN 0 521 58698 4 paperback
QC661.S65 1997 96-25107
539.2 – dc20 CIP

ISBN 978-0-521-58093-9 Hardback
ISBN 978-0-521-58698-6 Paperback

To the memory of my parents
Stanley *and* **Florence**,
to my wife
Linda,
and to the future of our children
Geoff *and* **Ellie**

Contents

Preface *xi*

1 Basic theory of classical electromagnetism **1**

1.1 Historical introduction: Maxwell's equations in integral form 1
1.2 Curl and divergence: Maxwell's equations in differential form 25
1.3 Surface densities, boundary conditions, and perfect conductors 34
1.3.1 Surface densities of charge and current 35
1.3.2 Electromagnetic boundary conditions 41
1.3.3 The concept of a perfect conductor 47
1.4 Energy of the electromagnetic field – Poynting's theorem 48
1.5 Electromagnetic boundary value problem – uniqueness theorem 65
1.6 Numerical solution of Maxwell's equations using finite differences: An example 71
1.7 Harmonic time dependence and the Fourier transform 88
1.7.1 Linear systems 88
1.7.2 Maxwell's equations 92
1.7.3 Poynting's theorem 94
1.7.4 Uniqueness theorem 98
1.7.5 Fourier transform 101
References 104
Problems 107

2 Electromagnetic plane waves in free space: Polarized waves **123**

2.1 General time dependence 123
2.2 Harmonic time dependence: Monochromatic plane waves 128
2.3 The polarization ellipse in the coordinate system of the principal axes 134
2.4 The Poincaré sphere and the Stokes parameters 140
2.5 Optical elements for processing polarized light 142
2.6 Transmission and reception of polarized waves with antennas 153

2.7 Historical note: The experiments of Hertz 161
References 168
Problems 169

3 Inhomogeneous plane waves and the plane-wave spectrum 177

3.1 Inhomogeneous plane waves 177
3.2 Two-dimensional, transverse electric and transverse magnetic fields 181
3.3 Plane-wave spectrum for two-dimensional electromagnetic fields 184
3.4 The uniformly illuminated slit 192
3.5 Asymptotic or radiated field – the method of stationary phase 199
3.6 Plane-wave spectrum for three-dimensional electromagnetic fields 207
3.6.1 General formulation 207
3.6.2 Asymptotic or radiated field 210
3.6.3 Wavefronts and rays 215
3.7 Examples of three-dimensional fields 218
3.7.1 Uniformly illuminated circular aperture 218
3.7.2 Circular aperture with tapered illumination – reflector antennas 232
3.7.3 Gaussian beam – paraxial approximation 240
References 250
Problems 251

4 Electromagnetic analogues of some optical principles 262

4.1 Huygens' principle: An alternate representation 262
4.1.1 General formulation 264
4.1.2 Radiated field 268
4.1.3 Discussion 272
4.2 Fresnel zones 273
4.3 The scattered field 277
4.4 Babinet's principle 281
4.4.1 Optical formulation 281
4.4.2 Electromagnetic formulation 283
4.5 Transmission coefficients and scattering cross sections 286
4.6 Complementary antennas 294
4.7 Images 299
4.8 General time dependence 305
4.9 Discussion 309
References 311
Problems 313

5 Radiation from distributions of charge and current: General formulation **320**

5.1 Electromagnetic potentials 320
5.1.1 Electrostatics 321
5.1.2 Magnetostatics 322
5.1.3 Electrodynamics 325
5.2 Dirac delta function – concept of a point charge 331
5.2.1 One-dimensional delta function 332
5.2.2 Three-dimensional delta function 335
5.2.3 Electrostatic point charge 337
5.3 Retarded potentials and electromagnetic field 341
5.4 Radiated field 347
5.5 Harmonic time dependence 352
References 355
Problems 356

6 Electromagnetic field of a moving point charge **358**

6.1 Derivation 358
6.1.1 Liénard-Wiechert potentials 359
6.1.2 Electromagnetic field 364
6.1.3 Radiated power 371
6.2 Special cases 375
6.2.1 Low velolcity: $v/c \ll 1$ 377
6.2.2 Relativistic velocity, $\vec{a}$ parallel to $\vec{v}$ 378
6.2.3 Relativistic velocity, $\vec{a}$ perpendicular to $\vec{v}$ 394
6.3 Synchrotron radiation 396
6.4 Cherenkov radiation 411
6.5 Self force 436
References 441
Problems 446

7 Dipole radiation **451**

7.1 Infinitesimal electric dipole or current element 452
7.1.1 General time dependence 452
7.1.2 Harmonic time dependence 465
7.2 Electrically short linear antennas 469
7.3 Duality and the infinitesimal magnetic dipole or current loop 477
7.4 Electrically small loop antennas 482
7.5 Simple arrays of electrically short linear antennas 485
7.5.1 General time dependence 486
7.5.2 Harmonic time dependence 493

7.6 Scattering by electrically small objects 506
7.7 The color and polarization of skylight 521
7.7.1 Natural light 524
7.7.2 Molecular scattering 528
References 537
Problems 541

8 Radiation from thin-wire antennas 546

8.1 Charge and current: Physical arguments 547
8.2 Basic traveling-wave element; dipole antennas 556
8.2.1 Description of element 556
8.2.2 Dipole antennas 562
8.3 Traveling-wave bends and loops 567
8.3.1 Simple bend 567
8.3.2 Circular loop 570
8.4 Other traveling-wave antennas 575
8.4.1 Dipole antenna with continuous resistive loading 575
8.4.2 Insulated linear antenna 579
8.5 Harmonic time dependence; examples 583
8.5.1 Basic traveling-wave element 584
8.5.2 Long-wire antennas 588
8.5.3 Standing-wave dipole 591
8.5.4 Resonance and the half-wave dipole 596
8.6 Perspective 601
References 602
Problems 604

Appendix A Units and dimensions 608

References 616

Appendix B Review of vector analysis 617

References 635
Problems 636

Supplemental references 638

Index 643

Preface

A glance at the literature on classical electromagnetism shows that there is an overwhelming amount of material associated with the subject. This is the product of over 300 years of study involving some of the greatest names in science (see the partial list of contributors in Table 1.1). Thus, when preparing a course or a book on classical electromagnetism, one is first faced with the question: What small subsection of the material will be addressed? This choice, of course, involves the background and interests of the intended audience.

This book was written as a text for use in a course in electrical engineering or applied physics populated by first-year graduate students (and possibly some advanced seniors). These students generally have come from many different undergraduate institutions, so their prior exposure to electromagnetism varies greatly. However, most have had at least a one-semester course on electromagnetism at the junior level that covers electrostatics, magnetostatics, and the propagation of plane waves. They also have some knowledge of the guiding of waves on transmission lines. Those that are the best prepared will have had at least a 'brush' with electromagnetic radiation (radiation from charged particles and antennas, electromagnetic scattering, etc.), but this cannot be assumed for the average student. Hence, the decision was made to make the focus of this book the fundamental aspects of electromagnetic radiation. Since the preparation of those in the intended audience is varied, a major objective for the first few chapters of the book is to normalize their knowledge of the subject. This mainly consists of reviewing the basic equations (e.g., Maxwell's equations), describing the basic elements of an electromagnetics problem, and providing common nomenclature and notation. Some review and consolidation of the concepts associated with electromagnetic plane waves is also included. The remainder of the book is concerned with the fundamental aspects of electromagnetic radiation illustrated by examples. A more detailed discussion of the topics covered is provided below.

A few special features of the book are worthy of mention. The emphasis throughout is on providing an understanding of the basic principles, with a balance between physical and mathematical explanation. I have tried to follow the advice of Maxwell:[1]

[1] From his address to the Mathematics and Physics Section, British Association for the Advancement of Science, 1870. T. Bartlett, *Familiar Quotations*, 14th Edition, Little-Brown, Boston, MA, 1968.

For the sake of persons of . . . different types, scientific truth should be presented in different forms, and should be regarded as equally scientific, whether it appears in robust form and the vivid coloring of a physical illustration, or in the tenuity and paleness of a symbolic expression.

A glance through the book shows that there are many more illustrations (over 300) than normally found in a book on this subject at this level. Experience has shown that students appreciate these pictures. A physical concept is often more easily grasped and remembered when presented with an illustration that supports the mathematics. Tables summarizing the important equations are included throughout the book. Their purpose is to keep the student from having to search through the text to find a useful equation, either when solving a homework problem or when using the text as a reference in the future.

Experimental results have been included in almost every chapter. These results serve several functions. For example, in Chapter 1 the numerical solution of Maxwell's partial differential equations using finite differences is described, and the theoretically determined radiation of a pulse by an antenna is presented. These theoretical results are compared with experimental measurements, and the excellent agreement shows the student the accuracy that can be expected from this beautiful theory. In Chapter 3 experimental results are used for a different purpose. The field in an electrically large aperture is assumed to be uniform when the aperture is illuminated by a normally incident plane wave. The experimental results are used to show the approximate nature of this assumption. And in Chapter 7, measurements for the color and polarization of skylight are used to motivate an examination of molecular scattering in the atmosphere.

Many texts at this level concentrate on electromagnetic fields that vary harmonically in time (usually the Fourier transform of the field). In this book, however, solutions in the time domain and the frequency domain are considered on an equal footing. The discussion of the former often precedes the discussion of the latter, because the radiation of time-varying signals (pulses) often is more easily understood from a physical point of view. An example is the radiation from thin-wire antennas presented in Chapter 8. A pulse of charge traveling along the antenna is considered first. Radiation occurs each time the pulse encounters a discontinuity or a bend, a situation analogous to a moving point charge undergoing acceleration. The infinite duration of time-harmonic excitation complicates this simple physical picture for the radiation.

Historical information is included in most chapters, with the dates of birth and death given at the first mention of most prominent scientists. This provides the student with some appreciation of the long and rather interesting development of the subject, as well as a rough idea of the period of time when various discoveries were made. When the students see the array of great scientists who developed this subject over a long period of time, they are often less exasperated by their struggle to become proficient in the subject during their first exposure.

The starting point for Chapter 1 is Maxwell's equations in integral form, which are familiar to most students. These equations are used to analyze a few of the basic experiments performed by early investigators – Coulomb, Oersted, Ampère, and Faraday. In addition to providing the student with a review, this material adds some historical perspective. The definitions for the vector operations of curl and divergence are introduced, and they are used to convert Maxwell's equations in integral form to those in differential form. Only electromagnetic fields in free space or "simple materials" are considered, that is, materials whose constitutive parameters are scalar constants in the time domain (linear, homogeneous, isotropic, and nondispersive materials). Next, the mathematical concept of a surface density of charge or current is described, and the electromagnetic boundary conditions are obtained.

The energy of the electromagnetic field is defined, and Poynting's theorem for the conservation of energy is obtained from Maxwell's equations. This theorem is applied to a problem studied by both Poynting and Heaviside, the transport of energy along a straight wire, and the results are used to illustrate the physical interpretation for the Poynting vector. Next, the elements of an electromagnetic boundary value problem are described, and a uniqueness theorem for the electromagnetic field (general time dependence) is obtained by manipulation of Poynting's theorem.

At this point, presentation of the basic equations of electromagnetism is complete, and the student is in a position to understand the formulation and solution of a practical problem. What is lacking is a technique for obtaining a solution to the equations. The technique presented is the finite-difference time-domain (FDTD) method, a conceptually simple, numerical method that involves the direct discretization of Maxwell's equations. After demonstrating the method for a one-dimensional problem, we use it to analyze a cylindrical monopole antenna with pulse excitation. The boundary value problem for the monopole is formulated based on the earlier discussion of the uniqueness theorem, the numerical solution for the electromagnetic field is obtained, and the theoretical results are compared with accurate measurements.

Chapter 1 ends with a discussion of harmonic time dependence and the Fourier transform. The special properties of linear systems are briefly stated, and the equations of electromagnetism are specialized to the case of harmonic time dependence. It is my experience that, even though most students are familiar with harmonic time dependence (phasors) from courses on linear circuits, they find this review helpful.

Chapter 2 is concerned with the propagation of homogeneous plane waves in free space. For many students this is partly a review. The Poincaré sphere and Stokes parameters are introduced to describe the state of polarization of a monochromatic wave. This material is illustrated with two practical examples: a discussion of optical elements for processing polarized light (the Jones calculus) and the reception of polarized waves by antennas. The chapter ends with a historical note describing Hertz's early experiments with electromagnetic waves, which confirmed Maxwell's equations.

Chapter 3 begins with a discussion of inhomogeneous plane waves in free space. A physical argument is then used to show that the complete set of homogeneous and inhomogeneous plane waves, i.e., the plane-wave spectrum, can be used to construct a solution to Maxwell's equations. Two-dimensional fields are treated first, with the problem of a uniformly illuminated slit as an example. The asymptotic or radiated field is obtained by a physical argument based on the method of stationary phase. The procedure is extended to three-dimensional fields and illustrated with additional examples: the uniformly illuminated circular aperture, the circular aperture with tapered illumination (reflector antennas), and the Gaussian beam (paraxial approximation).

Electromagnetic analogues of some of the famous principles from optics, such as Huygens' principle, Fresnel zones, Babinet's principle, and the method of images, are the focus of Chapter 4. Huygens' principle is obtained by directly manipulating results obtained in Chapter 3 for the plane-wave spectrum, and it is shown to be an alternate representation for an electromagnetic field. The principles are illustrated with examples. These examples introduce additional concepts, e.g., transmission coefficients and scattering cross sections for planar apertures and obstacles, the forward scattering theorem, Kirchhoff's approximation for the field in an aperture and the physical optics approximation for the current on an obstacle, and complementary antennas.

The remaining chapters, 5–8, are concerned with the radiation from distributions of current and charge. In Chapter 5, following discussions of the scalar electric and vector magnetic potentials and the Dirac delta function, the retarded potentials for a general distribution of charge and current are derived. The presentation is conventional, with the electromagnetic field being determined from the retarded potentials. The general results from Chapter 5 are applied to radiation from a moving point charge in Chapter 6, radiation from infinitesimal dipoles (electric and magnetic) in Chapter 7, and radiation from simple, thin-wire antennas in Chapter 8.

In Chapter 6, the electromagnetic field of a moving point charge is derived using the historically important approach based on the Liénard-Wiechert potentials. These results are then specialized for the cases of low velocity, general velocity with the acceleration parallel to the velocity (bremsstrahlung), and general velocity with the acceleration normal to the velocity. Graphics are used extensively to illustrate these cases. Sections of this chapter are devoted to synchrotron radiation (important for circular particle accelerators) and Cherenkov radiation (radiation from a particle whose velocity is greater than the speed of light in the surrounding dielectric). The chapter ends with a brief, qualitative discussion of the self force for a charged particle.

Chapter 7 deals with dipole radiation. After a derivation of the electromagnetic field for the infinitesimal electric dipole or current element (general time dependence), duality is introduced and used to obtain the field of the infinitesimal magnetic dipole or current loop. Brief discussions of the electrically short linear antenna and the electrically small loop antenna illustrate the use of these results.

Simple arrays of electrically short linear antennas (dipoles) are first dealt with in the time domain. Later, the case of harmonic time dependence is presented, and conventional topics associated with arrays are introduced: pattern multiplication, end-fire and broadside arrays, superdirectivity, etc.

The scattering from electrically small objects (Rayleigh scattering) is described in terms of induced electric and magnetic dipole moments. Polarizabilities and scattering cross sections are given for the perfectly conducting sphere, disc, and thin wire. Babinet's principle is used with the results for the disc to obtain the transmission coefficient for an electrically small, circular aperture. These results, together with those from Chapter 4, cover the two extremes for electromagnetic scattering and aperture transmission, i.e., scattering (transmission) for electrically small and electrically large objects (apertures).

Chapter 7 ends with a discussion of the color and polarization of skylight — the famous "blue sky problem." First, experimental results for these phenomena are presented, and then a reasonable explanation is sought based on the scattering of sunlight in the atmosphere. Brief discussions of natural light (unpolarized light) and molecular scattering accompany the explanation.

The last chapter, Chapter 8, is devoted to radiation from simple, thin-wire antennas. First there is a review of the numerical results from the accurate analysis of the monopole antenna presented in Chapter 1. These results are then used to obtain an approximation, based on an "educated guess," for the pulse of charge on a basic traveling-wave element. More complicated antennas (standing-wave dipole, traveling-wave loop, etc.) are viewed as a superposition of basic traveling-wave elements, and their radiated fields are easily obtained as the sum of the fields of the elements. Radiation is shown to occur when the pulse of charge encounters the discontinuities and bends in the wire. The analogy to the radiation from a moving point charge undergoing acceleration is stressed throughout. The chapter ends with the specialization of results to the practically important case of harmonic time dependence.

There are two appendices. Appendix A contains a discussion of units and dimensions, restricted to the International System (SI). It is mainly intended to show that the mysterious factors, such as ε_o, μ_o, and the 4π in Coulomb's law, have a rational explanation associated with measurement. Appendix B is a very brief discussion of vector analysis that includes a set of problems. Its purpose, apart from providing a list of vector relations, is to provide students with a review of a subject that is often "fuzzy" in their memories.

The maturity of this subject necessarily means that the source of all basic information is the writings of others. The references at the end of each chapter are the sources I consulted in preparing the material for the chapter. In most cases, these include the works of the original investigators of a subject as well as a number of interpretations of later writers. Of course, any errors in the presentation are my own, and I ask the readers to please bring these to my attention. A list of supplementary references with brief annotation, included at the end of the book, is intended to give the reader guidance in selecting material for further study of the subject. I made no attempt to make this list comprehensive, so I am sure that there are many more, excellent books with equivalent content that could have been included.

There is a set of problems (over 140 in all) at the end of each chapter. A solutions manual is available (to instructors) from the publisher.

I would like to thank the many people who have contributed to my understanding of this subject. Almost thirty years ago, I was privileged, as a graduate student at Harvard University, to receive the instruction of Professors Ronold W. P. King and Tai Tsun Wu. I hope that Professor King's insistence on a balance of theory with experiment is evident in this book.

My colleagues at the Georgia Institute of Technology have been most gracious in providing comments on the manuscript. Their suggestions have greatly improved the clarity of the presentation. Professor Waymond R. Scott, Jr. and Dr. William B. McFarland read the entire first draft of the manuscript. In addition, on a number of occasions, Professor Scott provided assistance with the computer-generated graphics. Others who have provided comments on selected chapters are: J. A. Buck, D. L. Brundrett, R. F. Fox, M. P. Kesler, and T. P. Montoya. Dr. James G. Maloney provided assistance with the material on the FDTD method and comments on the chapters associated with antenna analysis. Mrs. A. B. Powell expertly typed early versions of some chapters that originated as class notes.

My former colleague Professor John D. Nordgård (University of Colorado at Colorado Springs) read several chapters from the final draft of the manuscript. His comments were most helpful.

During the last eleven years, most of the time I spent writing this book was taken from the time I would normally spend with my family. I am very grateful to my wife Linda and children Geoff and Ellie for their patience and understanding during this period.

Writing this book and the associated teaching have been both enjoyable and rewarding experiences. In this regard, I can not do better than quote the physicist J. Robert Oppenheimer:[2]

> Whatever trouble life holds for you, that part of your lives which you spend finding out about things, things that you can tell others about, and that you can learn from them, that part will be essentially a gay, a sunny, a happy life ...

Glenn Stanley Smith
Atlanta, GA
May 1996

[2] From *Uncommon Sense / J. Robert Oppenheimer*, N. Metropolis, G.-C. Rota, and D. Sharp, Editors, p. 170, Birkhäuser Boston, Boston, MA, 1984.

1

Basic theory of classical electromagnetism

1.1 Historical introduction: Maxwell's equations in integral form

The development of electromagnetic theory has a long history, beginning perhaps with the ancients' experimentation with the electrical properties of amber and the magnetic properties of lodestone. The modern elements of the theory, however, had their origin in the investigations of 17th-, 18th-, and 19th-century scientists, or natural philosophers as they were called. The names of some of the more prominent of these scientists are listed in Table 1.1 in chronological order according to their dates of birth [1–3]. Prior to the early nineteenth century, the phenomena associated with electrostatics, magnetism, and optics were largely thought to be independent; hence, the names in Table 1.1 are divided into these categories.[1]

By the middle of the nineteenth century, Hans Christian Oersted (1777–1851) had discovered that an electric current produces a magnetic field, and the connection between steady currents and magnetism had been examined extensively by André Marie Ampère (1775–1836) and others. In addition, the experiments of Michael Faraday (1791–1867) had shown that a time-varying magnetic field could produce an electric current. In 1864, building on these earlier investigations, James Clerk Maxwell (1831–1879) presented a theory that provided a complete and unified explanation for electric, magnetic, and optical phenomena [4, 5]. A striking feature of Maxwell's theory was its prediction of electromagnetic waves with all the characteristics theretofore associated with optical waves.

The efforts during the remainder of the nineteenth century were devoted to obtaining additional experimental verification and refining Maxwell's theory. Heinrich Rudolf Hertz (1857–1894) demonstrated, unequivocally, the existence of electromagnetic waves, and Hertz, Oliver Heaviside (1850–1925), and Hendrik Antoon Lorentz (1853–1928), among others, put Maxwell's theory for classical electromagnetism in the form we use today.

In the twentieth century, classical electromagnetism has served as the origin for other fields of study, such as relativity and atomic physics. The unification accomplished by Maxwell for electromagnetism now serves as a model for the unification sought for all areas of physics.

[1] Many of the scientists listed in Table 1.1 made contributions to more than one area, although they are listed under a single area in the table. For example, Faraday experimented with electrostatics, magnetism, and optics.

OPTICS

Galileo Galilei (1564 - 1642)
Willebrord Snell (1591 - 1626)
René Descartes (1596 - 1650)
Pierre de Fermat (1601 - 1665)
Francesco Maria Grimaldi (1618 - 1663)
Christiaan Huygens (1629 - 1695)
Robert Hooke (1635 - 1703)
Isaac Newton (1642 - 1727)
Leonhard Euler (1707 - 1783)
Thomas Young (1773 - 1829)
Étienne Louis Malus (1775 - 1812)
Dominique François Jean Arago (1786 - 1853)
David Brewster (1781 - 1868)
Joseph von Fraunhofer (1787 - 1826)
Augustin Jean Fresnel (1788 - 1827)
Christian Johann Doppler (1803 - 1853)
William Rowan Hamilton (1805 - 1865)
Jean Bernard Léon Foucault (1819 - 1868)
Armand Hippolyte Fizeau (1819 - 1896)
George Gabriel Stokes (1819 - 1903)

TIME - VARYING ELECTROMAGNETIC FIELDS

Michael Faraday (1791 - 1867)
Joseph Henry (1797 - 1878)
Franz Ernst Neumann (1798 - 1895)
Wilhelm Eduard Weber (1804 - 1891)
Heinrich Friedrich Emil Lenz (1804 - 1865)
Hermann Ludwig Ferdinand von Helmholtz (1821 - 1894)
William Thomson (Lord Kelvin) (1824 - 1907)
Georg Friedrich Bernhard Riemann (1826 - 1866)
Ludwig Lorenz (1829 - 1891)

ELECTROSTATICS

Benjamin Franklin (1706 - 1790)
Henry Cavendish (1731 - 1810)
Joseph Priestley (1733 - 1804)
Charles Augustin de Coulomb (1736 - 1806)
Alessandro Giuseppe Antonio Anastasio Volta (1745 - 1827)
Pierre Simon de Laplace (1749 - 1827)
Johann Karl Friedrich Gauss (1777 - 1855)
Siméon Denis Poisson (1781 - 1840)
George Green (1793 - 1841)

MAGNETIC PHENOMENA ELECTRIC CURRENT

William Gilbert (1544 - 1603)
Luigi Galvani (1737 - 1798)
Jean Baptiste Biot (1774 - 1862)
André Marie Ampère (1775 - 1836)
Hans Christian Oersted (1777 - 1851)
Georg Simon Ohm (1787 - 1854)
Félix Savart (1791 - 1841)
Gustav Robert Kirchhoff (1824 - 1887)

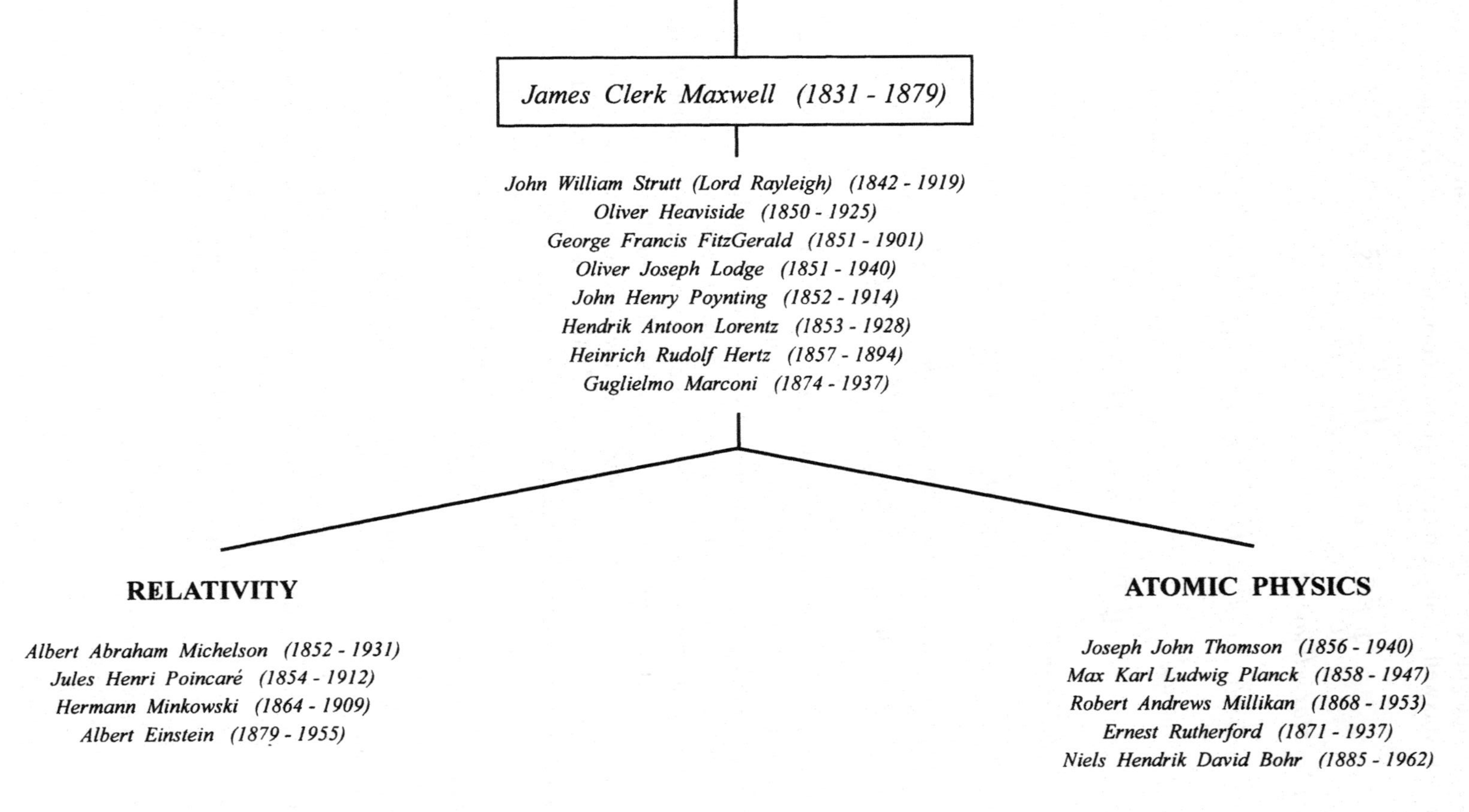

Table 1.1. *Contributors to the development of electromagnetic theory*

We will take as the starting point for our study of electromagnetism Maxwell's equations in integral form. They are the axioms on which the subsequent theory is developed. Maxwell's four field equations in modern notation are *Faraday's law*

$$\oint_C \vec{\mathcal{E}}(\vec{r},t)\cdot d\vec{\ell} = -\iint_S \frac{\partial \vec{\mathcal{B}}(\vec{r},t)}{\partial t}\cdot d\vec{S}, \tag{1.1}$$

the *Ampère-Maxwell law*

$$\oint_C \vec{\mathcal{H}}(\vec{r},t)\cdot d\vec{\ell} = \iint_S \vec{\mathcal{J}}(\vec{r},t)\cdot d\vec{S} + \iint_S \frac{\partial \vec{\mathcal{D}}(\vec{r},t)}{\partial t}\cdot d\vec{S}, \tag{1.2}$$

Gauss' electric law

$$\oint\!\!\oint_S \vec{\mathcal{D}}(\vec{r},t)\cdot d\vec{S} = \iiint_V \rho(\vec{r},t)dV, \tag{1.3}$$

and *Gauss' magnetic law*

$$\oint\!\!\oint_S \vec{\mathcal{B}}(\vec{r},t)\cdot d\vec{S} = 0. \tag{1.4}$$

To these equations we add the *equation of continuity for electric charge*

$$\oint\!\!\oint_S \vec{\mathcal{J}}(\vec{r},t)\cdot d\vec{S} = -\iiint_V \frac{\partial \rho(\vec{r},t)}{\partial t}dV. \tag{1.5}$$

All quantities in these equations are evaluated in the rest frame of the observer, and all surfaces and volumes are held fixed in that frame.

The integrals in Equations (1.1) and (1.2) are for the open surface S with contour C shown in Figure 1.1a. The differential surface area vector is $d\vec{S} = \hat{n}dS$, where $\hat{n}$ is the unit vector locally normal to S, and dS is the differential surface area. The vector differential length $d\vec{\ell}$ is an infinitesimal vector locally tangent to C. By convention, the direction of the normal and the direction in which the curve C is traversed are related by the right-hand rule: With the curled fingers of the right hand pointing in the positive direction for traversing C, the thumb points in the direction of $\hat{n}$. The integrals in Equations (1.3), (1.4), and (1.5) are for the volume V with closed surface S shown in Figure 1.1b. The differential volume is dV, and the differential surface area vector is $d\vec{S} = \hat{n}dS$, where $\hat{n}$ is the outward-pointing, unit vector normal to S.

The terminology we will use for each of the quantities in these equations is listed in Table 1.2. The terminology used for electromagnetic quantities is by no means standard. Here we take the suggestion of Sommerfeld (Arnold Johannes Wilhelm Sommerfeld, 1868–1951) and reserve the names *electric field strength* and *magnetic field strength* for the vectors $\vec{\mathcal{E}}$ and $\vec{\mathcal{B}}$, respectively; the vector $\vec{\mathcal{D}}$ will be called the *electric excitation* or *electric displacement*, and $\vec{\mathcal{H}}$ will be called the *magnetic excitation* [6]. Also listed in Table 1.2 are the units for each quantity in the International System of Units (Système International – SI). A discussion of units and dimensions can be found in Appendix A.

The state of the electric charge is described by the two functions, ρ and $\vec{\mathcal{J}}$. ρ is the *volume density of charge* – the charge per unit volume; $\vec{\mathcal{J}}$ is the *volume*

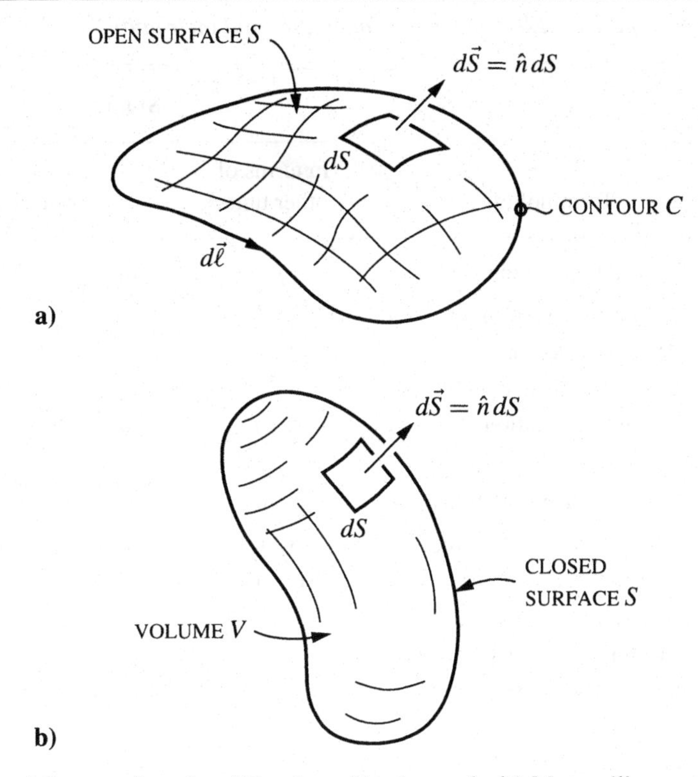

Fig. 1.1. a) Open surface S and b) volume V to be used with Maxwell's equations in integral form.

density of current – the charge passing per unit time through a unit area with normal $\hat{\mathcal{J}} = \vec{\mathcal{J}}/|\vec{\mathcal{J}}|$. The equation of continuity for electric charge (1.5) is a statement of charge conservation. Per unit time, the net charge passing out through the closed surface S (the left-hand side) must equal the net decrease in the charge within the volume V (the right-hand side).

In *free space* or vacuum the quantities $\vec{\mathcal{D}}$ and $\vec{\mathcal{H}}$ are simply related to $\vec{\mathcal{E}}$ and $\vec{\mathcal{B}}$:

$$\vec{\mathcal{D}}(\vec{r}, t) = \varepsilon_o \vec{\mathcal{E}}(\vec{r}, t), \tag{1.6}$$

$$\vec{\mathcal{H}}(\vec{r}, t) = \frac{1}{\mu_o} \vec{\mathcal{B}}(\vec{r}, t), \tag{1.7}$$

where $\varepsilon_o = 8.8541\ldots \times 10^{-12}$ F/m is the *permittivity of free space* and $\mu_o = 4\pi \times 10^{-7}$ H/m is the *permeability of free space*. The two constants ε_o and μ_o are related by the *speed of light c in a vacuum*:

$$c = 1/\sqrt{\varepsilon_o \mu_o} = 2.9979\ldots \times 10^8 \text{ m/s}. \tag{1.8}$$

In a material the connections between the vectors $\vec{\mathcal{E}}$, $\vec{\mathcal{B}}$, $\vec{\mathcal{D}}$, and $\vec{\mathcal{H}}$ are generally determined by experiment and are referred to as the *electromagnetic constitutive relations*. The simplest constitutive relations parallel the results for free

Table 1.2. *Electromagnetic quantities and their units*

Quantity	Terminology	SI unit: In terms of other units	SI unit: In terms of base units
$\vec{\mathcal{E}}$	Electric field strength	V/m	$m \cdot kg \cdot s^{-3} \cdot A^{-1}$
$\vec{\mathcal{B}}$	Magnetic field strength	T	$kg \cdot s^{-2} \cdot A^{-1}$
$\vec{\mathcal{D}}$	Electric excitation (electric displacement)	C/m^2	$m^{-2} \cdot s \cdot A$
$\vec{\mathcal{H}}$	Magnetic excitation	A/m	$m^{-1} \cdot A$
$\vec{\mathcal{J}}$	Volume density of current	A/m^2	$m^{-2} \cdot A$
ρ	Volume density of charge	C/m^3	$m^{-3} \cdot s \cdot A$
$\vec{\mathcal{J}}_s$	Surface density of current	A/m	$m^{-1} \cdot A$
ρ_s	Surface density of charge	C/m^2	$m^{-2} \cdot s \cdot A$
ε_o	Permittivity of free space ($8.8541\ldots \times 10^{-12}$)	F/m	$m^{-3} \cdot kg^{-1} \cdot s^4 \cdot A^2$
μ_o	Permeability of free space ($4\pi \times 10^{-7}$)	H/m	$m \cdot kg \cdot s^{-2} \cdot A^{-2}$
c	Speed of light in free space ($2.9979\ldots \times 10^8$)	–	$m \cdot s^{-1}$
$\vec{\mathcal{S}}$	Poynting vector	W/m^2	$kg \cdot s^{-3}$

space (1.6, 1.7):

$$\vec{\mathcal{D}}(\vec{r},t) = \varepsilon\vec{\mathcal{E}}(\vec{r},t) = \varepsilon_r\varepsilon_o\vec{\mathcal{E}}(\vec{r},t), \tag{1.9}$$

$$\vec{\mathcal{H}}(\vec{r},t) = \frac{1}{\mu}\vec{\mathcal{B}}(\vec{r},t) = \frac{1}{\mu_r\mu_o}\vec{\mathcal{B}}(\vec{r},t), \tag{1.10}$$

where $\varepsilon_r = \varepsilon/\varepsilon_o$ is the *relative permittivity* or dielectric constant and $\mu_r = \mu/\mu_o$ is the *relative permeability*. In a conductor, the volume density of current will also be related to the electric field:

$$\vec{\mathcal{J}}(\vec{r},t) = \vec{\mathcal{J}}_c(\vec{r},t) = \sigma\vec{\mathcal{E}}(\vec{r},t), \tag{1.11}$$

where σ is the *electrical conductivity* of the material. The subscript c indicates a conduction current; it will be omitted unless it is needed to resolve an ambiguity. Notice that ε, μ, and σ in these equations are independent of position ($\vec{r}$) and time (t) and that $\varepsilon_r \geq 1$, $\mu_r \geq 1$, and $\sigma \geq 0$.

Materials obeying the constitutive relations (1.9)–(1.11) will be called *simple materials*. We must emphasize that these simple constitutive relations are accurate representations for the electromagnetic behavior of actual materials only in restricted cases. More complicated relations are often required in practical applications. It is easy to understand one of the limitations of these relations: They require quantities like $\vec{\mathcal{E}}$ and $\vec{\mathcal{D}}$ to have the same temporal variation. Since the relationship between $\vec{\mathcal{E}}$ and $\vec{\mathcal{D}}$ is the result of physical mechanisms within the material, such as the orienting of microscopic dipoles, the simple constitutive relation (1.9) implies that the mechanisms respond instantaneously to any temporal variation in the field. Clearly, this will not be true for fields with extremely rapid temporal variations. In this book we will limit our discussion to electromagnetism in free space or in simple materials.

The equations we have presented so far, (1.1)–(1.11), provide a self-contained description of electromagnetism. The connection between electromagnetism and mechanics is provided by an additional relation – the *Lorentz force expression* [7]. For a volume V of free space at rest containing charge and current described by the volume densities ρ and $\vec{\mathcal{J}}$, the net mechanical force of electromagnetic origin acting on the charge and current is

$$\vec{\mathcal{F}} = \iiint_V (\rho\vec{\mathcal{E}} + \vec{\mathcal{J}} \times \vec{\mathcal{B}})dV. \tag{1.12}$$

The integrand of (1.12), which is a force per unit volume,

$$\vec{f} = \rho\vec{\mathcal{E}} + \vec{\mathcal{J}} \times \vec{\mathcal{B}}, \tag{1.13}$$

is often referred to as the *Lorentz force density*. If the volume V is a material body that carries the charge and the current, then the forces on the charge and the current are forces on the body, and (1.12) is the net mechanical force of electromagnetic origin acting on the body.

We can use Equation (1.12) to obtain an expression for the force on a charged particle moving in an applied electromagnetic field $\vec{\mathcal{E}}_a$, $\vec{\mathcal{B}}_a$ in free space. The charged particle is a small body with charge density ρ and total charge

$$q = \iiint_V \rho dV.$$

The velocity of the particle is $\vec{v}$, making the current density within the particle

$$\vec{\mathcal{J}} = \rho\vec{v}.$$

The total field at a point within the particle is the applied field $\vec{\mathcal{E}}_a$, $\vec{\mathcal{B}}_a$ plus the field due to the charge in the particle itself $\vec{\mathcal{E}}_s$, $\vec{\mathcal{B}}_s$, which is called the self field:

$$\vec{\mathcal{E}} = \vec{\mathcal{E}}_a + \vec{\mathcal{E}}_s, \qquad \vec{\mathcal{B}} = \vec{\mathcal{B}}_a + \vec{\mathcal{B}}_s.$$

From (1.12) the total force on the particle is

$$\vec{\mathcal{F}} = \vec{\mathcal{F}}_a + \vec{\mathcal{F}}_s, \tag{1.14}$$

with the applied force

$$\vec{\mathcal{F}}_a = \iiint_V (\rho\vec{\mathcal{E}}_a + \rho\vec{v} \times \vec{\mathcal{B}}_a)dV,$$

and the self force

$$\vec{\mathcal{F}}_s = \iiint_V (\rho\vec{\mathcal{E}}_s + \rho\vec{v} \times \vec{\mathcal{B}}_s)dV.$$

Now we will assume that the particle is so small that the applied field is essentially uniform over its extent. The force due to the applied field is then

$$\vec{\mathcal{F}}_a = (\vec{\mathcal{E}}_a + \vec{v} \times \vec{\mathcal{B}}_a) \iiint_V \rho \, dV,$$

which can be written

$$\vec{\mathcal{F}}_a = q(\vec{\mathcal{E}}_a + \vec{v} \times \vec{\mathcal{B}}_a). \tag{1.15}$$

This is the familiar *Lorentz force* for a moving particle in an applied electromagnetic field.

The total electromagnetic force on the particle (1.14) is the applied force $\vec{\mathcal{F}}_a$ plus the self force $\vec{\mathcal{F}}_s$.[2] In practical situations the self force is often negligible, and the total force on the particle is approximately the applied force

$$\vec{\mathcal{F}} = q(\vec{\mathcal{E}}_a + \vec{v} \times \vec{\mathcal{B}}_a). \tag{1.16}$$

This equation can be considered an operational definition for the electric and magnetic fields in free space. The applied electric field is defined by the force $\vec{\mathcal{F}}_e$ experienced by a stationary test charge q:

$$\vec{\mathcal{E}}_a = \vec{\mathcal{F}}_e / q.$$

The additional force $\vec{\mathcal{F}}_m$ that results from uniform motion of the test charge is used to define the applied magnetic field. The component of the magnetic field normal to the velocity $\vec{v}$ of the charge is

$$\vec{\mathcal{B}}_{a\perp} = \frac{-\vec{v} \times \vec{\mathcal{F}}_m}{q|\vec{v}\,|^2}.$$

The equations of electromagnetism, (1.1)–(1.12), are based on extensive experimental observation. Historically, deducing these equations from experiment was a long and arduous task. It involved many of the famous names in physics, as a glance at Table 1.1 shows. For our purposes, it would be of little benefit to reproduce the arguments that led from experimental observation to the equations of electromagnetism. What we will do instead is describe a few of the more important early experiments and show that the results of these experiments are predicted by the equations of electromagnetism. We will be using the theory to analyze the experiments rather than using the experiments to synthesize the theory, since the latter is a much more difficult process. Our approach provides an additional benefit: We will get to review the application of the integral form of Maxwell's equations to determine the fields, forces, etc. for a few simple geometries.

For the establishment of a physical law, which is expressed as a mathematical relationship, experimental observations of a quantitative nature are generally

[2] The self force first arose in the development of early classical models for the electron, in which the electron was assumed to be a small charged sphere. H. A. Lorentz was an important contributor to these early models. A brief qualitative argument for the origin of the self force is given in Section 6.5. The history of the classical models for the electron and detailed derivations of the self force are presented in References [8–10].

required. Until the middle of the eighteenth century, experimental observations of electric and magnetic phenomena were at best semiquantitative. For example, observations suggested that the electrostatic force between two charged objects increased as they came closer together, but the mathematical dependence of the force on the separation was not established.

In the latter half of the eighteenth century, several investigators, including Henry Cavendish (1731–1810) and Joseph Priestley (1733–1804), inferred from experiments that the electrostatic force between charged objects varies inversely as the square of the distance separating the objects [1]. In the modern literature, however, Charles Augustin de Coulomb (1736–1806) is generally credited with establishing this result, and it is Coulomb's experiment that we will describe [11, 12].

Coulomb had previously developed a torsion balance for measuring small forces, and he made use of it in his electrical studies. A sketch of his apparatus is shown in Figure 1.2.[3] The torsion balance is formed from a thin, vertical, silver wire held taut by a weight attached to its lower end. When the wire is twisted it produces a force of torsion. Coulomb had shown earlier that this force was directly proportional to the angle of twist ψ. Two small, gilded, elder wood pith balls are located in a plane normal to this wire. One is held fixed; the other is mounted on a horizontal arm attached to the wire and is free to move as the wire is twisted.

The balance was initially set so that there was no twist in the wire when the two uncharged balls were in contact. The pointer fixed to the upper end of the wire was zeroed. Then a small charged pin was brought in contact with the balls to give them equal charge of the same sign. The movable ball was repelled through the angle ψ_s (a counterclockwise rotation in Figure 1.2). The distance between the centers of the balls r and the angle ψ_s could be reduced by rotating the pointer through the angle ψ_p (a clockwise rotation in Figure 1.2), making the total angle of twist for the wire $\psi = \psi_s + \psi_p$.

Using two values of the angle of twist, ψ_1 and ψ_2, Coulomb was able to show that the electrostatic force of repulsion $\mathcal{F}$ behaves as

$$\frac{\mathcal{F}_1}{\mathcal{F}_2} = \frac{\psi_1}{\psi_2} = \left(\frac{r_2}{r_1}\right)^2;$$

that is, the force is inversely proportional to the square of the distance separating the balls. Additional experimentation led to the following result, known as Coulomb's law,

$$\mathcal{F} \propto \frac{qq'}{r^2},$$

where q and q' are the charges on the two balls.

We should mention that Coulomb also verified this relationship for attractive forces (i.e., for balls with charges of opposite sign). Initially he tried to use the apparatus in Figure 1.2 for this purpose but found that the mechanism was unstable,

[3] All of the drawings showing the apparatus used in the early experiments are this author's rendition, e.g., Figures 1.2, 1.4, 1.5a, 1.7, 1.9, and 1.12. In most cases details of the construction have been omitted to allow a clearer description of the mechanism.

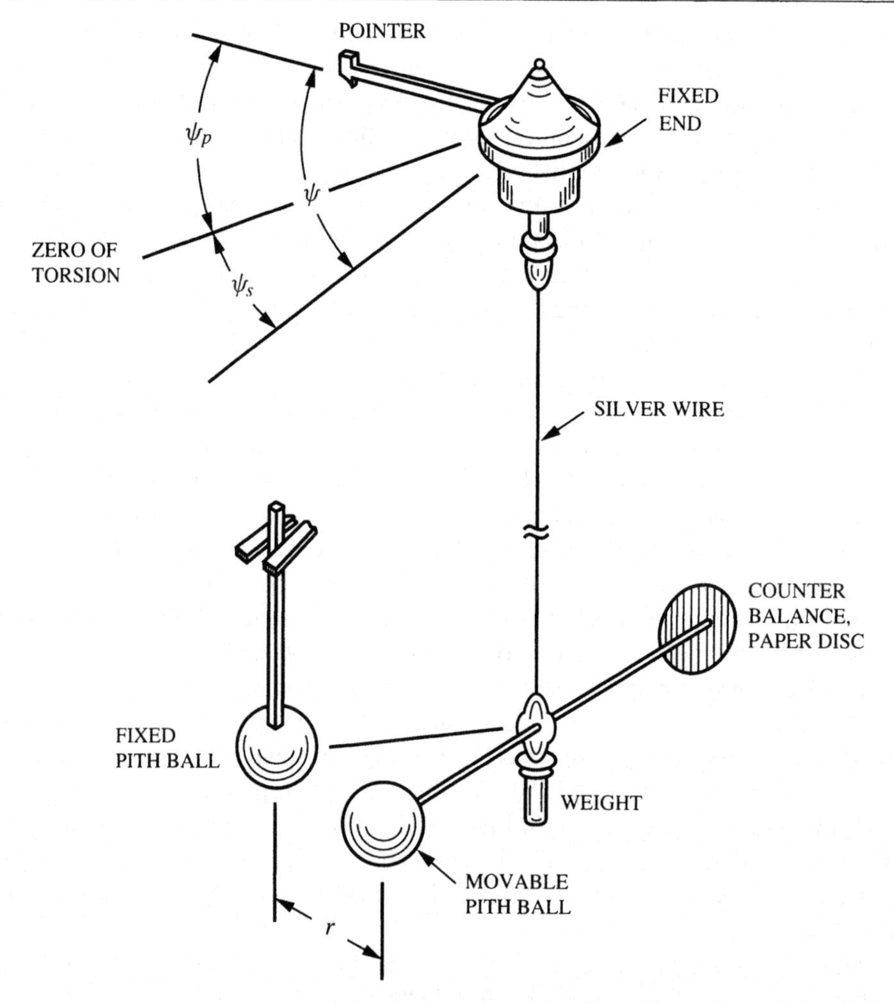

Fig. 1.2. Coulomb's apparatus for electrical studies. The apparatus was enclosed in glass tubes, not shown, to prevent air currents from disturbing the measurements.

with the pith balls often coming together. He then devised a new apparatus, which also used his torsion balance, to study the attractive forces [11, 12].

Coulomb's observations are predicted by the equations of electromagnetism, as we will show by applying these equations to the geometry in Figure 1.3. Here two spheres (1 and 2) in free space are separated by a distance r_{12} that is assumed to be much larger than the radius of either sphere. As a result, the charge on sphere 1 is approximately uniformly distributed over its surface, which means that its electric field is in the radial direction, $\hat{r}$, and is spherically symmetric. On applying Gauss' electric law (1.3) to a spherical surface of radius r surrounding sphere 1, we find that

$$\oiint_S \vec{\mathcal{D}}_1 \cdot d\vec{S} = \varepsilon_o \mathcal{E}_{r1} \oiint_S dS = \varepsilon_o \mathcal{E}_{r1}(4\pi r^2) = \iiint_V \rho_1 dV = q_1,$$

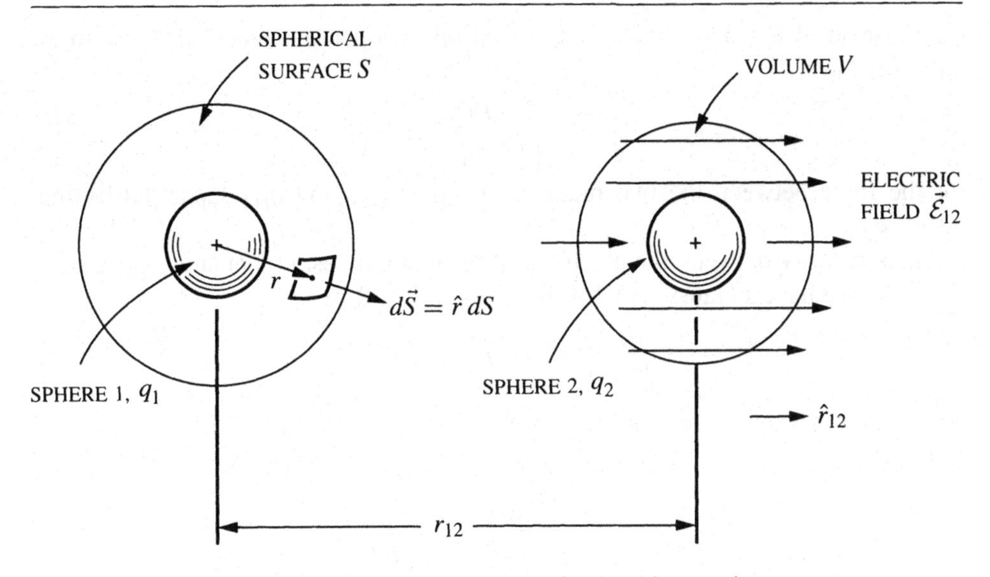

Fig. 1.3. Geometry for analyzing Coulomb's experiment.

or

$$\vec{\mathcal{E}}_1 = \frac{q_1}{4\pi\varepsilon_o r^2}\hat{r}.$$

Since the distance r_{12} between the spheres is much larger than the radius of sphere 2, the electric field of sphere 1 is approximately uniform over the volume of sphere 2:

$$\vec{\mathcal{E}}_{12} \approx \frac{q_1}{4\pi\varepsilon_o r_{12}^2}\hat{r}_{12}.$$

Now the force on sphere 2 due to the electric field of sphere 1 is obtained by applying the Lorentz force expression (1.12) to the volume V containing sphere 2:

$$\vec{\mathcal{F}} = \iiint_V \rho_2\vec{\mathcal{E}}_{12}\,dV \approx \vec{\mathcal{E}}_{12}\iiint_V \rho_2 dV = q_2\vec{\mathcal{E}}_{12},$$

or

$$\vec{\mathcal{F}} = \frac{q_1 q_2}{4\pi\varepsilon_o r_{12}^2}\hat{r}_{12}, \tag{1.17}$$

which is a mathematical statement of Coulomb's observations.

Coulomb also used his torsion balance to measure the force between magnets. By using a long, thin, magnetized needle, he was able to study the force near one pole of the magnet. From these measurements he concluded that the force between two magnetic poles of strength π_m and π'_m separated by the distance r was

$$\mathcal{F} \propto \frac{\pi_m \pi'_m}{r^2}.$$

Note the similarity of this expression to the one he obtained for the force between electric charges. When we write Coulomb's results in terms of magnetic charge

(q_m) instead of magnetic pole strength and take proper account of the system of units, we have

$$\vec{\mathcal{F}} = \frac{q_{m1} q_{m2}}{4\pi \mu_o r_{12}^2} \hat{r}_{12} \tag{1.18}$$

for the force between the two magnetic charges q_{m1} and q_{m2} separated by the distance r_{12}.

The similarity between the electric and magnetic cases, (1.17) and (1.18), suggests that we have a Gauss' law for the magnetic field

$$\oiint_S \vec{\mathcal{B}} \cdot d\vec{S} = \iiint_V \rho_m dV \tag{1.19}$$

and a Lorentz force expression for stationary magnetic charge

$$\vec{\mathcal{F}} = \iiint_V \rho_m \vec{\mathcal{H}} dV,$$

where ρ_m is the volume density of magnetic charge.[4]

Coulomb could not produce a body with a net magnetic pole strength (i.e., a net magnetic charge). When a magnet is broken into two pieces, two magnets (each with a north and a south pole) are formed, not two separate poles, one north and one south. All subsequent attempts to find a body with a net magnetic pole strength or a net magnetic charge have failed. Thus, the experimental evidence requires the right-hand side of (1.19) to be zero:

$$\oiint_S \vec{\mathcal{B}} \cdot d\vec{S} = 0.$$

This is just Equation (1.4) in our list of Maxwell's equations.

The qualitative similarities between electric and magnetic phenomena were noted long before Coulomb performed his experiments; for example, like electric charges and like magnetic poles were known to repel, whereas unlike electric charges and unlike magnetic poles were known to attract. In addition, other observations suggested a direct connection between electricity and magnetism; for example, steel objects located near lightning strikes, believed to be electric in origin, were found to be magnetized.

The first experiment that conclusively showed the connection between electricity and magnetism was performed by Hans Christian Oersted in 1820 [13, 14]. Oersted's experiment was extremely simple by today's standards; it involved no intricate or specially constructed apparatus. A sketch of his apparatus is in Figure 1.4. Oersted connected a piece of wire to the terminals of a battery. He then moved a compass needle under a straight section of the current-carrying wire and noticed that the needle was deflected to a position with its axis normal to the axis of the wire. The direction of the needle reversed when the needle was located above rather than below the wire or when the direction of the current in the wire was reversed.

[4] These equations depend somewhat on the convention used, that is, on how one defines magnetic charge. Other equally valid representations can be found in the literature (e.g., $\vec{\mathcal{H}}$ may appear in (1.19) instead of $\vec{\mathcal{B}}$). For additional discussion of magnetic charge see Problem 1.6 and Section 7.3.

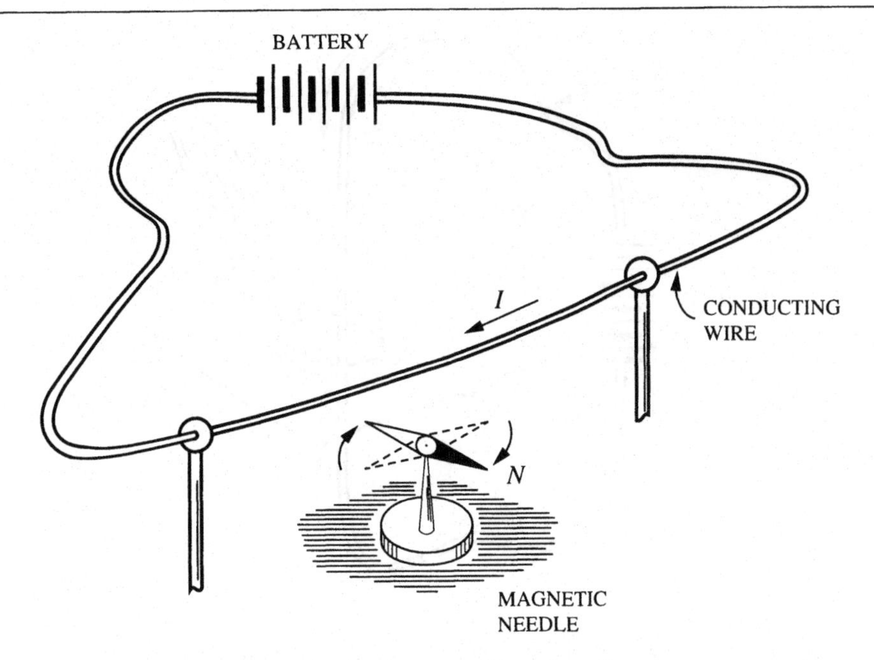

Fig. 1.4. Oersted's apparatus for showing the effect of an electric current on a magnetic needle.

These observations suggested the presence of a circular magnetic field around the wire. This point was graphically demonstrated some years later by Michael Faraday [15]. In Figure 1.5a we show Faraday's experiment. The straight section of the current-carrying wire pierces a piece of stiff paper. Iron filings sprinkled on the paper tend to align with the magnetic field. The circular magnetic field around the wire is clearly shown by the distribution of the filings in Figure 1.5b, which was made using an arrangement similar to Faraday's.[5]

The theoretical explanation for Oersted's observation is contained in the Ampère-Maxwell law (1.2). Because we are dealing with a steady current, all field quantities are time independent, and the second term on the right-hand side of (1.2) is zero:

$$\oint_C \vec{\mathcal{H}} \cdot d\vec{\ell} = \frac{1}{\mu_o} \oint_C \vec{\mathcal{B}} \cdot d\vec{\ell} = \iint_S \vec{\mathcal{J}} \cdot d\vec{S}. \tag{1.20}$$

We will now apply this equation to a surface S, which is a circular disc centered on the straight current-carrying wire in Figure 1.6. The right-hand side of (1.20) is then just the total current I passing through a cross section of the wire. Since the geometry is rotationally symmetric about the axis of the wire, $\vec{\mathcal{B}}$ is independent of the angle ϕ, and the left-hand side of (1.20) simplifies to become

$$\frac{1}{\mu_o} \int_{\phi=0}^{2\pi} \vec{\mathcal{B}} \cdot \hat{\phi} r d\phi = \frac{1}{\mu_o} r \mathcal{B}_\phi \int_{\phi=0}^{2\pi} d\phi = \frac{2\pi r}{\mu_o} \mathcal{B}_\phi .$$

[5] For this picture, a square coil of seven turns was connected to a source that produced a current of about 14 A. One side of the coil pierced the paper. Thus, the pattern is equivalent to what would be produced by a current of 7×14 A $= 98$ A in a single wire.

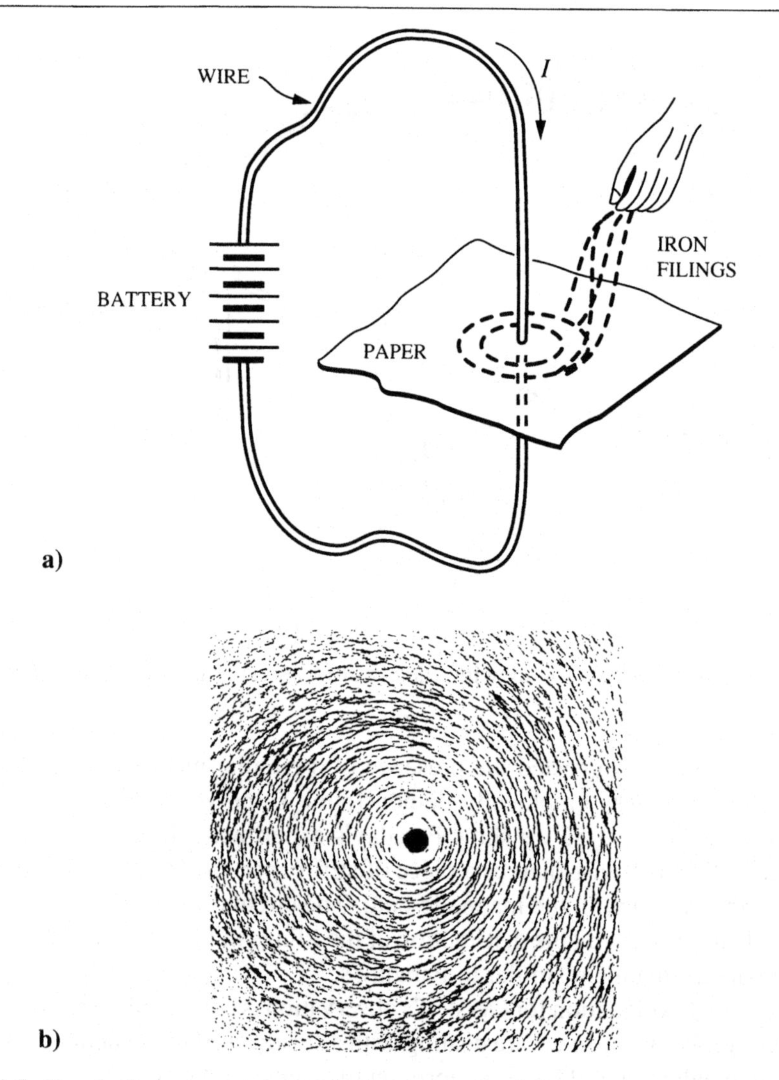

Fig. 1.5. Faraday's demonstration of the lines of magnetic force surrounding a long, straight conductor. a) Apparatus. b) Distribution of iron filings.

Thus,

$$\mathcal{B}_\phi = \frac{\mu_o I}{2\pi r},$$

so the magnetic field is

$$\vec{\mathcal{B}} = \frac{\mu_o I}{2\pi r}\hat{\phi}. \tag{1.21}$$

This is the circular magnetic field that Oersted observed around the wire.

The news of Oersted's discovery spread quickly throughout Europe, where several scientists tried to duplicate and extend his results. Jean Baptiste Biot (1774–1862) and Félix Savart (1791–1841) experimentally studied the force between a

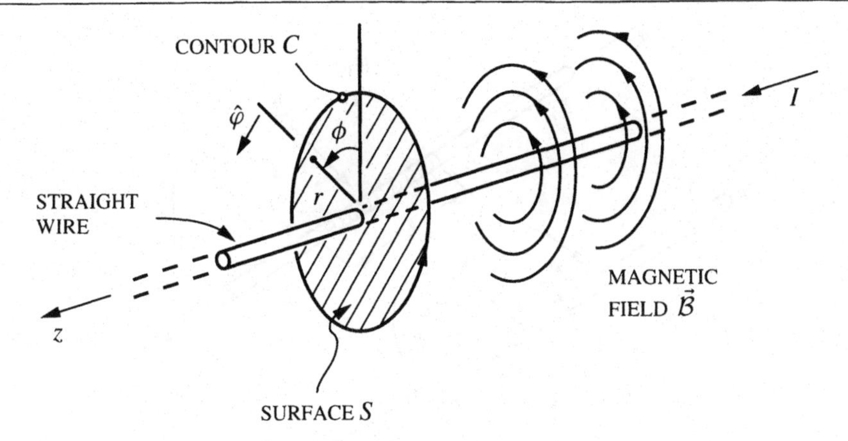

Fig. 1.6. Geometry for analyzing Oersted's experiment.

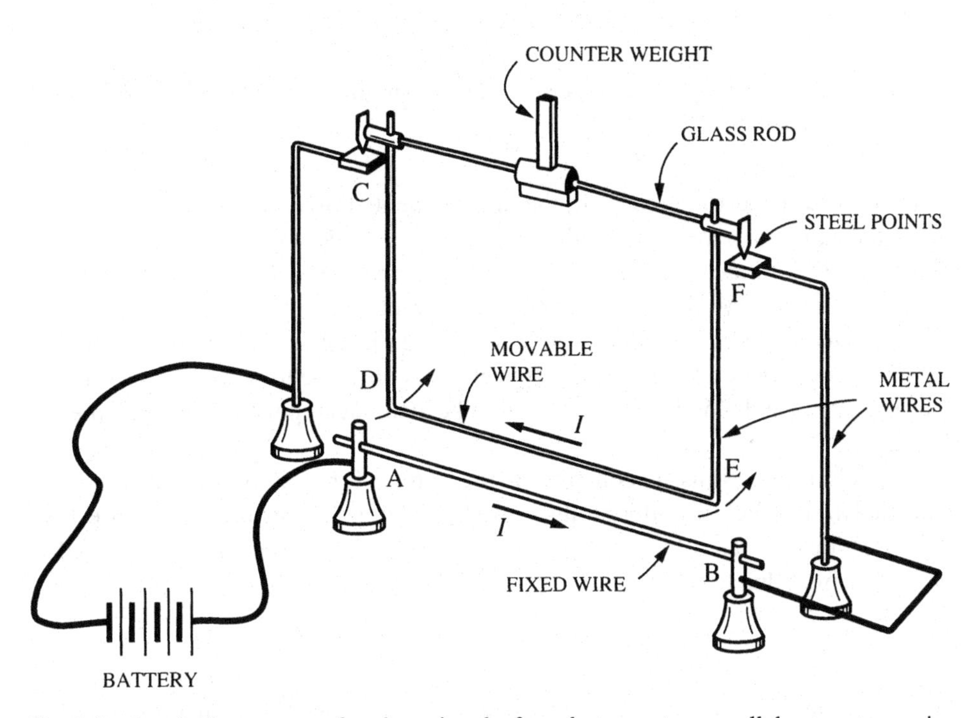

Fig. 1.7. Ampère's apparatus for observing the force between two parallel, current-carrying wires. The apparatus was enclosed in a glass case, not shown, to prevent air currents from disturbing the measurements.

current-carrying wire and a magnet. André Marie Ampère conjectured that electric currents were the cause of magnetism, and since there is a force between an electric current and a magnet, there must be a force between two electric currents. Ampère made extensive theoretical and experimental studies of this force.

Figure 1.7 shows the apparatus for one of Ampère's early experiments used to explore the force between two parallel, current-carrying wires [16]. The straight wire AB is held fixed, while the U-shaped wire CDEF is free to rotate on the

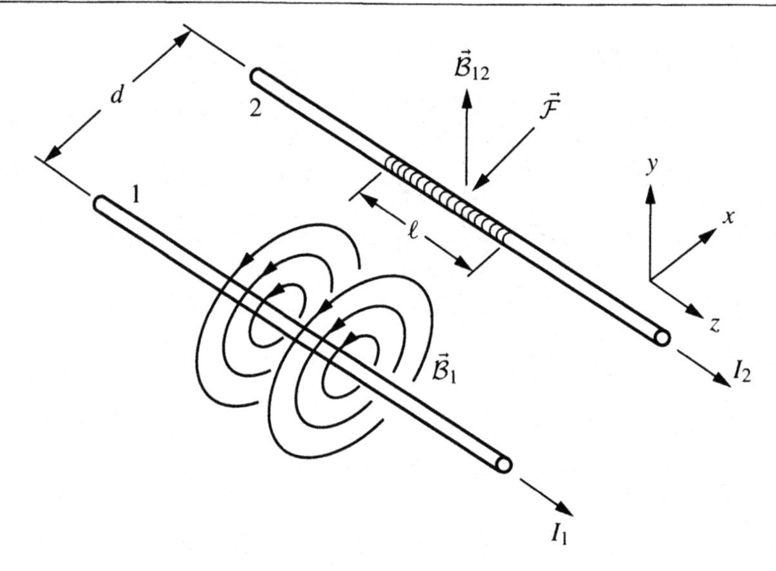

Fig. 1.8. Geometry for analyzing Ampère's experiment to observe the force on parallel, current-carrying wires.

steel points fastened to its ends. In the arrangement shown, the series connection causes the currents in the two parallel sections of wire AB and DE to be in opposite directions. The connections could easily be rearranged to make the currents in the same direction. With this apparatus Ampère observed that the parallel wires were attracted when the currents were in the same direction and repelled when the currents were in opposite directions. This effect is opposite to what is observed for electric charges and magnetic poles, where like elements repel and unlike elements attract.

Ampère's observations are contained in the Lorentz force expression (1.12). Consider the two parallel, infinitely long, current-carrying wires shown in Figure 1.8. The wires are separated by the distance d. From (1.21) the magnetic field at the center of wire 2 due to the current in wire 1 is

$$\vec{\mathcal{B}}_{12} = \frac{\mu_o I_1}{2\pi d}\hat{y}.$$

We will assume that the field is approximately uniform over the small cross section of wire 2; the force on a length ℓ of wire 2 due to the magnetic field of wire 1 is then

$$\begin{aligned}\vec{\mathcal{F}} &= \iiint_V \vec{\mathcal{J}}_2 \times \vec{\mathcal{B}}_{12}\, dV \approx \int_{z'=z}^{z+\ell} (I_2\hat{z}) \times \vec{\mathcal{B}}_{12}\, dz' \\ &= \frac{-\mu_o I_1 I_2 \ell}{2\pi d}\hat{x}. \end{aligned} \tag{1.22}$$

For equal currents in the same direction ($I_1 = I_2$), the wires are attracted; for equal currents in opposite directions ($I_1 = -I_2$), the wires are repelled, as Ampère observed.

In experiments like the one shown in Figure 1.7, the conductors essentially short circuit the battery. The battery then cannot maintain a constant voltage, so the current in the circuit varies with time, which makes quantitative measurements difficult. Ampère avoided this problem in his later experiments by basing them on the null principle. In these experiments, the same current passed through all conductors of the circuit. The physical arrangement of the conductors was chosen so that they produced no force – a null – at the location of a detector. Temporal variations in the current did not affect the null condition. By using different physical arrangements for the conductors, Ampère was able to infer the mathematical behavior for the force between current elements from these experiments, a feat described by Maxwell [5] as "one of the most brilliant achievements in science."

We will look at one of the four experiments, based on the null principle, that Ampère described in his famous memoir [17]. This particular experiment shows that the force of magnetic origin on a small element of current is always normal to the direction of the current. This result is obviously contained in the Lorentz force expression (1.12), since the cross product $\vec{\mathcal{J}} \times \vec{\mathcal{B}}$ is always normal to $\vec{\mathcal{J}}$.

Ampère's apparatus for this experiment is shown in Figure 1.9. A battery produces a current I in the conducting circuit AOBEB′OA′D, which rests on the top of a table. There are four radial conductors with common ends at O: AO, BO, A′O, and B′O. The circular arc of wire CC′ rests on the mercury-filled troughs at B and B′. A radial support arm EF is connected to this arc by a hinge at its midpoint E. The radial support arm is also attached to a vertical rod that is free to rotate about the axis OP.

The current-carrying conductors on the table are viewed as the two separate closed circuits shown in Figure 1.10. Circuit 1 is the loop AOA′DA; circuit 2 is the loop BOB′B. Circuit 1 produces a magnetic field $\vec{\mathcal{B}}_1$ at circuit 2, and it is the force that this field causes on the movable arc CC′ that is observed in the experiment. When the arc CC′ is positioned so that the circle of which it is a part has its center O′ coincident with O, as in Figure 1.11a, there is no movement of the arc. This null result implies that the force $\vec{\mathcal{F}}$ acting on the arc is normal to the arc and acts through O. The motion of the arc is thus prevented by the radial support arm EF. When the arc CC′ is positioned so that the circle of which it is a part has its center O′ away from O, as in Figure 1.11b, the arc moves. In this case the force $\vec{\mathcal{F}}$ is normal to the arc, but it does not act through O; thus, the radial support arm EF does not prevent it from moving.

Ampère obtained the same results for arcs of different length, and he concluded [18], "... that the action of a closed circuit, or an assembly of closed circuits (a magnetic field) on an infinitesimal element of an electric current is perpendicular to this element."

Note that the currents in the two segments BO and B′O of circuit 2 also produce a magnetic field at the arc CC′. For the case in Figure 1.11a, Ampère correctly argued that, due to symmetry, the forces caused by these currents would not move the arc. For the case in Figure 1.11b, the forces caused by these currents do move the arc. The additional effect due to the magnetic field of circuit 1 is then observed by changing the shape of the wire ADA′.

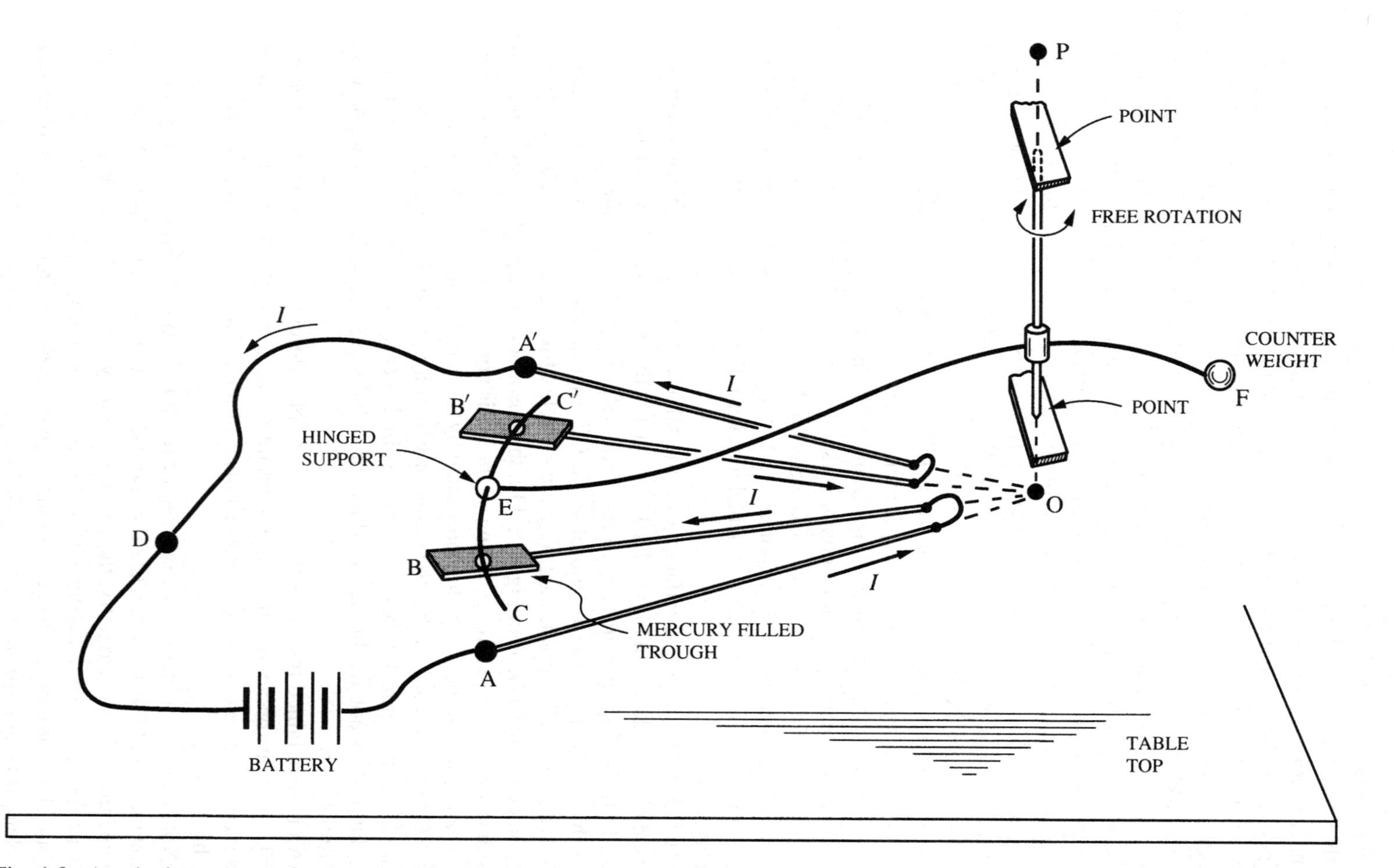

Fig. 1.9. Ampère's apparatus for demonstrating that the force of magnetic origin on a small element of current is always normal to the direction of the current.

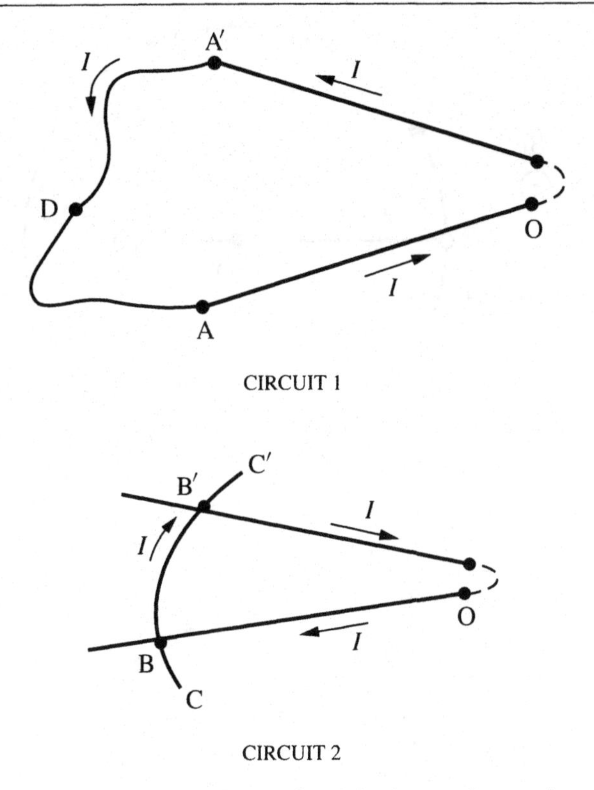

Fig. 1.10. The current-carrying conductors of Ampère's experiment viewed as forming two closed circuits.

After Oersted showed that an electric current produced a magnetic field, many scientists attempted to demonstrate the reciprocal effect, that is, an electric current produced by magnetism. There were a few near successes, but it was Michael Faraday in 1831 who finally obtained conclusive experimental proof for this effect. Faraday, like Ampère, based his theoretical reasoning on a series of experiments [19–21]. We will describe one of these experiments, the one with which Faraday first observed a magnetic field producing an electric current [22].

A drawing of Faraday's apparatus is shown in Figure 1.12. Two coils, A and B, are wound on an iron ring with a diameter of about 15 cm. The terminals of coil A are connected to a battery, while the terminals of coil B are connected to a loop of wire, a portion of which passes over a magnetic needle.

Faraday observed that the magnetic needle moved momentarily whenever the battery was connected to or disconnected from the circuit. The needle was deflected, oscillated, and eventually returned to its initial position. Faraday had discovered *electromagnetic induction*. When the battery was connected, it produced a rapidly increasing current $\mathcal{I}_A$ in coil A and, consequently, a rapidly increasing magnetic field in the iron ring. This time-varying magnetic field induced a transient current $\mathcal{I}_B$ in coil B and the connected wire loop. The current in the wire loop, in turn, produced a transient magnetic field that deflected the magnetic needle. Once the current in coil A was steady, there was no effect on the magnetic needle.

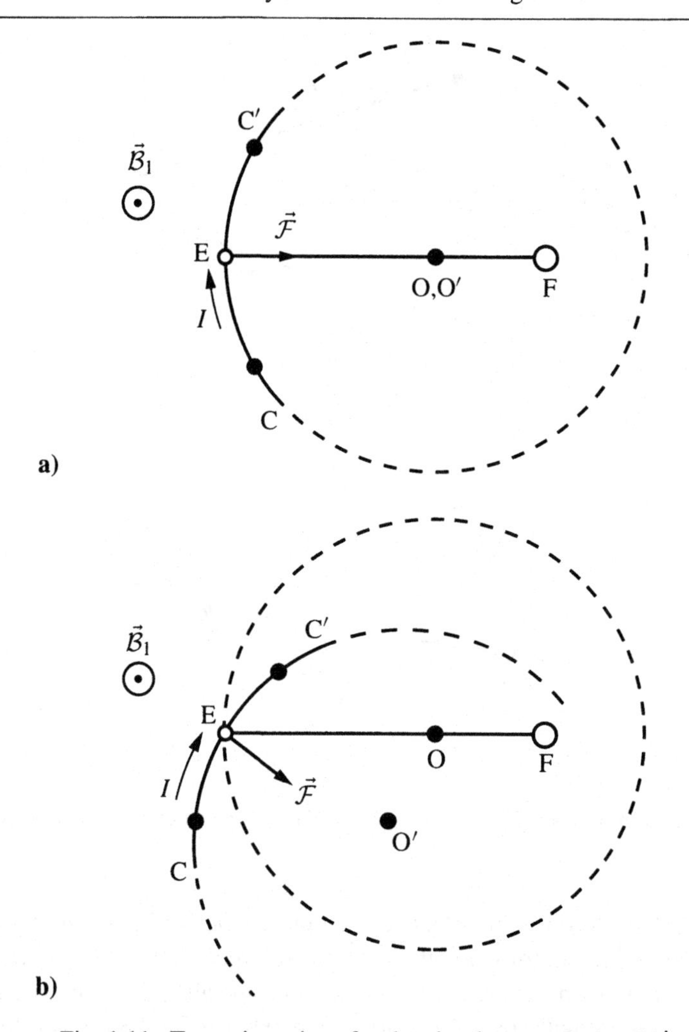

Fig. 1.11. Two orientations for the circular arc of wire CC′.

It is clear why the investigators before Faraday failed to discover electromagnetic induction. They used apparatus not significantly different from Faraday's, but they restricted their observations to the time when the current was steady and missed the small momentary effects produced when the battery was connected or disconnected.

The theoretical explanation for Faraday's observations rests on Equation (1.1), appropriately called Faraday's law. First, we will consider the cross section of the iron ring shown in Figure 1.13a. We will apply Equation (1.2) to the surface S_A on this cross section. The circular contour C_A, which is on the axis of the ring, bounds S_A. After neglecting the second term on the right-hand side of this equation, we have

$$\oint_{C_A} \vec{\mathcal{H}}_A \cdot d\vec{\ell} \approx \iint_{S_A} \vec{\mathcal{J}}_A \cdot d\vec{S}.$$

The magnetic excitation $\vec{\mathcal{H}}_A$ is approximately uniform on the axis of the ring,

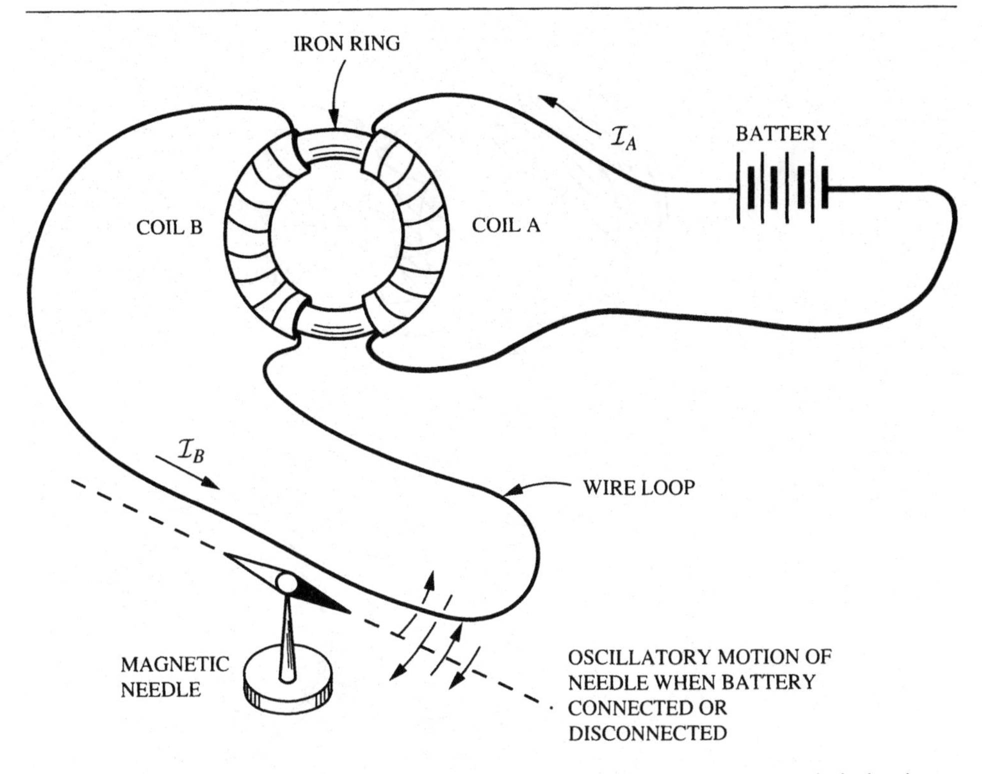

Fig. 1.12. The apparatus with which Faraday first observed electromagnetic induction.

$\vec{\mathcal{H}}_A \approx \mathcal{H}_{A\theta}\,\hat{\theta}$, and there are N_A turns, each carrying current $\mathcal{I}_A$ in coil A; thus

$$\begin{aligned}\oint_{C_A} \vec{\mathcal{H}}_A \cdot d\vec{\ell} &= \int_{\theta=0}^{2\pi} \vec{\mathcal{H}}_A \cdot \hat{\theta}\, b d\theta = 2\pi b \mathcal{H}_{A\theta} \\ &= \iint_{S_A} \vec{\mathcal{J}}_A \cdot d\vec{S} = N_A \mathcal{I}_A,\end{aligned}$$

or simply

$$\mathcal{H}_{A\theta} = \frac{N_A}{2\pi b}\mathcal{I}_A.$$

The magnetic field $\mathcal{B}_{A\theta}$ is related to $\mathcal{H}_{A\theta}$ by the permeability of the iron μ (1.10):

$$\mathcal{B}_{A\theta} = \mu \mathcal{H}_{A\theta} = \frac{\mu N_A}{2\pi b}\mathcal{I}_A.$$

Now we will apply Faraday's law (1.1) to the surface S_B, which is bounded by the contour C_B formed by the coil B and the connected wire loop:

$$\oint_{C_B} \vec{\mathcal{E}}_B \cdot d\vec{\ell} = -\iint_{S_B} \frac{\partial \vec{\mathcal{B}}_A}{\partial t} \cdot d\vec{S}. \tag{1.23}$$

For the right-hand side of this equation, we note that the portion of S_B that couples the magnetic field $\vec{\mathcal{B}}_A$ is the N_B turns, each of area $S_N = \pi a^2$ (see the lower drawing

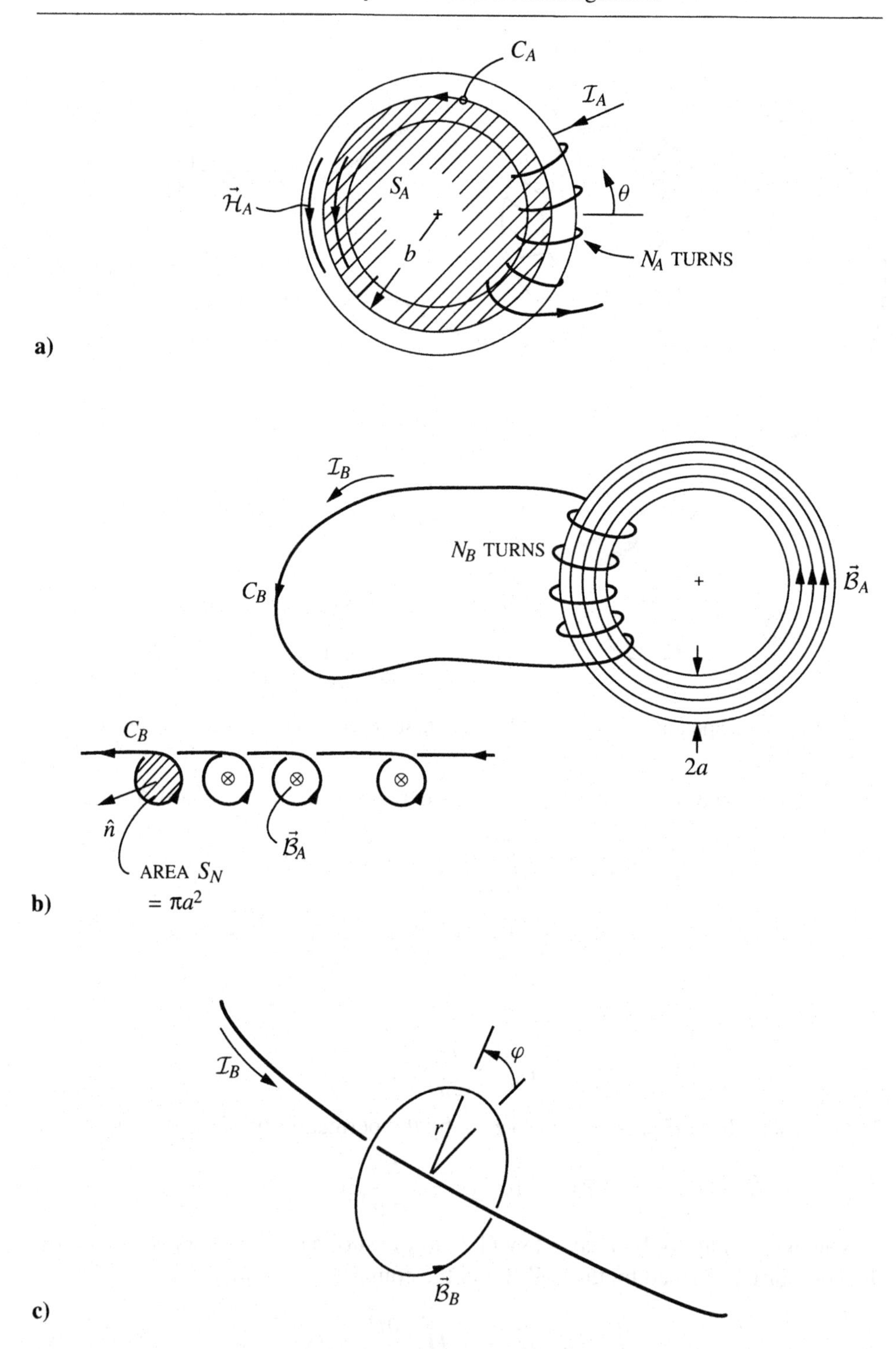

Fig. 1.13. Sketches for the analysis of Faraday's experiment. a) Magnetic field in iron ring. b) Current induced in coil B. c) Magnetic field of current in wire.

of Figure 1.13b). The magnetic field is approximately uniform over the area S_N; thus

$$-\iint_{S_B} \frac{\partial \vec{\mathcal{B}}_A}{\partial t} \cdot d\vec{S} = N_B \pi a^2 \frac{\partial \mathcal{B}_{A\theta}}{\partial t} = \frac{\mu N_A N_B a^2}{2b} \frac{\partial \mathcal{I}_A}{\partial t}. \tag{1.24}$$

The electric field $\vec{\mathcal{E}}_B$ in the wire of coil B is related to the volume current density by the conductivity of the wire σ_B (1.11):

$$\vec{\mathcal{E}}_B = \vec{\mathcal{J}}_B / \sigma_B.$$

If we assume that $\vec{\mathcal{J}}_B$ is in the axial direction and uniform throughout the wire,

$$\mathcal{I}_B = \pi a_B^2 |\vec{\mathcal{J}}_B|,$$

or

$$|\vec{\mathcal{E}}_B| = \mathcal{I}_B / \pi a_B^2 \sigma_B,$$

then

$$\oint_{C_B} \vec{\mathcal{E}}_B \cdot d\vec{\ell} = \frac{\mathcal{I}_B \ell_B}{\pi a_B^2 \sigma_B}, \tag{1.25}$$

where a_B and ℓ_B are the radius and length, respectively, of the wire forming coil B and the connected loop.

On inserting (1.24) and (1.25) into (1.23), we have

$$\mathcal{I}_B = \left(\frac{\pi a_B^2 \sigma_B}{\ell_B}\right)\left(\frac{\mu N_A N_B a^2}{2b}\right)\frac{\partial \mathcal{I}_A}{\partial t}.$$

The field $\vec{\mathcal{B}}_B$ produced by the straight piece of wire over the magnetic needle (Figure 1.13c) is from (1.21):

$$\vec{\mathcal{B}}_B = \frac{\mu_o \mathcal{I}_B}{2\pi r}\hat{\phi} = \left(\frac{\pi a_B^2 \sigma_B}{\ell_B}\right)\left(\frac{\mu_o \mu N_A N_B a^2}{4\pi b r}\right)\frac{\partial \mathcal{I}_A}{\partial t}\hat{\phi}.$$

Recognizing that $R_B = \ell_B / \sigma_B \pi a_B^2$ is the resistance of coil B and the connected wire, we get

$$\vec{\mathcal{B}}_B(t) = \left(\frac{\mu_o \mu N_A N_B a^2}{4\pi R_B b r}\right)\frac{\partial \mathcal{I}_A(t)}{\partial t}\hat{\phi}.$$

This equation is the mathematical statement of Faraday's observation: A time-varying current in coil A produces a magnetic field around the wire of coil B that deflects the magnetic needle. Once the current is steady, $\partial \mathcal{I}_A(t)/\partial t = 0$, this magnetic field ceases to exist and there is no effect on the magnetic needle.

Our theoretical explanations for the experiments we have examined have used all of Maxwell's equations and the Lorentz force expression. Coulomb's observations for static electric and magnetic fields are explained by Gauss' electric and magnetic laws, Equations (1.3) and (1.4), and the first term in the Lorentz force expression (1.12). Oersted's and Ampère's results for steady currents are explained by the Ampère-Maxwell law (1.2) (without the second term on the right-hand side) and the second term in the Lorentz force expression (1.12). Finally, Faraday's observation of electromagnetic induction is explained by his law (1.1).

The one term in Maxwell's equations that we have not used in our explanations is the second term on the right-hand side of the Ampère-Maxwell law, that is,

$$\oiint_S \frac{\partial \vec{\mathcal{D}}}{\partial t} \cdot d\vec{S}.$$

This is the famous term first introduced into the equations of electromagnetism by Maxwell. Maxwell believed that this term was due to a real displacement of a physical substance, hence the name electric displacement for the vector $\vec{\mathcal{D}}$.

The exact reasoning Maxwell used to arrive at this term is not known [23]. However, it is fairly certain that he did not make use of a definitive experiment of his own or another's design, for in describing the displacement within the dielectric of a condenser he states [24], "this part of the theory ... has not been verified by direct experiment. The experiment would be a very delicate and difficult one." And later in his treatise [5] he says, "We have very little experimental evidence relating to the direct electromagnetic action of currents due to the variation of electric displacement in dielectrics, ... " In fact, conclusive experimental proof for the displacement term in the Ampère-Maxwell law was not obtained until nearly fifteen years after the publication of Maxwell's treatise, when Hertz performed his famous experiments on electromagnetic waves [25]. We will delay the discussion of Hertz's experiments until we have developed the ideas associated with electromagnetic waves (Section 2.7).

There are several theoretical arguments that can be used to justify the displacement term in Equation (1.2). Perhaps the simplest argument shows that Equation (1.2) is inconsistent with Equations (1.3) and (1.5) unless this term is included. After differentiating (1.3) with respect to time and substituting into (1.5), we have

$$\oiint_S \vec{\mathcal{J}} \cdot d\vec{S} + \oiint_S \frac{\partial \vec{\mathcal{D}}}{\partial t} \cdot d\vec{S} = 0. \tag{1.26}$$

Now (1.2) is applied to the two portions S_1 and S_2 of the closed surface S shown in Figure 1.14:

$$\oint_{C_1} \vec{\mathcal{H}} \cdot d\vec{\ell} = \iint_{S_1} \vec{\mathcal{J}} \cdot d\vec{S} + \iint_{S_1} \frac{\partial \vec{\mathcal{D}}}{\partial t} \cdot d\vec{S}$$

and

$$\oint_{C_2} \vec{\mathcal{H}} \cdot d\vec{\ell} = \iint_{S_2} \vec{\mathcal{J}} \cdot d\vec{S} + \iint_{S_2} \frac{\partial \vec{\mathcal{D}}}{\partial t} \cdot d\vec{S}.$$

After combining these results, we have

$$\oint_{C_1} \vec{\mathcal{H}} \cdot d\vec{\ell} + \oint_{C_2} \vec{\mathcal{H}} \cdot d\vec{\ell} = \iint_{S_1} \vec{\mathcal{J}} \cdot d\vec{S} + \iint_{S_1} \frac{\partial \vec{\mathcal{D}}}{\partial t} \cdot d\vec{S} + \iint_{S_2} \vec{\mathcal{J}} \cdot d\vec{S} + \iint_{S_2} \frac{\partial \vec{\mathcal{D}}}{\partial t} \cdot d\vec{S}.$$

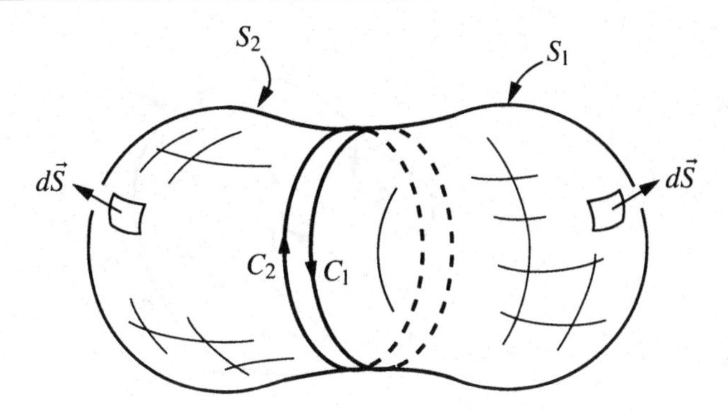

Fig. 1.14. Closed surface S composed of S_1 and S_2.

The two contour integrals on the left-hand side of this equation are equal except for their sense; therefore, they cancel leaving

$$\oiint_S \vec{\mathcal{J}} \cdot d\vec{S} + \oiint_S \frac{\partial \vec{\mathcal{D}}}{\partial t} \cdot d\vec{S} = 0. \tag{1.27}$$

The two results, (1.26) and (1.27), are the same, but they would not be if the displacement term had not been included in the Ampère-Maxwell law (1.2). An experimental demonstration of the displacement current is discussed in Problem 1.2.

1.2 Curl and divergence: Maxwell's equations in differential form

The integral form of Maxwell's equations, discussed in the previous section, applies to regions of space, either an open surface (Equations (1.1) and (1.2)), or a volume (Equations (1.3), (1.4), and (1.5)). These equations can be easily solved for the electromagnetic field, $\vec{\mathcal{E}}$ and $\vec{\mathcal{B}}$, only in cases where a high degree of physical symmetry exists. We saw examples of this when we analyzed the early experiments, which involved highly symmetric geometries, such as spherical pith balls; long, straight wires; etc.

The integral form of Maxwell's equations can be converted to a differential form, a system of partial differential equations, by the use of two vector operations – the curl and the divergence. The differential form of Maxwell's equations applies at a point in space rather than over a region of space. Systematic methods, using modern digital computers have been developed to solve these partial differential equations, even in cases where a high degree of physical symmetry is not present (we will discuss one of these methods in Section 1.6).

We will begin our derivation of the differential form of Maxwell's equations by considering the mathematical definitions for the curl and the divergence (see Appendix B and References [26] and [27]):

Curl The components of the vector field $\vec{A}$ are continuous in the neighborhood of the point P located by the position vector $\vec{r}$ shown in Figure 1.15a. The element of surface area ΔS on which P lies is bounded by the piecewise smooth, simple, closed curve C. At P the unit vector normal to the surface is $\hat{n}$. The circulation of

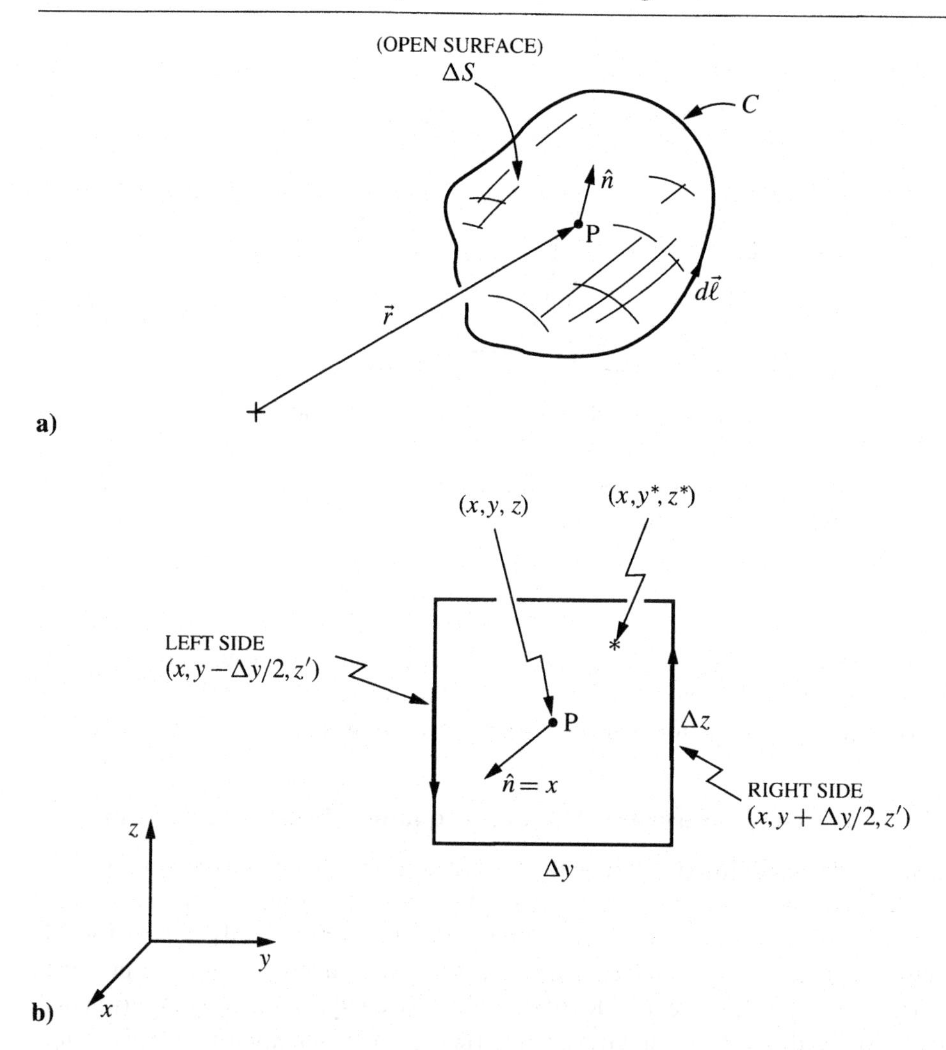

Fig. 1.15. Geometry used for defining the curl. a) Element of surface area ΔS bounded by curve C. b) Example for rectangular Cartesian coordinates.

the vector field $\vec{A}$ around the curve C is defined to be

$$\Gamma \equiv \oint_C \vec{A} \cdot d\vec{\ell}, \tag{1.28}$$

where $d\vec{\ell}$ is the vector differential length along C (an infinitesimal vector locally tangent to C). The sense in which the curve C is traversed relative to $\hat{n}$ is determined by the right-hand rule. The component in the direction $\hat{n}$ of the curl of the vector field at P is the limiting value of the circulation of $\vec{A}$ divided by the surface area ΔS, as the area approaches zero about P:

$$\hat{n} \cdot \operatorname{curl} \vec{A}(\vec{r}) \equiv \lim_{\Delta S \to 0} \left(\frac{\Gamma}{\Delta S} \right) = \lim_{\Delta S \to 0} \left(\frac{1}{\Delta S} \oint_C \vec{A} \cdot d\vec{\ell} \right). \tag{1.29}$$

This result is independent of the coordinate system.

Now we will consider the curl in a particular coordinate system – the rectangular Cartesian coordinate system (x, y, z). The component of curl $\vec{A}$ in the direction $\hat{x}$ is to be determined at the point P(x, y, z). The element of surface area is a rectangle with sides Δy and Δz and normal $\hat{x}$ centered at P. The geometry is shown in Figure 1.15b.

The line integral (1.29) now consists of four segments, one for each side of the rectangle. The integrals for the right and left sides,

$$I_R = \int_{z'=z-\Delta z/2}^{z+\Delta z/2} \vec{A}(x, y + \Delta y/2, z') \cdot \hat{z} dz' \tag{1.30a}$$

and

$$I_L = \int_{z'=z+\Delta z/2}^{z-\Delta z/2} \vec{A}(x, y - \Delta y/2, z') \cdot \hat{z} dz', \tag{1.30b}$$

combine to give

$$I_R + I_L = \int_{z'=z-\Delta z/2}^{z+\Delta z/2} \left[A_z(x, y + \Delta y/2, z') - A_z(x, y - \Delta y/2, z')\right] dz'. \tag{1.31}$$

If A_z is continuous over the interval $z - \Delta z/2 \leq z' \leq z + \Delta z/2$, the *mean value theorem for integrals* can be used to write (1.31) as[6]

$$I_R + I_L = \left[A_z(x, y + \Delta y/2, z^*) - A_z(x, y - \Delta y/2, z^*)\right] \Delta z, \tag{1.32}$$

where $z - \Delta z/2 < z^* < z + \Delta z/2$. Now if A_z is also continuous for the interval $y - \Delta y/2 \leq y' \leq y + \Delta y/2$ and has a first partial derivative $\partial A_z/\partial y$ for the interval $y - \Delta y/2 < y' < y + \Delta y/2$, the *mean value theorem* can be used to write (1.32) as

$$I_R + I_L = \left.\frac{\partial A_z}{\partial y}\right|_{(x, y^*, z^*)} \Delta y \Delta z, \tag{1.33}$$

where $y - \Delta y/2 < y^* < y + \Delta y/2$.

The line integrals over the other pair of sides (top and bottom) can be handled in a similar manner:

$$I_T + I_B = \left.\frac{-\partial A_y}{\partial z}\right|_{(x, y^{**}, z^{**})} \Delta y \Delta z. \tag{1.34}$$

[6] The mean value theorem for integrals states that if $f(\xi)$ is continuous for $a \leq \xi \leq b$, then

$$\int_a^b f(\xi) d\xi = f(\xi^*)(b - a), \qquad a < \xi^* < b.$$

The mean value theorem states that if $g(\xi)$ is continuous for $a \leq \xi \leq b$ and $g'(\xi)$ exists for $a < \xi < b$, then

$$g(b) - g(a) = g'(\xi^*)(b - a), \qquad a < \xi^* < b.$$

The relationship between these two theorems is seen on substituting $f(\xi) = g'(\xi)$ [28].

The component of the curl in the direction $\hat{x}$ is then

$$\begin{aligned}\hat{x}\cdot\operatorname{curl}\vec{A}(x,y,z) &= \lim_{\substack{\Delta y\to 0\\ \Delta z\to 0}}\left(\frac{I_R+I_L+I_T+I_B}{\Delta y\Delta z}\right)\\ &= \lim_{\substack{\Delta y\to 0\\ \Delta z\to 0}}\left(\frac{1}{\Delta y\Delta z}\left\{\left[\left.\frac{\partial A_z}{\partial y}\right|_{(x,y^*,z^*)} - \left.\frac{\partial A_y}{\partial z}\right|_{(x,y^{**},z^{**})}\right]\Delta y\Delta z\right\}\right).\end{aligned} \tag{1.35}$$

Assuming that the partial derivatives are continuous, in the limit $\Delta y \to 0$, we have $y^* \to y$, $y^{**} \to y$, etc., and

$$\hat{x}\cdot\operatorname{curl}\vec{A}(x,y,z) = \frac{\partial A_z}{\partial y} - \frac{\partial A_y}{\partial z}. \tag{1.36}$$

This is the x component of the vector curl $\vec{A}$; the y and z components, determined in a similar manner, are

$$\hat{y}\cdot\operatorname{curl}\vec{A}(x,y,z) = \frac{\partial A_x}{\partial z} - \frac{\partial A_z}{\partial x} \tag{1.37}$$

and

$$\hat{z}\cdot\operatorname{curl}\vec{A}(x,y,z) = \frac{\partial A_y}{\partial x} - \frac{\partial A_x}{\partial y}, \tag{1.38}$$

which makes

$$\begin{aligned}\operatorname{curl}\vec{A}(x,y,z) &= \hat{x}\left(\frac{\partial A_z}{\partial y} - \frac{\partial A_y}{\partial z}\right) + \hat{y}\left(\frac{\partial A_x}{\partial z} - \frac{\partial A_z}{\partial x}\right) + \hat{z}\left(\frac{\partial A_y}{\partial x} - \frac{\partial A_x}{\partial y}\right)\\ &= \sum_{i=x,y,z}\hat{i}\times\frac{\partial\vec{A}}{\partial i}.\end{aligned} \tag{1.39}$$

When the notation

$$\nabla\times\vec{A} = \sum_{i=x,y,z}\hat{i}\times\frac{\partial\vec{A}}{\partial i} \tag{1.40a}$$

is introduced, we have

$$\operatorname{curl}\vec{A}(x,y,z) = \nabla\times\vec{A}(x,y,z). \tag{1.40b}$$

Divergence The components of the vector field $\vec{A}$ are continuous in the neighborhood of the point P located by the position vector $\vec{r}$ shown in Figure 1.16a. The element of volume ΔV containing P has the piecewise smooth surface S with outward-pointing, unit normal vector $\hat{n}$. The divergence of the vector field at P is the limiting value of the net outward flux of $\vec{A}$ through the surface S divided by the volume ΔV, as the volume approaches zero about P:

$$\operatorname{div}\vec{A}(\vec{r}) \equiv \lim_{\Delta V\to 0}\left(\frac{1}{\Delta V}\oiint_S \hat{n}\cdot\vec{A}\,dS\right). \tag{1.41}$$

This result is independent of the coordinate system.

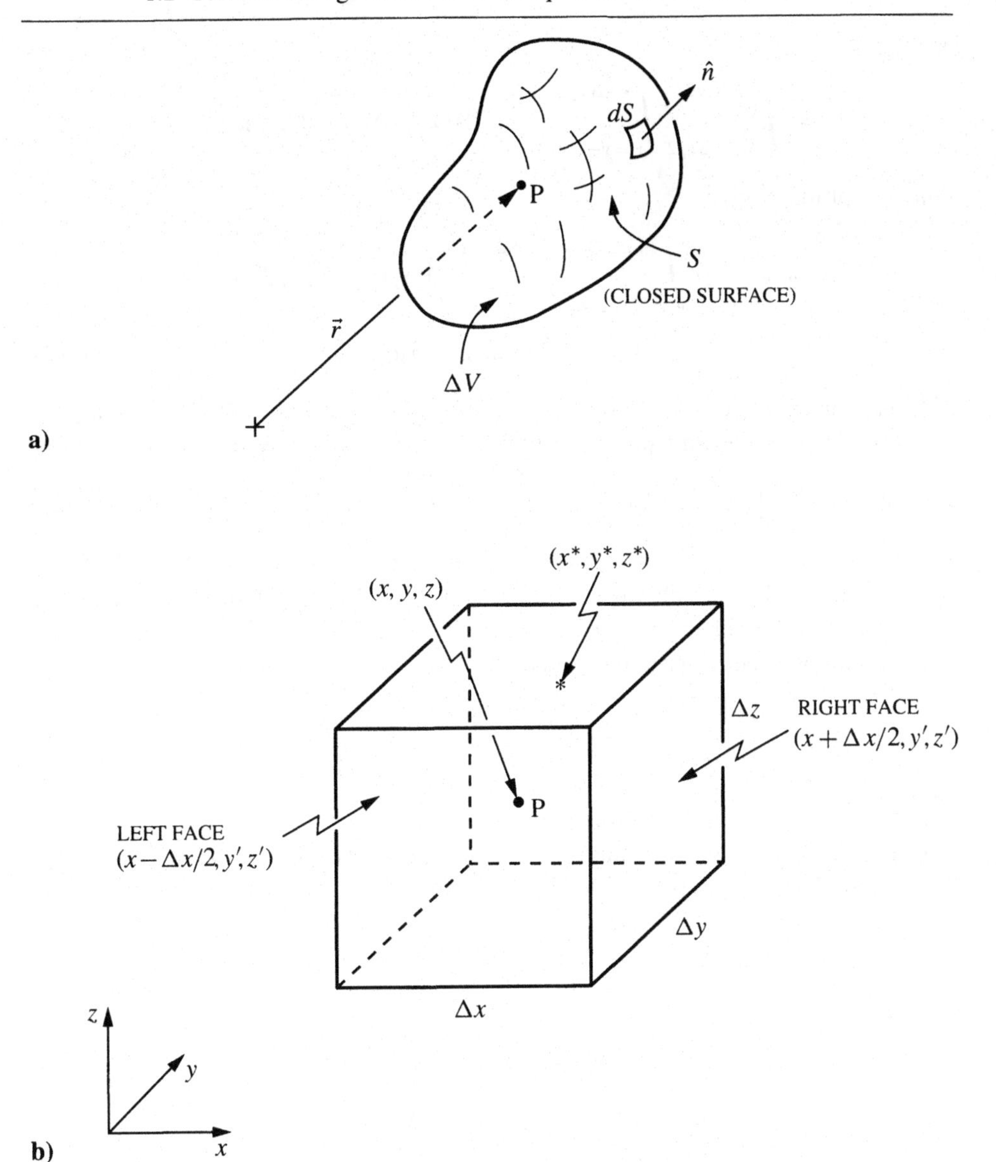

Fig. 1.16. Geometry used for defining the divergence and the gradient. a) Element of volume ΔV with surface S. b) Example for rectangular Cartesian coordinates.

Now we will consider the divergence in the rectangular Cartesian coordinate system (x, y, z). The volume element ΔV is a rectangular box with sides Δx, Δy, and Δz centered at P(x, y, z) (Figure 1.16b). The surface integral (1.41) consists of six integrals, one over each face of the box. For the right and left faces, the integrals are

$$I_R = \int_{z'=z-\Delta z/2}^{z+\Delta z/2} \int_{y'=y-\Delta y/2}^{y+\Delta y/2} \hat{x} \cdot \vec{A}(x + \Delta x/2, y', z') dy' dz'$$

and

$$I_L = \int_{z'=z-\Delta z/2}^{z+\Delta z/2} \int_{y'=y-\Delta y/2}^{y+\Delta y/2} -\hat{x} \cdot \vec{A}(x - \Delta x/2, y', z') dy' dz',$$

which combine to give

$$I_R + I_L = \int_{z'=z-\Delta z/2}^{z+\Delta z/2} \int_{y'=y-\Delta y/2}^{y+\Delta y/2} \big[A_x(x + \Delta x/2, y', z') - A_x(x - \Delta x/2, y', z')\big] dy' dz'. \tag{1.42}$$

If A_x is continuous for the intervals $y - \Delta y/2 \leq y' \leq y + \Delta y/2$ and $z - \Delta z/2 \leq z' \leq z + \Delta z/2$, the mean value theorem for integrals can be used to write (1.42) as

$$I_R + I_L = \big[A_x(x + \Delta x/2, y^*, z^*) - A_x(x - \Delta x/2, y^*, z^*)\big] \Delta y \Delta z, \tag{1.43}$$

where $y - \Delta y/2 < y^* < y + \Delta y/2$ and $z - \Delta z/2 < z^* < z + \Delta z/2$. If A_x is also continuous for the interval $x - \Delta x/2 \leq x' \leq x + \Delta x/2$ and has a first partial derivative $\partial A_x/\partial x$ for the interval $x - \Delta x/2 < x' < x + \Delta x/2$, the mean value theorem can be used to write (1.43) as

$$I_R + I_L = \left.\frac{\partial A_x}{\partial x}\right|_{(x^*, y^*, z^*)} \Delta x \Delta y \Delta z, \tag{1.44}$$

where $x - \Delta x/2 < x^* < x + \Delta x/2$.

The surface integrals over the other pairs of faces (front and back: I_F, I_B; top and bottom: I_T, I_V) can be handled in a similar manner:

$$I_F + I_B = \left.\frac{\partial A_y}{\partial y}\right|_{(x^{**}, y^{**}, z^{**})} \Delta x \Delta y \Delta z \tag{1.45}$$

$$I_T + I_V = \left.\frac{\partial A_z}{\partial z}\right|_{(x^{***}, y^{***}, z^{***})} \Delta x \Delta y \Delta z. \tag{1.46}$$

The divergence is then

$$\operatorname{div} \vec{A}(x, y, z,) = \lim_{\substack{\Delta x \to 0 \\ \Delta y \to 0 \\ \Delta z \to 0}} \left(\frac{I_R + I_L + I_F + I_B + I_T + I_V}{\Delta x \Delta y \Delta z} \right)$$

$$= \lim_{\substack{\Delta x \to 0 \\ \Delta y \to 0 \\ \Delta z \to 0}} \left(\frac{1}{\Delta x \Delta y \Delta z} \left\{ \left[\left.\frac{\partial A_x}{\partial x}\right|_{(x^*, y^*, z^*)} + \left.\frac{\partial A_y}{\partial y}\right|_{(x^{**}, y^{**}, z^{**})} + \left.\frac{\partial A_z}{\partial z}\right|_{(x^{***}, y^{***}, z^{***})} \right] \Delta x \Delta y \Delta z \right\} \right).$$

Assuming that the partial derivatives are continuous, in the limit $\Delta x \to 0$, we have

$x^* \to x, x^{**} \to x$, etc., and

$$\operatorname{div} \vec{A}(x, y, z) = \frac{\partial A_x}{\partial x} + \frac{\partial A_y}{\partial y} + \frac{\partial A_z}{\partial z} = \sum_{i=x,y,z} \hat{i} \cdot \frac{\partial \vec{A}}{\partial i}, \tag{1.47}$$

or, when the notation

$$\nabla \cdot \vec{A} = \sum_{i=x,y,z} \hat{i} \cdot \frac{\partial \vec{A}}{\partial i} \tag{1.48a}$$

is introduced, we have

$$\operatorname{div} \vec{A}(x, y, z) = \nabla \cdot \vec{A}(x, y, z). \tag{1.48b}$$

Note that the integral definitions we started with for the curl and divergence, (1.29) and (1.41), are more general than the differential forms that we subsequently obtained, (1.39) and (1.47). The use of the former requires the vector field to be continuous, whereas the use of the latter requires both the vector field and its first partial derivatives to be continuous.

We next apply the definitions of the curl and the divergence to convert Maxwell's equations in integral form to their differential form. Where required, quantities and their derivatives will be assumed continuous. First, we will consider Faraday's law (1.1) applied to the surface ΔS in Figure 1.15a:

$$\frac{1}{\Delta S} \oint_C \vec{\mathcal{E}} \cdot d\vec{\ell} = -\frac{1}{\Delta S} \iint_{\Delta S} \frac{\partial \vec{\mathcal{B}}}{\partial t} \cdot d\vec{S}.$$

After taking the limit $\Delta S \to 0$ and introducing the definition (1.29) for the curl, this equation becomes

$$\lim_{\Delta S \to 0} \left[\frac{1}{\Delta S} \oint_C \vec{\mathcal{E}} \cdot d\vec{\ell} \right] = \hat{n} \cdot \operatorname{curl} \vec{\mathcal{E}} = \lim_{\Delta S \to 0} \left[-\frac{1}{\Delta S} \iint_{\Delta S} \frac{\partial \vec{\mathcal{B}}}{\partial t} \cdot d\vec{S} \right] = -\hat{n} \cdot \frac{\partial \vec{\mathcal{B}}}{\partial t},$$

or simply

$$\hat{n} \cdot \operatorname{curl} \vec{\mathcal{E}} = -\hat{n} \cdot \frac{\partial \vec{\mathcal{B}}}{\partial t}. \tag{1.49}$$

The orientation for the surface ΔS (or the unit vector $\hat{n}$) is arbitrary in the definition for the curl; thus, (1.49) implies that

$$\operatorname{curl} \vec{\mathcal{E}}(\vec{r}, t) = -\frac{\partial \vec{\mathcal{B}}(\vec{r}, t)}{\partial t},$$

or using (1.40b) we obtain

$$\nabla \times \vec{\mathcal{E}}(\vec{r}, t) = -\frac{\partial \vec{\mathcal{B}}(\vec{r}, t)}{\partial t}. \tag{1.50}$$

This is the differential form of Faraday's law; the differential form for the Ampère-Maxwell law (1.2), obtained in a similar manner, is

$$\operatorname{curl} \vec{\mathcal{H}}(\vec{r}, t) = \vec{\mathcal{J}}(\vec{r}, t) + \frac{\partial \vec{\mathcal{D}}(\vec{r}, t)}{\partial t},$$

or

$$\nabla \times \vec{\mathcal{H}}(\vec{r}, t) = \vec{\mathcal{J}}(\vec{r}, t) + \frac{\partial \vec{\mathcal{D}}(\vec{r}, t)}{\partial t}. \tag{1.51}$$

When Gauss' electric law (1.3) is applied to the volume ΔV in Figure 1.16a, we have

$$\frac{1}{\Delta V} \oiint_S \vec{\mathcal{D}} \cdot d\vec{S} = \frac{1}{\Delta V} \iiint_{\Delta V} \rho dV.$$

After taking the limit $\Delta V \to 0$ and introducing the definition for the divergence (1.41), this equation becomes

$$\lim_{\Delta V \to 0} \left[\frac{1}{\Delta V} \oiint_S \vec{\mathcal{D}} \cdot d\vec{S} \right] = \operatorname{div} \vec{\mathcal{D}} = \lim_{\Delta V \to 0} \left[\frac{1}{\Delta V} \iiint_{\Delta V} \rho dV \right] = \rho,$$

or

$$\operatorname{div} \vec{\mathcal{D}}(\vec{r}, t) = \rho(\vec{r}, t),$$

which using (1.48b) becomes

$$\nabla \cdot \vec{\mathcal{D}}(\vec{r}, t) = \rho(\vec{r}, t). \tag{1.52}$$

This is the differential form of Gauss' electric law. The differential forms for Gauss' magnetic law (1.4) and for the equation of continuity for electric charge (1.5) are obtained in a similar manner:

$$\operatorname{div} \vec{\mathcal{B}}(\vec{r}, t) = 0,$$

or

$$\nabla \cdot \vec{\mathcal{B}}(\vec{r}, t) = 0; \tag{1.53}$$

and

$$\operatorname{div} \vec{\mathcal{J}}(\vec{r}, t) = -\frac{\partial \rho(\vec{r}, t)}{\partial t},$$

or

$$\nabla \cdot \vec{\mathcal{J}}(\vec{r}, t) = -\frac{\partial \rho(\vec{r}, t)}{\partial t}. \tag{1.54}$$

Equations (1.50)–(1.54) are the differential form of Maxwell's equations that we sought. These equations, together with the corresponding integral forms, are summarized in Table 1.3.

Let us examine these partial differential equations more closely [6, 29]. We will assume that the densities of charge and current, ρ and $\vec{\mathcal{J}}$, are known and that they satisfy the equation of continuity (1.54). The field vectors $\vec{\mathcal{E}}$, $\vec{\mathcal{B}}$, $\vec{\mathcal{D}}$, and $\vec{\mathcal{H}}$ are to be determined. For free space we can use Equations (1.6) and (1.7) to eliminate $\vec{\mathcal{D}}$ and $\vec{\mathcal{H}}$, leaving us with $\vec{\mathcal{E}}$ and $\vec{\mathcal{B}}$ to be determined. Each of the vectors $\vec{\mathcal{E}}$ and $\vec{\mathcal{B}}$ is composed of three scalar functions, one for each of the vector components. Thus, we have six scalar unknowns: $\mathcal{E}_x$, $\mathcal{E}_y$, $\mathcal{E}_z$; $\mathcal{B}_x$, $\mathcal{B}_y$, $\mathcal{B}_z$. Each of the curl equations, (1.50) and (1.51), is composed of three scalar equations; the divergence equations, (1.52) and (1.53), provide two additional scalar equations. Thus, with a total of eight scalar equations, we have more equations than unknowns.

Table 1.3. *The equations of electromagnetism*†

Name	Equation	
	Integral form	Differential form
Faraday's law	$\oint_C \vec{\mathcal{E}} \cdot d\vec{\ell} = -\iint_S \frac{\partial \vec{\mathcal{B}}}{\partial t} \cdot d\vec{S}$	$\nabla \times \vec{\mathcal{E}} = -\frac{\partial \vec{\mathcal{B}}}{\partial t}$
Ampère-Maxwell law	$\oint_C \vec{\mathcal{H}} \cdot d\vec{\ell} = \iint_S \vec{\mathcal{J}} \cdot d\vec{S} + \iint_S \frac{\partial \vec{\mathcal{D}}}{\partial t} \cdot d\vec{S}$	$\nabla \times \vec{\mathcal{H}} = \vec{\mathcal{J}} + \frac{\partial \vec{\mathcal{D}}}{\partial t}$
Gauss' electric law	$\oiint_S \vec{\mathcal{D}} \cdot d\vec{S} = \iiint_V \rho \, dV$	$\nabla \cdot \vec{\mathcal{D}} = \rho$
Gauss' magnetic law	$\oiint_S \vec{\mathcal{B}} \cdot d\vec{S} = 0$	$\nabla \cdot \vec{\mathcal{B}} = 0$
Equation of continuity for electric charge	$\oiint_S \vec{\mathcal{J}} \cdot d\vec{S} = -\iiint_V \frac{\partial \rho}{\partial t} dV$	$\nabla \cdot \vec{\mathcal{J}} = -\frac{\partial \rho}{\partial t}$
Lorentz force expression	$\vec{\mathcal{F}} = \iiint_V (\rho \vec{\mathcal{E}} + \vec{\mathcal{J}} \times \vec{\mathcal{B}}) dV$	
Poynting's theorem	$\iiint_V \vec{\mathcal{E}} \cdot \vec{\mathcal{J}} \, dV + \iiint_V \left(\vec{\mathcal{E}} \cdot \frac{\partial \vec{\mathcal{D}}}{\partial t} + \vec{\mathcal{H}} \cdot \frac{\partial \vec{\mathcal{B}}}{\partial t} \right) dV + \oiint_S \vec{\mathcal{S}} \cdot d\vec{S} = 0$	
Poynting vector	$\vec{\mathcal{S}} = \vec{\mathcal{E}} \times \vec{\mathcal{H}}$	

Note: †All quantities are evaluated in the rest frame of the observer, and all surfaces and volumes are held fixed in that frame.

This dilemma is the result of the five equations, (1.50)–(1.54), not being independent. To show this, consider the divergence of Equation (1.51) and recall that $\nabla \cdot (\nabla \times \vec{A}) = 0$ for any vector field $\vec{A}$ (Appendix B):

$$\nabla \cdot (\nabla \times \vec{\mathcal{H}}) = 0 = \nabla \cdot \vec{\mathcal{J}} + \nabla \cdot \frac{\partial \vec{\mathcal{D}}}{\partial t}.$$

After interchanging the order of differentiation with respect to the temporal and the spatial variables and substituting (1.54), we have

$$\frac{\partial}{\partial t}[\nabla \cdot \vec{\mathcal{D}} - \rho] = 0.$$

Now integrating this equation from an arbitrary earlier time t_o to t, we obtain

$$\begin{aligned} \nabla \cdot \vec{\mathcal{D}}(\vec{r}, t) - \rho(\vec{r}, t) &= \nabla \cdot \vec{\mathcal{D}}(\vec{r}, t_o) - \rho(\vec{r}, t_o) \\ &= \mathcal{C}(\vec{r}, t_o). \end{aligned}$$

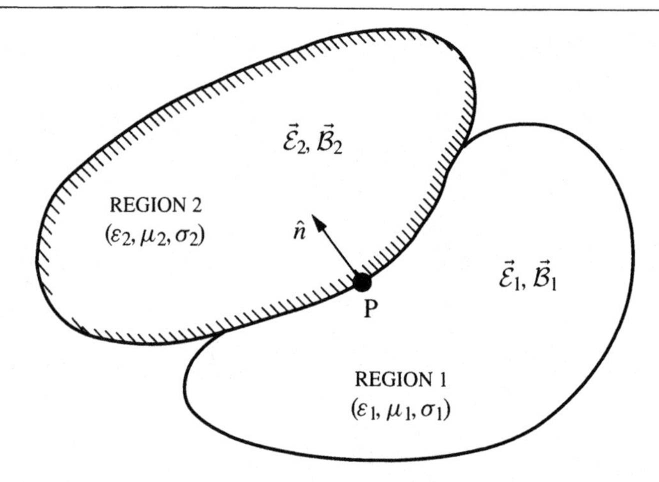

Fig. 1.17. Electromagnetic problem involving material regions with different electromagnetic properties. Illustration is for simple materials.

If the constant of integration $C(\vec{r}, t_o)$, which is determined at t_o, is chosen to be zero, then

$$\nabla \cdot \vec{\mathcal{D}}(\vec{r}, t) = \rho(\vec{r}, t),$$

which is Equation (1.52).[7]

We have shown that Gauss' electric law (1.52) is contained in the Ampère-Maxwell law (1.51) and the equation of continuity (1.54); in a similar way we can show that Gauss' magnetic law (1.53) is contained in Faraday's law (1.50). Thus, for time-varying electromagnetic fields, we really have two independent vector equations, (1.50) and (1.51), (six scalar equations) involving two vector unknowns, $\vec{\mathcal{E}}$ and $\vec{\mathcal{B}}$ (six scalar unknowns). The number of scalar equations equals the number of scalar unknowns.

Equations (1.50), (1.51), and (1.54) are sometimes referred to as the *independent Maxwell's equations* and (1.52) and (1.53) as the *dependent Maxwell's equations* [29].

1.3 Surface densities, boundary conditions, and perfect conductors

A typical problem in electromagnetics involves several material regions, each with different electromagnetic properties. For example, the regions may be simple materials with the electromagnetic constitutive parameters (1.9)–(1.11), as shown in Figure 1.17. Maxwell's equations in differential form (1.50)–(1.54) are solved in each material region to determine the electromagnetic field $\vec{\mathcal{E}}$, $\vec{\mathcal{B}}$. The field is assumed to be continuous and to have continuous first partial derivatives in each region.

[7] For a time-varying field, we can consider t_o to be a time in the distant past when both ρ and $\nabla \cdot \vec{\mathcal{D}}$ were zero throughout all space, making $C(\vec{r}, t_o) = 0$.

At the common surface boundary between two material regions, the electromagnetic properties of the materials often will be discontinuous: $\varepsilon_2 \neq \varepsilon_1$, $\mu_2 \neq \mu_1$, and $\sigma_2 \neq \sigma_1$. As a result, the components of the electromagnetic field may also be discontinuous at the boundary; for example, at the point P in Figure 1.17 we may have $\hat{n} \cdot \vec{\mathcal{E}}_1 \neq \hat{n} \cdot \vec{\mathcal{E}}_2$.

In this section we will establish relationships that the electromagnetic field must satisfy at a material boundary – the *electromagnetic boundary conditions*. We will apply Maxwell's equations in integral form (1.1)–(1.5) at the boundary to obtain these relationships. Here we are assuming that the integral form of the equations holds within regions where the field is discontinuous. Stated differently, we assume that the experimental evidence on which the integral form of Maxwell's equations is based shows that these equations hold when the field is discontinuous. Recall that the differential form for the equations was obtained from the integral form by assuming continuity for the field and its first partial derivatives. Thus, it would make no sense to apply the differential form of the equations at a boundary where we expect the components of the field to be discontinuous.[8]

1.3.1 Surface densities of charge and current

In Maxwell's equations, the state of the electric charge is given by the volume densities of charge and current, ρ and $\vec{\mathcal{J}}$. These densities are usually thought of as smooth functions of the position throughout a material body. However, it has been recognized since early times that, in certain situations, these densities may be significant only near the surface of a body. As an example, we cite the early electrostatic experiments of Stephen Gray (1666–1736) [31]. Gray constructed two cubes of oak, one solid and the other hollow. The two cubes were charged in a similar manner. A brass leaf was brought near the cubes, and its deflection was noted (a measure of the electric field). Gray found no difference in the deflections for the solid and the hollow cubes, and he concluded that only the surface of the cubes was involved in the attraction/repulsion (the electric charge was concentrated in a layer at the surface of both cubes).

More modern results also indicate that the charge on a body may be located in a very thin layer at the surface. Quantum-mechanical calculations can be used to estimate the charge distribution near the surface of a plane metal sheet (Figure 1.18a) [32–34]. The metal is viewed as a lattice of positive ions surrounded by a distribution of electrons. In the theoretical model, the ions are represented by a uniform positive charge density ρ_o^+ (solid line in Figure 18b), and the negative charge density ρ_o^- (dashed line in Figure 18b) for the electrons is computed [33]. The results in Figure 1.18b are for a neutral sheet with no applied electric field;

[8] The following questionable argument is sometimes used to obtain the boundary conditions: Maxwell's equations in differential form, which require continuity for the field and its first partial derivatives, are assumed to be obtained from experiment. Next the integral form of the equations is obtained from the differential form. The integral form is then used to obtain the boundary conditions, which show the discontinuity in the field. This point is discussed in References [29] and [30].

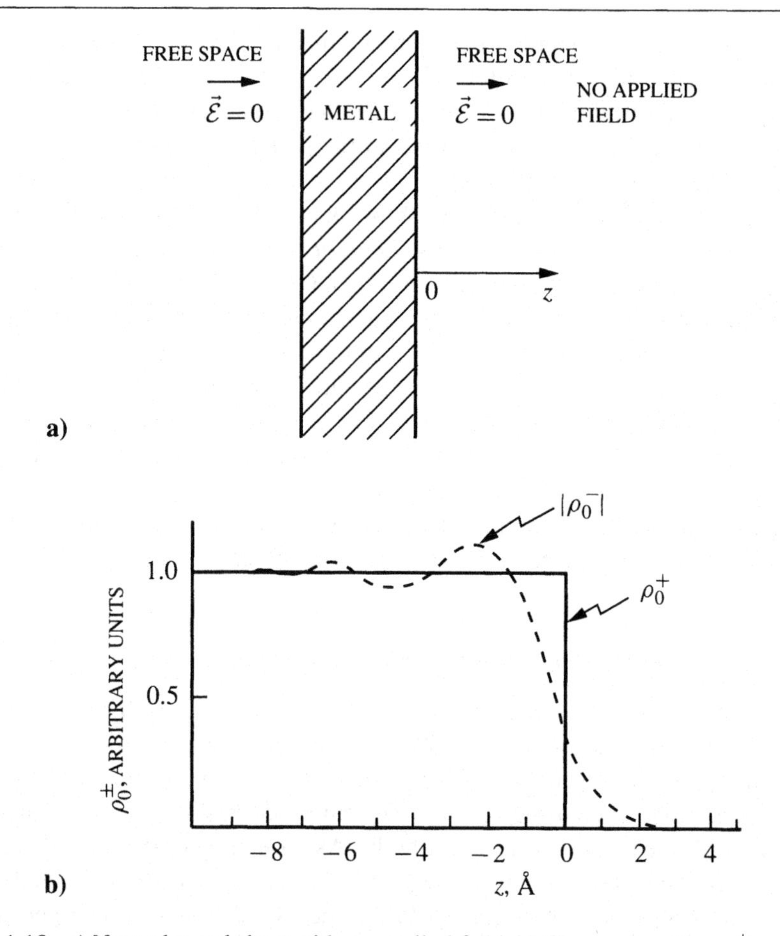

Fig. 1.18. a) Neutral metal sheet with no applied field. b) Charge densities ρ_o^+ and ρ_o^- near right-hand surface of sheet. (Densities are based on quantum-mechanical calculations from Lang and Kohn [33].)

hence, the net charge in the vicinity of each surface must be zero; for example, for the right-hand surface

$$\int_{z=-L}^{L} \rho_o(z)dz = \int_{z=-L}^{L} \left[\rho_o^+(z) + \rho_o^-(z)\right]dz = 0,$$

where L is large compared to the lattice spacing of the ions. The charge density ρ_o^- is seen to be nearly uniform in the metal away from the surface, to oscillate as the surface is approached, then to rapidly decay to zero.

When the metal sheet is placed in a uniform electrostatic field normal to its surface, as in Figure 1.19a, the uniform positive charge density is assumed to remain unchanged, $\rho^+(z) = \rho_o^+(z)$, and the electrons are allowed to redistribute to produce the new negative charge density $\rho^-(z)$ [34]. There is now an excess of positive charge at the right-hand surface of the sheet and an excess of negative charge at the left-hand surface of the sheet. The excess charge produces an electric field that cancels the applied field in the interior of the metal sheet.

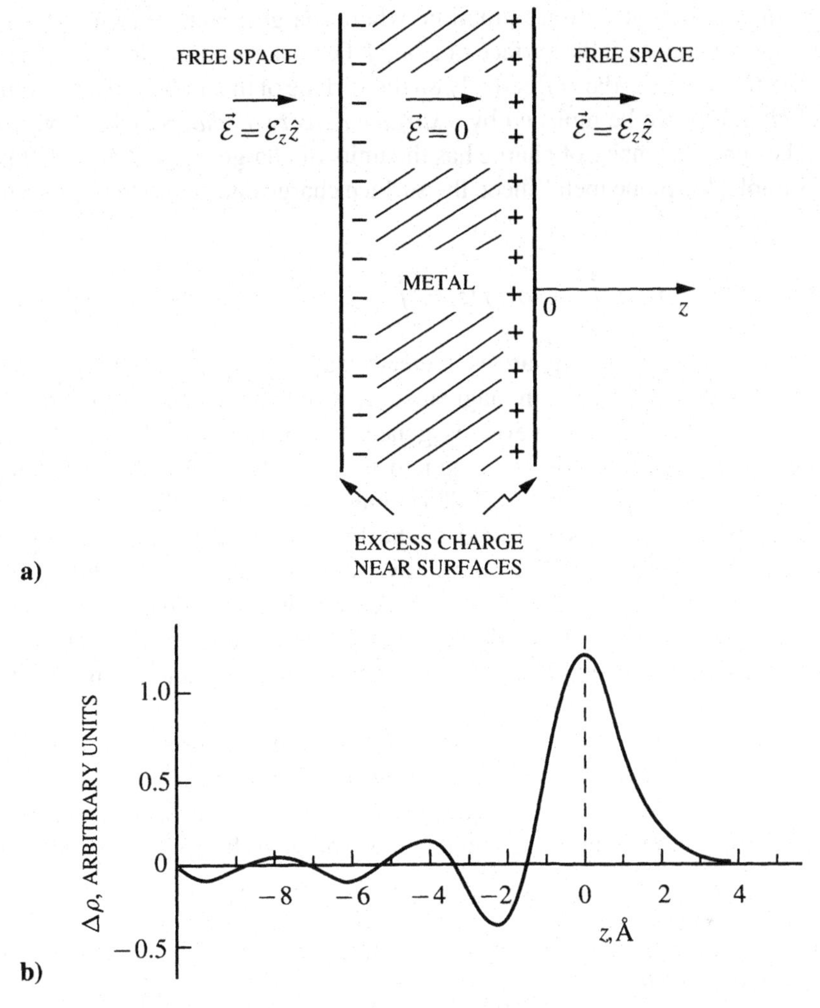

Fig. 1.19. a) Metal sheet placed in a uniform electrostatic field. b) Excess charge density $\Delta\rho$ near right-hand surface of sheet. (Based on quantum-mechanical calculations from Lang and Kohn [34].)

With the excess charge density $\Delta\rho(z)$ defined to be the difference in the charge densities with and without the applied field, we have

$$\Delta\rho(z) = \rho(z) - \rho_o(z) = \rho^+(z) + \rho^-(z) - \left[\rho_o^+(z) + \rho_o^-(z)\right] = \rho^-(z) - \rho_o^-(z),$$

and, for the right-hand surface,

$$\int_{z=-L}^{L} \rho(z)dz = \int_{z=-L}^{L} \Delta\rho(z)dz = \varepsilon_o\mathcal{E}_z,$$

where $\mathcal{E}_z\hat{z}$ is the electric field outside the sheet.

Figure 1.19b shows the excess charge density $\Delta\rho(z)$ for a sheet of metal like sodium placed in a weak external electric field [34]. The excess charge is seen to be located within a very small depth near the surface: -10 Å $\lesssim z \lesssim 10$ Å (1 Å $= 10^{-10}$ m).

These results show that there are situations where charge is confined within a layer of microscopic thickness at the surface of a body. For macroscopic electromagnetic calculations, the charge is then effectively on the surface of the body, and the volume density of charge ρ can be replaced by a *surface density of charge*, which we will call ρ_s.[9] The surface density of charge has the units of charge per unit area (C/m^2). For our example, the plane metal sheet, the surface charge density on the right-hand surface is simply

$$\rho_s = \int_{z=-L}^{L} \rho(z)dz = \int_{z=-L}^{L} \Delta\rho(z)dz.$$

Since a surface charge density arises in electrostatic problems, we might expect a surface current density to arise in magnetostatic problems (problems that involve only steady currents). This, however, is generally not the case. In Figure 1.20 we compare similar electrostatic and magnetostatic problems. A static line charge parallel to a metal half space (Figure 1.20a) induces a charge at the surface of the metal. The combined electrostatic field of the line charge and surface charge is zero within the metal. A steady line current over a metal half space (nonmagnetic, $\mu = \mu_o$), as shown in Figure 1.20b, induces no current within the metal. In fact, the magnetic field produced by the steady line current is the same as when the metal half space is absent. The metal is a shield for the electrostatic field but not for the magnetostatic field.

We must consider time-varying electromagnetic fields to observe current concentrated near the surface of a metal. If we let the line current of Figure 1.21a oscillate with a very high angular frequency $\omega = 2\pi f$, so that $\mathcal{I}(t) = I_0 \cos\omega t$, the volume density of current in the metal can be shown to be approximately [35]

$$\vec{\mathcal{J}}(y, z, t) = \hat{x}A(y)e^{-z/\delta}\cos\left[\omega t - z/\delta - \psi(y)\right].$$

The parameter δ is called the *skin depth*; it is simply related to the electrical constitutive parameters of the metal $(\varepsilon, \mu, \sigma)$ by

$$\delta = \sqrt{\frac{2}{\omega\mu\sigma}}. \tag{1.55}$$

Note that the current for this example is independent of the coordinate x. The terms $A(y)$ and $\psi(y)$ are the amplitude and phase of the current at the surface of the metal.

The behavior of the current is made clear when it is evaluated at the time $t = \psi(y)/\omega$ and normalized by dividing by the value at the surface $(z = 0)$:

$$\mathcal{J}_{xn}(z) = \frac{\mathcal{J}_x(y, z, t = \psi/\omega)}{\mathcal{J}_x(y, 0, t = \psi/\omega)} = e^{-z/\delta}\cos(z/\delta).$$

[9] We use the term *microscopic* to refer to observations made on a scale comparable to atomic dimensions. The contrasting term *macroscopic* is used to refer to observations made on a scale much larger than atomic dimensions. For example, the probe used in a microscopic measurement is capable of resolving the field between atoms, molecules, etc., whereas, the probe used in a macroscopic measurement is enormous compared to atomic dimensions and measures a spatial average of the microscopic field.

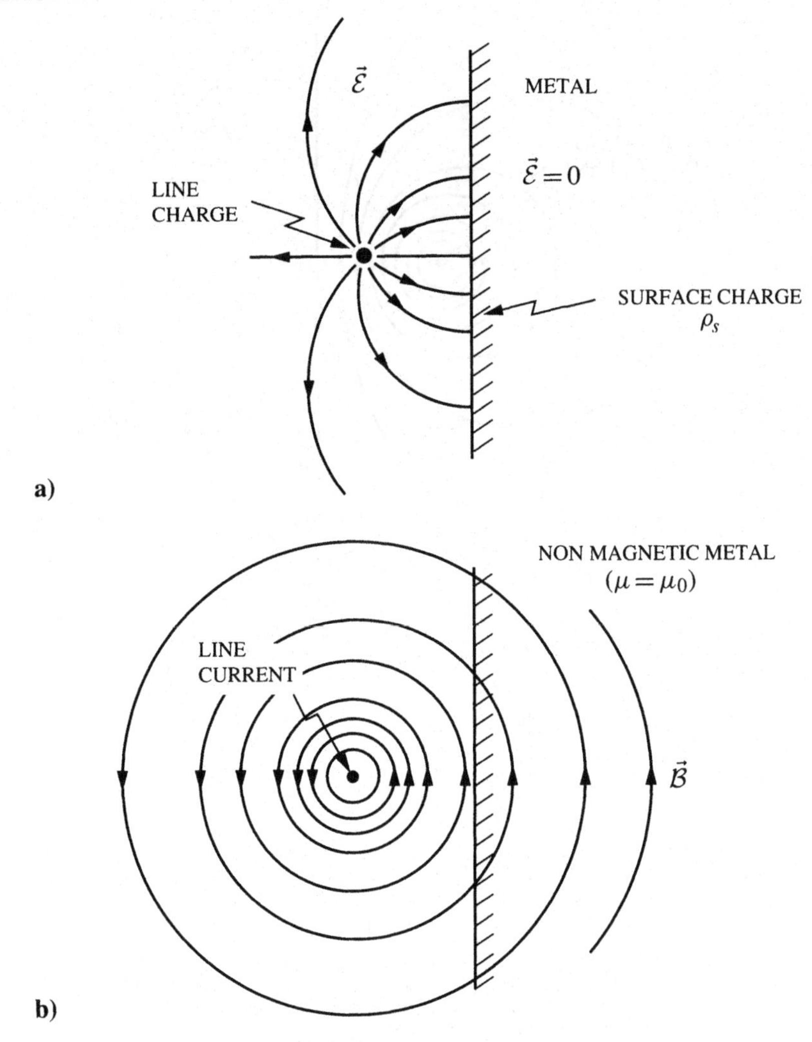

Fig. 1.20. a) Electrostatic problem: line charge over a half space of metal. b) Magnetostatic problem: line current over a half space of nonmagnetic metal ($\mu = \mu_0$).

The current in the metal (Figure 1.21b) is seen to decay exponentially from its value at the surface. The characteristic length for the decay is the skin depth δ. A detailed analysis shows that the electromagnetic field also decays with the same characteristic length; thus, the electromagnetic field ($\vec{\mathcal{E}}$, $\vec{\mathcal{B}}$) is essentially zero in the metal at a distance greater than a few skin depths from the surface. The metal is now a shield for the time-varying electromagnetic field; compare $\vec{\mathcal{B}}$ in Figures 1.20b and 1.21a.

The phenomenon we have described is known as the *skin effect approximation* [35, 36]. It is an approximation that holds when $\sigma/\omega\varepsilon \gg 1$ and the skin depth is small compared to all physical dimensions of the problem.[10] For our example

[10] For a metallic conductor of general shape, the skin depth must be small compared to the dimensions of the conductor and the radii of curvature of its surface. Also, the electromagnetic field must vary negligibly along the surface over distances comparable to a few skin depths [35].

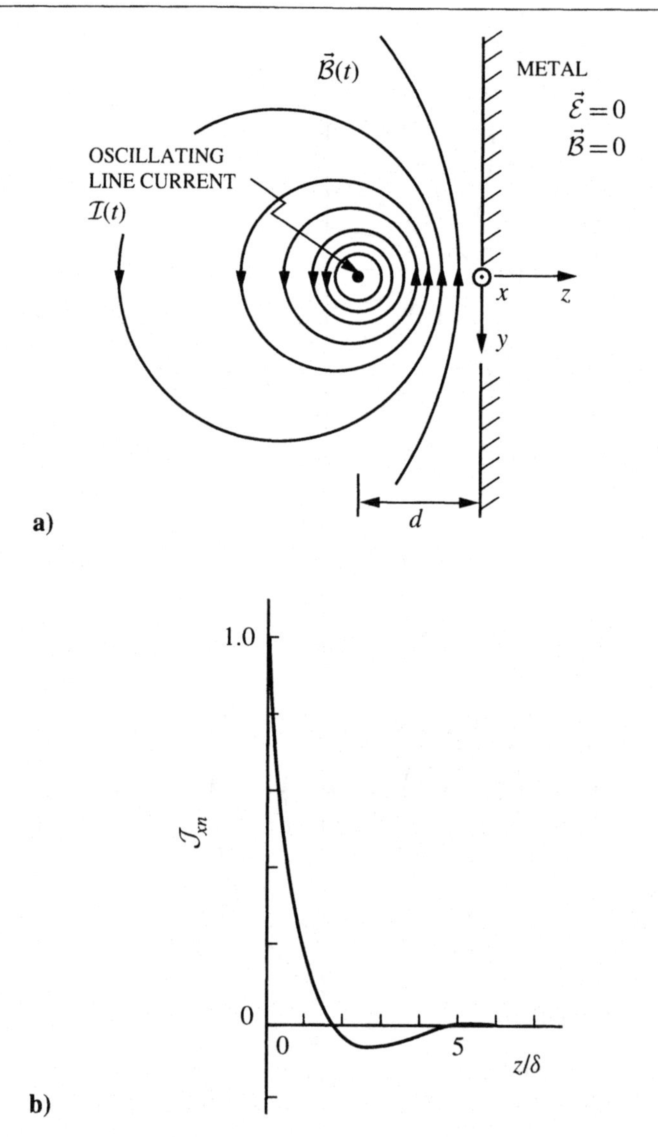

Fig. 1.21. a) Electrodynamic problem: rapidly oscillating line current over a half space of metal. b) Normalized volume density of current near surface of metal. Results are for $\delta^2 \ll d^2$, $k_o d = \omega d/c \ll 1$, where δ is the skin depth.

(Figure 1.21) the skin depth δ must be small compared to the spacing d between the line current and the metal half space ($\delta^2 \ll d^2$). For good conductors like copper ($\varepsilon = \varepsilon_o$, $\mu = \mu_o$, $\sigma = 5.8 \times 10^7$ S/m), the approximation is very useful, because the skin depth can be extremely small at practical frequencies; for example, at the radio frequency $f = 100$ kHz, $\sigma/\omega\varepsilon = 1.04 \times 10^{13}$ and $\delta = 2.09 \times 10^{-4}$ m, at the microwave frequency $f = 1$ GHz, $\sigma/\omega\varepsilon = 1.04 \times 10^9$ and $\delta = 2.09 \times 10^{-6}$ m. Of course, for a material that is not as good a conductor as copper and for low frequencies the approximation may be of little value; for example, for a typical soil

($\varepsilon \approx 20\varepsilon_o$, $\mu \approx \mu_o$, $\sigma \approx 5 \times 10^{-2}$ S/m) at the power frequency $f = 60$ Hz, $\sigma/\omega\varepsilon = 1.0 \times 10^5$ and $\delta = 2.9 \times 10^2$ m!

In situations where the skin effect approximation holds and the current is confined to a very thin layer at the surface of a body, the current is effectively on the surface of the body. For macroscopic electromagnetic calculations, the volume density of current $\vec{\mathcal{J}}$ can then be replaced by a *surface density of current*, which we will call $\vec{\mathcal{J}}_s$. The surface density of current has the units of current per unit length (A/m). For our example (Figure 1.21) the surface density of current on the metal is simply

$$\vec{\mathcal{J}}_s(y,t) = \int_{z=0}^{10\delta} \vec{\mathcal{J}}(y,z,t)dz,$$

where the upper limit is chosen large enough that $\vec{\mathcal{J}} \approx 0$ at that point.

It is important to remember that the surface densities of charge and current, ρ_s and $\vec{\mathcal{J}}_s$, are models – idealizations, used to represent charge and current that are actually confined to a thin layer near a surface. Simple mathematical statements for these models result when the surface is the plane $z = 0$, as in Figures 1.18–1.21. Let the actual volume densities of charge and current be $\rho(x, y, z)$ and $\vec{\mathcal{J}}(x, y, z)$. In the model these densities are replaced by the volume densities $\rho'(x, y, z)$ and $\vec{\mathcal{J}}'(x, y, z)$:

$$\rho'(x,y,z) = \rho_s(x,y)\delta(z) \tag{1.56}$$

$$\vec{\mathcal{J}}'(x,y,z) = \vec{\mathcal{J}}_s(x,y)\delta(z), \tag{1.57}$$

where $\delta(z)$ is the *Dirac delta function* (Paul Adrien Maurice Dirac, 1902–1984) or the *impulse function* with the following properties (see Section 5.2):

$$\delta(\xi) = 0, \qquad \xi \neq 0,$$

$$\int_{-\Delta\xi}^{\Delta\xi} f(\xi)\,\delta(\xi)d\xi = f(0). \tag{1.58}$$

The surface densities ρ_s and $\vec{\mathcal{J}}_s$ are chosen so that

$$\int_{-\Delta z}^{\Delta z} \rho'(x,y,z)dz = \rho_s(x,y) = \int_{-\Delta z}^{\Delta z} \rho(x,y,z)dz \tag{1.59}$$

and

$$\int_{-\Delta z}^{\Delta z} \vec{\mathcal{J}}'(x,y,z)dz = \vec{\mathcal{J}}_s(x,y) = \int_{-\Delta z}^{\Delta z} \vec{\mathcal{J}}(x,y,z)dz, \tag{1.60}$$

where Δz is chosen so that all of the charge or current is within the range of integration. The relationship between ρ and ρ_s is shown graphically in Figure 1.22.

1.3.2 Electromagnetic boundary conditions

We now return to our discussion of the behavior of the electromagnetic field at a material boundary, such as in Figure 1.17. In Figure 1.23 we have an enlarged

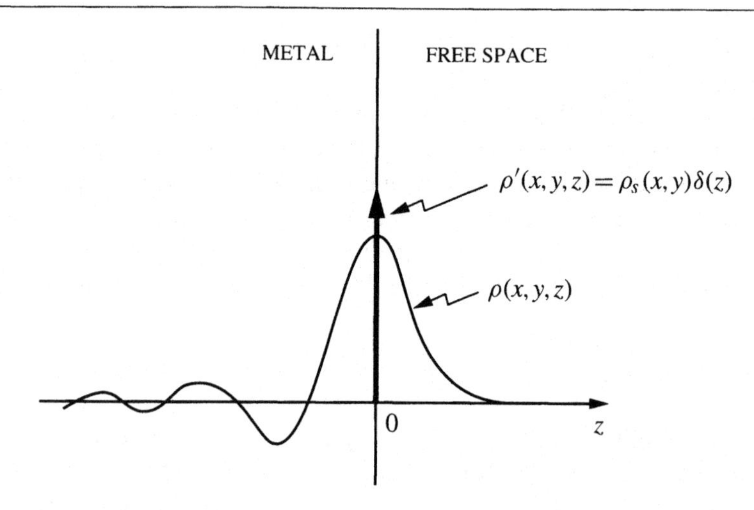

Fig. 1.22. Relationship between the volume densities of charge, ρ and ρ', and the surface density of charge, ρ_s.

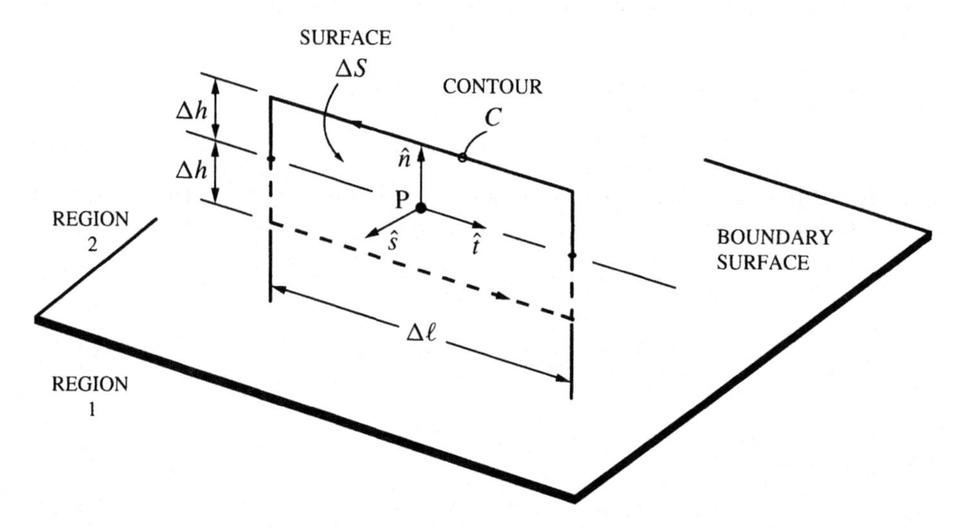

Fig. 1.23. Geometry used in obtaining the boundary conditions for the tangential components of the electromagnetic field.

picture of the boundary around the point P. The smooth boundary surface has the normal $\hat{n}$ at P pointing outward from region 1. The unit vectors $\hat{s}$ and $\hat{t}$ are tangent to the surface at P, and the vectors $\hat{n}$, $\hat{s}$, and $\hat{t}$ are mutually orthogonal and form a right-handed system.

The rectangular surface ΔS with contour C and normal $\hat{s}$ is centered on the boundary at P. The rectangle is small enough that the boundary surface can be assumed planar in the vicinity of P. The two sides of the rectangle of length $\Delta \ell$ are then parallel to the boundary.

Let's apply the Ampère-Maxwell law in integral form (1.2) to the rectangular surface:

$$\oint_C \vec{\mathcal{H}} \cdot d\vec{\ell} = \iint_{\Delta S} \vec{\mathcal{J}} \cdot d\vec{S} + \iint_{\Delta S} \frac{\partial \vec{\mathcal{D}}}{\partial t} \cdot d\vec{S}. \tag{1.61}$$

We will take the height of the contour $2\Delta h$ to be much smaller than its length $\Delta \ell$,

$$\Delta h / \Delta \ell \ll 1.$$

Assuming that $\vec{\mathcal{H}}$ is finite, the vertical portions of the contour make a negligible contribution to the integral. The left-hand side of (1.61) then becomes

$$\oint_C \vec{\mathcal{H}} \cdot d\vec{\ell} = \vec{\mathcal{H}}_1^* \cdot \hat{t} \Delta \ell - \vec{\mathcal{H}}_2^* \cdot \hat{t} \Delta \ell, \tag{1.62}$$

where, by the mean value theorem for integrals, $\vec{\mathcal{H}}_1^*$ ($\vec{\mathcal{H}}_2^*$) is $\vec{\mathcal{H}}_1$ ($\vec{\mathcal{H}}_2$) evaluated at a point on the horizontal portion of the contour.

Since the area ΔS is $2\Delta h \Delta \ell$, for small Δh and $\partial \vec{\mathcal{D}}/\partial t$ finite, the second integral on the right-hand side of (1.61) is negligible:

$$\iint_{\Delta S} \frac{\partial \vec{\mathcal{D}}}{\partial t} \cdot d\vec{S} = 0. \tag{1.63}$$

Now the first integral on the right-hand side of (1.61) will also be negligible unless there is a surface current on the boundary, and in that case

$$\iint_{\Delta S} \vec{\mathcal{J}} \cdot d\vec{S} = \vec{\mathcal{J}}_s^* \cdot \hat{s} \Delta \ell. \tag{1.64}$$

After substituting (1.62)–(1.64) into (1.61), we get

$$\vec{\mathcal{H}}_1^* \cdot \hat{t} \Delta \ell - \vec{\mathcal{H}}_2^* \cdot \hat{t} \Delta \ell = \vec{\mathcal{J}}_s^* \cdot \hat{s} \Delta \ell,$$

and in the limit $\Delta \ell \to 0$, $\vec{\mathcal{H}}_1^*$, $\vec{\mathcal{H}}_2^*$, and $\vec{\mathcal{J}}_s^*$ become the values $\vec{\mathcal{H}}_1$, $\vec{\mathcal{H}}_2$, and $\vec{\mathcal{J}}_s$ at the point P, so

$$\hat{t} \cdot (\vec{\mathcal{H}}_2 - \vec{\mathcal{H}}_1) = -\hat{s} \cdot \vec{\mathcal{J}}_s.$$

Since

$$\hat{t} = \hat{n} \times \hat{s},$$

we have

$$(\vec{\mathcal{H}}_2 - \vec{\mathcal{H}}_1) \cdot (\hat{n} \times \hat{s}) = -\hat{s} \cdot \vec{\mathcal{J}}_s,$$

or, on using the vector relation (Appendix B)

$$\vec{A} \cdot (\vec{B} \times \vec{C}) = -\vec{C} \cdot (\vec{B} \times \vec{A}),$$

we obtain

$$\hat{s} \cdot \left[\hat{n} \times (\vec{\mathcal{H}}_2 - \vec{\mathcal{H}}_1)\right] = \hat{s} \cdot \vec{\mathcal{J}}_s.$$

Now $\hat{s}$ is any unit vector tangent to the boundary surface at P, so we must have

$$\hat{n} \times (\vec{\mathcal{H}}_2 - \vec{\mathcal{H}}_1) = \vec{\mathcal{J}}_s, \tag{1.65}$$

which is our final boundary condition for the magnetic excitation.

If we applied the procedure used above to the integral form of Faraday's law (1.1), we would obtain the boundary condition for the electric field:

$$\hat{n} \times (\vec{\mathcal{E}}_2 - \vec{\mathcal{E}}_1) = 0. \tag{1.66}$$

Boundary conditions (1.65) and (1.66) apply to the components of the vectors that are tangent to the boundary surface. To see this, we take

$$\hat{n} \times \left[\hat{n} \times (\vec{\mathcal{H}}_2 - \vec{\mathcal{H}}_1)\right] = \hat{n} \times \vec{\mathcal{J}}_s$$

and use the vector relation (Appendix B)

$$\vec{A} \times (\vec{B} \times \vec{C}) = (\vec{A} \cdot \vec{C})\vec{B} - (\vec{A} \cdot \vec{B})\vec{C}$$

to obtain

$$\left[\vec{\mathcal{H}}_2 - (\hat{n} \cdot \vec{\mathcal{H}}_2)\hat{n}\right] - \left[\vec{\mathcal{H}}_1 - (\hat{n} \cdot \vec{\mathcal{H}}_1)\hat{n}\right] = -\hat{n} \times \vec{\mathcal{J}}_s.$$

Now $\vec{\mathcal{H}} - (\hat{n} \cdot \vec{\mathcal{H}})\hat{n}$ is the component of $\vec{\mathcal{H}}$ tangent to the boundary surface (it is the vector minus its normal component), so

$$\vec{\mathcal{H}}_{2t} - \vec{\mathcal{H}}_{1t} = -\hat{n} \times \vec{\mathcal{J}}_s, \tag{1.67}$$

where the subscript t signifies the component tangent to the boundary surface. In a similar manner, we have from (1.66)

$$\vec{\mathcal{E}}_{2t} - \vec{\mathcal{E}}_{1t} = 0. \tag{1.68}$$

In words, these boundary conditions state that the tangential component of $\vec{\mathcal{H}}$ is continuous on crossing a material boundary unless there is a surface current on the boundary; in that case, it is discontinuous by the amount $-\hat{n} \times \vec{\mathcal{J}}_s$. The tangential component of $\vec{\mathcal{E}}$ is always continuous on crossing a material boundary.

The boundary conditions for the normal components of the electromagnetic field are obtained by considering a pillbox-shaped volume centered on the boundary at the point P, as shown in Figure 1.24. The volume of the pillbox ΔV is small enough that the boundary surface can be assumed planar in the vicinity of P. The top and bottom surfaces ΔS of the pillbox are then parallel to the boundary. The side surface of the pillbox is ΔS_s, and the contour C is the intersection of the side surface with the boundary surface.

Now we will apply the equation of continuity for electric charge in integral form (1.5) to this volume:

$$\oint\!\!\!\oint_S \vec{\mathcal{J}} \cdot d\vec{S} = -\iiint_{\Delta V} \frac{\partial \rho}{\partial t} dV. \tag{1.69}$$

We will take the area of the side surface to be much smaller than the area of the top or bottom surfaces (Δh small):

$$\Delta S_s / \Delta S \propto \Delta h / \sqrt{\Delta S} \ll 1.$$

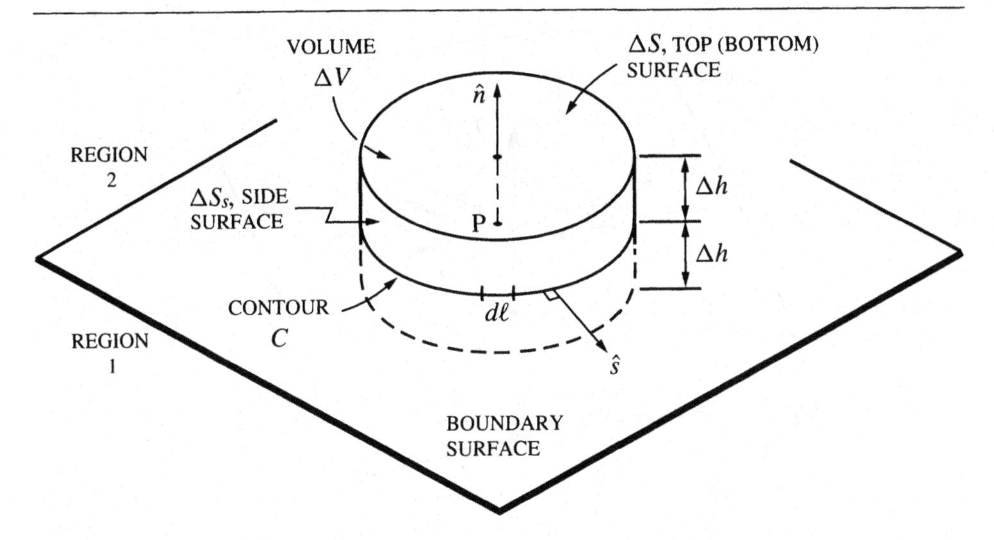

Fig. 1.24. Geometry used in obtaining the boundary conditions for the normal components of the electromagnetic field.

The integral over the side surface is then negligible unless there is a surface current on the boundary, and in which case,

$$\iint_{\Delta S_s} \vec{\mathcal{J}} \cdot d\vec{S} = \oint_C \vec{\mathcal{J}}_s \cdot \hat{s}\, d\ell. \tag{1.70}$$

The integrals over the top and bottom surfaces are simply

$$\iint_{\substack{\text{Top}\,+\\ \text{Bottom}}} \vec{\mathcal{J}} \cdot d\vec{S} = -\vec{\mathcal{J}}_1^* \cdot \hat{n}\Delta S + \vec{\mathcal{J}}_2^* \cdot \hat{n}\Delta S, \tag{1.71}$$

where, by the mean value theorem for integrals, $\vec{\mathcal{J}}_1^*$ ($\vec{\mathcal{J}}_2^*$) is $\vec{\mathcal{J}}_1$ ($\vec{\mathcal{J}}_2$) evaluated at a point on the area ΔS. Since the volume ΔV is $2\Delta h \Delta S$, for small Δh the integral on the right-hand side of (1.69) is zero unless there is a surface charge density on the boundary, in which case,

$$-\iiint_{\Delta V} \frac{\partial \rho}{\partial t} dV = -\frac{\partial \rho_s^*}{\partial t} \Delta S. \tag{1.72}$$

After substituting (1.70)–(1.72) into (1.69), we get

$$-\vec{\mathcal{J}}_1^* \cdot \hat{n}\Delta S + \vec{\mathcal{J}}_2^* \cdot \hat{n}\Delta S + \oint_C \vec{\mathcal{J}}_s \cdot \hat{s}\, d\ell = -\frac{\partial \rho_s^*}{\partial t} \Delta S,$$

and in the limit $\Delta S \to 0$, $\vec{\mathcal{J}}_1^*$, $\vec{\mathcal{J}}_2^*$, and ρ_s^* become the values $\vec{\mathcal{J}}_1$, $\vec{\mathcal{J}}_2$, and ρ_s at P:

$$\hat{n} \cdot (\vec{\mathcal{J}}_2 - \vec{\mathcal{J}}_1) + \lim_{\Delta S \to 0} \left(\frac{1}{\Delta S} \oint_C \hat{s} \cdot \vec{\mathcal{J}}_s d\ell \right) = -\frac{\partial \rho_s}{\partial t}. \tag{1.73}$$

The second term in this equation can be simplified by the introduction of the definition for the surface divergence or two-dimensional divergence [27], that is,

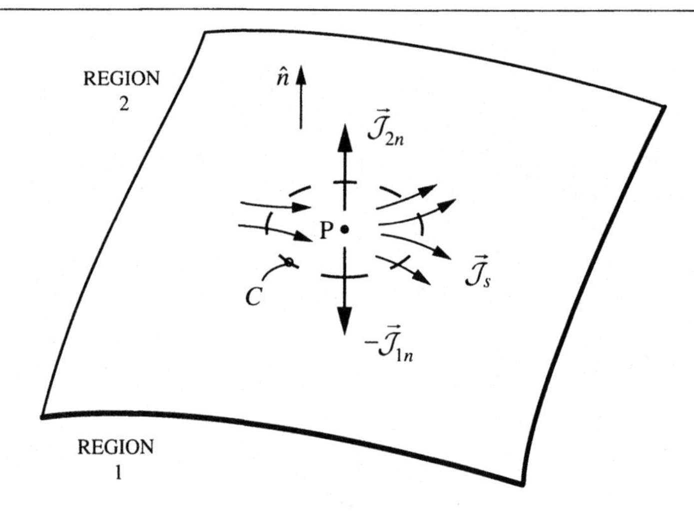

Fig. 1.25. Schematic drawing for explaining the surface equation of continuity for electric charge.

the specialization of (1.41) to a surface:

$$\mathrm{div}_s\, \vec{A}(\vec{r}\,) \equiv \lim_{\Delta S \to 0} \left(\frac{1}{\Delta S} \oint_C \hat{s} \cdot \vec{A}\, d\ell \right). \tag{1.74}$$

In the rectangular Cartesian coordinate system (x, y, z) the surface divergence can be evaluated using the same arguments as used in Section 1.2 for the regular divergence. For example, when the surface is the x-y plane with normal $\hat{z}$, and $\vec{A}$ is continuous and has continuous first partial derivatives, the surface divergence is

$$\mathrm{div}_s\, \vec{A}(x, y) = \frac{\partial A_x}{\partial x} + \frac{\partial A_y}{\partial y}, \tag{1.75}$$

or with the notation

$$\nabla_s \cdot \vec{A} = \sum_{i=x,y} \hat{i} \cdot \frac{\partial \vec{A}}{\partial i}, \tag{1.76}$$

we have

$$\mathrm{div}_s\, \vec{A}(x, y) = \nabla_s \cdot \vec{A}(x, y). \tag{1.77}$$

With the definition for the surface divergence, (1.73) becomes

$$\hat{n} \cdot (\vec{\mathcal{J}}_2 - \vec{\mathcal{J}}_1) + \nabla_s \cdot \vec{\mathcal{J}}_s = -\frac{\partial \rho_s}{\partial t}, \tag{1.78}$$

or

$$\mathcal{J}_{2n} - \mathcal{J}_{1n} + \nabla_s \cdot \vec{\mathcal{J}}_s = -\frac{\partial \rho_s}{\partial t}, \tag{1.79}$$

where the subscript n signifies the component normal to the boundary surface. This is the *surface equation of continuity for electric charge*; it is a statement of conservation of charge at a point on the surface and is illustrated in Figure 1.25. Charge can be taken away from the point P by the normal component of the volume

Table 1.4. *The electromagnetic boundary conditions*†

General boundary condition	Region 1: a perfect electric conductor
$\hat{n} \times (\vec{\mathcal{E}}_2 - \vec{\mathcal{E}}_1) = 0$	$\hat{n} \times \vec{\mathcal{E}}_2 = 0$
$\hat{n} \times (\vec{\mathcal{H}}_2 - \vec{\mathcal{H}}_1) = \vec{\mathcal{J}}_s$	$\hat{n} \times \vec{\mathcal{H}}_2 = \vec{\mathcal{J}}_s$
$\hat{n} \cdot (\vec{\mathcal{D}}_2 - \vec{\mathcal{D}}_1) = \rho_s$	$\hat{n} \cdot \vec{\mathcal{D}}_2 = \rho_s$
$\hat{n} \cdot (\vec{\mathcal{B}}_2 - \vec{\mathcal{B}}_1) = 0$	$\hat{n} \cdot \vec{\mathcal{B}}_2 = 0$
$\hat{n} \cdot (\vec{\mathcal{J}}_2 - \vec{\mathcal{J}}_1) + \nabla_s \cdot \vec{\mathcal{J}}_s = -\frac{\partial \rho_s}{\partial t}$	$\hat{n} \cdot \vec{\mathcal{J}}_2 + \nabla_s \cdot \vec{\mathcal{J}}_s = -\frac{\partial \rho_s}{\partial t}$

Note: †The normal $\hat{n}$ points outward from region 1 into region 2.

current in each region, $-\vec{\mathcal{J}}_{1n}$ and $\vec{\mathcal{J}}_{2n}$. The surface current $\vec{\mathcal{J}}_s$ can also remove charge from this point if it has a net outward pointing component on the curve C surrounding P (i.e., if $\nabla \cdot \vec{\mathcal{J}}_s$ is positive). The net charge removed by these currents per unit area, per unit time must equal $-\partial \rho_s / \partial t$ at P.

The procedure used above can also be applied to the integral forms of Gauss' electric law and Gauss' magnetic law, (1.3) and (1.4), to obtain boundary conditions for the normal components of $\vec{\mathcal{D}}$ and $\vec{\mathcal{B}}$, respectively:

$$\hat{n} \cdot (\vec{\mathcal{D}}_2 - \vec{\mathcal{D}}_1) = \rho_s \tag{1.80}$$

$$\hat{n} \cdot (\vec{\mathcal{B}}_2 - \vec{\mathcal{B}}_1) = 0, \tag{1.81}$$

or

$$\mathcal{D}_{2n} - \mathcal{D}_{1n} = \rho_s \tag{1.82}$$

$$\mathcal{B}_{2n} - \mathcal{B}_{1n} = 0. \tag{1.83}$$

In words, these boundary conditions state that the normal component of $\vec{\mathcal{D}}$ is continuous on crossing a material boundary unless there is a surface charge density on the boundary; in that case, it is discontinuous by the amount ρ_s. The normal component of $\vec{\mathcal{B}}$ is always continuous on crossing a material boundary.

These five boundary conditions [(1.65), (1.66), (1.78), (1.80), and (1.81)] are summarized in Table 1.4. Recall that for time-varying electromagnetic fields, only three of the five Maxwell's equations in differential form are independent. Similar arguments can be used to show that only three of the boundary conditions, (1.65), (1.66), and (1.78), are independent (Problem 1.8).

1.3.3 The concept of a perfect conductor

In our earlier discussion, we saw that the skin effect approximation holds for high-frequency oscillatory fields in good conductors. For this phenomenon, we must have $\sigma/\omega\varepsilon \gg 1$ and the skin depth $\delta = \sqrt{2/\omega\mu\sigma}$ must be small compared to all physical dimensions. The current in the conductor is then well approximated by a surface current density, and the electromagnetic field is essentially zero everywhere in the conductor.

Another way to view this approximation is to consider the limit as $\sigma \to \infty$, that is, the case in which the conductor becomes a *perfect conductor*. Then $\sigma/\omega\varepsilon \to \infty$, $\delta \to 0$, and the electromagnetic field is zero in the conductor for all frequencies $\omega > 0$. The perfect conductor is a model, an idealization that can be used whenever the skin effect approximation holds.

The electromagnetic boundary conditions simplify when one of the two regions is a perfect conductor. For example, when region 1 is a perfect conductor ($\vec{\mathcal{E}}_1 = 0$, $\vec{\mathcal{B}}_1 = 0$, $\vec{\mathcal{D}}_1 = 0$, and $\vec{\mathcal{H}}_1 = 0$), the boundary conditions (1.65), (1.66), (1.78), (1.80), and (1.81) become

$$\hat{n} \times \vec{\mathcal{E}}_2 = 0 \tag{1.84}$$

$$\hat{n} \times \vec{\mathcal{H}}_2 = \vec{\mathcal{J}}_s \tag{1.85}$$

$$\hat{n} \cdot \vec{\mathcal{J}}_2 + \nabla_s \cdot \vec{\mathcal{J}}_s = -\frac{\partial \rho_s}{\partial t} \tag{1.86}$$

$$\hat{n} \cdot \vec{\mathcal{D}}_2 = \rho_s \tag{1.87}$$

$$\hat{n} \cdot \vec{\mathcal{B}}_2 = 0. \tag{1.88}$$

These conditions are summarized in Table 1.4. Note that the tangential electric field $\mathcal{E}_{2t}$ and the normal magnetic field $\mathcal{B}_{2n}$ are always zero at the surface of a perfect conductor.

1.4 Energy of the electromagnetic field – Poynting's theorem

The Lorentz force expression (1.12) provides the connection between electromagnetism and classical mechanics. Through this expression, concepts such as energy, linear momentum, and angular momentum, which are familiar in mechanics, can be associated with the electromagnetic field.

In classical, nonrelativistic mechanics a particle of mass m, moving with velocity $\vec{v}$ at position $\vec{r}$ in an inertial reference frame, has *linear momentum* [37]

$$\vec{p} = m\frac{d\vec{r}}{dt} = m\vec{v}. \tag{1.89}$$

Newton's (Isaac Newton, 1642–1727) second law of motion states that the time rate of change of the linear momentum of a particle is equal to the total force applied to the particle:

$$\vec{\mathcal{F}} = \frac{d\vec{p}}{dt},$$

or, for a particle of constant mass,

$$\vec{\mathcal{F}} = m\frac{d\vec{v}}{dt} = m\frac{d^2\vec{r}}{dt^2} = m\vec{a}, \tag{1.90}$$

where $\vec{a}$ is the acceleration of the particle.

The *work* done by the applied force on the particle when it moves through the displacement $\Delta\vec{r}$ is defined to be

$$\Delta\mathcal{W} = \vec{\mathcal{F}} \cdot \Delta\vec{r}, \tag{1.91}$$

and the rate at which the work is done is the *power*:

$$\mathcal{P} = \lim_{\Delta t \to 0} \left(\frac{\Delta\mathcal{W}}{\Delta t} \right) = \lim_{\Delta t \to 0} \left(\vec{\mathcal{F}} \cdot \frac{\Delta\vec{r}}{\Delta t} \right) = \vec{\mathcal{F}} \cdot \vec{v}. \tag{1.92}$$

When we introduce (1.90), the power can be expressed in terms of the *kinetic energy* of the particle $\mathcal{K} = \frac{1}{2}m|\vec{v}\,|^2$:

$$\vec{\mathcal{F}} \cdot \vec{v} = m\frac{d\vec{v}}{dt} \cdot \vec{v} = \frac{d}{dt}\left(\frac{1}{2}m|\vec{v}|^2\right) = \frac{d\mathcal{K}}{dt},$$

or

$$\mathcal{P} = \vec{\mathcal{F}} \cdot \vec{v} = \frac{d\mathcal{K}}{dt}. \tag{1.93}$$

Thus, the rate at which work is done by the applied force – the power – is equal to the rate of increase in the kinetic energy of the particle.

Now let's examine a charged particle moving in an electromagnetic field $\vec{\mathcal{E}}, \vec{\mathcal{B}}$ in free space. The volume charge density in the particle is ρ, and its velocity is $\vec{v}$. From (1.12) the force of electromagnetic origin on the particle is

$$\vec{\mathcal{F}} = \iiint_V (\rho\vec{\mathcal{E}} + \vec{\mathcal{J}} \times \vec{\mathcal{B}})dV,$$

or, since

$$\vec{\mathcal{J}} = \rho\vec{v}, \tag{1.94}$$

$$\vec{\mathcal{F}} = \iiint_V (\rho\vec{\mathcal{E}} + \rho\vec{v} \times \vec{\mathcal{B}})dV. \tag{1.95}$$

The power (1.93) is then

$$\mathcal{P} = \vec{v} \cdot \iiint_V (\rho\vec{\mathcal{E}} + \rho\vec{v} \times \vec{\mathcal{B}})dV = \iiint_V \left[\rho\vec{v} \cdot \vec{\mathcal{E}} + \rho\vec{v} \cdot (\vec{v} \times \vec{\mathcal{B}})\right]dV,$$

where $\vec{v}$ was moved under the integral sign because the velocity is the same at all points in the particle. The second term in the integrand is zero: $\vec{v} \cdot (\vec{v} \times \vec{\mathcal{B}}) = 0$; the magnetic field does no work on the charged particle. After substituting (1.94), we have

$$\mathcal{P} = \iiint_V \vec{\mathcal{E}} \cdot \vec{\mathcal{J}} dV = \frac{d\mathcal{K}}{dt}. \tag{1.96}$$

We will assume that the electromagnetic field possesses energy and invoke the principle of conservation of energy. Equation (1.96) is then an expression for the rate at which energy is exchanged between the electromagnetic field and the mechanical motion of the charged particle. When $\mathcal{P}$ is positive, as in Figure 1.26a where $\vec{\mathcal{E}} \cdot \vec{\mathcal{J}}$ is positive throughout the particle, the kinetic energy of the particle is increasing; the field is instantaneously supplying energy to the mechanical motion of the particle.

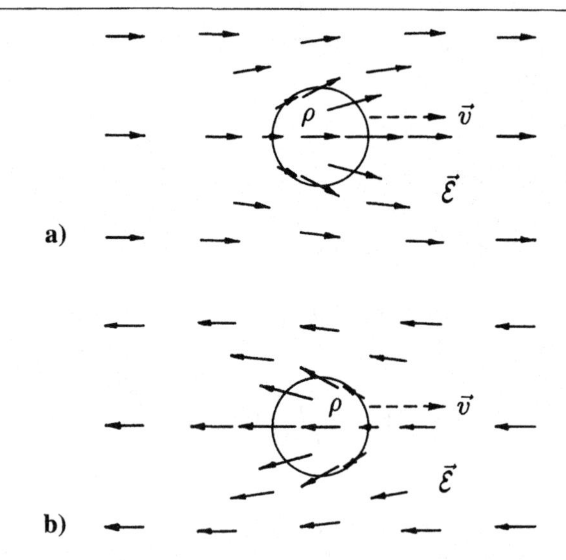

Fig. 1.26. Positively charged particle moving in an electromagnetic field. a) $\mathcal{P} = d\mathcal{K}/dt > 0$. b) $\mathcal{P} < 0$.

When $\mathcal{P}$ is negative, as in Figure 1.26b where $\vec{\mathcal{E}} \cdot \vec{\mathcal{J}}$ is negative throughout the particle, the kinetic energy of the particle is decreasing; the mechanical motion of the particle is instantaneously supplying energy to the field.

The expression we obtained above is for the energy associated with the motion of a charged particle. What we would really like to have is an expression for the energy that applies to general volume distributions of charge and current (ρ, $\vec{\mathcal{J}}$). We can obtain such an expression by starting with the integral from (1.96) applied to a general volume V and then substituting Maxwell's equations. First we will use the Ampère-Maxwell law (1.51) to replace $\vec{\mathcal{J}}$:

$$\iiint_V \vec{\mathcal{E}} \cdot \vec{\mathcal{J}} dV = \iiint_V \vec{\mathcal{E}} \cdot \nabla \times \vec{\mathcal{H}} dV - \iiint_V \vec{\mathcal{E}} \cdot \frac{\partial \vec{\mathcal{D}}}{\partial t} dV.$$

Next, the general vector relation (Appendix B)

$$\vec{A} \cdot (\nabla \times \vec{B}) = \vec{B} \cdot (\nabla \times \vec{A}) - \nabla \cdot (\vec{A} \times \vec{B})$$

is applied to the term $\vec{\mathcal{E}} \cdot (\nabla \times \vec{\mathcal{H}})$ to give

$$\iiint_V \vec{\mathcal{E}} \cdot \vec{\mathcal{J}} dV = \iiint_V \vec{\mathcal{H}} \cdot (\nabla \times \vec{\mathcal{E}}) dV - \iiint_V \nabla \cdot (\vec{\mathcal{E}} \times \vec{\mathcal{H}}) dV$$
$$- \iiint_V \vec{\mathcal{E}} \cdot \frac{\partial \vec{\mathcal{D}}}{\partial t} dV,$$

and Faraday's law (1.50) is used to replace $\nabla \times \vec{\mathcal{E}}$:

$$\iiint_V \vec{\mathcal{E}} \cdot \vec{\mathcal{J}} dV = - \iiint_V \vec{\mathcal{E}} \cdot \frac{\partial \vec{\mathcal{D}}}{\partial t} dV - \iiint_V \vec{\mathcal{H}} \cdot \frac{\partial \vec{\mathcal{B}}}{\partial t} dV - \iiint_V \nabla \cdot (\vec{\mathcal{E}} \times \vec{\mathcal{H}}) dV. \tag{1.97}$$

Now we introduce the divergence theorem or Gauss' theorem (Appendix B) [26]:

$$\iiint_V \nabla \cdot \vec{A} dV = \oiint_S \hat{n} \cdot \vec{A} dS. \tag{1.98}$$

After applying this theorem to the last integral in (1.97), we have

$$\iiint_V \vec{\mathcal{E}} \cdot \vec{\mathcal{J}} dV + \iiint_V \left(\vec{\mathcal{E}} \cdot \frac{\partial \vec{\mathcal{D}}}{\partial t} + \vec{\mathcal{H}} \cdot \frac{\partial \vec{\mathcal{B}}}{\partial t} \right) dV + \oiint_S (\vec{\mathcal{E}} \times \vec{\mathcal{H}}) \cdot d\vec{S} = 0. \tag{1.99}$$

It is customary to define the vector

$$\vec{\mathcal{S}}(\vec{r}, t) = \vec{\mathcal{E}}(\vec{r}, t) \times \vec{\mathcal{H}}(\vec{r}, t). \tag{1.100}$$

Then (1.99) becomes

$$\iiint_V \vec{\mathcal{E}} \cdot \vec{\mathcal{J}} dV + \iiint_V \left(\vec{\mathcal{E}} \cdot \frac{\partial \vec{\mathcal{D}}}{\partial t} + \vec{\mathcal{H}} \cdot \frac{\partial \vec{\mathcal{B}}}{\partial t} \right) dV + \oiint_S \vec{\mathcal{S}} \cdot d\vec{S} = 0. \tag{1.101}$$

This relation is known as *Poynting's theorem*, and the vector $\vec{\mathcal{S}}$ is called the *Poynting vector.*[11]

This relation was first derived by John Henry Poynting (1852–1914) in 1884 [38], and, as is the case with many scientific discoveries, it was obtained independently, at about the same time, by another investigator, Oliver Heaviside [39–41].

Before we examine the physical interpretation of this theorem, we should say a few words about the volume to which it applies. In the derivation of (1.101), we used Maxwell's equations in differential form and the divergence theorem; this usage implies that the field and its first partial derivatives are continuous, which may not be the case if material boundaries are included within the volume. Of course, we would like to apply the theorem to a large volume of space containing many subvolumes, each being a different material, such as a conductor, a dielectric, etc. Let's see how this can be done.

We will only consider a volume V composed of two subvolumes V_1 and V_2, as in Figure 1.27. Our argument can easily be extended to a volume containing any number of subvolumes. The surface of V_2 is S_2, and the surface of V_1 has an interior portion S_{1i} adjacent to S_2 plus an exterior portion S_{1e}, which is also the surface of V.[12] Poynting's theorem (1.101) can be applied to each of the volumes V_1 and V_2 separately, and the results can be added to give

$$\iiint_{V_1} \vec{\mathcal{E}} \cdot \vec{\mathcal{J}} dV + \iiint_{V_2} \vec{\mathcal{E}} \cdot \vec{\mathcal{J}} dV + \iiint_{V_1} \left(\vec{\mathcal{E}} \cdot \frac{\partial \vec{\mathcal{D}}}{\partial t} + \vec{\mathcal{H}} \cdot \frac{\partial \vec{\mathcal{B}}}{\partial t} \right) dV$$
$$+ \iiint_{V_2} \left(\vec{\mathcal{E}} \cdot \frac{\partial \vec{\mathcal{D}}}{\partial t} + \vec{\mathcal{H}} \cdot \frac{\partial \vec{\mathcal{B}}}{\partial t} \right) dV + \oiint_{S_1 = S_{1i} + S_{1e}} \vec{\mathcal{S}}_1 \cdot d\vec{S} + \oiint_{S_2} \vec{\mathcal{S}}_2 \cdot d\vec{S} = 0.$$

[11] The Poynting vector $\vec{\mathcal{S}}$ and the differential area vector $d\vec{S}$ should not be confused.

[12] In Figure 1.27 the surface S_1 is drawn slightly outside the boundary, and the surface S_2 is drawn slightly inside the boundary. The fields and their first partial derivatives are well behaved on these surfaces.

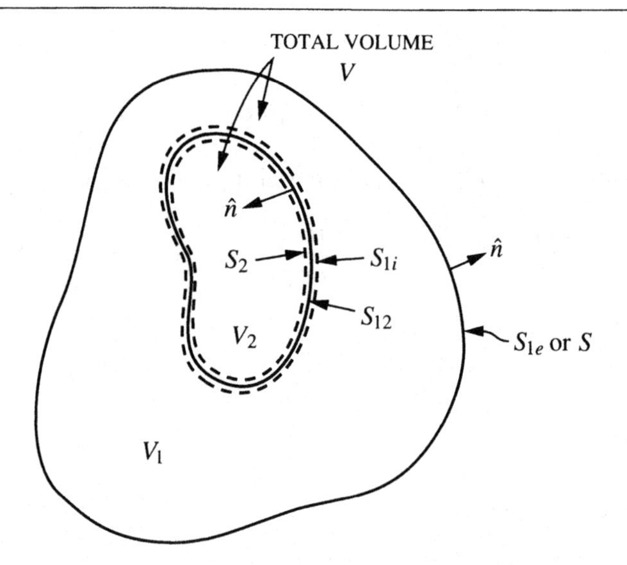

Fig. 1.27. Volume used in the discussion of Poynting's theorem.

After recognizing that on the interior surface $S_{1i} = S_2 = S_{12}$, $d\vec{S}_1 = \hat{n}dS$, and $d\vec{S}_2 = -\hat{n}dS$, and on the exterior surface $S_{1e} = S$ and $d\vec{S}_1 = d\vec{S}$, this result becomes

$$\iiint_{V_1+V_2} \vec{\mathcal{E}} \cdot \vec{\mathcal{J}} dV + \iiint_{V_1+V_2} \left(\vec{\mathcal{E}} \cdot \frac{\partial \vec{\mathcal{D}}}{\partial t} + \vec{\mathcal{H}} \cdot \frac{\partial \vec{\mathcal{B}}}{\partial t} \right) dV + \oiint_S \vec{\mathcal{S}} \cdot d\vec{S}$$
$$+ \oiint_{S_{12}} (\hat{n} \cdot \vec{\mathcal{S}}_1 - \hat{n} \cdot \vec{\mathcal{S}}_2) dS = 0. \tag{1.102}$$

The integrand of the last integral in (1.102) is

$$\begin{aligned} \hat{n} \cdot (\vec{\mathcal{S}}_1 - \vec{\mathcal{S}}_2) &= \hat{n} \cdot (\vec{\mathcal{E}}_1 \times \vec{\mathcal{H}}_1 - \vec{\mathcal{E}}_2 \times \vec{\mathcal{H}}_2) \\ &= \vec{\mathcal{H}}_1 \cdot (\hat{n} \times \vec{\mathcal{E}}_1) - \vec{\mathcal{H}}_2 \cdot (\hat{n} \times \vec{\mathcal{E}}_2), \end{aligned}$$

where the vector identity (Appendix B)

$$\vec{A} \cdot (\vec{B} \times \vec{C}) = \vec{C} \cdot (\vec{A} \times \vec{B})$$

was used. From the boundary condition (1.66), $\hat{n} \times \vec{\mathcal{E}}_1 = \hat{n} \times \vec{\mathcal{E}}_2$, which we will write as $\hat{n} \times \vec{\mathcal{E}}$, so

$$\begin{aligned} \hat{n} \cdot (\vec{\mathcal{S}}_1 - \vec{\mathcal{S}}_2) &= -(\vec{\mathcal{H}}_2 - \vec{\mathcal{H}}_1) \cdot (\hat{n} \times \vec{\mathcal{E}}) \\ &= \vec{\mathcal{E}} \cdot \left[\hat{n} \times (\vec{\mathcal{H}}_2 - \vec{\mathcal{H}}_1) \right], \end{aligned}$$

where the vector identity given above was used again. After using the boundary condition (1.65), we have

$$\hat{n} \cdot (\vec{\mathcal{S}}_1 - \vec{\mathcal{S}}_2) = \vec{\mathcal{E}} \cdot \vec{\mathcal{J}}_s,$$

or, on inserting this result into (1.102),

$$\iiint_{V_1+V_2} \vec{\mathcal{E}} \cdot \vec{\mathcal{J}} dV + \iiint_{V_1+V_2} \left(\vec{\mathcal{E}} \cdot \frac{\partial \vec{\mathcal{D}}}{\partial t} + \vec{\mathcal{H}} \cdot \frac{\partial \vec{\mathcal{B}}}{\partial t} \right) dV + \oiint_S \vec{\mathcal{S}} \cdot d\vec{S}$$
$$+ \oiint_{S_{12}} \vec{\mathcal{E}} \cdot \vec{\mathcal{J}}_s dS = 0. \tag{1.103}$$

A comparison of this equation with our original form for Poynting's theorem (1.101) shows that (1.103) contains the additional term

$$\oiint_{S_{12}} \vec{\mathcal{E}} \cdot \vec{\mathcal{J}}_s dS, \tag{1.104}$$

which involves the current on the internal surface S_{12}. Our conclusion then is that Poynting's theorem (1.101) is valid for volumes containing more than one material region if the first integral,

$$\iiint_V \vec{\mathcal{E}} \cdot \vec{\mathcal{J}} dV,$$

is interpreted as including the contributions (1.104) that arise due to surface currents on the boundaries between material regions. The term $\vec{\mathcal{E}} \cdot \vec{\mathcal{J}}_s$ is zero in most practical cases, since $\vec{\mathcal{J}}_s$ is zero unless one of the two regions is a perfect conductor, and, in that case, $\vec{\mathcal{E}} \cdot \vec{\mathcal{J}}_s = \vec{\mathcal{E}}_t \cdot \vec{\mathcal{J}}_s = 0$, because $\vec{\mathcal{E}}_t = 0$ at the surface of a perfect conductor.

Now let's return to the physical interpretation of Poynting's theorem (1.101). First we note that all of the terms (integrals) in this expression have the units of energy per unit time (J/s) or power (W); therefore, this relation is customarily viewed as a statement of conservation of energy. To better understand this interpretation, we will examine each of the terms in the theorem. Our discussion will be limited to a rigid, isothermal volume composed of regions that are simple materials obeying the electrical constitutive relations (1.9)–(1.11).

From our discussion of a charged particle moving in an electromagnetic field, we infer that the first term in (1.101) represents the rate at which energy is exchanged between the electromagnetic field and the mechanical motion of the charge within the volume V. In a *simple conductor* for which $\vec{\mathcal{J}} = \vec{\mathcal{J}}_c = \sigma \vec{\mathcal{E}}$, this term becomes

$$\iiint_V \vec{\mathcal{E}} \cdot \vec{\mathcal{J}}_c dV = \iiint_V |\vec{\mathcal{J}}_c|^2 / \sigma dV, \tag{1.105}$$

which is always positive; the electromagnetic field is always supplying energy to the mechanical motion of the charge within the conductor. For an isothermal conductor this energy is irreversibly transferred to a heat reservoir as Joule heat (James Prescott Joule, 1818–1889); thus, we can write this term as

$$\iiint_V \vec{\mathcal{E}} \cdot \vec{\mathcal{J}}_c dV = \frac{\partial \mathcal{Q}}{\partial t}, \tag{1.106}$$

where $\mathcal{Q}$ is the amount of energy transferred as heat.

This interpretation is consistent with the classical, microscopic model for a conductor. The electric field accelerates the charges (conduction electrons in a metal, ions in an electrolytic solution), increasing their kinetic energy. This energy, in turn, through "scattering processes" is converted to energy associated with the random motion of the microscopic particles (electrons and positive ions in a metal, ions and molecules in an electrolytic solution). For an isothermal conductor, the additional energy of random motion is removed from the conductor by contact with a heat reservoir.

In a *source region*, an impressed force of nonelectromagnetic origin, such as one due to mechanical or chemical action, can move positive electric charge in the direction opposite to the electric field. The volume density of current due to the motion of this charge is $\vec{\mathcal{J}}_i$, where the subscript i indicates an impressed current. When the motion of the charge is instantaneously supplying energy to the electromagnetic field throughout the volume V, we have $\vec{\mathcal{E}} \cdot \vec{\mathcal{J}}_i < 0$, and the first term in (1.101) becomes

$$\iiint_V \vec{\mathcal{E}} \cdot \vec{\mathcal{J}}_i dV = -\iiint_V |\vec{\mathcal{E}} \cdot \vec{\mathcal{J}}_i| dV, \tag{1.107}$$

which is negative. Since the source provides the energy to move the charge, we can write this term as

$$-\iiint_V \vec{\mathcal{E}} \cdot \vec{\mathcal{J}}_i dV = \frac{\partial \mathcal{W}_i}{\partial t}, \tag{1.108}$$

where $\mathcal{W}_i$ is the energy supplied by the nonelectromagnetic source to the electromagnetic field.

The simple model for a battery shown in Figure 1.28 includes both a conducting region and a source region. In the source region, which is generally at the surface of the plates in the battery, chemical processes produce a current $\vec{\mathcal{J}}_{is}$ in the direction opposite to the electric field $\vec{\mathcal{E}}_s$. In the electrolytic solution, a conduction current of ions $\vec{\mathcal{J}}_{ce}$ is in the same direction as the electric field $\vec{\mathcal{E}}_e$. In the steady state, $\vec{\mathcal{J}}_{is} = \vec{\mathcal{J}}_{ce}$, and the voltage across the terminals of the battery is

$$V = -\int_0^{x_s} \vec{\mathcal{E}}_s \cdot d\vec{x} - \int_{x_s}^{x_s+x_e} \vec{\mathcal{E}}_e \cdot d\vec{x} = \int_0^{x_s} |\vec{\mathcal{E}}_s| dx - \int_{x_s}^{x_s+x_e} |\vec{\mathcal{E}}_e| dx$$
$$\approx \int_0^{x_s} |\vec{\mathcal{E}}_s| dx,$$

where in the last step, the voltage drop across the electrolytic solution is assumed to be negligible.

The second term in Poynting's theorem (1.101) is the volume integral

$$\iiint_V \left(\vec{\mathcal{E}} \cdot \frac{\partial \vec{\mathcal{D}}}{\partial t} + \vec{\mathcal{H}} \cdot \frac{\partial \vec{\mathcal{B}}}{\partial t} \right) dV.$$

In simple materials with $\vec{\mathcal{D}} = \varepsilon \vec{\mathcal{E}}$ and $\vec{\mathcal{H}} = \vec{\mathcal{B}}/\mu$, the two terms in the integrand can be simplified. For the electric term we have

$$\vec{\mathcal{E}} \cdot \frac{\partial \vec{\mathcal{D}}}{\partial t} = \varepsilon \left(\vec{\mathcal{E}} \cdot \frac{\partial \vec{\mathcal{E}}}{\partial t} \right) = \varepsilon \left(\frac{1}{2} \frac{\partial |\vec{\mathcal{E}}|^2}{\partial t} \right) = \frac{\partial}{\partial t} \left(\frac{1}{2} \varepsilon |\vec{\mathcal{E}}|^2 \right), \tag{1.109}$$

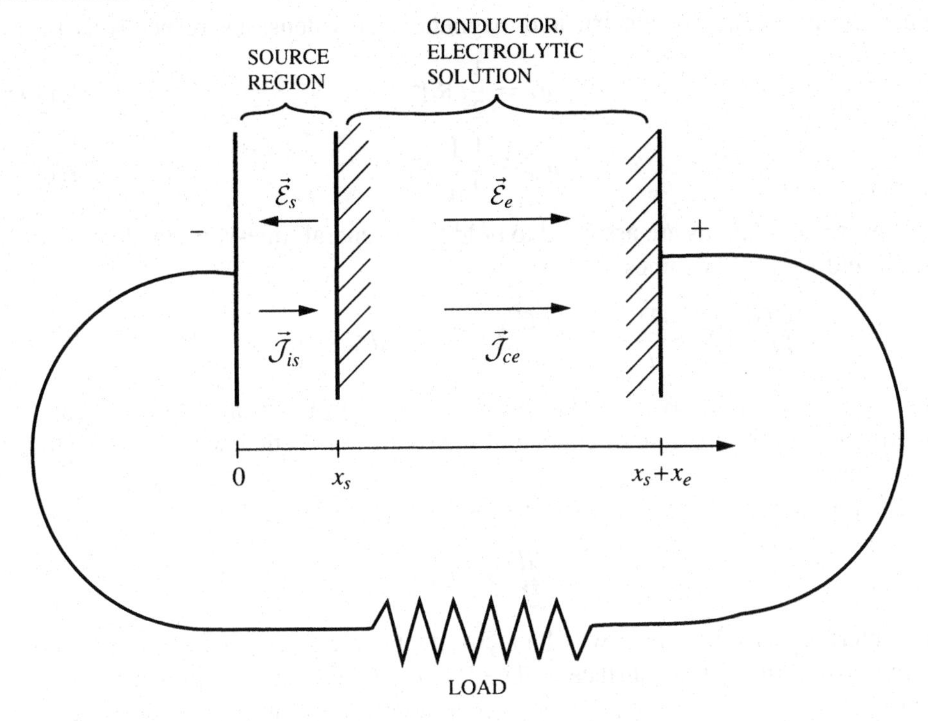

Fig. 1.28. Simple model for a battery.

and, similarly, for the magnetic term

$$\vec{\mathcal{H}} \cdot \frac{\partial \vec{\mathcal{B}}}{\partial t} = \frac{\partial}{\partial t}\left(\frac{1}{2}\frac{1}{\mu}|\vec{\mathcal{B}}|^2\right); \tag{1.110}$$

thus,

$$\iiint_V \left(\vec{\mathcal{E}} \cdot \frac{\partial \vec{\mathcal{D}}}{\partial t} + \vec{\mathcal{H}} \cdot \frac{\partial \vec{\mathcal{B}}}{\partial t}\right) dV = \frac{\partial}{\partial t}\left(\frac{1}{2}\iiint_V \varepsilon|\vec{\mathcal{E}}|^2 dV + \frac{1}{2}\iiint_V \frac{1}{\mu}|\vec{\mathcal{B}}|^2 dV\right). \tag{1.111}$$

For an *electrostatic field* in a simple material, the energy stored in the electric field is

$$\mathcal{U}_e = \frac{1}{2}\iiint_V \varepsilon|\vec{\mathcal{E}}|^2 dV = \iiint_V u_e dV, \tag{1.112}$$

and, similarly, for a *magnetostatic field* in a simple material, the energy stored in the magnetic field is

$$\mathcal{U}_m = \frac{1}{2}\iiint_V \frac{1}{\mu}|\vec{\mathcal{B}}|^2 dV = \iiint_V u_m dV, \tag{1.113}$$

where u_e and u_m are the electric and magnetic energy densities, respectively [42]:

$$u_e = \frac{1}{2}\varepsilon|\vec{\mathcal{E}}|^2 \tag{1.114}$$

$$u_m = \frac{1}{2}\frac{1}{\mu}|\vec{\mathcal{B}}|^2. \tag{1.115}$$

If these same relations are assumed to hold for a general time-varying electromagnetic field, (1.111) becomes

$$\iiint_V \left(\vec{\mathcal{E}}\cdot\frac{\partial\vec{\mathcal{D}}}{\partial t} + \vec{\mathcal{H}}\cdot\frac{\partial\vec{\mathcal{B}}}{\partial t}\right)dV = \frac{\partial}{\partial t}(\mathcal{U}_e + \mathcal{U}_m) = \frac{\partial}{\partial t}\mathcal{U}, \tag{1.116}$$

where $\mathcal{U} = \mathcal{U}_e + \mathcal{U}_m$. We see that the second term in Poynting's theorem can be interpreted as the time rate of change of the electromagnetic energy stored within the volume V.

The third term in Poynting's theorem is the surface integral

$$\oiint_S \vec{\mathcal{S}}\cdot d\vec{S}.$$

It is interpreted as the rate at which electromagnetic energy is leaving the volume V by passing through its surface S. The Poynting vector $\vec{\mathcal{S}}$, which appears in the integrand, has the units of energy per unit area per unit time (J/m^{2}s) or power per unit area (W/m^2). It can be regarded as the rate at which electromagnetic energy is passing through a unit area whose normal is in the direction of the vector $\vec{\mathcal{E}} \times \vec{\mathcal{H}}$. This interpretation proves particularly useful when dealing with the propagation of electromagnetic waves (Chapters 2 and 3). The direction of the Poynting vector is then normal to the wavefront, and the magnitude of the Poynting vector is the power per unit area of the wavefront (see Figure 3.18).

Now let's rewrite Poynting's theorem, substituting what we have learned from our discussion,

$$\begin{aligned}-\iiint_V \vec{\mathcal{E}}\cdot\vec{\mathcal{J}}_i dV = \iiint_V |\vec{\mathcal{J}}_c|^2/\sigma dV \\ + \frac{\partial}{\partial t}\left(\frac{1}{2}\iiint_V \varepsilon|\vec{\mathcal{E}}|^2 dV + \frac{1}{2}\iiint_V \frac{1}{\mu}|\vec{\mathcal{B}}|^2 dV\right) + \oiint_S \vec{\mathcal{S}}\cdot d\vec{S},\end{aligned} \tag{1.117}$$

or

$$\frac{\partial \mathcal{W}_i}{\partial t} = \frac{\partial \mathcal{Q}}{\partial t} + \frac{\partial \mathcal{U}}{\partial t} + \oiint_S \vec{\mathcal{S}}\cdot d\vec{S}. \tag{1.118}$$

The interpretation of this relation as a statement of conservation of energy within the volume V is now clear; it is summarized in Figure 1.29. The rate at which the nonelectromagnetic sources supply energy to the electromagnetic field must equal the rate at which energy is being transferred from the field as heat, plus the rate at which electromagnetic energy is being stored, plus the rate at which electromagnetic energy is leaving the volume by passing through its surfaces.

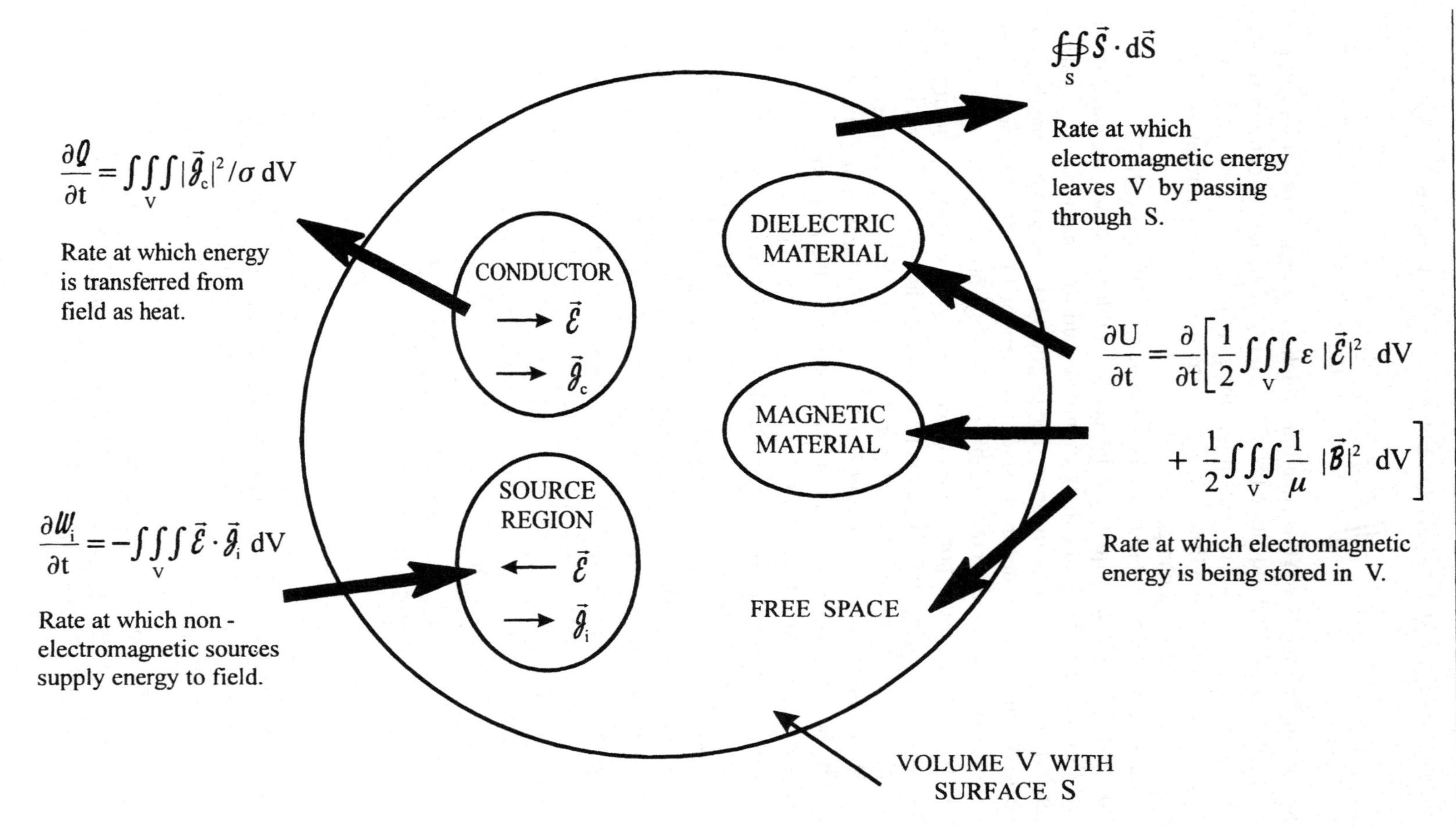

Fig. 1.29. Schematic drawing showing the interpretation for the terms in Poynting's theorem.

When $\partial \mathcal{W}_i/\partial t = 0$ and $\partial \mathcal{Q}/\partial t = 0$, Equation (1.118) becomes

$$\oiint_S \vec{\mathcal{S}} \cdot d\vec{S} = -\frac{\partial \mathcal{U}}{\partial t}.$$

Notice the similarity between this equation, which is a statement of conservation of energy, and the continuity Equation (1.5), which is a statement of conservation of charge. The energy density $\mathcal{U}$ corresponds to the charge density ρ, and the Poynting vector $\vec{\mathcal{S}}$, which represents a flow of energy, corresponds to the current density $\vec{\mathcal{J}}$, which represents a flow of charge.

It should be emphasized that Poynting's theorem (1.101) was obtained by simply combining Maxwell's equations; hence, its validity rests on the validity of Maxwell's equations. The same thing cannot be said for a particular physical interpretation of the terms in the theorem. The physical interpretation of the theorem has been a controversial issue almost since its inception [43]. It would not suit our purposes to give a detailed account of this controversy; however, consideration of a few of the more important points will increase our understanding of the theorem.

Any particular physical interpretation is, of course, tied to the physical makeup of the volume to which the theorem is applied. Recall that our interpretation is for a rigid isothermal volume composed of simple materials. We will not consider changes in the interpretation that result from changes in the properties of the materials. Instead, we will concentrate on the ambiguities that exist in the particular results we have already presented, Equations (1.117) and (1.118).

Let us accept that the surface integral in (1.117) represents the rate at which energy is leaving the volume V by passing through the surface S. Does it then necessarily follow that the Poynting vector $\vec{\mathcal{S}}$ can be interpreted as the flow of electromagnetic energy at a point? That is, does $\vec{\mathcal{S}}$ represent the rate at which energy is passing through a unit area whose normal is in the direction of $\vec{\mathcal{S}}$? The answer to this question is no, and we only need to consider the derivation of Poynting's theorem, namely Equation (1.97), to see this. A vector field $\vec{\mathcal{T}}$ whose divergence is zero can be added to the Poynting vector without affecting the theorem. In particular, if we let an *alternate Poynting vector* $\vec{\mathcal{S}}'$ be defined as

$$\vec{\mathcal{S}}' = \vec{\mathcal{S}} + \vec{\mathcal{T}}, \qquad (1.119a)$$

where

$$\nabla \cdot \vec{\mathcal{T}} = 0, \qquad (1.119b)$$

then the last integral in (1.97) is unchanged when $\vec{\mathcal{S}}'$ is substituted for $\vec{\mathcal{S}} = \vec{\mathcal{E}} \times \vec{\mathcal{H}}$. Which vector correctly describes the flow of electromagnetic energy at a point, $\vec{\mathcal{S}}$ or any one of the myriad choices for $\vec{\mathcal{S}}'$? There is no conclusive answer to this question, although additional constraints can often be imposed that exclude certain choices for $\vec{\mathcal{S}}'$ [44–47]. All we can really say is that in certain applications, particularly those involving wave propagation, the interpretation of the conventional Poynting vector $\vec{\mathcal{S}}$ as representing the flow of electromagnetic energy at a point gives physically meaningful results.

Our conclusion is that we can apply Poynting's theorem (1.101) to any electromagnetic problem without reservation, because its validity rests on the validity of Maxwell's equations. However, when we use a particular physical interpretation for the terms in the theorem, for quantities that appear in the integrands such as $\vec{S}$, u_e, and u_m, we are introducing an additional hypothesis whose usefulness will depend upon the particular problem being considered.

The concepts associated with Poynting's theorem and the Poynting vector are nicely illustrated by the simple example shown in Figure 1.30a. Here we have a pair of batteries producing a steady current I in a long, straight wire. The straight wire has finite conductivity σ, whereas the connecting wires are assumed to be perfectly conducting. We would like to know how the energy supplied by the batteries is transported to the wire where it is dissipated as heat. This is a problem that was discussed by both Poynting and Heaviside [38, 40, 48].

To simplify the analysis, the complex geometry of Figure 1.30a is replaced by the very idealized geometry of Figure 1.30b. The long, straight wire of radius a is surrounded by a perfectly conducting cylindrical shell of inner radius b. A battery is connected to each end of the structure. The batteries, which are shown schematically in Figure 1.30b, are assumed to be of radial design, so that the entire structure is rotationally symmetric about the z axis. This geometry and simple variations of it have been discussed many times in the literature [6, 49, 50].

We want to determine the electromagnetic field and the Poynting vector in a region near the center of the wire ($z = 0$) away from both ends. The interior of the wire is labeled 1, and the space between the conductors is labeled 2.

The time-independent electric field is expressed in terms of the scalar potential Φ, which is a solution to Laplace's partial differential equation (Pierre Simon de Laplace, 1749–1827):

$$\vec{\mathcal{E}} = -\nabla\Phi, \tag{1.120}$$

where

$$\nabla^2\Phi = 0. \tag{1.121}$$

In the wire, we find (Problem 1.18) that

$$\Phi_1 = \frac{-I}{\sigma\pi a}\left(\frac{z}{a}\right) \tag{1.122}$$

$$\vec{\mathcal{E}}_1 = \frac{I}{\sigma\pi a^2}\hat{z}, \tag{1.123}$$

and, in the space between the conductors,

$$\Phi_2 = \frac{-I}{\sigma\pi a\ln(b/a)}\left(\frac{z}{a}\right)\ln\left(\frac{b}{\rho}\right) \tag{1.124}$$

$$\vec{\mathcal{E}}_2 = \frac{-I}{\sigma\pi a^2\ln(b/a)}\left[\left(\frac{a}{\rho}\right)\left(\frac{z}{a}\right)\hat{\rho} - \ln\left(\frac{b}{\rho}\right)\hat{z}\right]. \tag{1.125}$$

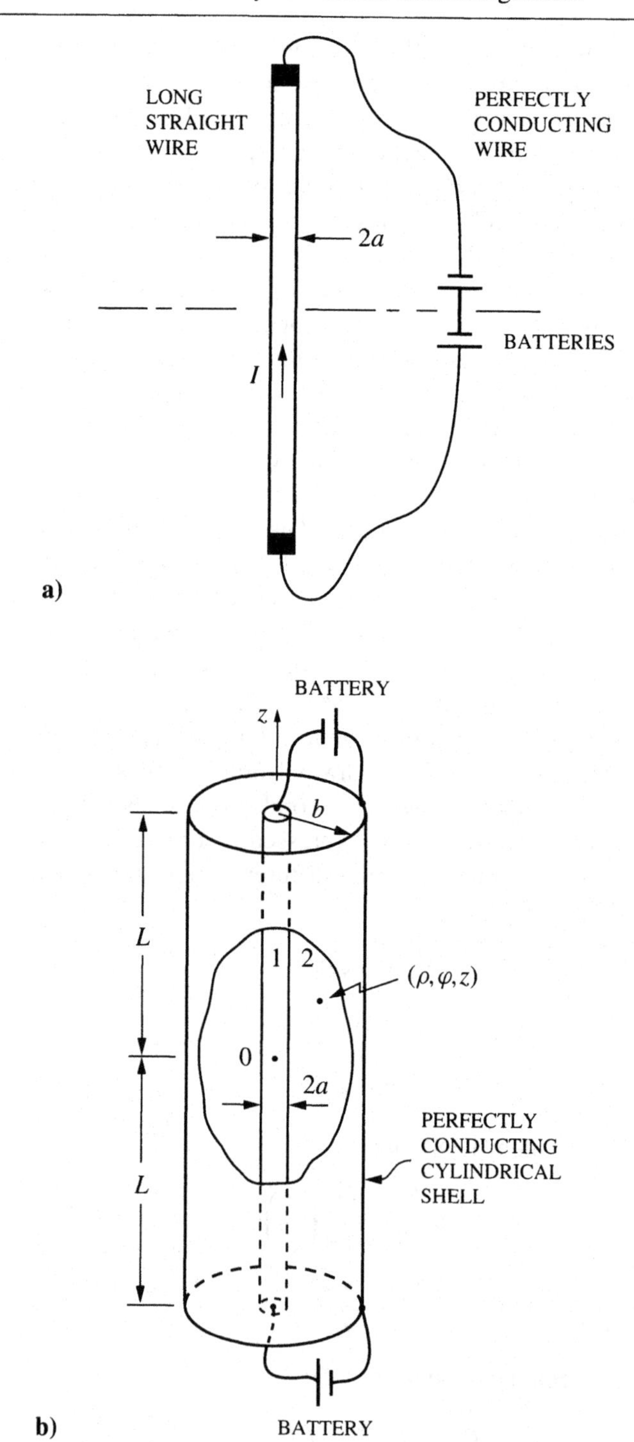

Fig. 1.30. a) Batteries producing current I in a long, straight wire. b) Idealized geometry for studying the electromagnetic field of (a).

The current density within the wire is then

$$\vec{\mathcal{J}} = \sigma \vec{\mathcal{E}}_1 = \frac{I}{\pi a^2}\hat{z}, \tag{1.126}$$

and the charge density on its surface is

$$\rho_s = \hat{n} \cdot (\vec{\mathcal{D}}_2 - \vec{\mathcal{D}}_1)_{\rho=a} = (\hat{\rho} \cdot \varepsilon_o \vec{\mathcal{E}}_2 - \hat{\rho} \cdot \varepsilon_o \vec{\mathcal{E}}_1)_{\rho=a} = \frac{-\varepsilon_o I}{\sigma \pi a^2 \ln(b/a)}\left(\frac{z}{a}\right), \tag{1.127}$$

where we have assumed $\varepsilon = \varepsilon_o$ in the wire. From the current (1.126), the magnetic excitation is found to be

$$\vec{\mathcal{H}}_1 = \frac{I}{2\pi a}\left(\frac{\rho}{a}\right)\hat{\phi}, \tag{1.128}$$

$$\vec{\mathcal{H}}_2 = \frac{I}{2\pi a}\left(\frac{a}{\rho}\right)\hat{\phi}. \tag{1.129}$$

The Poynting vector (1.100) is computed in region 1 from (1.123) and (1.128) and in region 2 from (1.125) and (1.129):

$$\vec{\mathcal{S}}_1 = -\left(\frac{I^2}{2\pi^2 a^3 \sigma}\right)\left(\frac{\rho}{a}\right)\hat{\rho} \tag{1.130}$$

$$\vec{\mathcal{S}}_2 = -\left(\frac{I^2}{2\pi^2 a^3 \sigma}\right)\frac{1}{\ln(b/a)}\left[\left(\frac{a}{\rho}\right)\ln\left(\frac{b}{\rho}\right)\hat{\rho} + \left(\frac{a}{\rho}\right)\left(\frac{z}{\rho}\right)\hat{z}\right]. \tag{1.131}$$

In Figures 1.31a and b the charge, current and electromagnetic field are sketched for a region near the center of the wire.[13] The volume current density $\vec{\mathcal{J}}$ is in the axial direction and is uniform throughout the wire. The surface charge density ρ_s is zero at the center of the wire ($z = 0$) and increases linearly with z; it is positive for $z < 0$ and negative for $z > 0$. The electric field $\vec{\mathcal{E}}$ has only an axial component within the wire. Outside the wire there are axial and radial components to the electric field. The magnetic excitation $\vec{\mathcal{H}}$ is in the azimuthal direction; it is zero on the axis of the wire, increases with increasing ρ until it reaches a maximum at the surface of the wire, then decreases with increasing ρ.

The Poynting vector $\vec{\mathcal{S}}$ is sketched in Figure 1.31c. *Outside* the wire, the Poynting vector has an axial component $\mathcal{S}_{2z}$, which is due to $\mathcal{E}_{2\rho}$ and $\mathcal{H}_{2\phi}$. This component of the Poynting vector represents an axial flow of energy in the space between the conductors. Since the surface charge density ρ_s produces $\mathcal{E}_{2\rho}$ (1.127), it is essential for the axial flow of energy.[14] There is also a radial component $\mathcal{S}_{2\rho}$ to the Poynting

[13] In Figures 1.31 and 1.32, the length of the arrow is used to indicate the relative magnitude of the current density, electric field, and Poynting vector; the number of symbols indicates the relative magnitude of the charge density and magnetic excitation.

[14] In Poynting's analysis of the wire, he did not include the radial component of the electric field, $\mathcal{E}_{2\rho}$; therefore, he obtained no axial component $\mathcal{S}_{2z}$ to the Poynting vector. He concluded that all energy flow was normal to the conductor [38]. This oversight was later pointed out by Heaviside [48].

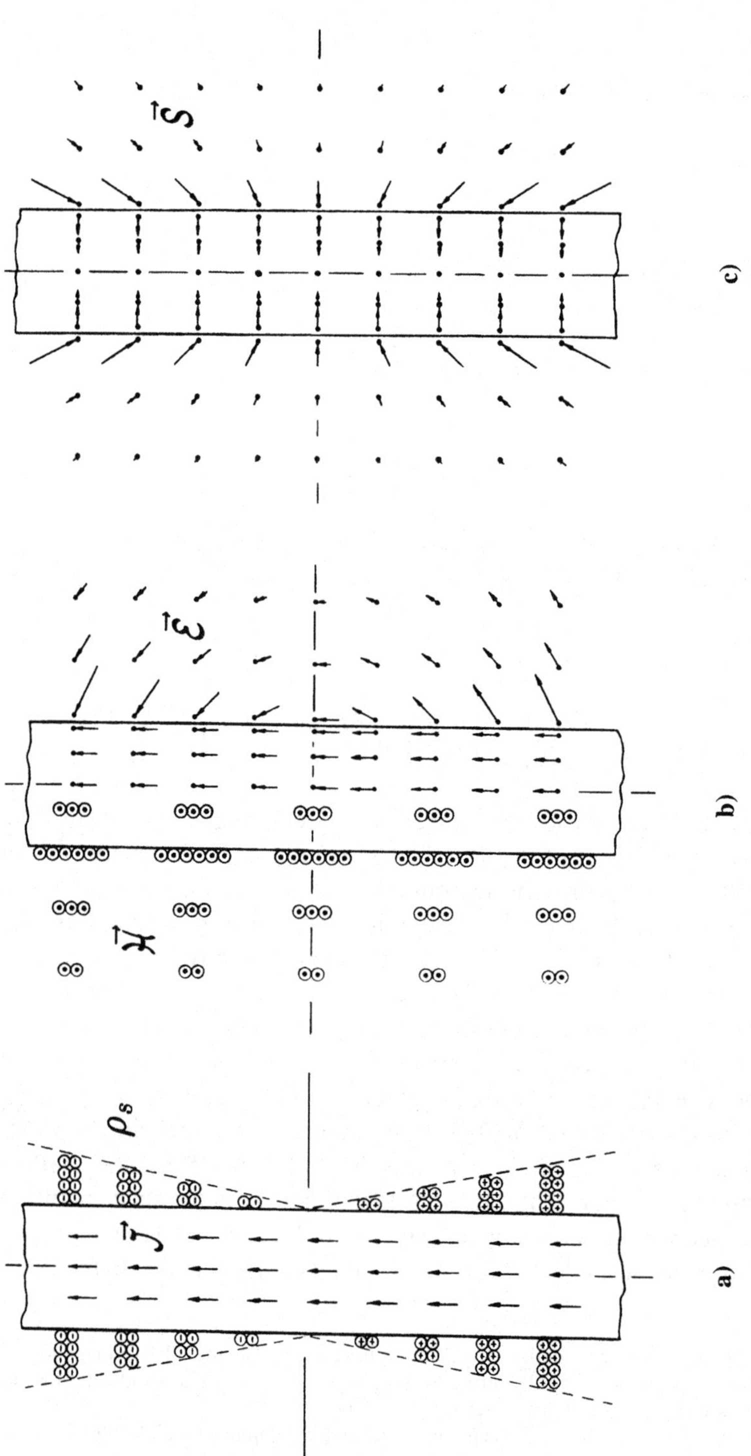

Fig. 1.31. Region near the center of a long, straight wire. a) Volume density of current and surface density of charge. b) Right-hand side: electric field; left-hand side: magnetic excitation. c) Poynting vector.

vector outside the wire, which is due to $\mathcal{E}_{2z}$ and $\mathcal{H}_{2\phi}$. At the surface of the wire, this component represents a radial flow of energy from space into the wire.

Inside the wire, the Poynting vector is in the radial direction, $\mathcal{S}_{1\rho}$; it represents a radial flow of energy from the surface of the wire to its interior, and it decreases with decreasing radius due to the dissipation of energy as heat within the conductor.

We now have a fairly clear picture for energy transport from the batteries to the wire. Outside the wire, energy flows from both ends of the structure (the batteries at $z = \pm L$) toward its center ($z = 0$), guided by the conductors (the wire and the perfectly conducting cylindrical shell). At the same time, energy is flowing into the wire through its surface. This flow is in the radial direction and uniform along the length of the wire. Inside the wire there exists only a radial flow of energy toward the axis. This flow decreases with decreasing radius due to dissipation of energy as heat.

We can show that energy is conserved in this process by applying Poynting's theorem to an annular volume of free space between the conductors. Let the bottom and top surfaces of this volume be at z and $z + \Delta z$, respectively. Because we are dealing with the steady state, the electromagnetic energy stored in this space is not changing with time, so Poynting's theorem (1.117) becomes simply

$$\oint\!\!\!\oint_S \vec{\mathcal{S}} \cdot d\vec{S} = 0.$$

With no energy flow into the perfect conductor ($\rho = b$), we only need to consider the surfaces at z, $z + \Delta z$, and $\rho = a$:

$$\begin{aligned}\oint\!\!\!\oint_S \vec{\mathcal{S}} \cdot d\vec{S} &= \int_{\rho=a}^{b} \int_{\phi=0}^{2\pi} (-\hat{z}) \cdot \vec{\mathcal{S}}_2\big|_z \rho d\phi d\rho + \int_{\rho=a}^{b} \int_{\phi=0}^{2\pi} \hat{z} \cdot \vec{\mathcal{S}}_2\big|_{z+\Delta z} \rho d\phi d\rho \\ &\quad + \int_{z'=z}^{z+\Delta z} \int_{\phi=0}^{2\pi} (-\hat{\rho}) \cdot \vec{\mathcal{S}}_2\big|_a a d\phi dz' \\ &= \frac{I^2}{\pi a^2 \sigma} z - \frac{I^2}{\pi a^2 \sigma}(z + \Delta z) + \frac{I^2}{\pi a^2 \sigma} \Delta z = 0.\end{aligned}$$

Before we leave this example, let us consider one alternative form for the Poynting vector. We will choose $\vec{\mathcal{S}}'$ (1.119a) to be

$$\vec{\mathcal{S}}' = \vec{\mathcal{S}} + \vec{\mathcal{T}} = \vec{\mathcal{S}} + \nabla \times (\Phi \vec{\mathcal{H}}), \tag{1.132}$$

where Φ is the scalar potential given by (1.122) and (1.124). Clearly $\nabla \cdot \vec{\mathcal{T}} = 0$, as required by (1.119b), since $\nabla \cdot (\nabla \times \vec{A}) = 0$ for any vector field $\vec{A}$ (Appendix B). In the two regions we then have

$$\vec{\mathcal{T}}_1 = \left(\frac{I^2}{2\pi^2 a^3 \sigma}\right)\left[\left(\frac{\rho}{a}\right)\hat{\rho} - 2\left(\frac{z}{a}\right)\hat{z}\right],$$

$$\vec{\mathcal{S}}'_1 = -2\left(\frac{I^2}{2\pi^2 a^3 \sigma}\right)\left(\frac{z}{a}\right)\hat{z},$$

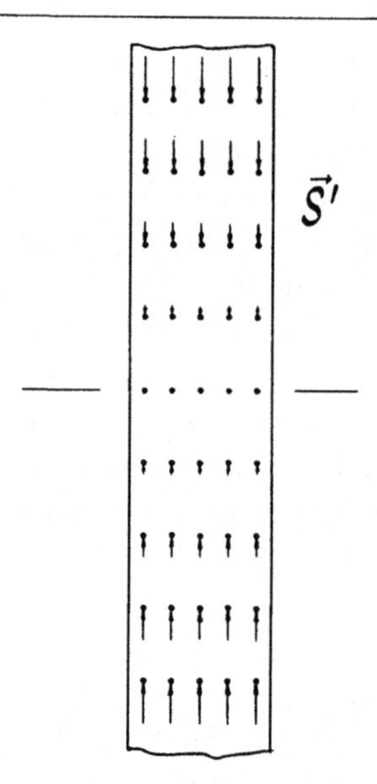

Fig. 1.32. Alternate Poynting vector $\vec{S}'$ in the region near the center of a long, straight wire.

and

$$\vec{T}_2 = \left(\frac{I^2}{2\pi^2 a^3 \sigma}\right)\frac{1}{\ln(b/a)}\left[\left(\frac{a}{\rho}\right)\ln\left(\frac{b}{\rho}\right)\hat{\rho} + \left(\frac{a}{\rho}\right)\left(\frac{z}{\rho}\right)\hat{z}\right],$$

$$\vec{S}_2' = 0.$$

The alternate Poynting vector $\vec{S}'$, sketched in Figure 1.32, gives a picture for the energy flow that is completely different from the one we obtained with the conventional Poynting vector $\vec{S}$ (Figure 1.31c). There is no energy flow in the space between the conductors; all energy flow is in the axial direction in the wire. This energy flow decreases linearly with axial position due to the dissipation of energy as heat in the conductor.

In this alternate physical interpretation, energy is no longer transported from the battery to the wire through the space surrounding the wire, but it is transported from the battery to different portions of the wire through the wire. Our original interpretation, based on the conventional Poynting vector $\vec{S}$, is the one usually accepted for this problem. However, this example illustrates the controversy we mentioned earlier associated with attaching a particular physical interpretation to the terms in Poynting's theorem.

At the beginning of this section, we used the Lorentz force expression to show that an electromagnetic field could cause a change in the kinetic energy of a charged particle. This fact together with the principle of conservation of energy was used to introduce the concept of energy for the electromagnetic field. In a similar manner, the mechanical concepts of linear momentum and angular momentum can be

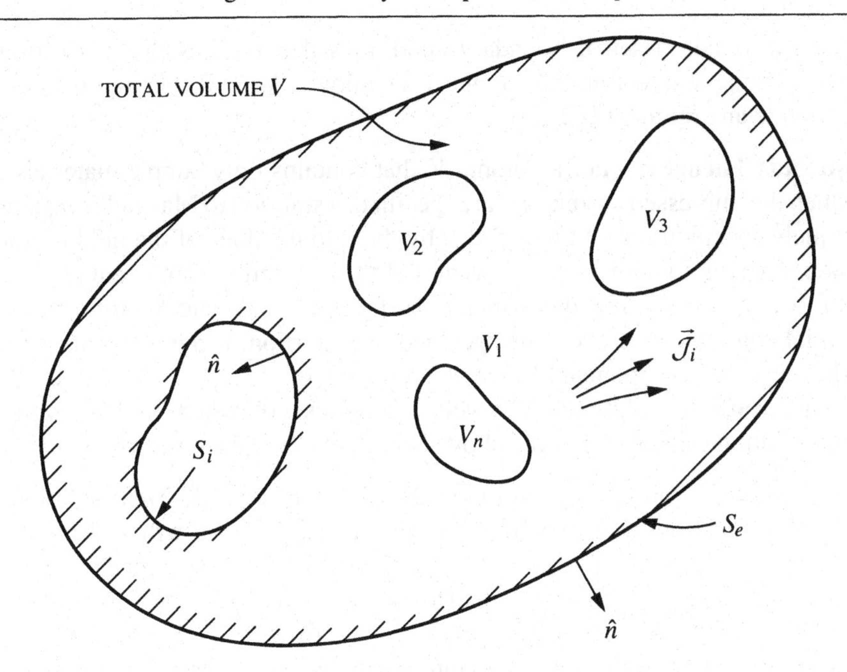

Fig. 1.33. Volume used in the discussion of uniqueness for solutions to Maxwell's equations.

introduced for the electromagnetic field. We will not pursue these topics here, but we refer the interested reader to the literature [42, 51].

1.5 Electromagnetic boundary value problem – uniqueness theorem

The geometry for a typical electromagnetic boundary value problem is sketched in Figure 1.33. Here we have a volume of space V with interior surface S_i and exterior surface S_e. Within this volume there may be regions occupied by different materials, such as dielectric or magnetic materials, conductors, and perfect conductors. These subvolumes are labeled $V_1, V_2, \ldots, V_n$. Maxwell's equations are to be solved for the electromagnetic field within each subvolume, and the field should satisfy the electromagnetic boundary conditions, given in Table 1.4, at each boundary surface. We should mention that the electromagnetic boundary conditions are not boundary conditions in the strict sense; they do not specify the field ($\vec{\mathcal{E}}$, $\vec{\mathcal{B}}$, etc.) on a boundary surface. They specify the jumps that must occur in components of the field when crossing a material boundary containing a surface charge or surface current. Perhaps it would be more appropriate to call these conditions the "electromagnetic jump conditions."

Before we can solve these partial differential equations, Maxwell's equations, we must supply supplementary information in the form of true boundary conditions.[15]

[15] Here we are using the term boundary condition in the general sense; it includes the specification of the field on spatial boundaries as well as the specification of the field at particular times – initial conditions.

We wish to obtain a set of boundary conditions that ensures that a solution to Maxwell's equations within the volume V is unique. We will state our results and then provide an argument [52, 53].

Uniqueness Theorem For a volume V that contains only simple materials and for which the impressed currents $\vec{\mathcal{J}}_i$ are specified, a solution to Maxwell's equations is uniquely specified for all times $t > 0$ by the initial values of the field ($\vec{\mathcal{E}}$ and $\vec{\mathcal{B}}$ at time $t = 0$) throughout V and the values of the tangential component of either $\vec{\mathcal{E}}$ or $\vec{\mathcal{B}}$ ($\hat{n} \times \vec{\mathcal{E}}$ or $\hat{n} \times \vec{\mathcal{B}}$) over the boundary surfaces of V (S_i and S_e) for $t \geq 0$. The tangential component of $\vec{\mathcal{E}}$ can be specified on a portion of the boundary surface and the tangential component of $\vec{\mathcal{B}}$ on the remainder.

In other words, if a solution to Maxwell's equations is obtained within the volume, and the solution satisfies all of the above conditions, it is the only solution.

We begin our argument by assuming that there are two different solutions to Maxwell's equations in the volume V for times $t > 0$. We will call these two solutions $\vec{\mathcal{E}}_a$, $\vec{\mathcal{B}}_a$ and $\vec{\mathcal{E}}_b$, $\vec{\mathcal{B}}_b$. At time $t = 0$ the field $\vec{\mathcal{E}}$, $\vec{\mathcal{B}}$ is specified everywhere in V, and both solutions a and b are equal to the specified field. Now we will consider the difference between the solutions: $\delta\vec{\mathcal{E}} = \vec{\mathcal{E}}_a - \vec{\mathcal{E}}_b$, $\delta\vec{\mathcal{B}} = \vec{\mathcal{B}}_a - \vec{\mathcal{B}}_b$, etc. Since Maxwell's equations are linear in all field quantities, the difference field must be a solution to Maxwell's equations

$$\nabla \times \delta\vec{\mathcal{E}} = -\frac{\partial(\delta\vec{\mathcal{B}})}{\partial t}$$

$$\nabla \times \delta\vec{\mathcal{H}} = \delta\vec{\mathcal{J}} + \frac{\partial(\delta\vec{\mathcal{D}})}{\partial t}.$$

This can be seen by simply writing Maxwell's equations separately for $\vec{\mathcal{E}}_a$, $\vec{\mathcal{B}}_a$ and $\vec{\mathcal{E}}_b$, $\vec{\mathcal{B}}_b$ and then subtracting the equations for b from those for a.

Maxwell's equations for the difference field can be combined as in the derivation of Poynting's theorem (Section 1.4) to obtain

$$\iiint_V \delta\vec{\mathcal{E}} \cdot \delta\vec{\mathcal{J}} dV + \iiint_V \left[\delta\vec{\mathcal{E}} \cdot \frac{\partial(\delta\vec{\mathcal{D}})}{\partial t} + \delta\vec{\mathcal{H}} \cdot \frac{\partial(\delta\vec{\mathcal{B}})}{\partial t}\right] dV$$

$$+ \oint\!\!\oint_S (\delta\vec{\mathcal{E}} \times \delta\vec{\mathcal{H}}) \cdot \hat{n} dS = 0. \tag{1.133}$$

Now we will assume that there are only simple materials or perfect conductors within V, so the constitutive relations (1.9)–(1.11) apply. In particular, for the volume density of current we have

$$\vec{\mathcal{J}}_a = \vec{\mathcal{J}}_{ca} + \vec{\mathcal{J}}_{ia} = \sigma\vec{\mathcal{E}}_a + \vec{\mathcal{J}}_{ia}$$

and

$$\vec{\mathcal{J}}_b = \vec{\mathcal{J}}_{cb} + \vec{\mathcal{J}}_{ib} = \sigma\vec{\mathcal{E}}_b + \vec{\mathcal{J}}_{ib},$$

where the first term in each expression applies within a conductor and the second term applies within a region where there are impressed currents. Since the impressed currents are specified, $\vec{\mathcal{J}}_{ia} = \vec{\mathcal{J}}_{ib}$ and

$$\delta\vec{\mathcal{J}} = \sigma(\vec{\mathcal{E}}_a - \vec{\mathcal{E}}_b) = \sigma(\delta\vec{\mathcal{E}}). \tag{1.134}$$

After inserting (1.134) and the other constitutive relations, (1.133) becomes

$$\iiint_V \sigma|\delta\vec{\mathcal{E}}|^2 dV + \frac{\partial}{\partial t}\left(\frac{1}{2}\iiint_V \varepsilon|\delta\vec{\mathcal{E}}|^2 dV + \frac{1}{2}\iiint_V \frac{1}{\mu}|\delta\vec{\mathcal{B}}|^2 dV\right)$$

$$+ \oint\!\!\oint_S \frac{1}{\mu}(\delta\vec{\mathcal{E}} \times \delta\vec{\mathcal{B}}) \cdot \hat{n}\, dS = 0. \tag{1.135}$$

The integrand of the last integral in (1.135) is

$$(\delta\vec{\mathcal{E}} \times \delta\vec{\mathcal{B}}) \cdot \hat{n} = \delta\vec{\mathcal{B}} \cdot (\hat{n} \times \delta\vec{\mathcal{E}}) = -\delta\vec{\mathcal{E}} \cdot (\hat{n} \times \delta\vec{\mathcal{B}}), \tag{1.136}$$

where we have used the vector identity (Appendix B)

$$\vec{A} \cdot (\vec{B} \times \vec{C}) = \vec{C} \cdot (\vec{A} \times \vec{B}) = -\vec{B} \cdot (\vec{A} \times \vec{C}).$$

We will assume that both solutions satisfy the boundary conditions on S stated above: $\hat{n} \times \vec{\mathcal{E}}$ or $\hat{n} \times \vec{\mathcal{B}}$ is specified on S for times $t > 0$. Thus $\hat{n} \times \delta\vec{\mathcal{E}}$ or $\hat{n} \times \delta\vec{\mathcal{B}}$ is zero at every point on S; this makes (1.136) and thus the last integral in (1.135) zero:

$$\frac{\partial}{\partial t}\left(\frac{1}{2}\iiint_V \varepsilon|\delta\vec{\mathcal{E}}|^2 dV + \frac{1}{2}\iiint_V \frac{1}{\mu}|\delta\vec{\mathcal{B}}|^2 dV\right) = -\iiint_V \sigma|\delta\vec{\mathcal{E}}|^2 dV,$$

or

$$\frac{\partial \mathcal{X}}{\partial t} = -\iiint_V \sigma|\delta\vec{\mathcal{E}}|^2 dV. \tag{1.137}$$

Here we have introduced the notation

$$\mathcal{X} = \frac{1}{2}\iiint_V \varepsilon|\delta\vec{\mathcal{E}}|^2 dV + \frac{1}{2}\iiint_V \frac{1}{\mu}|\delta\vec{\mathcal{B}}|^2 dV.$$

The term $\mathcal{X}$ is always a positive number or zero:

$$\mathcal{X} \geq 0, \qquad t > 0, \tag{1.138a}$$

and from (1.137) the partial derivative of $\mathcal{X}$ with respect to time must be a negative number or zero:

$$\frac{\partial \mathcal{X}}{\partial t} \leq 0, \qquad t > 0. \tag{1.138b}$$

The initial conditions we have assumed at $t = 0$ are $\vec{\mathcal{E}}_a = \vec{\mathcal{E}}_b$ or $\delta\vec{\mathcal{E}} = 0$, $\vec{\mathcal{B}}_a = \vec{\mathcal{B}}_b$ or $\delta\vec{\mathcal{B}} = 0$, which make

$$\mathcal{X} = 0, \qquad t = 0. \tag{1.138c}$$

These three statements (1.138a–c) are satisfied only if $\mathcal{X} = 0$ for $t > 0$.[16] This

[16] To see this, consider (1.138b), which we can write as

$$\frac{\partial \mathcal{X}}{\partial t} = \lim_{\Delta t \to 0}\left[\frac{\mathcal{X}(t + \Delta t) - \mathcal{X}(t)}{\Delta t}\right] \leq 0, \qquad t > 0.$$

means that $\delta\vec{\mathcal{E}} = 0$, $\delta\vec{\mathcal{B}} = 0$, and therefore $\vec{\mathcal{E}}_a = \vec{\mathcal{E}}_b$, $\vec{\mathcal{B}}_a = \vec{\mathcal{B}}_b$, for $t > 0$. Thus, there is only one solution, a unique solution, to Maxwell's equations within V for the stated conditions.

In the above argument we have ignored an important fact: The electromagnetic field propagates with a finite velocity. This fact can often be used to reduce the amount of information needed to obtain a unique solution. In other words, our original statement for uniqueness specifies information that is sufficient for obtaining a unique solution to Maxwell's equations, but some of this information may not be necessary.

This point is easily illustrated for a two-dimensional region of free space, the circular disc of radius ρ_o on the x-y plane shown in Figure 1.34a.[17] In two dimensions, our volume V with exterior surface S_e becomes the surface S bounded by the curve C_e. The electromagnetic field $\vec{\mathcal{E}}$, $\vec{\mathcal{B}}$ on S is assumed to be zero for times $t < 0$. The semi-infinite cylinder in Figure 1.34a shows this surface at times $t \geq 0$; that is, each cross section of the cylinder is the surface S at a different time.

In the next chapter, we will show that the electromagnetic field propagates with the speed of light in free space, c. Now consider the point P($x = 0$, $y = 0$) at time $t = t_1 < \rho_o/c$ in Figure 1.34b. Because of the finite velocity of propagation and causality, only the field at points within the right circular cone can influence the field at P.[18] The height of the cone is equal to the radius of its base, $ct_1 = \rho_1$. From this drawing we see that the initial values of the field ($\vec{\mathcal{E}}$ and $\vec{\mathcal{B}}$ at time $t = 0$) on the surface S_1 (disc of radius ρ_1) determine the field at P. The initial values of the field over the remainder of the surface S do not influence the field at P.

This means

$$\mathcal{X}(t + \Delta t) \leq \mathcal{X}(t),$$

and, with (1.138a),

$$0 \leq \mathcal{X}(t + \Delta t) \leq \mathcal{X}(t), \qquad t > 0.$$

Now we will assume that $\mathcal{X}$ is *continuous* at $t = 0$; then (1.138c) implies that

$$\mathcal{X} = 0, \qquad t = 0+,$$

and the above expression with $t = 0+$ gives

$$0 \leq \mathcal{X}(\Delta t) \leq \mathcal{X}(0+) = 0,$$

or

$$\mathcal{X}(\Delta t) = 0.$$

The same argument can be applied repeatedly to show that $\mathcal{X} = 0$ for all $t > 0$.

[17] Here we present a simple physical argument for two spatial dimensions; a rigorous uniqueness proof for three spatial dimensions, which includes the finite velocity of propagation, is presented in References [54] and [55].

[18] For a causal system, only events that have occurred earlier in time can influence an event at the present time. If causality were not invoked, Figure 1.34b would include an additional inverted cone with apex at P and base at $t > t_1$.

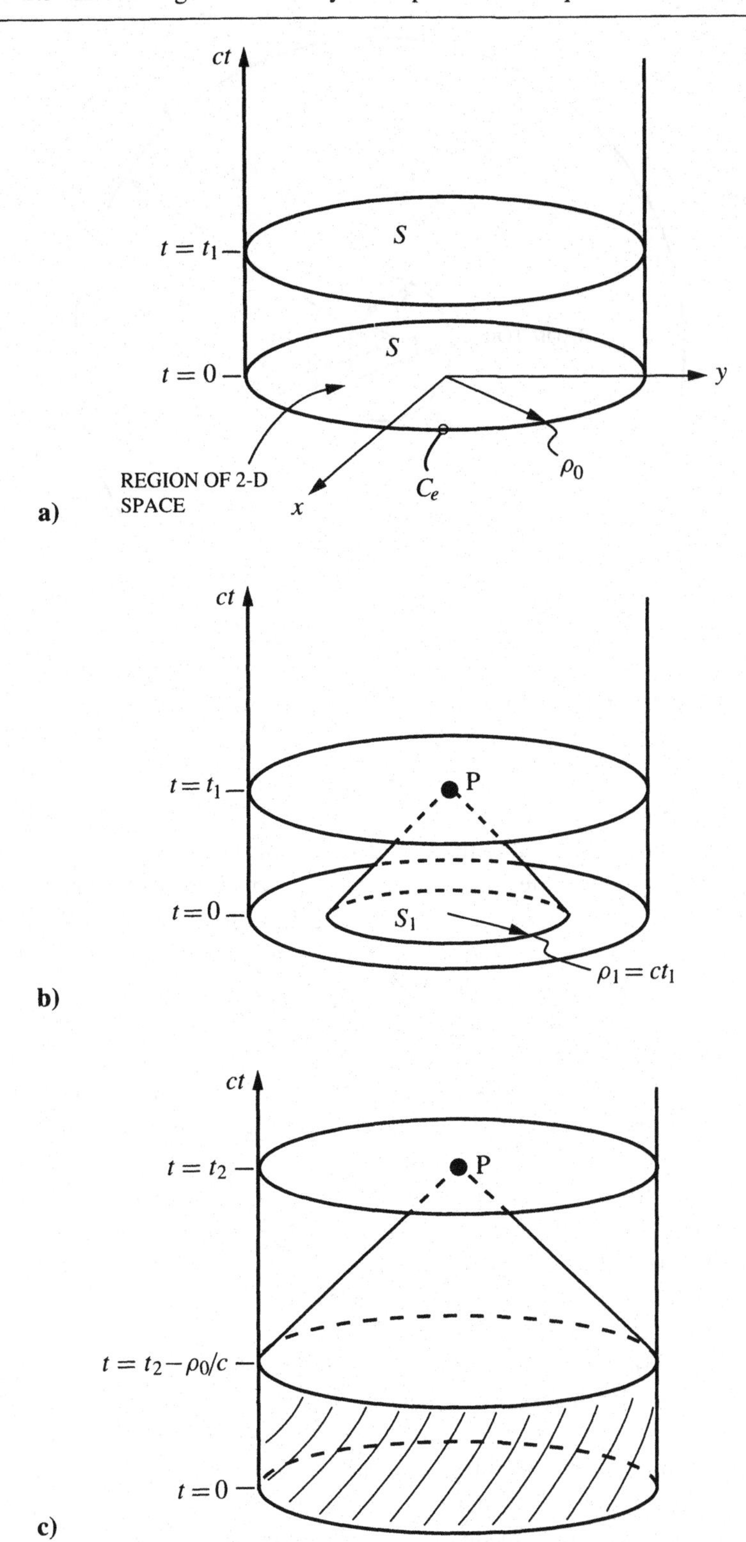

Fig. 1.34. Two-dimensional region in free space at various times.

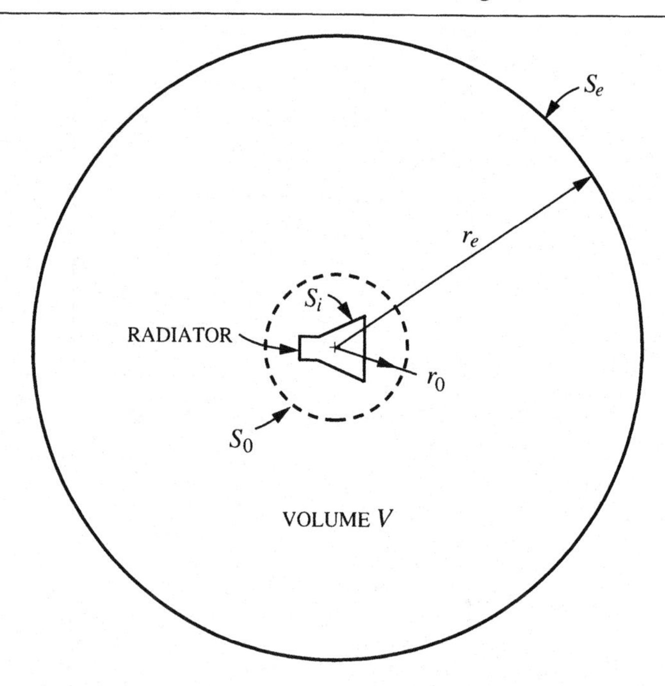

Fig. 1.35. Schematic drawing for a radiator in infinite space. The surface S_o encloses all sources.

At the later time $t = t_2 > \rho_o/c$ (Figure 1.34c) the surface of the cone intersects the surface of the cylinder. Now the initial values of the field ($t = 0$) on the entire surface S and the values of the tangential component of either $\vec{\mathcal{E}}$ or $\vec{\mathcal{B}}$ on the boundary curve C_e at times $0 \leq t \leq t_2 - \rho_o/c$ (shaded portion of the cylindrical surface) determine the field at P. The values of the tangential components of $\vec{\mathcal{E}}$ and $\vec{\mathcal{B}}$ on C_e for the times $t_2 - \rho_o/c < t \leq t_2$ do not influence the field at P.

In certain applications, the region in which Maxwell's equations are to be solved is infinite in extent; that is, in Figure 1.33 the external surface S_e recedes to infinity. The boundary condition on this surface, the specification of the tangential component of $\vec{\mathcal{E}}$ or $\vec{\mathcal{B}}$ for $t \geq 0$, then requires special consideration. This is best explained by discussing a simple example.

The example in Figure 1.35 is for a radiator (an antenna) in free space. The interior surface S_i of our volume V coincides with the surface of the radiator, and the exterior surface S_e is a sphere of radius r_e. The sphere of radius r_o encloses the radiator; it contains all of the sources for the field.

The electromagnetic field within V is zero for times $t < 0$, after which the radiator is turned on. Now the field is to be observed for times $0 \leq t \leq t_{\text{max}}$. If we choose the radius r_e of the exterior surface S_e so that $r_e > r_o + ct_{\text{max}}$, the physically relevant field $\vec{\mathcal{E}}, \vec{\mathcal{B}}$ (one that propagates with the speed of light and is causal) will be zero on this surface for the entire period of observation. As the exterior surface recedes to infinity, $r_e \to \infty$, the period of observation becomes infinite, $t_{\text{max}} \to \infty$. We conclude that a unique, physically relevant solution to Maxwell's equations is obtained by specifying the tangential component of either $\vec{\mathcal{E}}$ or $\vec{\mathcal{B}}$ on only the interior surface S_i. On the exterior surface S_e the tangential components of this field are zero due to the finite velocity of propagation for the electromagnetic field.

1.6 Numerical solution of Maxwell's equations using finite differences: An example

Up to this point, we have described the basic elements of the electromagnetic boundary value problem: the partial differential equations (Maxwell's equations), the constitutive relations, the boundary conditions, and the requirements for a unique solution. With these elements established, there are various mathematical methods that can be used to obtain a solution for a particular problem, that is, for a certain geometrical configuration of conductors, dielectrics, etc. These methods roughly fall into two categories: analytical methods and numerical methods.

Analytical methods provide solutions that are closed-form expressions (i.e., formulas) for the electromagnetic field. These solutions can be examined to obtain physical understanding of electromagnetic phenomena. For example, the behavior of the field at a large distance from a source can be found by evaluating the solution in the limit as the distance from the source becomes infinite (the radiated field, first discussed in Section 3.5). A particular analytical method is generally applicable to only a few geometrical configurations. Consequently, there are a limited number of analytical solutions available for study, and these are for rather simple geometries.

Numerical methods use appropriate algorithms with a digital computer to obtain solutions to Maxwell's equations. The solutions are a set of numbers, such as numerical values for the electric field at points on a grid in space for a particular set of times. Such solutions are not always useful for obtaining general physical understanding of electromagnetic phenomena. Nevertheless, numerical methods have an advantage in that they can be used to solve problems involving complicated geometries that are not solvable by standard analytical methods. Numerical methods are often the only alternative when an accurate solution is required to a practical electromagnetics problem.

In the remainder of this text, we will apply analytical methods to solve a few basic electromagnetic problems. Our objective will be to obtain a physical understanding of electromagnetic phenomena by examining the solutions.

In this section, we will describe a particularly straightforward, numerical method based on approximating partial derivatives by finite-difference quotients [56–61]. In the literature on electromagnetism, the method is often referred to as the finite-difference time-domain (FDTD) method. The approach, however, is not specific to electromagnetism, but is also used in other branches of physics, such as acoustics, and fluid mechanics. Our objective is simply to introduce numerical methods for solving electromagnetics problems. In this process we will see the solution of a practical electromagnetics problem from start to finish: the formulation of a theoretical model, a statement of the boundary value problem, the numerical solution, and the verification of the solution by comparison with experimental measurements.

To gain a basic understanding of the method, we will begin with a simple, one-dimensional problem in which the electromagnetic field is assumed to depend only

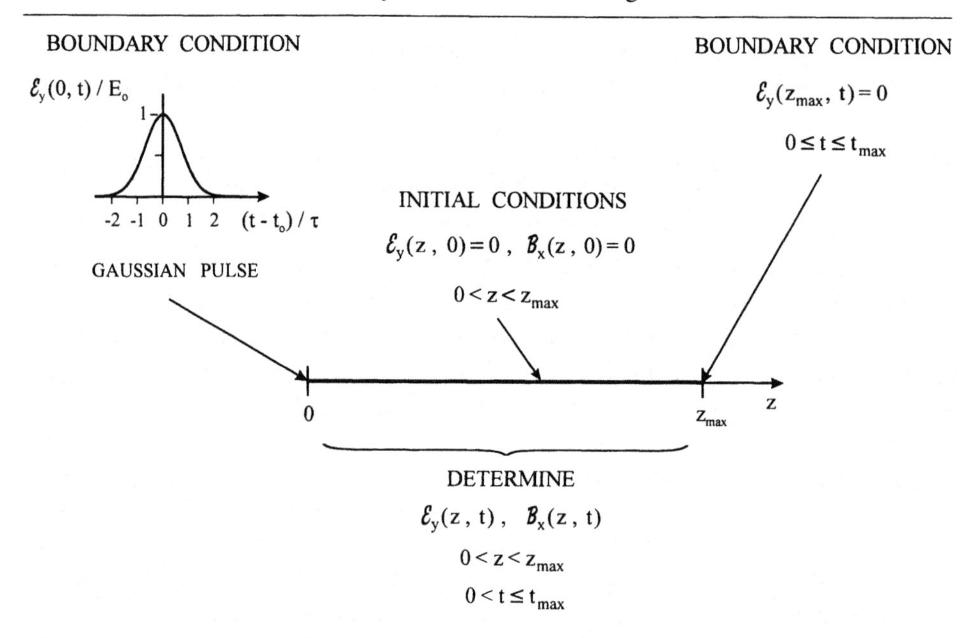

Fig. 1.36. Schematic drawing summarizing the elements of the one-dimensional problem.

on the z coordinate of the rectangular Cartesian system (x, y, z). In addition, the electric field is assumed to be in the y direction, $\vec{\mathcal{E}}(x, y, z, t) = \mathcal{E}_y(z, t)\hat{y}$. The two Maxwell's equations (1.50) and (1.51) for free space then become

$$\frac{\partial \mathcal{E}_y}{\partial z} = \frac{\partial \mathcal{B}_x}{\partial t} \tag{1.139}$$

and

$$\frac{\partial \mathcal{B}_x}{\partial z} = \varepsilon_o \mu_o \frac{\partial \mathcal{E}_y}{\partial t} = \frac{1}{c^2} \frac{\partial \mathcal{E}_y}{\partial t}. \tag{1.140}$$

Note that our assumptions have eliminated all but the x component of the magnetic field from these equations.

The geometry for the problem is sketched in Figure 1.36. The electromagnetic field is to be determined on the line $0 < z < z_{max}$ for times $0 < t \leq t_{max}$ (this line corresponds to the volume V in the uniqueness theorem). At time $t = 0$ the field is zero on this line $[\mathcal{E}_y(z, 0) = 0,\ \mathcal{B}_x(z, 0) = 0]$, and for times $t \geq 0$ the electric field $\mathcal{E}_y$ is specified at the boundary points $z = 0$ and $z = z_{max}$ (this corresponds to specifying $\hat{n} \times \vec{\mathcal{E}}$ on the surfaces S_i and S_e in the uniqueness theorem). We will let the electric field at the left-hand boundary be a Gaussian pulse in time:

$$\mathcal{E}_y(0, t) = E_o e^{-(t-t_o)^2/\tau^2}, \qquad 0 \leq t \leq t_{max}. \tag{1.141}$$

The parameter τ determines the width of the pulse (see Figure 1.36). The time shift t_o is chosen so that $t_o \gtrsim 2\tau$; the field (1.141) is then essentially zero at $t = 0$; this is compatible with our assumption that $\mathcal{E}_y(0+, 0) = 0$. The electric field at the right-hand boundary will be set to zero:

$$\mathcal{E}_y(z_{max}, t) = 0, \qquad 0 \leq t \leq t_{max}. \tag{1.142}$$

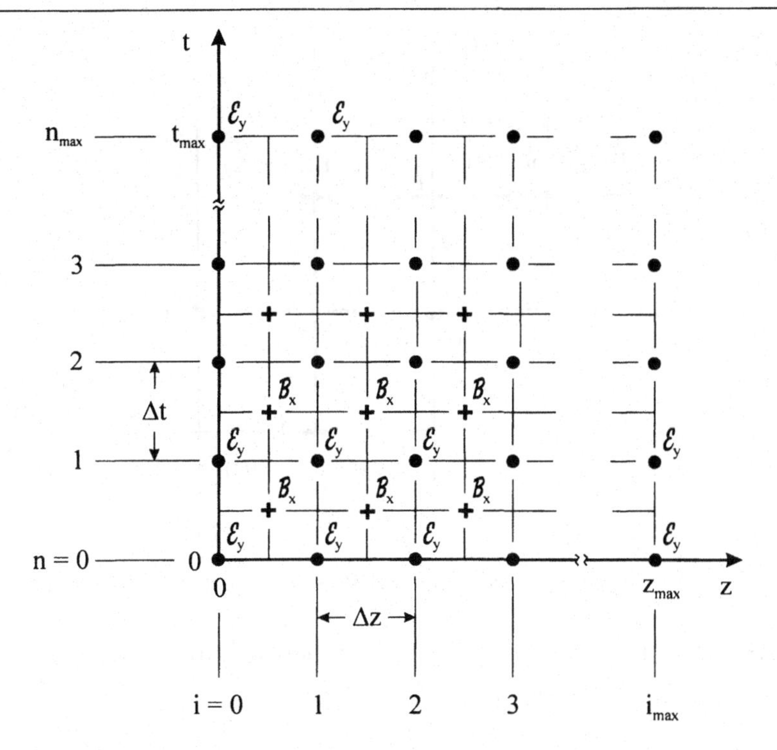

Fig. 1.37. Grid of points covering the region of spacetime for the one-dimensional problem.

The region of spacetime – the rectangle $0 \leq z \leq z_{\max}$, $0 \leq t \leq t_{\max}$ shown in Figure 1.37 – is covered by a grid of points. The grid spacings Δz and Δt are uniform and

$$\Delta z = z_{\max}/i_{\max}, \qquad \Delta t = t_{\max}/n_{\max}.$$

The following notation will be used for functions evaluated at points on the grid:

$$\mathcal{F}(i, n) = \mathcal{F}(i\Delta z, n\Delta t),$$

and similarly for points displaced by one half of a grid spacing in space or in time

$$\mathcal{F}\left(i + \frac{1}{2}, n + \frac{1}{2}\right) = \mathcal{F}\left[\left(i + \frac{1}{2}\right)\Delta z, \left(n + \frac{1}{2}\right)\Delta t\right],$$

with

$$i = 0, 1, 2, \ldots, i_{\max}, \qquad n = 0, 1, 2, \ldots, n_{\max}.$$

The electric field $\mathcal{E}_y$ is evaluated at each of the points (i, n) (solid dots); the magnetic field $\mathcal{B}_x$ is evaluated at each of the points $\left(i + \frac{1}{2}, n + \frac{1}{2}\right)$ (crosses) [60].

The region around the general point (i, n) is shown expanded in Figure 1.38. At the point $\left(i + \frac{1}{2}, n\right)$, which is marked A, the spatial derivative of $\mathcal{E}_y$ and the temporal

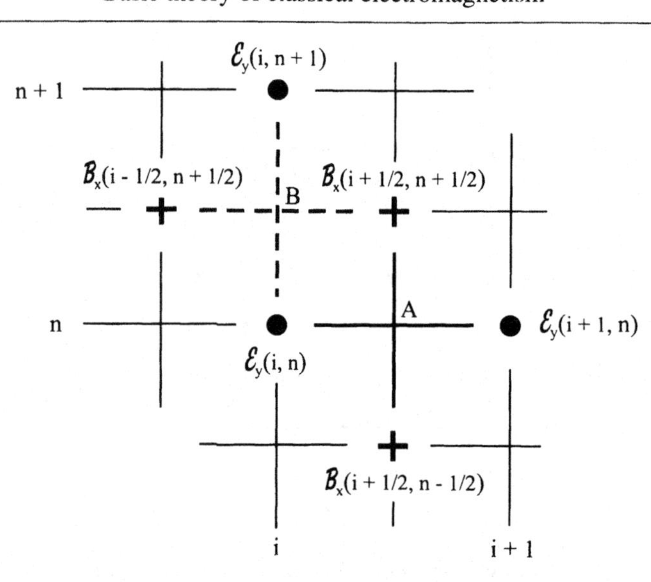

Fig. 1.38. Grid in the vicinity of the point (i, n).

derivative of $\mathcal{B}_x$ can be approximated by the following difference quotients:[19]

$$\left.\frac{\partial \mathcal{E}_y}{\partial z}\right|_{i+\frac{1}{2},n} \approx \frac{\mathcal{E}_y(i+1,n) - \mathcal{E}_y(i,n)}{\Delta z},$$

$$\left.\frac{\partial \mathcal{B}_x}{\partial t}\right|_{i+\frac{1}{2},n} \approx \frac{\mathcal{B}_x\left(i+\frac{1}{2}, n+\frac{1}{2}\right) - \mathcal{B}_x\left(i+\frac{1}{2}, n-\frac{1}{2}\right)}{\Delta t}.$$

When these results are substituted into (1.139), we get the following approximation for the first of Maxwell's equations:

$$\frac{\mathcal{E}_y(i+1,n) - \mathcal{E}_y(i,n)}{\Delta z} = \frac{\mathcal{B}_x\left(i+\frac{1}{2}, n+\frac{1}{2}\right) - \mathcal{B}_x\left(i+\frac{1}{2}, n-\frac{1}{2}\right)}{\Delta t},$$

or

$$c\mathcal{B}_x\left(i+\frac{1}{2}, n+\frac{1}{2}\right) = c\mathcal{B}_x\left(i+\frac{1}{2}, n-\frac{1}{2}\right) + \left(\frac{c\Delta t}{\Delta z}\right)\left[\mathcal{E}_y(i+1,n) - \mathcal{E}_y(i,n)\right]. \tag{1.143}$$

In a similar manner, the spatial derivative of $\mathcal{B}_x$ and the temporal derivative of $\mathcal{E}_y$ can be approximated at the point $\left(i, n+\frac{1}{2}\right)$, which is marked B:

$$\left.\frac{\partial \mathcal{B}_x}{\partial z}\right|_{i,n+\frac{1}{2}} \approx \frac{\mathcal{B}_x\left(i+\frac{1}{2}, n+\frac{1}{2}\right) - \mathcal{B}_x\left(i-\frac{1}{2}, n+\frac{1}{2}\right)}{\Delta z},$$

$$\left.\frac{\partial \mathcal{E}_y}{\partial t}\right|_{i,n+\frac{1}{2}} \approx \frac{\mathcal{E}_y(i,n+1) - \mathcal{E}_y(i,n)}{\Delta t}.$$

[19] For small Δz and Δt, these centered difference approximations differ from the actual derivatives by terms proportional to $(\Delta z)^2$ and $(\Delta t)^2$, respectively.

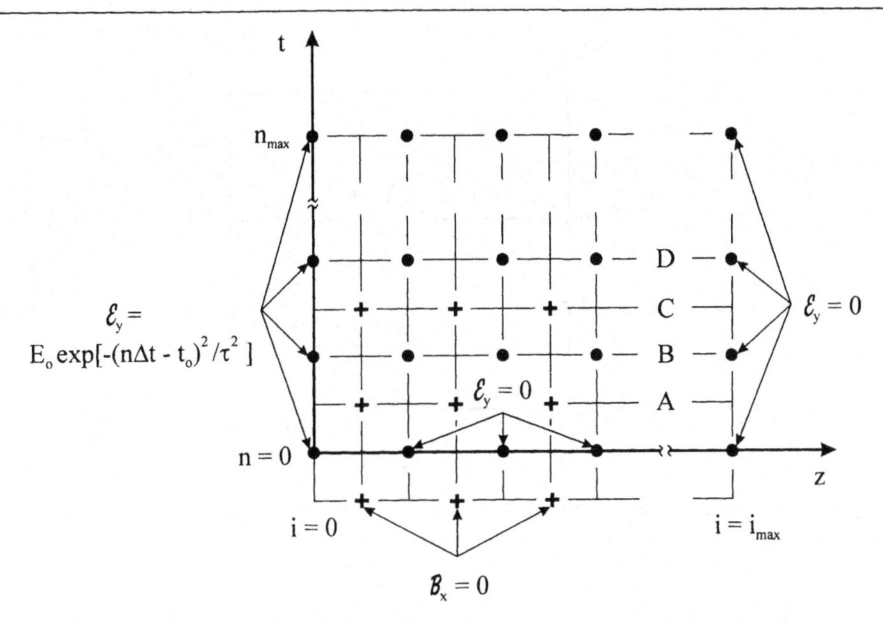

Fig. 1.39. Initial conditions and boundary conditions for the one-dimensional problem.

The approximation for the second of Maxwell's equations (1.140) is then

$$\frac{\mathcal{B}_x\left(i+\frac{1}{2}, n+\frac{1}{2}\right) - \mathcal{B}_x\left(i-\frac{1}{2}, n+\frac{1}{2}\right)}{\Delta z} = \varepsilon_o\mu_o \frac{\mathcal{E}_y(i, n+1) - \mathcal{E}_y(i, n)}{\Delta t},$$

or

$$\mathcal{E}_y(i, n+1) = \mathcal{E}_y(i, n) + \left(\frac{c\Delta t}{\Delta z}\right)\left[c\mathcal{B}_x\left(i+\frac{1}{2}, n+\frac{1}{2}\right) - c\mathcal{B}_x\left(i-\frac{1}{2}, n+\frac{1}{2}\right)\right]. \tag{1.144}$$

Note that Equations (1.143) and (1.144) allow $\mathcal{B}_x$ and $\mathcal{E}_y$, respectively, to be calculated at a particular time using only values of $\mathcal{B}_x$ and $\mathcal{E}_y$ from earlier times. For example, $\mathcal{B}_x$ at $t = \left(n+\frac{1}{2}\right)\Delta t$ is determined from $\mathcal{B}_x$ at $t = \left(n-\frac{1}{2}\right)\Delta t$ and $\mathcal{E}_y$ at $t = n\Delta t$.

The region of solution is redrawn in Figure 1.39. The initial conditions at $t = 0$ are $\mathcal{E}_y = 0$, $\mathcal{B}_x = 0$ for $0 < z < z_{\max}$; hence, $\mathcal{E}_y$ is set to zero at all points $i = 1, 2, \ldots, i_{\max} - 1$ in the row $n = 0$. Since $\mathcal{B}_x$ is not evaluated at $t = 0$, a row at $n = -1/2$ ($t = -\Delta t/2$) has been added, and $\mathcal{B}_x$ is set to zero at all points in this row. The electric field $\mathcal{E}_y$ is given by (1.141) at all points in the column $i = 0$, whereas it is set to zero (1.142) at all points in the column $i = i_{\max}$.

The electromagnetic field at the remaining grid points in Figure 1.39 is determined by alternately applying the finite-difference equations, that is, (1.143) and (1.144). First (1.143) is applied at each point in the row marked A to determine $\mathcal{B}_x$ at time $t = \Delta t/2$, that is, $\mathcal{B}_x\left(i+\frac{1}{2}, \frac{1}{2}\right)$, $i = 0, 1, 2, \ldots, i_{\max} - 1$. Next (1.144) is applied at each point in the row marked B to determine $\mathcal{E}_y$ at time $t = \Delta t$, that is, $\mathcal{E}_x(i, 1)$, $i = 1, 2, \ldots, i_{\max} - 1$. This process is then repeated for each of the other pairs of rows (C, D; E, F; etc.) until the electromagnetic field is determined at all

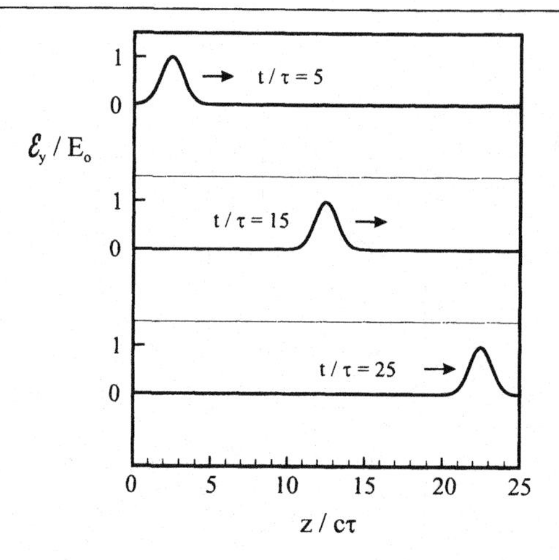

Fig. 1.40. Numerical results showing the distribution of the electric field in space at various times. The pulse has not reached the right-hand boundary, $t_o/\tau = 2.5$.

of the grid points. This procedure is often called "stepping-in-time" or "marching-in-time," because the solution is developed incrementally in time starting with the initial values at $t = 0$.

Numerical results calculated using this procedure are presented in Figure 1.40. The distribution of the electric field in space ($\mathcal{E}_y/E_o$ versus $z/c\tau$) is shown for three times: $t/\tau = 5$, 15, and 25. For this example $t_o/\tau = 2.5$. The field is seen to be a Gaussian pulse in space. A simple calculation shows that the pulse is propagating to the right with the speed of light c. From these purely numerical results, we infer that the electric field has the following form in space and time:

$$\mathcal{E}_y(z,t) = E_o e^{-[(t-z/c)-t_o]^2/\tau^2}, \tag{1.145}$$

and from similar numerical results for the magnetic field we find that

$$\mathcal{B}_x(z,t) = -\frac{1}{c}\mathcal{E}_y(z,t) = -\frac{1}{c}E_o e^{-[(t-z/c)-t_o]^2/\tau^2}. \tag{1.146}$$

These are the equations for an electromagnetic plane wave propagating in free space, as we will show by analytical means in Section 2.1.

For the times shown in Figure 1.40, the pulse has not yet reached the right-hand boundary at $z = z_{\max}$. Our assumption (1.142) that the electric field is zero at this point for the times of observation is consistent with these results. The pulse propagates as it would if the right-hand boundary were absent.

For the later times shown in Figure 1.41, the pulse has reached the boundary, where it is seen to be reflected in inverted form. Our assumption that the electric field is always zero at $z = z_{\max}$ physically corresponds to the region $z \geq z_{\max}$ being a perfect conductor: The tangential component of the electric field, which is $\mathcal{E}_y$ in this case, is always zero at the surface of a perfect conductor. From these purely numerical results, we conclude that the electric field of a plane wave (1.145) normally incident on a perfectly conducting half space is reflected in inverted form – a result that is easily confirmed by analytical means.

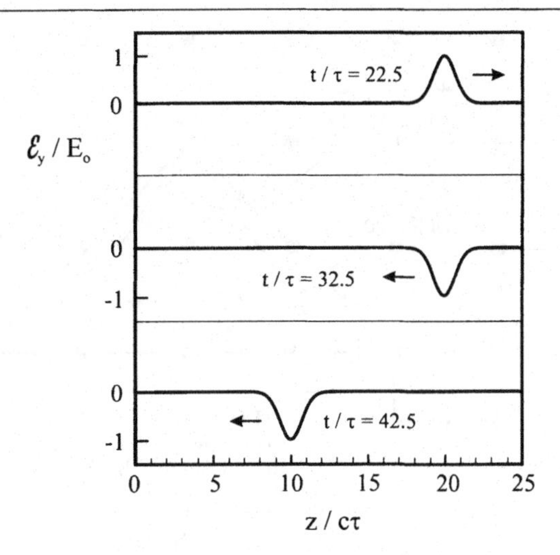

Fig. 1.41. Numerical results showing the distribution of the electric field in space at various times. The pulse has reflected from the right-hand boundary, $t_o/\tau = 2.5$.

We should say a few words about the size of the grid spacings Δz and Δt. Of course, we have assumed that our method of solution is convergent; that is, the solution to the difference equations, (1.143) and (1.144), tends to the solution of the differential equations, (1.139) and (1.140), as Δz and Δt tend to zero. Thus, we expect our results to differ negligibly from the actual solution when Δz and Δt are chosen reasonably small, that is, small enough to resolve the spatial and temporal variations of the field.

A mathematical argument shows that Δz and Δt cannot be chosen independently if our method of solution is to be convergent [56–59]. In fact, for convergence the grid spacings must satisfy the *domain of dependence condition* (Courant–Friedrichs–Lewy condition), which, for a one-dimensional problem, is

$$\frac{c\Delta t}{\Delta z} \leq 1. \tag{1.147}$$

Below, we will offer a simple argument for this condition.

Consider a point at which the field is to be determined, like the point marked A in Figure 1.42. Examination of the difference equations, (1.143) and (1.144), shows that the field at this point is determined solely by the values of the field at the points within the triangle shaded dark gray. The sides of this triangle have slopes $\pm\Delta t/\Delta z$. The interior of the triangle is called the *domain of dependence for the difference equations.* We know from Maxwell's differential equations that the electromagnetic field propagates with a maximum speed equal to the speed of light c. Therefore, the field at any point within the triangle shaded light gray in Figure 1.42 can influence the field at point A. The sides of this triangle have slopes $\pm 1/c$, and the interior of the triangle is called the *domain of dependence for the differential equations.*

This domain of dependence for the difference equations must include the domain of dependence for the differential equations. Otherwise, there would be points that could theoretically influence the result at A that would not be included in our

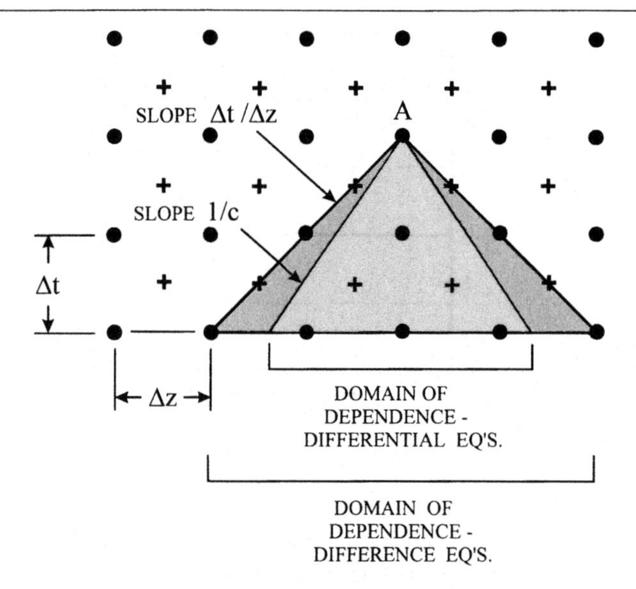

Fig. 1.42. Domains of dependence for the differential equations (triangle shaded light gray) and the difference equations (triangle shaded dark gray).

numerical scheme. So we must choose

$$\frac{\Delta t}{\Delta z} \leq \frac{1}{c},$$

or

$$\frac{c\Delta t}{\Delta z} \leq 1,$$

which is Equation (1.147).

The numerical method of solution we have described for one spatial dimension can be extended to two and three spatial dimensions. The mathematics are no more complicated; however, it is a bit more difficult to visualize the spatial and temporal arrangement of the field components for the higher dimensions. For three spatial dimensions and the rectangular Cartesian coordinate system (x, y, z), the grid in space (called the *Yee lattice*) is composed of cuboids with sides Δx, Δy, and Δz. The six components of the electromagnetic field are evaluated at staggered positions in space as shown in Figure 1.43, and the finite-difference equations corresponding to the two Maxwell's equations, (1.50) and (1.51), in free space are [60, 61]:

$$c\mathcal{B}_x^{n+\frac{1}{2}}\left(i, j+\frac{1}{2}, k+\frac{1}{2}\right) =$$

$$c\mathcal{B}_x^{n-\frac{1}{2}}\left(i, j+\frac{1}{2}, k+\frac{1}{2}\right) + c\Delta t\left\{\frac{1}{\Delta z}\left[\mathcal{E}_y^n\left(i, j+\frac{1}{2}, k+1\right)\right.\right.$$

$$\left.\left. - \mathcal{E}_y^n\left(i, j+\frac{1}{2}, k\right)\right] - \frac{1}{\Delta y}\left[\mathcal{E}_z^n\left(i, j+1, k+\frac{1}{2}\right) - \mathcal{E}_z^n\left(i, j, k+\frac{1}{2}\right)\right]\right\},$$

(1.148a)

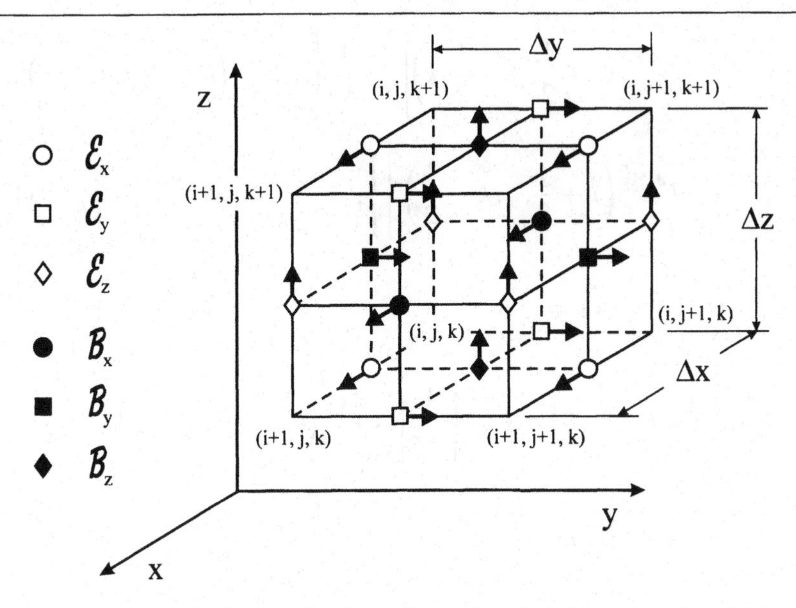

Fig. 1.43. Spatial grid (*Yee lattice*) for three-dimensional problem showing the location of the field components about the point $(i,\ j,\ k)$.

$$c\mathcal{B}_y^{n+\frac{1}{2}}\left(i+\frac{1}{2},j,k+\frac{1}{2}\right)=$$

$$c\mathcal{B}_y^{n-\frac{1}{2}}\left(i+\frac{1}{2},j,k+\frac{1}{2}\right)+c\Delta t\left\{\frac{1}{\Delta x}\left[\mathcal{E}_z^n\left(i+1,j,k+\frac{1}{2}\right)\right.\right.$$

$$\left.\left.-\mathcal{E}_z^n\left(i,j,k+\frac{1}{2}\right)\right]-\frac{1}{\Delta z}\left[\mathcal{E}_x^n\left(i+\frac{1}{2},j,k+1\right)-\mathcal{E}_x^n\left(i+\frac{1}{2},j,k\right)\right]\right\},$$

(1.148b)

$$c\mathcal{B}_z^{n+\frac{1}{2}}\left(i+\frac{1}{2},j+\frac{1}{2},k\right)=$$

$$c\mathcal{B}_z^{n-\frac{1}{2}}\left(i+\frac{1}{2},j+\frac{1}{2},k\right)+c\Delta t\left\{\frac{1}{\Delta y}\left[\mathcal{E}_x^n\left(i+\frac{1}{2},j+1,k\right)\right.\right.$$

$$\left.\left.-\mathcal{E}_x^n(i+\frac{1}{2},j,k)\right]-\frac{1}{\Delta x}\left[\mathcal{E}_y^n\left(i+1,j+\frac{1}{2},k\right)-\mathcal{E}_y^n\left(i,j+\frac{1}{2},k\right)\right]\right\},$$

(1.148c)

and

$$\mathcal{E}_x^{n+1}\left(i+\frac{1}{2},j,k\right)=$$

$$\mathcal{E}_x^n\left(i+\frac{1}{2},j,k\right)+c\Delta t\left\{\frac{1}{\Delta y}\left[c\mathcal{B}_z^{n+\frac{1}{2}}\left(i+\frac{1}{2},j+\frac{1}{2},k\right)\right.\right.$$

$$- c\mathcal{B}_z^{n+\frac{1}{2}}\left(i+\frac{1}{2}, j-\frac{1}{2}, k\right)\Bigg] - \frac{1}{\Delta z}\Bigg[c\mathcal{B}_y^{n+\frac{1}{2}}\left(i+\frac{1}{2}, j, k+\frac{1}{2}\right)$$

$$- c\mathcal{B}_y^{n+\frac{1}{2}}\left(i+\frac{1}{2}, j, k-\frac{1}{2}\right)\Bigg]\Bigg\}, \tag{1.149a}$$

$$\mathcal{E}_y^{n+1}\left(i, j+\frac{1}{2}, k\right) =$$

$$\mathcal{E}_y^{n}\left(i, j+\frac{1}{2}, k\right) + c\Delta t\Bigg\{\frac{1}{\Delta z}\Bigg[c\mathcal{B}_x^{n+\frac{1}{2}}\left(i, j+\frac{1}{2}, k+\frac{1}{2}\right)$$

$$- c\mathcal{B}_x^{n+\frac{1}{2}}\left(i, j+\frac{1}{2}, k-\frac{1}{2}\right)\Bigg] - \frac{1}{\Delta x}\Bigg[c\mathcal{B}_z^{n+\frac{1}{2}}\left(i+\frac{1}{2}, j+\frac{1}{2}, k\right)$$

$$- c\mathcal{B}_z^{n+\frac{1}{2}}\left(i-\frac{1}{2}, j+\frac{1}{2}, k\right)\Bigg]\Bigg\}, \tag{1.149b}$$

$$\mathcal{E}_z^{n+1}\left(i, j, k+\frac{1}{2}\right) =$$

$$\mathcal{E}_z^{n}\left(i, j, k+\frac{1}{2}\right) + c\Delta t\Bigg\{\frac{1}{\Delta x}\Bigg[c\mathcal{B}_y^{n+\frac{1}{2}}\left(i+\frac{1}{2}, j, k+\frac{1}{2}\right)$$

$$- c\mathcal{B}_y^{n+\frac{1}{2}}\left(i-\frac{1}{2}, j, k+\frac{1}{2}\right)\Bigg] - \frac{1}{\Delta y}\Bigg[c\mathcal{B}_x^{n+\frac{1}{2}}\left(i, j+\frac{1}{2}, k+\frac{1}{2}\right)$$

$$- c\mathcal{B}_x^{n+\frac{1}{2}}\left(i, j-\frac{1}{2}, k+\frac{1}{2}\right)\Bigg]\Bigg\}. \tag{1.149c}$$

Here the conventional notation

$$\mathcal{F}^n(i, j, k) = \mathcal{F}(i\Delta x, j\Delta y, k\Delta z, n\Delta t)$$

has been used.

These finite-difference equations are solved in the same manner as for one dimension: The equations are applied alternately to advance the solution in time. The grid spacings must satisfy the three-dimensional version of the domain of dependence condition (1.147):

$$c\Delta t\sqrt{\frac{1}{\Delta x^2} + \frac{1}{\Delta y^2} + \frac{1}{\Delta y^2}} \leq 1, \tag{1.150}$$

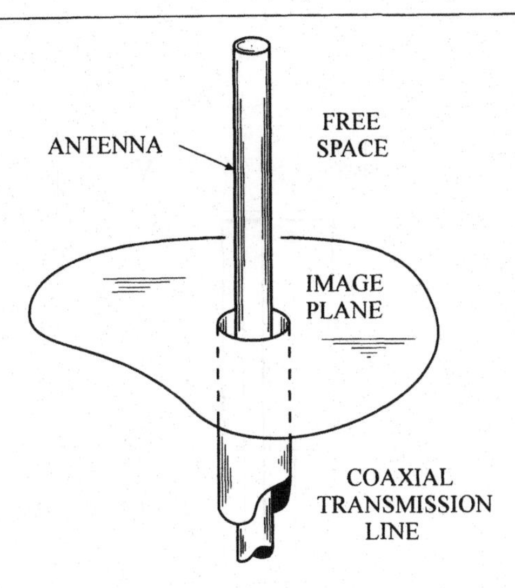

Fig. 1.44. Cylindrical monopole antenna fed through an image plane from a coaxial transmission line.

which for $\Delta x = \Delta y = \Delta z$ is simply[20]

$$\frac{c\Delta t}{\Delta z} \leq \frac{1}{\sqrt{3}}.$$

The utility and accuracy of the finite-difference time-domain method are best demonstrated by a practical example. Consider the cylindrical monopole antenna shown in Figure 1.44 [62]. The antenna is a vertical metal rod placed over a metal image plane; it is fed through the image plane by a coaxial transmission line. This is an idealized theoretical model for a practical antenna; for example, the vertical rod might represent the radio antenna on an automobile, with the image plane representing the sheet metal (roof, fender, etc.) on which the antenna is mounted.

A cross section of the antenna is shown in Figure 1.45. The image plane is assumed to be infinite in extent, and all conductors are assumed to be perfect. The volume of free space V in which the electromagnetic field is to be determined surrounds the antenna and extends into the coaxial line to the depth $z = -\ell_A$. The boundary surface of this region is indicated by the dashed line in Figure 1.45.

To obtain a unique solution to Maxwell's equations within V for times $0 < t \leq t_{\text{max}}$, we must specify $\vec{\mathcal{E}}$ and $\vec{\mathcal{B}}$ within V at time $t = 0$. In addition $\hat{n} \times \vec{\mathcal{E}}$ or $\hat{n} \times \vec{\mathcal{B}}$ must be specified on the boundary surface of V for all times $0 < t \leq t_{\text{max}}$.

We will assume that the electromagnetic field ($\vec{\mathcal{E}}$ and $\vec{\mathcal{B}}$) is zero within V at time $t = 0$. On the cross section of the coaxial line at A ($z = -\ell_A$), the tangential

[20] When the grid spacings are equal, say all spacings equal Δz, the domain of dependence condition for n-dimensional space is simply

$$\frac{c\Delta t}{\Delta z} \leq \frac{1}{\sqrt{n}},$$

which agrees with our results for one and three dimensions.

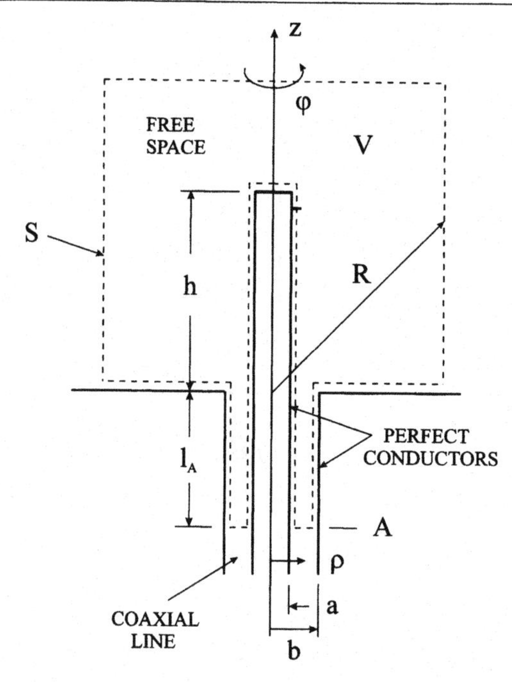

Fig. 1.45. Cross section of cylindrical monopole antenna showing region of solution.

component of the incident field $(-\hat{z} \times \vec{\mathcal{E}}^i)$ is specified for times $0 < t \leq t_{max}$:[21]

$$\vec{\mathcal{E}}^i(\rho, -\ell_A, t) = \frac{\mathcal{V}^i(t)}{\ln(b/a)\rho}\hat{\rho}, \tag{1.151}$$

where $\mathcal{V}^i(t)$ is the time-varying, incident voltage. This will be the only electric field at this cross section if we choose $2\ell_A > ct_{max}$, because the field reflected from the end of the line will not reach this cross section during the observation time. Since we have perfect conductors, we have $\hat{n} \times \vec{\mathcal{E}} = 0$ on the surface of the coaxial line, antenna, and image plane. Above the image plane, the minimum distance to the surface S that surrounds the antenna is R_{min}. When $R_{min} > ct_{max} - \ell_A$, the electromagnetic field ($\hat{n} \times \vec{\mathcal{E}}$ and $\hat{n} \times \vec{\mathcal{B}}$) will be zero on this portion of the surface for all observation times. A review of the above specifications shows that all of the requirements have been met for ensuring a unique solution to Maxwell's equations within the volume V for times $0 < t \leq t_{max}$.[22]

The application of the FDTD method to this problem is described in detail in Reference [62]; here, we will summarize the procedure and discuss the numerical results. The geometry for the monopole and the incident electric field (1.151) are both rotationally symmetric [independent of the coordinate ϕ in the circular cylindrical coordinate system (ρ, ϕ, z)], and the incident field has only the components, $\vec{\mathcal{E}}_\rho$ and $\vec{\mathcal{B}}_\phi$. As a result, the electromagnetic field for this problem has only three

[21] This is the electric field of a transverse electromagnetic (TEM) mode in the coaxial line.

[22] The conditions on the locations of the surfaces A (in the coaxial line) and S (above the image plane) can be relaxed by the use of absorbing boundary conditions on these surfaces. This point is discussed in Reference [62] but is not important to the argument presented here.

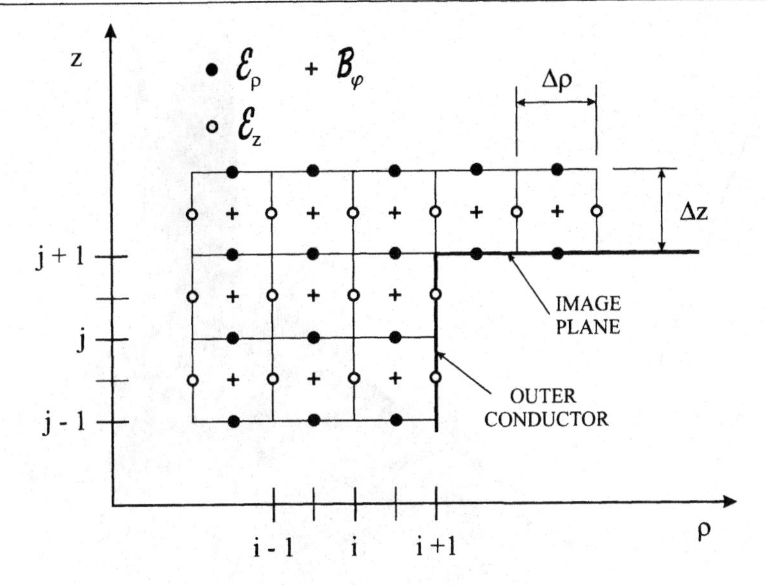

Fig. 1.46. Spatial grid showing location of the field components about the point (i, j) for application of the finite-difference time-domain method to the cylindrical monopole antenna.

components, $\mathcal{E}_\rho$, $\mathcal{E}_z$, and $\mathcal{B}_\phi$, and these are only a function of the two coordinates ρ and z. The two-dimensional, spatial grid of points (ρ, z) and the field components evaluated at these points are shown in Figure 1.46. Note that the grid is arranged so that the electric field component tangential to the surface of a perfect conductor is evaluated at the surface where it is zero.

The finite-difference equations that correspond to the two Maxwell's equations, (1.50) and (1.51), are (see Problem 1.15)

$$c\mathcal{B}_\phi^{n+\frac{1}{2}}\left(i+\frac{1}{2}, j+\frac{1}{2}\right) = c\mathcal{B}_\phi^{n-\frac{1}{2}}\left(i+\frac{1}{2}, j+\frac{1}{2}\right) + c\Delta t\left\{\frac{1}{\Delta\rho}\left[\mathcal{E}_z^n\left(i+1, j+\frac{1}{2}\right) - \mathcal{E}_z^n\left(i, j+\frac{1}{2}\right)\right] - \frac{1}{\Delta z}\left[\mathcal{E}_\rho^n\left(i+\frac{1}{2}, j+1\right) - \mathcal{E}_\rho^n\left(i+\frac{1}{2}, j\right)\right]\right\}, \tag{1.152}$$

$$\mathcal{E}_\rho^{n+1}\left(i+\frac{1}{2}, j\right) = \mathcal{E}_\rho^n\left(i+\frac{1}{2}, j\right) - \left(\frac{c\Delta t}{\Delta z}\right) \left[c\mathcal{B}_\phi^{n+\frac{1}{2}}\left(i+\frac{1}{2}, j+\frac{1}{2}\right) - c\mathcal{B}_\phi^{n+\frac{1}{2}}\left(i+\frac{1}{2}, j-\frac{1}{2}\right)\right], \tag{1.153}$$

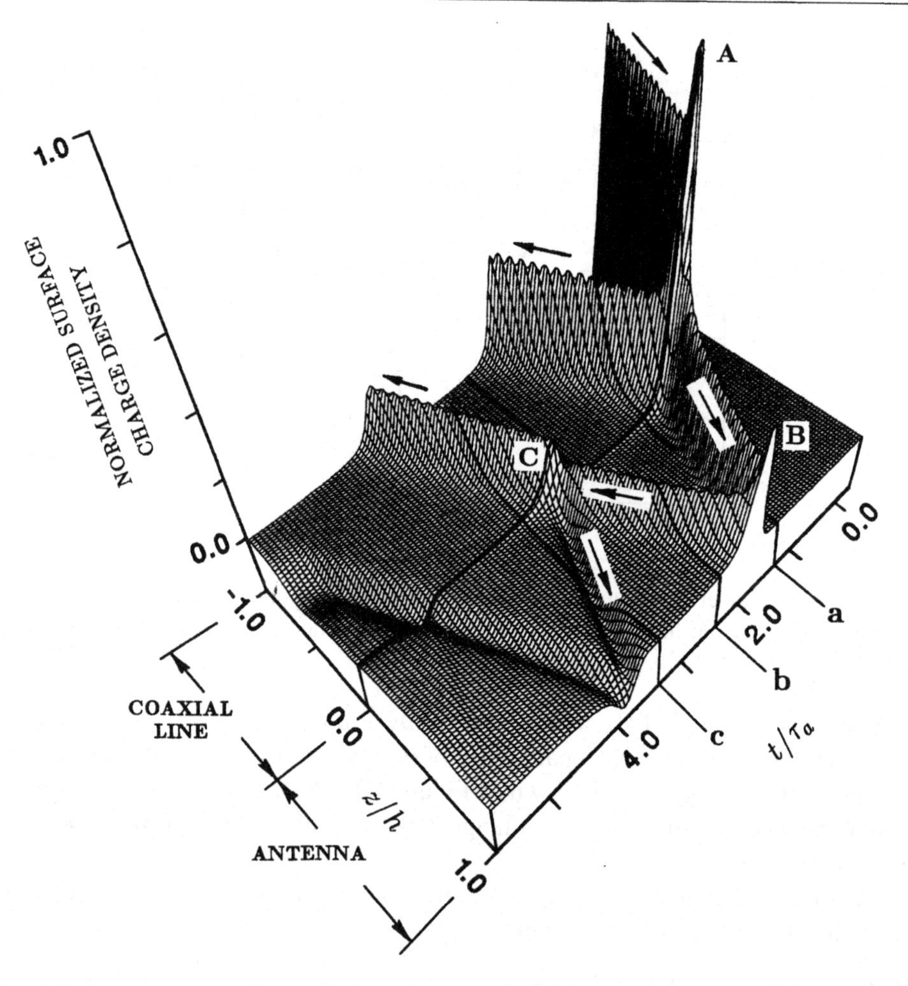

Fig. 1.47. Normalized surface charge density on the cylindrical monopole antenna as a function of the normalized position, z/h, and the normalized time, t/τ_a : $b/a = 2.30$, $h/a = 65.8$, $\tau/\tau_a = 0.114$. (After Maloney et al. [62], © 1990 IEEE.)

and

$$\mathcal{E}_z^{n+1}\left(i, j+\frac{1}{2}\right) = \mathcal{E}_z^{n}\left(i, j+\frac{1}{2}\right) + \left(\frac{c\Delta t}{\Delta\rho}\right)\frac{1}{\rho_i}$$

$$\left[\rho_{i+\frac{1}{2}} c\mathcal{B}_\phi^{n+\frac{1}{2}}\left(i+\frac{1}{2}, j+\frac{1}{2}\right) - \rho_{i-\frac{1}{2}} c\mathcal{B}_\phi^{n+\frac{1}{2}}\left(i-\frac{1}{2}, j+\frac{1}{2}\right)\right]. \tag{1.154}$$

These equations are used with the time-stepping procedure to obtain the electromagnetic field in the volume V for times $0 < t \leq t_{\max}$.

The parameters that describe the monopole antenna are the height h and the radii a and b of the conductors of the coaxial line. We will let $b/a = 2.30$, which corresponds to a characteristic impedance of 50 Ω for the air-filled coaxial line. For the results shown in Figures 1.47 and 1.48, the antenna is excited by a Gaussian

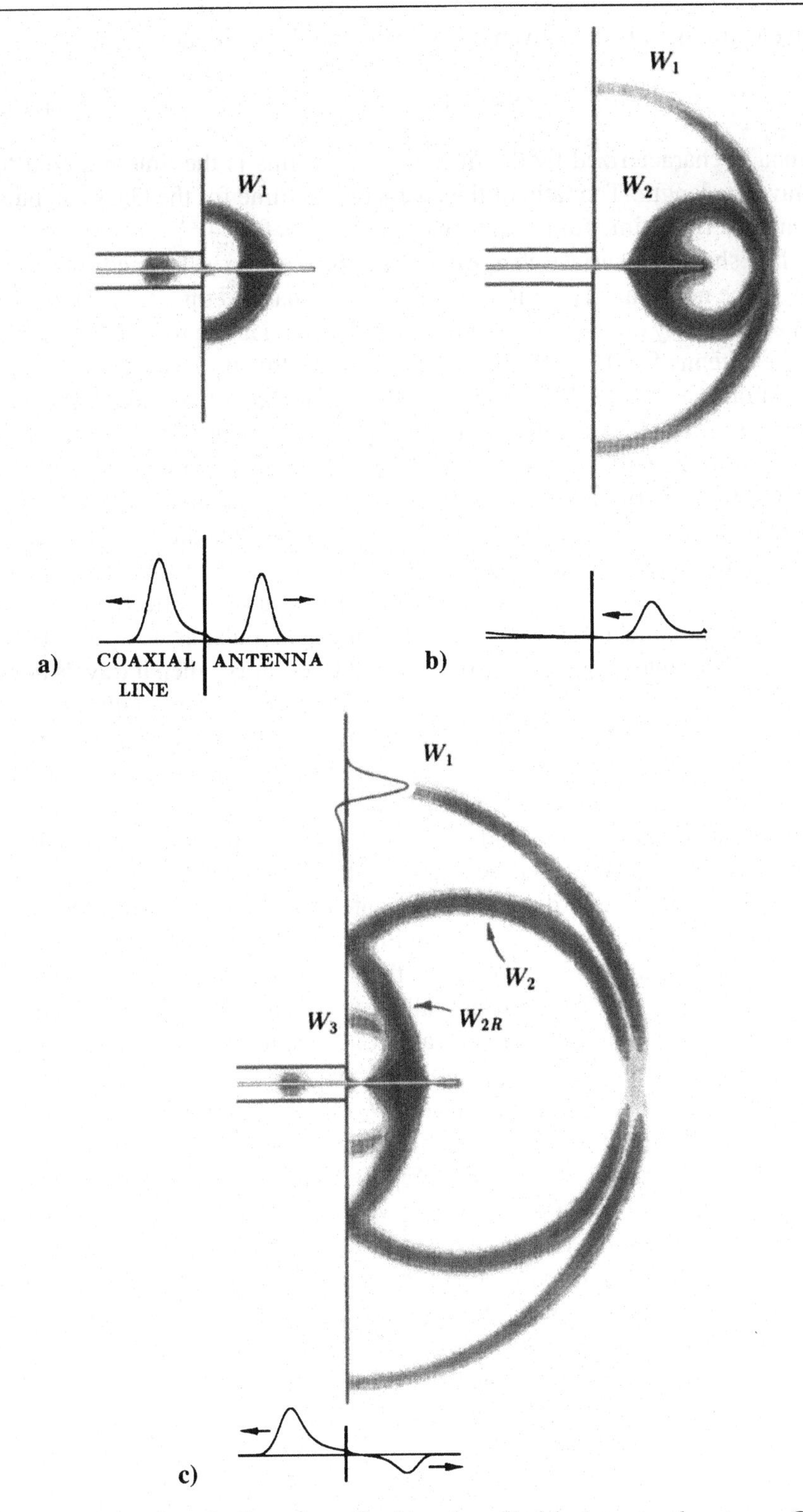

Fig. 1.48. Radiation of a Gaussian pulse from the cylindrical monopole antenna. Gray scale plots show the magnitude of the electric field, while line drawings below show the surface charge density on the antenna: $b/a = 2.30$, $h/a = 65.8$, $\tau/\tau_a = 0.114$. (After Maloney et al. [62], ©1990 IEEE.)

pulse; the electric field is then given by (1.151) with the incident voltage

$$\mathcal{V}^i(t) = V_o e^{-(t-t_o)^2/\tau^2}. \tag{1.155}$$

The antenna is characterized by the time $\tau_a = h/c$; this is the time required for light to travel its length. The ratio of the characteristic time for the Gaussian pulse τ to the characteristic time for the antenna τ_a is $\tau/\tau_a = 0.114$.[23]

Figure 1.47 shows the surface charge density ρ_s on the coaxial line and antenna as a function of the normalized position z/h and the normalized time t/τ_a. This is the surface charge density on the inner conductor of the coaxial line for $-1.0 \leq z/h < 0$ and on the antenna for $0 \leq z/h \leq 1.0$. Each slice of this figure (with t/τ_a fixed) is similar to one of the graphs in Figure 1.40 for our one-dimensional problem. At the surface of a perfect conductor, the normal component of the electric field is proportional to the surface charge density, so this is also a graph for the radial component of the electric field at the surface of the inner conductor and antenna.

This diagram, called a "bounce diagram," clearly shows that the pulse of electric field travels up the coaxial line until it reaches the aperture ($z/h \approx 0.0$, $t/\tau_a \approx 1.0$; point A in Figure 1.47). At this point, a portion of the pulse is reflected back into the line and the remainder emerges on the antenna. The pulse is next reflected at the open end of the antenna ($z/h \approx 1.0$, $t/\tau_a \approx 2.0$; point B); then it travels down to the aperture ($z/h \approx 0.0$, $t/\tau_a \approx 3.0$; point C) where it is partially reflected and enters the coaxial line. This process is repeated continuously, with the pulse alternately reflected from the open end of the antenna and from the coaxial aperture until there is no longer any charge on the antenna.

The magnitude of the electric field, $|\vec{\mathcal{E}}|$, in the space immediately surrounding the monopole antenna is displayed on a gray scale in Figure 1.48 for three times. Graphs of the surface charge density on the antenna are below each plot; these correspond to the slices marked a, b, and c in Figure 1.47. The spacing between the conductors of the coaxial line is expanded in these plots to clarify the presentation.

In Figure 1.48a the pulse has been partially reflected at the aperture and is traveling out along the antenna. A spherical wavefront is centered on the aperture (W_1 in Figure 1.48a); it terminates at the packets of charge on the antenna and on the image plane. A second spherical wavefront (W_2) is produced when this pulse is reflected at the open end of the antenna. This spherical wavefront, centered on the open end of the antenna, is clearly shown in Figure 1.48b. The pulse, after reflection from the open end, travels down the antenna, eventually being partially reflected at the aperture and entering the coaxial line. A third spherical wavefront (W_3), centered at the aperture, is produced on this reflection (Figure 1.48c). The spherical wavefront W_{2R}, which is the reflection of wavefront W_2 at the image plane, is also clearly shown in this figure. For times beyond those shown in Figure 1.48, similar wavefronts are produced on each reflection of the pulse from the open end of the antenna and the coaxial aperture.

[23] The pulse duration was chosen short enough to resolve the reflections from different points on the antenna.

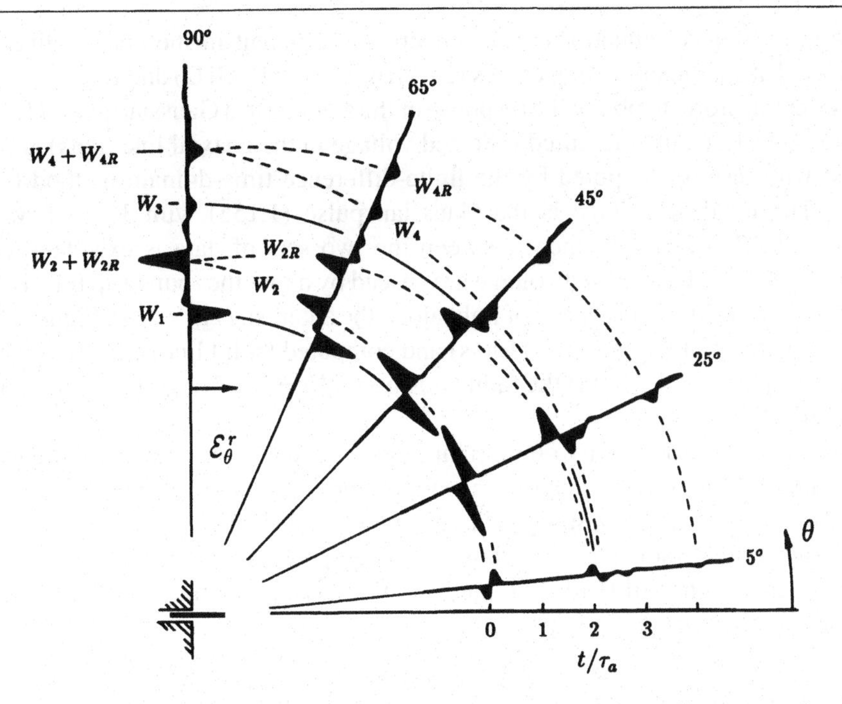

Fig. 1.49. Electric field as seen on a spherical surface of large radius centered on the monopole antenna. Each graph shows the field at a fixed polar angle θ as a function of the normalized time, t/τ_a: $b/a = 2.30$, $h/a = 65.8$, $\tau/\tau_a = 0.114$. (After Maloney et al. [62], © 1990 IEEE.)

To complete the picture for the radiation, in Figure 1.49 we show the electric field as seen by a distant observer, one situated on a sphere of large radius centered on the antenna. This is the radiated or far-zone field of the antenna, which we will discuss in more detail in Chapter 8. Each graph in this figure shows the electric field at a fixed polar angle θ (measured from the axis of the antenna) as a function of the normalized time t/τ_a. The origin for the time, $t/\tau_a = 0$, and the amplitude of the field were selected to clarify the presentation. Notice that spherical wavefronts centered at the same point on the antenna are always separated by a time interval that is a multiple of $2\tau_a$, the round-trip transit time for the pulse on the antenna. For example, wavefronts W_1 and W_3, which are centered on the drive point, are separated by the time $2\tau_a$, as are wavefronts W_2 and W_4, which are centered on the open end. However, the relative times of arrival of the wavefront pairs, such as W_1, W_3 and W_2, W_4, change with the viewing angle θ. For example, at $\theta = 90°$ the wavefronts W_1, W_2, W_3, and W_4 are all separated, but at $\theta = 5°$, W_1 and W_2 are superposed, as are W_3 and W_4. The radiated field off the end of the antenna ($\theta = 0°$) is zero. Additional insight into the radiation from this antenna is given in Section 8.1.

An experimental model was constructed for the cylindrical monopole antenna with the dimensions $b/a = 2.30$ and $h/a = 32.8$. This model was mounted on an aluminum image plane (120 cm × 155 cm) that was surrounded by absorbing material to reduce reflections. The reflection coefficient in the coaxial line was measured at a number of frequencies (50 MHz $\le f \le$ 18 GHz) and then used with a Fourier transform to obtain the response of the antenna to a Gaussian pulse (1.155).

In Figure 1.50a the measured reflected voltage in the coaxial line (dots) is compared with results computed by the finite-difference time-domain method (solid line). The incident voltage is the Gaussian pulse (1.155) with $V_o = 1$ V and $\tau/\tau_a = 0.228$. The agreement between the two sets of data is excellent. Note that the peaks in the reflected voltage are spaced by about the round trip transit time for a pulse traveling at the speed of light along the antenna, $t/\tau_a = 2.0$. Figure 1.50b is a comparison of the measured (dots) and computed (solid line) electric fields on the image plane ($\theta = 90°$) at the radial position $\rho/h = 12.7$. Again, the agreement is excellent.

These results illustrate two important points: 1) that electromagnetic theory, as embodied in Maxwell's equations, accurately predicts experimental observations and 2) that a particularly straightforward numerical method can be used with a digital computer to solve these equations. As the speed and memory of digital computers increase, the utility of such methods for solving practical electromagnetics problems will certainly increase.

1.7 Harmonic time dependence and the Fourier transform

Maxwell's equations as written in Table 1.3 are for an electromagnetic field with general time dependence; that is, no restrictions are placed on the temporal behavior of quantities such as the electric field $\vec{\mathcal{E}}(\vec{r},t)$. In certain applications, these equations can be simplified by specifying a field with harmonic time dependence: A field for which all quantities are cosinusoidal in time, for example, $\vec{\mathcal{E}}(\vec{r},t) = \hat{x}A_x \cos(\omega t + \phi_x)$, where the parameter ω is the angular frequency. The simplification that results for harmonic time dependence is a consequence of the special properties of linear systems. Thus, before we consider Maxwell's equations for harmonic time dependence, it will be helpful to review a few of the basic properties of linear systems [63].

1.7.1 Linear systems

The input to a system, which we will call $f(t)$, produces a unique output, which we will call $g(t)$, as illustrated in Figure 1.51a. The relationship between $f(t)$ and $g(t)$ may be specified by a set of equations; we will represent this relationship by the operator L:

$$L\big[f(t)\big] = g(t). \tag{1.156}$$

We might think of $f(t)$ as being the volume density of electric current $\vec{\mathcal{J}}(\vec{r},t)$ and $g(t)$ as being the resulting electromagnetic field $\vec{\mathcal{E}}(\vec{r},t)$, $\vec{\mathcal{B}}(\vec{r},t)$; the two are related through Maxwell's equations.

A *linear system* (Figure 1.51b) has the special property that if

$$L\big[f_1(t)\big] = g_1(t)$$

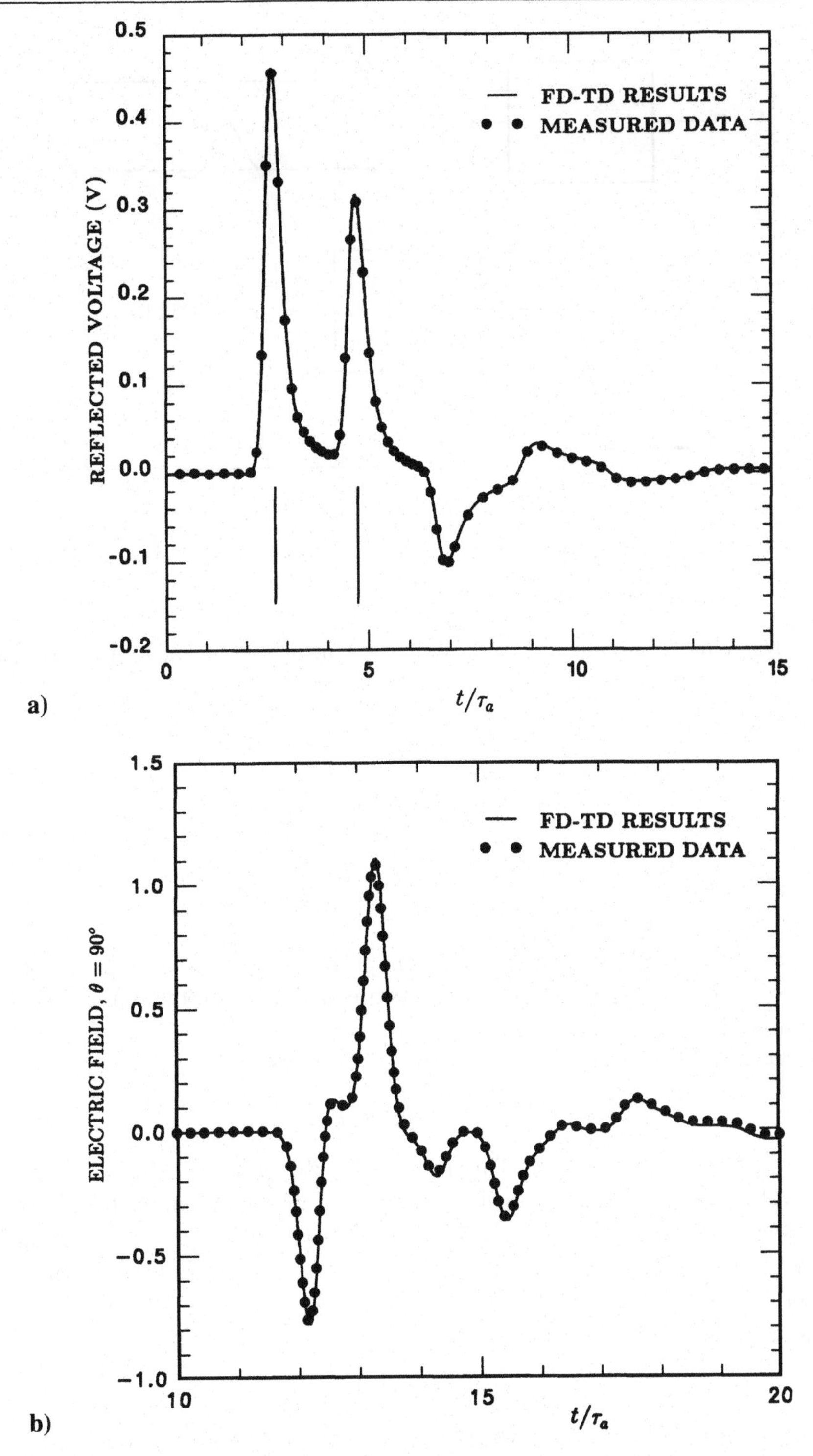

Fig. 1.50. Comparison of calculated and measured results for the cylindrical monopole antenna excited by a 1 V Gaussian pulse: $b/a = 2.30$, $h/a = 32.8$, $\tau/\tau_a = 0.228$. a) Reflected voltage in the coaxial line. b) Electric field on image plane ($\theta = 90°$) at $\rho/h = 12.7$. (After Maloney et al. [62], © 1990 IEEE.)

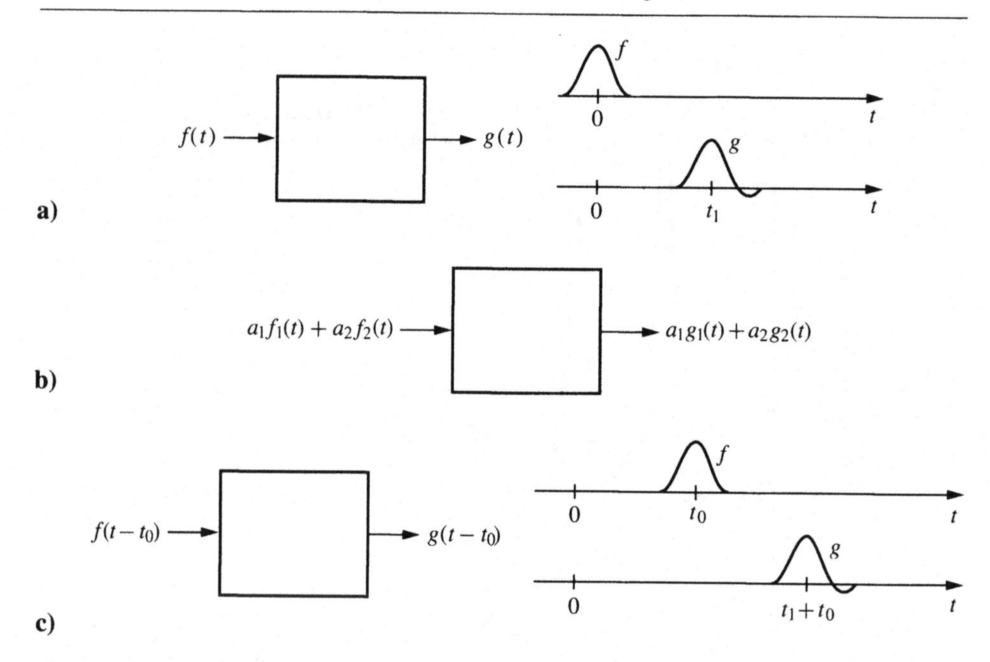

Fig. 1.51. Diagrams illustrating a) general system, b) linear system, and c) time-invariant system.

and

$$L\big[f_2(t)\big] = g_2(t),$$

then

$$L\big[a_1 f_1(t) + a_2 f_2(t)\big] = a_1 g_1(t) + a_2 g_2(t), \qquad (1.157)$$

where a_1 and a_2 are arbitrary constants. This property is called the "principle of superposition." Examples of linear operations are addition, subtraction, differentiation, and integration.

A *time-invariant system* is one for which

$$L\big[f(t - t_o)\big] = g(t - t_o), \qquad (1.158)$$

where t_o is an arbitrary constant. This property is illustrated by the drawings in Figures 1.51a and 1.51c, where a shift in the time at which the input starts is seen to cause a shift in the time at which the output starts, the shape of the output is unchanged by the shift. A system with this property is often called "stationary" or "non–time varying."

A function that satisfies the relationship

$$L\big[f(t)\big] = Cf(t)$$

(the output is proportional to the input) is an *eigenfunction* of the operator. For a *time-invariant, linear system* (operator) the eigenfunctions are exponentials:

$$L[e^{j\omega t}] = H(\omega)e^{j\omega t}. \qquad (1.159)$$

The time independent factor $H(\omega)$, which is a function of the parameter ω, is called the "system function."[24]

For a *real system*, a real input produces a real output. We will be dealing with *real, time-invariant, linear systems*. From our discussion, we see that for such systems the input[25]

$$f(t) = C(\omega)\cos\left[\omega t + \psi(\omega)\right] = \mathrm{Re}\left[F(\omega)e^{j\omega t}\right] \tag{1.160}$$

with

$$F(\omega) = C(\omega)e^{j\psi(\omega)}$$

produces the output

$$g(t) = A(\omega)\cos\left[\omega t + \phi(\omega)\right] = \mathrm{Re}\left[G(\omega)e^{j\omega t}\right] \tag{1.161}$$

with

$$G(\omega) = H(\omega)F(\omega) = A(\omega)e^{j\phi(\omega)}.$$

In words, this important result states that a cosinusoidal input to a real, time-invariant, linear system produces a cosinusoidal output.

[24] The following simple argument verifies (1.159) [64]. For a general system with exponential input, we have

$$f(t) = e^{j\omega t}, \qquad g(t) = \mathcal{W}(t,\omega)e^{j\omega t}.$$

Here we have factored the exponential from the output, leaving the function $\mathcal{W}$ that depends on both t and the parameter ω. The system is time-invariant so

$$f(t-t_o) = e^{j\omega(t-t_o)}, \qquad g(t-t_o) = \mathcal{W}(t-t_o,\omega)e^{j\omega(t-t_o)}.$$

This input also can be written as

$$f(t-t_o) = e^{-j\omega t_o}e^{j\omega t} = e^{-j\omega t_o}f(t),$$

where $e^{-j\omega t_o}$ can be viewed as a constant. Because the system is linear, the output for this input is

$$g(t-t_o) = e^{-j\omega t_o}g(t) = e^{-j\omega t_o}\mathcal{W}(t,\omega)e^{j\omega t}$$
$$= \mathcal{W}(t,\omega)e^{j\omega(t-t_o)}.$$

Equating the two expressions above for the output, we find that

$$\mathcal{W}(t-t_o,\omega) = \mathcal{W}(t,\omega)$$

for arbitrary t_o. Hence, $\mathcal{W}$ must be independent of t:

$$\mathcal{W}(t,\omega) = H(\omega).$$

[25] The complex input

$$f(t) = Fe^{j\omega t} = \mathrm{Re}(Fe^{j\omega t}) + j\,\mathrm{Im}(Fe^{j\omega t})$$

produces the complex output

$$g(t) = HFe^{j\omega t} = \mathrm{Re}(HFe^{j\omega t}) + j\,\mathrm{Im}(HFe^{j\omega t}),$$

where Re indicates the real part and Im the imaginary part of the complex number. The input can be considered as the superposition of the two terms on the right-hand side of the equation for $f(t)$. Because the system is linear, the output is the superposition of the outputs due to each of these two inputs. A real input produces a real output; hence, the component of the output $\mathrm{Re}(HFe^{j\omega t})$ must result from the component of the input $\mathrm{Re}(Fe^{j\omega t})$.

1.7.2 Maxwell's equations

Now let us return to Maxwell's equations (Table 1.3). Maxwell's equations involve only time-invariant linear operations on the quantities $\vec{\mathcal{E}}$, $\vec{\mathcal{B}}$, $\vec{\mathcal{D}}$, $\vec{\mathcal{H}}$, etc., all of which are real. However, the complete description of an electromagnetic problem requires not only Maxwell's equations but also the electromagnetic constitutive relations. The constitutive relations we have discussed so far, those for simple materials (1.9)–(1.11), are time invariant and linear. Therefore, Maxwell's equations together with these constitutive relations form a real, time-invariant, linear system. The constitutive relations are approximately time-invariant and linear for many practical materials; however, we must emphasize that the constitutive relations do not necessarily exhibit these properties. For example, a nonlinear relationship might exist between $\vec{\mathcal{D}}$ and $\vec{\mathcal{E}}$, such as $\mathcal{D}_x \propto \mathcal{E}_x^2$.

From our general discussion of real, time-invariant, linear systems, we know that if one of the quantities in Maxwell's equations, the input, is specified to be cosinusoidal in time, then all of the other quantities, the outputs, will also be cosinusoidal in time. For example, the volume density of electric current may be specified as cosinusoidal:

$$\vec{\mathcal{J}}(\vec{r},t) = C_x(\vec{r},\omega)\cos\left[\omega t + \psi_x(\vec{r},\omega)\right]\hat{x} + C_y(\vec{r},\omega)\cos\left[\omega t + \psi_y(\vec{r},\omega)\right]\hat{y}$$
$$+ C_z(\vec{r},\omega)\cos\left[\omega t + \psi_z(\vec{r},\omega)\right]\hat{z}, \tag{1.162}$$

or in more compact form

$$\vec{\mathcal{J}}(\vec{r},t) = \mathrm{Re}\left[\vec{J}(\vec{r},\omega)e^{j\omega t}\right], \tag{1.163}$$

where $\vec{J}(\vec{r},\omega)$ is the *complex vector phasor*

$$\vec{J}(\vec{r},\omega) = \sum_{i=x,y,z} \vec{J}_i(\vec{r},\omega) = \sum_{i=x,y,z} J_i(\vec{r},\omega)\hat{i} = \sum_{i=x,y,z} C_i(\vec{r},\omega)e^{j\psi_i(\vec{r},\omega)}\hat{i}. \tag{1.164}$$

The electromagnetic field produced by this current is also cosinusoidal:

$$\vec{\mathcal{E}}(\vec{r},t) = A_x(\vec{r},\omega)\cos\left[\omega t + \phi_x(\vec{r},\omega)\right]\hat{x} + A_y(\vec{r},\omega)\cos\left[\omega t + \phi_y(\vec{r},\omega)\right]\hat{y}$$
$$+ A_z(\vec{r},\omega)\cos\left[\omega t + \phi_z(\vec{r},\omega)\right]\hat{z},$$

or

$$\vec{\mathcal{E}}(\vec{r},t) = \mathrm{Re}\left[\vec{E}(\vec{r},\omega)e^{j\omega t}\right], \tag{1.165}$$

where $\vec{E}(\vec{r},\omega)$ is the complex vector phasor

$$\vec{E}(\vec{r},\omega) = \sum_{i=x,y,z} \vec{E}_i(\vec{r},\omega) = \sum_{i=x,y,z} E_i(\vec{r},\omega)\hat{i} = \sum_{i=x,y,z} A_i(\vec{r},\omega)e^{j\phi_i(\vec{r},\omega)}\hat{i}, \tag{1.166}$$

with similar expressions for $\vec{\mathcal{B}}$, $\vec{\mathcal{D}}$, and $\vec{\mathcal{H}}$. Notice the notation we have introduced: The time-varying field is indicated by the typeface $\vec{\mathcal{E}}$, whereas the corresponding vector phasor is indicated by the typeface $\vec{E}$. Each component of the vector phasor

$\vec{E}$ has a modulus A_i and argument ϕ_i that depend on the position $\vec{r}$ and the parameter ω.[26]

When the electromagnetic field is expressed as in (1.165), the Maxwell equation (1.50) becomes

$$\nabla \times \mathrm{Re}\left[\vec{E}(\vec{r},\omega)e^{j\omega t}\right] = -\frac{\partial}{\partial t}\mathrm{Re}\left[\vec{B}(\vec{r},\omega)e^{j\omega t}\right].$$

The operations of taking the real part and differentiation with respect to the spatial or temporal variables can be interchanged without affecting this equation; thus,

$$\mathrm{Re}\left[\nabla \times \vec{E}(\vec{r},\omega)e^{j\omega t}\right] = \mathrm{Re}\left[-j\omega\vec{B}(\vec{r},\omega)e^{j\omega t}\right]. \tag{1.167}$$

It then follows that[27]

$$\nabla \times \vec{E}(\vec{r},\omega)e^{j\omega t} = -j\omega\vec{B}(\vec{r},\omega)e^{j\omega t}, \tag{1.168}$$

or

$$\nabla \times \vec{E}(\vec{r},\omega) = -j\omega\vec{B}(\vec{r},\omega). \tag{1.169}$$

The other Maxwell's equations, (1.51)–(1.54), when treated in a similar manner, become

$$\nabla \times \vec{H}(\vec{r},\omega) = \vec{J}(\vec{r},\omega) + j\omega\vec{D}(\vec{r},\omega) \tag{1.170}$$

$$\nabla \cdot \vec{D}(\vec{r},\omega) = \rho(\vec{r},\omega) \tag{1.171}$$

$$\nabla \cdot \vec{B}(\vec{r},\omega) = 0 \tag{1.172}$$

$$\nabla \cdot \vec{J}(\vec{r},\omega) = -j\omega\rho(\vec{r},\omega). \tag{1.173}$$

These are Maxwell's equations for the complex vector phasors $\vec{E}(\vec{r},\omega)$, $\vec{B}(\vec{r},\omega)$, etc.; for future reference they are listed in Table 1.5. They apply to electromagnetic systems in which the constitutive relations for all materials are time invariant and linear, and they are generally used in the following manner to solve a practical problem. A cosinusoidal excitation, such as the current $\vec{\mathcal{J}}(\vec{r},t)$, is specified; from this a vector phasor, $\vec{J}(\vec{r},\omega)$, is determined. The Maxwell's equations with this excitation are then solved to obtain the vector phasors for the electromagnetic field, $\vec{E}(\vec{r},\omega)$, $\vec{B}(\vec{r},\omega)$. Finally, the cosinusoidal field is determined from $\vec{\mathcal{E}}(\vec{r},t) = \mathrm{Re}[\vec{E}(\vec{r},\omega)e^{j\omega t}]$, $\vec{\mathcal{B}}(\vec{r},t) = \mathrm{Re}[\vec{B}(\vec{r},\omega)e^{j\omega t}]$.

These equations are simpler than the general equations in Table 1.3 because they contain one less variable, the time t. For this reason they are often called the *time-*

[26] For now, we have included the parameter ω in the argument of the phasors [e.g., $\vec{E}(\vec{r},\omega)$] to emphasize that we are dealing with a field that varies harmonically in time. Later, we will drop the ω and simply write $\vec{E}(\vec{r})$.

[27] Equation (1.167) holds for all time. When we let $t \to t + (\pi/2\omega)$ we have

$$\mathrm{Re}[\nabla \times \vec{E}e^{j\omega t}e^{j\pi/2}] = \mathrm{Re}[-j\omega\vec{B}e^{j\omega t}e^{j\pi/2}],$$

or

$$\mathrm{Re}[j\nabla \times \vec{E}e^{j\omega t}] = \mathrm{Re}[j(-j\omega\vec{B}e^{j\omega t})].$$

Thus,

$$\mathrm{Im}[\nabla \times \vec{E}e^{j\omega t}] = \mathrm{Im}[-j\omega\vec{B}e^{j\omega t}],$$

which when combined with (1.167) gives (1.168).

Table 1.5. *The equations of electromagnetism for harmonic time dependence*

Name	Equation: Integral form	Equation: Differential form
Faraday's law	$\oint_C \vec{E}\cdot d\vec{\ell} = -j\omega \iint_S \vec{B}\cdot d\vec{S}$	$\nabla\times\vec{E} = -j\omega\vec{B}$
Ampère-Maxwell law	$\oint_C \vec{H}\cdot d\vec{\ell} = \iint_S \vec{J}\cdot d\vec{S} + j\omega \iint_S \vec{D}\cdot d\vec{S}$	$\nabla\times\vec{H} = \vec{J} + j\omega\vec{D}$
Gauss' electric law	$\oiint_S \vec{D}\cdot d\vec{S} = \iiint_V \rho\, dV$	$\nabla\cdot\vec{D} = \rho$
Gauss' magnetic law	$\oiint_S \vec{B}\cdot d\vec{S} = 0$	$\nabla\cdot\vec{B} = 0$
Equation of continuity for electric charge	$\oiint_S \vec{J}\cdot d\vec{S} = -j\omega \iiint_V \rho\, dV$	$\nabla\cdot\vec{J} = -j\omega\rho$
Complex Poynting's theorem	$\iiint_V \left(\frac{1}{2}\vec{E}\cdot\vec{J}^*\right)dV - j\omega \iiint_V \left(\frac{1}{2}\vec{E}\cdot\vec{D}^* - \frac{1}{2}\vec{H}^*\cdot\vec{B}\right)dV + \oiint_S \vec{S}_c\cdot d\vec{S} = 0$	
Complex Poynting vector	$\vec{S}_c = \frac{1}{2}\vec{E}\times\vec{H}^*$	

independent Maxwell's equations. The parameter ω will often be omitted from the argument of the vector phasor when this introduces no ambiguity: $\vec{E}(\vec{r}\,) = \vec{E}(\vec{r}, \omega)$.

The other equations of electromagnetism can be specialized for the case of harmonic time dependence. For equations that are *linear* in the field quantities, this usually involves simply replacing the first partial derivatives with respect to time by $j\omega$ and changing $\vec{\mathcal{E}}(\vec{r}, t)$ to $\vec{E}(\vec{r}, \omega)$, etc. This is the case for the electromagnetic boundary conditions, which are given for harmonic time dependence in Table 1.6. Equations that are *nonlinear* in the field quantities, such as Poynting's theorem, which involves products of field quantities, require additional consideration.

1.7.3 Poynting's theorem

Our derivation of Poynting's theorem for a field with general time dependence (Section 1.4) is based on the integral

$$\iiint_V \vec{\mathcal{E}}(\vec{r}, t)\cdot\vec{\mathcal{J}}(\vec{r}, t) dV.$$

Table 1.6. *The electromagnetic boundary conditions for harmonic time dependence*†

General boundary condition	Region 1: a perfect electric conductor
$\hat{n} \times (\vec{E}_2 - \vec{E}_1) = 0$	$\hat{n} \times \vec{E}_2 = 0$
$\hat{n} \times (\vec{H}_2 - \vec{H}_1) = \vec{J}_s$	$\hat{n} \times \vec{H}_2 = \vec{J}_s$
$\hat{n} \cdot (\vec{D}_2 - \vec{D}_1) = \rho_s$	$\hat{n} \cdot \vec{D}_2 = \rho_s$
$\hat{n} \cdot (\vec{B}_2 - \vec{B}_1) = 0$	$\hat{n} \cdot \vec{B}_2 = 0$
$\hat{n} \cdot (\vec{J}_2 - \vec{J}_1) + \nabla_s \cdot \vec{J}_s = -j\omega\rho_s$	$\hat{n} \cdot \vec{J}_2 + \nabla_s \cdot \vec{J}_s = -j\omega\rho_s$

Note: †The normal $\hat{n}$ points outward from region 1 into region 2.

When the field is harmonic in time, the product $\vec{\mathcal{E}} \cdot \vec{\mathcal{J}}$ in the integrand becomes

$$\begin{aligned}\vec{\mathcal{E}}(\vec{r},t)\cdot\vec{\mathcal{J}}(\vec{r},t) &= \mathrm{Re}\left[\vec{E}(\vec{r},\omega)e^{j\omega t}\right]\cdot\mathrm{Re}\left[\vec{J}(\vec{r},\omega)e^{j\omega t}\right]\\ &= \frac{1}{4}\left[\vec{E}(\vec{r},\omega)e^{j\omega t} + \vec{E}^*(\vec{r},\omega)e^{-j\omega t}\right]\cdot\left[\vec{J}(\vec{r},\omega)e^{j\omega t}\right.\\ &\quad\left. + \vec{J}^*(\vec{r},\omega)e^{-j\omega t}\right]\\ &= \frac{1}{4}\left[\vec{E}(\vec{r},\omega)\cdot\vec{J}^*(\vec{r},\omega) + \vec{E}^*(\vec{r},\omega)\cdot\vec{J}(\vec{r},\omega)\right.\\ &\quad\left. + \vec{E}(\vec{r},\omega)\cdot\vec{J}(\vec{r},\omega)e^{j2\omega t} + \vec{E}^*(\vec{r},\omega)\cdot\vec{J}^*(\vec{r},\omega)e^{-j2\omega t}\right]\\ &= \frac{1}{2}\left\{\mathrm{Re}\left[\vec{E}(\vec{r},\omega)\cdot\vec{J}^*(\vec{r},\omega)\right] + \mathrm{Re}\left[\vec{E}(\vec{r},\omega)\cdot\vec{J}(\vec{r},\omega)e^{j2\omega t}\right]\right\},\end{aligned} \tag{1.174}$$

where the $*$ indicates the complex conjugate. This is the instantaneous rate at which energy is exchanged between the electromagnetic field and the mechanical motion of the charge, per unit volume. For quantities that are cosinusoidal in time, we are generally interested in the time-average value of products like (1.174) rather than their instantaneous value. The average of (1.174) over one period of oscillation, $T = 2\pi/\omega$, is (from Problem 1.19)[28]

$$\left\langle\vec{\mathcal{E}}(\vec{r},t)\cdot\vec{\mathcal{J}}(\vec{r},t)\right\rangle = \frac{1}{T}\int_0^T \vec{\mathcal{E}}(\vec{r},t)\cdot\vec{\mathcal{J}}(\vec{r},t)dt = \mathrm{Re}\left[\frac{1}{2}\vec{E}(\vec{r},\omega)\cdot\vec{J}^*(\vec{r},\omega)\right]. \tag{1.175}$$

[28] This result holds for the time-average value of the scalar and vector products of any two quantities that are harmonic in time. For example, when $\vec{\mathcal{A}}(\vec{r},t)$ and $\vec{\mathcal{B}}(\vec{r},t)$ are harmonic in time,

$$\langle\vec{\mathcal{A}}(\vec{r},t)\cdot\vec{\mathcal{B}}(\vec{r},t)\rangle = \mathrm{Re}\left[\frac{1}{2}\vec{A}(\vec{r},\omega)\cdot\vec{B}^*(\vec{r},\omega)\right]$$

and

$$\langle\vec{\mathcal{A}}(\vec{r},t)\times\vec{\mathcal{B}}(\vec{r},t)\rangle = \mathrm{Re}\left[\frac{1}{2}\vec{A}(\vec{r},\omega)\times\vec{B}^*(\vec{r},\omega)\right].$$

This result suggests that we should examine the integral

$$\iiint_V \left[\frac{1}{2}\vec{E}(\vec{r},\omega)\cdot\vec{J}^*(\vec{r},\omega)\right]dV \tag{1.176}$$

for fields with harmonic time dependence.

On substituting Maxwell's equations (1.169) and (1.170) into (1.176) and using the divergence theorem (1.98), we find that[29]

$$\iiint_V \left(\frac{1}{2}\vec{E}\cdot\vec{J}^*\right)dV - j\omega\iiint_V \left(\frac{1}{2}\vec{E}\cdot\vec{D}^* - \frac{1}{2}\vec{H}^*\cdot\vec{B}\right)dV + \oint\!\!\oint_S \left(\frac{1}{2}\vec{E}\times\vec{H}^*\right)\cdot d\vec{S} = 0. \tag{1.177}$$

This formula is known as the *complex Poynting's theorem*. When we introduce the *complex Poynting vector*

$$\vec{S}_c(\vec{r},\omega) = \frac{1}{2}\vec{E}(\vec{r},\omega)\times\vec{H}^*(\vec{r},\omega), \tag{1.178}$$

it becomes

$$\iiint_V \left(\frac{1}{2}\vec{E}\cdot\vec{J}^*\right)dV - j\omega\iiint_V \left(\frac{1}{2}\vec{E}\cdot\vec{D}^* - \frac{1}{2}\vec{H}^*\cdot\vec{B}\right)dV + \oint\!\!\oint_S \vec{S}_c\cdot d\vec{S} = 0. \tag{1.179}$$

From our discussion of time-average values, it is clear that

$$\mathrm{Re}\left[\vec{S}_c(\vec{r},\omega)\right] = \langle\vec{S}(\vec{r},t)\rangle; \tag{1.180}$$

that is, the real part of the complex Poynting vector is the time average of the Poynting vector (1.100). Notice that $\vec{S}_c$ is not a phasor in the sense of $\vec{\mathcal{J}}$ and $\vec{\mathcal{E}}$. It cannot be used as in (1.163) to obtain the time-varying Poynting vector.

A physical interpretation of this theorem requires the specification of constitutive relations; we will assume simple materials (1.9)–(1.11):

$$\vec{D}(\vec{r},\omega) = \varepsilon\vec{E}(\vec{r},\omega), \tag{1.181}$$

$$\vec{H}(\vec{r},\omega) = \frac{1}{\mu}\vec{B}(\vec{r},\omega), \tag{1.182}$$

and

$$\vec{J}(\vec{r},\omega) = \vec{J}_i(\vec{r},\omega) + \vec{J}_c(\vec{r},\omega), \tag{1.183}$$

with

$$\vec{J}_c(\vec{r},\omega) = \sigma\vec{E}(\vec{r},\omega). \tag{1.184}$$

Here we have separated the volume density of current into an impressed current $\vec{J}_i$ and a conduction current $\vec{J}_c$.

[29] These are the same mathematical steps that we used in obtaining Poynting's theorem for fields with general time dependence (1.99).

With (1.181)–(1.184), the complex Poynting's theorem (1.179) becomes

$$-\iiint_V \left(\frac{1}{2}\vec{E}\cdot\vec{J}_i^*\right)dV = \iiint_V \left(\frac{1}{2}|\vec{J}_c|^2/\sigma\right)dV$$
$$- j\omega \iiint_V \left(\frac{1}{2}\varepsilon|\vec{E}|^2 - \frac{1}{2}\frac{1}{\mu}|\vec{B}|^2\right)dV + \oiint_S \vec{S}_c\cdot d\vec{S}. \tag{1.185}$$

On equating the real and imaginary parts of this equation, we obtain the two equations:

$$-\iiint_V \mathrm{Re}\left(\frac{1}{2}\vec{E}\cdot\vec{J}_i^*\right)dV = \iiint_V \left(\frac{1}{2}|\vec{J}_c|^2/\sigma\right)dV + \oiint_S \mathrm{Re}(\vec{S}_c)\cdot d\vec{S} \tag{1.186}$$

and

$$-\iiint_V \mathrm{Im}\left(\frac{1}{2}\vec{E}\cdot\vec{J}_i^*\right)dV = -\omega \iiint_V \left(\frac{1}{2}\varepsilon|\vec{E}|^2 - \frac{1}{2}\frac{1}{\mu}|\vec{B}|^2\right)dV$$
$$+ \oiint_S \mathrm{Im}(\vec{S}_c)\cdot d\vec{S}, \tag{1.187}$$

where we have indicated the Hermitian magnitude of a complex vector by

$$|\vec{A}| = \sqrt{\vec{A}\cdot\vec{A}^*}. \tag{1.188}$$

The first of these equations, (1.186), is an expression for the time-average power within the volume V. To see this, we must identify each term in the equation as the time average of a quantity; from (1.106)

$$\iiint_V \left(\frac{1}{2}|\vec{J}_c|^2/\sigma\right)dV = \iiint_V \langle|\vec{\mathcal{J}}_c|^2/\sigma\rangle dV = \left\langle\frac{\partial\mathcal{Q}}{\partial t}\right\rangle, \tag{1.189}$$

and similarly from (1.108)

$$-\iiint_V \mathrm{Re}\left(\frac{1}{2}\vec{E}\cdot\vec{J}_i^*\right)dV = -\iiint_V \langle\vec{\mathcal{E}}\cdot\vec{\mathcal{J}}_i\rangle dV = \left\langle\frac{\partial\mathcal{W}_i}{\partial t}\right\rangle. \tag{1.190}$$

With these results and (1.180), (1.186) becomes

$$\left\langle\frac{\partial\mathcal{W}_i}{\partial t}\right\rangle = \left\langle\frac{\partial\mathcal{Q}}{\partial t}\right\rangle + \oiint_S \langle\vec{\mathcal{S}}\rangle\cdot d\vec{S}, \tag{1.191}$$

which states that the time-average power supplied by nonelectromagnetic sources to the electromagnetic field must equal the time-average power transferred from the field as heat plus the time-average power leaving the volume by passing through its surface.

The second equation, (1.187), concerns the time-average energy stored in the electromagnetic field within the volume. From (1.112) and (1.113), we have

$$\iiint_V \left(\frac{1}{2}\varepsilon|\vec{E}|^2\right)dV = 2\iiint_V \left\langle\frac{1}{2}\varepsilon|\vec{\mathcal{E}}|^2\right\rangle dV = 2\langle\mathcal{U}_e\rangle \tag{1.192}$$

and

$$\iiint_V \left(\frac{1}{2}\frac{1}{\mu}|\vec{B}|^2\right) dV = 2 \iiint_V \left\langle\frac{1}{2}\frac{1}{\mu}|\vec{\mathcal{B}}|^2\right\rangle dV = 2\langle\mathcal{U}_m\rangle. \quad (1.193)$$

Thus, (1.187) becomes

$$\langle\mathcal{U}_e\rangle - \langle\mathcal{U}_m\rangle = \frac{1}{2\omega}\left[\iiint_V \mathrm{Im}\left(\frac{1}{2}\vec{E}\cdot\vec{J}_i^*\right) dV + \oiint_S \mathrm{Im}(\vec{S}_c)\cdot d\vec{S}\right], \quad (1.194)$$

which is an equation for the difference in the time-average energies stored in the electric and magnetic fields within the volume. When the right-hand side of this equation is zero, we have the interesting result

$$\langle\mathcal{U}_e\rangle = \langle\mathcal{U}_m\rangle,$$

or, for the total energy $\mathcal{U} = \mathcal{U}_e + \mathcal{U}_m$,

$$\langle\mathcal{U}\rangle = 2\langle\mathcal{U}_e\rangle = 2\langle\mathcal{U}_m\rangle.$$

Half of the total time-average energy is stored in the electric field and half is stored in the magnetic field.

1.7.4 Uniqueness theorem

Earlier we presented a uniqueness theorem for electromagnetic fields with general time dependence (Section 1.5). We will now develop a separate uniqueness theorem that applies specifically to fields with harmonic time dependence; it will involve the complex vector phasors $\vec{E}(\vec{r},\omega)$, $\vec{B}(\vec{r},\omega)$ rather than the time-varying field $\vec{\mathcal{E}}(\vec{r},t)$, $\vec{\mathcal{B}}(\vec{r},t)$. The procedure we will use is essentially the same as before, except that Maxwell's equations and Poynting's theorem for general time dependence (Table 1.3) are replaced by their counterparts for harmonic time dependence (Table 1.5) [65]. As before, we will state our result and then present an argument.

Uniqueness Theorem For a volume V that contains only simple materials and for which the impressed currents $\vec{J}_i$ are specified, a time-harmonic solution to Maxwell's equations is uniquely specified by the values of the tangential component of either $\vec{E}$ or $\vec{B}$ ($\hat{n}\times\vec{E}$ or $\hat{n}\times\vec{B}$) over the boundary surface of V (S_i and S_e in Figure 1.33). The tangential component of $\vec{E}$ can be specified on a portion of the boundary surface and the tangential component of $\vec{B}$ on the remainder. Each region within V, with the exception of perfect conductors, is assumed to have finite conductivity. The case for lossless materials ($\sigma = 0$) is viewed as the case for lossy materials ($\sigma \neq 0$) in the limit as the loss goes to zero ($\sigma \to 0$).

In other words, if a solution to Maxwell's equations is obtained within the volume, and the solution satisfies all of the above conditions, it is the only solution.

We will assume that there are two different solutions to Maxwell's equations described by the complex vector phasors $\vec{E}_a$, $\vec{B}_a$ and $\vec{E}_b$, $\vec{B}_b$. The difference between these solutions, $\delta\vec{E} = \vec{E}_a - \vec{E}_b$, $\delta\vec{B} = \vec{B}_a - \vec{B}_b$, is also a solution to Maxwell's equations; therefore, it can be used in the complex Poynting's theorem (1.185):

$$j\omega \iiint_V \left(\frac{1}{2}\varepsilon|\delta\vec{E}|^2 - \frac{1}{2}\frac{1}{\mu}|\delta\vec{B}|^2\right)dV - \iiint_V \left(\frac{1}{2}\sigma|\delta\vec{E}|^2\right)dV$$
$$-\iiint_V \left(\frac{1}{2}\delta\vec{E}\cdot\delta\vec{J}_i^*\right)dV = \oiint_S \frac{1}{2}\frac{1}{\mu}\delta\vec{B}^*\cdot(\hat{n}\times\delta\vec{E})dS$$
$$= -\oiint_S \frac{1}{2}\frac{1}{\mu}\delta\vec{E}\cdot(\hat{n}\times\delta\vec{B}^*)dS. \qquad (1.195)$$

We have assumed that both solutions arise from the same impressed sources and have the same tangential components for $\vec{E}$ or $\vec{B}$ on the surface S. Hence, $\delta\vec{J}_i^* = 0$ in the last integral on the left, and $\hat{n}\times\delta\vec{E} = 0$ or $\hat{n}\times\delta\vec{B}^* = 0$ in the integral on the right, making (1.195)

$$j\omega \iiint_V \left(\frac{1}{2}\varepsilon|\delta\vec{E}|^2 - \frac{1}{2}\frac{1}{\mu}|\delta\vec{B}|^2\right)dV = \iiint_V \left(\frac{1}{2}\sigma|\delta\vec{E}|^2\right)dV. \qquad (1.196)$$

On separating the real and imaginary parts, we get

$$\iiint_V \left(\frac{1}{2}\sigma|\delta\vec{E}|^2\right)dV = 0, \qquad (1.197)$$

and

$$\iiint_V \left(\frac{1}{2}\varepsilon|\delta\vec{E}|^2 - \frac{1}{2}\frac{1}{\mu}|\delta\vec{B}|^2\right)dV = 0. \qquad (1.198)$$

For $\sigma \neq 0$, Equation (1.197) can only be satisfied by $|\delta\vec{E}| = 0$ or $\vec{E}_a = \vec{E}_b$; then, from (1.198), it follows that $|\delta\vec{B}| = 0$ or $\vec{B}_a = \vec{B}_b$. Thus, there is only one solution, a unique solution, to Maxwell's equations within V for the stated conditions.

For lossless materials ($\sigma = 0$) our argument fails; we are left with the inconclusive Equation (1.198). The case for lossless materials can be viewed as the case for lossy materials in the limit as the loss goes to zero. The physically relevant solution for the lossless case ($\sigma = 0$) is taken to be the physically relevant solution for the lossy case ($\sigma \neq 0$) in the limit as the conductivity goes to zero ($\sigma \to 0$). This method is sometimes called the "principle of limiting absorption."[30]

As for fields with general time dependence, problems that involve regions of infinite extent require special consideration. We will again restrict our discussion to the problem described in Figure 1.35: a source of finite extent in infinite space. When the surrounding material has loss ($\sigma \neq 0$), the physically relevant solution to this problem is a field that decays with increasing distance from the source. For this solution, the integral over the surface S_e in (1.195) goes to zero as $r_e \to \infty$; all of the energy supplied by the source is dissipated before it reaches the boundary

[30] For uniqueness theorems that do not require lossy materials see References [27] and [55].

surface at infinity.[31] We conclude that a unique, physically relevant solution to Maxwell's equations is obtained by specifying the tangential component of either $\vec{E}$ or $\vec{B}$ on only the interior surface S_i. On the exterior surface S_e the tangential components of this field are zero due to the dissipation in the surrounding material. To obtain the physically relevant solution for the lossless case ($\sigma = 0$), we take the limit of this solution as $\sigma \to 0$.

It seems disconcerting that the boundary conditions differ in the uniqueness theorems for general time dependence (Section 1.5) and for harmonic time dependence: The former requires specification of $\hat{n} \times \vec{\mathcal{E}}$ or $\hat{n} \times \vec{\mathcal{B}}$ on the surface S as well as the initial values of $\vec{\mathcal{E}}$ and $\vec{\mathcal{B}}$ throughout the volume V, whereas the latter requires only specification of $\hat{n} \times \vec{E}$ or $\hat{n} \times \vec{B}$ on the surface S. This difference is explained by a simple physical argument. Consider the following situation: The field is initially zero within V, and at time $t = 0$ a source is turned on. The source gradually builds up to a cosinusoidal oscillation that continues for an indefinite time. For example, the source may be an impressed current of the form

$$\mathcal{J}_i \propto \mathcal{R}(t) \cos(\omega t), \tag{1.199}$$

where $\mathcal{R}(t)$ is a ramp function that goes from 0 at time $t = 0$ to 1.0 after a time equal to several periods ($2\pi/\omega$) of the cosinusoid. This is essentially the form that all oscillatory signals have in practice. The volume V is finite in extent and contains only lossy materials ($\sigma \neq 0$). Now the field is observed after a very long period of time; it is oscillating cosinusoidally everywhere within V; the transient field associated with turning on the source has long since died out due to the dissipation within V. The field has reached the steady state; it is harmonic in time, and it contains no information about the initial conditions at time $t = 0$. Since the initial values of the field do not affect the steady state solution, it is reasonable that they are not required in the uniqueness theorem for harmonic time dependence.[32]

[31] The physically relevant solution behaves as an outward-going wave that decays exponentially with increasing radial distance r_e:

$$\lim_{r_e \to \infty} \left\{ \begin{matrix} \vec{E} \\ \vec{B} \end{matrix} \right\} \propto e^{-jkr_e} = e^{-\alpha r_e} e^{-j\beta r_e}$$

with $k = \beta - j\alpha$, where

$$\alpha = \omega \sqrt{\frac{\varepsilon\mu}{2} \left[\sqrt{1 + (\sigma/\omega\varepsilon)^2} - 1 \right]}$$

and

$$\beta = \omega \sqrt{\frac{\varepsilon\mu}{2} \left[\sqrt{1 + (\sigma/\omega\varepsilon)^2} + 1 \right]}$$

[32] A mathematical argument that supports this conclusion is presented in Reference [66].

1.7.5 Fourier transform

So far our discussion in this section has been confined to fields with harmonic time dependence. However, a field with general time dependence can be thought of as a linear superposition of fields that vary harmonically in time at different frequencies. The formal mechanism that describes this relationship is the *Fourier transformation* (Jean Baptiste Joseph Fourier, 1768–1830):

$$f(t) = \frac{1}{2\pi} \int_{-\infty}^{\infty} \tilde{F}(\omega) e^{j\omega t} d\omega, \tag{1.200}$$

where the weight associated with the harmonic function of frequency ω is

$$\tilde{F}(\omega) = \int_{-\infty}^{\infty} f(t) e^{-j\omega t} dt. \tag{1.201}$$

Equation (1.201) is called the *Fourier transform* of the function $f(t)$; Equation (1.200) is called the *Fourier inversion formula*. The following shorthand notation will be used for a pair of functions that satisfy these relations:

$$f(t) \leftrightarrow \tilde{F}(\omega). \tag{1.202}$$

We will not consider the conditions required for the existence of the transform; they are discussed in detail in References [63, 64] and [67]. We simply note that the transform always exists for a physically realizable function $f(t)$, such as one obtained from a measurement.

It is easy to show that the transform of a real function $f(t)$ has

$$\tilde{F}(-\omega) = \tilde{F}^*(\omega), \tag{1.203}$$

and that the transform of the n-th temporal derivative of a function is

$$\frac{d^n f(t)}{dt^n} \leftrightarrow (j\omega)^n \tilde{F}(\omega). \tag{1.204}$$

Now we will consider the Fourier transform of the electric field

$$\tilde{\vec{E}}(\vec{r}, \omega) = \int_{-\infty}^{\infty} \vec{\mathcal{E}}(\vec{r}, t) e^{-j\omega t} dt. \tag{1.205}$$

Taking the curl of this equation and using the Maxwell's Equation (1.50), we have

$$\begin{aligned} \nabla \times \tilde{\vec{E}}(\vec{r}, \omega) &= \int_{-\infty}^{\infty} \nabla \times \vec{\mathcal{E}}(\vec{r}, t) e^{-j\omega t} dt \\ &= -\int_{-\infty}^{\infty} \left[\frac{\partial}{\partial t} \vec{\mathcal{B}}(\vec{r}, t) \right] e^{-j\omega t} dt. \end{aligned} \tag{1.206}$$

After introducing the Fourier transform of the magnetic field,

$$\tilde{\vec{B}}(\vec{r}, \omega) = \int_{-\infty}^{\infty} \vec{\mathcal{B}}(\vec{r}, t) e^{-j\omega t} dt, \tag{1.207}$$

and employing (1.204), Equation (1.206) becomes

$$\nabla \times \tilde{\vec{E}}(\vec{r}, \omega) = -j\omega \tilde{\vec{B}}(\vec{r}, \omega). \tag{1.208}$$

Similar operations can be used with the other Maxwell's equations (1.51)–(1.54) to obtain

$$\nabla \times \tilde{\vec{H}}(\vec{r}, \omega) = \tilde{\vec{J}}(\vec{r}, \omega) + j\omega\tilde{\vec{D}}(\vec{r}, \omega) \tag{1.209}$$

$$\nabla \cdot \tilde{\vec{D}}(\vec{r}, \omega) = \tilde{\rho}(\vec{r}, \omega) \tag{1.210}$$

$$\nabla \cdot \tilde{\vec{B}}(\vec{r}, \omega) = 0 \tag{1.211}$$

$$\nabla \cdot \tilde{\vec{J}}(\vec{r}, \omega) = -j\omega\tilde{\rho}(\vec{r}, \omega). \tag{1.212}$$

These are Maxwell's equations for the Fourier transform of the electromagnetic field $\tilde{\vec{E}}(\vec{r}, \omega)$, $\tilde{\vec{B}}(\vec{r}, \omega)$, etc. Notice that they have the same form as Maxwell's equations for a time-harmonic field (Table 1.5). However, the quantities in these two sets of equations are not the same. This is easily seen by examining their units. For example, the vector phasor $\vec{E}(\vec{r}, \omega)$ for harmonic time dependence has the units of the electric field (V/m), whereas the Fourier transform $\tilde{\vec{E}}(\vec{r}, \omega)$ has the units of the electric field per unit frequency [(V/m)/Hz]. In addition, the vector phasor $\vec{E}(\vec{r}, \omega)$ is a complex number with the parameter ω fixed, whereas the Fourier transform $\tilde{\vec{E}}(\vec{r}, \omega)$ is a complex function of the variable ω (Problem 1.14).

Equations (1.208)–(1.212) with the Fourier transform of the appropriate constitutive relations are generally used in the following manner to solve a practical problem. An excitation, such as the current $\vec{\mathcal{J}}(\vec{r}, t)$, is specified. The Fourier transform of the excitation, $\tilde{\vec{J}}(\vec{r}, \omega)$, is obtained from (1.201). Maxwell's equations are then solved with this excitation to obtain the Fourier transform of the electromagnetic field $\tilde{\vec{E}}(\vec{r}, \omega)$, $\tilde{\vec{B}}(\vec{r}, \omega)$. Finally, the time-varying electromagnetic field $\vec{\mathcal{E}}(\vec{r}, t)$, $\vec{\mathcal{B}}(\vec{r}, t)$ is obtained using the Fourier inversion formula (1.200).

From what we have shown above, we can conclude that any direct manipulation of Maxwell's equations for the Fourier transform of the field will produce a result that is symbolically the same as for a time-harmonic field. For example, Poynting's theorem is given by Equation (1.179) or (1.186) with all field quantities specified as Fourier transforms rather than vector phasors [e.g., $\tilde{\vec{E}}(\vec{r}, \omega)$ rather than $\vec{E}(\vec{r}, \omega)$]. However, we must provide a new interpretation for the quantities appearing in this equation. To this end, we will examine the following integral of the complex Poynting vector:

$$\frac{2}{\pi}\int_0^\infty \hat{n} \cdot \mathrm{Re}\left[\vec{S}_c(\omega)\right]d\omega = \frac{2}{\pi}\int_0^\infty \hat{n} \cdot \left\{ \mathrm{Re}\left[\frac{1}{2}\,\tilde{\vec{E}}(\omega) \times \tilde{\vec{H}}^{*}(\omega)\right]\right\}d\omega. \tag{1.213}$$

Parseval's formula for the real functions $f_1(t)$ and $f_2(t)$ states that [63, 64]

$$\int_{-\infty}^{\infty} f_1(t) f_2(t) dt = \frac{1}{\pi}\int_0^\infty \mathrm{Re}\left[\tilde{F}_1(\omega)\tilde{F}_2^*(\omega)\right]d\omega, \tag{1.214}$$

where

$$f_1(t) \leftrightarrow \tilde{F}_1(\omega)$$

Table 1.7. *Notation for field quantities*

General time dependence	$\vec{\mathcal{E}}(\vec{r},t)=\sum_{i=x,y,z}\mathcal{E}_i(\vec{r},t)\,\hat{i}$	$\lvert\vec{\mathcal{E}}\rvert=\sqrt{\vec{\mathcal{E}}\cdot\vec{\mathcal{E}}}$	(Magnitude of a real vector)
		$\lvert\mathcal{E}_i\rvert$	(Absolute value of a real number)
Harmonic time dependence	$\vec{\mathcal{E}}(\vec{r},t)=Re\left[\vec{E}(\vec{r},\omega)e^{j\omega t}\right]$	$\lvert\vec{E}\rvert=\sqrt{\vec{E}\cdot\vec{E}^*}$	(Hermitian magnitude of a complex vector)
	$\vec{E}(\vec{r},\omega)=\sum_{i=x,y,z}\vec{E}_i(\vec{r},\omega)$	$\lvert E_i\rvert=\sqrt{E_iE_i^*}$	(Modulus of a complex number)
		$=A_i$	
	$=\sum_{i=x,y,z}E_i(\vec{r},\omega)\,\hat{i}$		
	$=\sum_{i=x,y,z}A_i(\vec{r},\omega)e^{j\phi_i(\vec{r},\omega)}\,\hat{i}$		
Fourier transform	$\vec{\mathcal{E}}(\vec{r},t)=\frac{1}{2\pi}\int_{-\infty}^{\infty}\tilde{\tilde{\vec{E}}}(\vec{r},\omega)e^{j\omega t}d\omega$		
	$\tilde{\tilde{\vec{E}}}(\vec{r},\omega)=\int_{-\infty}^{\infty}\vec{\mathcal{E}}(\vec{r},t)e^{-j\omega t}dt$		

and

$$f_2(t)\leftrightarrow\tilde{F}_2(\omega).$$

Using this formula with (1.213), we find

$$\int_{-\infty}^{\infty}\hat{n}\cdot\left[\vec{\mathcal{E}}(t)\times\vec{\mathcal{H}}(t)\right]dt=\frac{2}{\pi}\int_{0}^{\infty}\hat{n}\cdot\left\{\mathrm{Re}\left[\frac{1}{2}\tilde{\tilde{\vec{E}}}(\omega)\times\tilde{\tilde{\vec{H}}}^{*}(\omega)\right]\right\}d\omega,$$

or

$$\int_{-\infty}^{\infty}\hat{n}\cdot\vec{\mathcal{S}}(t)dt=\int_{0}^{\infty}\frac{2}{\pi}\hat{n}\cdot\left\{\mathrm{Re}\left[\vec{S}_c(\omega)\right]\right\}d\omega. \qquad (1.215)$$

The left-hand side of this equation is just the total electromagnetic energy passing through a unit area of surface with the unit normal $\hat{n}$. Therefore, the integrand on the right-hand side of the equation,

$$\frac{2}{\pi}\hat{n}\cdot\left\{\mathrm{Re}\left[\vec{S}_c(\omega)\right]\right\}, \qquad (1.216)$$

must be the energy passing through a unit area of this surface, per unit frequency $[(\mathrm{J/m^2})/\mathrm{Hz}]$. It is called the energy spectral density.[33] All of the quantities appearing in Poynting's theorem (1.186) have this interpretation. For example,

$$\frac{2}{\pi}\left(\frac{1}{2}|\tilde{\vec{J}}_c(\omega)|^2/\sigma\right)$$

is the energy per unit volume, per unit frequency transferred from the field as heat.

We have used the tilde ($\tilde{}$) with the Fourier transform $\tilde{\vec{E}}(\vec{r}, \omega)$ to distinguish it from the vector phasor $\vec{E}(\vec{r}, \omega)$ for the time-harmonic field. The tilde will be omitted when this causes no confusion. The equations in Tables 1.5 and 1.6 can then be thought of as applying to the vector phasors for a field with harmonic time dependence or to the Fourier transform of a field with general time dependence. The equations in Table 1.5 are often referred to as *Maxwell's equations for the frequency domain*, whereas our original equations in Table 1.3 are referred to as *Maxwell's equations for the time domain*.

For future reference the notation we have introduced in this section is summarized in Table 1.7.

References

[1] E. Whittaker, *A History of the Theories of Aether and Electricity*, Volumes I and II, Tomash Publ./American Institute of Physics, New York, 1987. Originally published, Nelson, London, 1951, 1953.

[2] V. Ronchi, *The Nature of Light*, Harvard University Press, Cambridge, MA, 1970.

[3] P. F. Mottelay, *Bibliographical History of Electricity and Magnetism*, Charles Griffin, London, 1922.

[4] J. C. Maxwell, "A Dynamical Theory of the Electromagnetic Field," *Philos. Trans. Roy. Soc. London*, Vol. 155, pp. 459–512, 1865. Paper read December 8, 1864. Also reprinted in book form, *A Dynamical Theory of the Electromagnetic Field*, Scottish Academic Press, Edinburgh, 1982.

[5] —, *A Treatise on Electricity and Magnetism*, Clarendon Press, Oxford, 1873. Republication of 3rd Edition (1891), Dover Publications, New York, 1954.

[6] A. Sommerfeld, *Electrodynamics*, Lectures on Theoretical Physics, Vol. II, Academic Press, New York, 1952.

[7] H. A. Lorentz, *The Theory of Electrons and Its Application to the Phenomena of Light and Radiant Heat*, 1st Edition, Leipzig, 1909. Reprint of Second Edition, Dover Publications, New York, 1952.

[8] T. Erber, "The Classical Theories of Radiation Reaction," *Fortschritte der Physik*, Vol. 9, pp. 343–92, 1961.

[9] F. Rohrlich, *Classical Charged Particles*, Addison-Wesley, Reading, MA, 1965.

[10] P. Pearle, "Classical Electron Models," Ch. 7 in *Electromagnetism Paths to Research*, D. Teplitz, editor, pp. 211–85, Plenum, New York, 1982.

[33] The factor $2/\pi$ that appears in the energy spectral density is a consequence of the definition used for the Fourier transform pair (1.200), (1.201). For other definitions of this pair, the factor will be different.

Notice that we have not placed a $\tilde{}$ over $\vec{S}_c(\omega)$. This is to avoid the *incorrect* interpretation that $\vec{S}_c(\omega)$ is the Fourier transform of $\vec{S}(t)$.

[11] C. A. Coulomb, First and second memoirs on electricity and magnetism, *Mémoires de l'Académie Royale des Sciences,* for 1785, pp. 569–77 and pp. 578–611, published 1788.

[12] C. S. Gillmor, *Coulomb and the Evolution of Physics and Engineering in Eighteenth-Century France*, Princeton University Press, Princeton, NJ, 1971.

[13] J. C. Oersted, "Experiments on the Effect of a Current of Electricity on the Magnetic Needle," *Annals of Philosophy,* Vol. 16, pp. 274–75, October 1820. An English translation of a tract in Latin first printed and circulated by Oersted.

[14] B. Dibner, *Oersted and the Discovery of Electromagnetism,* Burndy Library, Norwalk, CT, 1961.

[15] T. Martin, editor, *Faraday's Diary,* Vol. VI, Entry for December 11, 1851, G. Bell, London, 1935.

[16] A.-M. Ampère, "Mémoire sur l'action des courans voltaiques," *Annales de Chimie et de Physique,* Vol. 15, pp. 59–76, 170–218, 1820.

[17] —, *Mémoire sur la théorie mathématique des phénomènes électrodynamiques, uniquement déduite de l'expérience*, 1827. Reprinted by Librairie Scientifique Albert Blanchard, Paris, 1958.

[18] R. A. R. Tricker, *Early Electrodynamics, The First Law of Circulation,* Pergamon Press, Oxford, 1965.

[19] S. P. Thompson, *Michael Faraday, His Life and Works,* Cassell, London, 1901.

[20] L. P. Williams, *Michael Faraday,* Basic Books, New York, 1965.

[21] R. A. R. Tricker, *The Contributions of Faraday and Maxwell to Electrical Science,* Pergamon Press, Oxford, 1966.

[22] T. Martin, editor, *Faraday's Diary,* Vol. VI, G. Bell, London, 1935.

[23] A. M. Bork, "Maxwell, Displacement Current, and Symmetry," *Am. J. Phys.*, Vol. 31, pp. 854–59, November 1963.

[24] J. C. Maxwell, "On a Method of Making a Direct Comparison of Electrostatic with Electromagnetic Force; with a Note on the Electromagnetic Theory of Light," *Philos. Trans. Roy. Soc. London,* Vol. 158, pp. 643–57, 1868.

[25] H. Hertz, *Electric Waves,* Macmillan, London, 1893. Republication, Dover Publications, New York, 1962.

[26] O. D. Kellogg, *Foundations of Potential Theory*, Springer-Verlag, Berlin, 1929.

[27] C. Müller, *Foundations of the Mathematical Theory of Electromagnetic Waves*, Springer-Verlag, Berlin, 1969.

[28] A. Schwartz, *Analytic Geometry and Calculus*, Holt, Rinehart, and Winston, New York, 1960.

[29] C. T. Tai, "On the Presentation of Maxwell's Theory," *Proc. IEEE*, Vol. 60, pp. 936–45, August 1972.

[30] S. A. Schelkunoff, "On Teaching the Undergraduate Electromagnetic Theory," *IEEE Trans. on Education*, Vol. E-15, pp. 15–25, February 1972.

[31] S. Gray, "A Letter to Cromwell Mortimer, M. D. Secr. R. S. Containing Several Experiments Concerning Electricity," *Philos. Trans. Roy. Soc. London*, Vol. 37, pp. 18–44, January–February 1731.

[32] N. D. Lang and W. Kohn, "Theory of Metal Surfaces: Charge Density and Surface Energy," *Phys. Rev. B*, Vol. 1, pp. 4,555–68, June 1970.

[33] —, "Theory of Metal Surfaces: Work Functions," *Phys. Rev. B*, Vol. 3, pp. 1,215–23, February 1971.

[34] —, "Theory of Metal Surfaces: Induced Surface Charge and Image Potential," *Phys. Rev. B*, Vol. 7, pp. 3,541–50, April 1973.

[35] G. S. Smith, "On the Skin Effect Approximation," *Am. J. Phys.*, Vol. 58, pp. 996–1,002, October 1990.

[36] H. B. G. Casimir and J. Ubbink, "The Skin Effect," *Philips Tech. Rev.*, Vol. 28, Part I, No. 9, pp. 271–83; Part II, No. 10, pp. 300–15; Part III, No. 12, pp. 366–81, 1967.

[37] H. Goldstein, *Classical Mechanics*, 2nd Edition, Addison-Wesley, Reading, MA, 1980.

[38] J. H. Poynting, "On the Transfer of Energy in the Electromagnetic Field," *Philos. Trans. Roy. Soc. London*, Vol. 175, Part II, pp. 343–61, 1884.

[39] O. Heaviside, "The Induction of Currents in Cores," *The Electrician*, Vol. 13, pp. 133–4, June 21, 1884.

[40] —, "Electromagnetic Induction and Its Propagation," *The Electrician*, Vol. 14, pp. 178–80, January 10, 1885; pp. 306–7, February 21, 1885.

[41] P. J. Nahin, *Oliver Heaviside: Sage in Solitude*, IEEE Press, New York, 1988.

[42] J. D. Jackson, *Classical Electrodynamics*, 2nd Edition, Wiley, New York, 1975.

[43] M. Mason and W. Weaver, *The Electromagnetic Field*, University of Chicago Press, Chicago, 1929.

[44] J. Slepian, "Energy and Energy Flow in the Electromagnetic Field," *J. Appl. Phys.*, Vol. 13, pp. 512–8, August 1942.

[45] W. H. Furry, "Examples of Momentum Distributions in the Electromagnetic Field in Matter," *Am. J. Phys.*, Vol. 37, pp. 621–36, June 1969.

[46] C. S. Lai, "Alternative Choice for the Energy Flow Vector of the Electromagnetic Field," *Am. J. Phys.*, Vol. 49, pp. 841–3, September 1981. Comments on this paper: P. Lorrain, Vol. 50, p. 492, June 1982; D. H. Kobe, Vol. 50, pp. 1,162–4, December 1982; P. C. Peters, Vol. 50, pp. 1,165–6, December 1982; R. H. Romer, Vol. 50, pp. 1,166–8, December 1982.

[47] U. Backhaus and K. Schäfer, "On the Uniqueness of the Vector for Energy Flow Density in Electromagnetic Fields," *Am. J. Phys.*, Vol. 54, pp. 279–80, March 1986.

[48] O. Heaviside, "Electromagnetic Induction and its Propagation, XXXV. The Transfer of Energy and its Application to Wires. Energy-Current," *The Electrician*, Vol. 18, pp. 211–3, January 14, 1887.

[49] C. Schaefer, *Einführung in die theoretische Physik*, Vol. III, Part 1, Walter de Gruyter, Berlin, 1932.

[50] A. Marcus, "The Electric Field Associated with a Steady Current in Long Cylindrical Conductor," *Am. J. Phys.*, Vol. 9, pp. 225–6, August 1941.

[51] L. D. Landau and E. M. Lifshitz, *The Classical Theory of Fields*, Course of Theoretical Physics, Vol. 2, 4th Edition, Pergamon Press, New York, 1975.

[52] J. A. Stratton, *Electromagnetic Theory*, McGraw-Hill, New York, 1941.

[53] I. E. Tamm, *Fundamentals of the Theory of Electricity*, Mir Publishers, Moscow, USSR, 1979. First Russian Edition, 1929.

[54] R. Courant and D. Hilbert, *Methods of Mathematical Physics, Volume II, Partial Differential Equations*, Wiley, New York, 1962.

[55] D. S. Jones, *The Theory of Electromagnetism*, Pergamon Press, Oxford, 1964.

[56] R. Courant, K. Friedrichs, and H. Lewy, "Über die partiellen Differenzengleichungen der mathematischen Physik," *Math. Ann.*, Vol. 100, pp. 32–74, 1928. English translation, P. Fox, *On the Partial Difference Equations of Mathematical Physics*, Report NYO-7689, Courant Institute of Mathematical Sciences, New York University, New York, September 1956; also *IBM. Res. Dev.*, Vol. 11, pp. 215–34, March 1967.

[57] P. Fox, "The Solution of Hyperbolic Partial Differential Equations by Difference

Methods," in *Mathematical Methods for Digital Computers*, A. Ralston and H. S. Wilf, editors, Ch. 16, Wiley, New York, 1964.

[58] E. Isaacson and H. B. Keller, *Analysis of Numerical Methods*, Wiley, New York, 1966.

[59] W. F. Ames, *Numerical Methods for Partial Differential Equations*, 2nd Edition, Academic Press, New York, 1977.

[60] K. S. Yee, "Numerical Solution of Initial Boundary Value Problems Involving Maxwell's Equations in Isotropic Media," *IEEE Trans. Antennas Propagat.*, Vol. AP-14, pp. 302–7, May 1966.

[61] A. Taflove, *Computational Electrodynamics, the Finite-Difference Time-Domain Method,* Artech House, Norwood, MA 1995.

[62] J. G. Maloney, G. S. Smith, and W. R. Scott, Jr., "Accurate Computation of the Radiation from Simple Antennas Using the Finite-Difference Time-Domain Method," *IEEE Trans. Antennas Propagat.*, Vol. AP-38, pp. 1,059–68, July 1990.

[63] A. Papoulis, *The Fourier Integral and Its Applications*, McGraw-Hill, New York, 1962.

[64] R. N. Bracewell, *The Fourier Transform and Its Applications*, 2nd Edition, McGraw-Hill, New York, 1978.

[65] R. F. Harrington, *Time-Harmonic Electromagnetic Fields*, McGraw-Hill, New York, 1961.

[66] I. Stakgold, *Boundary Value Problems of Mathematical Physics*, Volume II, Macmillan, New York, 1968.

[67] E. C. Titchmarsh, *Introduction to the Theory of Fourier Integrals*, 2nd Edition, Oxford University Press, London, 1948. Republication, Chelsea, New York, 1986.

[68] T. R. Carver and J. Rajhel, "Direct 'Literal' Demonstration of the Effect of a Displacement Current," *Am. J. Phys.*, Vol. 42, pp. 246–9, March 1974.

[69] D. F. Bartlett and T. R. Corle, "Measuring Maxwell's Displacement Current Inside a Capacitor," *Phys. Rev. Lett.*, Vol. 55, pp. 59–62, July 1, 1985.

[70] E. Katz, "Concerning the Number of Independent Variables of the Classical Electromagnetic Field," *Am. J. Phys.*, Vol. 33, pp. 306–12, April 1965.

[71] P. G. H. Sandars, "Magnetic Charge," *Contemporary Physics*, Vol. 7, pp. 419–29, November 1966.

[72] D. R. Moorcroft, "Faraday's Law – Demonstration of a Teaser," *Am. J. Phys.*, Vol. 37, p. 221, February 1969; also, Vol. 38, pp. 376–7, March 1970.

[73] R. H. Romer, "What do 'Voltmeters' Measure?: Faraday's Law in a Multiply Connected Region," *Am. J. Phys.*, Vol. 50, pp. 1,089–93, December 1982.

[74] P. C. Peters, "The Role of Induced EMF's in Simple Circuits," *Am. J. Phys.*, Vol. 52, pp. 208–11, March 1984.

[75] E. M. Purcell, *Electricity and Magnetism*, Berkeley Physics Course Vol. 2, 2nd Edition, McGraw-Hill, New York, 1985.

[76] M. A. Heald, "Electric Fields and Charges in Elementary Circuits," *Am. J. Phys.*, Vol. 52, pp. 522–6, June 1984.

[77] J. Van Bladel, *Singular Electromagnetic Fields and Sources*, Oxford University Press, Oxford, 1991.

Problems

1.1 Ampère invented a device known as the *astatic combination* for use in his experiments. This device is basically a detector for a spatially nonuniform magnetic field. Its construction is shown in Figure P1.1. The wire forming the detector lies in the plane of the page and is bent to form two identical loops, L and R. The current I_d is in opposite directions in the two loops. The detector is free to rotate about the vertical axis C—C′.

a) This device is called astatic (having no tendency toward a change in direction) because it is not affected by the presence of the Earth's magnetic field. Assume that the Earth's magnetic field $\vec{B}_e$ has the orientation shown in Figure P1.2a and is uniform over the area of the detector. Show that the Earth's magnetic field does not cause the detector to rotate.
b) The detector does rotate when it is placed in a magnetic field that is not uniform over its area. Show that the detector rotates when placed near the long, straight current-carrying wire with the orientation shown in Figure P1.2b.
c) Ampère used this detector in some of his experiments based on the null principle. One of these experiments is shown in Figure P1.2c. The detector is placed midway between two parallel wires carrying the same current. The plane of the detector is normal to the plane of the two wires, and one

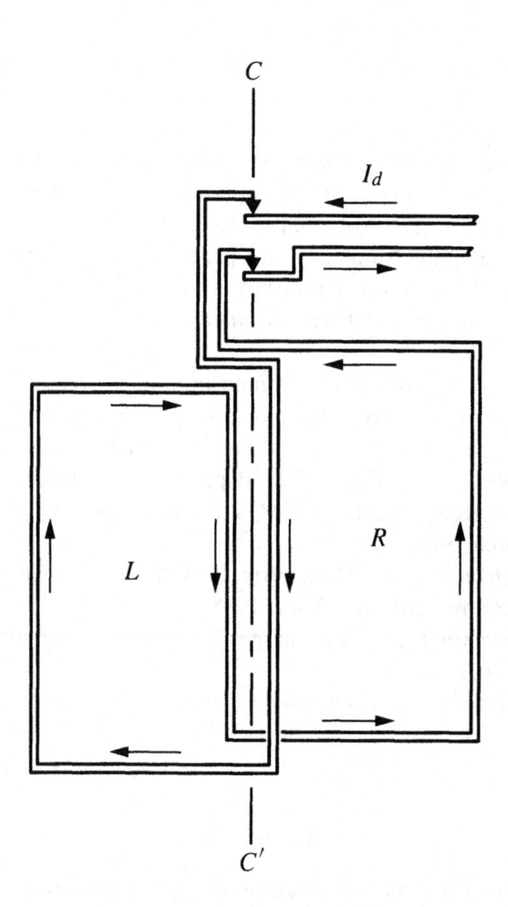

Fig. P1.1. Ampère's astatic combination.

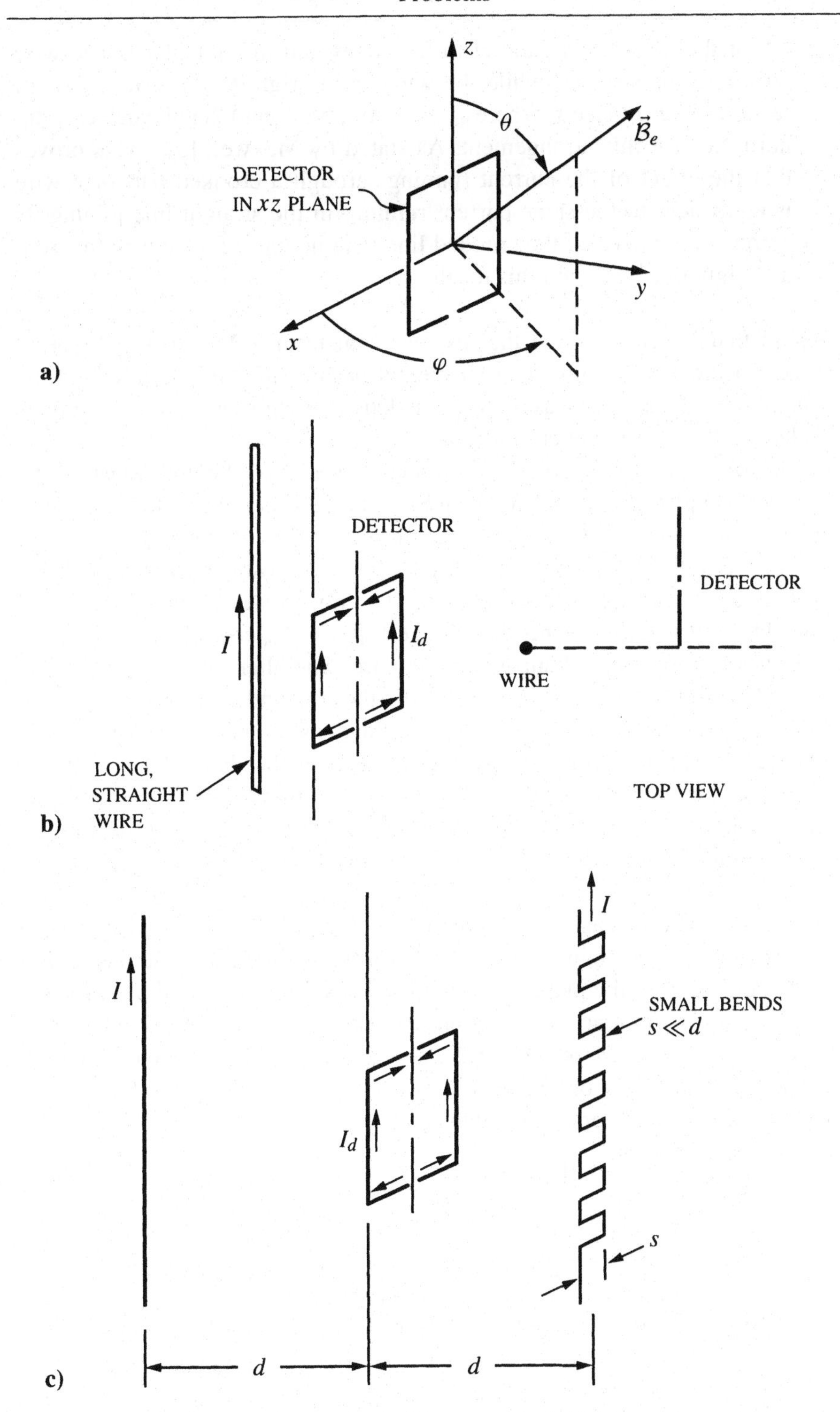

Fig. P1.2. a) Astatic combination in Earth's magnetic field. b) Astatic combination near a long, straight, current-carrying wire. c) Ampère's experiment.

leg of the detector is parallel to the wires and in their plane. The wire on the left is straight, while the wire on the right has a series of small regular bends of size s, where $s \ll d$. Ampère found that the detector did not rotate for this arrangement. As stated by Maxwell [5], "This proves that the effect of the current running through a crooked part of a wire is equivalent to the same current running in the straight line joining its extremities, provided the crooked line is in no part of its course far from a straight one." Explain this result.

1.2 Recall that Maxwell added the displacement term $\partial\vec{\mathcal{D}}/\partial t$ to the Ampère-Maxwell law without direct experimental proof. Since this term enters the equation in the same way as the volume density of current $\vec{\mathcal{J}}$, it is sometimes called the "displacement current."

The following simple experiment supports the inclusion of this term [68, 69]. The parallel plate capacitor shown in Figure P1.3 is connected to an alternating current source. This produces the current $\mathcal{I}(t) = I_o \sin(\omega t)$ in the wire connected to the plates and the voltage $\mathcal{V}(t) = V_o \cos(\omega t)$ across the plates. A voltmeter is connected to the terminals of a tightly wound, circular toroidal coil. The voltmeter measures the voltage $\mathcal{V}_m(t) = V_m \cos(\omega t)$.

In position A, the coil surrounds the wire. The alternating current in the wire ($\vec{\mathcal{J}}$) produces a magnetic field within the coil, and by Faraday's law, the time-varying magnetic field produces a voltage (electric field) at the terminals of the coil. In position B, the coil is inserted between the plates of the capacitor, and a voltage is again observed at the terminals of the coil. However, it is now the alternating displacement current ($\partial\vec{\mathcal{D}}/\partial t$) that produces the time-varying magnetic field within the coil and the resulting voltage across its terminals.

a) Assume that the capacitor is ideal: The electric field exists only between the plates (no fringing), and it is uniform and normal to the plates. Let the diameter of the plates be d and their spacing s. The coil has radius b ($b < d/2$), circular cross section of area $S_N = \pi a^2$, and N turns. The coil is very thin ($a \ll b$). Obtain an expression for V_m when the coil is between the plates of the capacitor.

b) For the following practical values:

$$V_o = 100 \text{ V} \qquad s = 10 \text{ cm}$$
$$f = 20 \text{ kHz} \qquad b = 35 \text{ cm}$$
$$d = 80 \text{ cm} \qquad a = 1.3 \text{ cm},$$

how many turns are needed on the coil to produce a voltage $V_m = 10^{-3}$ V?

1.3 An infinitesimal current element is shown in Figure 7.1a. It is a current $\mathcal{I}(t)$ acting over the length Δl in the limit as $\Delta l \to 0$. When the current varies slowly with time, the electric field of the element, given in Table 7.1, is

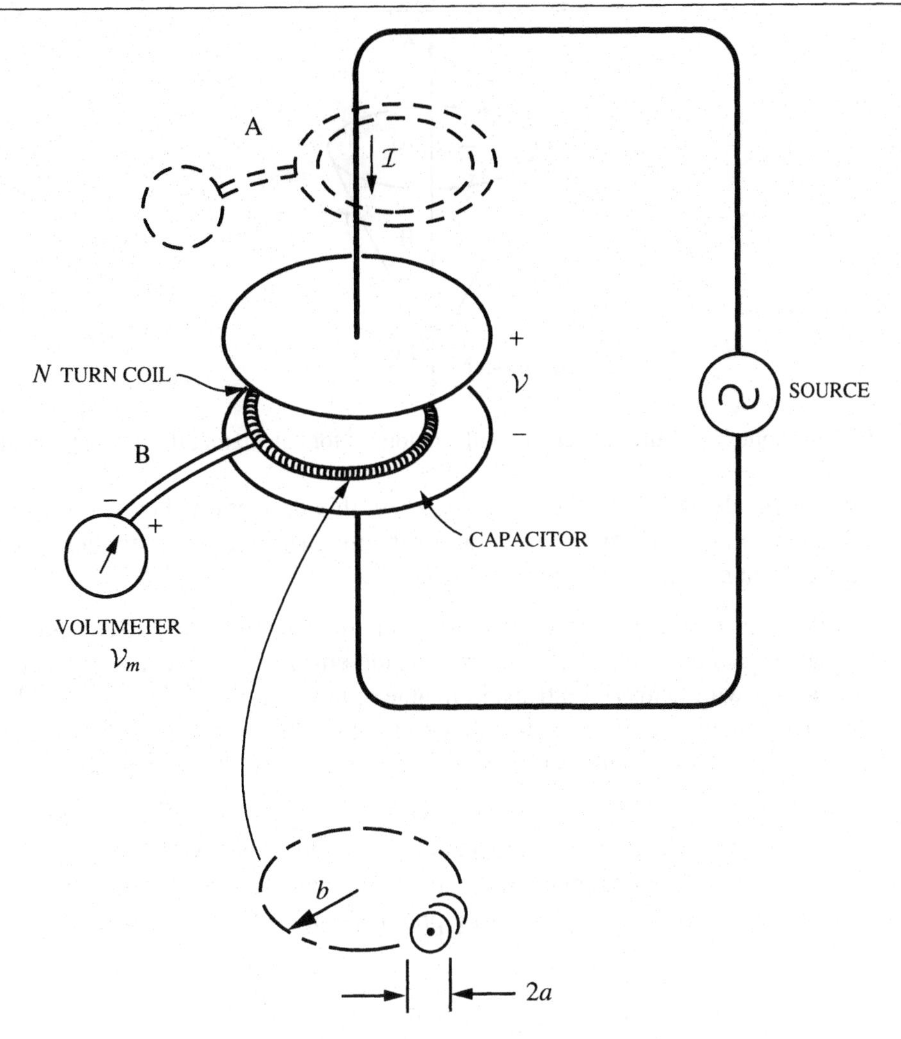

Fig. P1.3. Experiment for observing the "displacement current."

approximately

$$\vec{\mathcal{E}}(\vec{r}, t) \approx \frac{1}{4\pi\varepsilon_o r^3}(2\cos\theta\,\hat{r} + \sin\theta\,\hat{\theta}\,) \int_{-\infty}^{t} \mathcal{I}(t')\Delta l dt'.$$

Here we have assumed that the element is aligned with the z axis, and we have used the spherical coordinates shown in Figure 7.1b.

Obtain the magnetic field $\mathcal{B}_\phi$ of the current element by applying the Ampère-Maxwell law in integral form to the open surface S whose boundary is the circular contour C shown in Figure P1.4. Show that your result is essentially the *Biot-Savart formula*, which states that the magnetic field of a current element is

$$\vec{\mathcal{B}}(\vec{r}, t) = \frac{\mu_o \mathcal{I}(t)\Delta l}{4\pi r^2}\hat{z} \times \hat{r},$$

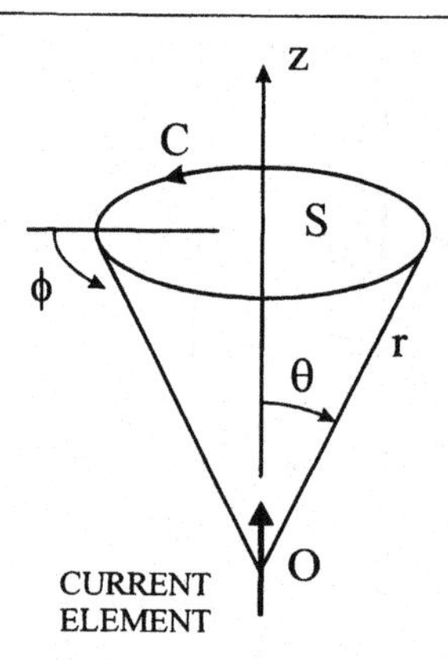

Fig. P1.4. Infinitesimal current element with contour C for calculating the magnetic field.

when the element is in the $\hat{z}$ direction. As this derivation shows, the Biot-Savart formula is an approximation for the magnetic field that applies when the current element is slowly varying in time.

1.4 In the text, Faraday's law (1.50), the Ampère-Maxwell law (1.51), and the equation of continuity (1.54) were considered the independent Maxwell's equations with Gauss' electric and magnetic laws, (1.52) and (1.53), the dependent equations. Examine the alternate point of view where (1.50), (1.51), and (1.52) are considered independent with (1.53) and (1.54) dependent.

1.5 For low-frequency circuits, the charge on the wires connecting the components is generally insignificant. Use the model developed in the text, Figure 1.30 and Equation (1.127), to estimate the total charge on the upper half ($0 \leq z \leq L$) of a straight wire. Use the following practical values for the parameters:

$$I = 1 \text{ mA} \qquad L = 30 \text{ cm}$$
$$\sigma = 5.8 \times 10^7 \text{ S/m} \qquad b = 10 \text{ cm}$$
$$a = 1 \text{ mm}.$$

Compare the charge you obtain to that of a single electron: -1.602×10^{-19} C.

1.6 To date, magnetic charge has not been observed in any experiment. However, the failure to observe magnetic charge may simply be a result of the convention used in defining electromagnetic quantities, as the following argument shows [42, 70, 71].

a) The equations of electromagnetism (Table 1.3) can be modified to include the effects of magnetic charge. After introducing the volume densities of magnetic charge and current, ρ_m and $\vec{\mathcal{J}}_m$, Maxwell's equations become

$$\nabla \times \vec{\mathcal{E}} = -\vec{\mathcal{J}}_m - \frac{\partial \vec{\mathcal{B}}}{\partial t}, \qquad \nabla \times \vec{\mathcal{H}} = \vec{\mathcal{J}}_e + \frac{\partial \vec{\mathcal{D}}}{\partial t},$$
$$\nabla \cdot \vec{\mathcal{B}} = \rho_m, \qquad \nabla \cdot \vec{\mathcal{D}} = \rho_e,$$
$$\nabla \cdot \vec{\mathcal{J}}_m = -\frac{\partial \rho_m}{\partial t}, \qquad \nabla \cdot \vec{\mathcal{J}}_e = -\frac{\partial \rho_e}{\partial t},$$

with the Lorentz force density

$$\vec{f} = \rho_e \vec{\mathcal{E}} + \rho_m \vec{\mathcal{H}} + \vec{\mathcal{J}}_e \times \vec{\mathcal{B}} - \vec{\mathcal{J}}_m \times \vec{\mathcal{D}}.$$

Here we have used the subscripts e and m to distinguish electric and magnetic sources. The three equations on the left involve ρ_m and $\vec{\mathcal{J}}_m$. The second of these is the differential form of (1.19), and the third is the equation of continuity for magnetic charge. Show that the first equation, Faraday's law modified to include magnetic current, is consistent with the other two.

b) Let the sources be in free space so that (1.6) and (1.7) apply. Show that Maxwell's equations are invariant under the following transformation:

$$\vec{\mathcal{E}} = \vec{\mathcal{E}}' \cos\psi + c\vec{\mathcal{B}}' \sin\psi$$

$$\vec{\mathcal{B}} = -\frac{1}{c}\vec{\mathcal{E}}' \sin\psi + \vec{\mathcal{B}}' \cos\psi$$

$$\rho_e = \rho_e' \cos\psi + \frac{1}{\zeta_o}\rho_m' \sin\psi$$

$$\vec{\mathcal{J}}_e = \vec{\mathcal{J}}_e{}' \cos\psi + \frac{1}{\zeta_o}\vec{\mathcal{J}}_m{}' \sin\psi$$

$$\rho_m = -\zeta_o \rho_e' \sin\psi + \rho_m' \cos\psi$$

$$\vec{\mathcal{J}}_m = -\zeta_o \vec{\mathcal{J}}_e{}' \sin\psi + \vec{\mathcal{J}}_m{}' \cos\psi,$$

where $\zeta_o = \sqrt{\mu_o/\varepsilon_o}$. That is, substitute the expressions for the unprimed quantities $\vec{\mathcal{E}}$, ρ_e, etc., and show that Maxwell's equations are obtained for the primed quantities $\vec{\mathcal{E}}'$, ρ_e', etc. Also show that the Lorentz force density, the Poynting vector $\vec{\mathcal{S}}$, and the total energy density $u = u_e + u_m$ are invariant under this transformation.

c) The results from part b show that both representations for the electromagnetic field and sources (the primed quantities and the unprimed quantities) satisfy Maxwell's equations. Both representations also predict the same experimental results, since they give the same values for quantities directly related to observations, such as the Lorentz force density, the Poynting vector, and the energy density.

Assume that the magnetic sources are set to zero for the unprimed representation ($\rho_m = 0$, $\vec{\mathcal{J}}_m = 0$). This is just the conventional electromagnetism summarized in Table 1.3. We can redefine the electromagnetic field and sources using the transformation in part b. The magnetic sources will no longer be zero ($\rho_m' \neq 0$, $\vec{\mathcal{J}}_m{}' \neq 0$) in the new primed representation. Because the same experimental results are predicted by both representations, we see that the exclusion of magnetic charge may be a matter of convention. Show that the ratio of the magnetic charge density to the electric charge density in the new representation is fixed at the value

$$\frac{1}{\zeta_o}\frac{\rho'_m}{\rho'_e} = \frac{1}{\zeta_o}\frac{\vec{\mathcal{J}}_m{}'}{\vec{\mathcal{J}}_e{}'} = \tan(\psi),$$

where ψ is arbitrary.

If we assume that matter is made up of charged particles, in the new representation, all particles would have the same ratio of magnetic to electric charge

$$\frac{1}{\zeta_o}\frac{q'_m}{q'_e} = \tan(\psi).$$

d) Now assume that matter is made up of two types of particles, 1 and 2. Each has electric and magnetic charge with the following ratios:

$$\frac{1}{\zeta_o}\frac{q_{m1}}{q_{e1}} = \tan(\psi_1)$$

$$\frac{1}{\zeta_o}\frac{q_{m2}}{q_{e2}} = \tan(\psi_2).$$

When $\psi_1 \neq \psi_2$, can an electromagnetic field and sources be defined so that the conventional equations of electromagnetism (those without magnetic sources) accurately predict experimental observations? Show that ψ_2 must equal ψ_1 for electrically and magnetically neutral matter ($q_{e2} = -q_{e1}$, $q_{m2} = -q_{m1}$).

1.7 Several variations of the following problem have been discussed in the literature [72–74]. It is a valuable problem, because it encourages a clear understanding of Faraday's law. Consider the two devices shown in Figure P1.5a: a resistor and a voltmeter. Each device is characterized by the line integral of the electric field over the path (*a*–*b*) through its terminals: For the resistor

$$-\int_a^b \vec{\mathcal{E}} \cdot d\vec{\ell} = \mathcal{I}_R R = \mathcal{V}_R,$$

where R is the resistance, $\mathcal{I}_R$ the current through the resistor, and $\mathcal{V}_R$ the voltage across the resistor; and for the voltmeter

$$-\int_a^b \vec{\mathcal{E}} \cdot d\vec{\ell} = \mathcal{V}_m,$$

where $\mathcal{V}_m$ is the voltmeter reading. The current through the voltmeter is assumed to be negligible.

a) A closed, planar circuit is formed from the two resistors, R_1 and R_2, and perfectly conducting wire. A uniform magnetic field is normal to the plane of the circuit and confined to the shaded region of area S in Figure 1.5b. This field increases linearly with time:

$$\vec{\mathcal{B}}(t) = C_o t \hat{z},$$

where $\hat{z}$ is pointing out from the page. This field could be produced by a long solenoid, whose axis is normal to the plane of the circuitry, carrying a current that varied linearly with time. Two identical voltmeters are connected to the circuit at points A and B; one voltmeter is to the right of the circuit; the other is to the left. What are the readings on the voltmeters, $\mathcal{V}_{m1}$ and $\mathcal{V}_{m2}$? Under what conditions are the two meter readings the same?

b) A "short circuit," comprised of a perfectly conducting wire, is connected between points A and B on the right side of the circuit, as in Figure 1.5c. What are the readings on the two voltmeters?

c) The connections to the voltmeters are rearranged, as in Figure 1.5d. What are the readings on the two voltmeters?

1.8 Show that the two boundary conditions for the normal components of the electromagnetic field (1.80) and (1.81), can be obtained from the three other boundary conditions, (1.65), (1.66), and (1.78), and the Maxwell's equations

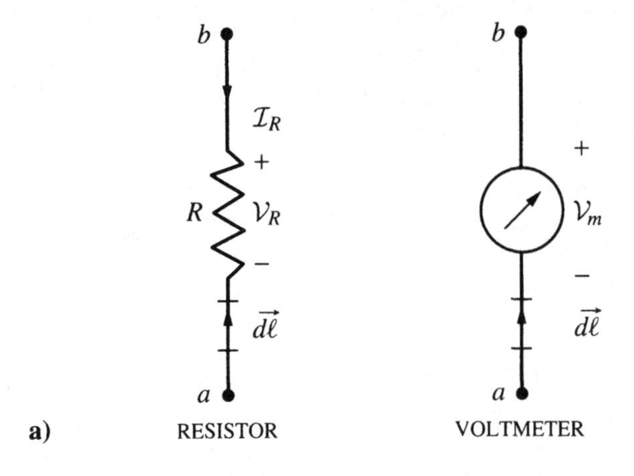

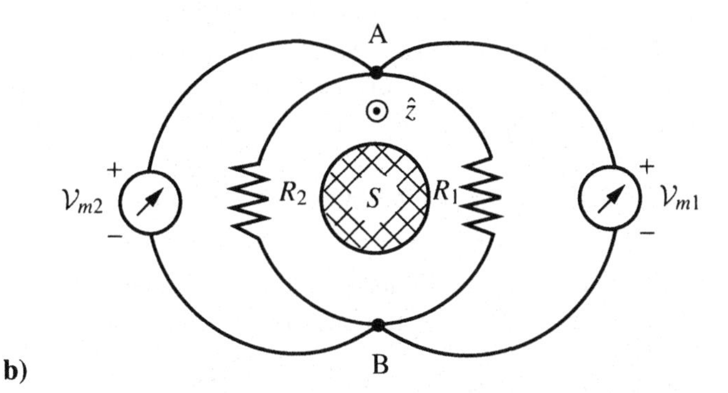

Fig. P1.5. *(continued on the next page)*

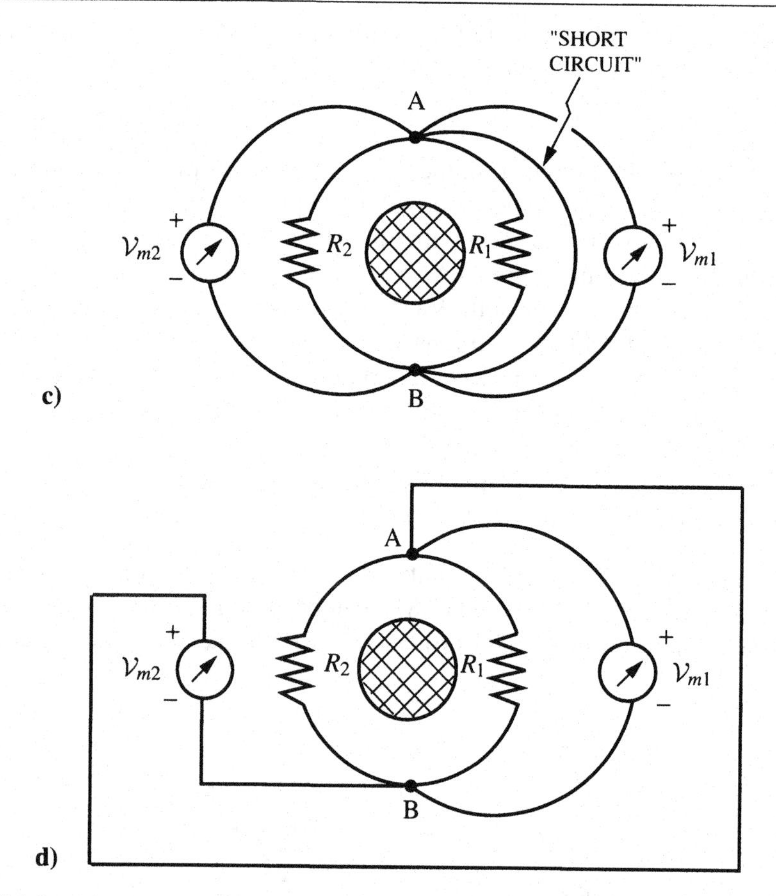

Fig. P1.5. a) Models for a resistor and a voltmeter. b), c), and d) Connections for various measurements.

(1.50) and (1.51). Use the rectangular Cartesian coordinate system (x, y, z) and assume that the boundary surface is the x-y plane with the unit normal $\hat{n} = \hat{z}$, as in Figure P1.6.

1.9 Assume that the plane boundary in Figure P1.6 is between a perfect conductor, region 1, and free space, region 2. A small probe is used to measure the magnetic field in free space at the surface of the conductor, and it is found to be a simple function of position:

$$\vec{\mathcal{B}}_2(x, y, 0+, t) = C_o(y\hat{x} - x\hat{y})\cos(\omega t).$$

a) Use the electromagnetic boundary conditions to determine the surface densities of charge $\rho_s(x, y, t)$ and current $\vec{\mathcal{J}}_s(x, y, t)$ on the perfect conductor.

b) What is the electric field $\vec{\mathcal{E}}_2(x, y, 0+, t)$ in free space at the surface of the conductor?

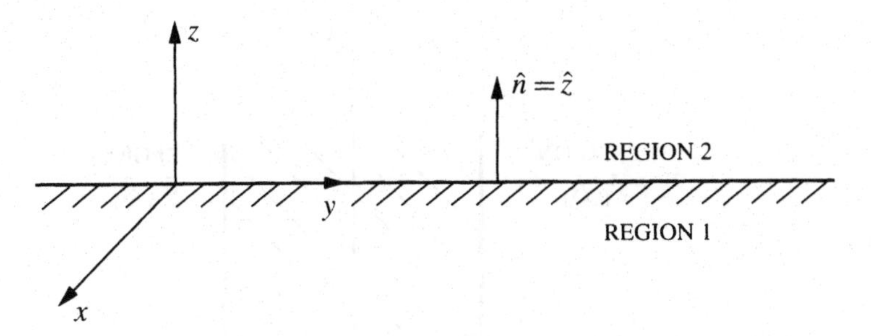

Fig. P1.6. Plane boundary with accompanying rectangular Cartesian coordinates.

1.10 The plane boundary in Figure P1.6 is between two simple materials with the constitutive parameters $\varepsilon_1, \sigma_1, \mu_1 = \mu_o$ and $\varepsilon_2, \sigma_2, \mu_2 = \mu_o$. The electromagnetic field is harmonic in time with frequency ω. At the surface, the phasor for the normal component of the electric field in region 1, $E_{1z}(x, y, 0-, \omega)$, is known. Determine the phasor for the normal component of the electric field in region 2, $E_{2z}(x, y, 0+, \omega)$, and the surface density of charge $\rho_s(x, y, \omega)$ on the boundary.

1.11 For a time harmonic field in a simple material $(\varepsilon, \mu, \sigma)$, show that the volume density of charge ρ must be zero and, as a result, $\nabla \cdot \vec{E} = 0$.

1.12 a) For a perfectly conducting body of volume V with surface S, show that the Lorentz force expression (1.12) becomes

$$\vec{\mathcal{F}} = \oint\!\!\!\oint_S (\rho_s \vec{\mathcal{E}} + \vec{\mathcal{J}}_s \times \vec{\mathcal{B}}) dS.$$

b) Although the electromagnetic field on the surface of the conductor appears in this expression, it is not well defined. Just inside the surface, all of the components of both $\vec{\mathcal{E}}$ and $\vec{\mathcal{B}}$ are zero, while just outside the surface some of the components of $\vec{\mathcal{E}}$ and $\vec{\mathcal{B}}$ are not zero. What value should be used in the expression for the force? The following argument shows that the correct value is the average of the fields just inside and just outside the surface [75]. The surface of the body is approximated locally by the planar geometry in Figure P1.7. The charge and current are assumed to be distributed in a thin layer $(-L < z < L)$ at the surface, rather than in an infinitesimal layer as for the perfect conductor. Within this layer, the volume densities of charge and current are functions of only the normal coordinate z, and the current is in the y direction: $\rho(z)$ and $\vec{\mathcal{J}}(z) = \mathcal{J}_y(z)\hat{y}$. To the left of the layer, the field is $\vec{\mathcal{E}}_1, \vec{\mathcal{B}}_1$, and to the right $\vec{\mathcal{E}}_2, \vec{\mathcal{B}}_2$. For both regions, 1 and 2, assume that $\varepsilon = \varepsilon_o$ and $\mu = \mu_o$. Show that the electrostatic and magnetostatic forces per unit area of the surface are

$$d\vec{\mathcal{F}}_e = \rho_s \left[\frac{1}{2}(\vec{\mathcal{E}}_1 + \vec{\mathcal{E}}_2) \right]$$

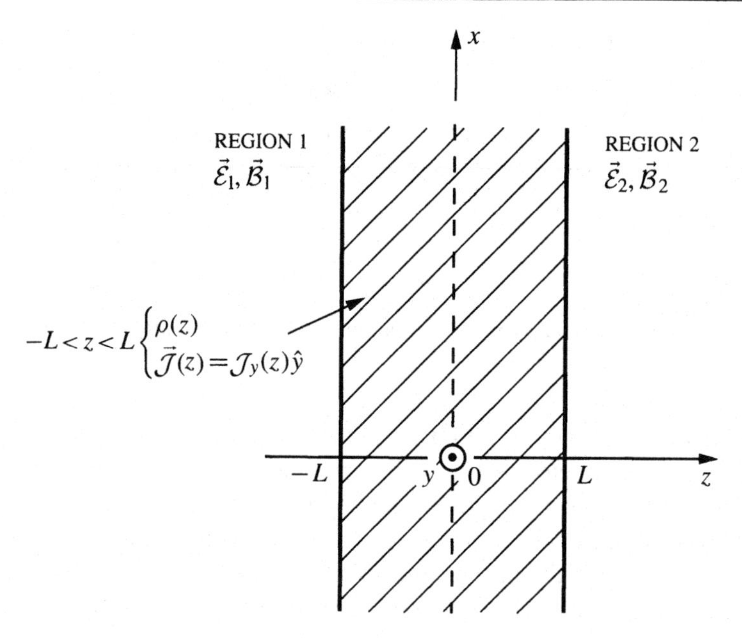

Fig. P1.7. Electric charge and current distributed within a thin layer at the surface.

and

$$d\vec{\mathcal{F}}_m = \vec{\mathcal{J}}_s \times \left[\frac{1}{2}(\vec{\mathcal{B}}_1 + \vec{\mathcal{B}}_2)\right],$$

where

$$\rho_s = \int_{-L}^{L} \rho(z)dz, \qquad \vec{\mathcal{J}}_s = \int_{-L}^{L} \vec{\mathcal{J}}(z)dz.$$

c) Use the results from part b and the boundary conditions at the surface of a perfect conductor to show that

$$\vec{\mathcal{F}} = \frac{1}{2} \oint\!\!\!\oint_S \left(\frac{1}{\varepsilon_o}|\rho_s|^2 - \mu_o|\vec{\mathcal{J}}_s|^2\right)\hat{n}dS.$$

1.13 The constitutive relation between $\vec{\mathcal{D}}$ and $\vec{\mathcal{E}}$ is written as an operator L (1.156),

$$L[\vec{\mathcal{E}}(\vec{r}, t)] = \vec{\mathcal{D}}(\vec{r}, t).$$

Which of the following relations are linear and which are time invariant?

$$\vec{\mathcal{D}}(\vec{r}, t) = \alpha\vec{\mathcal{E}}(\vec{r}, t) + \beta\frac{\partial\vec{\mathcal{E}}(\vec{r}, t)}{\partial t} + \gamma\frac{\partial^2\vec{\mathcal{E}}(\vec{r}, t)}{\partial t^2}$$

$$\mathcal{D}_x(\vec{r}, t) = \alpha\mathcal{E}_x(\vec{r}, t) + \beta\mathcal{E}_x^2(\vec{r}, t)$$

$$\vec{\mathcal{D}}(\vec{r}, t) = \alpha t\vec{\mathcal{E}}(\vec{r}, t)$$

$$\vec{\mathcal{D}}(\vec{r}, t) = \alpha \vec{\mathcal{E}}(\vec{r}, t - t')$$

$$\vec{\mathcal{D}}(\vec{r}, t) = \int_{t'=0}^{\infty} f(t')\vec{\mathcal{E}}(\vec{r}, t - t')dt'.$$

1.14 Consider the time-harmonic electric field (1.165)

$$\vec{\mathcal{E}}(\vec{r}, t) = \sum_{i=x,y,z} A_i(\vec{r}, \omega_o) \cos\left[\omega_o t + \phi_i(\vec{r}, \omega_o)\right] \hat{i},$$

where ω_o is used to emphasize that the frequency is a fixed parameter. Examine the relationship between the complex vector phasor for this field (1.166)

$$\vec{E}(\vec{r}, \omega_o) = \sum_{i=x,y,z} A_i(\vec{r}, \omega_o) e^{j\phi_i(\vec{r}, \omega_o)} \hat{i}$$

and the Fourier transform of the field (1.201)

$$\vec{\tilde{E}}(\vec{r}, \omega) = \int_{-\infty}^{\infty} \vec{\mathcal{E}}(\vec{r}, t) e^{-j\omega t} dt.$$

The following Fourier transform pair will be of use [63]:

$$e^{j\omega_o t} \leftrightarrow 2\pi\,\delta(\omega - \omega_o),$$

where δ is the Dirac delta function (1.58), which is discussed in detail in Section 5.2.

1.15 Derive the finite-difference equations (1.152), (1.153), and (1.154) for the electromagnetic field with the components $\mathcal{E}_\rho$, $\mathcal{E}_z$, and $\mathcal{B}_\phi$ in the circular cylindrical coordinate system (ρ, ϕ, z). Each of the field components is independent of ϕ. Use the spatial grid in Figure 1.46.

1.16 a) Write a computer program to solve the one-dimensional, finite-difference equations (1.143) and (1.144). Check your program by obtaining the results in Figures 1.40 and 1.41.

b) Violate the domain of dependence condition by letting $c\Delta t/\Delta z$ be slightly greater than one and make a graph as in Figure 1.40. What effect does this have on the solution?

1.17 An ideal battery is connected to a resistor using perfectly conducting wire, as in Figure P1.8a. A two-dimensional analogue of this circuit is shown in Figure P1.8b. This is a simpler geometry to analyze because the electromagnetic field is invariant in the axial direction. Make a sketch showing the Poynting vector on the cross section of this geometry. Your sketch should have the same form as Figure 1.31c. Assume that the magnetic field is uniform and in the axial direction inside the cylindrical tube, while it is insignificant outside the tube. Give a physical interpretation for what you observe [76].

1.18 Show that the scalar potential and the electric field for the long, straight wire (Figure 1.30b), are given by Equations (1.122)–(1.125). Let the constitutive parameters for the wire be ε_o, μ_o, and σ. Hint: Assume that the

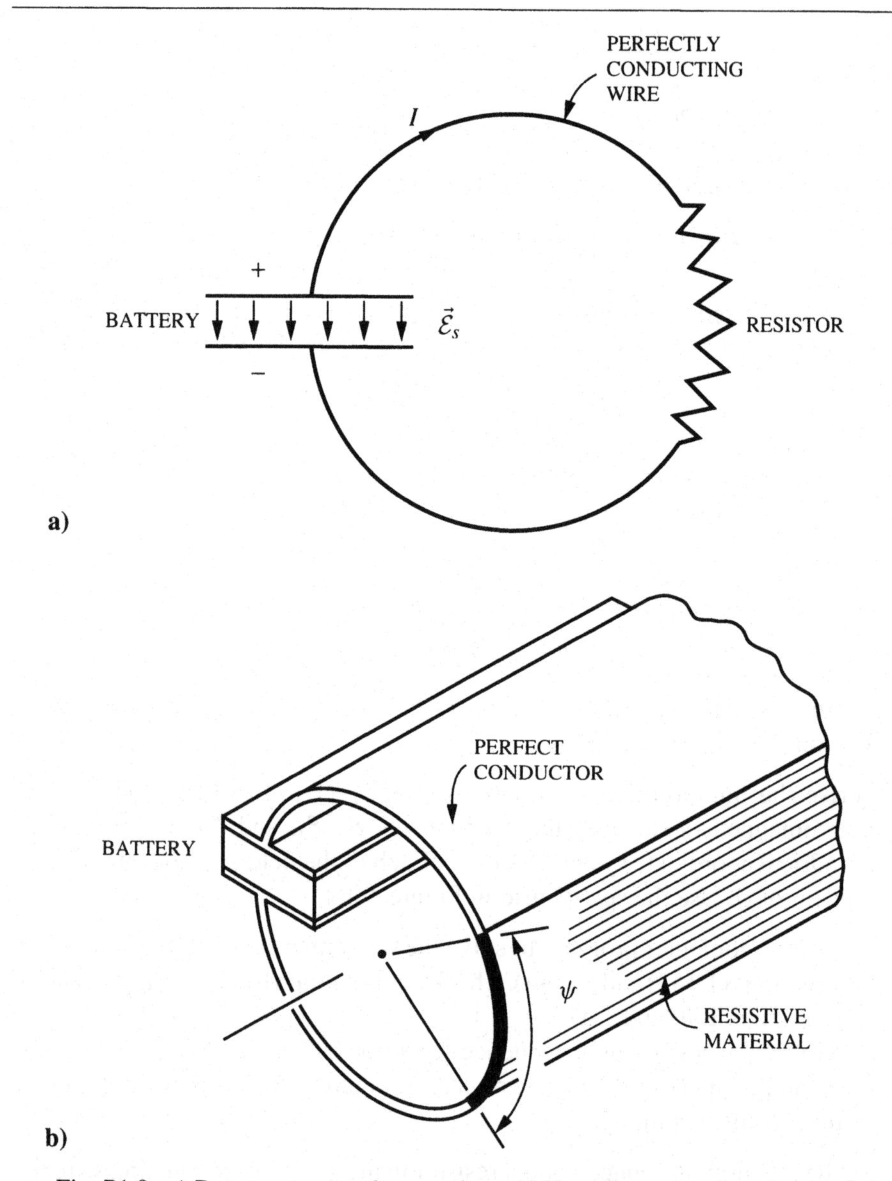

Fig. P1.8. a) Battery connected to a resistor. b) Two-dimensional analogue of a).

solution to Laplace's equation (1.121) in cylindrical coordinates is separable, $\Phi(\rho, z) = f(\rho)g(z)$. Solve for Φ in regions 1 and 2, and use the symmetry of the geometry and the electromagnetic boundary conditions to determine any unknown constants in the solutions.

1.19 Obtain the expression for the time average of a product of two time-harmonic quantities (1.175).

1.20 We have examined boundary conditions for the cases where single layers of charge or current are present on the interface between two materials, causing discontinuities in the normal component of $\vec{\mathcal{D}}$ and the tangential component

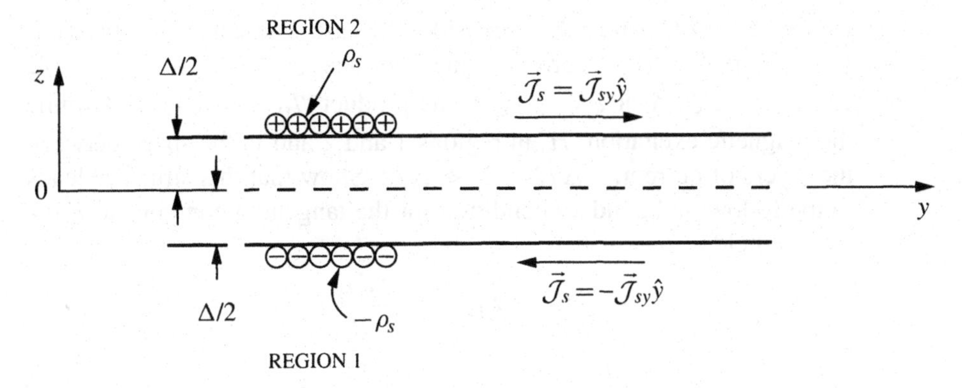

Fig. P1.9. Double layers of charge and current at boundary.

of $\vec{\mathcal{H}}$. For some materials, the physical structure at the interface may be more complex and require a model containing more than one layer of charge or current. For example, when the interface is between a metal and a polar liquid of electrolytic solution, a "double layer" of charge may be present at the interface [77].

In this problem we will examine the boundary conditions that result for the simple case of a plane surface with a uniform double layer of charge or current.

a) *Double layer of charge*: Let the plane surface separating regions 1 and 2 have the normal $\hat{n} = \hat{z}$. A uniform layer of positive surface charge, ρ_s, lies just above this surface at $z = \Delta/2$, and a uniform layer of negative surface charge, $-\rho_s$, lies just below this surface at $z = -\Delta/2$ (see Figure P1.9). The product $\rho_s \Delta = C_q$.

   Determine the electric field, $\vec{\mathcal{E}}$, in regions 1 and 2 and in the space between the layers of charge, $-\Delta/2 < z < \Delta/2$. Assume that the permittivity and permeability of all regions are those of free space. Show that this structure leads to the following boundary condition for the electrostatic potential Φ:

$$\Phi_2 - \Phi_1 = C_q/\varepsilon_o,$$

   where the potential at point b with respect to that at point a is

$$\Phi = -\int_a^b \vec{\mathcal{E}} \cdot d\vec{\ell}.$$

   Let C_q remain constant in the limit as $\Delta \to 0$ and show that the volume density of charge for the structure can be expressed as

$$\rho = -C_q \frac{\partial \delta(z)}{\partial z} = -C_q \frac{\partial \delta}{\partial n},$$

   where δ is the Dirac delta function.

b) *Double layer of current*: Now consider the case where there are two uniform layers of surface current at the interface: $\vec{\mathcal{J}}_s = \mathcal{J}_{sy}\hat{y}$ at $z = \Delta/2$ and $\vec{\mathcal{J}}_s = -\mathcal{J}_{sy}\hat{y}$ at $z = -\Delta/2$; the product $\mathcal{J}_{sy}\Delta = \mathcal{C}_{Jy}$. Determine the magnetic excitation, $\vec{\mathcal{H}}$, in regions 1 and 2 and in the space between the layers of current, $-\Delta/2 < z < \Delta/2$. Show that this structure leads to the following boundary condition for the tangential component of the electric field:

$$\mathcal{E}_{2y} - \mathcal{E}_{1y} = \mu_o \frac{\partial \mathcal{C}_{Jy}}{\partial t},$$

or

$$\vec{\mathcal{E}}_2 - \vec{\mathcal{E}}_1 = \mu_o \frac{\partial \vec{\mathcal{C}}_J}{\partial t}, \qquad \vec{\mathcal{C}}_J = \mathcal{C}_{Jy}\hat{y}.$$

Let $\mathcal{C}_{Jy}$ remain constant in the limit as $\Delta \to 0$, and show that the volume density of current for the structure can be expressed as

$$\vec{\mathcal{J}} = -\mathcal{C}_{Jy}\hat{y}\frac{\partial \delta(z)}{\partial z} = -\vec{\mathcal{C}}_J \frac{\partial \delta}{\partial n}.$$

2

Electromagnetic plane waves in free space: Polarized waves

When we encounter a new physical problem, we often try to predict or interpret its solution based on our knowledge of a similar but simpler problem. The simpler problems, for which we know the solutions, form a catalog we can turn to for insight when we encounter a new problem. In electrostatics the catalog contains problems such as the uniformly charged sphere and the parallel plate capacitor; in magnetostatics the catalog contains problems such as the infinitely long, current-carrying wire and the ideal solenoidal coil.

In electrodynamics the most important problem/solution in the catalog is probably the electromagnetic plane wave. We use it again and again to predict or interpret the solutions to new, more complex problems. Thus, it is fitting that the first analytical solution to Maxwell's equations we will examine in detail will be the electromagnetic plane wave. Another reason for this choice is that we are most likely familiar with the pictorial concepts used in describing waves. Perhaps this is because of our practical experience with mechanical waves such as those on a string or on the ocean's surface or because most of us have seen light waves used to simply describe common optical phenomena.

2.1 General time dependence

The simplest electromagnetic system we can use to study wave propagation is a region of free space or vacuum containing no sources. The electromagnetic field in this region is described by the four Maxwell's equations [(1.50)–(1.53) with the constitutive relations (1.6)–(1.8)]:

$$\nabla \times \vec{\mathcal{E}}(\vec{r}, t) = -\frac{\partial \vec{\mathcal{B}}(\vec{r}, t)}{\partial t} \tag{2.1}$$

$$\nabla \times \vec{\mathcal{B}}(\vec{r}, t) = \frac{1}{c^2}\frac{\partial \vec{\mathcal{E}}(\vec{r}, t)}{\partial t} \tag{2.2}$$

$$\nabla \cdot \vec{\mathcal{E}}(\vec{r}, t) = 0 \tag{2.3}$$

$$\nabla \cdot \vec{\mathcal{B}}(\vec{r}, t) = 0. \tag{2.4}$$

After taking the curl of (2.1) and the derivative with respect to time of (2.2), these equations can be combined to obtain the *vector wave equation* for the electric field:

$$\nabla^2 \vec{\mathcal{E}} - \frac{1}{c^2}\frac{\partial^2 \vec{\mathcal{E}}}{\partial t^2} = 0, \tag{2.5}$$

where the vector relation $\nabla^2 \vec{A} \equiv \nabla(\nabla \cdot \vec{A}) - \nabla \times (\nabla \times \vec{A})$ (Appendix B) has been used. The wave equation for the magnetic field is obtained in a similar manner, first using (2.2) and then (2.1):

$$\nabla^2 \vec{\mathcal{B}} - \frac{1}{c^2}\frac{\partial^2 \vec{\mathcal{B}}}{\partial t^2} = 0. \tag{2.6}$$

We now seek a solution to the wave equation (2.5) for which the electric field depends on only one of the coordinates of the rectangular Cartesian coordinate system (x, y, z). Without loss of generality, we can take this to be the z coordinate. The electric field is independent of x and y, that is, it has the same value on any plane for which z is a constant. Equation (2.5) then reduces to the one-dimensional wave equation

$$\left[\frac{\partial^2}{\partial z^2} - \frac{1}{c^2}\frac{\partial^2}{\partial t^2}\right]\vec{\mathcal{E}} = 0,$$

or

$$\left[\frac{\partial}{\partial t} - c\frac{\partial}{\partial z}\right]\left[\frac{\partial}{\partial t} + c\frac{\partial}{\partial z}\right]\vec{\mathcal{E}} = 0. \tag{2.7}$$

After introducing the new variables p and q, defined as

$$p = t - z/c, \qquad q = t + z/c,$$

or

$$t = (q+p)/2, \qquad z = c(q-p)/2, \tag{2.8}$$

and the resulting relationships for the derivatives

$$\begin{aligned}\frac{\partial}{\partial p} &= \frac{\partial t}{\partial p}\frac{\partial}{\partial t} + \frac{\partial z}{\partial p}\frac{\partial}{\partial z} = \frac{1}{2}\left[\frac{\partial}{\partial t} - c\frac{\partial}{\partial z}\right]\\ \frac{\partial}{\partial q} &= \frac{\partial t}{\partial q}\frac{\partial}{\partial t} + \frac{\partial z}{\partial q}\frac{\partial}{\partial z} = \frac{1}{2}\left[\frac{\partial}{\partial t} + c\frac{\partial}{\partial z}\right],\end{aligned} \tag{2.9}$$

the one-dimensional wave equation (2.7) becomes simply

$$4\frac{\partial^2 \vec{\mathcal{E}}}{\partial p \partial q} = 0. \tag{2.10}$$

There are two solutions to this equation; one is a function of only the variable p:

$$\vec{\mathcal{E}}_+(z,t) = \vec{\mathcal{F}}_+(p) = \vec{\mathcal{F}}_+(t - z/c), \tag{2.11a}$$

and the other is a function of only the variable q:

$$\vec{\mathcal{E}}_-(z,t) = \vec{\mathcal{F}}_-(q) = \vec{\mathcal{F}}_-(t + z/c). \tag{2.11b}$$

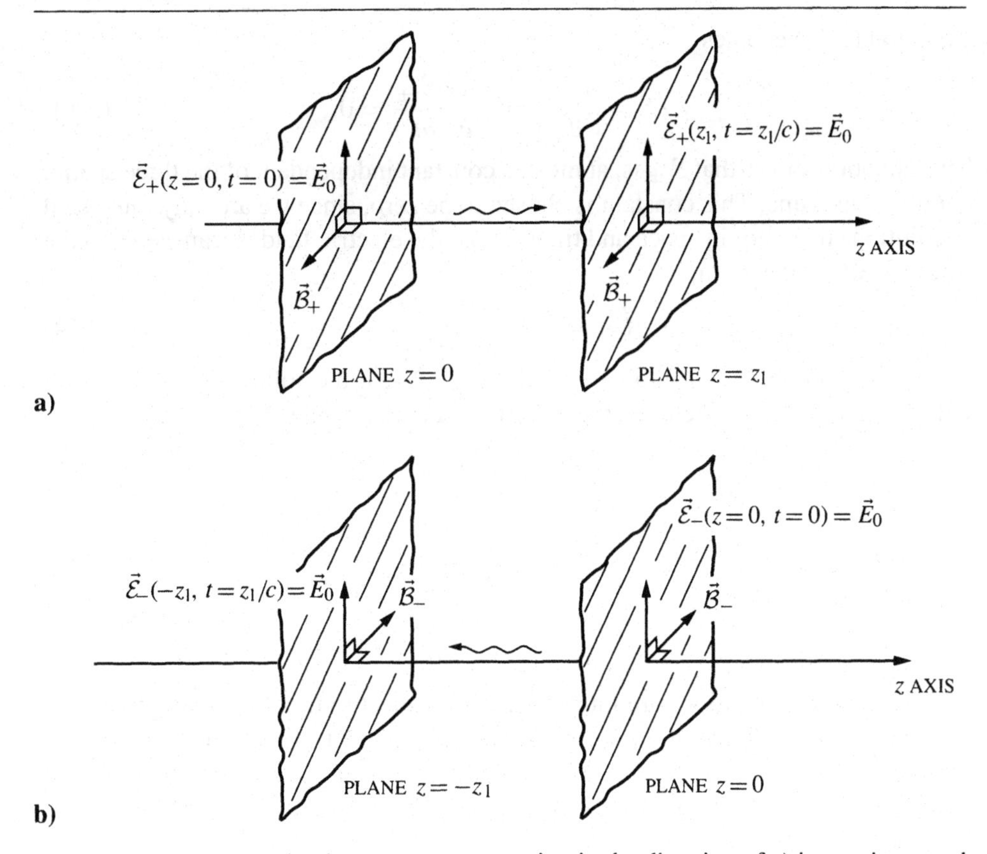

Fig. 2.1. Electromagnetic plane waves propagating in the direction of a) increasing z and b) decreasing z.

The schematic drawing in Figure 2.1 shows the physical situations corresponding to these two solutions. The solution $\vec{\mathcal{E}}_+$ is a plane wave propagating in the direction of increasing z, whereas the solution $\vec{\mathcal{E}}_-$ is a plane wave propagating in the direction of decreasing z. Both solutions propagate with a velocity equal to c, the velocity of light in free space. With reference to Figure 2.1, for the solution $\vec{\mathcal{E}}_+$, the electric field is $\vec{E}_o$ on the plane $z = 0$ at time $t = 0$, and it assumes the same value on the plane $z = z_1$ at the later time $t = z_1/c$. For the solution $\vec{\mathcal{E}}_-$, the electric field is $\vec{E}_o$ on the plane $z = 0$ at time $t = 0$, and it assumes the same value on the plane $z = -z_1$ at the later time $t = z_1/c$. For these solutions, the derivatives with respect to z and t are simply related by

$$\frac{\partial \mathcal{E}_{i\pm}}{\partial z} = \mp \frac{1}{c} \frac{\partial \mathcal{E}_{i\pm}}{\partial t}, \qquad i = x, y, z. \tag{2.12}$$

Other properties of the plane-wave solution are obtained by inserting (2.11a), (2.11b), and (2.12) into Maxwell's equations. Since

$$\frac{\partial \mathcal{E}_{x\pm}}{\partial x} = 0, \qquad \frac{\partial \mathcal{E}_{y\pm}}{\partial y} = 0,$$

Equation (2.3) becomes

$$\nabla \cdot \vec{\mathcal{E}} = \frac{\partial \mathcal{E}_{z\pm}}{\partial z} = \mp \frac{1}{c} \frac{\partial \mathcal{E}_{z\pm}}{\partial t} = 0. \tag{2.13}$$

This equation shows that $\mathcal{E}_{z\pm}$ is, at most, a constant independent of the three spatial variables and time. This constant is chosen to be zero, since we are only interested in solutions that vary in space and time. Thus, the electric field is transverse to the direction of propagation:

$$\vec{\mathcal{E}}_{\pm} = \mathcal{E}_{x\pm}\hat{x} + \mathcal{E}_{y\pm}\hat{y}. \tag{2.14}$$

On substituting (2.14) into (2.1) and using (2.12), the following relations are obtained for the components of the electric and magnetic fields:

$$\frac{\partial \mathcal{B}_{z\pm}}{\partial t} = 0 \tag{2.15a}$$

$$\frac{\partial \mathcal{B}_{x\pm}}{\partial t} = \mp \frac{1}{c} \frac{\partial \mathcal{E}_{y\pm}}{\partial t} \tag{2.15b}$$

$$\frac{\partial \mathcal{B}_{y\pm}}{\partial t} = \pm \frac{1}{c} \frac{\partial \mathcal{E}_{x\pm}}{\partial t}. \tag{2.15c}$$

After these three equations are integrated with respect to time, each equation contains a constant of integration representing a component of the field that is independent of time. Again, because we are only interested in solutions that vary in space and time, these constants are chosen to be zero, and (2.15a) through (2.15c) become

$$\mathcal{B}_{z\pm} = 0 \tag{2.16a}$$

$$\mathcal{B}_{x\pm} = \mp \frac{1}{c} \mathcal{E}_{y\pm} \tag{2.16b}$$

$$\mathcal{B}_{y\pm} = \pm \frac{1}{c} \mathcal{E}_{x\pm}. \tag{2.16c}$$

In vector notation, these three equations are simply

$$\vec{\mathcal{B}}_{\pm} = \pm \frac{1}{c} \hat{z} \times \vec{\mathcal{E}}_{\pm}. \tag{2.17}$$

From (2.17) we see that the magnetic field, as well as the electric field, is transverse to the direction of propagation. The magnetic field is also orthogonal to and simply related to the electric field. This is a *TEM (transverse electromagnetic) plane wave.*

All that remains to complete the description of a particular plane wave is to specify the function $\vec{\mathcal{F}}_{\pm}$ in Equation (2.11a) or Equation (2.11b). As an example, we will consider the case where $\vec{\mathcal{F}}_{+}$ has only a y component, and this is a Gaussian function:

$$\vec{\mathcal{E}}_{+}(z,t) = \vec{\mathcal{F}}_{+}(t - z/c) = E_o e^{-[(t-z/c)-t_o]^2/\tau^2}\,\hat{y}$$

$$\vec{\mathcal{B}}_{+}(z,t) = -\frac{1}{c} E_o e^{-[(t-z/c)-t_o]^2/\tau^2}\,\hat{x}.$$

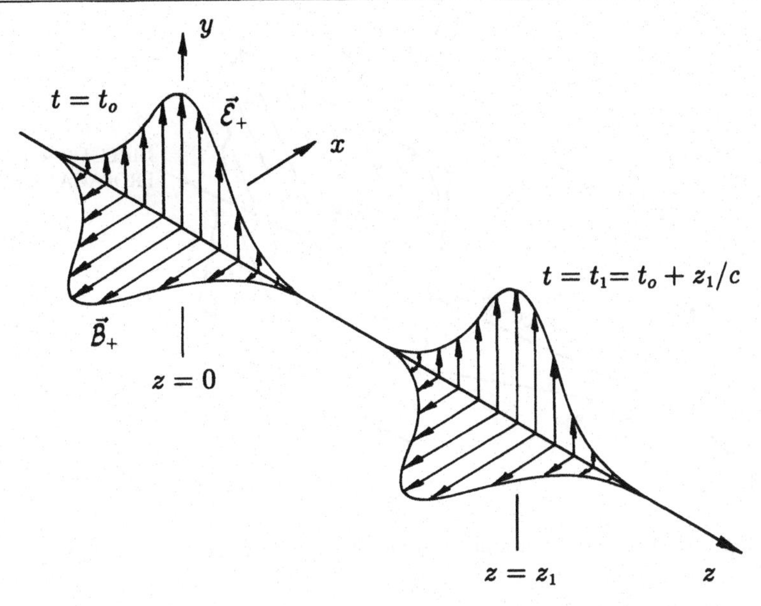

Fig. 2.2. Propagation of a Gaussian pulse. Drawing shows spatial dependence of the field for the two times, t_o and t_1.

The spatial distribution for this field is shown for two instants of time in Figure 2.2. This is a Gaussian pulse propagating in the direction of increasing z with a velocity equal to c. Recall that we encountered this electromagnetic field earlier in Section 1.6, where the analytical description given above was inferred from purely numerical results obtained by directly solving Maxwell's equations with a computer.

Our discussion to this point has been for a plane wave propagating in the $\pm z$ direction. We will now remove this restriction and summarize the properties of the plane wave by considering a wave propagating in the direction of the arbitrary unit vector $\hat{k}$. This vector together with the unit vectors $\hat{u}$ and $\hat{v}$ are the basis vectors for the orthogonal, right-handed system (u, v, k) shown in Figure 2.3. The electromagnetic field is now

$$\vec{\mathcal{E}}(\vec{r}, t) = \vec{\mathcal{F}}(t - \hat{k} \cdot \vec{r}/c) \tag{2.18a}$$

$$\vec{\mathcal{B}}(\vec{r}, t) = \frac{1}{c}\hat{k} \times \vec{\mathcal{E}}(\vec{r}, t), \tag{2.18b}$$

with the transverse nature of the wave specified by the relations

$$\hat{k} \cdot \vec{\mathcal{E}} = 0 \tag{2.19a}$$

and

$$\hat{k} \cdot \vec{\mathcal{B}} = 0. \tag{2.19b}$$

Notice that it is no longer necessary to include the $\pm$ sign to distinguish the two directions of propagation for the wave. The unit vector $\hat{k}$ includes this information; when the direction of $\hat{k}$ is reversed, the scalar product $\hat{k} \cdot \vec{r}$ automatically changes sign. At a fixed time t the electromagnetic field, (2.18a) and (2.18b), has the same

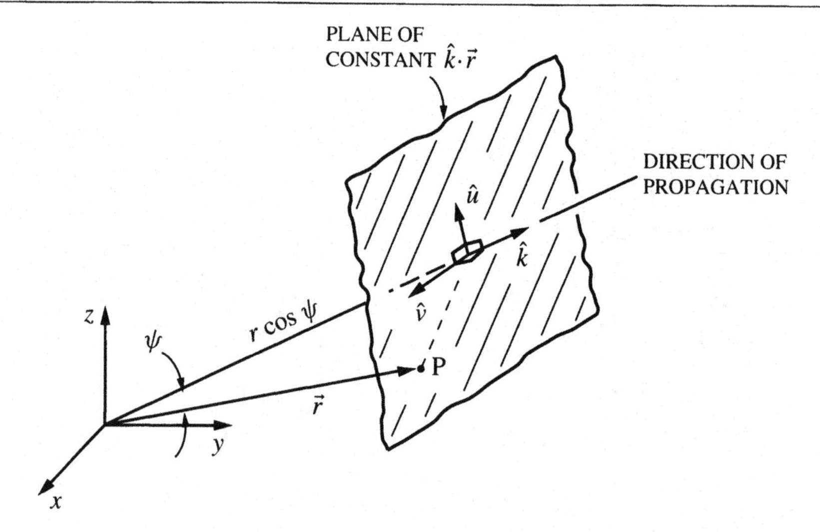

Fig. 2.3. Geometry for a plane wave propagating in a general direction.

value at every point P on a plane for which $\hat{k}\cdot\vec{r} = r\cos\psi = \text{constant}$; this is indicated schematically in Figure 2.3.

The magnetic excitation $\vec{\mathcal{H}}$ is

$$\vec{\mathcal{H}} = \frac{1}{\mu_o}\vec{\mathcal{B}} = \frac{1}{\zeta_o}\hat{k}\times\vec{\mathcal{E}}; \tag{2.20}$$

thus, the Poynting vector (1.100) for the plane wave becomes

$$\vec{\mathcal{S}} = \vec{\mathcal{E}}\times\vec{\mathcal{H}} = \frac{1}{\zeta_o}|\vec{\mathcal{E}}|^2\hat{k}, \tag{2.21}$$

where

$$\zeta_o = \sqrt{\mu_o/\varepsilon_o} = 376.7\ldots\,\Omega \tag{2.22}$$

is the wave impedance of free space. The Poynting vector is uniform over any plane surface with normal $\hat{k}$ and represents the instantaneous rate at which electromagnetic energy is passing through a unit area of the plane (the power per unit area).

2.2 Harmonic time dependence: Monochromatic plane waves

We will now consider a case of practical importance in which the source of the electromagnetic plane wave varies harmonically in time with the frequency f (angular frequency $\omega = 2\pi f$) [1–3]. The components of $\vec{\mathcal{F}}_\pm$ in (2.11a) or (2.11b) are then cosinusoidal functions, and the electric field for the plane wave propagating in the direction of the $\pm z$ axis (2.14) becomes

$$\begin{aligned}\vec{\mathcal{E}}_\pm(z,t) &= \mathcal{E}_{x\pm}(z,t)\hat{x} + \mathcal{E}_{y\pm}(z,t)\hat{y}\\ &= A_x\cos\left[\omega(t\mp z/c)+\phi_x\right]\hat{x} + A_y\cos\left[\omega(t\mp z/c)+\phi_y\right]\hat{y}.\end{aligned} \tag{2.23}$$

Note that the amplitudes and the phases of the field components $\mathcal{E}_x$ and $\mathcal{E}_y$ are different. There is no required interrelationship between these two field components for the plane wave.

The wave number for propagation in free space is defined as

$$k_o = 2\pi/\lambda_o = \omega/c, \tag{2.24}$$

where λ_o is the wavelength in free space. The relative amplitude and the relative phase of the components $\mathcal{E}_x$ and $\mathcal{E}_y$ of the field are described by the angles γ and δ:

$$\tan(\gamma) = A_y/A_x, \qquad 0 \le \gamma \le \pi/2 \tag{2.25}$$

$$\delta = \phi_y - \phi_x, \qquad -\pi < \delta \le \pi. \tag{2.26}$$

After using (2.24)–(2.26), the electric field (2.23) becomes

$$\vec{\mathcal{E}}_\pm(z,t) = A_x\Big\{\cos\big[s_\pm(z,t)\big]\hat{x} + \tan(\gamma)\cos\big[s_\pm(z,t)+\delta\big]\hat{y}\Big\}, \tag{2.27}$$

with

$$s_\pm(z,t) = \omega t \mp k_o z + \phi_x, \tag{2.28}$$

and the accompanying magnetic field is

$$\vec{\mathcal{B}}_\pm(z,t) = \pm\frac{A_x}{c}\Big\{\cos\big[s_\pm(z,t)\big]\hat{y} - \tan(\gamma)\cos\big[s_\pm(z,t)+\delta\big]\hat{x}\Big\}. \tag{2.29}$$

Equations (2.27)–(2.29) completely describe a monochromatic plane wave propagating in free space. The term monochromatic indicates that the electromagnetic field oscillates at a single frequency; a wave of visible light would be a single (mono) color (chromatic).

The electric field for a monochromatic plane wave propagating in the direction of the positive z axis is shown in Figure 2.4. In Figure 2.4a the spatial dependences of the two components of the field, $\mathcal{E}_{x+}$ and $\mathcal{E}_{y+}$, are shown at a fixed time t_o. The two components are seen to vary periodically along the z axis with a period equal to the wavelength λ_o. The total electric field $\vec{\mathcal{E}}_+ = \mathcal{E}_{x+}\hat{x} + \mathcal{E}_{y+}\hat{y}$, shown in Figure 2.4b, rotates around the z axis, making one complete revolution per wavelength in z. For this example, the rotation of $\vec{\mathcal{E}}_+$ about the z axis has the same sense as a right-handed screw.

At a fixed position z_o, the electric field vector rotates in time with the angular frequency ω, and the tip of the vector traces out a closed curve, the *polarization ellipse*. For the example in Figure 2.4b, the ellipse is traced out in a counterclockwise sense when viewed by an observer looking in the direction of propagation (i.e., looking in the direction of the positive z axis).

The equation describing the polarization ellipse is obtained by combining the expressions for the field components $\mathcal{E}_{x\pm}$ and $\mathcal{E}_{y\pm}$ to eliminate the function $s_\pm(z,t)$:

$$\frac{\mathcal{E}_{x\pm}(z,t)}{A_x} = \cos\big[s_\pm(z,t)\big]$$

$$\frac{\mathcal{E}_{y\pm}(z,t)}{A_x\tan(\gamma)} = \cos\big[s_\pm(z,t)+\delta\big] = \cos\big[s_\pm(z,t)\big]\cos(\delta) - \sin\big[s_\pm(z,t)\big]\sin(\delta),$$

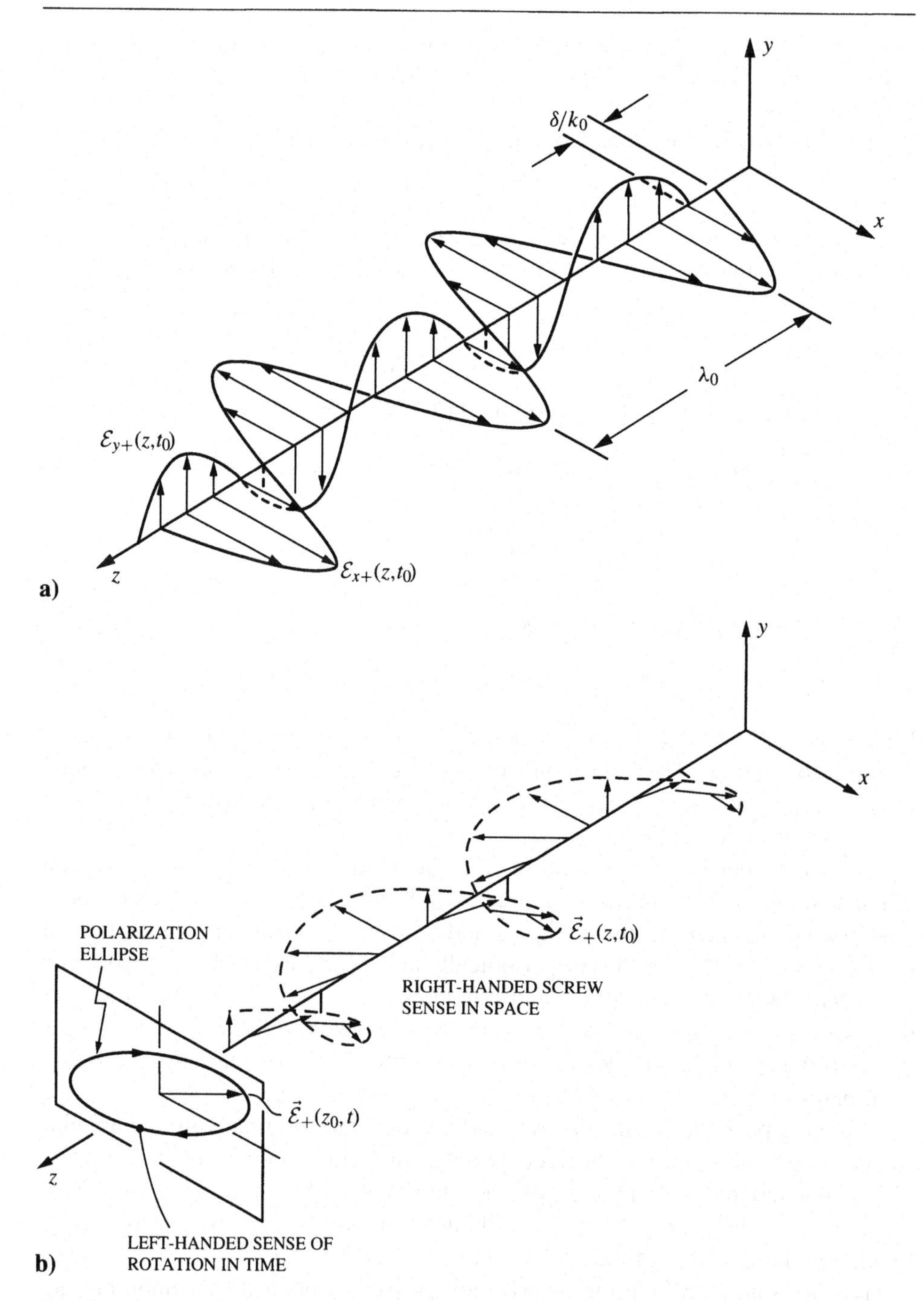

Fig. 2.4. Monochromatic plane wave. a) Spatial dependence for the components of the electric field, $\mathcal{E}_{x+}$, $\mathcal{E}_{y+}$, at the fixed time t_o. b) Spatial dependence for the total field, $\vec{\mathcal{E}}_+$, at the fixed time t_o, and the curve (polarization ellipse) traced out in time by the electric field at the fixed position z_o.

or

$$\frac{\mathcal{E}_{y\pm}(z,t)}{A_x \tan(\gamma)} - \frac{\mathcal{E}_{x\pm}(z,t)}{A_x}\cos(\delta) = -\left\{1 - \left[\frac{\mathcal{E}_{x\pm}(z,t)}{A_x}\right]^2\right\}^{\frac{1}{2}} \sin(\delta). \tag{2.30}$$

After taking the square of (2.30) and rearranging terms, the equation for the polarization ellipse results:

$$\left[\frac{\mathcal{E}_{x\pm}(z,t)}{A_x}\right]^2 - \frac{2\cos(\delta)}{\tan(\gamma)}\left[\frac{\mathcal{E}_{x\pm}(z,t)}{A_x}\frac{\mathcal{E}_{y\pm}(z,t)}{A_x}\right] + \frac{1}{\tan^2(\gamma)}\left[\frac{\mathcal{E}_{y\pm}(z,t)}{A_x}\right]^2 = \sin^2(\delta). \tag{2.31}$$

The terminology used to describe the state of polarization for the electric field of a plane wave is a matter of convention. Here we will adopt the convention used by the Institute of Electrical and Electronics Engineers (IEEE) [4]. The sense in which the polarization ellipse is traced out is termed right-handed (left-handed) if an observer looking in the direction of propagation for the wave sees a clockwise (counterclockwise) rotation of the electric field vector as time advances.[1] Thus, the sense of the polarization for the plane wave described in Figure 2.4 is left-handed. Note that this is opposite to the right-handed screw sense for the rotation of the electric field vector in space (time fixed at t_o). In the optical literature, the screw sense in space is sometimes used to describe the state of polarization. Thus, in the optical terminology the sense of the polarization for the plane wave in Figure 2.4 would be right-handed.

The shape and the orientation of the polarization ellipse and the sense in which the polarization ellipse is traced out are determined by the angles γ and δ. The size of the ellipse is determined by the amplitude of the electric field. Any quantity indicative of the amplitude, such as A_x, can be used to set the size of the ellipse; this is evident in (2.31) where both field components $\mathcal{E}_{x\pm}$ and $\mathcal{E}_{y\pm}$ are divided by A_x.

When the polarization ellipse is circumscribed in a rectangle whose sides are parallel to the x and y axes (Figure 2.5a), the angle between the diagonal of the rectangle and the x axis is γ. The effect of the angle δ on the polarization ellipse is illustrated in Figure 2.6a, where polarization ellipses are shown for various values of δ with the angle γ fixed at $\tan(\gamma) = 0.5$ ($\gamma = 26.57°$). Propagation is in the direction of the positive z axis. For $\delta = 0$ we have *linear polarization*; the field components $\mathcal{E}_x$ and $\mathcal{E}_y$ are in phase and the polarization ellipse reduces to a straight line. For $0 < \delta < \pi$ the polarization ellipse is traced with a left-handed sense; the component $\mathcal{E}_y$ leads the component $\mathcal{E}_x$ in time. When $\delta = \pi$ we again have linear polarization, and the polarization ellipse reduces to a straight line. For $-\pi < \delta < 0$ the polarization ellipse is traced with a right-handed sense; the component $\mathcal{E}_y$ lags the component $\mathcal{E}_x$ in time.

For the special cases $\tan(\gamma) = 1$ ($A_x = A_y$) and $\delta = \pi/2,\ -\pi/2$, which are shown in Figure 2.6b, the polarization ellipse reduces to a circle. These two cases

[1] A simple procedure for remembering this terminology is to place the thumb of the right (left) hand in the direction of propagation. If the curled fingers of the right (left) hand point in the direction in which the electric field vector is rotating, the sense of the polarization is right- (left-) handed.

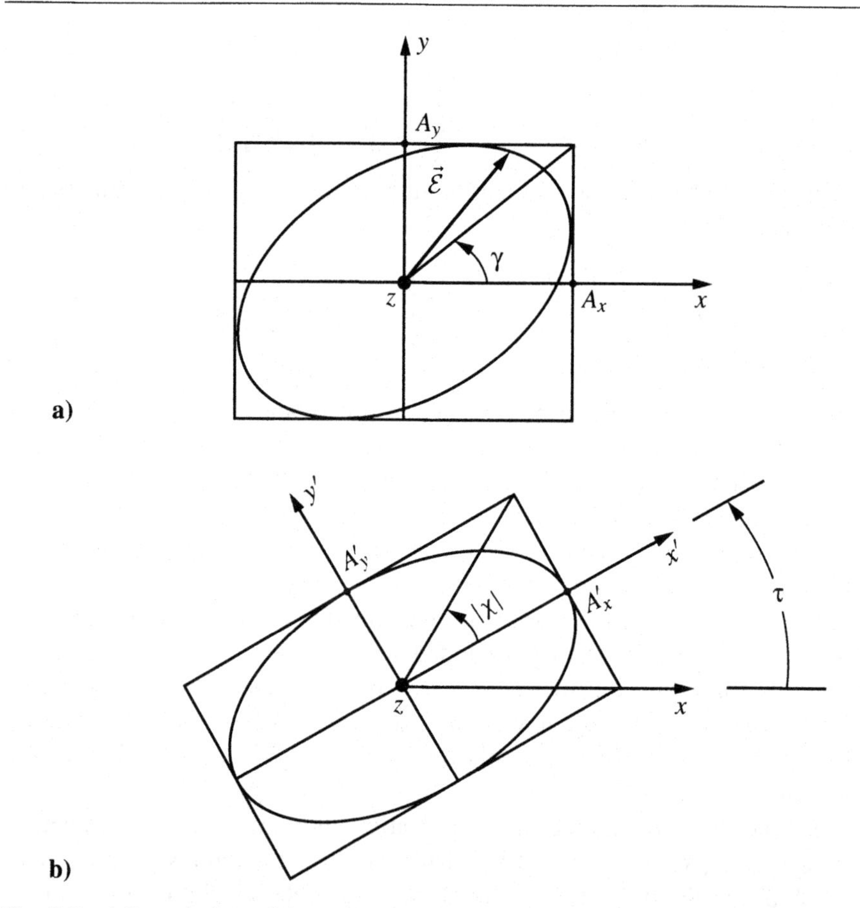

Fig. 2.5. a) Description of the polarization ellipse in terms of the coordinate system x, y. b) Description of the polarization ellipse in terms of the coordinate system x', y' (the coordinates of the principal axes of the ellipse). Propagation is in the direction of the positive z axis.

are referred to as *left-handed circular polarization* ($\delta = \pi/2$) and *right-handed circular polarization* ($\delta = -\pi/2$).

The electromagnetic field of the monochromatic plane wave can also be expressed in terms of the complex vector phasors $\vec{E}(\vec{r})$ and $\vec{B}(\vec{r})$ (1.165):

$$\vec{\mathcal{E}}(\vec{r}, t) = \text{Re}\big[\vec{E}(\vec{r})e^{j\omega t}\big], \tag{2.32a}$$

$$\vec{\mathcal{B}}(\vec{r}, t) = \text{Re}\big[\vec{B}(\vec{r})e^{j\omega t}\big]. \tag{2.32b}$$

For a plane wave propagating in the direction of the arbitrary unit vector $\hat{k}$, as in Figure 2.3, these phasors are

$$\vec{E}(\vec{r}) = (A_u e^{j\phi_u}\hat{u} + A_v e^{j\phi_v}\hat{v})e^{-j\vec{k}\cdot\vec{r}} = A_u\big[\hat{u} + \tan(\gamma)e^{j\delta}\hat{v}\big]e^{-j(\vec{k}\cdot\vec{r}-\phi_u)}, \tag{2.33a}$$

$$\vec{B}(\vec{r}) = \frac{1}{c}\hat{k} \times \vec{E}(\vec{r}), \tag{2.33b}$$

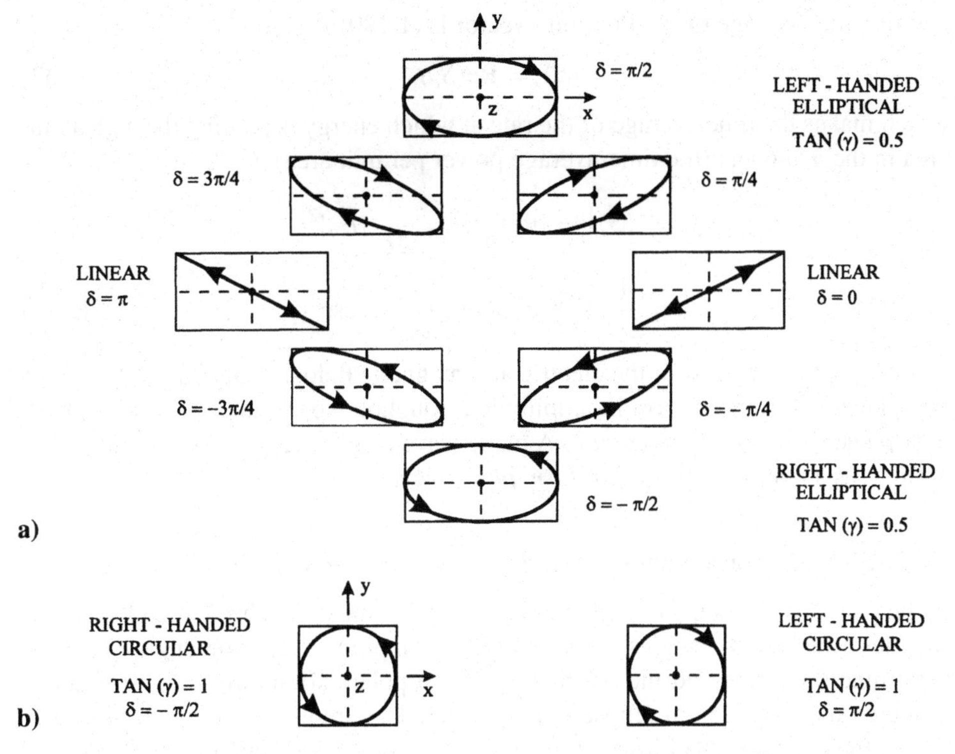

Fig. 2.6. a) The polarization ellipse for various values of the phase difference $\delta = \phi_y - \phi_x$, with $\tan(\gamma) = A_y/A_x = 0.5$. b) Circular polarization: $\delta = \pm\pi/2$, $\tan(\gamma) = 1.0$. Propagation is in the direction of the positive z axis.

where the vector wave number is

$$\vec{k} = k_o\hat{k}. \tag{2.34}$$

The angles γ and δ now describe the relative amplitude and the relative phase, respectively, of the electric field components in the $\hat{u}$ and $\hat{v}$ directions.

Sometimes it is convenient to write the phasors for the field of a plane wave with the propagating factor $\exp(-j\vec{k}\cdot\vec{r})$ removed. Thus, we introduce the following notation, for (2.33a)

$$\vec{E}(\vec{r}) = \vec{E}_o e^{-j\vec{k}\cdot\vec{r}}, \tag{2.35a}$$

with

$$\vec{E}_o = \vec{E}_{ou} + \vec{E}_{ov} = E_{ou}\hat{u} + E_{ov}\hat{v}. \tag{2.35b}$$

Here the subscript o indicates that $\vec{E}_o$ is $\vec{E}(\vec{r})$ evaluated for $\vec{r} = 0$.

The complex Poynting vector for the plane wave is

$$\vec{S}_c(\vec{r}) = \frac{1}{2}\vec{E}(\vec{r}) \times \vec{H}^*(\vec{r}) = \frac{1}{2\zeta_o}\left|\vec{E}(\vec{r})\right|^2\hat{k} = \frac{1}{2\zeta_o}|\vec{E}_o|^2\hat{k}$$

$$= \frac{1}{2\zeta_o}A_u^2\sec^2(\gamma)\hat{k} = \frac{1}{2\zeta_o}A_v^2\csc^2(\gamma)\hat{k}, \tag{2.36}$$

and the time average of the Poynting vector is (1.180)

$$\langle \vec{S} \rangle = \text{Re}(\vec{S}_c), \tag{2.37}$$

which makes the time average of the rate at which energy is passing through a unit area in the u-v plane (the time-average power per unit area)

$$S_0 = \left|\langle \vec{S} \rangle\right| = \left|\text{Re}(\vec{S}_c)\right| = \frac{1}{2\zeta_o}|\vec{E}_o|^2$$
$$= \frac{1}{2\zeta_o}A_u^2 \sec^2(\gamma) = \frac{1}{2\zeta_o}A_v^2 \csc^2(\gamma). \tag{2.38}$$

Notice that the phasors for the electric and magnetic fields, (2.33a) and (2.33b), of this plane wave are uniform in amplitude throughout space and uniform in phase over plane surfaces with normal $\hat{k}$. A plane wave exhibiting these properties is referred to as a *homogeneous* or *uniform plane wave*.

2.3 The polarization ellipse in the coordinate system of the principal axes

In the previous section, the state of polarization for the monochromatic plane wave was described by the angles γ and δ and the amplitude A_x. The angles γ and δ determine the shape and the orientation of the polarization ellipse and the sense in which the polarization ellipse is traced out. The amplitude A_x determines the size of the polarization ellipse. The time-average power per unit area S_0 (2.38) can also be used to specify the size of the polarization ellipse. The power S_0 has the additional advantage that it is independent of the particular coordinate system used to describe the ellipse.

The principal axes of the polarization ellipse are along the coordinate axes x', y' shown in Figure 2.5b. When the electric field is expressed in terms of this new coordinate system (x', y'), the equation describing the polarization ellipse (2.31) reduces to the "central equation" for the ellipse. Letting $\gamma \to \chi$, $\delta \to \pm\pi/2$, and $\mathcal{E}_x \to \mathcal{E}'_x$, $A_x \to A'_x$, etc., in Equation (2.31), we get

$$\left[\frac{\mathcal{E}'_{x\pm}(z,t)}{A'_x}\right]^2 + \frac{1}{\tan^2(\chi)}\left[\frac{\mathcal{E}'_{y\pm}(z,t)}{A'_x}\right]^2 = 1. \tag{2.39}$$

Note that the new coordinate system (x', y') is obtained by rotating the original system (x, y) through the angle τ about the z axis. The axis x' is along the major axis of the ellipse, and the angle τ is measured in the right-handed sense from the x axis to the x' axis such that $0 \le \tau < \pi$.

The state of polarization for the monochromatic plane wave now can be described by the auxiliary angle χ, the tilt angle τ, and S_0. The shape of the ellipse is determined by the auxiliary angle χ with

$$\tan(\chi) = \pm\frac{A'_y}{A'_x}, \qquad -\pi/4 \le \chi \le \pi/4, \tag{2.40}$$

where A'_y/A'_x is the "axial ratio" (minor axis/major axis) of the ellipse (see Figure 2.5b). The sign in (2.40) is introduced to indicate the sense in which the polarization ellipse is traced out: + (−) for left-handed (right-handed) sense. Notice that this sign does not affect the equation for the polarization ellipse (2.39), since $\tan^2(\chi)$ appears there. The orientation of the polarization ellipse is determined by the tilt angle τ, and the size of the polarization ellipse is again determined by S_0.

The relationships between the two sets of angles (γ, δ) and (χ, τ) used to describe the state of polarization for the wave in the two coordinate systems (x, y) and (x', y') will be determined next.[2] The time-average power per unit area (2.38) for the wave can be expressed in terms of the components of the field in either coordinate system; thus,

$$S_0 = \frac{1}{2\zeta_o}(A_x)^2\sec^2(\gamma) = \frac{1}{2\zeta_o}(A'_x)^2\sec^2(\chi),$$

which makes

$$A'_x = \frac{\sec(\gamma)}{\sec(\chi)}A_x. \tag{2.41}$$

After using (2.41) to normalize all of the field components with respect to A_x (i.e., $\mathcal{E}_{xn} = \mathcal{E}_x/A_x$, $\mathcal{E}'_{xn} = \mathcal{E}'_x/A_x$, etc.) the equations for the polarization ellipse, (2.31) and (2.39), become

$$\frac{1}{\sin^2(\delta)}\left[\mathcal{E}_{xn}^2 - \frac{2\cos(\delta)}{\tan(\gamma)}\mathcal{E}_{xn}\,\mathcal{E}_{yn} + \frac{1}{\tan^2(\gamma)}\mathcal{E}_{yn}^2\right] = 1 \tag{2.42}$$

and

$$\frac{\sec^2(\chi)}{\sec^2(\gamma)}\left[(\mathcal{E}'_{xn})^2 + \frac{1}{\tan^2(\chi)}(\mathcal{E}'_{yn})^2\right] = 1. \tag{2.43}$$

These equations are quadratic forms and can be expressed as the following matrix products:

$$E_n^T Q E_n = 1 \tag{2.44}$$

$$(E'_n)^T Q' E'_n = 1, \tag{2.45}$$

where

$$E_n = \begin{bmatrix} \mathcal{E}_{xn} \\ \mathcal{E}_{yn} \end{bmatrix}, \qquad E'_n = \begin{bmatrix} \mathcal{E}'_{xn} \\ \mathcal{E}'_{yn} \end{bmatrix}, \tag{2.46}$$

$$Q = \frac{1}{\sin^2(\delta)}\begin{bmatrix} 1 & \dfrac{-\cos(\delta)}{\tan(\gamma)} \\ \dfrac{-\cos(\delta)}{\tan(\gamma)} & \dfrac{1}{\tan^2(\gamma)} \end{bmatrix}, \tag{2.47}$$

[2] The reader not interested in the mathematical details of this procedure can proceed to Equation (2.64).

and

$$Q' = \frac{\sec^2(\chi)}{\sec^2(\gamma)} \begin{bmatrix} 1 & 0 \\ 0 & \dfrac{1}{\tan^2(\chi)} \end{bmatrix}. \tag{2.48}$$

The superscript T indicates the transpose of the matrix.

The column vectors E_n and E_n' are related by the rotation matrix R_ℓ:

$$E_n = R_\ell E_n', \tag{2.49}$$

where

$$R_\ell = \begin{bmatrix} \cos\tau & -\sin\tau \\ \sin\tau & \cos\tau \end{bmatrix}. \tag{2.50}$$

The subscript ℓ indicates a left-handed rotation about the z axis. After substituting (2.49), Equation (2.44) becomes

$$(E_n')^T R_\ell^T Q R_\ell E_n' = 1, \tag{2.51}$$

which on comparison with (2.45) shows that $R_\ell^T Q R_\ell = Q'$. The inverse of the rotation matrix is equal to its transpose, $R_\ell^{-1} = R_\ell^T$, and thus,

$$R_\ell^{-1} Q R_\ell = Q'. \tag{2.52}$$

Notice that this operation transforms the real, symmetric matrix Q into the real, diagonal matrix Q'. A transformation of this form is called a "similarity transformation."

The matrix equation (2.52) is equivalent to four nonlinear equations involving the angles γ, δ, χ, and τ. These equations will be used to obtain the desired expressions for χ and τ in terms of γ and δ. This can be accomplished by a "brute force" manipulation of (2.52), or it can be accomplished by a somewhat more elegant application of linear algebra in which the matrix R_ℓ is determined in the similarity transformation (2.52) that diagonalizes a *general*, real, symmetric matrix Q. We will outline the latter approach [5].

If the column vectors

$$X_i = \begin{bmatrix} X_{xi} \\ X_{yi} \end{bmatrix}, \qquad i = 1, 2 \tag{2.53}$$

are the eigenvectors of the matrix Q, then

$$Q X_i = \lambda_i X_i. \tag{2.54}$$

Here the λ_i are the eigenvalues of the matrix Q; that is, they are roots of the equation

$$|Q - \lambda_i I| = 0, \tag{2.55}$$

where $|\ |$ indicates the determinant of the matrix and I is the identity matrix. Now consider the matrix X whose column vectors are the eigenvectors X_i of Q:

$$X = \begin{bmatrix} X_{x1} & X_{x2} \\ X_{y1} & X_{y2} \end{bmatrix}. \tag{2.56}$$

Equation (2.54) shows that for this matrix

$$QX = X\Lambda, \tag{2.57}$$

where the diagonal matrix Λ is

$$\Lambda = \begin{bmatrix} \lambda_1 & 0 \\ 0 & \lambda_2 \end{bmatrix}. \tag{2.58}$$

From (2.57), we see that

$$X^{-1}QX = \Lambda. \tag{2.59}$$

A comparison of Equations (2.52) and (2.59) shows that the diagonal elements of the matrix Q' must be the eigenvalues of the matrix Q, that is, $Q' = \Lambda$, and that the matrix R_ℓ must be the matrix of the eigenvectors of Q, that is, $R_\ell = X$.

After applying the general results presented above to our specific problem, we find that the eigenvalues and eigenvectors of the matrix Q are

$$\lambda_i = \left[1 \mp \sqrt{1 - \sin^2(\delta)\sin^2(2\gamma)}\right] \Big/ 2\sin^2(\gamma) \tag{2.60}$$

and

$$X_i = C_i \begin{bmatrix} 1 \\ \dfrac{(1 - \lambda_i)\tan(\gamma)}{\cos(\delta)} \end{bmatrix}, \tag{2.61}$$

where the $-(+)$ sign in (2.60) is for $i = 1(2)$ and the C_i are arbitrary constants. Here λ_1 is chosen to be the smallest eigenvalue, because $Q'_{11} \le Q'_{22}$ for $-\pi/4 < \chi \le \pi/4$.

On equating the matrices Q' and Λ, we find that

$$\sec^2(\chi)/\sec^2(\gamma) = \left[1 - \sqrt{1 - \sin^2(\delta)\sin^2(2\gamma)}\right] \Big/ 2\sin^2(\gamma)$$

and

$$\left[\sec^2(\chi)/\sec^2(\gamma)\right] \Big/ \tan^2(\chi) = \left[1 + \sqrt{1 - \sin^2(\delta)\sin^2(2\gamma)}\right] \Big/ 2\sin^2(\gamma);$$

thus,

$$\tan(\chi) = \sqrt{\frac{1 - \sqrt{1 - \sin^2(\delta)\sin^2(2\gamma)}}{1 + \sqrt{1 - \sin^2(\delta)\sin^2(2\gamma)}}}.$$

After making the substitution $\sin(\alpha) = \sin(\delta)\sin(2\gamma)$, this equation becomes

$$\tan(\chi) = \sqrt{\frac{1 - \cos(\alpha)}{1 + \cos(\alpha)}} = \tan(\alpha/2),$$

and so it follows that $\alpha = 2\chi$ and

$$\sin(2\chi) = \sin(\delta)\sin(2\gamma). \tag{2.62}$$

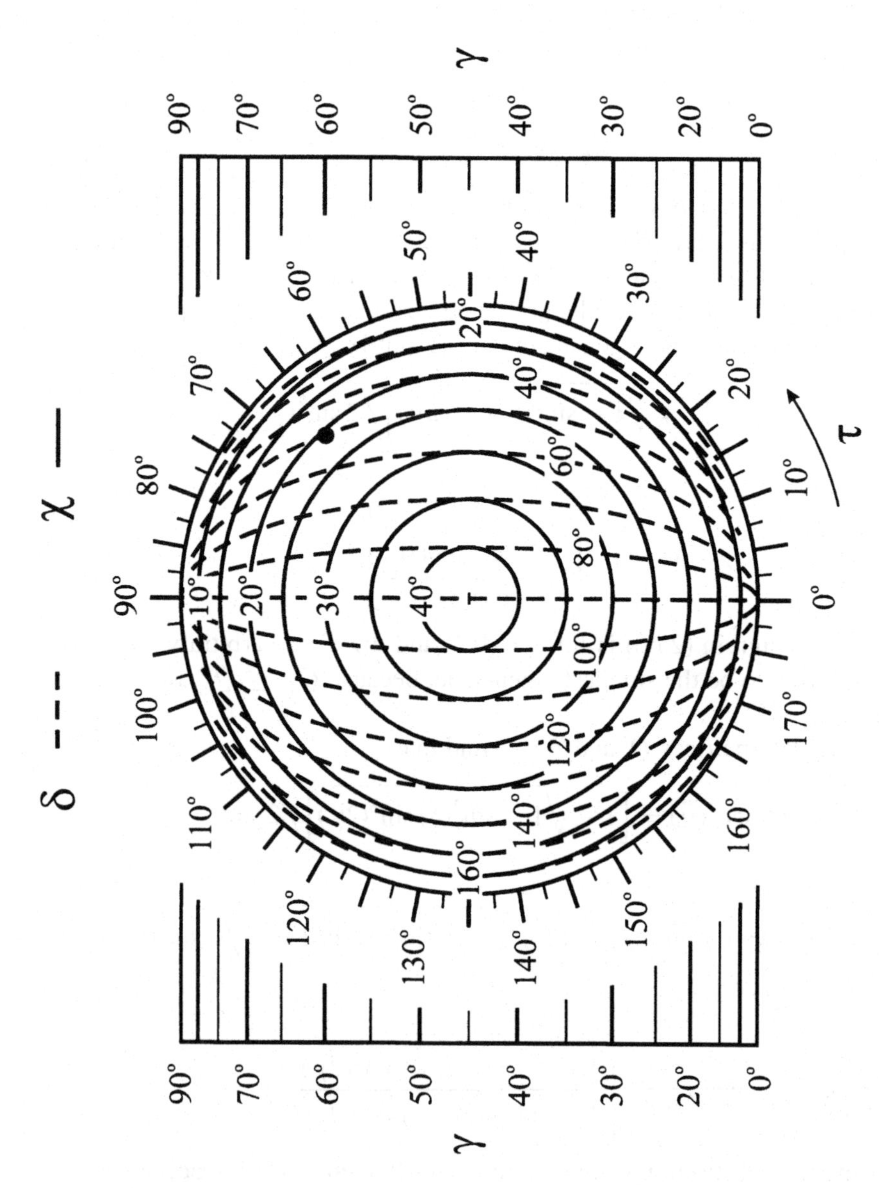

Fig. 2.7. Polarization chart for graphically converting between the angles γ, δ and χ, τ. This chart is for left-handed polarization. For right-handed polarization, $\delta \rightarrow -\delta$ and $\chi \rightarrow -\chi$.

From the eigenvector X_1 and the first column of the matrix R_ℓ we have

$$\tan(\tau) = \tan(\gamma)(1 - \lambda_1)/\cos(\delta),$$

or, on inserting λ_1 and rearranging,

$$\tan(\tau) = \frac{\sqrt{1 + \cos^2(\delta)\tan^2(2\gamma)} - 1}{\cos(\delta)\tan(2\gamma)}.$$

After making the substitution $\tan(\beta) = \cos(\delta)\tan(2\gamma)$, we get

$$\tan(\tau) = \frac{1 - \cos(\beta)}{\sin(\beta)} = \tan(\beta/2).$$

It therefore follows that $\beta = 2\tau$ and

$$\tan(2\tau) = \cos(\delta)\tan(2\gamma). \tag{2.63}$$

We now have the desired relationships for determining the angles χ and τ from the angles γ and δ:

$$\sin(2\chi) = \sin(\delta)\sin(2\gamma) \tag{2.64}$$

$$\tan(2\tau) = \cos(\delta)\tan(2\gamma). \tag{2.65}$$

These results can also be used to express γ and δ in terms of χ and τ:

$$\cos(2\gamma) = \cos(2\chi)\cos(2\tau) \tag{2.66}$$

$$\tan(\delta) = \tan(2\chi)/\sin(2\tau). \tag{2.67}$$

When any one of these expressions is solved for a given angle, an inverse trigonometric function occurs, and some care must be exercised to ensure that the correct values of these multivalued functions are used. For example, from (2.64)

$$\chi = \frac{1}{2}\sin^{-1}\left[\sin(\delta)\sin(2\gamma)\right], \tag{2.68}$$

and from (2.65)

$$\tau = \begin{cases} \frac{1}{2}\tan^{-1}\left[\cos(\delta)\tan(2\gamma)\right], & \gamma \le \pi/4, \quad |\delta| \le \pi/2 \\ \frac{1}{2}\tan^{-1}\left[\cos(\delta)\tan(2\gamma)\right] + \pi, & \gamma \le \pi/4, \quad |\delta| > \pi/2\,. \\ \frac{1}{2}\tan^{-1}\left[\cos(\delta)\tan(2\gamma)\right] + \pi/2, & \gamma > \pi/4 \end{cases} \tag{2.69}$$

The principal values of the functions $\sin^{-1}$ and $\tan^{-1}$ are assumed in these equations (i.e., $-\pi/2 \le \sin^{-1}(\,) \le \pi/2$ and $-\pi/2 \le \tan^{-1}(\,) \le \pi/2$).

Figure 2.7 is a chart for graphically converting between the angles γ, δ and χ, τ (Problem 2.6).[3] On this chart the curves of constant γ are horizontal straight lines, and the curves of constant δ are segments of ellipses, whereas the curves of constant χ are circles, and the curves of constant τ are radial lines. The chart in Figure 2.7 is drawn for left-handed states of polarization; for right-handed states the substitutions

[3] The use of graphical aids, such as this chart and the Poincaré sphere described in the next section, make it easier to determine the proper values for the inverse trigonometric functions, such as those in (2.68) and (2.69).

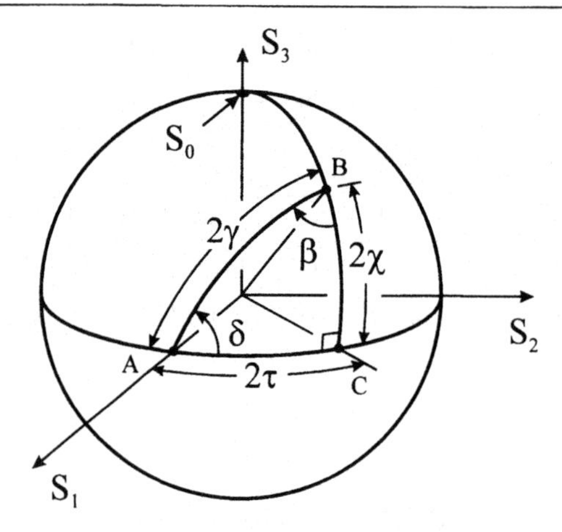

Fig. 2.8. Coordinates for the Poincaré sphere: $0 \le \gamma \le \pi/2$, $-\pi < \delta \le \pi$; $-\pi/4 \le \chi \le \pi/4$, $0 \le \tau < \pi$.

$\delta \to -\delta$ and $\chi \to -\chi$ should be made. As an example, we can use the chart to find that, for $\gamma = 60°$, $\delta = 50°$ (this point is marked by a dot on the chart), the transformed angles are $\chi \approx 21°$, $\tau \approx 66°$; from Equations (2.68) and (2.69), the precise values are $\chi = 20.8°$, $\tau = 66.0°$.

2.4 The Poincaré sphere and the Stokes parameters

In the preceding section, the state of polarization for the monochromatic plane wave was described in the two coordinate systems (x, y) and (x', y') by the parameters S_0, γ, δ and S_0, χ, τ, respectively. We will now consider a valuable graphical representation due to Jules Henri Poincaré (1854–1912) for visualizing the state of polarization as described by either set of parameters [6]. In Poincaré's representation, shown in Figure 2.8, the state of polarization is indicated by the location of a point on the surface of a sphere – the *Poincaré sphere*. The radius of the sphere is equal to the time-average power per unit area S_0, and the spherical angular coordinates of the point are 2τ and 2χ, the longitude and the latitude, respectively. It is easy to show from Equations (2.66) and (2.67) and Napier's rules for a right-angled spherical triangle that the angles 2γ and δ are as indicated in Figure 2.8 [7]. An application of the *cosine rule for sides* to the triangle ABC yields

$$\cos(2\gamma) = \cos(2\chi)\cos(2\tau),$$

which is Equation (2.66). From the *sine rule*

$$\sin(\delta) = \sin(\beta)\sin(2\chi)/\sin(2\tau),$$

and from the *cosine rule for angles*

$$\cos(\delta) = \sin(\beta)\cos(2\chi);$$

thus,

$$\tan(\delta) = \tan(2\chi)/\sin(2\tau),$$

which is Equation (2.67).

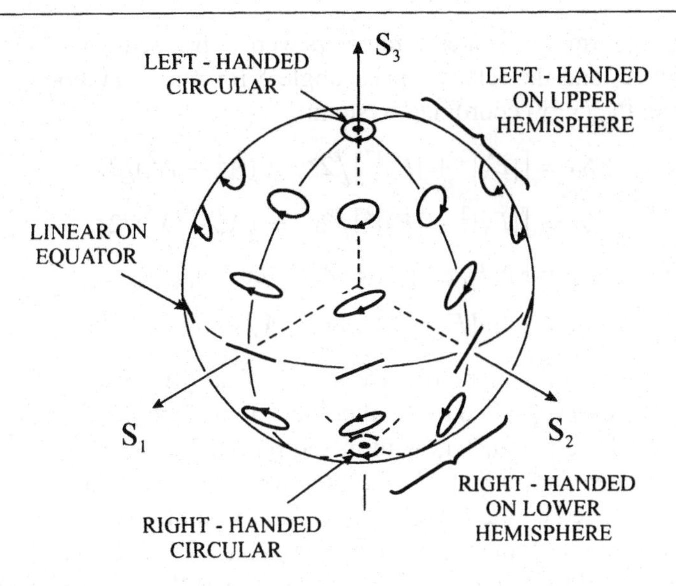

Fig. 2.9. Various states of polarization on the Poincaré sphere.

There is a one-to-one correspondence between the states of polarization and the points on the Poincaré sphere. The various states of polarization on the sphere are shown in Figure 2.9. On the equator the polarization is linear. States with left-handed polarization are on the upper hemisphere, and states with right-handed polarization are on the lower hemisphere. Left-handed circular polarization is at the north pole, and right-handed circular polarization is at the south pole.

The Poincaré sphere provides a convenient graphical tool for visualizing changes in the state of polarization. For example, an optical system may contain several devices that alter the state of polarization of a light wave passing through the system. The state of polarization as the wave passes through the system can be graphed as a curve on the Poincaré sphere. Any alteration in the system is then visualized as a change in the curve on the Poincaré sphere. We will examine this application in more detail in the next section.

The three rectangular coordinates (S_1, S_2, S_3) of a point (S_0, χ, τ) on the Poincaré sphere in Figure 2.8 are

$$S_1 = S_0 \cos(2\chi) \cos(2\tau), \tag{2.70a}$$

$$S_2 = S_0 \cos(2\chi) \sin(2\tau), \tag{2.70b}$$

and

$$S_3 = S_0 \sin(2\chi). \tag{2.70c}$$

The four parameters S_0, S_1, S_2, and S_3 are the *Stokes parameters* (George Gabriel Stokes, 1819–1903) for the monochromatic plane wave [8]. Note that the four parameters are not independent, since

$$S_0^2 = S_1^2 + S_2^2 + S_3^2. \tag{2.71}$$

In their original form, the Stokes parameters were expressed in terms of the amplitudes A_x and A_y and the relative phase angle δ for representation of the state of polarization in the (x, y) coordinate system:

$$S_0 = \left[|E_x|^2 + |E_y|^2\right]/2\zeta_o = (A_x^2 + A_y^2)/2\zeta_o \tag{2.72a}$$

$$S_1 = \left[|E_x|^2 - |E_y|^2\right]/2\zeta_o = (A_x^2 - A_y^2)/2\zeta_o \tag{2.72b}$$

$$S_2 = \mathrm{Re}[E_x E_y^*]/\zeta_o = A_x A_y \cos(\delta)/\zeta_o \tag{2.72c}$$

$$S_3 = -\mathrm{Im}[E_x E_y^*]/\zeta_o = A_x A_y \sin(\delta)/\zeta_o. \tag{2.72d}$$

The four Stokes parameters all have the physical dimensions of time-average power per unit area as opposed to the alternative representations for the wave (e.g., S_0, γ, δ), which have parameters with mixed physical dimensions – time-average power per unit area and radians. In the literature, the four Stokes parameters are sometimes designated by the letters I, Q, U, and V, and they may be defined with a common factor multiplying the parameters given in (2.72a–d). The column vector formed by the four Stokes parameters is referred to as the Stokes vector:

$$S = \begin{bmatrix} S_0 \\ S_1 \\ S_2 \\ S_3 \end{bmatrix}. \tag{2.73}$$

The Poincaré sphere and the Stokes parameters are often used in normalized form. The radius of the sphere is then unity, and the normalized Stokes parameters are $S_{in} = S_i/S_0, i = 0, 1, 2, 3$. The normalized Stokes vectors are given in Table 2.1 for several states of polarization of a monochromatic plane wave.

2.5 Optical elements for processing polarized light

We have examined the propagation of plane electromagnetic waves in free space and shown that polarization is an important characteristic of a monochromatic plane wave. Our analysis to this point has been completely independent of the particular frequency (optical, radio, etc.) of the electromagnetic field and also independent of the methods used to generate and detect the waves. We will now complete our discussion by briefly describing the use of polarized waves in two specific examples. In this section, we will describe optical elements for processing polarized light, and, in the following section, we will consider the transmission and reception of polarized waves with antennas.

We will assume that the discrete optical elements are located in free space and that a beam of light propagates between the elements as a completely polarized monochromatic plane wave. Each optical element is represented by a plate, shown schematically in Figure 2.10a, with the direction of propagation of the plane wave $(\hat{z})$ normal to the plane of the plate. The optical elements change the state of polarization of the wave, but they do not affect the direction of propagation.

Table 2.1. *Normalized Stokes and Jones vectors for various states of polarization of a monochromatic plane wave*

State	Angles† (γ, δ) (χ, τ)	Stokes vector* S_n	Jones vector E_n
Horizontal linear	$(0, -)$ $(0, 0)$	$\{1, 1, 0, 0\}$	$\begin{bmatrix} 1 \\ 0 \end{bmatrix}$
Vertical linear	$(\pi/2, -)$ $(0, \pi/2)$	$\{1, -1, 0, 0\}$	$\begin{bmatrix} 0 \\ 1 \end{bmatrix}$
45° linear	$(\pi/4, 0)$ $(0, \pi/4)$	$\{1, 0, 1, 0\}$	$\frac{1}{\sqrt{2}}\begin{bmatrix} 1 \\ 1 \end{bmatrix}$
−45° linear	$(\pi/4, \pi)$ $(0, 3\pi/4)$	$\{1, 0, -1, 0\}$	$\frac{1}{\sqrt{2}}\begin{bmatrix} 1 \\ -1 \end{bmatrix}$
Left-handed circular	$(\pi/4, \pi/2)$ $(\pi/4, -)$	$\{1, 0, 0, 1\}$	$\frac{1}{\sqrt{2}}\begin{bmatrix} 1 \\ j \end{bmatrix}$
Right-handed circular	$(\pi/4, -\pi/2)$ $(-\pi/4, -)$	$\{1, 0, 0, -1\}$	$\frac{1}{\sqrt{2}}\begin{bmatrix} 1 \\ -j \end{bmatrix}$

Note: *The column vector for S_n has been written as $\{S_{0n}, S_{1n}, S_{2n}, S_{3n}\}$ to conserve space.
†A dash indicates that the angle is undefined.

The electric field of the plane wave is expressed in terms of the complex vector phasor $\vec{E}$, that is,

$$\vec{\mathcal{E}}(t) = \mathrm{Re}(\vec{E} e^{j\omega t}). \tag{2.74}$$

At the input to the optical element, the vector phasor for the incident field is

$$\vec{E}^i = E^i_x \hat{x} + E^i_y \hat{y} = A^i_x \big[\hat{x} + \tan(\gamma_i) e^{j\delta_i} \hat{y}\big] e^{j\phi_{xi}}, \tag{2.75}$$

and at the output of the optical element, the vector phasor for the transmitted field is

$$\vec{E}^t = E^t_x \hat{x} + E^t_y \hat{y} = A^t_x \big[\hat{x} + \tan(\gamma_t) e^{j\delta_t} \hat{y}\big] e^{j\phi_{xt}}. \tag{2.76}$$

We will represent these fields by the column vectors

$$E^i = \begin{bmatrix} E^i_x \\ E^i_y \end{bmatrix} = A^i_x e^{j\phi_{xi}} \begin{bmatrix} 1 \\ \tan(\gamma_i) e^{j\delta_i} \end{bmatrix}$$

and

$$E^t = \begin{bmatrix} E^t_x \\ E^t_y \end{bmatrix} = A^t_x e^{j\phi_{xt}} \begin{bmatrix} 1 \\ \tan(\gamma_t) e^{j\delta_t} \end{bmatrix}. \tag{2.77}$$

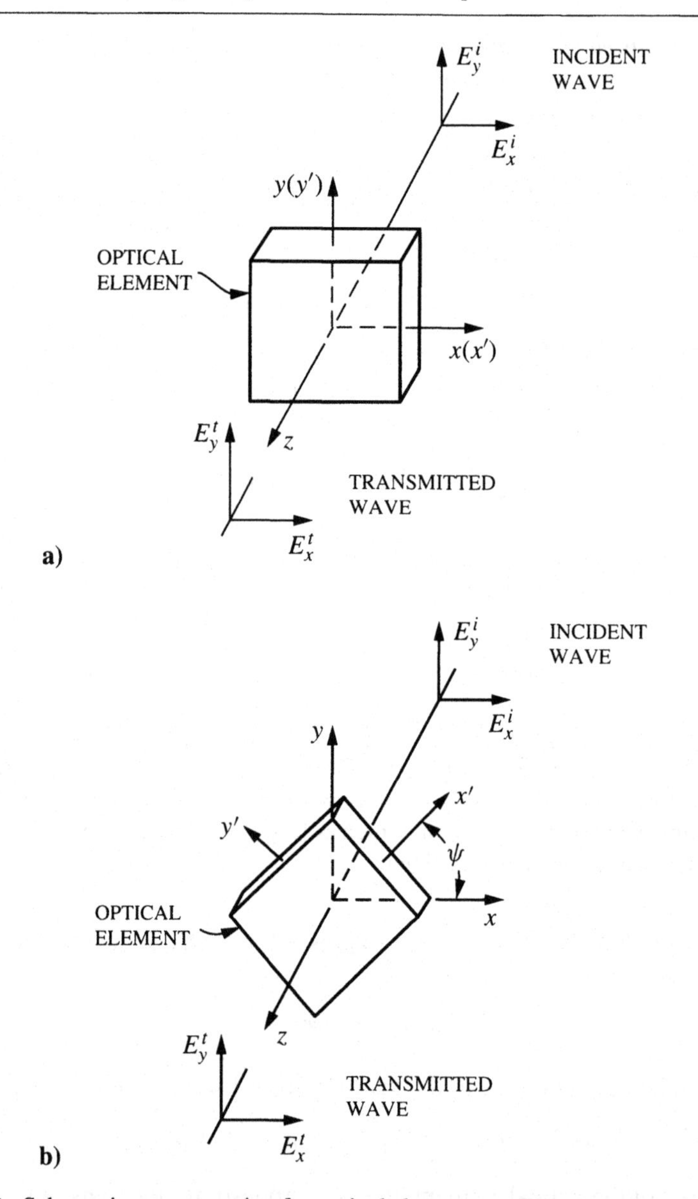

Fig. 2.10. Schematic representation for optical elements. a) Element aligned with axes x, y. b) Element rotated through angle ψ about z axis (right-handed rotation).

For a linear optical element, the incident and transmitted fields are simply related by the transmission matrix T:

$$E^t = TE^i,$$

or

$$\begin{bmatrix} E_x^t \\ E_y^t \end{bmatrix} = \begin{bmatrix} T_{xx} & T_{xy} \\ T_{yx} & T_{yy} \end{bmatrix} \begin{bmatrix} E_x^i \\ E_y^i \end{bmatrix}. \tag{2.78}$$

In the optical literature, the column vectors (2.77) are called *Jones vectors*, and the transmission matrices are called *Jones matrices*, after Robert Clark Jones who introduced this formulation [9, 10]. The Jones vectors are given in Table 2.1 for several states of polarization; they are normalized so that $E_n^T E_n^* = 1.0$:

$$E_n = \frac{E}{A_x e^{j\phi_x} \sec\gamma} = \begin{bmatrix} \cos\gamma \\ \sin\gamma e^{j\delta} \end{bmatrix}.$$

Now consider the arrangement shown in Figure 2.10b. The optical element has been rotated through the angle ψ about the z axis. Different coordinate axes are used in the description of the field (x, y) and in the description of the transmission matrix (x', y'). The coordinates x', y' are obtained by a right-handed rotation of the coordinates x, y through the angle ψ. The transmission matrix T for the unprimed coordinates is simply related to the transmission matrix for the primed coordinates T' via

$$T = R_r^T T' R_r, \tag{2.79}$$

where the rotation matrix for a right-handed rotation is

$$R_r = \begin{bmatrix} \cos\psi & \sin\psi \\ -\sin\psi & \cos\psi \end{bmatrix} \tag{2.80}$$

and $R_r^{-1} = R_r^T$. Note that $R_r = R_\ell^T$, where R_ℓ is the matrix introduced in (2.50) for a left-handed rotation.

When several optical elements are used in series, the transmission matrix for the combination is the product of the transmission matrices for the individual elements:

$$T = T_m T_{m-1} \ldots T_3 T_2 T_1. \tag{2.81}$$

Here the field is assumed to be incident first on element 1. Notice that this simple formulation neglects reflections from the surfaces of the elements, which can cause interference effects when two or more elements are used.

We will now describe a few optical elements in terms of their Jones matrices. An *ideal linear polarizer* is an element that produces a linearly polarized transmitted field from an incident field of arbitrary polarization. The direction of the linearly polarized field is along the transmission axis, which we will assume to be the x axis. The form of the Jones matrix is then

$$T = \begin{bmatrix} T_{xx} & T_{xy} \\ 0 & 0 \end{bmatrix}.$$

If we add the additional constraints that an incident wave linearly polarized along the transmission axis (x) is transmitted unattenuated, and that an incident wave linearly polarized along the axis (y) normal to the transmission axis is completely absorbed, the coefficients T_{xx} and T_{xy} are easily determined:

$$T = e^{-j\zeta} \begin{bmatrix} 1 & 0 \\ 0 & 0 \end{bmatrix},$$

where $-\zeta$ is the phase shift on transmission. It is customary to normalize the Jones matrix by removing common phase factors such as $\exp(-j\zeta)$:

$$T_n = \begin{bmatrix} 1 & 0 \\ 0 & 0 \end{bmatrix}. \tag{2.82}$$

When the transmission axis (x') of the linear polarizer is at the angle ψ, as in Figure 2.10b, the normalized Jones matrix, obtained by applying (2.79) to (2.82), is

$$T_n = \begin{bmatrix} \cos^2\psi & \cos\psi\sin\psi \\ \cos\psi\sin\psi & \sin^2\psi \end{bmatrix}. \tag{2.83}$$

Now consider a field linearly polarized in the direction of the x axis incident on an ideal linear polarizer with transmission axis at angle ψ. The Jones vector for the transmitted wave is

$$E^t = TE^i = E^i_x e^{-j\zeta}\cos\psi \begin{bmatrix} \cos\psi \\ \sin\psi \end{bmatrix}.$$

Thus, the ratio of the time-average powers per unit area for the transmitted and incident waves is

$$\frac{S_{0t}}{S_{0i}} = \frac{|\vec{E}^t|^2}{|\vec{E}^i|^2} = \frac{(E^t)^T(E^t)^*}{(E^i)^T(E^i)^*} = \cos^2\psi.$$

This is a statement of the law of Malus (Étienne Louis Malus, 1775–1812) for linear polarizers.

Retarders are optical elements used to change the state of polarization. The retarder works by dividing the incident field into two components, such as E^i_x and E^i_y, and retarding the phase of one of these components relative to the other. When the two components of the wave are reunited to form the transmitted field, the state of polarization is changed. We will consider an *ideal linear retarder* in which the x axis is the fast axis and the y axis is the slow axis, that is, the phase of the component E_y is retarded relative to the phase of the component E_x. In addition, we will assume that the retarder does not attenuate the wave. The Jones matrix for an ideal linear retarder with retardance Δ is then

$$T = e^{-j\zeta}\begin{bmatrix} 1 & 0 \\ 0 & e^{-j\Delta} \end{bmatrix},$$

or, in normalized form,

$$T_n = \begin{bmatrix} e^{j\Delta/2} & 0 \\ 0 & e^{-j\Delta/2} \end{bmatrix}. \tag{2.84}$$

When the fast axis (x') of the linear retarder is at the angle ψ, as in Figure 2.10b, the normalized Jones matrix, obtained by applying (2.79) to (2.84), is

$$T_n = \begin{bmatrix} \cos(\Delta/2) + j\cos 2\psi\sin(\Delta/2) & j\sin 2\psi\sin(\Delta/2) \\ j\sin 2\psi\sin(\Delta/2) & \cos(\Delta/2) - j\cos 2\psi\sin(\Delta/2) \end{bmatrix}. \tag{2.85}$$

A *half-wave plate* is a retarder with $\Delta = \pi$, and a *quarter-wave plate* is a retarder with $\Delta = \pi/2$. When the fast axis is along the x axis, the normalized Jones matrices for these wave plates are

$$T_n(\Delta = \pi) = \begin{bmatrix} j & 0 \\ 0 & -j \end{bmatrix} \quad \text{(half-wave plate)} \tag{2.86}$$

and

$$T_n(\Delta = \pi/2) = \frac{1}{\sqrt{2}} \begin{bmatrix} 1+j & 0 \\ 0 & 1-j \end{bmatrix} \quad \text{(quarter-wave plate).} \tag{2.87}$$

There are many different physical constructions for linear polarizers and linear retarders; these are discussed in detail in the optical literature. Here we will only give a qualitative description of representative sheet-type linear polarizers and retarders, which have many commercial applications. The modern sheet-type polarizer was invented by Edwin Herbert Land and is referred to as Polaroid sheet [11, 12]. A simple, commercial polarizer called an H-sheet polarizer is constructed from a polymeric material such as polyvinyl alcohol. A sheet of this material is heated and stretched unidirectionally, which causes the long-chain hydrocarbon molecules to orient parallel to the stretched direction. The sheet is then stained with a solution containing iodine. The iodine atoms form long strings that lie parallel to the molecules of the polymer. Conduction electrons can move along the string of iodine atoms, and each string can be thought of crudely as a straight resistive conductor (wire) with its axis parallel to the stretched direction.

To see how the sheet linearly polarizes light, consider a wave incident on the sheet, as in Figure 2.10a, with the electric field lying in the plane of the sheet. Let the field be decomposed into components parallel to and perpendicular to the stretched direction (y). The field component parallel to the stretched direction produces an axial electric current in each string of iodine atoms. The currents dissipate power in these resistive conductors; thus, this component of the field is severely attenuated as the wave passes through the sheet. The field component perpendicular to the stretched direction produces no significant axial currents in the strings of iodine atoms, and it suffers little attenuation as the wave passes through the sheet. Consequently, the wave that emerges from the sheet is linearly polarized with the electric field perpendicular to the stretched direction. For polarizers with this construction, the ratio $|T_{yy}/T_{xx}|$ is typically in the range 5×10^{-3} to 2×10^{-1} at optical wavelengths (4,000 Å$\leq \lambda_o \leq$ 7,000 Å); for an ideal polarizer, of course, this ratio is zero. The sheet-type polarizer is an example of a *dichroic* linear polarizer, a polarizer that works on the principle of differential absorption for two states of polarization of the incident wave.

Clear sheets of stretched polymeric material are also used to construct linear retarders. Within the sheet, a wave with electric field parallel to the stretched direction experiences a higher effective permittivity, therefore a lower phase velocity, than a wave with electric field perpendicular to the stretched direction. The thickness of the sheet is adjusted to obtain the desired relative phase retardance for waves with the two orientations of electric field.

These two simple optical elements – the linear polarizer and the linear retarder – can be used to fabricate more complicated optical devices. As an example, we will describe the construction of an *ideal circular polarizer.* The polarizer has the following property: It produces either a left-handed or right-handed circularly polarized transmitted field from an incident field of arbitrary polarization. Thus, the Jones matrix performs the following transformation:

$$\begin{bmatrix} T_{xx} & T_{xy} \\ T_{yx} & T_{yy} \end{bmatrix} \begin{bmatrix} A \\ B \end{bmatrix} = C \begin{bmatrix} 1 \\ \pm j \end{bmatrix}, \tag{2.88}$$

where the upper (lower) sign refers to left-handed (right-handed) circular polarization. The complex constants A, B, and C are arbitrary, provided $2|C|^2 \leq |A|^2+|B|^2$ so that the polarizer does not produce power. If we specify that the polarizer transmits a left-handed (right-handed) circularly polarized wave unattenuated and completely absorbs a right-handed (left-handed) circularly polarized wave, then (2.88) can be used to write four equations in the four unknown transmission coefficients, $T_{xx}, T_{yy}, \ldots$. After solving these equations, the normalized Jones matrix for the ideal left-handed (right-handed) circular polarizer becomes

$$T_n = \frac{1}{2} \begin{bmatrix} 1 & \mp j \\ \pm j & 1 \end{bmatrix}. \tag{2.89}$$

A circular polarizer with the Jones matrix (2.89) can be constructed from a linear polarizer sandwiched between two quarter-wave plates. The fast axes of the two quarter-wave plates are orthogonal, and the transmission axis of the linear polarizer makes an angle of 45° with the fast axis of either plate. The arrangement is shown in Figure 2.11. Note that the transmission axis of the linear polarizer is at the angle $\psi = 45^\circ$ for a left-handed circular polarizer (LCP) and at the angle $\psi = -45^\circ$ for a right-handed circular polarizer (RCP). It is easy to verify that the arrangement shown in Figure 2.11 has the Jones matrix (2.89); we simply multiply the Jones matrices for the three elements:

$$T_n = T_{n3}T_{n2}T_{n1} = \begin{bmatrix} e^{-j\pi/4} & 0 \\ 0 & e^{j\pi/4} \end{bmatrix} \frac{1}{2} \begin{bmatrix} 1 & \pm 1 \\ \pm 1 & 1 \end{bmatrix} \begin{bmatrix} e^{j\pi/4} & 0 \\ 0 & e^{-j\pi/4} \end{bmatrix}$$

$$= \frac{1}{2} \begin{bmatrix} 1 & \mp j \\ \pm j & 1 \end{bmatrix}.$$

The changes in the state of polarization for a wave passing through the ideal circular polarizer can be illustrated by use of the normalized Poincaré sphere. Let the arbitrary state of polarization for the incident wave be represented by point A on the Poincaré sphere in Figure 2.12. On passing through the first quarter-wave plate, the relative phase angle δ between the two components (E_x and E_y) of the wave is decreased by $\pi/2$, while the relative amplitude, indicated by the angle γ, remains fixed. This moves the state of polarization from point A to point B on the sphere. The linear polarizer, with transmission axis at $\psi = 45^\circ$, moves the state of polarization on the sphere from point B to point C. The wave is now linearly polarized with its

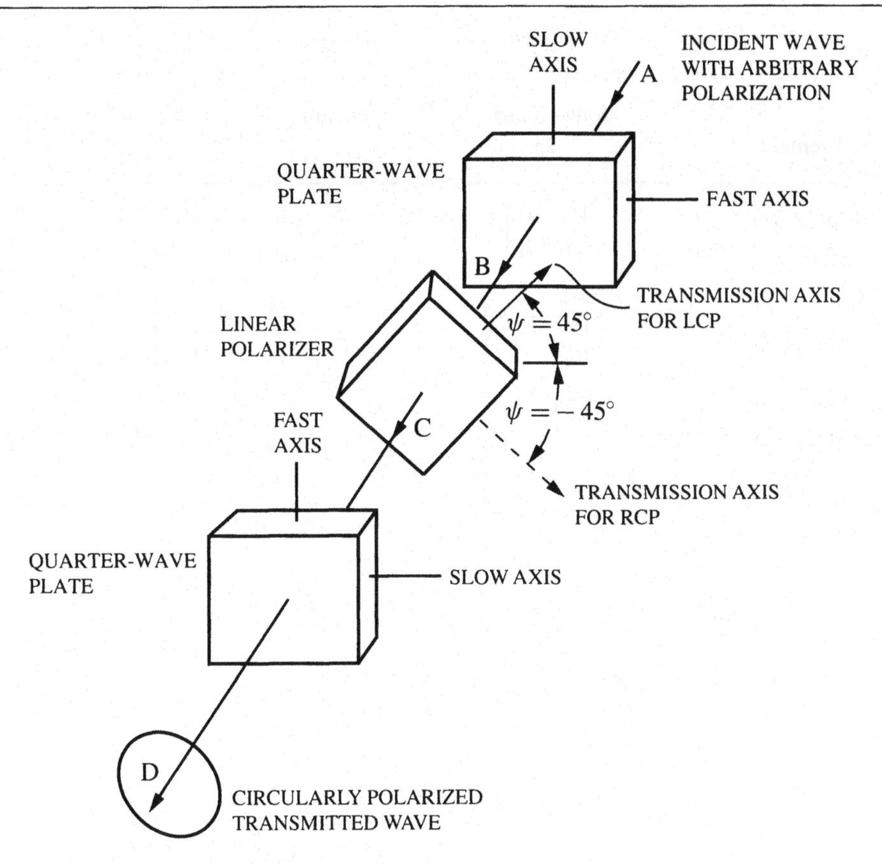

Fig. 2.11. Three-element, ideal circular polarizer formed from a linear polarizer and two quarter-wave plates. Angle of linear polarizer: $\psi = 45°$ for left-handed circular polarizer, $\psi = -45°$ for right-handed circular polarizer.

axis at 45° ($\chi = 0$, $\tau = \pi/4$, or $\gamma = \pi/4$, $\delta = 0$). The second quarter-wave plate increases the relative phase angle δ by $\pi/2$, while the relative amplitude remains fixed at $\gamma = \pi/4$. This moves the state of polarization on the sphere from point C to the north pole, point D, where the wave is left-handed circularly polarized.

We could have eliminated the first quarter-wave plate in the above construction and still have obtained a left-handed circularly polarized wave from an incident wave with arbitrary polarization. For this two-element polarizer, the states of polarization at points A and C on the Poincaré sphere in Figure 2.12 are connected by the dashed curve. The extra element (the first quarter-wave plate) in the three-element polarizer is added to make the polarizer pass unattenuated left-handed circularly polarized waves and completely absorb right-handed circularly polarized waves. The two-element polarizer does not have these properties.

The normalized Jones matrices for the ideal elements we have discussed are summarized in Table 2.2. The eigenvalues and eigenvectors (λ_i and X_i, $i = 1, 2$) for the matrices are also given in the table. The eigenvectors are normalized so that $X^T X^* = 1$. Recall from Section 2.3 that the eigenvectors and eigenvalues of

Table 2.2. *Normalized Jones matrices for ideal elements*

Element	Jones matrix T_n	Eigenvalues λ_1, λ_2	Eigenvectors X_1, X_2
Linear polarizer, transmission axis x	$\begin{bmatrix} 1 & 0 \\ 0 & 0 \end{bmatrix}$	1 0	$\begin{bmatrix} 1 \\ 0 \end{bmatrix}$ $\begin{bmatrix} 0 \\ 1 \end{bmatrix}$
Left-handed circular polarizer	$\frac{1}{2}\begin{bmatrix} 1 & -j \\ j & 1 \end{bmatrix}$	1 0	$\frac{1}{\sqrt{2}}\begin{bmatrix} 1 \\ j \end{bmatrix}$ $\frac{1}{\sqrt{2}}\begin{bmatrix} 1 \\ -j \end{bmatrix}$
Right-handed circular polarizer	$\frac{1}{2}\begin{bmatrix} 1 & j \\ -j & 1 \end{bmatrix}$	1 0	$\frac{1}{\sqrt{2}}\begin{bmatrix} 1 \\ -j \end{bmatrix}$ $\frac{1}{\sqrt{2}}\begin{bmatrix} 1 \\ j \end{bmatrix}$
Half-wave plate, fast axis x	$\begin{bmatrix} j & 0 \\ 0 & -j \end{bmatrix}$	j $-j$	$\begin{bmatrix} 1 \\ 0 \end{bmatrix}$ $\begin{bmatrix} 0 \\ 1 \end{bmatrix}$
Quarter-wave plate, fast axis x	$\frac{1}{\sqrt{2}}\begin{bmatrix} 1+j & 0 \\ 0 & 1-j \end{bmatrix}$	$\frac{(1+j)}{\sqrt{2}}$ $\frac{(1-j)}{\sqrt{2}}$	$\begin{bmatrix} 1 \\ 0 \end{bmatrix}$ $\begin{bmatrix} 0 \\ 1 \end{bmatrix}$

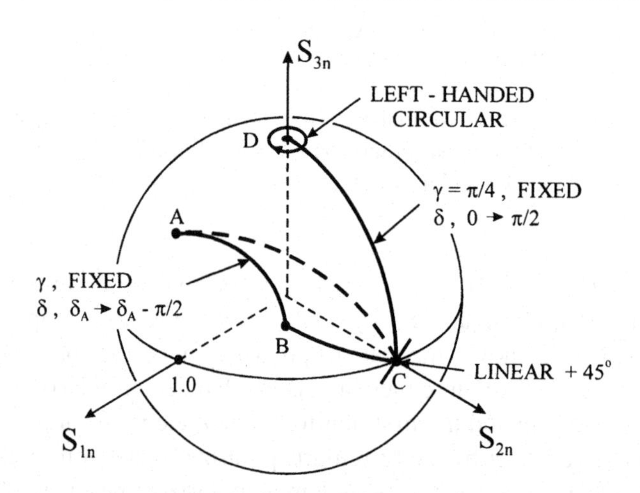

Fig. 2.12. The state of polarization on the normalized Poincaré sphere as a wave passes through the three-element, left-handed circular polarizer. The state of polarization at A is arbitrary, and the solid line traces the state of polarization as the wave passes through the various elements.

a matrix satisfy the relation

$$T_n X_i = \lambda_i X_i.$$

From this result we see that an eigenvector is the normalized Jones vector for a particular incident field and that for this incident field the state of polarization is not changed when the wave passes through the optical element; that is, the incident and transmitted fields have the same state of polarization. For example, for the linear polarizer with transmission axis along the x axis, the eigenvectors are linearly polarized fields in the x and y directions. The eigenvalue for the x-directed field is one, so this field passes unattenuated through the polarizer. The eigenvalue for the y-directed field is zero, so this field is completely absorbed by the polarizer. Similar arguments apply to the left- and right-handed circular polarizers. For half- and quarter-wave plates whose fast axes lie along the x axis, the eigenvectors are linearly polarized fields in the x and y directions. The eigenvalues are the relative phase shifts for each of these fields on passing through the plate.

In the above discussion, we have considered optical elements arranged to form *polarizers* – devices that produce a wave with a known state of polarization from a wave with an arbitrary state of polarization. These same optical elements can be arranged to form devices for determining the state of polarization of a wave.[4] We will illustrate this point with a simple *polarimeter* used to determine the Stokes parameters of an arbitrarily polarized incident wave. The elements of the polarimeter, shown in Figure 2.13, are a rotatable, ideal quarter-wave plate; a rotatable, ideal linear polarizer, referred to as an analyzer; and a detector. The fast axis of the quarter-wave plate and the transmission axis of the polarizer are, respectively, at the angles ψ_r and ψ_p from the x axis. The normalized Jones matrix for the wave plate–analyzer combination is the product of the matrices for the individual elements $T_n = T_{n2}T_{n1}$; the elements in this matrix are

$$\begin{aligned}
T_{nxx} &= \cos\psi_p\big[\cos\psi_p + j\cos(\psi_p - 2\psi_r)\big]/\sqrt{2}\\
T_{nxy} &= \cos\psi_p\big[\sin\psi_p - j\sin(\psi_p - 2\psi_r)\big]/\sqrt{2}\\
T_{nyx} &= \sin\psi_p\big[\cos\psi_p + j\cos(\psi_p - 2\psi_r)\big]/\sqrt{2}\\
T_{nyy} &= \sin\psi_p\big[\sin\psi_p - j\sin(\psi_p - 2\psi_r)\big]/\sqrt{2}.
\end{aligned} \tag{2.90}$$

The detector senses the time-average power per unit area of the transmitted wave (2.38):

$$S_0^t(\psi_r, \psi_p) = (E^t)^T(E^t)^*/2\zeta_o = (T_n E^i)^T(T_n E^i)^*/2\zeta_o. \tag{2.91}$$

After some algebra, we find from (2.91) that

$$\begin{aligned}
S_0^t(\psi_r, \psi_p) = {} & \frac{1}{2}\Big[\big(|E_x^i|^2 + |E_y^i|^2\big)/2\zeta_o\Big]\big(|T_{nxx}|^2 + |T_{nyy}|^2 + |T_{nxy}|^2 + |T_{nyx}|^2\big)\\
& + \frac{1}{2}\Big[\big(|E_x^i|^2 - |E_y^i|^2\big)/2\zeta_o\Big]\big(|T_{nxx}|^2 - |T_{nyy}|^2 - |T_{nxy}|^2 + |T_{nyx}|^2\big)\\
& + \big[\mathrm{Re}(E_x^i E_y^{i*})/\zeta_o\big]\mathrm{Re}(T_{nxx}T_{nxy}^* + T_{nyy}^* T_{nyx})\\
& + \big[-\mathrm{Im}(E_x^i E_y^{i*})/\zeta_o\big]\,\mathrm{Im}(T_{nxx}T_{nxy}^* + T_{nyy}^* T_{nyx}).
\end{aligned} \tag{2.92}$$

[4] Section 7.7 contains additional discussion of the use of optical elements to determine the state of polarization for light.

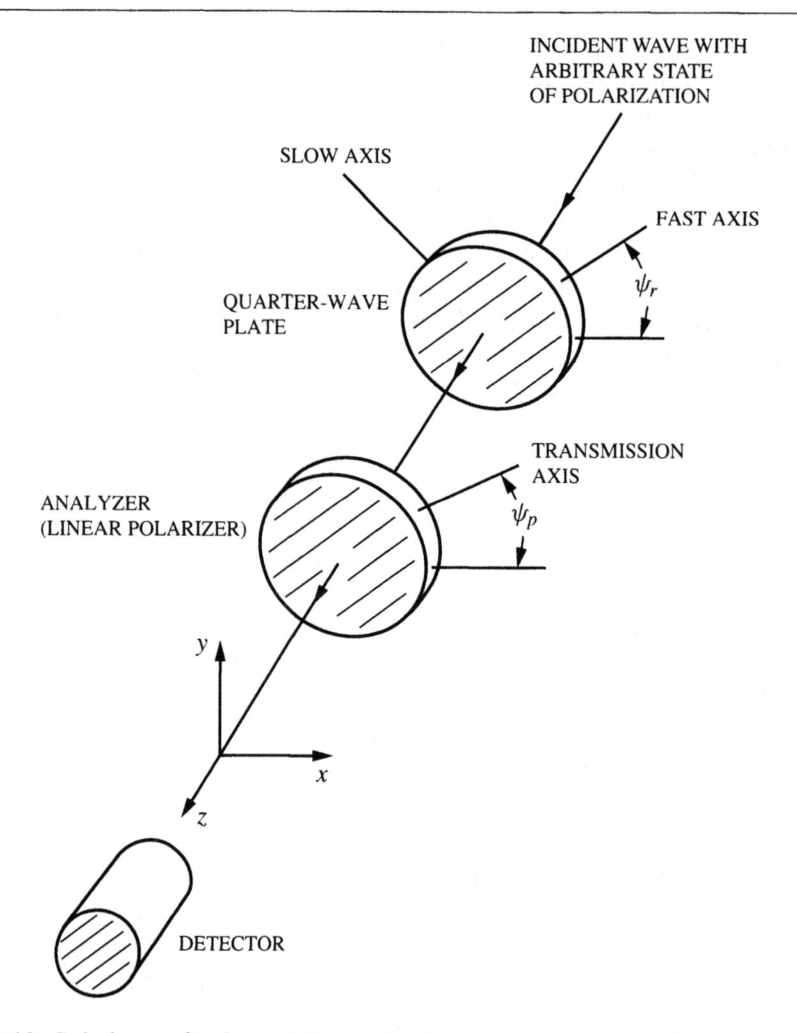

Fig. 2.13. Polarimeter for determining the Stokes parameters of an incident plane wave.

On comparison with (2.72a–d), the four bracketed terms in (2.92) are seen to be the Stokes parameters of the incident field [e.g., $(|E_x^i|^2 + |E_y^i|^2)/2\zeta_o = S_0^i$]. Our final result is obtained by inserting the Stokes parameters and the elements of the Jones matrix (2.90) into (2.92):

$$S_0^t(\psi_r, \psi_p) = \frac{1}{2}\Big\{S_0^i + \big[S_1^i \cos(2\psi_r) + S_2^i \sin(2\psi_r)\big]\cos\big[2(\psi_p - \psi_r)\big] + S_3^i \sin\big[2(\psi_p - \psi_r)\big]\Big\}. \tag{2.93}$$

It is now clear that the four Stokes parameters for the incident wave can be determined by combining detector readings made with various orientations of the quarter-wave plate and analyzer in Figure 2.13, that is, with various combinations of the angles ψ_r and ψ_p:

$$\begin{aligned}
S_0^i &= S_0^t(0^\circ, 0^\circ) + S_0^t(90^\circ, 90^\circ) \\
S_1^i &= S_0^t(0^\circ, 0^\circ) - S_0^t(90^\circ, 90^\circ) \\
S_2^i &= S_0^t(45^\circ, 45^\circ) - S_0^t(-45^\circ, -45^\circ) \\
S_3^i &= S_0^t(0^\circ, 45^\circ) - S_0^t(0^\circ, -45^\circ).
\end{aligned} \tag{2.94}$$

These equations represent what are referred to as the phenomenological definitions of the Stokes parameters: S_0^i is the overall intensity of the light, S_1^i is the excess of the light passed by a horizontal (x) linear polarizer over that passed by a vertical (y) linear polarizer, S_2^i is the excess of the light passed by a linear polarizer at $+45^\circ$ over that passed by a linear polarizer at -45°, and S_3^i is the excess of the light passed by a left-handed circular polarizer over that passed by a right-handed circular polarizer.

2.6 Transmission and reception of polarized waves with antennas

Antennas are devices used to produce an electromagnetic wave in space from energy provided by a source (transmitting antenna) or to recover energy from an electromagnetic wave in space and supply it to a detector (receiving antenna). The electromagnetic wave propagating through space is polarized, and the state of polarization is important in determining the efficiency of the energy transfer between a transmitting antenna and a receiving antenna.

The general properties of antennas will be considered in more detail in later chapters. For now we will just summarize a few of these results, so that we can illustrate with a simple example the important role polarization plays in the transmission and reception of waves with antennas.

Figure 2.14 shows a transmitting antenna located at the origin of a spherical coordinate system (r, θ, ϕ). We will only be concerned with the monochromatic electromagnetic field at a large radial distance from the antenna. This is the radiated or far-zone field, and it is described by the complex vector phasors[5]

$$\begin{aligned}
\vec{E}^r(\vec{r}) &= E_\theta^r(r, \theta, \phi)\hat{\theta} + E_\phi^r(r, \theta, \phi)\hat{\phi} \\
&= \left[F_\theta(\theta, \phi)\hat{\theta} + F_\phi(\theta, \phi)\hat{\phi}\right]e^{-j\vec{k}\cdot\vec{r}}/r, \\
\vec{B}^r(\vec{r}) &= \frac{1}{c}\hat{r} \times \vec{E}^r(\vec{r}),
\end{aligned} \tag{2.95}$$

where

$$\vec{k} = k_o\hat{r}.$$

In Section 3.6 we will derive these expressions and give a more precise definition for the radiated field.

[5] The superscript r is used to indicate the radiated field.

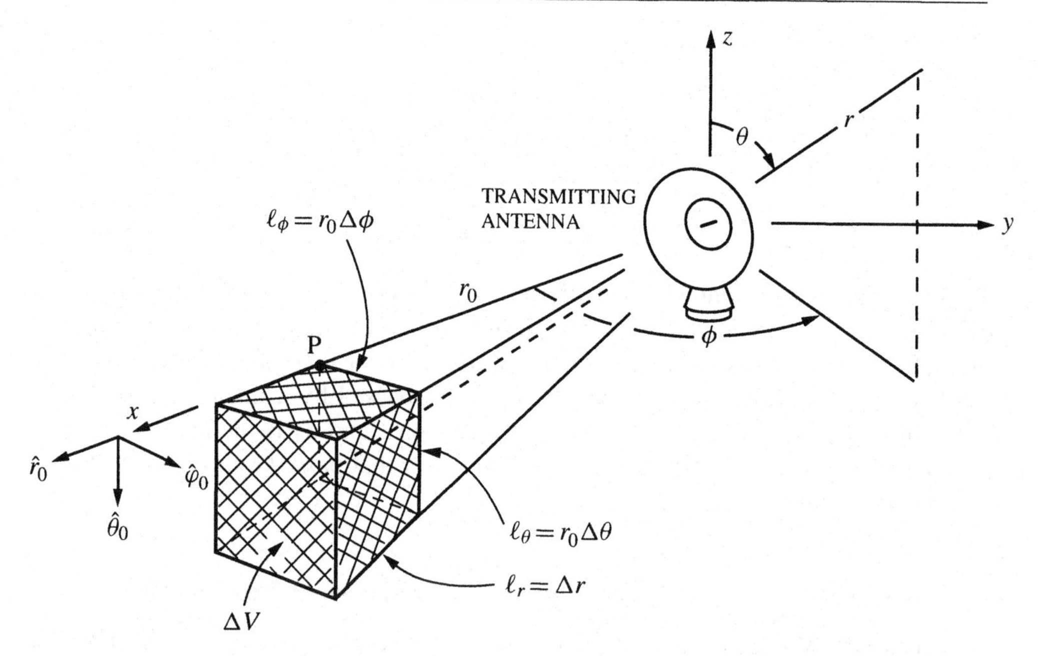

Fig. 2.14. Coordinates used to describe the radiated electromagnetic field of a transmitting antenna.

This field is a wave propagating in the radial direction away from the antenna. The phase factor $\exp(-j\vec{k}\cdot\vec{r})$ is constant on a spherical surface, and the components of the field, $\vec{E}^r$ and $\vec{B}^r$, are transverse to the direction of propagation ($\hat{r}$). The functions $F_\theta(\theta, \phi)$ and $F_\phi(\theta, \phi)$ determine the directional characteristics of the antenna.

We will now show that the electromagnetic field (2.95) behaves locally as a plane wave within a limited volume ΔV of space located at an arbitrary point P. For convenience of representation, this point is taken to be on the x axis (r_o, $\theta_o = \pi/2$, $\phi_o = 0$), as shown in Figure 2.14. The volume located at P has sides $\ell_r = \Delta r \ll r_o$, $\ell_\theta = r_o\Delta\theta \ll r_o$, and $\ell_\phi = r_o\Delta\phi \ll r_o$. When these inequalities are satisfied, the following approximations can be used in (2.95): $1/r \approx 1/r_o$, $\hat{r} \approx \hat{r}_o$, $\hat{\theta} \approx \hat{\theta}_o$, $\hat{\phi} \approx \hat{\phi}_o$, and since $\Delta\theta$ and $\Delta\phi$ are small, $F_\theta(\theta, \phi) \approx F_\theta(\theta_o, \phi_o)$ and $F_\phi(\theta, \phi) \approx F_\phi(\theta_o, \phi_o)$. In addition, we will require $k_o\ell_\theta(\ell_\theta/r_o) \ll 1$ and $k_o\ell_\phi(\ell_\phi/r_o) \ll 1$, which permits the approximation $\vec{k}\cdot\vec{r} \approx k_o\hat{r}_o\cdot\vec{r}$ to be used in the phase factor.

Subject to the above inequalities, the electromagnetic field (2.95) within the volume ΔV of space at the point P is approximately

$$\vec{E}^r(\vec{r}) \approx \frac{1}{r_o}\left[F_\theta(\theta_o, \phi_o)\hat{\theta}_o + F_\phi(\theta_o, \phi_o)\hat{\phi}_o\right]e^{-jk_o\hat{r}_o\cdot\vec{r}},$$

$$\vec{B}^r(\vec{r}) \approx \frac{1}{c}\hat{r}_o \times \vec{E}^r(\vec{r}). \qquad (2.96)$$

A comparison with Equations (2.33a), (2.33b), and (2.34) shows that this is the electromagnetic field for a monochromatic plane wave propagating in the direction $\hat{r}_o$. Thus, all of our methods for describing the state of polarization of a monochromatic

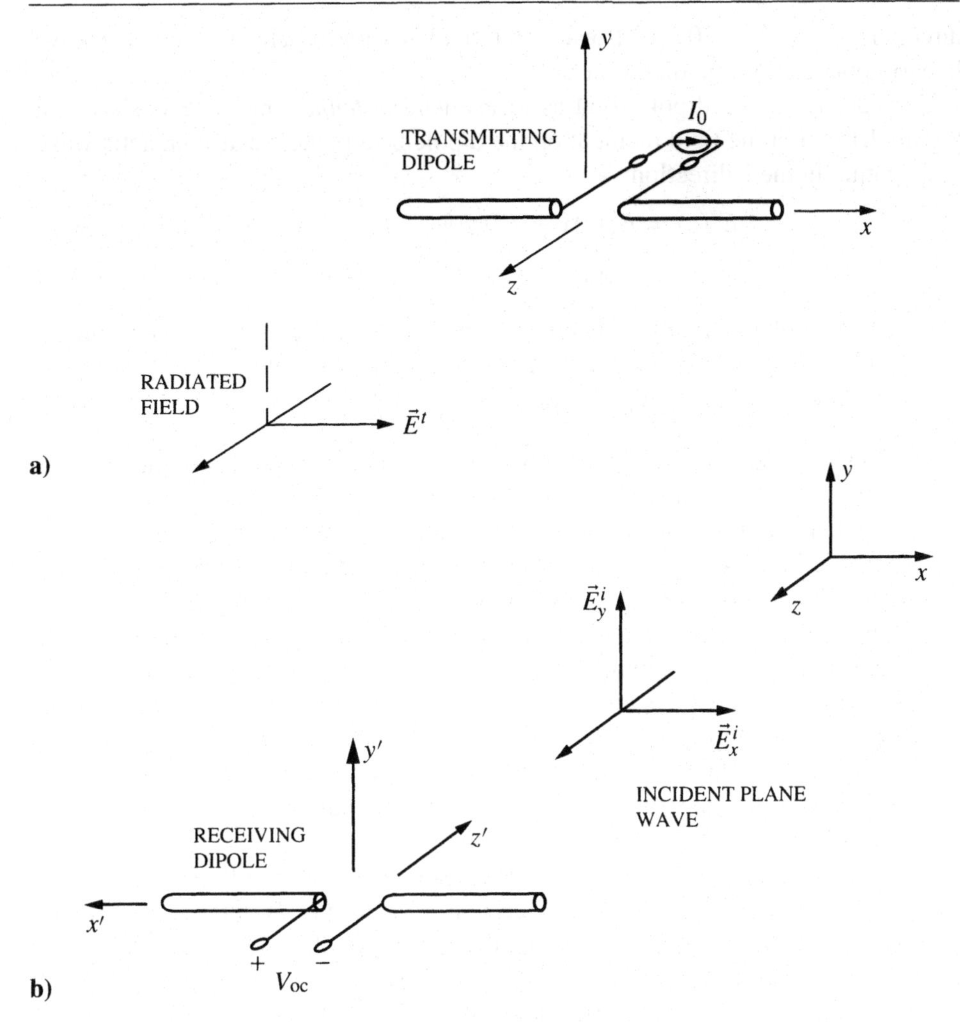

Fig. 2.15. a) Dipole transmitting antenna. b) Dipole receiving antenna.

plane wave (Poincaré sphere, Stokes parameters, etc.) apply for the state of polarization of the radiated field of an antenna. In fact, for experimental purposes, it is common to use the radiated field of an antenna to approximate a plane wave with the same state of polarization.

Now we will consider a specific antenna, the simple linear dipole formed from a thin metallic wire with terminals at its midpoint. The wire lies along the x axis in Figure 2.15a, and we will only be concerned with the field at points along the z axis. When the dipole is used as a *transmitting antenna* and is driven at its terminals with the current I_o, the radiated electric field along the z axis ($x = 0$, $y = 0$, z) can be shown to be

$$\vec{E}^t(z) = \frac{-j\zeta_o h_e I_o}{2\lambda_o z} e^{-jk_o z}\hat{x}, \tag{2.97}$$

where the constant h_e is the "effective length" of the antenna for radiation in the direction ($x = 0, y = 0, z$).[6] Locally, the field is seen to behave as a plane wave linearly polarized in the direction $\hat{x}$.

Consider the same dipole used as a *receiving antenna*; this case is shown in Figure 2.15b. Let the field incident on the dipole be a plane electromagnetic wave propagating in the $\hat{z}$ direction:

$$\begin{aligned} \vec{E}^i(z) &= E_x^i(z)\hat{x} + E_y^i(z)\hat{y} \\ &= A_x^i\big[\hat{x} + \tan(\gamma_i)e^{j\delta_i}\hat{y}\big]e^{-j(k_o z - \phi_x)}, \end{aligned} \tag{2.98}$$

with the state of polarization determined by the angles γ_i and δ_i. The voltage produced at the open-circuited terminals of the receiving dipole can be shown to be

$$V_{\text{oc}} = -h_e\hat{x}' \cdot \vec{E}^i(z) = -h_e E_x^i(z)\hat{x}' \cdot \hat{x}. \tag{2.99}$$

From (2.99), we see that the dipole only responds to the component of the incident electric field that is parallel to its axis. Notice that there are two separate right-handed rectangular Cartesian coordinate systems in Figure 2.15b. The coordinates (x, y, z) are used to describe the incident wave, and the coordinates (x', y', z') are used to describe the orientation of the antenna.

A slightly more complicated antenna is shown in Figure 2.16a. It is formed from two identical dipoles, one aligned with the x axis and the other aligned with the y axis. An additional device is connected to the terminals of the y-axis dipole. This device changes the amplitude by the factor $\tan(\gamma_t)$ and the phase by the angle δ_t for the current on transmission and for the open-circuit voltage on reception. We will not consider the construction of this device here, but an example is discussed in Problem 2.15.

When this antenna is *transmitting*, a current source is connected in series with the dipoles, as shown in Figure 2.16b. The current at the terminals of the x-axis dipole is then I_o, and that at the terminals of the y-axis dipole is $I_o \tan(\gamma_t)\exp(j\delta_t)$. The radiated electric field along the z axis ($x = 0, y = 0, z$) is obtained by superimposing the fields for the two dipoles (2.97):

$$\vec{E}^t(z) = \frac{-j\zeta_o h_e I_o}{2\lambda_o z}\big[\hat{x} + \tan(\gamma_t)e^{j\delta_t}\hat{y}\big]e^{-jk_o z}. \tag{2.100}$$

With this antenna, any desired state of polarization for the field can be obtained by adjusting γ_t and δ_t. For example, when $\tan(\gamma_t) = 1$ and $\delta_t = \pi/2$ $(-\pi/2)$ the field is left-handed (right-handed) circularly polarized.

Now consider this same antenna (γ_t and δ_t fixed) as a *receiving* antenna placed in the path of an incident plane wave (2.98) propagating in the $\hat{z}$ direction. The dipoles are connected in series as shown in Figure 2.16b, and the sum of their open-circuit voltages (2.99) is

$$\begin{aligned} V_{\text{oc}} &= -h_e\big[\hat{x}' + \tan(\gamma_t)e^{j\delta_t}\hat{y}'\big] \cdot \vec{E}^i(z) \\ &= -h_e A_x^i\big[-1 + \tan(\gamma_t)\tan(\gamma_i)e^{j(\delta_t + \delta_i)}\big]e^{-j(k_o z - \phi_x)}. \end{aligned} \tag{2.101}$$

Again, the coordinates (x, y, z) are used for the incident wave and the coordinates

[6] In Section 7.2 we will examine the electrically short, linear antenna or dipole. The effective height for that antenna is discussed in Footnote 6 of Chapter 7.

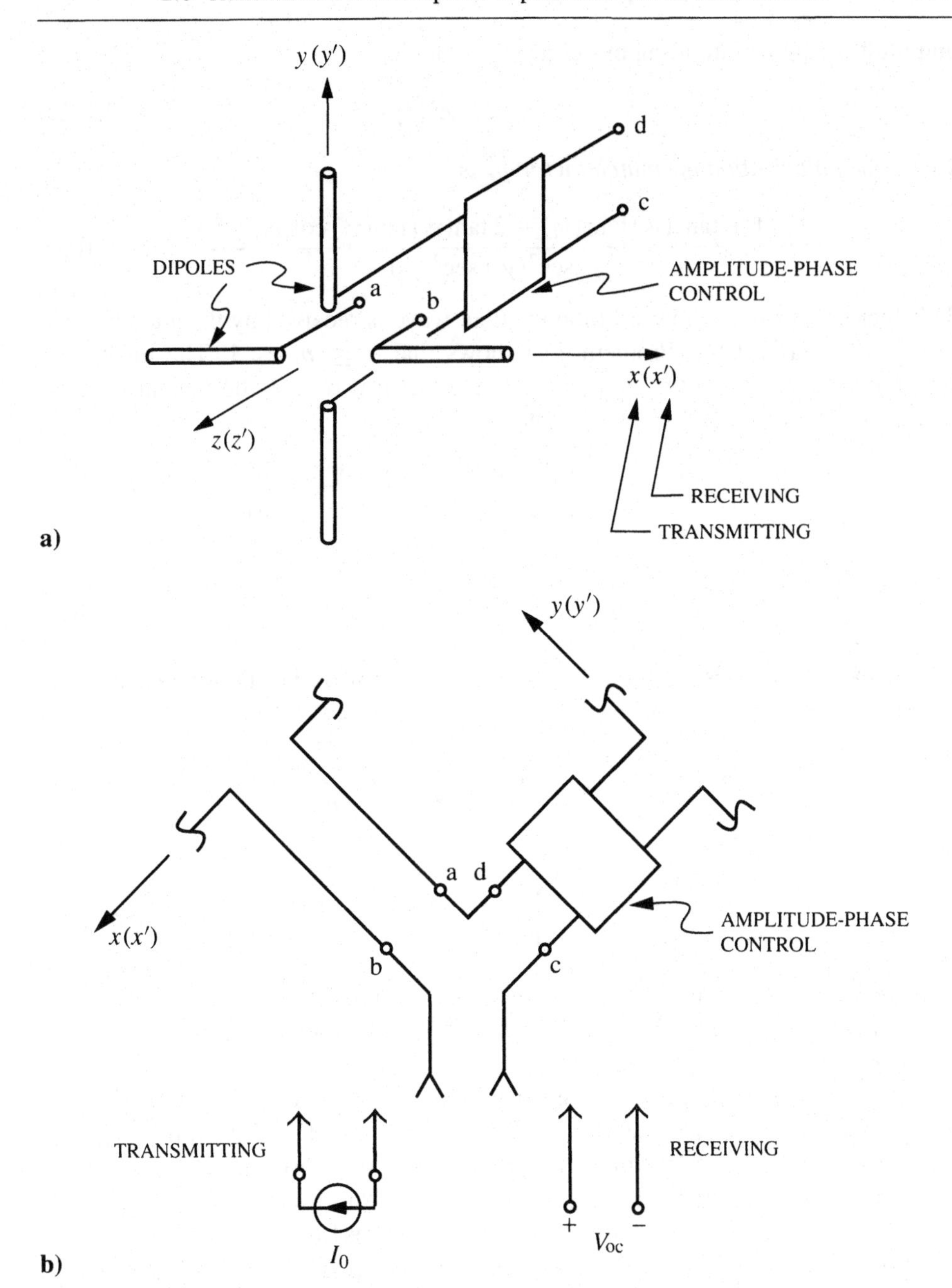

Fig. 2.16. a) Antenna formed from a pair of dipoles. b) Methods for connecting the antenna when transmitting and receiving.

(x', y', z') for the orientation of the antenna. We will examine the squared magnitude of this voltage. After introducing the time-average power per unit area of the incident wave (2.38),

$$S_0^i = \frac{|A_x^i|^2 \sec^2(\gamma_i)}{2\zeta_o},$$

and multiplying and dividing by $\sec^2(\gamma_t)$, we obtain

$$|V_{oc}|^2 = 2\zeta_o |h_e|^2 S_0^i \sec^2(\gamma_t) p_r, \tag{2.102}$$

where the *polarization mismatch factor* p_r is

$$p_r = \frac{1 + \tan^2(\gamma_t)\tan^2(\gamma_i) - 2\tan(\gamma_t)\tan(\gamma_i)\cos(\delta_t + \delta_i)}{\sec^2(\gamma_t)\sec^2(\gamma_i)}. \tag{2.103}$$

This factor determines how well the incident wave is received by the antenna. An examination of (2.103) (Problem 2.9) shows that $0 \le p_r \le 1$. The mismatch factor p_r is a *maximum* ($p_r = 1$) when the incident wave has the special state of polarization $\gamma_i = \gamma_r$, $\delta_i = \delta_r$, such that

$$\gamma_r = \gamma_t, \qquad \delta_r = \begin{cases} \pi - \delta_t, & 0 \le \delta_t \le \pi \\ -\pi - \delta_t, & -\pi < \delta_t < 0 \end{cases},$$

or

$$\chi_r = \chi_t, \qquad \tau_r = \pi - \tau_t. \tag{2.104}$$

The mismatch factor is *zero* (null) when the incident wave has the special state of polarization $\gamma_i = \gamma_n$, $\delta_i = \delta_n$, such that

$$\gamma_n = \pi/2 - \gamma_r, \qquad \delta_n = \begin{cases} \delta_r - \pi, & 0 < \delta_r \le \pi \\ \delta_r + \pi, & -\pi < \delta_r \le 0 \end{cases},$$

or

$$\chi_n = -\chi_r, \qquad \tau_n = \begin{cases} \tau_r + \pi/2, & \tau_r < \pi/2 \\ \tau_r - \pi/2, & \tau_r \ge \pi/2 \end{cases}. \tag{2.105}$$

Here we used (2.68) and (2.69) to obtain χ and τ from γ and δ, and we have been careful to keep the angles within the ranges we originally established: $0 \le \gamma \le \pi/2$, $-\pi < \delta \le \pi$, $-\pi/4 \le \chi \le \pi/4$, and $0 \le \tau < \pi$.

These are interesting results. They show that the states of polarization for the incident field that produce a maximum reception (γ_r, δ_r) and a null reception (γ_n, δ_n) for the receiving antenna are simply related to the state of polarization of the field that is radiated when the antenna is transmitting (γ_t, δ_t).[7] This relationship is illustrated in Figure 2.17, where representative polarization ellipses for three fields (t, r, and n) are shown. For maximum reception the polarization ellipse of the incident field (r) has the same shape, orientation, and sense of rotation with respect to the direction of propagation (both are left-handed in the figure) as the polarization ellipse of the transmitted field (t). For null reception the polarization ellipse of the incident field (n) has the same shape, an orthogonal orientation (the major axes are orthogonal),

[7] The subscript t is used to indicate the state of polarization of the radiated field for the transmitting antenna. The subscripts r and n are used to indicate the states of polarization of the incident field for maximum reception and null reception, respectively, for the receiving antenna.

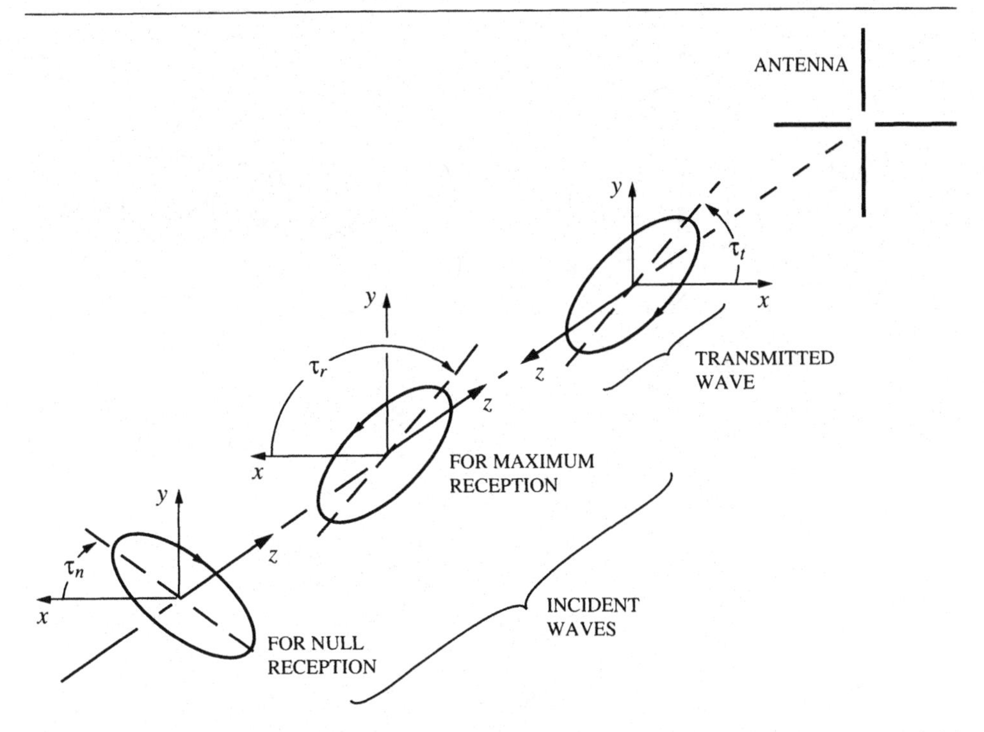

Fig. 2.17. Polarization ellipses for the field of the transmitting antenna (t), the incident field for maximum reception (r), and the incident field for null reception (n).

and an opposite sense of rotation (right-handed in the figure) as the polarization ellipse of the field for maximum reception (r).

The normalized Poincaré sphere is convenient for visualizing the states of polarization for an antenna. The states of polarization for transmission (t) and for maximum reception (r) are shown on the Poincaré sphere in Figure 2.18a, and the states of polarization for maximum reception (r) and null reception (n) are shown on the Poincaré sphere in Figure 2.18b. Note that the points r and n are diametrically opposite on the Poincaré sphere. Thus, we see that an antenna that has maximum reception for a horizontally (vertically) polarized incident field will have null reception for a vertically (horizontally) polarized incident field. Similarly, if maximum reception is for a left-handed (right-handed) circularly polarized wave, there will be null reception for a right-handed (left-handed) circularly polarized wave.

Consider an incident wave with an arbitrary state of polarization (γ_i, δ_i) represented by the point i on the normalized Poincaré sphere in Figure 2.18b. The points r and i are joined by the great-circle arc of length s. We will now show that the polarization mismatch factor p_r (2.103) is a function of the arc length s, decreasing with increasing s. Thus, the polarization mismatch factor is the same for all of the states of polarization represented by points on a circle centered at r on the Poincaré sphere. This is illustrated in Figure 2.18b. Note that as s increases and p_r decreases, the circle moves away from the point r. Eventually, when $s = \pi$ the circle reduces to the point n and $p_r = 0$.

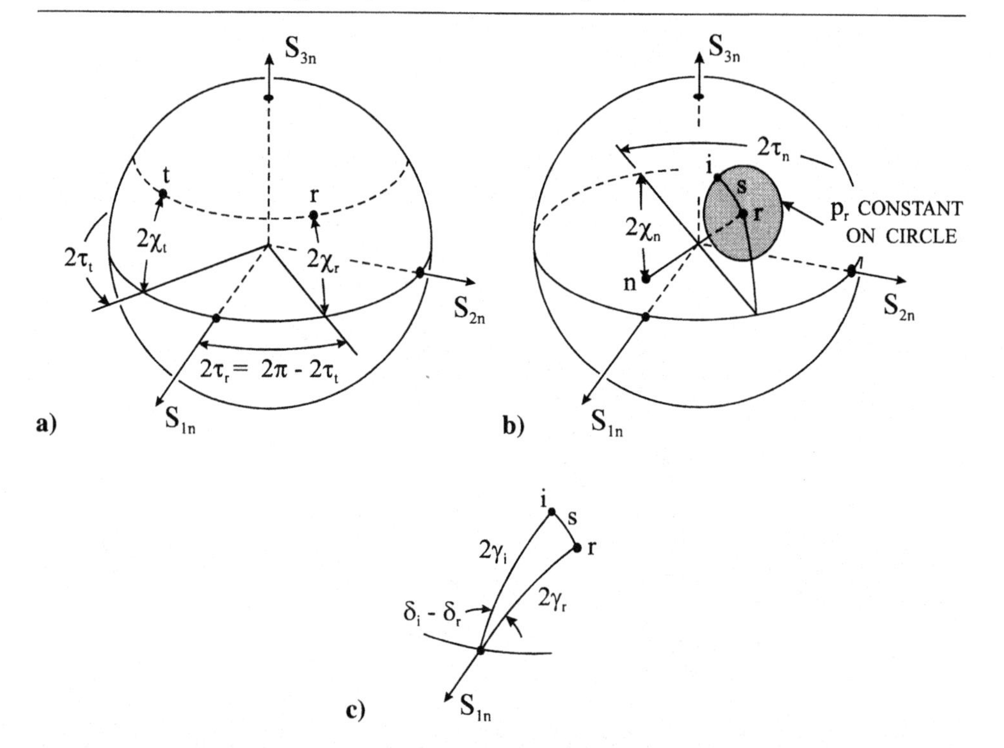

Fig. 2.18. a) and b) States of polarization on the normalized Poincaré sphere: transmitting antenna (t), receiving antenna – maximum reception (r), receiving antenna – null reception (n), and general incident wave (i). c) Detail of spherical triangle.

After introducing the angles γ_r and δ_r and using trigonometric identities, (2.103) becomes

$$p_r = \frac{1}{2}\big[1 + \cos(2\gamma_r)\cos(2\gamma_i) + \sin(2\gamma_r)\sin(2\gamma_i)\cos(\delta_i - \delta_r)\big]. \quad (2.106)$$

For the spherical triangle shown in Figure 2.18c, we find from the cosine rule for sides that [7]

$$\cos(s) = \cos(2\gamma_r)\cos(2\gamma_i) + \sin(2\gamma_r)\sin(2\gamma_i)\cos(\delta_i - \delta_r). \quad (2.107)$$

A comparison of (2.106) with (2.107) shows that

$$p_r = \frac{1}{2}\big[1 + \cos(s)\big] = \cos^2(s/2), \quad (2.108)$$

proving our assertion that the polarization mismatch factor is a function of the arc length s.[8]

We have examined a very simple antenna, the pair of dipoles shown in Figure 2.16a, for a specific direction of propagation ($x = 0$, $y = 0$, z), and we have developed interesting relationships between the states of polarization for the fields transmitted and received by this antenna. These results are not restricted to this

[8] Because we are dealing with a sphere of unit radius, the arc length s is the same as the angle subtended by the arc at the center of the sphere. Some authors use this angle in the expression for p_r.

antenna or to a particular direction of propagation; they apply for any antenna and for a general direction of propagation [13]. We will now summarize these results for a general antenna. The radiated field of a transmitting antenna has the state of polarization γ_t, δ_t in the direction determined by the angles θ, ϕ. When the antenna is receiving an incident plane wave from the same direction θ, ϕ, there will be maximum response, $p_r = 1$, for the state of polarization of the incident field γ_r, δ_r (2.104), and there will be a null response, $p_r = 0$, for the state of polarization γ_n, δ_n (2.105). These two states of polarization (r and n) are at diametrically opposite points on the Poincaré sphere. For a general state of polarization γ_i, δ_i of the incident field, the polarization mismatch factor is $p_r = \cos^2(s/2)$, where s is the length of the great-circle arc joining the points that represent the states of polarization γ_i, δ_i and γ_r, δ_r on the normalized Poincaré sphere.

It follows that for two antennas (1 and 2) used in a communications link, their states of polarization must be related by (2.104) for maximum reception on the link ($\gamma_{t2} = \gamma_{t1}$; $\delta_{t2} = \pi - \delta_{t1}, 0 \le \delta_{t1} \le \pi$; $\delta_{t2} = -\pi - \delta_{t1}, -\pi < \delta_{t1} < 0$). A few examples will illustrate this relationship. If antenna 1 transmits a linearly polarized wave with $\gamma_{t1} = \pi/4$, $\delta_{t1} = 0$ in the direction of antenna 2, then antenna 2 should transmit a linearly polarized wave with $\gamma_{t2} = \gamma_{t1} = \pi/4$, $\delta_{t2} = \pi - \delta_{t1} = \pi$ in the direction of antenna 1. If antenna 1 transmits a left-handed circularly polarized wave ($\gamma_{t1} = \pi/4, \delta_{t1} = \pi/2$) in the direction of antenna 2, then antenna 2 should transmit a left-handed circularly polarized wave ($\gamma_{t2} = \gamma_{t1} = \pi/4, \delta_{t2} = \pi - \delta_{t1} = \pi/2$) in the direction of antenna 1.

2.7 Historical note: The experiments of Hertz

In this chapter, starting with only Maxwell's equations, we showed that the electromagnetic field in free space can propagate as a plane wave with a velocity equal to the speed of light. The vectors for the electric and magnetic fields of the plane wave are transverse to the direction of propagation, and for the monochromatic plane wave, the field can assume various states of polarization.

One of the great and immediate achievements of Maxwell's theory was that it predicted the wave nature of the electromagnetic field, which we have described above. The theory showed that optics was a branch of electromagnetism and that many of the phenomena that were observed for light waves should be present for electromagnetic waves of lower frequency. Maxwell's theory and its predictions were not accepted by all of his contemporaries, and indisputable experimental proof for the theory did not come until some fifteen years after the publication of Maxwell's celebrated treatise (1873). This proof was provided by a series of ingenious experiments performed by the German physicist Heinrich R. Hertz during the period 1887–1888. These experiments represent one of the greatest achievements in the development of electrodynamics, and we will now give a brief description of a few of the more important experiments that deal with electromagnetic wave propagation in free space [14–18].

At the time of Hertz, there was no standard instrumentation for generating or detecting high-frequency electromagnetic signals, and he had to develop suitable

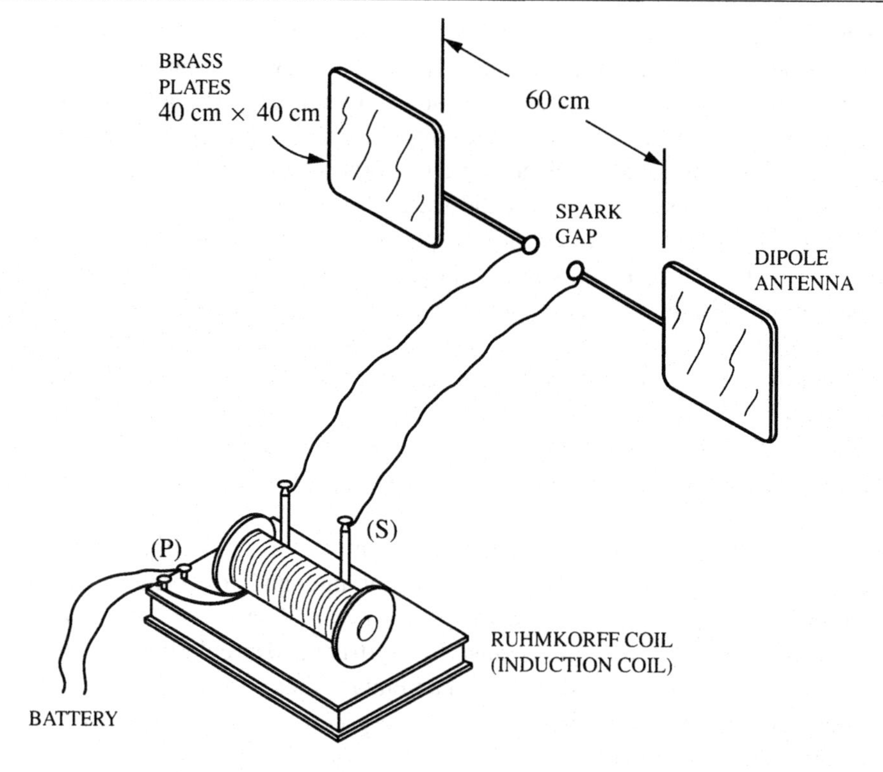

Fig. 2.19. Hertz's spark-gap transmitter.

transmitters and detectors before he could make observations on electromagnetic waves. A sketch of Hertz's spark-gap transmitter is shown in Figure 2.19. The Ruhmkorff coil is a large induction coil, the primary of which is connected to a set of batteries. The secondary of the coil is connected to the terminals of a dipole antenna. The antenna is formed from two square (40 cm × 40 cm) brass plates connected by a 60-cm copper wire with the terminals at its center. The operation of the transmitter is outlined in Figure 2.20. A vibrator (S) in the primary of the coil alternately opens and closes the primary circuit. With the vibrator closed, a large current I_p develops in the primary of the coil. When the vibrator opens, this current ceases, the magnetic field in the coil collapses, and the voltage V_s across the secondary rapidly increases. The charge on the plates of the antenna also rapidly builds. When the voltage across the secondary is sufficient to break down the air, a spark appears across the spark gap at the terminals of the antenna. The spark has low resistance, and it essentially short circuits the terminals of the antenna. The antenna is now the only active element in the transmitter circuit. The charge oscillates between the two plates of the antenna, producing an oscillatory current I_a in the wire connecting the plates, as shown in Figure 2.20d. The oscillatory current is damped by the loss due to radiation. The frequency of the oscillation is roughly $\omega_a \approx 1/\sqrt{L_a C_a}$, where L_a is the inductance of the antenna (primarily that of the wire) and C_a is the capacitance of the antenna (primarily due to the plates).

The sequence described above is repeated each time the vibrator completes one cycle (closes then opens). The radiation from the antenna is not monochromatic, but contains a distribution of frequencies about ω_a.

Hertz used a second antenna, a thin-wire, circular loop, to detect the electromagnetic signal. The loop was 35 cm in radius and had a spark gap at its terminals. The size of the loop was chosen to produce the strongest spark in the gap when the loop had the orientation with respect to the applied electric field shown in Figure 2.21a. With this orientation, the loop can be approximately represented by the equivalent circuit in Figure 2.21b. The current source I_ℓ in this circuit is proportional to the magnitude of the applied electric field $\vec{\mathcal{E}}$, and L_ℓ and C_ℓ are the inductance and the capacitance of the loop, respectively. The voltage V_g across the spark gap

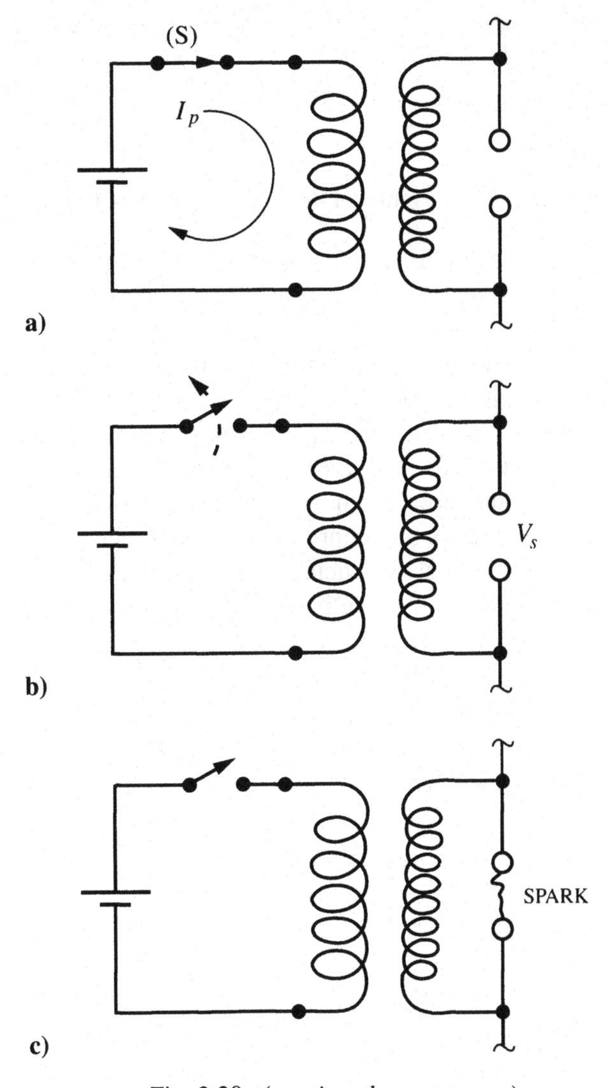

Fig. 2.20. (*continued on next page*)

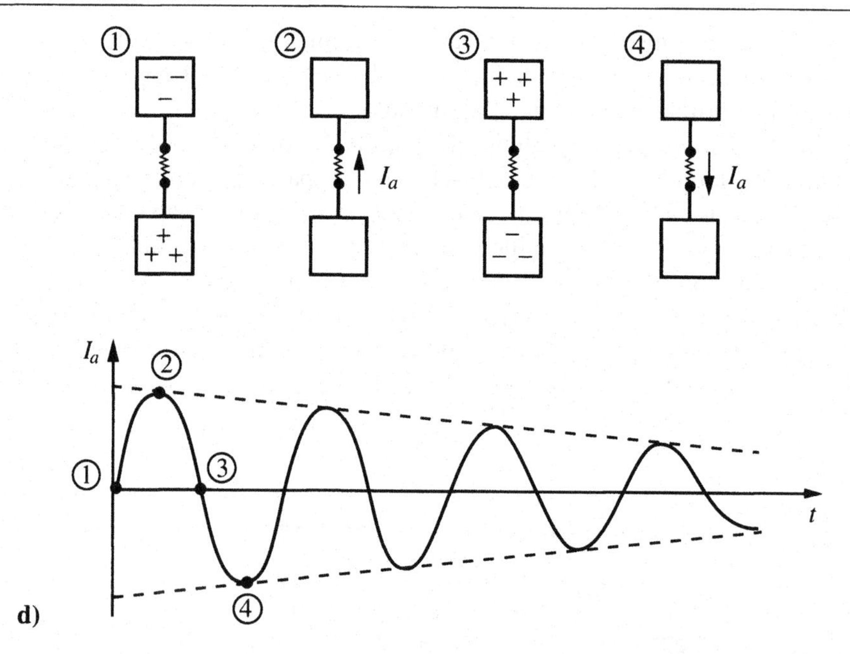

Fig. 2.20. The operation of the spark-gap transmitter.

is maximum at the resonant frequency for the loop, $\omega_\ell \approx 1/\sqrt{L_\ell C_\ell}$.[9] The relative magnitude of the applied field was estimated from the strength of the spark at the gap.

Hertz used the phenomenon of interference to demonstrate the existence of the electromagnetic wave. Figure 2.22a shows a sketch of his arrangement. The dipole antenna was placed about 13 meters from a metal sheet mounted on the wall of the room. The loop detector was used to observe the pattern produced by the interference of the incident wave from the dipole with its reflection from the metal sheet.

Let's assume that the source is monochromatic and the metal sheet is at a large distance from the dipole, so that the incident field is approximately a linearly polarized, normally incident, plane wave in the vicinity of the sheet. We will also assume that the metal sheet is perfectly conducting and of infinite extent. The reflected field is then also a linearly polarized plane wave, but traveling in the direction opposite to the incident wave and with the electric field pointing in the opposite direction to that of the incident wave at the sheet. For the coordinates shown in Figure 2.22a, the superposition of the electric fields for the incident and reflected waves is

$$\begin{aligned}\vec{\mathcal{E}}(z,t) = \mathcal{E}_x(z,t)\hat{x} &= A_o \cos\left[\omega(t - z/c)\right]\hat{x} - A_o \cos\left[\omega(t + z/c)\right]\hat{x} \\ &= 2A_o \sin(\omega t)\sin(k_o z)\hat{x}. \qquad (2.109)\end{aligned}$$

[9] The loops used by Hertz were nearly self resonant, which means their circumference was about one half of a wavelength. Loops this large have properties quite different from those of electrically small loops; for example, their impedance (Figure 2.21b) is not a simple inductance. The reception for loops this large is sensitive to the orientation of the loop with respect to both the electric and the magnetic fields, and it can be quite complicated [19].

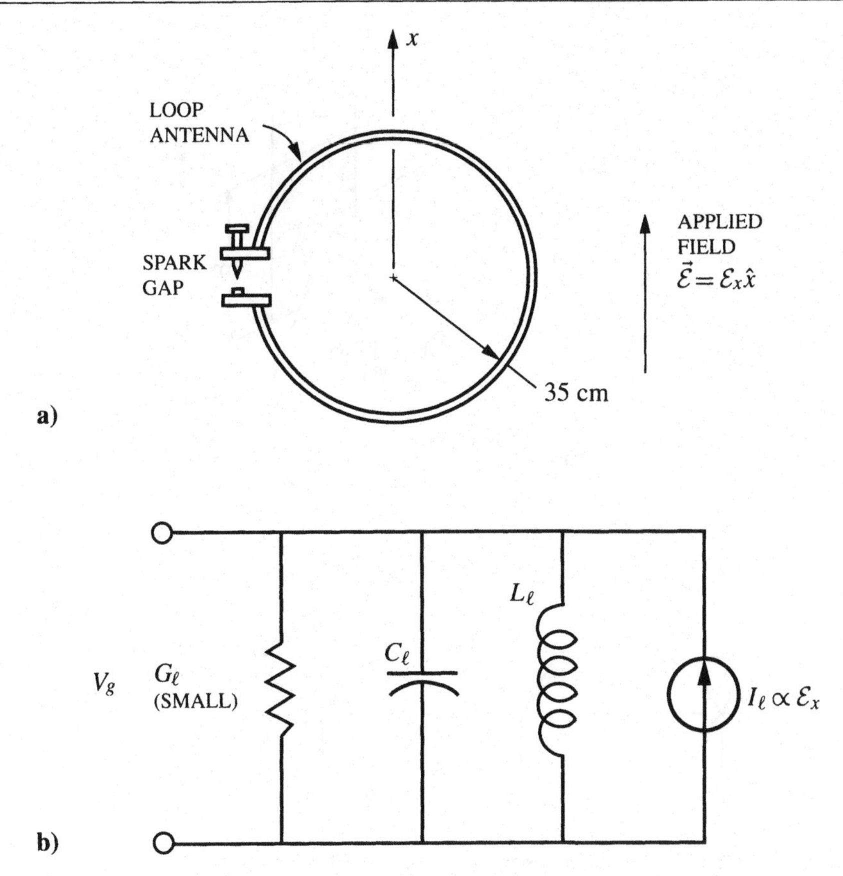

Fig. 2.21. a) Hertz's receiving loop. b) Approximate equivalent circuit for the receiving loop.

This is a *standing wave* whose spatial distribution for various times is sketched in Figure 2.22b. The magnitude of the distribution is seen to vary periodically in time, but the form of the distribution $\sin(k_o z)$ is fixed. In particular, the nodes of the distribution are spaced at half-wavelength intervals starting at the sheet.

In Hertz's experiment, the source was not monochromatic, the waves were not plane, and the metal sheet was not infinite in extent. Nevertheless, the interference patterns he determined from his measurements were very similar to the one described above. A sketch of one of Hertz's patterns is shown in Figure 2.22c. Notice that a null does not occur precisely at the surface of the sheet. From these patterns, Hertz estimated the wavelength of the radiation he measured to be $\lambda_o = 9.6$ m.

After Hertz had successfully demonstrated the existence of electromagnetic waves with the apparatus just described, he constructed improved apparatus for observing electromagnetic waves of shorter wavelength – $\lambda_o \approx 66$ cm ($f \approx 455$ MHz). Both the transmitting and receiving antennas were dipoles positioned on the focal lines of parabolic reflectors to concentrate the waves (see Figure 2.23). With this higher frequency apparatus, Hertz showed that electromagnetic waves produce many of the phenomena associated with light.

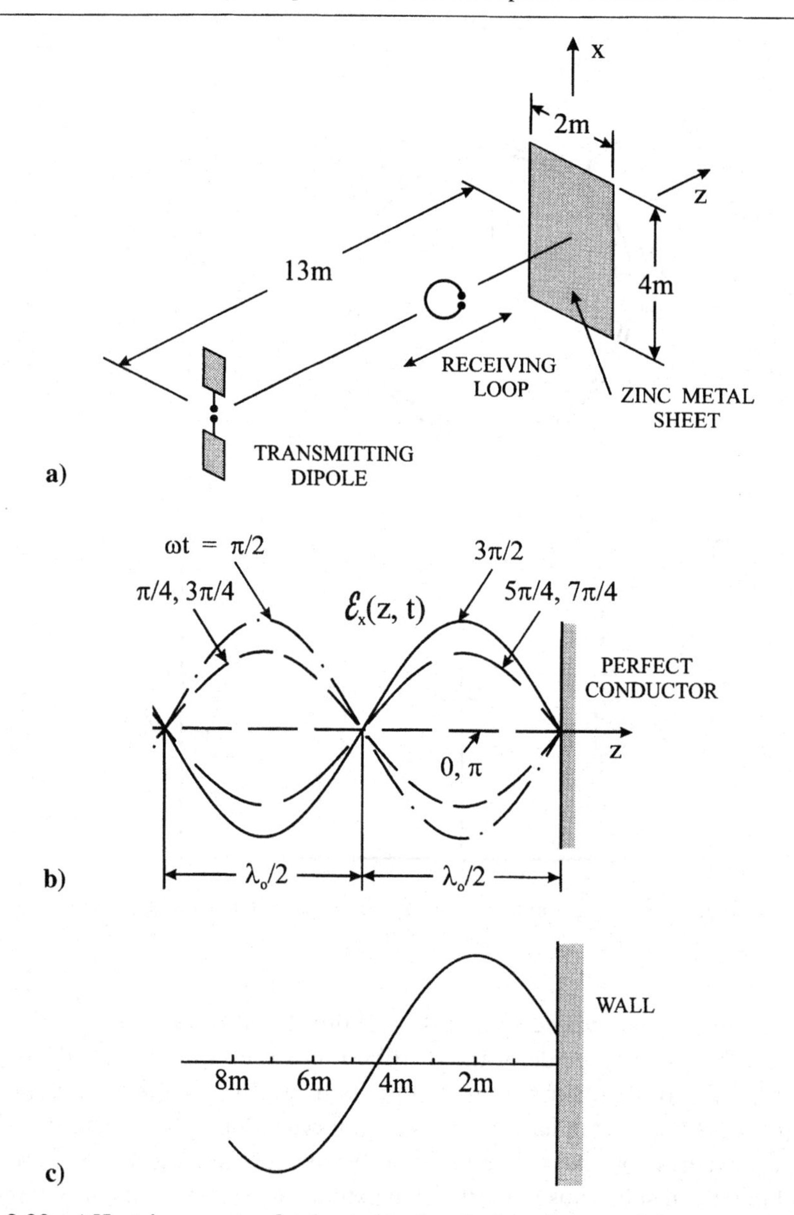

Fig. 2.22. a) Hertz's apparatus for demonstrating the interference of electromagnetic waves. b) Standing wave formed when a plane wave is reflected from a perfectly conducting plane of infinite extent. c) Example of the interference patterns observed by Hertz.

Hertz formed a slit using two large coplanar metal sheets, and with the transmitting and receiving antennas on opposite sides of the slit, he observed diffraction phenomena, such as we will discuss in Section 3.4. He constructed a large prism from hard pitch and showed that electromagnetic waves were refracted by the prism, just as light waves are refracted by a glass prism.

Hertz used a grating of parallel wires to demonstrate the polarization of electromagnetic waves. The wires of the grating were 1 mm in diameter and spaced 3 cm

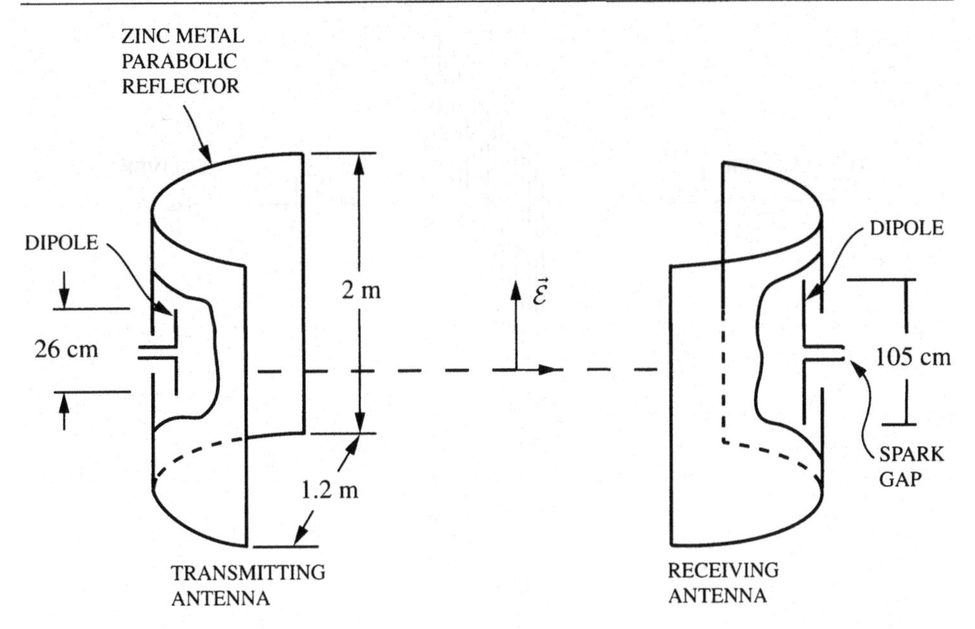

Fig. 2.23. Hertz's apparatus for observing electromagnetic waves of shorter wavelength ($\lambda_o \approx 66$ cm).

apart. When the grating was placed between the transmitting and receiving antennas, as in Figure 2.24, the reception was a function of the orientation of the grating. With the wires of the grating parallel to the electric field of the wave (Figure 2.24a) there is no reception. In this case the incident electric field produces axial electrical currents in the wires. These currents radiate a field that nearly cancels the incident field in the forward direction (to the right of the grating in Figure 2.24a). The grating effectively reflects the incident wave back toward the transmitting antenna. With the wires of the grating perpendicular to the electric field of the wave (Figure 2.24b) the field produces no significant axial electric currents in the wires, and the grating has very little effect on the reception.

Note the similarity between Hertz's grating of wires and the H-sheet polarizer discussed in Section 2.5. Both devices work on the same principle, but at very different frequencies (radio versus optical frequencies). The only difference is that the conducting wires of the grating reflect the incident wave, whereas the resistive atomic strings in the polarizer absorb the incident wave.

Perhaps the best statement of what Hertz's experiments accomplished is his own conclusion [16]:

> Casting now a glance backwards we see that by the experiments above sketched the propagation in time of a supposed action-at-a-distance is for the first time proved. This fact forms the philosophic result of the experiments; and, indeed, in a certain sense the most important result. The proof includes a recognition of the fact that the electric forces can disentangle themselves from material bodies, and can continue to subsist as conditions or changes in the state of space. The details of the experiments further prove that the particular manner in which the electric force is propagated exhibits the closest analogy with the propagation

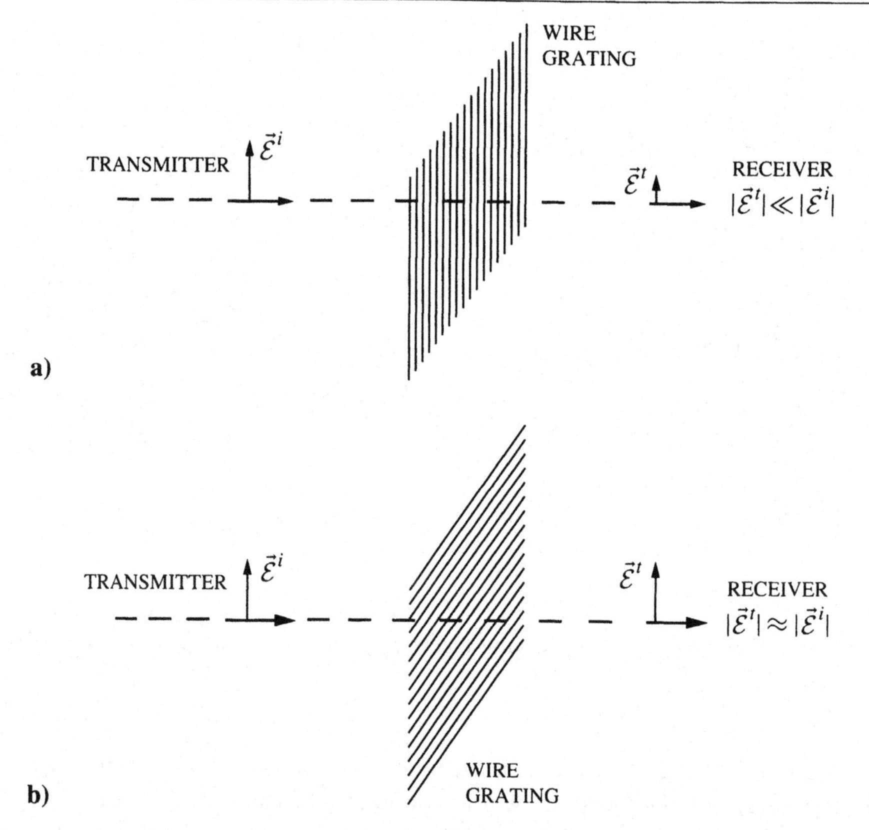

Fig. 2.24. Wire grating for demonstrating the polarization of electromagnetic waves. a) Almost complete blockage of wave by grating. b) Very little blockage of wave by grating.

of light; indeed, that it corresponds almost completely to it. The hypothesis that light is an electrical phenomenon is thus made highly probable. To give a strict proof of this hypothesis would logically require experiments upon light itself.

References

[1] M. Born and W. Wolf, *Principles of Optics*, 3rd Edition, Sec. 1.4, Pergamon Press, New York, 1965.

[2] W. Swindell, editor, *Polarized Light*, Benchmark Papers in Optics/1, Halstead Press, New York, 1975. This reference contains papers of historical interest; included are [6, 8, 10, 11], and [12].

[3] J. D. Kraus, *Radio Astronomy*, Ch. 4, McGraw-Hill, New York, 1966.

[4] *IEEE Standard Definitions of Terms for Radio Wave Propagation*, The Institute of Electrical and Electronics Engineers, Inc., Std. 211–1977, 1977.

[5] A. J. Pettrofrezzo, *Matrices and Transformations*, Ch. 4, Prentice-Hall, Englewood Cliffs, NJ, 1966. Republication, Dover Publications, New York, 1978.

[6] H. Poincaré, *Théorie Mathématique de la Lumière*, Vol. 2, Ch. 12, Gauthier-Villars, Paris, 1892.

[7] W. Gellert, H. Kustner, M. Hellwich, and H. Kastner, *The VNR Concise Encyclopedia of Mathematics*, Ch. 12, Van Nostrand Reinhold, New York, 1977.

[8] G. G. Stokes, "On the Composition and Resolution of Streams of Polarized Light from Different Sources," *Trans. Cambridge. Philos. Soc.*, Vol. 9, pp. 399–416, 1852. Reprinted in *Mathematical and Physical Papers*, Volume III, pp. 233–58, Johnson Reprint Corp., New York, 1966.
[9] W. A. Shurcliff, *Polarized Light: Production and Use*, Harvard University Press, Cambridge, MA, 1966.
[10] R. C. Jones, "A New Calculus for the Treatment of Optical Systems," Parts I through VIII, *J. Opt. Soc. Am.*, I–III, Vol. 31, pp. 488–503, July 1941; IV, Vol. 32, pp. 486–93, August 1942; V–VI, Vol. 37, pp. 107–12, February 1947; VII, Vol. 38, pp. 671–85, August 1948; VIII, Vol. 46, pp. 126–31, February 1956. Part II is coauthored with H. Hurwitz, J.
[11] E. H. Land and C. D. West, "Dichroism and Dichroic Polarizers," in *Colloid Chemistry*, J. Alexander, editor, Vol. 6, pp. 160–90, Reinhold, New York, 1946.
[12] E. H. Land, "Some Aspects of the Development of Sheet Polarizers," *J. Opt. Soc. Am.*, Vol. 41, pp. 957–63, December 1951.
[13] G. A. Deschamps, "Techniques for Handling Elliptically Polarized Waves with Special Reference to Antennas, Part II – Geometrical Representation of the Polarization of a Plane Electromagnetic Wave," *Proc. IRE*, Vol. 39, pp. 540–4, May 1951.
[14] H. Hertz, "Ueber electrodynamische Wellen im Luftraume und deren Reflexion," *Ann. Physik*, Vol. 34, pp. 609–23, 1888.
[15] — "Ueber Strahlen electrischer Kraft," *Ann. Physik*, Vol. 36, pp. 764–83, 1889.
[16] — *Electric Waves*, Macmillan, London, 1893. Republication, Dover Publications, New York, 1962.
[17] G. W. Pierce, *Principles of Wireless Telegraphy*, Ch. VIII, McGraw-Hill, New York, 1910.
[18] J. A. Fleming, *The Principles of Electric Wave Telegraphy*, 3rd Edition, Longmans Green, London, 1916.
[19] R. W. P. King, "The Loop Antenna for Transmission and Reception," in *Antenna Theory*, R. E. Collin and F. J. Zucker, editors, Part I, Ch. 11, McGraw-Hill, New York, 1969.
[20] J. W. Evans, "The Birefringent Filter," *J. Opt. Soc. Am.*, Vol. 39, pp. 229–42, March 1949.

Problems

2.1 Consider the electric field of a plane wave with elliptical polarization:

$$\vec{E}(z) = [A_x e^{j\phi_x}\hat{x} + A_y e^{j\phi_y}\hat{y}]e^{-jk_o z}.$$

Show that this field can be written as the sum of a right-handed circularly polarized wave and a left-handed circularly polarized wave.

2.2 Show that the expression for $\vec{B}(\vec{r})$ given in (2.33b) follows from Maxwell's equations and the expression for $\vec{E}(\vec{r})$ given in (2.33a).

2.3 Verify that the matrix product in Equation (2.44) is the equation for the polarization ellipse (2.42).

2.4 Two electromagnetic waves, one with frequency ω_1 and the other with frequency ω_2 ($\omega_1 \neq \omega_2$), propagate in the direction of the positive z axis. The electric field for each wave is

$$\vec{\mathcal{E}}_i(z,t) = A_{xi}\Big\{\cos\big[\omega_i t - k_o(\omega_i)z + \phi_{xi}\big]\hat{x}$$

$$+ \tan(\gamma_i)\cos\big[\omega_i t - k_o(\omega_i)z + \phi_{xi} + \delta_i\big]\hat{y}\Big\}, \qquad i = 1, 2.$$

a) Determine the real Poynting vectors $\vec{S}(z,t)$ for the field $\vec{\mathcal{E}}_1$ alone, for the field $\vec{\mathcal{E}}_2$ alone, and for the combination $\vec{\mathcal{E}}_c = \vec{\mathcal{E}}_1 + \vec{\mathcal{E}}_2$.

b) An instrument measures the intensity $I = |\langle\vec{S}(\vec{r},t)\rangle|$ of the wave:

$$\big\langle\vec{S}(\vec{r},t)\big\rangle = \frac{1}{T}\int_0^T \vec{S}(\vec{r},t)dt.$$

The time T is very large compared to the period of either wave: $T \gg T_1 = 2\pi/\omega_1$, $T \gg T_2 = 2\pi/\omega_2$, and $T \gg T_1T_2/(T_2 - T_1)$. Show that

$$I_c \approx I_1 + I_2,$$

that is, the response of the instrument to the combination of fields is the same as the sum of the responses to the individual fields.

c) If n plane-wave fields with different frequencies are combined, and the above inequalities are satisfied for all waves, will

$$I_c \approx \sum_{i=1}^{n} I_i \ ?$$

2.5 Obtain the expressions for the Stokes parameters S_1, S_2, and S_3 in Equations (2.72b–d) from the corresponding expressions in Equations (2.70a–c). Hint: Use the relations (2.64)–(2.67) and the following trigonometric formulas:

$$\sin(2\psi) = \frac{2\tan(\psi)}{1 + \tan^2(\psi)}$$

$$\cos(2\psi) = \frac{1 - \tan^2(\psi)}{1 + \tan^2(\psi)}$$

$$\tan(2\psi) = \frac{2}{\cot(\psi) - \tan(\psi)}.$$

2.6 Consider the plane that contains the equator of the normalized Poincaré sphere, the S_{1n}-S_{2n} plane. Show that the polarization chart in Figure 2.7 is an orthographic projection of the lines of constant γ, δ, χ, and τ from the upper hemisphere of the Poincaré sphere onto this plane. For the orthographic projection, a line perpendicular to the plane and passing through the point (γ, δ) on the sphere locates the point (γ, δ) on the plane. Obtain the equations that describe lines of constant γ, δ, χ, and τ on this plane.

2.7 A commercial, two-element, left-handed circular polarizer is formed from a sheet of linear polarizer and a sheet that is a quarter-wave plate. The construction is the same as that for the three-element polarizer in Figure 2.11, only the first quarter-wave plate, the one that is the closest to the source, is absent.

a) Determine the Jones matrix for the circular polarizer and verify that the device produces a left-handed circularly polarized wave for an incident wave with general polarization.
b) Now assume that the orientation of the polarizer is reversed, so that the incident field first encounters the quarter-wave plate (that is, both elements of the device are rotated through 180° about the fast axis of the quarter-wave plate). Determine the Jones matrix for this device. How does this device now affect the state of polarization of the incident wave?

2.8 Two linear polarizers are placed in series. The first polarizer has its transmission axis at an angle of $\psi = 45°$ with respect to the x axis, and the second polarizer has its transmission axis in the direction of the y axis.

a) Determine the normalized Jones matrix (normalized transmission matrix) for the combination.
b) The electric field of the incident light is linearly polarized in the direction of the x axis, with the normalized Jones vector

$$E_n^i = \begin{bmatrix} 1 \\ 0 \end{bmatrix}.$$

Show that the electric field of the transmitted light is linearly polarized in the direction of the y axis. Thus, this combination of linear polarizers has changed the linear polarization of the light from horizontal to vertical.

2.9 Starting from Equation (2.103) obtain the values of γ and δ, given in (2.104) and (2.105), for which p_r is a maximum ($p_r = 1$) and a minimum ($p_r = 0$). Only consider the case for $0 \leq \delta_t \leq \pi$.

2.10 a) The state of polarization of an incident plane wave is to be determined using a single dipole receiving antenna. The dipole and wave have the orientation shown in Figure P2.1a. A device placed at the terminals of the dipole measures the modulus of the received voltage $|V_{oc}|$. The dipole is in the x'-y' plane, and it is rotated slowly about the z' axis. Let ψ be the angle the dipole makes with the x' axis. Obtain an expression for $|V_{oc}|$ as a function of the angle ψ.
b) Assume that the incident wave has the state of polarization given by the following parameters: $A_x^i = 1.0$, $\gamma_i = 33.21°$, $\delta_i = 40.90°$. Also let $|h_e| = 1.0$ m. Use the expression obtained in part a to make a polar graph with $|V_{oc}|$ the radial variable and ψ the angular variable, as indicated in Figure P2.1b.
c) Determine the angles χ_i and τ_i for the incident wave of part b. On your graph, draw the polarization ellipse for the incident wave so that it is circumscribed by the curve of part b.

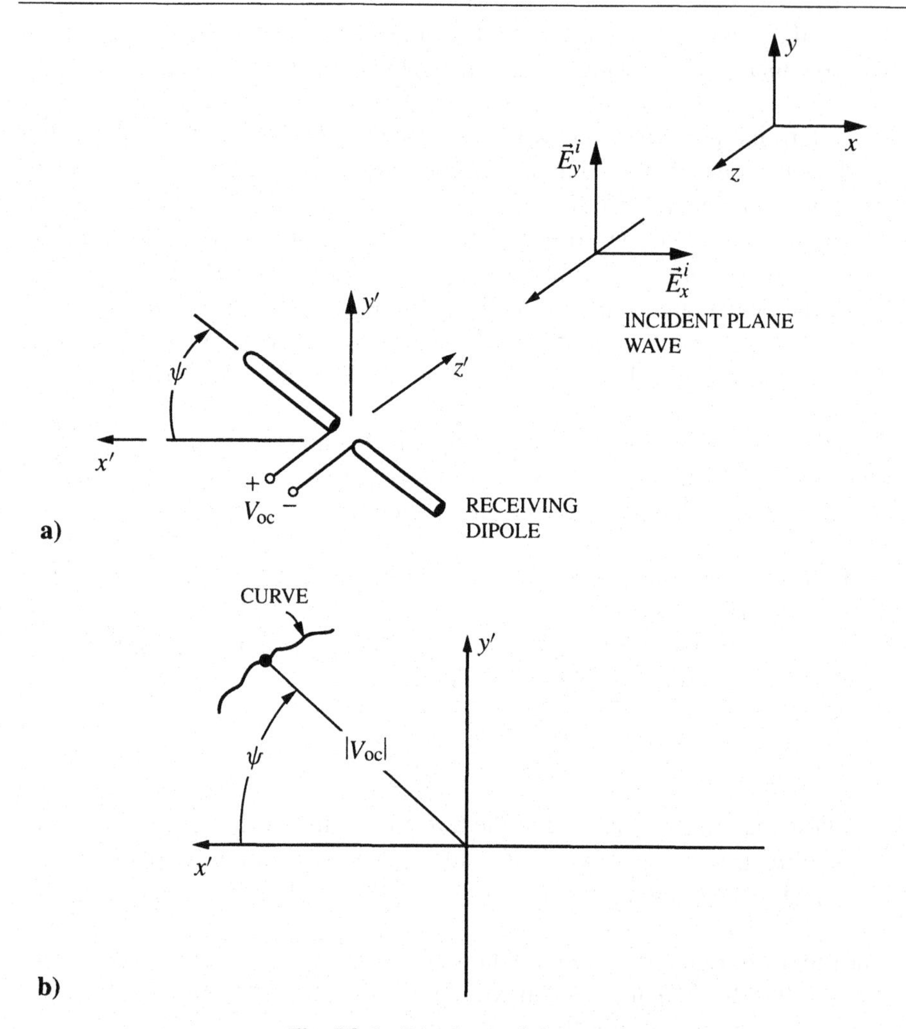

Fig. P2.1. Dipole receiving antenna.

d) Explain how the angles χ_i and τ_i for the incident wave could be determined from the polar graph of $|V_{oc}|$ versus ψ.

e) Make a sketch of the normalized Poincaré sphere and locate the state of polarization for the incident wave of part b on the sphere. Show how the state of polarization for the receiving dipole changes position on the sphere as the angle ψ changes. Indicate how the polarization mismatch factor p_r changes as ψ changes, and use this to explain the shape of the curve obtained in part b.

2.11 An incident plane wave has the state of polarization specified by the angles $\chi_i = 0$, $\tau_i = \pi/8$. Two antennas are available for use in receiving this signal. The first is a linearly polarized antenna when transmitting with $\chi_{t1} = 0$, $\tau_{t1} = 0$, and the second is a left-handed circularly polarized antenna when transmitting with $\chi_{t2} = \pi/4$.

a) On the Poincaré sphere locate the state of polarization for the incident wave and the states of polarization for transmission (χ_t, τ_t) and maximum reception (χ_r, τ_r) for the two antennas.
b) Use the Poincaré sphere to determine which of the two antennas will provide the best reception (the largest p_r) for the specified incident wave.

2.12 Consider a communications link with one antenna on the Earth and the second antenna on a satellite. The wave propagating between the antennas must pass through the ionosphere of the Earth. This is a plasma composed of free electrons, ions, and neutral particles. The ionosphere when biased by the Earth's magnetic field forms an anisotropic medium. As the wave passes through this medium, its state of polarization is changed. This change may be as simple as a rotation of the polarization ellipse for the wave (i.e., a change in the tilt angle of the ellipse from τ to $\tau + \Delta\tau$) on passing through the ionosphere. This phenomenon, known as *Faraday rotation*, is most important for propagation through the ionosphere at frequencies below about 7 GHz. The effect is nonreciprocal in that the sense of rotation (right-handed or left-handed) for waves propagating from the Earth to the satellite is opposite to that for waves propagating from the satellite to the Earth.

a) Assume that the two antennas are linearly polarized and aligned so that the polarization mismatch factor is maximum, $p_r = 1$, in the absence of the ionosphere. What will the polarization mismatch factor be when the ionosphere produces a Faraday rotation of $\Delta\tau$?
b) Let the two antennas of part a be circularly polarized, either left- or right-handed. What effect does the Faraday rotation now have on the polarization mismatch factor?

2.13 An antenna transmits a wave with the state of polarization γ_t, δ_t. The radiated field is

$$\vec{E}^t = A_x e^{j\phi_x}[\hat{x} + \tan\gamma_t e^{j\delta_t}\hat{y}].$$

Maximum reception for the same antenna occurs for an incident wave with the state of polarization γ_r, δ_r. The electric field for this wave is

$$\vec{E}^r = A'_x e^{j\phi'_x}[\hat{\underline{x}} + \tan\gamma_r e^{j\delta_r}\hat{\underline{y}}].$$

The relationships between γ_t, δ_t and γ_r, δ_r are given in Equation (2.104). Note that the two fields $\vec{E}^t$ and $\vec{E}^r$ are described by the two different right-handed coordinate systems (x, y, z) and ($\underline{x}$, $\underline{y}$, $\underline{z}$) shown in Figure P2.2, with $\underline{x} = -x$, $\underline{y} = y$, and $\underline{z} = -z$. Show that

$$\vec{E}^r = C(\vec{E}^t)^*,$$

where $*$ indicates the complex conjugate and C is a complex constant. This equation constitutes another way of expressing the relationship between the state of polarization on transmission and the state of polarization for an incident wave that produces maximum reception.

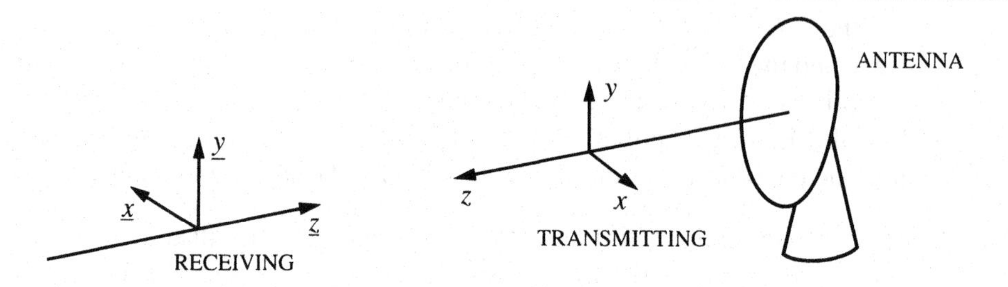

Fig. P2.2. Coordinates for the antenna on transmission and reception.

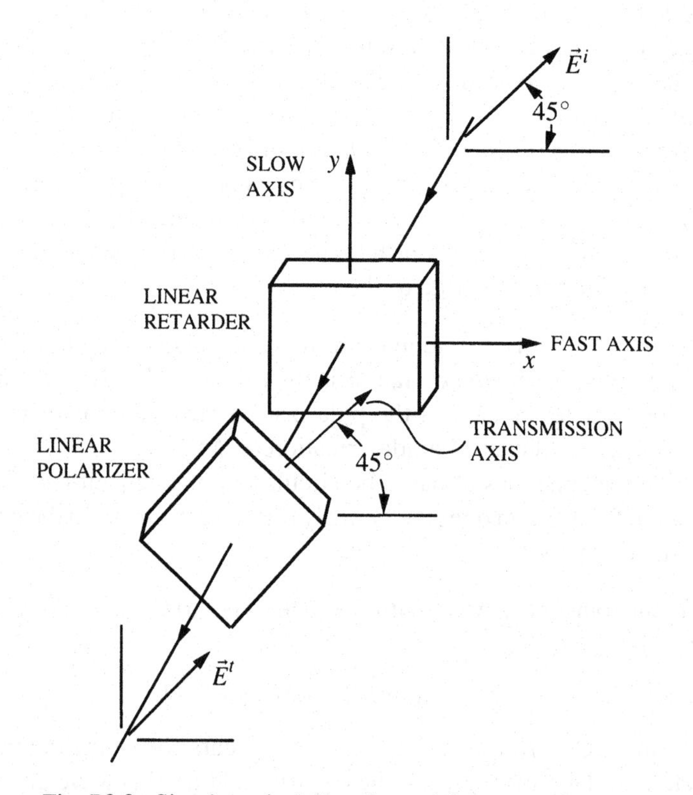

Fig. P2.3. Simple optical filter formed from two elements.

2.14 *Lyot filter* The simple optical filter shown in Figure P2.3 is formed by a linear retarder followed by a linear polarizer [20]. The x and y axes are the fast and slow axes, respectively, of the retarder, and the transmission axis of the polarizer makes an angle of 45° with the x axis.

a) Assume that the electric field $\vec{E}^i$ of the linearly polarized incident wave makes an angle of 45° with the x axis; it has the Jones vector

$$E^i = \frac{1}{\sqrt{2}}\begin{bmatrix} 1 \\ 1 \end{bmatrix}.$$

Use the Jones matrices for the retarder and polarizer to determine the Jones vector for the electric field $\vec{E}^t$ of the transmitted wave. Obtain an expression for the transmission ratio

$$S_0^t/S_0^i$$

that involves only the retardance Δ.

b) The retardance is

$$\Delta = 2\pi f \cdot \delta n \cdot \ell/c,$$

where ℓ is the physical length of the retarder and δn is the difference in the indices of refraction for fields oriented along the slow and the fast axes. Make a graph of the ratio S_0^t/S_0^i versus the frequency f for frequencies in the range

$$m - \frac{1}{2} \le f/\left[c/(\delta n \cdot \ell)\right] \le m + \frac{3}{2},$$

where m is an integer.

Identify the center frequencies f_1 and f_2 for two adjacent pass bands of the filter (i.e., points at which $S_0^t/S_0^i = 1$). What parameters determine the relative spacing $(f_2 - f_1)/f_1$ between these pass bands?

c) The performance of the filter can be improved by using several of the simple filters shown in Figure P2.3 in series. Let there be N of these simple, two-element filters in series, with the retardance of the i-th filter being

$$\Delta_i = 2^{i-1}(2\pi f \cdot \delta n \cdot \ell/c).$$

Determine the transmission ratio S_0^t/S_0^i for this composite. Make a graph of this ratio for $N = 3$ and the same range of frequencies used in part b. What is the effect of adding more elements to the filter?

2.15 In this problem we will consider the antenna shown in Figure 2.16a, which is an orthogonal arrangement of dipoles.

a) When the antennas are *transmitting*, the input impedance to each dipole is Z_A. Let the device that controls the amplitude and the phase of the current in the y-axis dipole be simply an impedance Z_p in parallel with that dipole. For the transmitting arrangement in Figure 2.16b, show that the factor in Equation (2.100) is

$$\tan(\gamma_t)e^{j\delta_t} = Z_p/(Z_A + Z_p).$$

b) Now consider the same arrangement when the antennas are *receiving*. Each dipole can be represented by a Thévenin equivalent circuit that consists of the open-circuit voltage V_{oc} in series with the impedance Z_A. For the receiving arrangement in Figure 2.16b, show that the factor in Equation (2.101) is

$$\tan(\gamma_t)e^{j\delta_t} = Z_p/(Z_A + Z_p).$$

Notice that this is the same result found in part a, so the device (impedance Z_p) produces the same change (amplitude and phase) in the current on transmission as in the open-circuit voltage on reception.

c) The relation obtained in parts a and b must be solved to determine the impedance Z_p that will produce the desired values of γ_t and δ_t. Show that

$$Z_p = Z_A/\left[\cot(\gamma_t)e^{-j\delta_t} - 1\right].$$

Assume that the dipoles are resonant and have the input impedance $Z_A \approx R_A \approx 73\ \Omega$. What should Z_p be for the antenna in Figure 2.16a to produce the following states of polarization on transmission:

right-handed elliptical: $\gamma_t = \pi/8, \quad \delta_t = -\pi/4,$
linear: $\gamma_t = \pi/4, \quad \delta_t = 0,$
right-handed circular: $\gamma_t = \pi/4, \quad \delta_t = -\pi/2.$

Do you see any problems with these answers?

2.16 The following plane waves propagate in the direction of the z axis in free space. For each wave, determine the angles δ, γ, χ, and τ. Make a sketch of the polarization ellipse for each wave, and locate the state of polarization on the normalized Poincaré sphere.

a) $\vec{\mathcal{E}}(z,t) = A_x\{\cos[\omega(t - z/c)]\hat{x} + \cos[\omega(t - z/c) + \pi/4]\hat{y}\}$
b) $\vec{\mathcal{E}}(z,t) = A_x\{\cos[\omega(t - z/c)]\hat{x} + 0.5774\cos[\omega(t - z/c) - 0.9553]\hat{y}\}$
c) $\vec{\mathcal{E}}(z,t) = A_x\{\cos[\omega(t - z/c) + \pi/4]\hat{x} + \sin[\omega(t - z/c) + \pi/4]\hat{y}\}.$

3

Inhomogeneous plane waves and the plane-wave spectrum

3.1 Inhomogeneous plane waves

In the previous chapter, we obtained a wave equation (2.5) for the electromagnetic field in free space from Maxwell's equations expressed in the time domain. Solutions to this equation that are uniform over plane surfaces at any instant of time are plane waves. For the special case of a harmonic time dependence (a monochromatic plane wave), the phasors for the electromagnetic field are uniform in amplitude throughout space and uniform in phase over plane surfaces, and the wave is referred to as a homogeneous or uniform plane wave.

The case for harmonic time dependence could also have been developed by starting with Maxwell's equations for the phasors $\vec{E}(\vec{r})$ and $\vec{B}(\vec{r})$ [(1.169)–(1.172) with (1.6)–(1.8)]:

$$\nabla \times \vec{E}(\vec{r}) = -j\omega\vec{B}(\vec{r}) \tag{3.1}$$

$$\nabla \times \vec{B}(\vec{r}) = \frac{j\omega}{c^2}\vec{E}(\vec{r}) \tag{3.2}$$

$$\nabla \cdot \vec{E}(\vec{r}) = 0 \tag{3.3}$$

$$\nabla \cdot \vec{B}(\vec{r}) = 0. \tag{3.4}$$

These equations can be combined to obtain *vector Helmholtz equations* (Hermann Ludwig Ferdinand von Helmholtz, 1821–1894) for the electric and magnetic fields:

$$\nabla^2\vec{E}(\vec{r}) + k_o^2\vec{E}(\vec{r}) = 0 \tag{3.5}$$

$$\nabla^2\vec{B}(\vec{r}) + k_o^2\vec{B}(\vec{r}) = 0, \tag{3.6}$$

with $k_o = \omega/c$. Alternatively, Equations (3.5) and (3.6) can be obtained by assuming harmonic time dependence in the wave equations (2.5) and (2.6); hence, (3.5) and (3.6) are sometimes referred to as reduced wave equations.

It is easy to show that the previously obtained complex electric field (2.35a) for the monochromatic plane wave,

$$\vec{E} = \vec{E}_o e^{-j\vec{k}\cdot\vec{r}}, \tag{3.7}$$

with

$$\vec{k} \cdot \vec{k} = k_o^2, \tag{3.8}$$

satisfies the Helmholtz equation (3.5) (Problem 3.1). The relationship between $\vec{B}$ and $\vec{E}$ for this wave follows from (3.1):

$$\vec{B} = \frac{1}{\omega}\vec{k} \times \vec{E}, \tag{3.9}$$

and the transverse nature of the wave follows from (3.3) and (3.4):

$$\vec{k} \cdot \vec{E} = 0 \tag{3.10}$$

$$\vec{k} \cdot \vec{B} = 0. \tag{3.11}$$

Consider the case in which the vector wave number $\vec{k}$ is *real*. The planes of constant phase for this wave (3.7) are determined from $\vec{k} \cdot \vec{r} =$ constant and are also planes of constant amplitude. This is the field of the *homogeneous* or *uniform plane wave* that we described in the previous chapter. The time average of the Poynting vector is (1.180)

$$\langle \vec{S} \rangle = \mathrm{Re}(\vec{S}_c) = \frac{1}{2\zeta_o}|\vec{E}_o|^2\hat{k}. \tag{3.12}$$

Thus, the direction for maximum time-average energy flow for the wave is $\hat{k} = \vec{k}/k_o$.

Now consider the case where the vector wave number $\vec{k}$ is *complex* with real and imaginary parts $\vec{k}_r$ and $\vec{k}_i$, respectively:

$$\vec{k} = \vec{k}_r + j\vec{k}_i. \tag{3.13}$$

Since (3.8) still must be satisfied, we have

$$\vec{k} \cdot \vec{k} = (\vec{k}_r \cdot \vec{k}_r - \vec{k}_i \cdot \vec{k}_i) + 2j(\vec{k}_r \cdot \vec{k}_i) = k_o^2, \tag{3.14}$$

which on equating the real and imaginary parts becomes

$$\vec{k}_r \cdot \vec{k}_r - \vec{k}_i \cdot \vec{k}_i = k_o^2 \tag{3.15a}$$

and

$$\vec{k}_r \cdot \vec{k}_i = 0. \tag{3.15b}$$

This last condition (3.15b) requires the real vectors $\vec{k}_r$ and $\vec{k}_i$ to be orthogonal. After introducing the notation

$$k_r = |\vec{k}_r|, \qquad k_i = |\vec{k}_i|, \tag{3.16}$$

(3.15a) becomes

$$k_o^2 = k_r^2 - k_i^2. \tag{3.17}$$

All of the relations that describe the electromagnetic wave for harmonic time dependence (3.5)–(3.11) are valid when $\vec{k}$ is a complex vector. The electric field (3.7) is then

$$\vec{E} = \vec{E}_o e^{\vec{k}_i \cdot \vec{r}} e^{-j\vec{k}_r \cdot \vec{r}}. \tag{3.18}$$

The planes of constant phase for this wave are determined from $\vec{k}_r \cdot \vec{r} =$ constant, and those of constant amplitude are determined from $\vec{k}_i \cdot \vec{r} =$ constant. These two sets of planes are clearly orthogonal in space as a result of (3.15b). A plane wave of this type, one whose planes of constant amplitude and planes of constant phase are not coincident, is called an *inhomogeneous* or *nonuniform plane wave*.[1] The complex Poynting vector for the wave is (see Problem 3.12)

$$\vec{S}_c = \frac{1}{2\zeta_o}\left\{|\vec{E}|^2\frac{\vec{k}_r}{k_o} + \frac{\vec{k}_i}{k_o} \times \mathrm{Im}(\vec{E} \times \vec{E}^*) - j\left[|\vec{E}|^2\frac{\vec{k}_i}{k_o} - \frac{\vec{k}_r}{k_o} \times \mathrm{Im}(\vec{E} \times \vec{E}^*)\right]\right\}, \tag{3.19}$$

and the time average of the Poynting vector is

$$\langle\vec{\mathcal{S}}\rangle = \mathrm{Re}(\vec{S}_c) = \frac{1}{2\zeta_o}e^{2\vec{k}_i\cdot\vec{r}}\left[|\vec{E}_o|^2\frac{\vec{k}_r}{k_o} + \frac{\vec{k}_i}{k_o} \times \mathrm{Im}(\vec{E}_o \times \vec{E}_o^*)\right]. \tag{3.20}$$

From (3.20), we see that there is no time-average energy flow in the direction of $\vec{k}_i$ [$\vec{k}_r$ and $\vec{k}_i \times (\ )$ are both normal to $\vec{k}_i$]. Thus, the maximum time-average energy flow is in a direction normal to $\vec{k}_i$, but not necessarily in the direction of $\vec{k}_r$.

For graphical illustration and comparison of the homogeneous and inhomogeneous plane waves, we will consider the simple case where $\vec{E}_o = A_y\hat{y}$, with $\vec{k} = k_o\hat{z}$ for the homogeneous wave, and $\vec{k} = k_o[(\sqrt{37}/6)\hat{x} - j(1/6)\hat{z}]$ for the inhomogeneous wave. The electromagnetic field (phasor) for the *homogeneous wave* is then

$$\vec{E}(z) = A_y e^{-jk_oz}\hat{y}$$

$$\vec{B}(z) = \frac{-A_y}{c}e^{-jk_oz}\hat{x},$$

or in the time domain

$$\vec{\mathcal{E}}(z,t) = A_y\cos(\omega t - k_oz)\hat{y}$$

$$\vec{\mathcal{B}}(z,t) = \frac{-A_y}{c}\cos(\omega t - k_oz)\hat{x},$$

with

$$\langle\vec{\mathcal{S}}\rangle = \frac{1}{2\zeta_o}|A_y|^2\hat{z}. \tag{3.21}$$

The electromagnetic field (phasor) for the *inhomogeneous wave* is

$$\vec{E}(x,z) = A_y e^{-(1/6)k_oz}e^{-j(\sqrt{37}/6)k_ox}\hat{y}$$

$$\vec{B}(x,z) = \frac{A_y}{c}e^{-(1/6)k_oz}e^{-j(\sqrt{37}/6)k_ox}\{j(1/6)\hat{x} + (\sqrt{37}/6)\hat{z}\},$$

[1] The orthogonality of these planes in the present case is a consequence of the medium, free space, being lossless (k_o real). For an inhomogeneous plane wave in a lossy medium, the planes are not orthogonal.

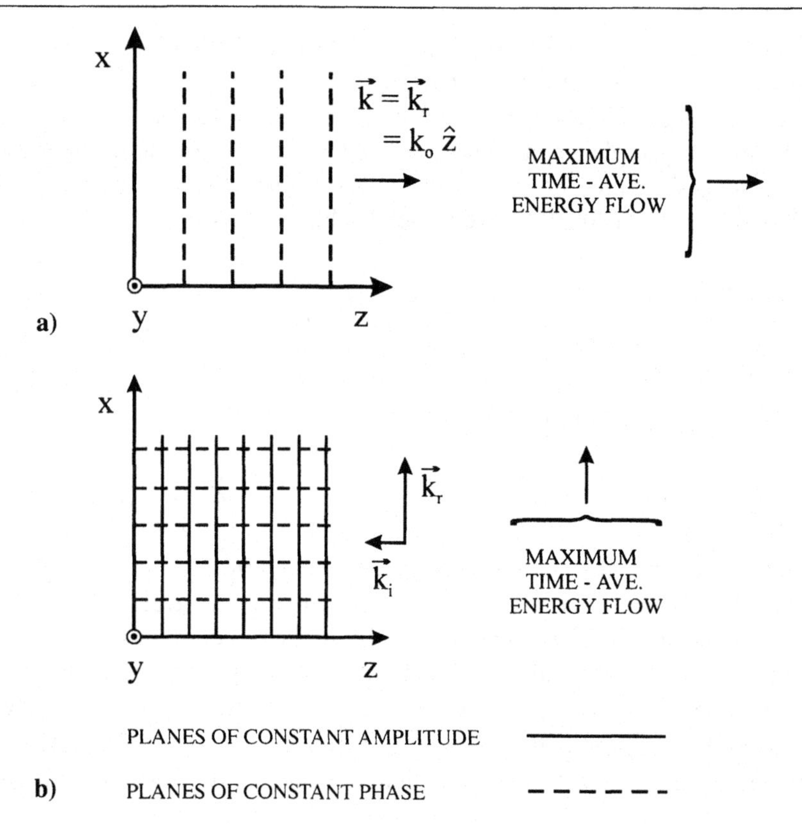

Fig. 3.1. a) Homogeneous plane wave. b) Inhomogeneous plane wave.

or in the time domain

$$\vec{\mathcal{E}}(x, z, t) = A_y e^{-(1/6)k_o z} \cos\left[\omega t - (\sqrt{37}/6)k_o x\right]\hat{y}$$

$$\vec{\mathcal{B}}(x, z, t) = \frac{A_y}{c} e^{-(1/6)k_o z}\Big\{-(1/6)\sin\left[\omega t - (\sqrt{37}/6)k_o x\right]\hat{x}$$

$$+ (\sqrt{37}/6)\cos\left[\omega t - (\sqrt{37}/6)k_o x\right]\hat{z}\Big\},$$

with

$$\langle\vec{\mathcal{S}}\rangle = \frac{1}{2\zeta_o}\frac{\sqrt{37}}{6}|A_y|^2 e^{-(1/3)k_o z}\hat{x}. \tag{3.22}$$

The relationships between the planes of constant amplitude and the planes of constant phase are shown for both waves in Figure 3.1, and the spatial distributions for the electric fields at the time $t = 0$ are graphed in Figure 3.2.

The homogeneous plane wave propagates and has maximum time-average energy flow in the direction $\vec{k}_r = k_o\hat{z}$. The phase velocity for propagation is $\omega/k_o = c$, the speed of light in free space. The inhomogeneous plane wave has an exponential

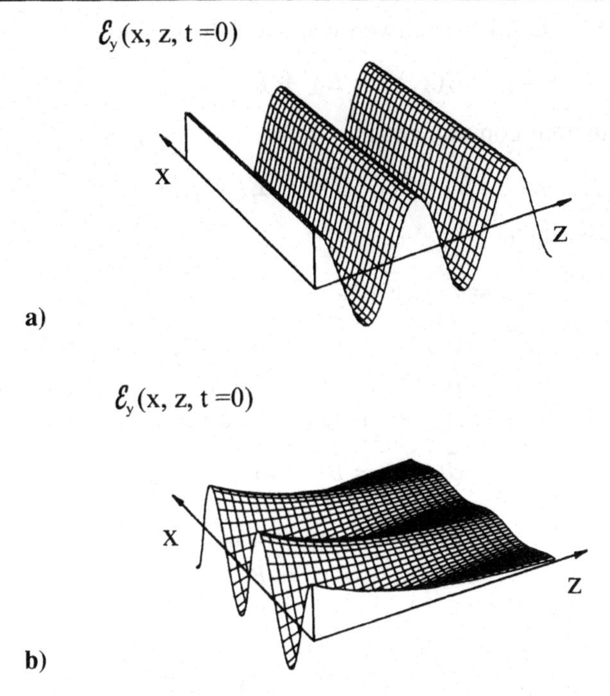

Fig. 3.2. Instantaneous electric field $\mathcal{E}_y(x, z, t = 0)$ for a) homogeneous plane wave and b) inhomogeneous plane wave.

amplitude variation and no time-average energy flow in the direction of $\vec{k}_i = -(1/6)k_o\hat{z}$. Propagation is in the direction of $\vec{k}_r = (\sqrt{37}/6)k_o\hat{x}$, and for this example, the maximum time-average energy flow is also in this direction. The phase velocity for propagation in this direction is $\omega/k_r = (6/\sqrt{37})c$. Notice that this phase velocity is less than the speed of light, as it always must be for the inhomogeneous wave because

$$k_r = \sqrt{k_o^2 + k_i^2} \geq k_o. \tag{3.23}$$

The field of the homogeneous plane wave at any instant of time is uniform over every plane in space with unit normal $\hat{k}$ (the planes $z =$ constant in Figure 3.2a). No similar planes exist for the field of the inhomogeneous plane wave; this is clear from Figure 3.2b. If we had insisted on defining a plane wave as one whose field is *instantaneously* uniform over plane surfaces in space, then the term plane wave would be a misnomer for the inhomogeneous wave.

3.2 Two-dimensional, transverse electric and transverse magnetic fields

In the next few sections, we will be dealing with electromagnetic fields that depend upon only two of the three spatial coordinates of the rectangular Cartesian system (x, y, z). These two-dimensional fields will be chosen to be independent of the coordinate y [e.g., $\vec{E}(x, z)$]. Thus, the partial derivative $\partial/\partial y$ in Maxwell's equations will be zero.

The electric field is divided into two components:

$$\vec{E}(x,z) = \vec{E}_\perp + \vec{E}_\parallel, \tag{3.24a}$$

where the perpendicular component is

$$\vec{E}_\perp = (\hat{y}\cdot\vec{E})\hat{y} = E_\perp\hat{y} \tag{3.24b}$$

and the parallel component is

$$\vec{E}_\parallel = \vec{E} - \vec{E}_\perp = -\hat{y}\times(\hat{y}\times\vec{E}). \tag{3.24c}$$

The reason for this choice of notation becomes clear when we consider $\vec{E}_\perp$ and $\vec{E}_\parallel$ to be real vectors; $\vec{E}_\perp$ is then perpendicular to the x-z plane, and $\vec{E}_\parallel$ is parallel to the x-z plane. Similarly, the magnetic field can be divided into components:

$$\vec{B}(x,z) = \vec{B}_\perp + \vec{B}_\parallel, \tag{3.25a}$$

where

$$\vec{B}_\perp = (\hat{y}\cdot\vec{B})\hat{y} = B_\perp\hat{y} \tag{3.25b}$$

and

$$\vec{B}_\parallel = \vec{B} - \vec{B}_\perp = -\hat{y}\times(\hat{y}\times\vec{B}). \tag{3.25c}$$

On substitution of (3.24a) and (3.25a) into Maxwell's equations, (3.1) and (3.2), we obtain two sets of equations (Problem 3.3),

$$\nabla\times\vec{E}_\perp = -j\omega\vec{B}_\parallel \tag{3.26a}$$

$$\nabla\times\vec{B}_\parallel = \frac{j\omega}{c^2}\vec{E}_\perp \tag{3.26b}$$

$$\nabla^2 E_\perp + k_o^2 E_\perp = 0 \tag{3.26c}$$

and

$$\nabla\times\vec{B}_\perp = \frac{j\omega}{c^2}\vec{E}_\parallel \tag{3.27a}$$

$$\nabla\times\vec{E}_\parallel = -j\omega\vec{B}_\perp \tag{3.27b}$$

$$\nabla^2 B_\perp + k_o^2 B_\perp = 0. \tag{3.27c}$$

Here the first two equations in each set were used to obtain the two-dimensional, scalar Helmholtz equations for $E_\perp$ and $B_\perp$, (3.26c) and (3.27c).

From these two sets of equations, (3.26a–c) and (3.27a–c), we see that there are two independent fields, each of which satisfies the source-free Maxwell's equations in free space. The first is a two-dimensional field composed of $\vec{E}_\perp$ and $\vec{B}_\parallel$; the second is a two-dimensional field composed of $\vec{E}_\parallel$ and $\vec{B}_\perp$. We will refer to the former as the *transverse electric* (*TE*) field, and use the alternate representation $\vec{E}^{TE}$, $\vec{B}^{TE}$ for $\vec{E}_\perp$, $\vec{B}_\parallel$; the latter will be referred to as the *transverse magnetic* (*TM*) field, with the alternate representation $\vec{E}^{TM}$, $\vec{B}^{TM}$ for $\vec{E}_\parallel$, $\vec{B}_\perp$. For the TE (TM) field the electric (magnetic) field is transverse to the x-z plane.

The TE problem is solved by first obtaining a solution E^{TE} to the scalar Helmholtz equation

$$\nabla^2 E^{TE} + k_o^2 E^{TE} = 0,$$

and then obtaining $\vec{B}^{TE}$ from

$$\vec{B}^{TE} = \frac{j}{\omega}\nabla \times \vec{E}^{TE}.$$

Similarly, the TM problem is solved by first obtaining a solution B^{TM} to the equation

$$\nabla^2 B^{TM} + k_o^2 B^{TM} = 0,$$

and then obtaining $\vec{E}^{TM}$ from

$$\vec{E}^{TM} = \frac{-jc^2}{\omega}\nabla \times \vec{B}^{TM}.$$

Any two-dimensional electromagnetic field in free space can be expressed as a linear combination of a TE field and a TM field.

As examples of two-dimensional TE and TM fields, we will consider plane waves in free space whose vector wave numbers have components in the x-z plane:

$$\vec{k} = k_x\hat{x} + k_z\hat{z}. \tag{3.28}$$

Thus, as required, these fields are independent of the coordinate y. With k_x chosen to be real, k_z from (3.14) becomes

$$k_z = \sqrt{k_o^2 - k_x^2}, \qquad 0 \le k_x^2 \le k_o^2 \tag{3.29a}$$

$$k_z = -j\sqrt{k_x^2 - k_o^2}, \qquad k_x^2 > k_o^2, \tag{3.29b}$$

where the positive square root is assumed.

Values of k_x in the first range (3.29a) represent homogeneous plane waves with

$$\vec{k}_r = k_x\hat{x} + k_z\hat{z}, \qquad \vec{k}_i = 0,$$

whereas those in the second range (3.29b) represent inhomogeneous plane waves with

$$\vec{k}_r = k_x\hat{x}, \qquad j\vec{k}_i = k_z\hat{z}.$$

The signs in front of the square roots in (3.29a) and (3.29b) were chosen so that the homogeneous waves propagate in the direction of increasing z and the inhomogeneous waves decay with increasing z.

The electromagnetic field for the *TE plane wave* is

$$\vec{E}^{TE} = E_{oy}e^{-j(k_x x + k_z z)}\hat{y} \tag{3.30a}$$

$$\vec{B}^{TE} = \frac{1}{\omega}\vec{k} \times \vec{E}^{TE} = -\frac{1}{c}E_{oy}\left[\left(\frac{k_z}{k_o}\right)\hat{x} - \left(\frac{k_x}{k_o}\right)\hat{z}\right]e^{-j(k_x x + k_z z)}, \tag{3.30b}$$

with

$$\langle \vec{S} \rangle = \frac{1}{2\zeta_o} |E_{oy}|^2 e^{-2|\mathrm{Im}(k_z)|z} \left[\left(\frac{k_x}{k_o} \right) \hat{x} + \mathrm{Re}\left(\frac{k_z}{k_o} \right) \hat{z} \right]. \tag{3.30c}$$

For the *TM plane wave* the electromagnetic field is

$$\vec{B}^{TM} = B_{oy} e^{-j(k_x x + k_z z)} \hat{y} \tag{3.31a}$$

$$\vec{E}^{TM} = -\frac{c^2}{\omega} \vec{k} \times \vec{B}^{TM} = c B_{oy} \left[\left(\frac{k_z}{k_o} \right) \hat{x} - \left(\frac{k_x}{k_o} \right) \hat{z} \right] e^{-j(k_x x + k_z z)}. \tag{3.31b}$$

For later use, it will be convenient to have the TM field expressed in terms of the x component of the electric field rather than the y component of the magnetic field (Problem 3.4):

$$\vec{E}^{TM} = E_{ox} \left[\hat{x} - \left(\frac{k_x}{k_z} \right) \hat{z} \right] e^{-j(k_x x + k_z z)} \tag{3.32a}$$

$$\vec{B}^{TM} = \frac{1}{\omega} \vec{k} \times \vec{E}^{TM} = \frac{1}{c} \left(\frac{k_o}{k_z} \right) E_{ox} e^{-j(k_x x + k_z z)} \hat{y}, \tag{3.32b}$$

with

$$\langle \vec{S} \rangle = \frac{1}{2\zeta_o} |E_{ox}|^2 e^{-2|\mathrm{Im}(k_z)|z} \left(\frac{k_o}{|k_z|} \right)^2 \left[\left(\frac{k_x}{k_o} \right) \hat{x} + \mathrm{Re}\left(\frac{k_z}{k_o} \right) \hat{z} \right]. \tag{3.32c}$$

Notice that the maximum time-average energy flow for individual TE and TM waves is always in the direction $\vec{k}_r$: For homogeneous waves $\vec{k}_r = k_x \hat{x} + k_z \hat{z}$, $0 \leq k_x^2 \leq k_o^2$; for inhomogeneous waves $\vec{k}_r = k_x \hat{x}$, $k_x^2 > k_o^2$. The plane waves used for the illustration in the last section are TE plane waves (see Figures 3.1 and 3.2).

3.3 Plane-wave spectrum for two-dimensional electromagnetic fields

Figure 3.3 is a schematic representation for the problem we will now address [1, 2]. All sources of the electromagnetic field are in the left half space, $z < 0$, and we wish to determine the two-dimensional electromagnetic field everywhere in the right half space, $z > 0$. On the plane $z = 0$, the tangential components of the electric field or the tangential components of the magnetic field are specified. For many problems of practical interest, the field will be nonzero over only a finite range for the variable x on this plane. In addition, we will assume that the field is zero everywhere on the cylindrical surface of infinite radius that closes the right half space. We can think of this as being the result of the medium having a very small amount of loss (infinitesimal conductivity σ). From the uniqueness theorem for time-harmonic fields, discussed in Section 1.7, we know that this information ensures a unique solution to Maxwell's equations in the right half space.

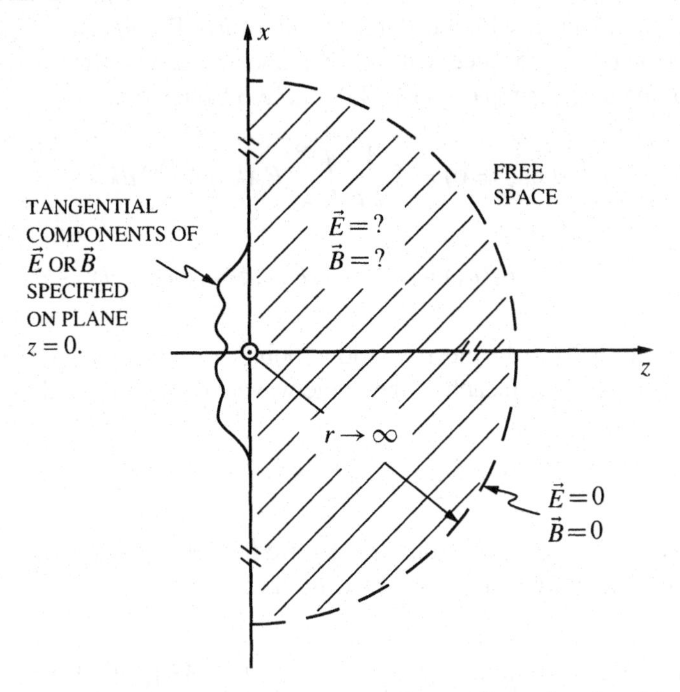

Fig. 3.3. Schematic representation for the two-dimensional problem.

Initially we will assume that the field is a two-dimensional TE field with the tangential component of the electric field specified on the plane $z = 0$:

$$\vec{E}^{TE}(x, z = 0) = f_y(x)\hat{y}. \tag{3.33}$$

The function $f_y(x)$ can be expressed as an integral through the use of the Fourier transform

$$f_y(x) = \frac{1}{2\pi}\int_{-\infty}^{\infty} F_y(\xi)e^{-j\xi x}d\xi, \tag{3.34a}$$

where the spectral function $F_y(\xi)$ is

$$F_y(\xi) = \int_{-\infty}^{\infty} f_y(x)e^{j\xi x}dx. \tag{3.34b}$$

With a change of notation, this is the same Fourier transform pair introduced in Section 1.7 for representation of time-varying fields [Equations (1.200) and (1.201)]. Here we assume that the function $f_y(x)$ satisfies the conditions sufficient for the validity of the transform [3].

Now consider the electric field of the TE plane wave described by (3.30a). On the plane $z = 0$, this field is

$$\vec{E}^{TE}(x, z = 0) = E_{oy}e^{-jk_x x}\hat{y}. \tag{3.35}$$

The integrand of (3.34a) has the same form (dependence on x) as (3.35) when $\xi = k_x$. Thus, (3.34a) can be viewed as the field on the plane $z = 0$ of a superposition of TE plane waves, each with a different value of k_x. The spectral function F_y is the complex weight associated with each plane wave. On setting $\xi = k_x$ and using (3.33), the transform pair in (3.34a) and (3.34b) becomes

$$\vec{E}^{TE}(x, z = 0) = \hat{y}\frac{1}{2\pi}\int_{-\infty}^{\infty} F_y(k_x)e^{-jk_x x}dk_x \tag{3.36}$$

$$F_y(k_x) = \int_{-\infty}^{\infty} \hat{y} \cdot \vec{E}(x, z = 0)e^{jk_x x}dx. \tag{3.37}$$

By analogy with (3.30a), the electric field at points $z > 0$ is obtained by including the factor $\exp(-jk_z z)$ with each of the plane waves in the integrand of (3.36):[2]

$$\vec{E}^{TE}(x, z) = \hat{y}\frac{1}{2\pi}\int_{-\infty}^{\infty} F_y(k_x)e^{-j(k_x x+k_z z)}dk_x. \tag{3.38}$$

The magnetic field is obtained from the electric field by applying the relationship (3.30b) to each plane wave in the superposition (3.38):

$$\vec{B}^{TE}(x, z) = \frac{-1}{2\pi c}\int_{-\infty}^{\infty} F_y(k_x)\left[\left(\frac{k_z}{k_o}\right)\hat{x} - \left(\frac{k_x}{k_o}\right)\hat{z}\right]e^{-j(k_x x+k_z z)}dk_x. \tag{3.39}$$

In these equations, k_z is related to k_x by (3.29a) and (3.29b); all of the plane waves propagate or decay exponentially in the direction pointing away from the source (increasing z).

This electromagnetic field, Equations (3.38) and (3.39) with (3.37), is the solution to our TE problem: On the plane $z = 0$ the tangential component of the electric field is equal to the specified field (3.33), and the field is a solution to Maxwell's equations everywhere in the half space $z > 0$, since each of the plane waves in the superposition is a solution to Maxwell's equations.

It is now a simple matter to obtain the solution for the TM problem by analogy with the TE problem. Let the tangential component of the electric field be specified on the plane $z = 0$:

$$\vec{E}_x(x, z = 0) = f_x(x)\hat{x}. \tag{3.40}$$

The Fourier transform pair is now

$$\vec{E}_x^{TM}(x, z = 0) = \hat{x}\frac{1}{2\pi}\int_{-\infty}^{\infty} F_x(k_x)e^{-jk_x x}dk_x \tag{3.41}$$

$$F_x(k_x) = \int_{-\infty}^{\infty} \hat{x} \cdot \vec{E}(x, z = 0)e^{jk_x x}dx. \tag{3.42}$$

[2] For a different, less intuitive argument for obtaining $\vec{E}^{TE}$ see Problem 3.5.

After consideration of (3.32a) and (3.32b), we obtain the superposition of TM plane waves that comprises the electromagnetic field in the half space $z > 0$:

$$\vec{E}^{TM}(x, z) = \frac{1}{2\pi} \int_{-\infty}^{\infty} F_x(k_x) \left[\hat{x} - \left(\frac{k_x}{k_z} \right) \hat{z} \right] e^{-j(k_x x + k_z z)} dk_x \qquad (3.43)$$

$$\vec{B}^{TM}(x, z) = \hat{y} \frac{1}{2\pi c} \int_{-\infty}^{\infty} \left(\frac{k_o}{k_z} \right) F_x(k_x) e^{-j(k_x x + k_z z)} dk_x. \qquad (3.44)$$

For the general two-dimensional problem, where the specified tangential electric field on the plane $z = 0$ has both $\hat{x}$ and $\hat{y}$ components, the two spectral functions F_y and F_x are determined from (3.37) and (3.42), and the complete electromagnetic field in the half space $z > 0$ is the combination of the TE and TM fields given by Equations (3.38), (3.39), (3.43), and (3.44). The results for the TE and TM cases are summarized in Table 3.1. In the table we have also included expressions for the spectral functions F_y and F_x for the case where the tangential components of the magnetic field are specified on the plane $z = 0$.

A simple physical picture helps with the visualization of the procedure developed above. The two-dimensional plane waves all have the following propagating factor (dependence on x and z):

$$\exp\left[-j(k_x x + k_z z)\right] = \begin{cases} \exp(-jk_x x - j\sqrt{k_o^2 - k_x^2}\, z), & 0 \le k_x^2 \le k_o^2 \\ \exp(-jk_x x - \sqrt{k_x^2 - k_o^2}\, z), & k_x^2 > k_o^2. \end{cases} \qquad (3.45)$$

The waves in the first range for k_x are homogeneous plane waves propagating in the direction of $\vec{k}$. The direction of these *propagating waves* is also specified by the angle ψ between $\vec{k}$ and the z axis, which is given by

$$\psi = \tan^{-1}(k_x/k_z). \qquad (3.46)$$

Schematic representations for propagating waves with $\psi = 0°$, $30°$, and $90°$ are shown in Figures 3.4a, b, and c. Each propagating wave produces a field that is a cosinusoidal function of x on the plane $z = 0$. This is illustrated in Figure 3.4, where the real part of the propagating factor $\exp[-j(k_x x + k_z z)]$ is graphed for points along each of the axes. As the direction of the propagating wave changes from $\psi = 0°$ to $\psi = 90°$ $(0 \le k_x \le k_o)$, the spatial period $\lambda_x = 2\pi/|k_x| = \lambda_o/|\sin\psi|$ of the oscillations along the x axis changes from $\lambda_x = \infty$ to $\lambda_x = \lambda_o$.

The waves in the second range for k_x in (3.45) are inhomogeneous plane waves that decay or evanesce in amplitude with increasing z. Schematic representations for these *evanescent waves* are in Figures 3.4d and e. Along the x axis, the oscillations for these waves are more rapid than for the propagating waves, with the spatial period ranging from $\lambda_x = \lambda_o$ to $\lambda_x = 0$ $(k_o < k_x \le \infty)$. The faster the oscillation with x, the more rapid the decay with increasing z.

The spectral representations in Table 3.1 superimpose plane wave fields (propagating and evanescent) that oscillate in x with all spatial periods $0 \le \lambda_x \le \infty$ to represent the specified field on the plane $z = 0$. This is the fundamental idea used with all Fourier series and Fourier integral representations.

Table 3.1. *Spectral representations for two-dimensional fields*

Transverse electric fields (TE)

$$\vec{E}^{TE}(\vec{r}) = \hat{y}\,\frac{1}{2\pi}\int_{-\infty}^{\infty} F_y(k_x)\,e^{-j(k_x x + k_z z)}dk_x$$

$$= \hat{y}\,\frac{1}{2\pi}\int_C F_y(k_o \sin\psi)\,e^{-jk_o r\cos(\theta-\psi)}k_o\cos\psi\,d\psi$$

$$\vec{B}^{TE}(\vec{r}) = \frac{-1}{2\pi c}\int_{-\infty}^{\infty} F_y(k_x)\left[\left(\frac{k_z}{k_o}\right)\hat{x} - \left(\frac{k_x}{k_o}\right)\hat{z}\right]e^{-j(k_x x + k_z z)}dk_x$$

$$= \frac{-1}{2\pi c}\int_C F_y(k_o\sin\psi)[\sin(\theta-\psi)\hat{r} + \cos(\theta-\psi)\hat{\theta}]e^{-jk_o r\cos(\theta-\psi)}k_o\cos\psi\,d\psi$$

$$F_y(k_x) = \int_{-\infty}^{\infty}\hat{y}\cdot\vec{E}(x, z=0)\,e^{jk_x x}dx = -c\left(\frac{k_o}{k_z}\right)\int_{-\infty}^{\infty}\hat{x}\cdot\vec{B}(x, z=0)\,e^{jk_x x}dx$$

Transverse magnetic fields (TM)

$$\vec{E}^{TM}(\vec{r}) = \frac{1}{2\pi}\int_{-\infty}^{\infty} F_x(k_x)\left[\hat{x} - \left(\frac{k_x}{k_z}\right)\hat{z}\right]e^{-j(k_x x + k_z z)}dk_x$$

$$= \frac{1}{2\pi}\int_C F_x(k_o\sin\psi)\left[\sin(\theta-\psi)\hat{r} + \cos(\theta-\psi)\hat{\theta}\,\right]e^{-jk_o r\cos(\theta-\psi)}k_o d\psi$$

$$\vec{B}^{TM}(\vec{r}) = \hat{y}\,\frac{1}{2\pi c}\int_{-\infty}^{\infty}\left(\frac{k_o}{k_z}\right)F_x(k_x)\,e^{-j(k_x x + k_z z)}dk_x$$

$$= \hat{y}\,\frac{1}{2\pi c}\int_C F_x(k_o\sin\psi)\,e^{-jk_o r\cos(\theta-\psi)}k_o d\psi$$

$$F_x(k_x) = \int_{-\infty}^{\infty}\hat{x}\cdot\vec{E}(x, z=0)\,e^{jk_x x}dx = c\left(\frac{k_z}{k_o}\right)\int_{-\infty}^{\infty}\hat{y}\cdot\vec{B}(x, z=0)\,e^{jk_x x}dx$$

The complex Poynting vector for the general two-dimensional field composed of TE and TM components is

$$\vec{S}_c = \frac{1}{2\mu_o}(\vec{E}^{TE} + \vec{E}^{TM}) \times (\vec{B}^{TE} + \vec{B}^{TM})^*, \tag{3.47}$$

and the net time-average power transmitted per unit length in y through a plane of constant z is

$$P_z = \int_{x=-\infty}^{\infty}\langle\vec{S}\rangle\cdot\hat{z}dx = \int_{x=-\infty}^{\infty}\mathrm{Re}(\vec{S}_c)\cdot\hat{z}dx. \tag{3.48}$$

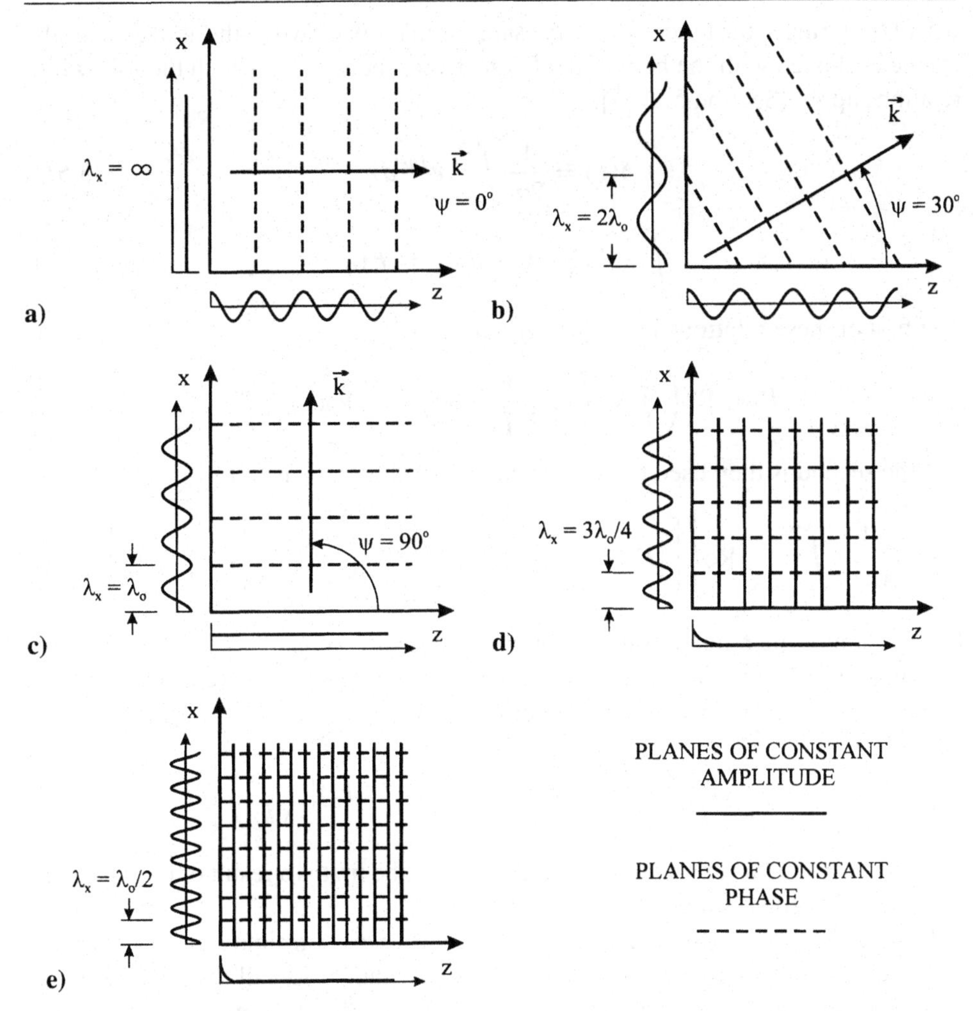

Fig. 3.4. a)–c) Propagating plane waves, $0 \leq k_x \leq k_o$. d), e) Evanescent plane waves, $k_o < k_x \leq \infty$. The real part of the propagating factor $\exp[-j(k_x x + k_z z)]$ is graphed along the x and z axes.

After inserting the expressions for the fields from Table 3.1 into (3.48), we have

$$P_z = \text{Re}\Bigg(\int_{x=-\infty}^{\infty} \int_{k_x'=-\infty}^{\infty} \int_{k_x=-\infty}^{\infty} \Bigg\{ \frac{1}{8\pi^2 \zeta_o} \Bigg[F_y(k_x) F_y^*(k_x') \left(\frac{k_z'}{k_o} \right)^* + F_x(k_x) F_x^*(k_x') \left(\frac{k_o}{k_z'} \right)^* \Bigg] e^{-j(k_z - k_z'^*)z} \Bigg\} e^{-j(k_x - k_x')x} dk_x dk_x' dx \Bigg), \tag{3.49}$$

and, with an interchange of the order of the integrations, we can write

$$P_z = \text{Re}\Bigg(\int_{-\infty}^{\infty} \int_{-\infty}^{\infty} \{\,\} \int_{x=-\infty}^{\infty} e^{-j(k_x - k_x')x} dx dk_x dk_x' \Bigg), \tag{3.50}$$

where the terms in the braces { } are the same as in (3.49). Two of the above integrals can be evaluated with the help of the Dirac delta function, for which the following relations apply (Section 5.2) [3]:

$$\delta(\xi) = \frac{1}{2\pi}\int_{-\infty}^{\infty} e^{j\eta\xi}\,d\eta \tag{3.51}$$

$$\int_{-\infty}^{\infty} F(\xi)\delta(\xi-\eta)d\xi = F(\eta). \tag{3.52}$$

The first of these relations is used to write (3.50) as

$$P_z = \mathrm{Re}\left(\int_{-\infty}^{\infty}\int_{-\infty}^{\infty}\left\{\ \right\}2\pi\delta(k_x' - k_x)dk_x dk_x'\right),$$

and the second is then used to evaluate the integral with respect to k_x':

$$P_z = \frac{1}{4\pi\zeta_o}\int_{k_x=-\infty}^{\infty}\mathrm{Re}\left\{\left[\left|F_y(k_x)\right|^2\left(\frac{k_z}{k_o}\right)^* + \left|F_x(k_x)\right|^2\left(\frac{k_o}{k_z}\right)^*\right]e^{-j(k_z-k_z^*)z}\right\}dk_x.$$

On division of the range of integration into the two regions $0 \le k_x^2 \le k_o^2$ (k_z real; propagating waves) and $k_x^2 > k_o^2$ (k_z imaginary; evanescent waves), we see that only the former contributes to the integral, and so

$$P_z = \frac{1}{4\pi\zeta_o}\int_{k_x=-k_o}^{k_o}\left[\left|F_y(k_x)\right|^2\left(\frac{k_z}{k_o}\right) + \left|F_x(k_x)\right|^2\left(\frac{k_o}{k_z}\right)\right]dk_x. \tag{3.53}$$

The two terms in this expression are independent; the first is a result of the TE field, whereas the second is a result of the TM field.

From (3.53) we see that only the propagating waves contribute to the net time-average power transmitted through a plane of constant z. Recall that these are the waves in the spectrum that have spatial periods in x within the range $\lambda_o \le \lambda_x \le \infty$. The evanescent waves make no direct contribution to the net time-average power transmitted through a plane of constant z. The evanescent waves in the spectrum, with their shorter spatial periods $0 \le \lambda_x < \lambda_o$, are, however, required to represent the fine details of the field near the plane $z = 0$.

The spectral representations we obtained for the electromagnetic field are expressed in terms of the rectangular coordinates of the position (x, z) and the rectangular components of the vector wave number (k_x, k_z). Often it is convenient to express these quantities in terms of their components in a cylindrical system. In the cylindrical coordinates (r, θ) of Figure 3.5a the position is simply[3]

$$\left.\begin{aligned} x &= r\sin\theta \\ z &= r\cos\theta \end{aligned}\right\} \qquad -\pi/2 \le \theta \le \pi/2. \tag{3.54}$$

[3] This is not the cylindrical coordinate system (ρ, ϕ, z) discussed in Appendix B, but the cylindrical coordinate system formed from the spherical coordinates r and θ in the x-z plane and y. We use this system so that the notation for the two-dimensional and the three-dimensional fields will be similar.

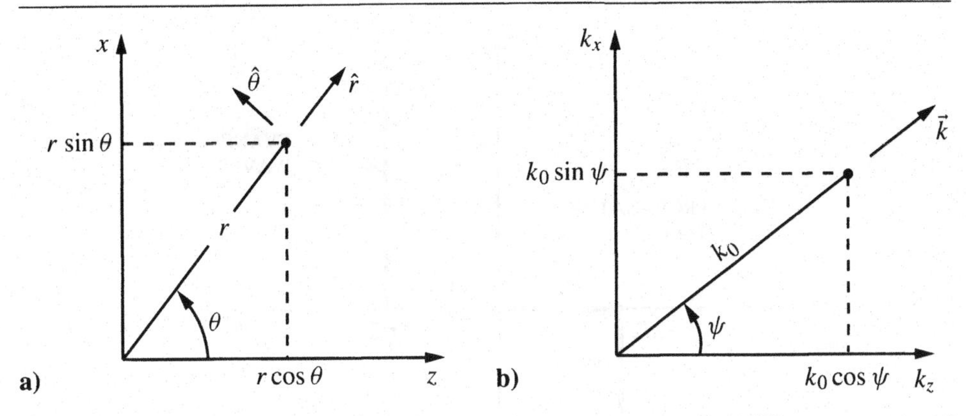

Fig. 3.5. a) The position expressed in terms of the coordinates r, θ. b) The components of the vector wave number expressed in terms of k_o and ψ. The drawing is for a wave number with k_z real.

When the wave number represents a *propagating wave* ($0 \leq k_x^2 \leq k_o^2$), the real components of the wave number are easily expressed in terms of k_o and the real angle ψ as shown in Figure 3.5b:

$$\left.\begin{aligned} k_x &= k_o \sin\psi \\ k_z &= k_o \cos\psi \end{aligned}\right\} \qquad -\pi/2 \leq \psi \leq \pi/2. \tag{3.55}$$

When the wave number represents an *evanescent wave* ($k_x^2 > k_o^2$), the components of the wave number are complex, and the angle ψ must also be complex, $\psi = \psi_r + j\psi_i$:

$$k_x = k_o \sin(\psi_r + j\psi_i) = k_o[\sin\psi_r \cosh\psi_i + j\cos\psi_r \sinh\psi_i]$$

$$k_z = k_o \cos(\psi_r + j\psi_i) = k_o[\cos\psi_r \cosh\psi_i - j\sin\psi_r \sinh\psi_i].$$

Since k_x is always real, we must have $\psi_r = \pm\pi/2$, which makes

$$k_x = k_o \sin(\pm\pi/2)\cosh\psi_i$$

$$k_z = -jk_o \sin(\pm\pi/2)\sinh\psi_i.$$

For the evanescent wave to decrease in amplitude with increasing z, k_z must be negative imaginary. Thus, when $\psi_r = -\pi/2$, ψ_i must be negative ($-\infty \leq \psi_i < 0$); and when $\psi_r = \pi/2$, ψ_i must be positive ($0 < \psi_i \leq \infty$). The components of the wave number for the evanescent wave are then

$$\left.\begin{aligned} k_x &= k_o \sin\psi = -k_o \cosh\psi_i \\ k_z &= k_o \cos\psi = jk_o \sinh\psi_i \end{aligned}\right\} \qquad \begin{aligned} &\psi_r = -\pi/2 \\ &-\infty \leq \psi_i < 0 \end{aligned}$$

and

$$\left.\begin{aligned} k_x &= k_o \sin\psi = k_o \cosh\psi_i \\ k_z &= k_o \cos\psi = -jk_o \sinh\psi_i \end{aligned}\right\} \qquad \begin{aligned} &\psi_r = \pi/2 \\ &0 < \psi_i \leq \infty. \end{aligned} \tag{3.56}$$

Equations (3.55) and (3.56) describe the path the angle ψ traces in the complex plane as the real variable k_x goes from $-\infty$ to $+\infty$. This path is the contour C in

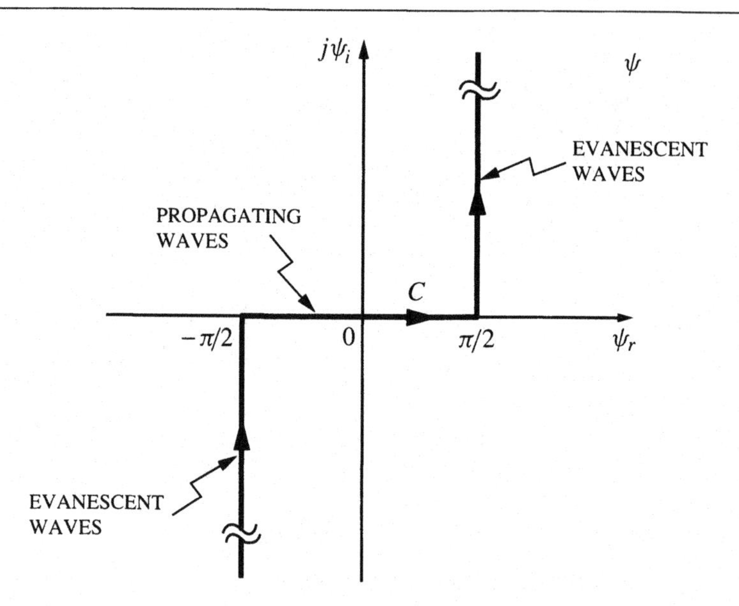

Fig. 3.6. The contour C in the complex ψ plane. The contour $C_{1/2}$ is the right half of C.

Figure 3.6; it starts at $\psi = -\pi/2 - j\infty$ ($k_x = -\infty$), goes to $\psi = -\pi/2$ ($k_x = -k_o$), then to $\psi = \pi/2$ ($k_x = k_o$), and finally to $\psi = \pi/2 + j\infty$ ($k_x = \infty$). The horizontal portion of the contour corresponds to propagating waves; the vertical portions of the contour correspond to evanescent waves.

The spectral representations for the electromagnetic field are now easily expressed in terms of the coordinates r, θ and k_o, ψ. For example, the field $\vec{E}^{TE}$ (3.38) on substitution of (3.54) and (3.55) becomes

$$\begin{aligned}\vec{E}^{TE} &= \hat{y}\frac{1}{2\pi}\int_C F_y(k_o \sin\psi) e^{-jk_o r(\sin\psi\sin\theta + \cos\psi\cos\theta)} k_o \cos\psi\, d\psi \\ &= \hat{y}\frac{1}{2\pi}\int_C F_y(k_o \sin\psi) e^{-jk_o r\cos(\theta-\psi)} k_o \cos\psi\, d\psi. \qquad (3.57)\end{aligned}$$

The representations for the other components of the field are listed in Table 3.1.

3.4 The uniformly illuminated slit

We will now illustrate the use of the two-dimensional plane-wave spectrum with a simple example. The physical geometry to be considered is shown in Figure 3.7. A plane wave, either TE or TM, is normally incident on an infinitesimally thin, perfectly conducting screen containing a slit of width d. The electric field for the incident wave is

$$\vec{E}^i = \begin{cases} E_o^i e^{-jk_o z}\hat{y}, & \text{(TE)} \\ E_o^i e^{-jk_o z}\hat{x}, & \text{(TM)}. \end{cases}$$

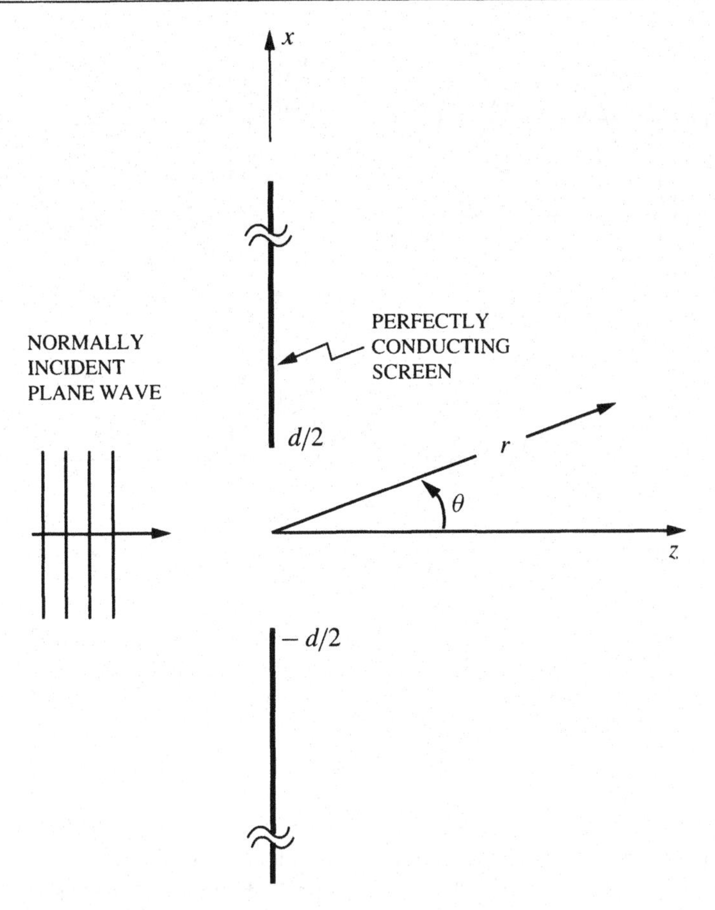

Fig. 3.7. Plane wave normally incident on an infinitesimally thin, perfectly conducting screen containing a slit of width d.

The screen is in the plane $z = 0$, and we wish to determine the field in the half space $z > 0$. This is a classical problem in the theory of diffraction.

We must know the tangential component of the electric or magnetic field on the plane $z = 0+$ to apply our procedure; we will assume knowledge of the former. Because the screen is perfectly conducting, the tangential component of the electric field is zero for $z = 0+$, $|x| \geq d/2$. What is the electric field on the remainder of the plane, in the aperture $z = 0+$, $|x| < d/2$? For sufficiently wide slits ($d/\lambda_o \gg 1$), we might assume that the electric field in the aperture is equal to the incident field; that is, the screen has little effect on the field in the aperture. This assumption is sometimes referred to as *Kirchhoff's approximation* after Gustav Robert Kirchhoff (1824–1887), who used a similar assumption in optical calculations [4].

To check the validity of the assumption, very good estimates for the electric field in the aperture are graphed in Figure 3.8.[4] Here results are shown for TE and TM

[4] The results presented in Figure 3.8 were computed using Babinet's principle (Section 4.4) and formulas for electrically wide strips given in Reference [5].

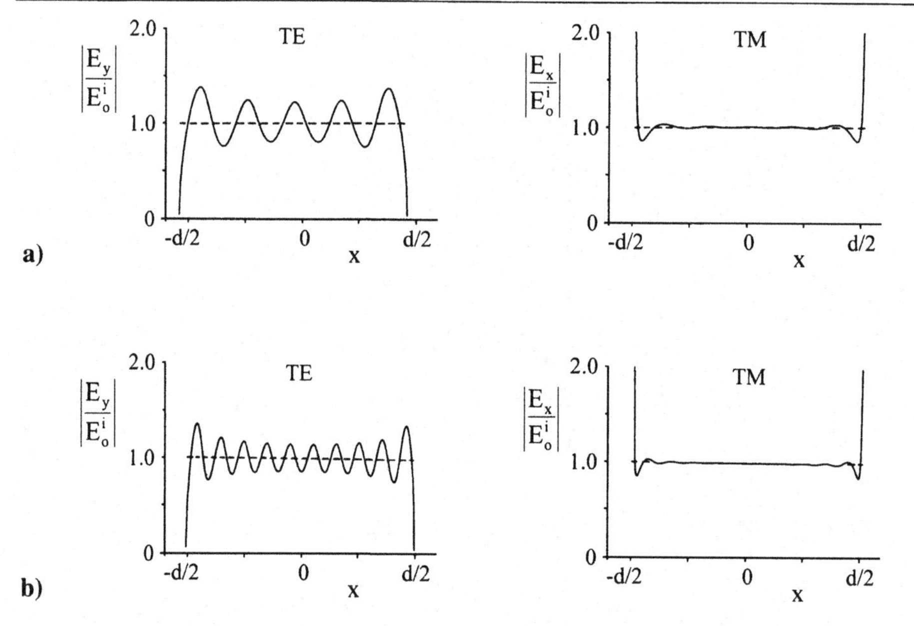

Fig. 3.8. Tangential component of the electric field in the slit ($z = 0$) for a normally incident plane wave (TE and TM). Solid line: good estimate for electrically wide slit. Dashed line: uniformly illuminated slit. a) $d/\lambda_o = 5.0$. b) $d/\lambda_o = 10.0$.

incident waves and two widths for the slit, $d/\lambda_o = 5$ and 10. From these graphs it is clear that the tangential component of the electric field is only approximately uniform in the aperture and that the approximation becomes better as the electrical width of the slit, d/λ_o, becomes larger. The field for the TE case shows a pronounced ripple; this is caused by the interaction of the fields produced by the two edges of the slit. The amplitude of the ripple decreases with increasing slit width. The field for the TM case is fairly uniform in the aperture except near the edges, where there are large peaks. In fact, at the edges the field becomes infinite. This is a consequence of assuming infinitely sharp edges for the perfectly conducting screen.

Let us now proceed with our analysis, subject to the assumption that the slit is electrically wide ($d/\lambda_o \gg 1$) and the tangential component of the electric field is approximately uniform in the aperture and zero elsewhere on the plane $z = 0$, as shown in Figure 3.9a. For the TE case

$$\vec{E}_y(x, z = 0) = \begin{cases} E_o^i \hat{y}, & |x| < d/2 \\ 0, & |x| \geq d/2, \end{cases} \tag{3.58}$$

and for the TM case

$$\vec{E}_x(x, z = 0) = \begin{cases} E_o^i \hat{x}, & |x| < d/2 \\ 0, & |x| \geq d/2. \end{cases} \tag{3.59}$$

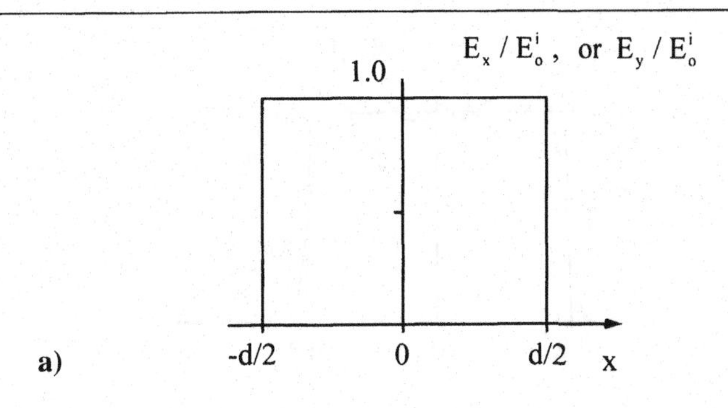

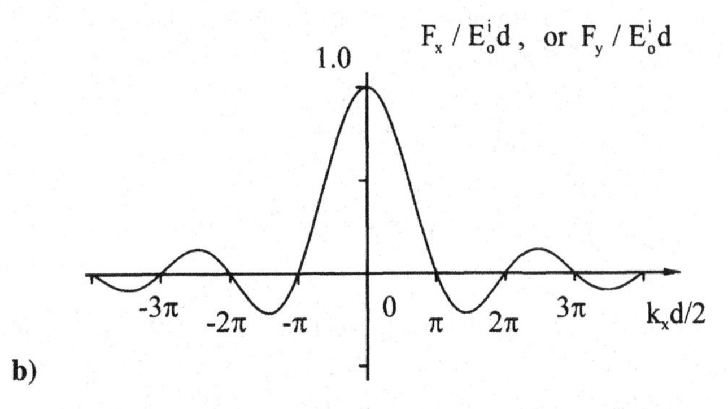

Fig. 3.9. Uniformly illuminated slit. a) Tangential component of the electric field on the plane $z = 0$. b) Spectral function for the field of a).

The spectral functions for these fields are obtained from the integrals in Table 3.1. For the TE case

$$\begin{aligned} F_y(k_x) &= \int_{-\infty}^{\infty} \hat{y} \cdot \vec{E}(x, z = 0) e^{jk_x x}\, dx \\ &= E_o^i \int_{-d/2}^{d/2} e^{jk_x x}\, dx \\ &= E_o^i d \,\mathrm{sinc}(k_x d/2), \end{aligned} \tag{3.60}$$

and similarly for the TM case

$$F_x(k_x) = E_o^i d \,\mathrm{sinc}(k_x d/2). \tag{3.61}$$

Here we have introduced the "sinc" function

$$\mathrm{sinc}(z) = \sin(z)/z. \tag{3.62}$$

The spectral function, $F_y(k_x)$ or $F_x(k_x)$, is graphed in Figure 3.9b.

The electromagnetic field at points in the half space $z > 0$ can now be obtained using the spectral functions with the appropriate integrals from Table 3.1. For

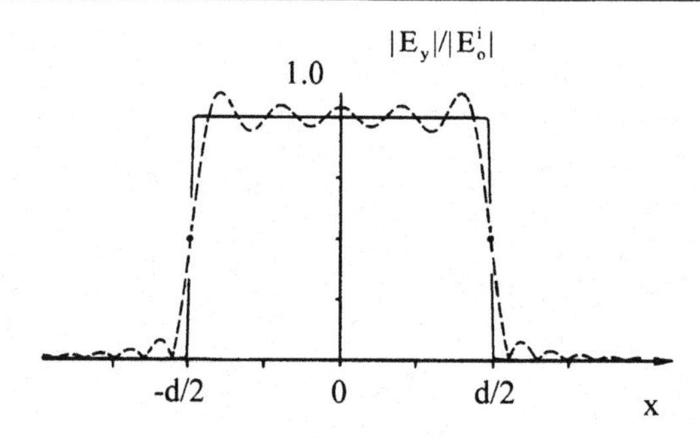

Fig. 3.10. Tangential component of the TE electric field on the plane $z = 0$ for a slit of width $d/\lambda_o = 5.0$. Solid line: field of the total spectrum. Dashed line: field of only the propagating spectrum ($0 \le k_x^2 \le k_o^2$).

example, the electric field for the TE case is

$$\begin{aligned} \vec{E}^{TE} &= \hat{y}\frac{E_o^i d}{2\pi}\int_{-\infty}^{\infty}\text{sinc}\left(\frac{k_x d}{2}\right)e^{-j(k_x x + k_z z)}dk_x \\ &= \hat{y}\frac{E_o^i k_o d}{2\pi}\int_C \text{sinc}\left(\frac{k_o d}{2}\sin\psi\right)e^{-jk_o r\cos(\theta-\psi)}\cos\psi\, d\psi. \end{aligned} \tag{3.63}$$

At this point we should emphasize that the field given by (3.63) is the *exact* TE field in the half space $z > 0$ for the specified field (3.58) on the plane $z = 0$. However, it is only the *approximate* field for the physical geometry described in Figure 3.7. If we could construct another physical geometry that produced the specified field (3.58) on the plane $z = 0$, then (3.63) would be the exact field in the half space $z > 0$ for that geometry.

Let us now study the electric field produced by the uniformly illuminated slit. We will only consider the TE case, which involves the evaluation of the integral in (3.63). For general values of $k_o d$, $k_o r$, and θ, the integration cannot be performed in closed form, and numerical evaluation must be used. It is instructive to evaluate this integral first on the plane $z = 0$. If the theory is self-consistent, we should obtain the tangential electric field we started with, (3.58) and Figure 3.9a. The result for a slit of width $d/\lambda_o = 5$ is shown as a solid line in Figure 3.10; as expected, the tangential field is recovered. Note that at the points $x = \pm d/2$ the field is $E_y = E_o^i/2$, which is the mean of the assumed values for the field at $|x| = d/2+$ and $|x| = d/2-$. This illustrates a well known property of the Fourier inversion formula: At a discontinuity it returns the mean of the values on the two sides [3].

Recall that the contour of integration C in (3.63) consists of two parts, one corresponding to the propagating portion and the other to the evanescent portion of the spectrum. The dashed line in Figure 3.10 shows the tangential electric field on the plane $z = 0$ that is due only to the propagating portion of the spectrum ($0 \le k_x^2 \le k_o^2$). It is clear that the field of the propagating spectrum has the general shape of the aperture field, but it cannot produce the rapid variation in the field near the points $|x| = d/2$. The shorter spatial periods ($\lambda_x < \lambda_o$) of the evanescent spectrum are required to produce this variation. The period of the ripple on the dashed line in Figure 3.10 is roughly equal to the minimum period for a wave in the propagating spectrum (i.e., $\lambda_x = \lambda_o$, or $\lambda_x = d/5$ since $d/\lambda_o = 5$).

The development of the field in the half space $z > 0$ is shown in Figure 3.11a for a slit of width $d/\lambda_o = 5$. Here the normalized amplitude of the field is graphed as a function of position, x/λ_o and z/λ_o. The field is seen to spread out as it travels away from the slit and to extend into the region behind the screen, the shadow region. This phenomenon is known as diffraction.[5] The oscillations with x/λ_o seen in the

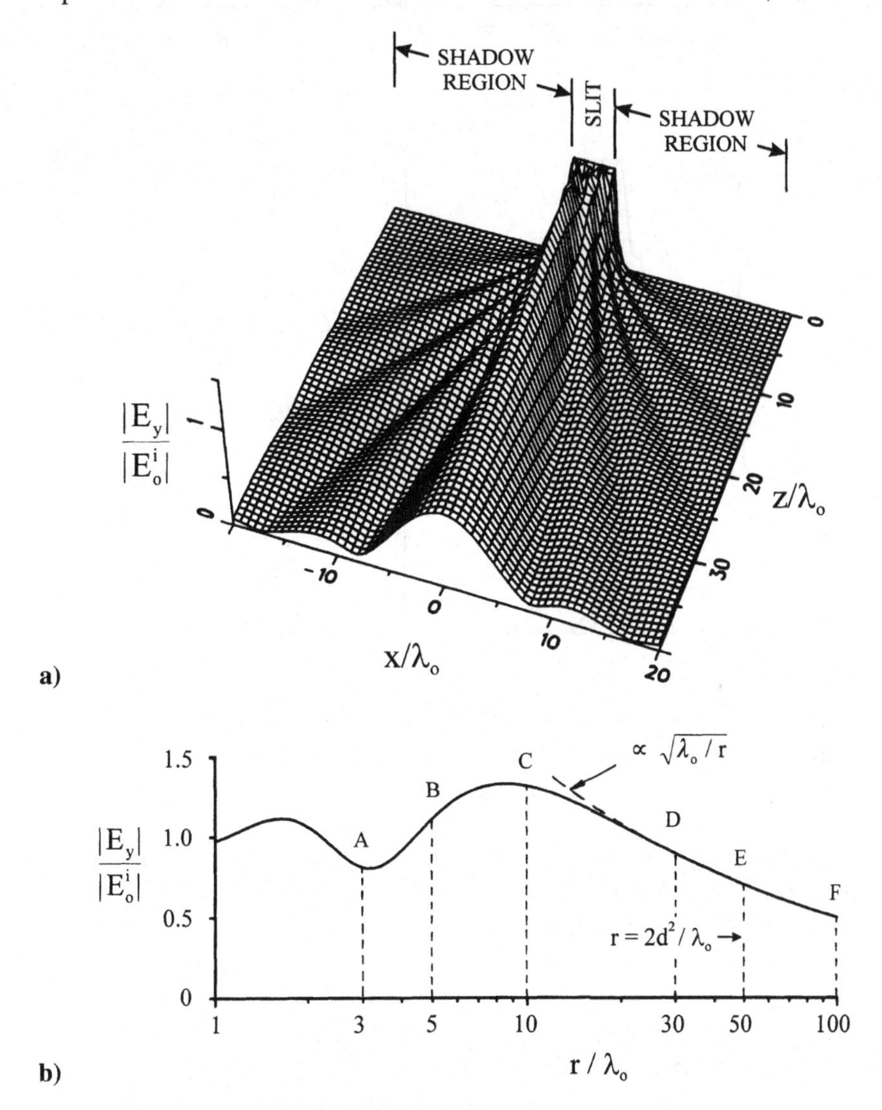

Fig. 3.11. *(continued on next page)*

[5] The crudest approximation for the field behind the slit is that of *geometrical optics*, for which the region $|x| < d/2$, $z > 0$ would be uniformly illuminated, and the remainder would be in shadow. That is, on each cross section z/λ_o in Figure 3.11, the plot would be the same rectangular pulse as in the aperture plane, $z/\lambda_o = 0$. The *diffracted field* is the difference between the actual field and the field of geometrical optics.

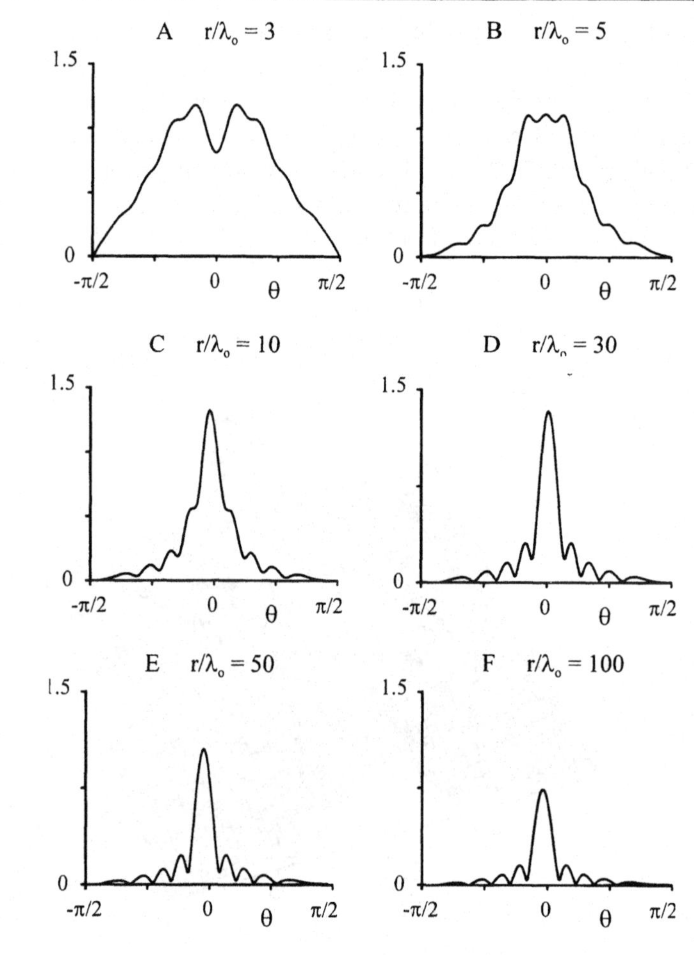

Fig. 3.11. The electric field in the half space $z > 0$ for a slit of width $d/\lambda_o = 5.0$. a) Field as a function of x/λ_o and z/λ_o. b) Field as a function of r/λ_o for points on the z axis ($r = z$, $\theta = 0$). c) Field as a function of θ for various values of r/λ_o.

amplitude of the field appear as light and dark regions at optical frequencies and are called diffraction bands.

Additional insight is gained when the field is graphed in the cylindrical coordinates (r, θ) of Figure 3.7. In Figure 3.11b the normalized amplitude of the field along the z axis ($r = z, \theta = 0$) is graphed as a function of the radius r/λ_o, and in Figure 3.11c the field is shown as a function of the angle θ for various radii. The field is seen to exhibit the rectangular shape of the aperture for the smaller values of r/λ_o (A, B). This is true even though the field is graphed on a cylindrical surface (r = constant) rather than on a planar surface (z = constant).

The amplitude of the field oscillates as we move along the z axis away from the aperture plane; however, at large distances (D, E, F) it becomes a monotonically decreasing function of r/λ_o with the dependence $\sqrt{\lambda_o/r}$ or $1/\sqrt{k_o r}$. In this region the distribution for the field (i.e., the field as a function of θ) is roughly the same for all values of r/λ_o (D, E, F). A comparison with Figure 3.9b shows that this distribution is approximately the same as the spectral function (3.60) with $k_x = k_o \sin\theta$. Thus, from these numerical results, we infer that at very large distances from the aperture the electric field will have the simple form

$$|\vec{E}^{TE}| \propto \left|F_y(k_o \sin\theta)\right| / \sqrt{k_o r}. \tag{3.64}$$

In the next section, we will obtain this result by an approximation of the integral in Equation (3.57).

3.5 Asymptotic or radiated field – the method of stationary phase

We now wish to establish by analytical means the simple relationship (3.64) we inferred from the numerical results for the field at large radial distances from the uniformly illuminated slit. We will not restrict our argument to the uniformly illuminated slit, but we will consider general field distributions on the plane $z = 0$. To be more precise, we seek an approximation to the integral representation for the TE field (3.57)

$$\vec{E}^{TE} = \hat{y}\frac{1}{2\pi}\int_C F_y(k_o \sin\psi)e^{-jk_o r\cos(\theta-\psi)}k_o \cos\psi\, d\psi \tag{3.65}$$

that is valid in the limit as $k_o r = 2\pi r/\lambda_o \to \infty$, for $-\pi/2 < \theta < \pi/2$.

The branch of mathematics that deals with such approximations is called asymptotic methods [6, 7]. Briefly, the problem is often one of obtaining an approximation $h(x)$ to a function $f(x)$ that applies for large x; the basic idea is that $f(x)/h(x)$ tends to unity as $x \to \infty$. We then say that $h(x)$ is an asymptotic approximation to $f(x)$ for large x, and we indicate this by

$$f(x) \sim h(x), \qquad x \to \infty,$$

where the symbol $\sim$ is read "is asymptotic to." The larger is x, the better $h(x)$ approximates $f(x)$. A familiar example of an asymptotic approximation is Stirling's formula for estimating $n!$ when n is large [8]:

$$n! \sim \left(\frac{n}{e}\right)^n \sqrt{2\pi n}, \qquad n \to \infty.$$

When $n = 5$, this formula is in error by 1.7%; however, when $n = 50$, it is in error by only 0.17%.

Returning to our problem and using the above notation, we seek an approximation to the field $\vec{E}^{TE}(\vec{r})$, which we will call $\vec{E}^{r}(\vec{r})$, such that

$$\vec{E}^{TE}(\vec{r}) \sim \vec{E}^{r}(\vec{r}), \qquad k_o r \to \infty.$$

The asymptotic field is often referred to as the *radiated field* (hence the superscript r) or the far-zone field in antenna analysis, and the Fraunhofer field (Joseph von Fraunhofer, 1787–1826) in optics.

The procedure we will use to asymptotically evaluate our integral (3.65) is generally called the *method of stationary phase*. It was first used by G. G. Stokes [9]; however, our treatment will follow more closely that given later by Lord Kelvin (William Thomson, 1824–1907) [10].

We begin by noting from (3.56) and (3.57) that along the two vertical portions of the contour C in Figure 3.6 the amplitude of the exponential term in the integrand is

$$e^{-k_o r \cos\theta \sinh|\psi_i|}.$$

Thus, in the limit $k_o r \to \infty$, these portions of the contour make a negligible contribution to the integral and can be ignored. Stated differently, the evanescent portion of the spectrum does not contribute to the field when $k_o r$ is large. We are now left with the following integral for evaluation:

$$I = \frac{1}{2\pi}\int_{\psi=-\pi/2}^{\pi/2} F_y(k_o \sin\psi) e^{-jk_o r\cos(\psi-\theta)} k_o \cos\psi \, d\psi. \tag{3.66}$$

Let us again concentrate on the exponential term. The "phase" of the exponential,

$$g(\psi) = -k_o r \cos(\psi - \theta),$$

is graphed as a function of the angle $\psi - \theta$ in Figure 3.12a; the real and imaginary parts of the exponential are shown in Figure 3.12b. Both graphs are for the numerical value $k_o r = 20\pi$. Notice that the vertical axis in Figure 3.12a was made discontinuous so that the scales on the vertical and horizontal axes could be equal. The phase of the exponential is seen to vary very rapidly with a change in the angle $\psi - \theta$, except near the point $\psi - \theta = 0$. At this point the derivative of the phase with respect to the angle ψ is zero:

$$\frac{\partial g}{\partial \psi} = \frac{\partial}{\partial \psi}\left[-k_o r \cos(\psi - \theta)\right] = k_o r \sin(\psi - \theta) = 0,$$

and this point is referred to as "a point of stationary phase." The real and imaginary parts of the integrand are seen to oscillate rapidly with a change in the angle $\psi - \theta$, except in the vicinity of the point of stationary phase. An increase in $k_o r$ would cause more rapid oscillation, and the oscillations would start closer to the point of stationary phase.

This behavior suggests the following argument. Consider $F_y(k_o \sin\psi)$ to be a smooth function of the angle ψ. Then away from the stationary point, each oscillation in the exponential produces positive and negative contributions to the integral (3.66) that are of equal area and, therefore, cancel. Thus, the value of the integral is determined mainly by the behavior of the integrand near the stationary point. A simple numerical example will illustrate this argument. Let $F_y(k_o \sin\psi) = 1.0$ and $\theta = 0$, making the stationary point $\psi = 0$. Now the integral (3.66) will be evaluated only over a panel of width 2χ centered on the stationary point. After suitable normalization the integral becomes

$$I_n = \frac{\int_{-\chi}^{\chi} e^{-jk_o r\cos\psi} k_o \cos\psi \, d\psi}{\int_{-\pi/2}^{\pi/2} e^{-jk_o r\cos\psi} k_o \cos\psi \, d\psi}.$$

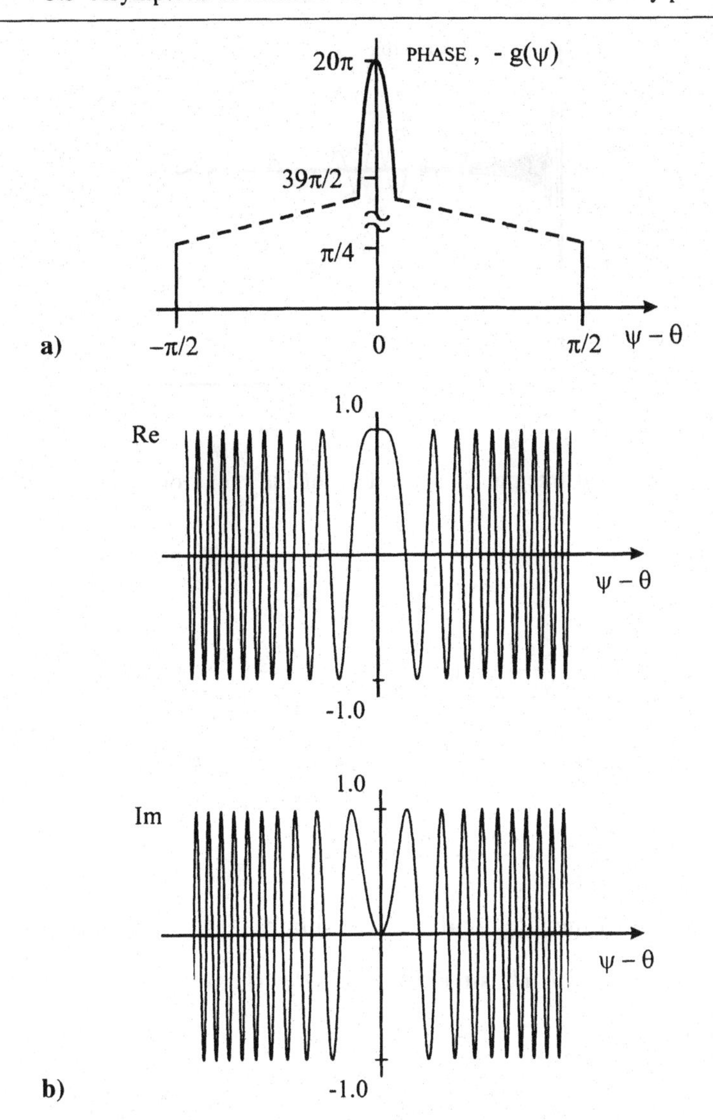

Fig. 3.12. a) Phase of exponential, $g(\psi)$. b) Real and imaginary parts of exponential, $\exp[jg(\psi)]$. For both graphs $k_o r = 20\pi$.

This integral shows us what fraction of the total integration is coming from a panel of width 2χ about the stationary point. In Figure 3.13 the magnitude of the integral is plotted as a function of the angle χ for two values of $k_o r$. The curves are seen to rise to a value close to 1.0 and then oscillate about this value with increasing χ. The larger $k_o r$, the more rapidly the integral settles to the value 1.0, indicating that only a small region around the stationary point is contributing to the integration.

Since the major contribution to the integral comes from a small region around the stationary point, $\psi = \theta$, we can replace the elements in the integrand of (3.66) by approximations valid near this point. For the exponent we will use the first three terms in a Taylor series expansion (where the third term is the first nonzero term to

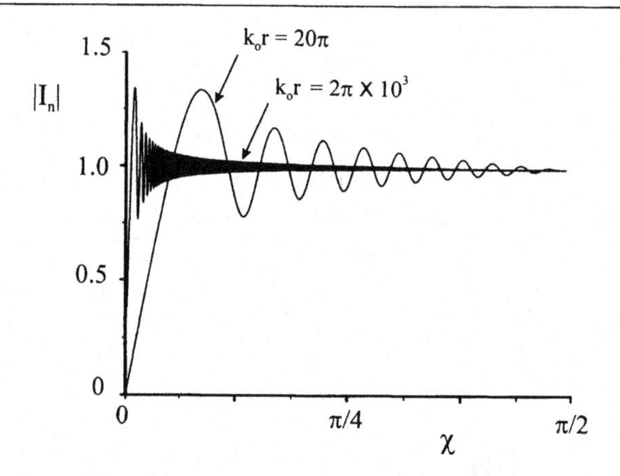

Fig. 3.13. The normalized integral as a function of the range of integration about the stationary point.

depend on ψ):

$$g(\psi) \approx g(\theta) + \left.\frac{\partial g}{\partial \psi}\right|_{\psi=\theta}(\psi - \theta) + \frac{1}{2!}\left.\frac{\partial^2 g}{\partial \psi^2}\right|_{\psi=\theta}(\psi - \theta)^2$$

$$= -k_o r\left[\cos(0) - \sin(0)(\psi - \theta) - \frac{1}{2}\cos(0)(\psi - \theta)^2\right]$$

$$= -k_o r\left[1 - \frac{(\psi - \theta)^2}{2}\right],$$

and in the remainder of the integrand, we will simply set $\psi = \theta$, namely,

$$k_o \cos\psi\, F_y(k_o \sin\psi) \approx k_o \cos\theta\, F_y(k_o \sin\theta).$$

The integral is then

$$I \approx \frac{1}{2\pi} k_o \cos\theta\, F_y(k_o \sin\theta) e^{-jk_o r} \int_{\theta-\Delta}^{\theta+\Delta} e^{jk_o r(\psi-\theta)^2/2} d\psi,$$

where the limits $(\theta - \Delta,\ \theta + \Delta)$ define a small region centered on the stationary point. Notice that the limits of integration can be changed to $(-\infty,\ \infty)$ without affecting the value of the integral; no new stationary points are introduced, and the additional oscillations added by extending the limits cancel on integration. With the change of variable $\xi = \sqrt{k_o r/2}(\psi - \theta)$, the integral becomes

$$I \approx \frac{1}{2\pi} k_o \cos\theta\, F_y(k_o \sin\theta) \sqrt{\frac{2}{k_o r}} e^{-jk_o r} \int_{-\infty}^{\infty} e^{j\xi^2} d\xi.$$

The definite integral above has the value $\sqrt{\pi}\exp(j\pi/4)$ [8]; thus,

$$I \approx \frac{k_o}{\sqrt{2\pi}} \cos\theta\, F_y(k_o \sin\theta) \frac{1}{\sqrt{k_o r}} e^{-j(k_o r - \pi/4)},$$

Table 3.2. *Spectral representations for two-dimensional radiated fields*

Transverse electric fields (TE)

$$\vec{E}^r(\vec{r}) = \hat{y}\,\frac{k_o}{\sqrt{2\pi}}\cos\theta\, F_y(k_o\sin\theta)\frac{1}{\sqrt{k_o r}}\,e^{-j(k_o r-\pi/4)}$$

$$\vec{B}^r(\vec{r}) = -\hat{\theta}\,\frac{k_o}{\sqrt{2\pi}\,c}\cos\theta\, F_y(k_o\sin\theta)\frac{1}{\sqrt{k_o r}}\,e^{-j(k_o r-\pi/4)}$$

$$= \frac{1}{c}\hat{r}\times\vec{E}^r(\vec{r})$$

Transverse magnetic fields (TM)

$$\vec{E}^r(\vec{r}) = \hat{\theta}\,\frac{k_o}{\sqrt{2\pi}}\, F_x(k_o\sin\theta)\frac{1}{\sqrt{k_o r}}\,e^{-j(k_o r-\pi/4)}$$

$$\vec{B}^r(\vec{r}) = \hat{y}\,\frac{k_o}{\sqrt{2\pi}\,c}\, F_x(k_o\sin\theta)\frac{1}{\sqrt{k_o r}}\,e^{-j(k_o r-\pi/4)}$$

$$= \frac{1}{c}\hat{r}\times\vec{E}^r(\vec{r})$$

and the radiated field we seek is

$$\vec{E}^r = \hat{y}\frac{k_o}{\sqrt{2\pi}}\cos\theta\, F_y(k_o\sin\theta)\frac{1}{\sqrt{k_o r}}e^{-j(k_o r-\pi/4)}. \tag{3.67}$$

A comparison with (3.64) shows that this field has the same dependence as we deduced from our earlier numerical results for the slit. Asymptotic approximations can also be obtained for the TE magnetic field and for the field components of the TM problem; these are summarized in Table 3.2.

The results we have obtained for the radiated field are most interesting and deserve some further discussion. The field (3.67) is the product of a function of θ with a function of r. This means that the angular dependence of the field is the same at different radii, as we observed for the slit in Figure 3.11c (panels D, E, and F). Furthermore, it is the angular dependence of the radiated field that characterizes different distributions for the field on the plane $z = 0$; the radial dependence is the same for all distributions. The radiated field has many of the same features as a homogeneous plane wave, with the vector $k_o\hat{r}$ assuming the role of the vector wave number $\vec{k} = k_o\hat{k}$: a transverse field with an exponential variation of the phase with distance, $\exp(-jk_o r)$; a simple relationship between $\vec{B}^r$ and $\vec{E}^r$,

$$\vec{B}^r = \frac{1}{c}\hat{r}\times\vec{E}^r; \tag{3.68}$$

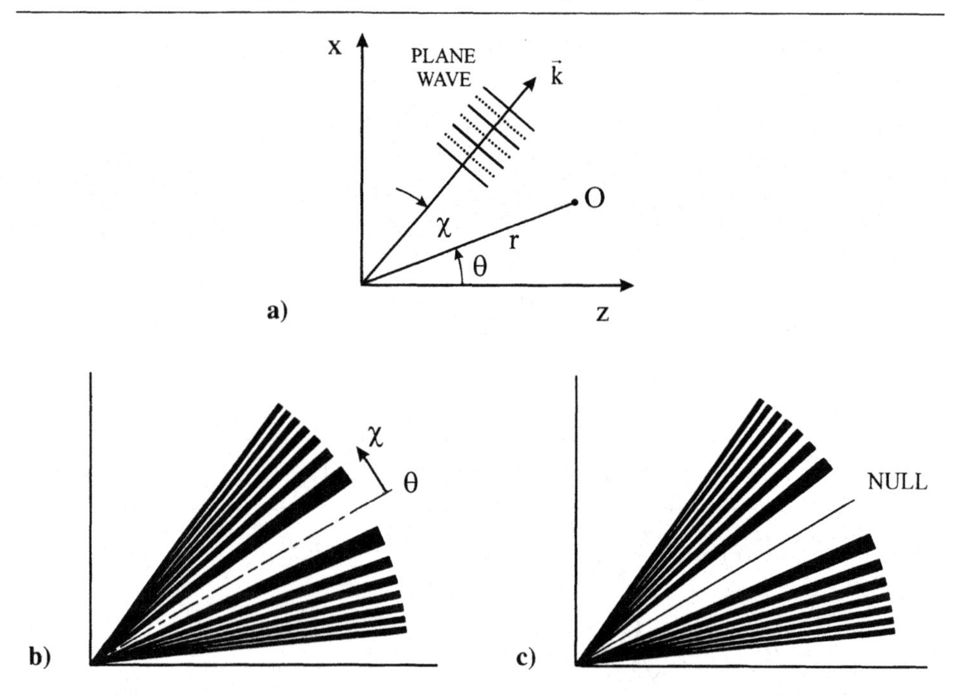

Fig. 3.14. a) Coordinates for the plane wave and observation point O. b) Regions where $\vec{k}$ lies for positive (white) and negative (black) contributions to the real part of $\exp[jk_o r(1 - \cos\chi)]$. c) Same as b), except for the imaginary part. Drawings are for $k_o r = 160\pi$.

and a complex Poynting vector in the direction of propagation, $\hat{r}$:

$$\vec{S}_c^r = \frac{1}{2\zeta_o}|\vec{E}^r|^2\hat{r} = \hat{r}\frac{k_o}{4\pi\zeta_o r}\begin{cases}\cos^2\theta\left|F_y(k_o\sin\theta)\right|^2, & \text{(TE)}\\ \left|F_x(k_o\sin\theta)\right|^2, & \text{(TM).}\end{cases} \tag{3.69}$$

Probably the most striking feature of the radiated field (3.67) is that its angular dependence is simply related to the spectral function $F_y(k_o\sin\psi)$ evaluated at the angle of observation $\psi = \theta$. In other words, the radiated field depends only on the plane wave in the spectrum that is propagating in the direction of observation. There is a simple physical argument that elucidates this result; it is closely related to the mathematical argument presented above.

Consider the plane wave in the continuous spectrum propagating at a small angle χ with respect to the direction of observation θ, as sketched in Figure 3.14a. The field of this wave at the observation point O is proportional to

$$F_y\left[k_o\sin(\theta+\chi)\right]e^{-jk_o r}e^{jk_o r(1-\cos\chi)}. \tag{3.70}$$

We will assume that the spectral function is a slowly varying function of the angle χ and concentrate on the last factor:

$$e^{jk_o r(1-\cos\chi)} = \cos\left[k_o r(1-\cos\chi)\right] + j\sin\left[k_o r(1-\cos\chi)\right].$$

The real and imaginary parts of this function may be positive or negative depending on the angle χ. In Figure 3.14b, we show the angular sectors where the angle χ, or the vector wave number $\vec{k}$, lies for the real part of this function to be positive (white areas) and negative (black areas). Figure 3.14c is a similar drawing for the imaginary part. From these pictures it is clear that away from the direction of observation (θ), adjacent regions representing positive and negative contributions at O have equal areas and, therefore, they approximately cancel. Only the plane waves with vector wave numbers lying within a narrow angular sector surrounding the direction of observation contribute to the field; hence, the factor $F_y(k_o \sin\theta)$ appears in the expression (3.67). Stated differently, all of the plane waves in the spectrum destructively interfere at the observation point, except for a small group of waves propagating in the direction of observation. The portion of the spectrum – the angular sector surrounding θ – that contributes to the field decreases with increasing $k_o r$. To see this, assume that the contribution comes roughly from the angular sector for which $|k_o r(1-\cos\chi)| \leq \pi/2$; in Figure 3.14b this is the white sector centered about θ. For $k_o r \gg 1$, the angular extent of this sector is approximately $2\chi \approx 2\sqrt{\pi/k_o r}$. The portion of the spectrum that contributes to the field decreases as $1/\sqrt{k_o r}$ with increasing $k_o r$; hence, the factor $1/\sqrt{k_o r}$ appears in the expression (3.67).

Let us now apply the formulas for the radiated field to our example – the uniformly illuminated slit. From the spectral functions (3.60) and (3.61), we can obtain the radiated electric fields. For the TE field we get

$$\vec{E}^r = \hat{y}\,\frac{E_o^i k_o d}{\sqrt{2\pi}}\cos\theta\left[\frac{\sin\left(\frac{k_o d}{2}\sin\theta\right)}{\frac{k_o d}{2}\sin\theta}\right]\frac{1}{\sqrt{k_o r}}e^{-j(k_o r-\pi/4)}, \tag{3.71}$$

and the TM field is found to be

$$\vec{E}^r = \hat{\theta}\,\frac{E_o^i k_o d}{\sqrt{2\pi}}\left[\frac{\sin\left(\frac{k_o d}{2}\sin\theta\right)}{\frac{k_o d}{2}\sin\theta}\right]\frac{1}{\sqrt{k_o r}}e^{-j(k_o r-\pi/4)}. \tag{3.72}$$

These fields, in normalized form $|\vec{E}^r(\theta)|/|\vec{E}^r(0)|$, are graphed as a function of θ in Figure 3.15 (dashed lines). Results are shown for two widths of the slit ($d/\lambda_o = 5$ and 10); these are the same widths used for Figure 3.8.

Recall that the uniformly illuminated slit is an approximation to the physical problem shown in Figure 3.7. Very good estimates for the radiated field for this physical problem are also graphed in Figure 3.15 (solid lines).[6] The two sets of data are seen to be in good agreement, except at angles well away from the normal where the levels are low. As expected, the agreement improves with increasing slit width. The reason for the good agreement observed in Figure 3.15 is fairly simple. The difference between the actual field in the slit (Figure 3.8) and our approximation of uniform illumination is a rapidly varying field that is mostly a result of the evanescent portion of the plane-wave spectrum. Because the evanescent spectrum does not contribute to the radiated field, the radiated fields for the two cases are nearly the same.

The asymptotic expressions we have developed apply in the limit $k_o r \to \infty$; however, we expect these expressions to be reasonable approximations to the actual

[6] The results presented in Figure 3.15 were computed using Babinet's principle (Section 4.4) and formulas for electrically wide strips given in Reference [11].

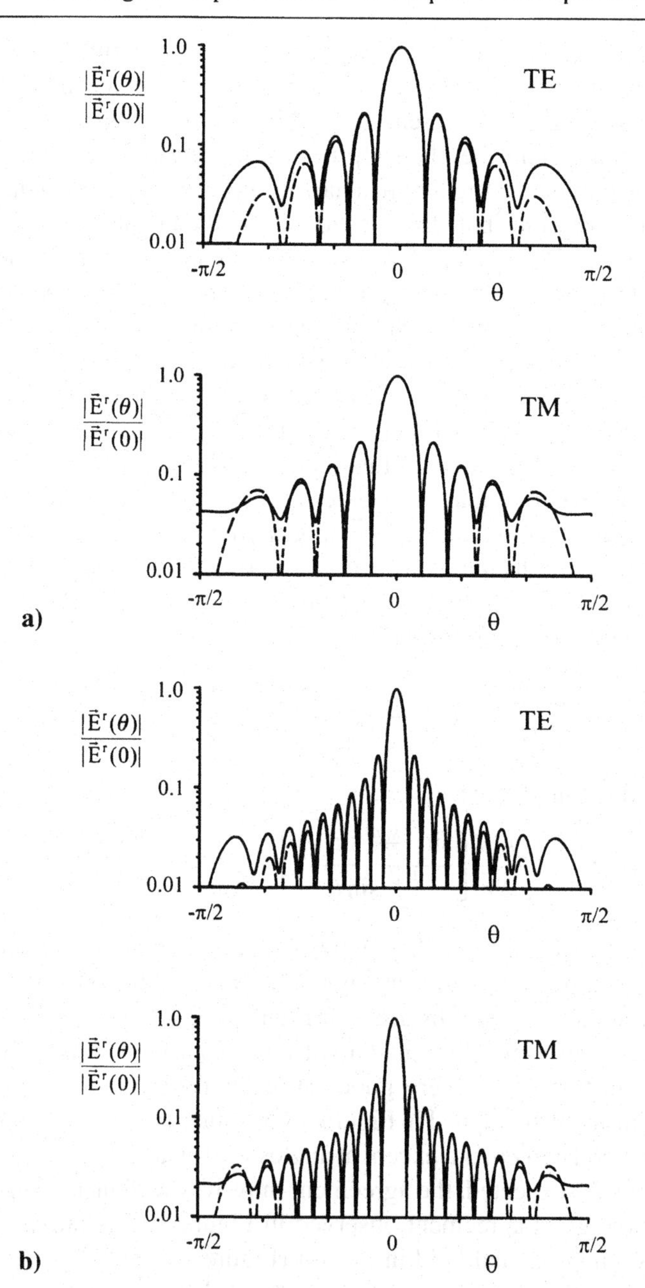

Fig. 3.15. Comparison of normalized radiated electric fields. Solid line: good estimate for electrically wide slit. Dashed line: uniformly illuminated slit. a) $d/\lambda_o = 5.0$. b) $d/\lambda_o = 10.0$.

fields when $k_o r$ is simply large. The question then is how large must $k_o r$ be for the expressions to be useful? A precise answer to this question, of course, will depend on the specific field we started with on the plane $z = 0$. However, we can make an estimate by examining the uniformly illuminated slit.

Recall that the contribution to the integral (3.66) comes from the region surrounding the stationary point. We will again assume that $k_o r \gg 1$, so that this region is roughly of angular extent $2\chi \approx 2\sqrt{\pi/k_o r}$ (the white sector centered about θ in Figure 3.14b). For the asymptotic evaluation, the spectral function, F_y or F_x, must be approximately constant over this angle.

Now consider the spectral function for the uniformly illuminated slit shown in Figure 3.9b. We see that the spectral function is fairly constant within a region $|\Delta k_x(d/2)| \ll \pi$, that is, well within a region bounded by two consecutive nulls. In terms of the angles θ and χ, this inequality becomes

$$\left|k_o \sin(\theta + \chi)d/2 - k_o \sin(\theta - \chi)d/2\right| = k_o d|\cos\theta \sin\chi| \le k_o d\chi \ll \pi, \quad (3.73)$$

or

$$2\chi \ll 2\pi/k_o d.$$

Our criterion for the asymptotic evaluation now becomes

$$2\chi \approx 2\sqrt{\pi/k_o r} \ll 2\pi/k_o d,$$

or

$$k_o r \gg (k_o d)^2/\pi$$

$$r \gg 2d^2/\lambda_o. \quad (3.74)$$

This inequality together with the inequality $k_o r \gg 1$ determines the region where the radiated field is applicable for the uniformly illuminated slit. As an illustration, we have marked the radius $r = 2d^2/\lambda_o$ ($k_o r = 314$) in Figure 3.11b. The graphical results (E,F) indicate that the electric field has approximately the asymptotic form beyond this point.

3.6 Plane-wave spectrum for three-dimensional electromagnetic fields

The simplicity associated with working in two dimensions has allowed us to easily examine the phenomena associated with the plane-wave spectral representation of electromagnetic fields. Most practical problems, however, involve three-dimensional fields, and fortunately the results for three dimensions can be readily obtained using arguments that parallel those for the two-dimensional case.

3.6.1 General formulation

Our problem is again one of determining the electromagnetic field everywhere in the half space $z > 0$ when the tangential components of the electric or magnetic field are specified on the plane $z = 0$. Now the tangential field on this plane will be a function of two variables, x and y, rather than one variable as for the two-dimensional case:

$$\vec{E}(x, y, z = 0) = \vec{E}_x(x, y, z = 0) + \vec{E}_y(x, y, z = 0) = f_x(x, y)\hat{x} + f_y(x, y)\hat{y}. \quad (3.75)$$

The electromagnetic field in the half space $z > 0$ will be represented by a superposition of plane waves, each with an electric field of the form

$$\vec{E}(\vec{r}) = (E_{ox}\hat{x} + E_{oy}\hat{y} + E_{oz}\hat{z})e^{-j\vec{k}\cdot\vec{r}}, \tag{3.76}$$

where the vector wave number is

$$\vec{k} = k_x\hat{x} + k_y\hat{y} + k_z\hat{z}. \tag{3.77}$$

The plane waves are transverse, so

$$\vec{k} \cdot \vec{E}(\vec{r}) = 0, \tag{3.78}$$

which on substitution of (3.76) and (3.77) gives

$$E_{oz} = -\frac{1}{k_z}(k_x E_{ox} + k_y E_{oy}). \tag{3.79}$$

The vector wave number must satisfy (3.8),

$$\vec{k} \cdot \vec{k} = k_x^2 + k_y^2 + k_z^2 = k_o^2, \tag{3.80}$$

and with k_x and k_y taken to be real, we have

$$k_z = \sqrt{k_o^2 - k_x^2 - k_y^2}, \qquad 0 \le k_x^2 + k_y^2 \le k_o^2 \tag{3.81a}$$

$$k_z = -j\sqrt{k_x^2 + k_y^2 - k_o^2}, \qquad k_x^2 + k_y^2 > k_o^2. \tag{3.81b}$$

The values of k_z in the first range (3.81a) represent propagating waves,

$$e^{-j\vec{k}\cdot\vec{r}} = e^{-j(k_x x + k_y y + k_z z)},$$

whereas those in the second range (3.81b) represent evanescent waves that decay with increasing z,

$$e^{-j\vec{k}\cdot\vec{r}} = e^{-|k_z|z}\, e^{-j(k_x x + k_y y)}.$$

On substitution of (3.77) and (3.79), the electric field for the plane wave (3.76) becomes

$$\vec{E}(\vec{r}) = \left\{E_{ox}\hat{x} + E_{oy}\hat{y} - \left(\frac{k_o}{k_z}\right)\left[\left(\frac{k_x}{k_o}\right)E_{ox} + \left(\frac{k_y}{k_o}\right)E_{oy}\right]\hat{z}\right\}e^{-j(k_x x + k_y y + k_z z)}, \tag{3.82a}$$

with the accompanying magnetic field obtained from (3.9):

$$\begin{aligned}\vec{B}(\vec{r}) = \frac{-1}{c}\Bigg[\left(\frac{k_o}{k_z}\right)\Bigg(&\left\{\left(\frac{k_x}{k_o}\right)\left(\frac{k_y}{k_o}\right)E_{ox} + \left[1 - \left(\frac{k_x}{k_o}\right)^2\right]E_{oy}\right\}\hat{x} \\ &- \left\{\left(\frac{k_x}{k_o}\right)\left(\frac{k_y}{k_o}\right)E_{oy} + \left[1 - \left(\frac{k_y}{k_o}\right)^2\right]E_{ox}\right\}\hat{y}\Bigg) \\ &- \left[\left(\frac{k_x}{k_o}\right)E_{oy} - \left(\frac{k_y}{k_o}\right)E_{ox}\right]\hat{z}\Bigg]e^{-j(k_x x + k_y y + k_z z)}.\end{aligned} \tag{3.82b}$$

In the half space $z > 0$, the following electromagnetic field, which is a superposition of plane waves of the form (3.82a), (3.82b), is a solution to Maxwell's equations:

$$\vec{E}(\vec{r}) = \frac{1}{(2\pi)^2} \int_{-\infty}^{\infty} \int_{-\infty}^{\infty} \left\{ F_x(k_x, k_y)\hat{x} + F_y(k_x, k_y)\hat{y} - \left(\frac{k_o}{k_z}\right)\left[\left(\frac{k_x}{k_o}\right) F_x(k_x, k_y) + \left(\frac{k_y}{k_o}\right) F_y(k_x, k_y)\right]\hat{z} \right\} e^{-j(k_x x + k_y y + k_z z)} dk_x dk_y, \tag{3.83a}$$

$$\begin{aligned}\vec{B}(\vec{r}) = \frac{-1}{(2\pi)^2 c} \int_{-\infty}^{\infty} \int_{-\infty}^{\infty} \Bigg[\left(\frac{k_o}{k_z}\right)\Bigg(\Bigg\{\left(\frac{k_x}{k_o}\right)\left(\frac{k_y}{k_o}\right) F_x(k_x, k_y) \\ + \left[1 - \left(\frac{k_x}{k_o}\right)^2\right] F_y(k_x, k_y)\Bigg\}\hat{x} - \Bigg\{\left(\frac{k_x}{k_o}\right)\left(\frac{k_y}{k_o}\right) F_y(k_x, k_y) \\ + \left[1 - \left(\frac{k_y}{k_o}\right)^2\right] F_x(k_x, k_y)\Bigg\}\hat{y}\Bigg) - \Bigg[\left(\frac{k_x}{k_o}\right) F_y(k_x, k_y) \\ - \left(\frac{k_y}{k_o}\right) F_x(k_x, k_y)\Bigg]\hat{z}\Bigg] e^{-j(k_x x + k_y y + k_z z)} dk_x dk_y.\end{aligned} \tag{3.83b}$$

Here the spectral functions F_x and F_y play the same role as E_{ox} and E_{oy} in Equations (3.82a) and (3.82b). On the plane $z = 0$, the tangential component of the electric field (3.83a) must equal the specified electric field (3.75); thus,

$$f_x(x, y)\hat{x} + f_y(x, y)\hat{y} = \frac{1}{(2\pi)^2} \int_{-\infty}^{\infty} \int_{-\infty}^{\infty} \left[F_x(k_x, k_y)\hat{x} + F_y(k_x, k_y)\hat{y}\right] e^{-j(k_x x + k_y y)} dk_x dk_y. \tag{3.84}$$

After introducing the two-dimensional Fourier transform pair

$$f(x, y) = \frac{1}{(2\pi)^2} \int_{-\infty}^{\infty} \int_{-\infty}^{\infty} F(\xi, \eta) e^{-j(\xi x + \eta y)} d\xi d\eta \tag{3.85a}$$

$$F(\xi, \eta) = \int_{-\infty}^{\infty} \int_{-\infty}^{\infty} f(x, y) e^{j(\xi x + \eta y)} dx dy, \tag{3.85b}$$

it is evident that the spectral functions F_x and F_y are determined by the following Fourier integrals:

$$\begin{aligned} F_i(k_x, k_y) &= \int_{-\infty}^{\infty} \int_{-\infty}^{\infty} f_i(x, y) e^{j(k_x x + k_y y)} dx dy \\ &= \int_{-\infty}^{\infty} \int_{-\infty}^{\infty} \hat{i} \cdot \vec{E}(x, y, z = 0) e^{j(k_x x + k_y y)} dx dy, \end{aligned} \tag{3.86}$$

with $i = x, y$.

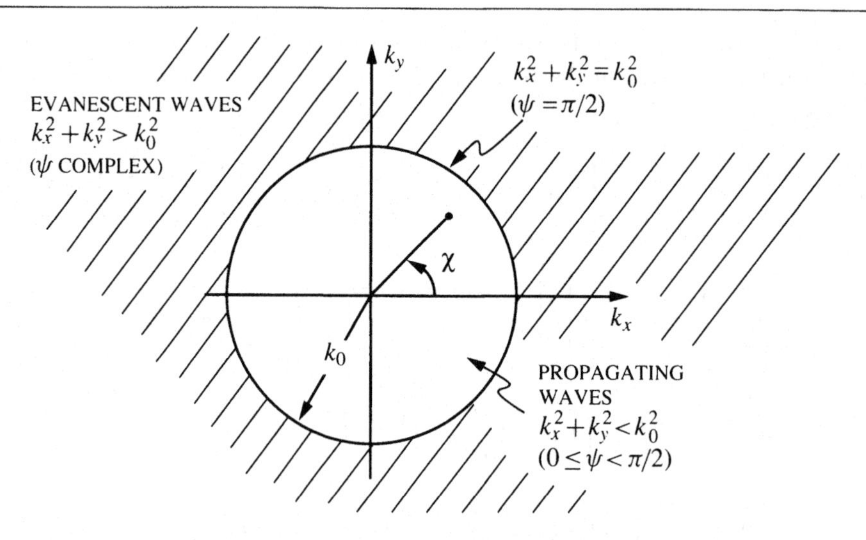

Fig. 3.16. The k_x-k_y plane showing regions for propagating and evanescent waves.

Each of the plane waves in the superposition (3.83a) is represented by a point in the $k_x - k_y$ plane of Figure 3.16. The circle $k_x^2 + k_y^2 = k_o^2$ is the boundary between the region where the waves are propagating (interior) and the region where the waves are evanescent (exterior). The point at the origin (center of the circle) is a wave propagating normal to the plane $z = 0$ ($\vec{k} = k_o\hat{z}$); points on the circle are for waves propagating along the plane $z = 0$ ($\vec{k} = k_x\hat{x} + k_y\hat{y}$).

Equations (3.83a) and (3.83b) together with (3.86) describe the three-dimensional electromagnetic field we sought; it is a solution to Maxwell's equations in the half space $z > 0$ and is equal to the tangential electric field on the plane $z = 0$. These results are summarized in Table 3.3 together with those for the case where the tangential components of the magnetic field, rather than the tangential components of the electric field, are specified on the plane $z = 0$.

3.6.2 *Asymptotic or radiated field*

The asymptotic forms for the three-dimensional field are obtained using the same procedures as for the two-dimensional field. First, we will represent the position (x, y, z) and vector wave number $\vec{k}$ in terms of the spherical coordinates shown in Figure 3.17.[7] The position in the coordinate system (r, θ, ϕ) is

$$x = r\sin\theta\cos\phi, \qquad y = r\sin\theta\sin\phi, \qquad z = r\cos\theta, \tag{3.87}$$

with $0 \leq \theta \leq \pi/2$, $-\pi \leq \phi \leq \pi$, and the components of the vector wave number in the coordinate system (k_o, ψ, χ) are

$$k_x = k_o\cos\chi\sin\psi, \qquad k_y = k_o\sin\chi\sin\psi, \qquad k_z = k_o\cos\psi. \tag{3.88}$$

[7] Note that these coordinates are consistent with those used earlier in Figure 3.5 for the two-dimensional case. The latter are obtained from the former by setting $\phi = 0$ and $\chi = 0$.

Table 3.3. *Spectral representations for three-dimensional fields*

$$\vec{E}(\vec{r}) = \frac{1}{(2\pi)^2} \int_{-\infty}^{\infty} \int_{-\infty}^{\infty} \left\{ F_x(k_x, k_y)\,\hat{x} + F_y(k_x, k_y)\,\hat{y} - \left(\frac{k_o}{k_z}\right) \left[\left(\frac{k_x}{k_o}\right) F_x(k_x, k_y) + \left(\frac{k_y}{k_o}\right) F_y(k_x, k_y) \right] \hat{z} \right\} e^{-j(k_x x + k_y y + k_z z)} dk_x dk_y$$

$$\vec{B}(\vec{r}) = \frac{-1}{(2\pi)^2 c} \int_{-\infty}^{\infty} \int_{-\infty}^{\infty} \left[\left(\frac{k_o}{k_z}\right) \left(\left\{ \left(\frac{k_x}{k_o}\right) \left(\frac{k_y}{k_o}\right) F_x(k_x, k_y) + \left[1 - \left(\frac{k_x}{k_o}\right)^2 \right] F_y(k_x, k_y) \right\} \hat{x} - \left\{ \left(\frac{k_x}{k_o}\right) \left(\frac{k_y}{k_o}\right) F_y(k_x, k_y) + \left[1 - \left(\frac{k_y}{k_o}\right)^2 \right] F_x(k_x, k_y) \right\} \hat{y} \right) - \left[\left(\frac{k_x}{k_o}\right) F_y(k_x, k_y) - \left(\frac{k_y}{k_o}\right) F_x(k_x, k_y) \right] \hat{z} \right] e^{-j(k_x x + k_y y + k_z z)} dk_x dk_y$$

$$F_i(k_x, k_y) = \int_{-\infty}^{\infty} \int_{-\infty}^{\infty} \hat{\imath} \cdot \vec{E}(x, y, z = 0)\, e^{j(k_x x + k_y y)} dx dy$$

$$F_i(k_x, k_y) = \pm c \left(\frac{k_o}{k_z}\right) \int_{-\infty}^{\infty} \int_{-\infty}^{\infty} \left\{ \left(\frac{k_x}{k_o}\right) \left(\frac{k_y}{k_o}\right) \hat{\imath} \cdot \vec{B}(x, y, z = 0) \mp \left[1 - \left(\frac{k_i}{k_o}\right)^2 \right] (\hat{\imath} \times \hat{z}) \cdot \vec{B}(x, y, z = 0) \right\} e^{j(k_x x + k_y y)} dx dy$$

$i = x$, top sign
$i = y$, bottom sign

For the *propagating waves* ($0 \le k_x^2 + k_y^2 \le k_o^2$), the components of the wave number are real and the angles are in the ranges

$$0 \le \psi \le \pi/2$$
$$-\pi \le \chi \le \pi. \tag{3.89}$$

For the *evanescent waves* ($k_x^2 + k_y^2 > k_o^2$), the component k_z of the wave number is complex, as is the angle $\psi = \psi_r + j\psi_i$; thus, (3.88) becomes

$$k_x = k_o \cos\chi(\sin\psi_r \cosh\psi_i + j\cos\psi_r \sinh\psi_i)$$
$$k_y = k_o \sin\chi(\sin\psi_r \cosh\psi_i + j\cos\psi_r \sinh\psi_i) \tag{3.90}$$
$$k_z = k_o(\cos\psi_r \cosh\psi_i - j\sin\psi_r \sinh\psi_i).$$

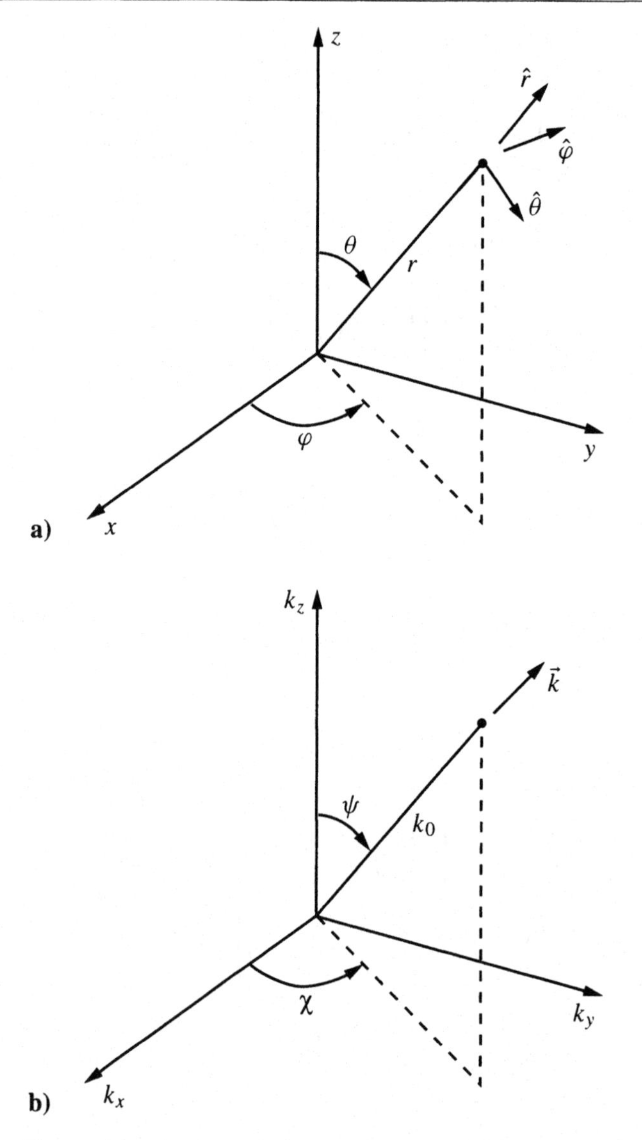

Fig. 3.17. a) The position in the spherical coordinate system r, θ, ϕ. b) The components of the vector wave number expressed in terms of k_o, ψ, and χ. The drawing is for a wave number with k_z real.

Because k_x and k_y are always real, $\psi_r = \pi/2$. Since k_z is chosen negative imaginary for evanescent waves (3.81b), we must have $0 < \psi_i \leq \infty$ when $\psi_r = \pi/2$. The angles for the evanescent waves are, therefore,

$$\psi = \pi/2 + j\psi_i, \qquad 0 < \psi_i \leq \infty$$
$$-\pi \leq \chi \leq \pi. \tag{3.91}$$

After introducing the changes in coordinates (3.87) and (3.88), the electric field (3.83a) becomes

$$\vec{E}(\vec{r}) = \frac{k_o^2}{(2\pi)^2} e^{-jk_o r} \int_{C_{1/2}} \int_{\chi=-\pi}^{\pi} \Big\{ F_x(k_o \cos\chi \sin\psi,\ k_o \sin\chi \sin\psi)\hat{x}$$

$$+ F_y(k_o \cos\chi \sin\psi,\ k_o \sin\chi \sin\psi)\hat{y}$$

$$- \frac{1}{\cos\psi}\big[\cos\chi \sin\psi\, F_x(k_o \cos\chi \sin\psi,\ k_o \sin\chi \sin\psi)$$

$$+ \sin\chi \sin\psi\, F_y(k_o \cos\chi \sin\psi,\ k_o \sin\chi \sin\psi)\big]\hat{z}\Big\}$$

$$e^{-jk_o r\{\sin\theta \sin\psi[\cos(\phi-\chi)-1]+[\cos(\theta-\psi)-1]\}} \cos\psi \sin\psi\, d\chi\, d\psi. \tag{3.92}$$

Here we have used the following relationship for the elemental areas:[8]

$$dk_x dk_y = k_o^2 \cos\psi \sin\psi\, d\chi\, d\psi. \tag{3.93}$$

The contour $C_{1/2}$ is in the right half of the complex ψ plane, and it is the same as the right half of the contour C for the two-dimensional case (Figure 3.6).

In the limit as $k_o r \rightarrow \infty$, the evanescent portion of the spectrum does not contribute to the integral (3.92); thus, the contour $C_{1/2}$ becomes the segment $0 \leq \psi \leq \pi/2$, and we are left with integrals of the following form to evaluate for each component of the field:

$$I = \int_{\psi=0}^{\pi/2} \int_{\chi=-\pi}^{\pi} W(\psi, \chi) e^{-jk_o r\{\sin\theta \sin\psi[\cos(\chi-\phi)-1]+[\cos(\psi-\theta)-1]\}} d\chi\, d\psi. \tag{3.94}$$

The asymptotic approximation for this double integral is obtained by applying the method of stationary phase.

The phase of the exponential

$$g(\psi, \chi) = -k_o r\Big\{\sin\theta \sin\psi\big[\cos(\chi - \phi) - 1\big] + \big[\cos(\psi - \theta) - 1\big]\Big\}$$

has a stationary point whenever

$$\frac{\partial g}{\partial \psi} = -k_o r\Big\{\sin\theta \cos\psi\big[\cos(\chi - \phi) - 1\big] - \sin(\psi - \theta)\Big\} = 0$$

and

$$\frac{\partial g}{\partial \chi} = k_o r\big[\sin\theta \sin\psi \sin(\chi - \phi)\big] = 0,$$

[8] This relationship is obtained by projecting the element of surface area, $k_o^2 \sin\psi\, d\chi\, d\psi$, on the sphere of radius k_o onto the k_x–k_y plane: $dk_x dk_y = (\hat{z} \cdot \hat{k}) k_o^2 \sin\psi\, d\chi\, d\psi = k_o^2 \cos\psi \sin\psi\, d\chi\, d\psi$.

which occurs for $\psi = \theta$, $\chi = \phi$. The Taylor series expansion that approximates the phase near this point is

$$g(\psi,\chi) \approx g(\theta,\phi) + \frac{\partial g}{\partial \psi}\bigg|_{\substack{\psi=\theta\\ \chi=\phi}}(\psi-\theta) + \frac{\partial g}{\partial \chi}\bigg|_{\substack{\psi=\theta\\ \chi=\phi}}(\chi-\phi) + \frac{1}{2}\left[\frac{\partial^2 g}{\partial \psi^2}\bigg|_{\substack{\psi=\theta\\ \chi=\phi}}(\psi-\theta)^2\right.$$

$$\left. + 2\frac{\partial^2 g}{\partial \psi \partial \chi}\bigg|_{\substack{\psi=\theta\\ \chi=\phi}}(\psi-\theta)(\chi-\phi) + \frac{\partial^2 g}{\partial \chi^2}\bigg|_{\substack{\psi=\theta\\ \chi=\phi}}(\chi-\phi)^2\right]$$

$$= \frac{1}{2}k_o r\left[(\psi-\theta)^2 + \sin^2\theta(\chi-\phi)^2\right].$$

After substituting this result, the integral (3.94) becomes

$$I \approx W(\theta,\phi)\int_{\psi=0}^{\pi/2}\int_{\chi=-\pi}^{\pi} e^{jk_o r(\psi-\theta)^2/2} e^{jk_o r\sin^2\theta(\chi-\phi)^2/2}\, d\chi\, d\psi$$

$$\approx \frac{2W(\theta,\phi)}{k_o r\sin\theta}\int_{-\infty}^{\infty}\int_{-\infty}^{\infty} e^{j\xi^2} e^{j\eta^2}\, d\eta\, d\xi$$

$$= \frac{j2\pi}{k_o r\sin\theta}W(\theta,\phi). \tag{3.95}$$

Now using (3.95) and (3.92) and changing to the unit vectors $\hat{r}$, $\hat{\theta}$, and $\hat{\phi}$, we obtain the radiated electric field

$$\vec{E}^r(\vec{r}) = \frac{jk_o}{2\pi}\Big\{\big[\cos\phi F_x(k_o\cos\phi\sin\theta, k_o\sin\phi\sin\theta)$$

$$+ \sin\phi F_y(k_o\cos\phi\sin\theta,\ k_o\sin\phi\sin\theta)\big]\hat{\theta}$$

$$+ \cos\theta\big[-\sin\phi F_x(k_o\cos\phi\sin\theta,\ k_o\sin\phi\sin\theta)$$

$$+ \cos\phi\ F_y(k_o\cos\phi\sin\theta,\ k_o\sin\phi\sin\theta)\big]\hat{\phi}\Big\}e^{-jk_o r}/r, \tag{3.96}$$

or

$$\vec{E}^r(\vec{r}) = \big[F_\theta(\theta,\phi)\hat{\theta} + F_\phi(\theta,\phi)\hat{\phi}\big]e^{-jk_o r}/r, \tag{3.97}$$

where we have introduced the pattern functions F_θ and F_ϕ. The accompanying magnetic field, obtained in a similar manner, is

$$\vec{B}^r(\vec{r}) = \frac{1}{c}\big[-F_\phi(\theta,\phi)\hat{\theta} + F_\theta(\theta,\phi)\hat{\phi}\big]e^{-jk_o r}/r. \tag{3.98}$$

Note that the radial and angular dependences for the radiated field are separate, with the angular dependence for the two components of the field determined by the pattern functions F_θ and F_ϕ. These results are summarized in Table 3.4.

As in the case for two-dimensional fields, the three-dimensional radiated field has many of the same features as a homogeneous plane wave, with the vector $k_o\hat{r}$ assuming the role of the vector wave number $\vec{k} = k_o\hat{k}$: a transverse field with an exponential variation of the phase with radial distance, $\exp(-jk_o r)$; a simple

Table 3.4. *Spectral representations for three-dimensional radiated fields*

$$\vec{E}^r(\vec{r}) = \left[F_\theta(\theta,\phi)\hat{\theta} + F_\phi(\theta,\phi)\hat{\phi}\right]e^{-jk_o r}/r$$

$$\vec{B}^r(\vec{r}) = \frac{1}{c}\left[-F_\phi(\theta,\phi)\hat{\theta} + F_\theta(\theta,\phi)\hat{\phi}\right]e^{-jk_o r}/r$$

$$= \frac{1}{c}\hat{r} \times \vec{E}^r(\vec{r})$$

$$F_\theta(\theta,\phi) = \frac{jk_o}{2\pi}\big[\cos\phi\, F_x(k_o\cos\phi\sin\theta, k_o\sin\phi\sin\theta)$$

$$+ \sin\phi\, F_y(k_o\cos\phi\sin\theta, k_o\sin\phi\sin\theta)\big]$$

$$F_\phi(\theta,\phi) = \frac{jk_o}{2\pi}\big\{\cos\theta\big[-\sin\phi\, F_x(k_o\cos\phi\sin\theta, k_o\sin\phi\sin\theta)$$

$$+ \cos\phi\, F_y(k_o\cos\phi\sin\theta, k_o\sin\phi\sin\theta)\big]\big\}$$

relationship between $\vec{B}^r$ and $\vec{E}^r$,

$$\vec{B}^r = \frac{1}{c}\hat{r} \times \vec{E}^r; \tag{3.99}$$

and a complex Poynting vector in the direction of propagation, $\hat{r}$:

$$\vec{S}_c^r = \frac{1}{2\zeta_o}|\vec{E}^r|^2\hat{r}$$

$$= \frac{1}{2\zeta_o r^2}\left[\left|F_\theta(\theta,\phi)\right|^2 + \left|F_\phi(\theta,\phi)\right|^2\right]\hat{r}. \tag{3.100}$$

Application of our results for three-dimensional fields to a uniformly illuminated aperture on the plane $z = 0$ would show that the inequality (3.74)

$$r \gg 2d^2/\lambda_o \tag{3.101}$$

together with the inequality $k_o r \gg 1$ again determines the region where the radiated field is applicable. Now, however, d is the maximum dimension of the aperture.

3.6.3 Wavefronts and rays

The structure we have observed for the radiated field (3.97)–(3.100), in particular its similarity to the field of a homogeneous plane wave, makes possible a simple geometrical picture for this field.

First we will consider the homogeneous plane wave shown in Figure 3.18a. The *geometrical wavefronts* are surfaces of constant phase. For the plane wave they are plane surfaces normal to the direction of propagation $\hat{k}$. The *geometrical rays*

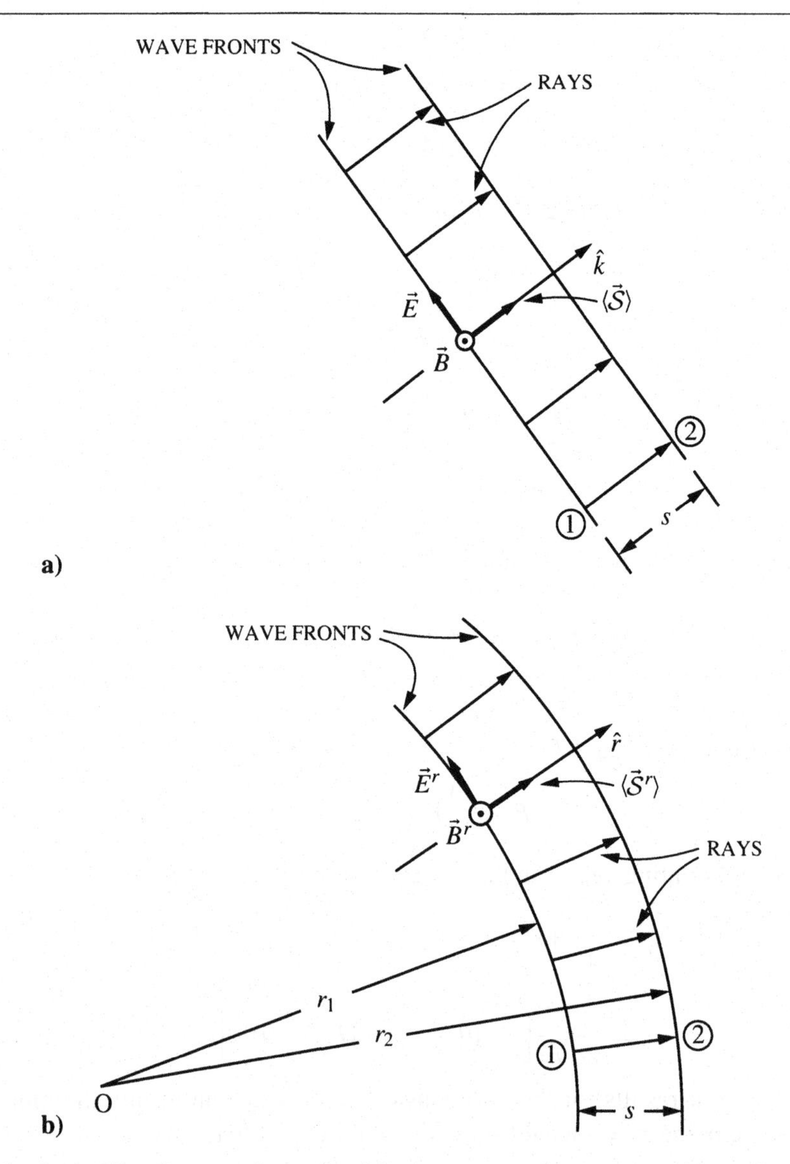

Fig. 3.18. Wavefronts and rays for a) homogeneous plane wave and b) radiated field.

are vectors that indicate the direction of the time-average energy flow. This is the direction of the real part of the complex Poynting vector $\langle \vec{S} \rangle = \mathrm{Re}(\vec{S}_c)$. For the plane wave the rays are in the direction of $\hat{k}$ and are orthogonal to the wavefronts. Although the rays specify the direction of the time-average energy flow, they tell us nothing about its magnitude. For the plane wave, of course, the time-average power per unit area of wavefront is a constant. Thus, the transport of energy along the trajectory of a ray from a point on wavefront 1 to a point on wavefront 2 is simply expressed:

$$\langle \vec{S}_2 \rangle = \langle \vec{S}_1 \rangle. \tag{3.102a}$$

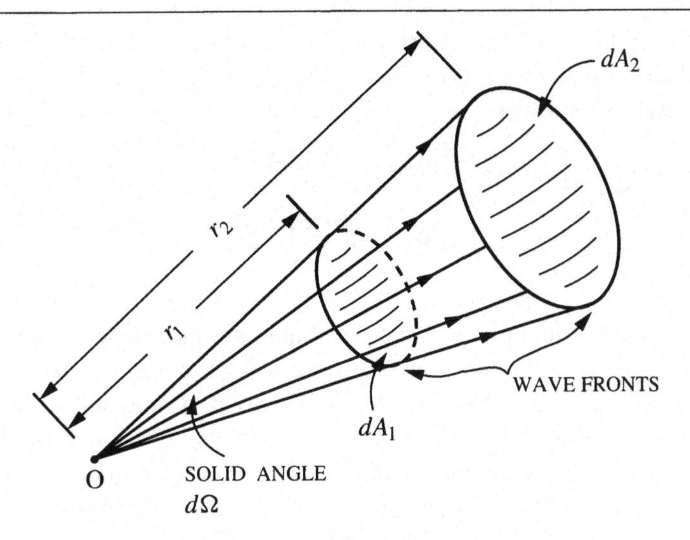

Fig. 3.19. Conical tube of rays.

The relationship for the electric fields at these points is

$$\vec{E}_2 = e^{-jk_o s}\vec{E}_1, \tag{3.102b}$$

where s is the length of the ray joining these points.

Now let us return to the radiated field, which is shown in Figure 3.18b. The electric field (3.97) is the product of two factors, $[F_\theta(\theta, \phi)\hat{\theta} + F_\phi(\theta, \phi)\hat{\phi}]/r$ and the exponential $\exp(-jk_o r)$. For small relative changes in position, only the exponential factor changes significantly, and this only affects the phase of the field.[9] The surfaces of constant phase, the *geometrical wavefronts*, are therefore locally determined by $k_o r$ = constant and are spherical. The time-average energy flow for the radiated field – the real part of the complex Poynting vector (3.100) – is in the radial direction. Thus, the *geometrical rays* are in the radial direction and are orthogonal to the wavefronts.

The transport equation relating the time-average powers per unit area of wavefront at two points along the trajectory of a ray is obtained from (3.100):

$$\langle \vec{S}_2^r \rangle = \left(\frac{r_1}{r_2}\right)^2 \langle \vec{S}_1^r \rangle = \left(\frac{r_1}{r_1 + s}\right)^2 \langle \vec{S}_1^r \rangle, \tag{3.103a}$$

and the relationship for the electric fields at these points is, from (3.97),

$$\vec{E}_2^r = \left(\frac{r_1}{r_1 + s}\right) e^{-jk_o s}\vec{E}_1^r. \tag{3.103b}$$

The geometrical interpretation of Equation (3.103a) is obtained from the narrow "tube" or "pencil" of rays shown in Figure 3.19. Each ray is a radial line from the origin O, and the conical tube of rays subtends the solid angle $d\Omega$. The tube cuts out the areas $dA_1 = r_1^2 d\Omega$ and $dA_2 = r_2^2 d\Omega$ from the wavefronts 1 and 2, respectively.

[9] The change in position is much less than r; however, the change in position may be several wavelengths.

Since the energy flow is in the direction of the rays, no energy flows through the sides of the tube. Conservation of energy then dictates that the energies flowing outward through the surfaces dA_1 and dA_2 must be the same:

$$\langle \vec{S}_2^r \rangle dA_2 = \langle \vec{S}_1^r \rangle dA_1,$$

$$\langle \vec{S}_2^r \rangle = \left(\frac{r_1}{r_2} \right)^2 \langle \vec{S}_1^r \rangle,$$

which is (3.103a).

Our simple geometrical picture for the radiated field is now complete: The geometrical wavefronts are the local surfaces of constant phase, the geometrical rays are in the direction of the local time-average energy flow, and energy conservation within a tube of rays is expressed by the transport equation (3.103a). We should mention that two observed features of these rays are a consequence of the electrical properties of the medium, free space: The rays are straight lines because the medium is homogeneous, and the rays are orthogonal to the wavefronts because the medium is isotropic.

The terminology we have introduced for the radiated field has its roots in classical *geometrical optics*, a science that was well developed before the advent of Maxwell's equations. Geometrical optics uses the concepts of wavefronts and rays to describe the propagation, reflection, and refraction of light, such as in the familiar analysis of a lens based on ray tracing. The connection between geometrical optics and electromagnetic theory, as expressed by Maxwell's equations, is examined in detail in References [12] and [13].

3.7 Examples of three-dimensional fields

As a starting point for the application of our results for the plane-wave spectrum, we must know the tangential components of the electric or magnetic field on a plane surface. The location of this reference plane will depend on the particular problem being considered; a few examples are given in Figure 3.20. The plane may be on the far side of an illuminated aperture in a plane screen, as in Figure 3.20a. We have already discussed this case for two-dimensional fields (the uniformly illuminated slit). Other possibilities are the aperture plane of a reflector antenna, such as the paraboloidal reflector shown in Figure 3.20b, or the plane at the center of a cavity resonator for a gas laser (Figure 3.20c). Of course, the details of the electromagnetic fields for devices like these will be different. However, many of the general features of the fields are similar, and these can be deduced by considering a few simple distributions for the field on the reference plane.

3.7.1 Uniformly illuminated circular aperture

We will first consider a uniformly illuminated circular aperture of diameter d. The tangential component of the electric field is assumed to be in the y direction:

$$\vec{E}_y(x, y, z = 0) = \begin{cases} E_o^i \hat{y}, & \rho = \sqrt{x^2 + y^2} < d/2 \\ 0, & \rho \geq d/2. \end{cases} \tag{3.104}$$

This is probably the most important aperture distribution for practical use. It illustrates many of the features that are common to all aperture distributions, and it is often used as a benchmark against which other distributions are compared. This field is Kirchhoff's approximation for the physical arrangement in Figure 3.21. Here a plane wave of the form

$$\vec{E}^i(\vec{r}) = E_o^i e^{-jk_o z}\hat{y}$$

is normally incident on a perfectly conducting plane screen containing a circular aperture of diameter d. From what we learned in our discussion of the slit, we expect (3.104) to be a reasonable approximation to the actual field when the aperture is electrically large ($d/\lambda_o \gg 1$).

From Table 3.3, the spectral function corresponding to the field (3.104) is

$$F_y(k_x, k_y) = E_o^i \int_{\phi=0}^{2\pi} \int_{\rho=0}^{d/2} e^{j\rho(k_x \cos\phi + k_y \sin\phi)} \rho d\rho d\phi, \tag{3.105}$$

where the change of variables

$$x = \rho\cos\phi, \qquad y = \rho\sin\phi$$

was used in representing a point on the plane $z = 0$. The term in the exponent of (3.105) will now be rewritten as

$$k_x \cos\phi + k_y \sin\phi = A(\cos\xi\cos\phi + \sin\xi\sin\phi) = A\cos(\phi - \xi),$$

from which we find

$$\xi = \tan^{-1}(k_y/k_x)$$

and

$$A = \sqrt{k_x^2 + k_y^2};$$

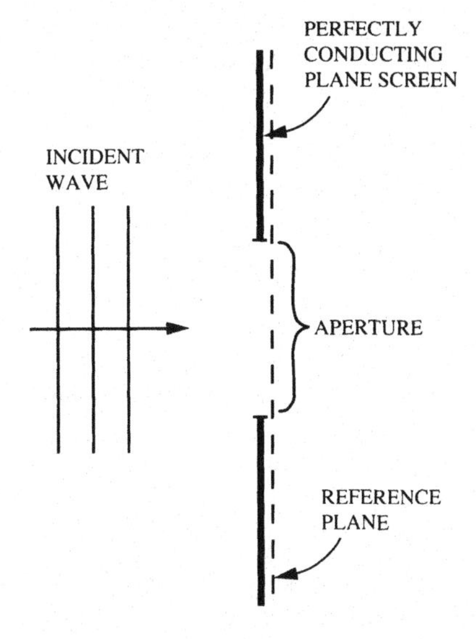

Fig. 3.20. *(continued on next page)*

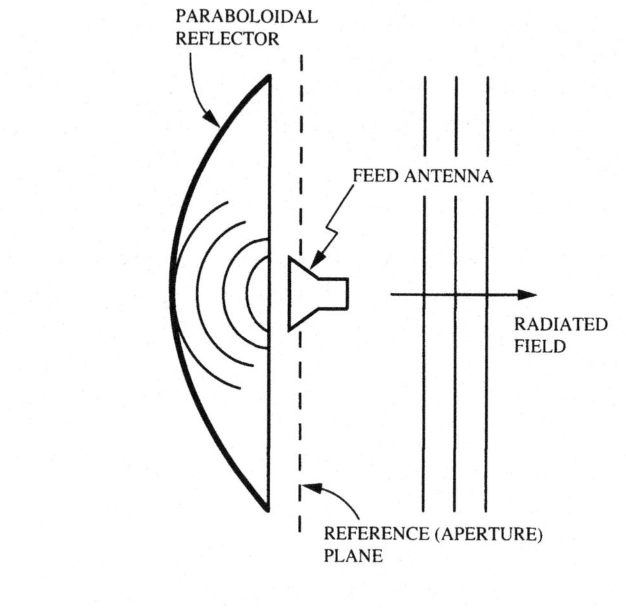

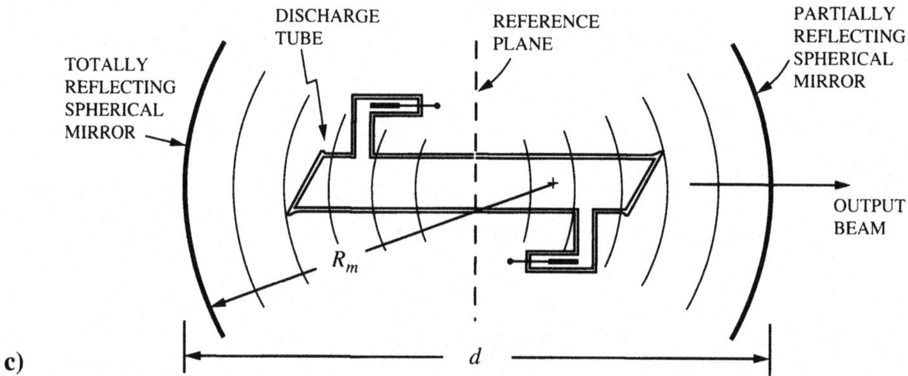

Fig. 3.20. Devices showing the reference planes for the application of the plane-wave spectrum. a) Aperture in plane screen. b) Paraboloidal reflector antenna. c) Gas laser with symmetric cavity.

thus,

$$F_y(k_x, k_y) = E_o^i \int_{\phi=0}^{2\pi} \int_{\rho=0}^{d/2} e^{j\sqrt{k_x^2+k_y^2}\,\rho\cos(\phi-\xi)} \rho d\rho d\phi. \tag{3.106}$$

With the change of variable $\eta = \phi - \xi$, (3.106) becomes

$$F_y(k_x, k_y) = E_o^i \int_{\eta=-\xi}^{2\pi-\xi} \int_{\rho=0}^{d/2} e^{j\sqrt{k_x^2+k_y^2}\,\rho\cos(\eta)} \rho d\rho d\eta. \tag{3.107}$$

Because we are integrating over one period of the function $\cos(\eta)$, the value of the integral (3.107) is independent of ξ. Letting $\xi = 0$ and using the symmetry of the

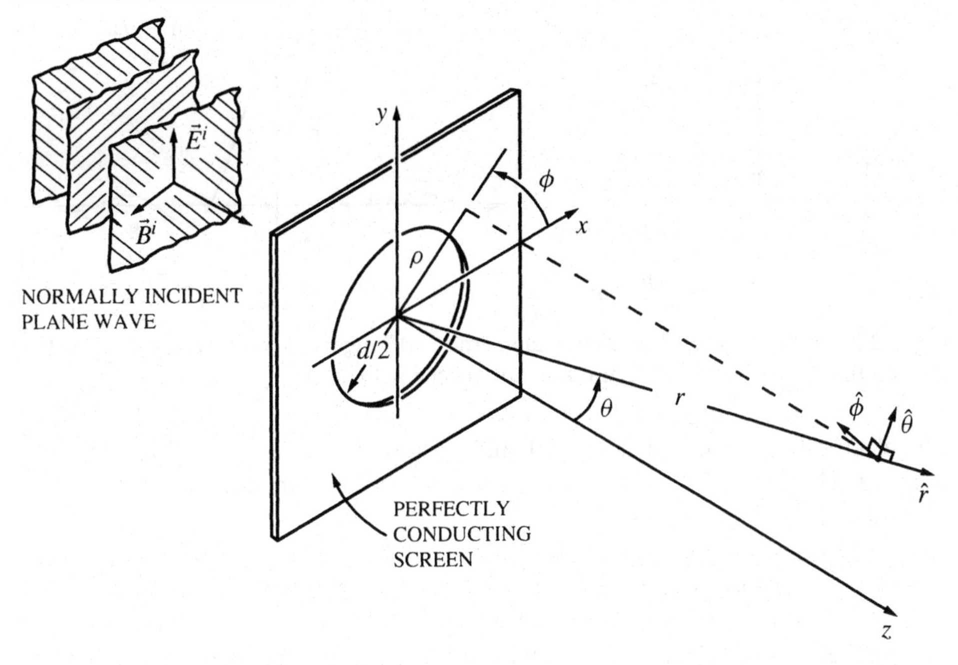

Fig. 3.21. Plane wave normally incident on an infinitesimally thin, perfectly conducting screen containing a circular aperture of diameter d.

integrand, we have

$$F_y(k_x, k_y) = 2E_o^i \int_{\eta=0}^{\pi} \int_{\rho=0}^{d/2} \cos\left[\sqrt{k_x^2 + k_y^2}\, \rho \cos(\eta)\right] \rho d\rho d\eta. \quad (3.108)$$

Now the integral with respect to η is identified as a standard form – an integral representation for the Bessel function [8]:[10]

$$J_o(z) = \frac{1}{\pi} \int_0^{\pi} \cos\left[z \cos(u)\right] du, \quad (3.109)$$

which makes (3.108)

$$F_y(k_x, k_y) = 2\pi E_o^i \int_{\rho=0}^{d/2} J_o\left(\sqrt{k_x^2 + k_y^2}\, \rho\right) \rho d\rho. \quad (3.110)$$

After using the following property of the Bessel function [8]:

$$\int_0^z u J_o(u) du = z J_1(z), \quad (3.111)$$

the spectral function becomes

$$F_y(k_x, k_y) = \frac{\pi d^2 E_o^i}{4} \left[\frac{2J_1\left(\sqrt{k_x^2 + k_y^2}\, d/2\right)}{\sqrt{k_x^2 + k_y^2}\, d/2}\right]. \quad (3.112)$$

[10] $J_n(z)$ is the notation for the Bessel function of the first kind and order n.

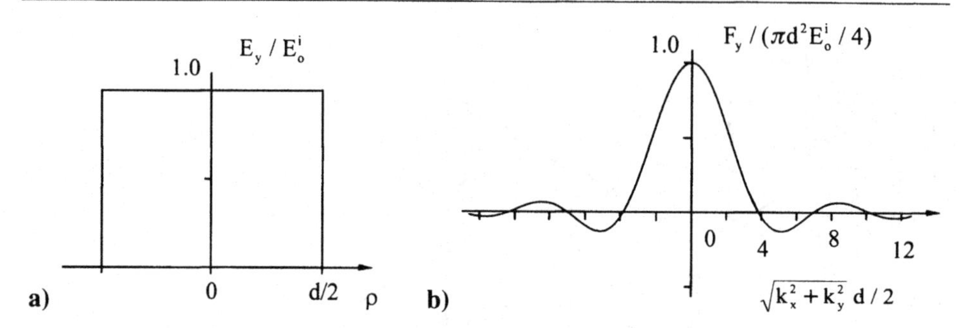

Fig. 3.22. Uniformly illuminated circular aperture. a) Tangential component of the electric field on the plane $z = 0$. b) Spectral function for the field of a).

The electric field on the plane $z = 0$ and the accompanying spectral function are graphed in Figure 3.22. Note the similarity to the spectral function for the uniformly illuminated slit (Figure 3.9b).

The electromagnetic field in the half space $z > 0$ is obtained by inserting (3.112) into the expressions in Table 3.3. For a general position $\vec{r}$, the resulting integrals cannot be performed in closed form. The radiated field, however, is readily obtained by substituting (3.112) into the formulas in Table 3.4; the radiated electric field is

$$\vec{E}^r(\vec{r}) = \left[F_\theta(\theta,\phi)\hat{\theta} + F_\phi(\theta,\phi)\hat{\phi}\right]e^{-jk_o r}/r, \tag{3.113a}$$

with

$$F_\theta(\theta,\phi) = \frac{j}{2k_o}\left(\frac{k_o d}{2}\right)^2 E_o^i \left[\frac{2J_1\left(\frac{k_o d}{2}\sin\theta\right)}{\frac{k_o d}{2}\sin\theta}\right]\sin\phi, \tag{3.113b}$$

$$F_\phi(\theta,\phi) = \frac{j}{2k_o}\left(\frac{k_o d}{2}\right)^2 E_o^i \left[\frac{2J_1\left(\frac{k_o d}{2}\sin\theta\right)}{\frac{k_o d}{2}\sin\theta}\right]\cos\phi\cos\theta. \tag{3.113c}$$

Graphs of this field for an aperture of diameter $d/\lambda_o = 5$ are shown in Figures 3.23 and 3.24.[11] Figure 3.23a is for the normalized field in the plane $\phi = 0\ (\pi)$:

$$\left|\vec{E}^r(\theta,0)\right| / \left|\vec{E}^r(0,0)\right| = \left|F_\phi(\theta,0)\right| / \left|F_\phi(0,0)\right|, \tag{3.114a}$$

and Figure 3.23b is for the normalized field in the plane $\phi = \pi/2\ (3\pi/2)$:

$$\left|\vec{E}^r(\theta,\pi/2)\right| / \left|\vec{E}^r(0,\pi/2)\right| = \left|F_\theta(\theta,\pi/2)\right| / \left|F_\theta(0,\pi/2)\right|. \tag{3.114b}$$

The former is referred to as the "principal H-plane pattern"; the latter is referred to as the "principal E-plane pattern." The principal H plane (E plane) is the plane that contains the maximum of the field and for which the $\vec{H}^r$ ($\vec{E}^r$) field lies entirely within the plane. In the three-dimensional representation (Figure 3.24) the magnitude of the field (vertical axis) is graphed as a function of θ (radial variable) and ϕ (angular variable). The principal H- and E-plane patterns are shown as sections of this figure.

[11] The three-dimensional graphs, such as Figure 3.24, that appear throughout the book were drawn using the program described in Reference [14].

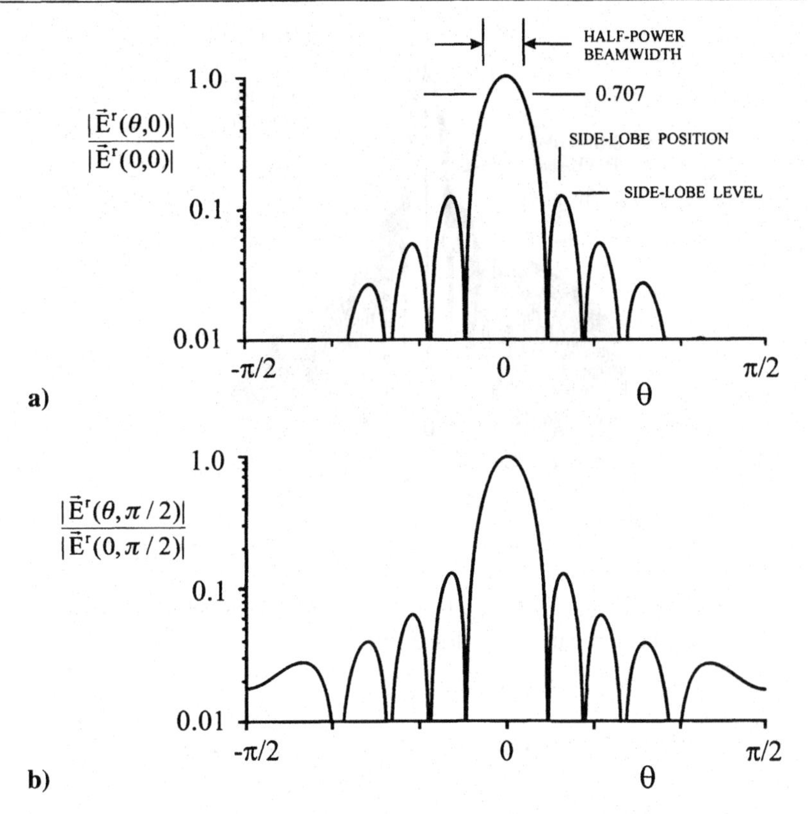

Fig. 3.23. Normalized electric field patterns for the uniformly illuminated circular aperture, where $d/\lambda_o = 5.0$. a) Principal H-plane pattern, $\phi = 0\ (\pi)$. b) Principal E-plane pattern, $\phi = \pi/2\ (3\pi/2)$.

The radiated field is seen to consist of a *main lobe* or beam with maximum field in the direction of $\theta = 0$ and several *side lobes* of lower level. There are a number of fairly standard measures that can be used to describe the structure of this field. The ability of a radiator (antenna) to concentrate radiation in a particular direction θ, ϕ is expressed by the *directivity*:

$$D(\theta, \phi) = \frac{\text{time-average power radiated per unit area in the direction } \theta, \phi}{\text{time-average power radiated per unit area averaged over all directions}}$$

$$= \frac{\hat{r} \cdot \mathrm{Re}[\vec{S}_c^r(\theta, \phi)]}{\left\{\int_{\phi'=0}^{2\pi} \int_{\theta'=0}^{\pi} \hat{r} \cdot \mathrm{Re}[\vec{S}_c^r(\theta', \phi')] r^2 \sin\theta' \, d\theta' \, d\phi'\right\} / 4\pi r^2}$$

$$= \frac{4\pi [|F_\theta(\theta, \phi)|^2 + |F_\phi(\theta, \phi)|^2]}{\int_{\phi'=0}^{2\pi} \int_{\theta'=0}^{\pi} [|F_\theta(\theta', \phi')|^2 + |F_\phi(\theta', \phi')|^2] \sin\theta' \, d\theta' \, d\phi'}. \tag{3.115}$$

The direction θ, ϕ is usually chosen to be that for the maximum field of the main lobe. For an aperture that radiates only into the half space $z > 0$, the limits on the θ'

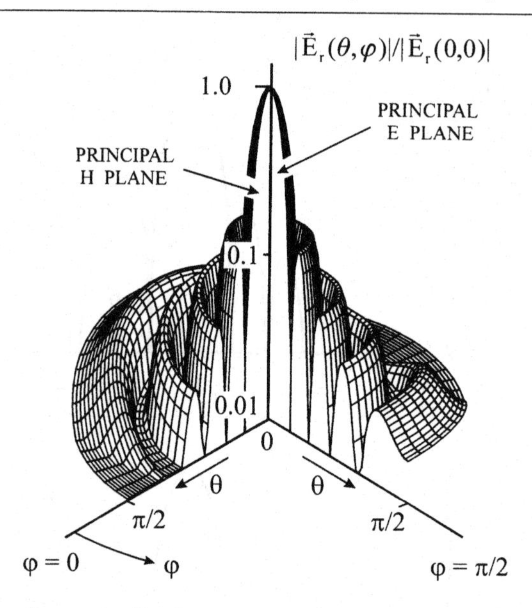

Fig. 3.24. Normalized magnitude of the electric field versus the angles θ and ϕ, where $d/\lambda_o = 5.0$. Note that the vertical scale is logarithmic.

integration in (3.115) become 0 and $\pi/2$, and (3.115) is referred to as the aperture directivity D_A.

If we substitute (3.113a–c) into (3.115) and note that in the limit as z goes to zero $2J_1(z)/z = 1$, the directivity of the uniformly illuminated circular aperture in the direction $\theta = 0$ becomes

$$D_A(\theta=0) = \left\{ \frac{1}{2} \int_{\theta'=0}^{\pi/2} \left[\frac{2J_1\left(\frac{k_o d}{2}\sin\theta'\right)}{\frac{k_o d}{2}\sin\theta'} \right]^2 \left(1 - \frac{1}{2}\sin^2\theta'\right) \sin\theta' d\theta' \right\}^{-1}. \tag{3.116}$$

The width of the main lobe in the principal H or E plane (see Figure 3.23) is characterized by the *half-power beamwidth*. This is the angular separation between the points at which the power per unit area has dropped to one-half of its maximum value (the field has dropped to $1/\sqrt{2} = 0.707\ldots$ of its maximum value). For the uniformly illuminated circular aperture, the half-power beamwidths $2\theta_{\mathrm{BW}}$ are determined from the smallest solutions of the following equations:
in the **H plane**,

$$\left[\frac{2J_1\left(\frac{k_o d}{2}\sin\theta_{\mathrm{BWH}}\right)}{\frac{k_o d}{2}\sin\theta_{\mathrm{BWH}}} \right] \cos\theta_{\mathrm{BWH}} = \frac{1}{\sqrt{2}}, \tag{3.117}$$

and in the **E plane**,

$$\left[\frac{2J_1\left(\frac{k_o d}{2}\sin\theta_{\mathrm{BWE}}\right)}{\frac{k_o d}{2}\sin\theta_{\mathrm{BWE}}} \right] = \frac{1}{\sqrt{2}}. \tag{3.118}$$

The side lobes adjacent to the main lobe (the first side lobes) are generally of most concern. For the uniformly illuminated circular aperture, the positions of these lobes θ_{SL} are given by the appropriate solutions to the following equations:
in the **H plane**,

$$\frac{d}{d\theta} E^r_\phi(\theta, 0)\Big|_{\theta_{SLH}} = 0,$$

or

$$J_o\left(\frac{k_o d}{2}\sin\theta_{SLH}\right)\left(\frac{k_o d}{2}\sin\theta_{SLH}\right)\cos^2\theta_{SLH} - J_1\left(\frac{k_o d}{2}\sin\theta_{SLH}\right)(1+\cos^2\theta_{SLH}) = 0, \tag{3.119}$$

and in the **E plane**,

$$\frac{d}{d\theta} E^r_\theta(\theta, \pi/2)\Big|_{\theta_{SLE}} = 0,$$

or

$$J_o\left(\frac{k_o d}{2}\sin\theta_{SLE}\right)\left(\frac{k_o d}{2}\sin\theta_{SLE}\right) - 2J_1\left(\frac{k_o d}{2}\sin\theta_{SLE}\right) = 0. \tag{3.120}$$

The levels of the side lobes relative to the main lobe are determined by inserting θ_{SLH} into (3.114a) and θ_{SLE} into (3.114b).

The four quantities – directivity, half-power beamwidth, side-lobe level, and side-lobe position – are shown as a function of the electrical size of the aperture $k_o d = 2\pi d/\lambda_o$ in Figure 3.25.[12] As the size of the aperture increases, the directivity is seen to increase and the beamwidths narrow. Side lobes appear on the patterns when $k_o d > 7.66\ldots$ [the first zero of $J_1(k_o d/2)$]. The positions of the first side lobes move toward the axis ($\theta = 0$), and their levels become fairly constant as the size of the aperture increases.

For *electrically large apertures* ($k_o d \gg 1$), a case of practical importance, the radiation patterns simplify. The function

$$2J_1\left(\frac{k_o d}{2}\sin\theta\right)\Big/\left(\frac{k_o d}{2}\sin\theta\right),$$

which appears in the pattern functions (3.113b) and (3.113c), is negligibly small when the argument $(k_o d/2)\sin\theta$ is large. This can be seen from the graph of this function in Figure 3.22b. Thus, when $k_o d \gg 1$ this function will be significant only for small values of the angle θ. After introducing the small angle approximations

[12] The unit used for the directivity and relative side-lobe level is the decibel (dB): $D_A(\text{dB}) = 10\log_{10}(D_A)$; side-lobe level (dB) $= 20\log_{10}[|\vec{E}^r(\theta_{SL}, 0)|/|\vec{E}^r(0, 0)|]$.

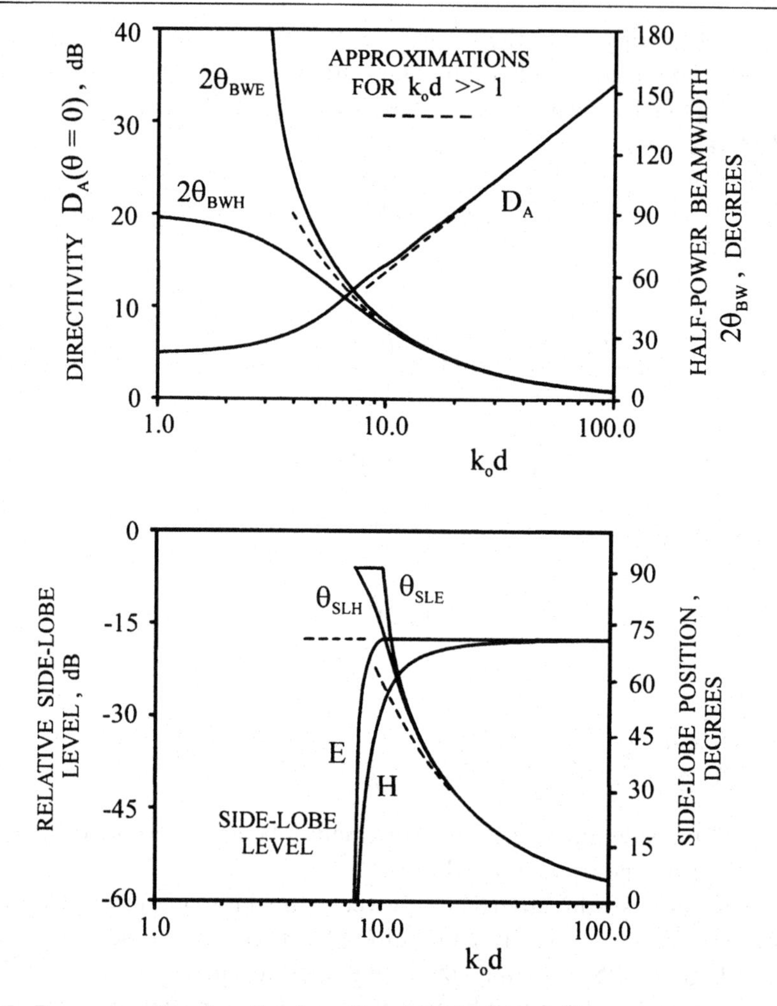

Fig. 3.25. Parameters that describe the radiated field of the uniformly illuminated circular aperture as functions of the electrical size of the aperture, $k_o d = 2\pi d/\lambda_o$. Dashed lines show approximations for electrically large apertures, $k_o d \gg 1$.

$\sin\theta \approx \theta$ and $\cos\theta \approx 1$, the pattern functions (3.113b) and (3.113c) become

$$F_\theta(\theta, \phi) = \frac{j}{2k_o}\left(\frac{k_o d}{2}\right)^2 E_o^i \left[\frac{2J_1\left(\frac{k_o d}{2}\theta\right)}{\frac{k_o d}{2}\theta}\right] \sin\phi \tag{3.121a}$$

and

$$F_\phi(\theta, \phi) = \frac{j}{2k_o}\left(\frac{k_o d}{2}\right)^2 E_o^i \left[\frac{2J_1\left(\frac{k_o d}{2}\theta\right)}{\frac{k_o d}{2}\theta}\right] \cos\phi. \tag{3.121b}$$

Note, to this degree of approximation, the principal H- and E-plane patterns,

(3.114a) and (3.114b), are identical and equal to

$$\frac{|\vec{E}^r(\theta,0)|}{|\vec{E}^r(0,0)|} = \frac{|\vec{E}^r(\theta,\pi/2)|}{|\vec{E}^r(0,\pi/2)|} = \left|\frac{2J_1\left(\frac{k_o d}{2}\theta\right)}{\frac{k_o d}{2}\theta}\right|. \tag{3.122}$$

With the small-angle approximations, the directivity of the aperture (3.116) becomes

$$D_A(\theta=0) = \frac{1}{2}\left(\frac{k_o d}{2}\right)^2\left\{\int_0^\infty \frac{[J_1(u)]^2}{u}du\right\}^{-1}, \tag{3.123}$$

where we have used the fact that the integrand is negligible for large arguments to replace the upper limit of the integration by ∞. The integral in (3.123) is a standard form with the value 1/2 [8]; thus,

$$D_A(\theta=0) = \left(\frac{k_o d}{2}\right)^2 = \left(\frac{\pi d}{\lambda_o}\right)^2. \tag{3.124}$$

The half-power beamwidths and the side-lobe positions and levels are the same for the principal H and E planes. The equations for the half-power beamwidths, (3.117) and (3.118), become

$$\frac{2J_1\left(\frac{k_o d}{2}\theta_{BW}\right)}{\frac{k_o d}{2}\theta_{BW}} = \frac{1}{\sqrt{2}},$$

or

$$2\theta_{BW} = 6.46/k_o d = 1.03(\lambda_o/d), \tag{3.125}$$

and the equations for the side-lobe positions, (3.119) and (3.120), become

$$\left(\frac{k_o d}{2}\theta_{SL}\right)J_o\left(\frac{k_o d}{2}\theta_{SL}\right) - 2J_1\left(\frac{k_o d}{2}\theta_{SL}\right) = 0, \tag{3.126}$$

which is equivalent to [8]

$$J_2\left(\frac{k_o d}{2}\theta_{SL}\right) = 0,$$

or

$$\theta_{SL} = 10.3/k_o d = 1.63(\lambda_o/d). \tag{3.127}$$

The side-lobe levels are

$$\left|\vec{E}^r(\theta_{SL},0)\right|/\left|\vec{E}^r(0,0)\right| = \left|\vec{E}^r(\theta_{SL},\pi/2)\right|/\left|\vec{E}^r(0,\pi/2)\right| = 0.132 = -17.6 \text{ dB}. \tag{3.128}$$

The approximate values for the directivity, half-power beamwidth, and side-lobe position and level that apply for electrically large apertures are shown as dashed lines in Figure 3.25.

In terms of the area of the aperture ($A = \pi d^2/4$) the directivity (3.124) is

$$D_A(\theta=0) = 4\pi A/\lambda_o^2. \tag{3.129a}$$

This important relationship can be shown to apply approximately to uniformly illuminated, electrically large apertures of arbitrary shape [15].

The main lobe or beam covers a solid angle that is roughly $\Omega_{\mathrm{BW}} = 2\pi[1 - \cos(\theta_{\mathrm{BW}})] \approx \pi(\theta_{\mathrm{BW}})^2$. The fraction of the total power radiated that is contained in the main beam is proportional to the product $D_A(\theta = 0)\Omega_{\mathrm{BW}}/4\pi$. On substitution of (3.124) and (3.125), this is seen to be a constant, independent of $k_o d$:

$$D_A(\theta = 0)\Omega_{\mathrm{BW}}/4\pi \approx (\pi/4)^2,$$

or

$$D_A(\theta = 0)(2\theta_{\mathrm{BW}})^2 \approx \pi^2. \tag{3.129b}$$

Thus, increasing the electrical size of the aperture, $k_o d$, increases the power density on the axis of the beam but decreases the beamwidth in a manner that keeps the total power in the main beam constant. The relationships (3.129a) and (3.129b) are worth committing to memory, since they are very useful for estimating the directivity (D_A) and beamwidth ($2\theta_{\mathrm{BW}}$) that can be expected from electrically large apertures.

Our description of the electromagnetic field of the uniformly illuminated circular aperture can be compared to the results of experimental measurements. These experiments generally attempt to duplicate the physical arrangement shown in Figure 3.21: a plane wave normally incident on a perfectly conducting screen containing the circular aperture. The field is measured in the half space $z > 0$. We must stress again that the assumption of uniform illumination in the aperture is an approximation for this arrangement, only applicable for electrically large apertures. This point is nicely illustrated by measurements made at a microwave frequency ($\lambda_o = 8.0$ cm) with the apparatus shown in Figure 3.26a [16, 17]. Here an antenna (parabolic reflector) produces an approximation to a plane wave in the vicinity of an aperture in a thin metallic screen. The electric field is measured with a small dipole probe. Figure 3.26b shows the electric field measured within ($z = 0$) an aperture of diameter $d/\lambda_o = 8.0$. The field is seen to be only approximately uniform, with oscillations about the value $|E_y|/|E_o^i| = 1.0$. Note the similarity of these measured results to the theoretical computations for the slit (Figure 3.8). The H-plane results for the circular aperture are similar to the TE results for the slit; both are for a plane in which the incident electric field is parallel to edges of the aperture. The E-plane results for the circular aperture are similar to the TM results for the slit; both are for a plane in which the incident electric field is normal to edges of the aperture.

The electric field along the axis of an aperture of the same electrical size $d/\lambda_o = 8.0$ is shown in Figure 3.26c. A slightly different experimental apparatus was used in measuring these results: $\lambda_o = 6.0$ cm and the metallic screen was coated with a material that reduced reflections [18]. The theory for the uniformly illuminated aperture (dashed line) is seen to be in good agreement with the measured results except at points very close to the screen. Note that there is a succession of alternating "light" and "dark" regions along the axis directly behind the aperture, with the maximum electric field being about twice that for the incident wave.

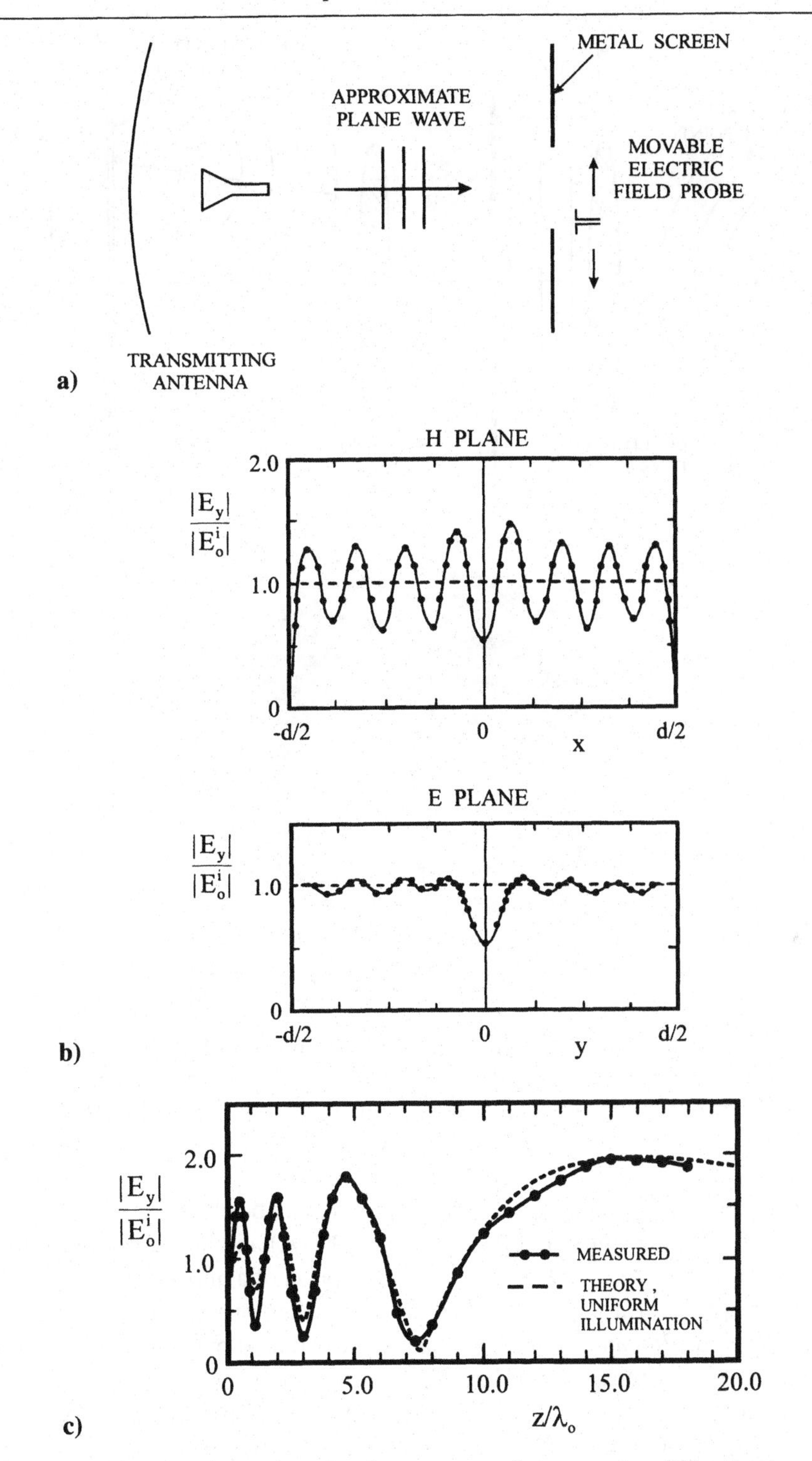

Fig. 3.26. a) Schematic drawing showing apparatus for measuring diffraction by a circular aperture at microwave frequencies. b) The normalized electric field in the aperture, $d/\lambda_o = 8.0$. (Measured data from Andrews [17].) c) The normalized electric field along the z axis, $d/\lambda_o = 8.0$. (Measured data from Severin [18].)

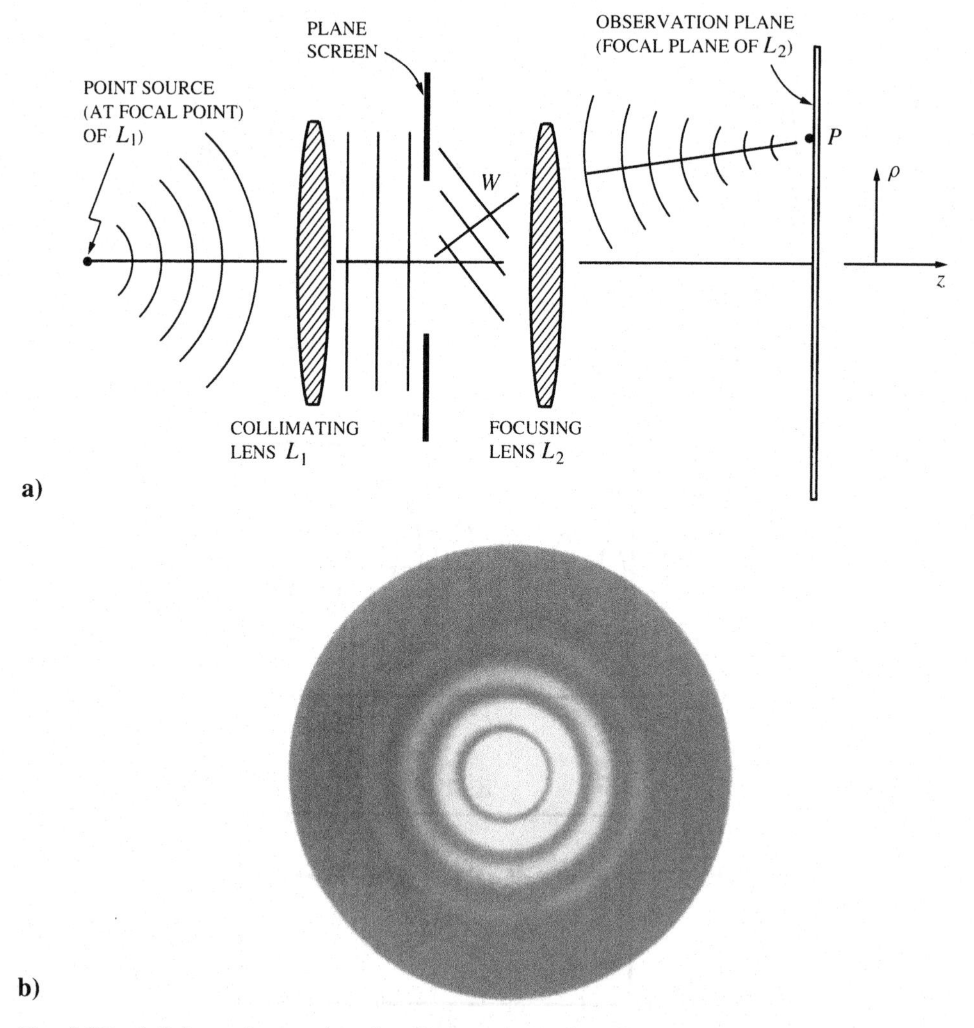

Fig. 3.27. a) Schematic drawing showing apparatus for observing Fraunhofer diffraction at optical frequencies. b) Fraunhofer diffraction pattern of a circular aperture, $d/\lambda_o = 1.04 \times 10^4$. (Reproduced by permission of E. Wolf from Born and Wolf, *Principles of Optics* [12].)

At optical frequencies, a physically small aperture can be electrically very large; for example, for visible light (4,000 Å $\leq \lambda_o \leq$ 7,000 Å) and an aperture 6 mm in diameter, $8.6 \times 10^3 \lesssim d/\lambda_o \lesssim 1.5 \times 10^4$. The assumption of a uniformly illuminated aperture for the physical arrangement in Figure 3.21 is then very good. The apparatus shown in Figure 3.27a can be used to measure the radiated or Fraunhofer field of an aperture at optical frequencies. Here light from a point source is collimated by a lens to produce a plane wave incident on the aperture. For direct observation of the Fraunhofer diffraction pattern, we would have to be at a distance $r \gg 2d^2/\lambda_o$ (3.101) from the aperture (the Fraunhofer region). This distance can be quite large; for example, for yellow light with λ_o = 5,790 Å and an aperture 6 mm in diameter, $r \gg$ 124 m! An alternative arrangement for observing the Fraunhofer diffraction

Table 3.5. *Numerical values for Airy pattern*

$u = \left(\frac{k_o d}{2z}\right)\rho$	$\left[\frac{2J_1(u)}{u}\right]^2$	Percent of energy within disc of radius ρ
0	1 (max)	
3.832	0 (min)	83.8
5.136	0.0175 (max)	
7.016	0 (min)	91.0
8.417	0.0042 (max)	
10.17	0 (min)	93.8
11.62	0.0016 (max)	

pattern is to place a focusing lens on the far side of the aperture, as in Figure 3.27a. This lens causes each propagating plane wave W in the spectrum of the aperture to be focused to a separate point P on the observation plane. Thus, the lens performs the same operation as moving the observation point to the Fraunhofer region; it produces a field at each point in space that is simply related to the field of a particular plane wave in the spectrum (a plane wave propagating in a particular direction).[13] The pattern on the observation plane is a scaled version of the Fraunhofer diffraction pattern.

The complex Poynting vector in the Fraunhofer region is obtained by inserting (3.113a–c) into (3.100), which gives

$$\vec{S}^r_c = \frac{1}{8\zeta_o(k_o r)^2}\left(\frac{k_o d}{2}\right)^4 |E^i_o|^2 \left[\frac{2J_1\left(\frac{k_o d}{2}\sin\theta\right)}{\frac{k_o d}{2}\sin\theta}\right]^2 \hat{r}.$$

The irradiance or intensity at a point on the observation plane (a plane of constant z) is

$$I = \left|\langle \hat{z}\cdot\vec{\mathcal{S}}^r\rangle\right| = \left|\mathrm{Re}(\hat{z}\cdot\vec{S}^r_c)\right|. \tag{3.130}$$

Recall that for electrically large apertures only small values of θ are of importance, and note that $\hat{z}\cdot\hat{r} = \cos\theta \approx 1$ and $\sin\theta = \rho/r \approx \rho/z$, where ρ is the radial distance on the observation plane measured from the z axis (see Figure 3.27a). The normalized irradiance on an observation plane in the Fraunhofer region is then

$$\frac{I(\rho)}{I(0)} \approx \left[\frac{2J_1\left(\frac{k_o d}{2z}\rho\right)}{\frac{k_o d}{2z}\rho}\right]^2. \tag{3.131}$$

This distribution, referred to as the *Airy pattern*, is graphed in Figure 3.28, and specific values are given in Table 3.5. It was first derived by George Biddell Airy (1801–1892) in a quite different form, without employing Bessel functions [20].

[13] Recall that the radiated or Fraunhofer field is simply related to the two-dimensional Fourier transform (spectral function) of the aperture distribution. Thus, the lens optically performs a two-dimensional Fourier transform of the aperture distribution; this is a general property of converging lenses [19].

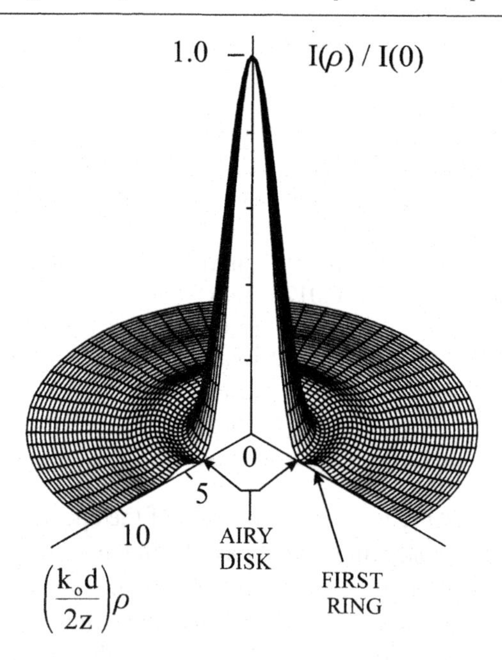

Fig. 3.28. Airy pattern for the uniformly illuminated circular aperture: normalized irradiance on an observation plane in the Fraunhofer region.

For comparison with these theoretical results, a Fraunhofer diffraction pattern, measured in the manner of Figure 3.27a, is shown in Figure 3.27b [12]. The circular aperture is 6 mm in diameter and the wavelength of the yellow light is $\lambda_o = 5{,}790$ Å, making $d/\lambda_o = 1.04 \times 10^4$. The central portion of the pattern was overexposed to show the presence of the weak outer rings. The circular spot at the center of the pattern, known as the *Airy disc*, contains about 84% of the total energy in the pattern.

3.7.2 Circular aperture with tapered illumination – reflector antennas

The radiated field for antennas, like the one shown in Figure 3.20b, is determined by the field on the aperture plane, which, in turn, is determined by the feed antenna and the reflector.

At microwave and millimeter-wave frequencies (3×10^8 Hz $\lesssim f \lesssim 3 \times 10^{11}$ Hz) reflector antennas can have apertures that are electrically very large; an example is shown in Figure 3.29. This is the NASA (National Aeronautics and Space Administration) Goldstone antenna, which is used for deep-space communications. The diameter of the main reflector or "dish" is 64 m; this is several thousands of wavelengths at the highest frequencies of operation. The great size of this antenna can be seen by noting that three of the feed cones shown in Figure 3.29b are located at the center of the reflector shown in Figure 3.29a. The feed system for this antenna is discussed in more detail in Problem 3.6. Large reflector antennas like this one are generally used to concentrate radiation in a single direction – to produce a "pencil beam."

Now let us consider the aperture field for the simple paraboloidal reflector (parabola of revolution) in Figure 3.20b, which is shown in more detail in Figure 3.30. When the reflector is electrically large ($k_o d \gg 1$), the same arguments (geometrical optics) as used for optical instruments (mirrors, lens, etc.) can be used to approximately analyze this antenna. The paraboloid is assumed to be in the radiated field of the feed antenna, which is at the focal point of the paraboloid, a "prime focus feed." Each ray (AB) from the feed, after reflection from the paraboloid, is parallel to the z axis (BC). Recalling that, by definition, AB = DB or AB + BC = DC = a constant for a point on the paraboloid, we see that all rays from the feed travel the same distance to arrive at the aperture plane. Thus, all of the rays accumulate the same phase change on propagating from the feed to the aperture plane.

a)

Fig. 3.29. *(continued on next page)*

b)

Fig. 3.29. a) The NASA 210 ft (64 m) Goldstone reflector antenna. b) Feed cones for the NASA Goldstone reflector antenna. (Photographs by the author.)

We will assume that all of the rays leave the feed with the same phase, which, according to the argument above, makes the phase of the aperture field uniform. In addition, we will assume that the feed produces a linearly polarized, rotationally symmetric aperture field. The amplitude of the aperture field will generally not be uniform, but will taper toward the edge of the aperture. Although some of the tapering is a natural result of the geometry of the reflector, additional tapering may be introduced by the radiation pattern of the feed, (i.e., a reduction of the radiation in the direction of the reflector's edge). To see how the former arises, consider the two tubes of rays of equal solid angle in Figure 3.30: Tube 1 is near the center of the reflector, and tube 2 is near the edge. The distance from the focus to the reflector is longer for tube 2 than for tube 1. Consequently, for equal net power in the two tubes, the power per unit area and the electric field will be smaller at the reflector for tube 2, leading to a radial taper in the amplitude of the aperture field.

We conclude that a linearly polarized, rotationally symmetric field with uniform phase and a radial taper in amplitude is a reasonable approximation to the aperture field of our paraboloidal reflector antenna. This field is often represented by the

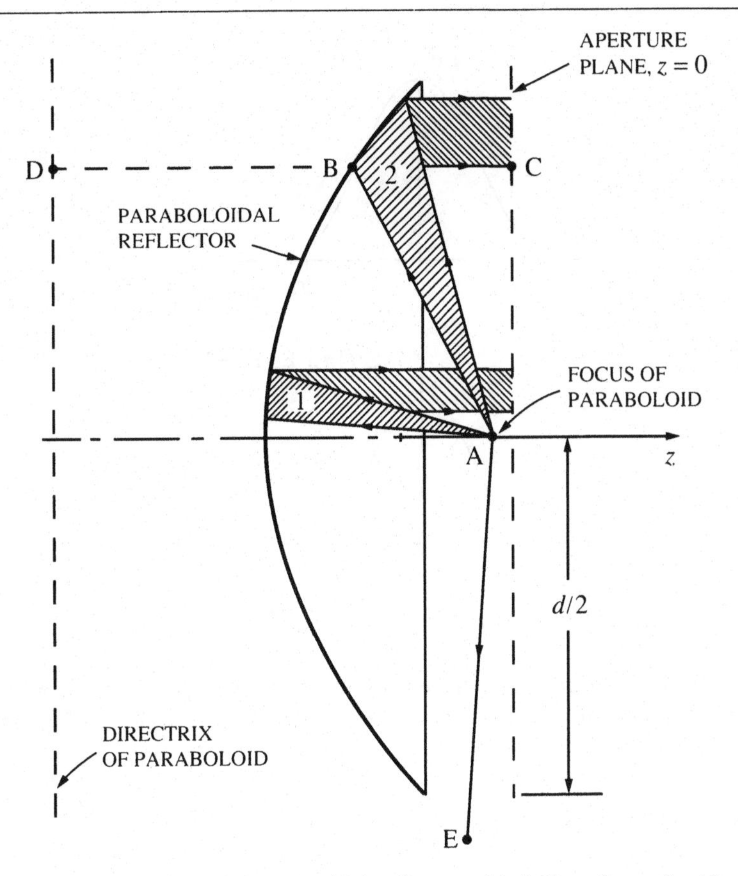

Fig. 3.30. Detail for paraboloidal reflector with "prime focus feed."

empirical formula

$$\vec{E}_y(x, y, z = 0) = \begin{cases} E_o\{p + (1 - p)[1 - (2\rho/d)^2]^n\}\hat{y}, & \rho = \sqrt{x^2 + y^2} \leq d/2 \\ 0, & \rho > d/2, \end{cases} \tag{3.132}$$

which is graphed for the case $p = 0.1, n = 2$ in Figure 3.31a [21]. In this expression the pedestal height p determines the level of the field at the edge of the aperture, and the integer n controls the rate at which the field decreases with the radial position ρ on the aperture plane.

The spectral function corresponding to the aperture field (3.132) is

$$F_y(k_x, k_y) = E_o \int_{\phi=0}^{\pi/2} \int_{\rho=0}^{d/2} \{p + (1 - p)[1 - (2\rho/d)^2]^n\} e^{j\rho(k_x \cos\phi + k_y \sin\phi)} \rho d\rho d\phi. \tag{3.133}$$

After following the same procedures as used with the integral for the uniformly

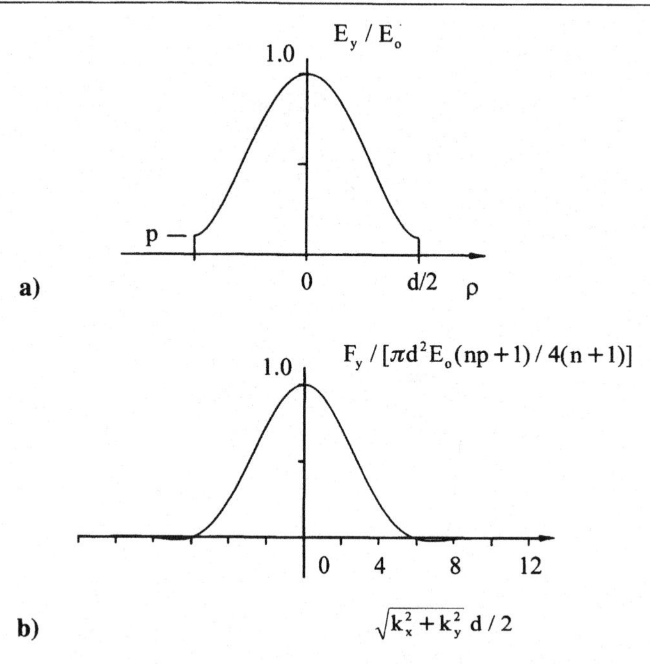

Fig. 3.31. Circular aperture with tapered illumination: $p = 0.1$, $n = 2$. a) Tangential component of the electric field on the aperture plane, $z = 0$. b) Spectral function for the field of a).

illuminated aperture (3.105), (3.133) becomes

$$F_y(k_x, k_y) = \frac{\pi d^2 E_o}{4}\left\{p\left[\frac{2J_1\left(\sqrt{k_x^2+k_y^2}\,d/2\right)}{\sqrt{k_x^2+k_y^2}\,d/2}\right]\right.$$

$$\left. + 2(1-p)\int_0^1 u(1-u^2)^n J_o\left[\sqrt{k_x^2+k_y^2}\,(d/2)u\right]du\right\}. \qquad (3.134)$$

The remaining integral is evaluated in Reference [22]:

$$F_y(k_x, k_y) = \frac{\pi d^2 E_o}{4}\left\{p\left[\frac{2J_1\left(\sqrt{k_x^2+k_y^2}\,d/2\right)}{\sqrt{k_x^2+k_y^2}\,d/2}\right]\right.$$

$$\left. + \left(\frac{1-p}{n+1}\right)\left[\frac{2^{n+1}(n+1)!\,J_{n+1}\left(\sqrt{k_x^2+k_y^2}\,d/2\right)}{\left(\sqrt{k_x^2+k_y^2}\,d/2\right)^{n+1}}\right]\right\}. \qquad (3.135)$$

This spectral function is graphed for the case $p = 0.1$ and $n = 2$ in Figure 3.31b.

The radiated electric field of the aperture is obtained by substituting (3.135) into the formulas in Table 3.4:

$$\vec{E}^r(\vec{r}) = \left[F_\theta(\theta,\phi)\hat{\theta} + F_\phi(\theta,\phi)\hat{\phi}\right]e^{-jk_o r}/r, \qquad (3.136a)$$

with

$$F_\theta(\theta,\phi) = \frac{j}{k_o}\left(\frac{k_o d}{2}\right)^2 E_o \left\{ p \left[\frac{J_1\left(\frac{k_o d}{2}\sin\theta\right)}{\frac{k_o d}{2}\sin\theta} \right] + (1-p)(2)^n n! \left[\frac{J_{n+1}\left(\frac{k_o d}{2}\sin\theta\right)}{\left(\frac{k_o d}{2}\sin\theta\right)^{n+1}} \right] \right\} \sin\phi, \tag{3.136b}$$

$$F_\phi(\theta,\phi) = \frac{j}{k_o}\left(\frac{k_o d}{2}\right)^2 E_o \left\{ p \left[\frac{J_1\left(\frac{k_o d}{2}\sin\theta\right)}{\frac{k_o d}{2}\sin\theta} \right] + (1-p)(2)^n n! \left[\frac{J_{n+1}\left(\frac{k_o d}{2}\sin\theta\right)}{\left(\frac{k_o d}{2}\sin\theta\right)^{n+1}} \right] \right\} \cos\phi\cos\theta. \tag{3.136c}$$

This field is compared with the radiated field of a uniformly illuminated circular aperture in Figure 3.32. The diameter of the aperture is $d/\lambda_o = 5$, and the total power radiated by both aperture distributions is the same. On the right side of the figures, the field for an aperture with tapered illumination $n = 2$ and no pedestal $p = 0$ (dashed line) is compared with the field of the uniformly illuminated aperture $p = 1$ (solid line). The taper is seen to decrease the field and therefore the directivity on axis ($\theta = 0$), widen the main lobe, and decrease the level of the side lobes. For the graphs on the left side of the figures (dashed line) a small pedestal has been added to the tapered illumination ($n = 2$, $p = 0.1$). The addition of the pedestal slightly improves the patterns: It increases the directivity, narrows the main lobe, and decreases the level of the first side lobes. These effects are the result of the radiation from the tapered illumination and the radiation from the pedestal being out of phase at angles near the first side lobe; they therefore cancel. The pedestal height $p = 0.1$ used for Figure 3.32 is the value that minimizes the level of the first side lobes for $n = 2$ [21].

When the aperture is electrically large ($k_o d \gg 1$), the directivity for the field (3.132) is determined using the same approximations as for the uniformly illuminated aperture (Problem 3.10):

$$\frac{D_A(\theta = 0,\ n,\ p)}{D_A(\theta = 0,\ p = 1)} = \left[1 + \frac{n^2(1-p)^2}{(2n+1)(1+np)^2}\right]^{-1}. \tag{3.137}$$

It is clear from this result that the directivity with tapered illumination will always be less than with uniform illumination, $D_A(\theta = 0,\ p = 1)$, no matter what values of n and p are selected.

For a reflector antenna the *gain* $G(\theta, \phi)$ is generally a more useful measure of the performance than the directivity of the aperture; the two quantities, however,

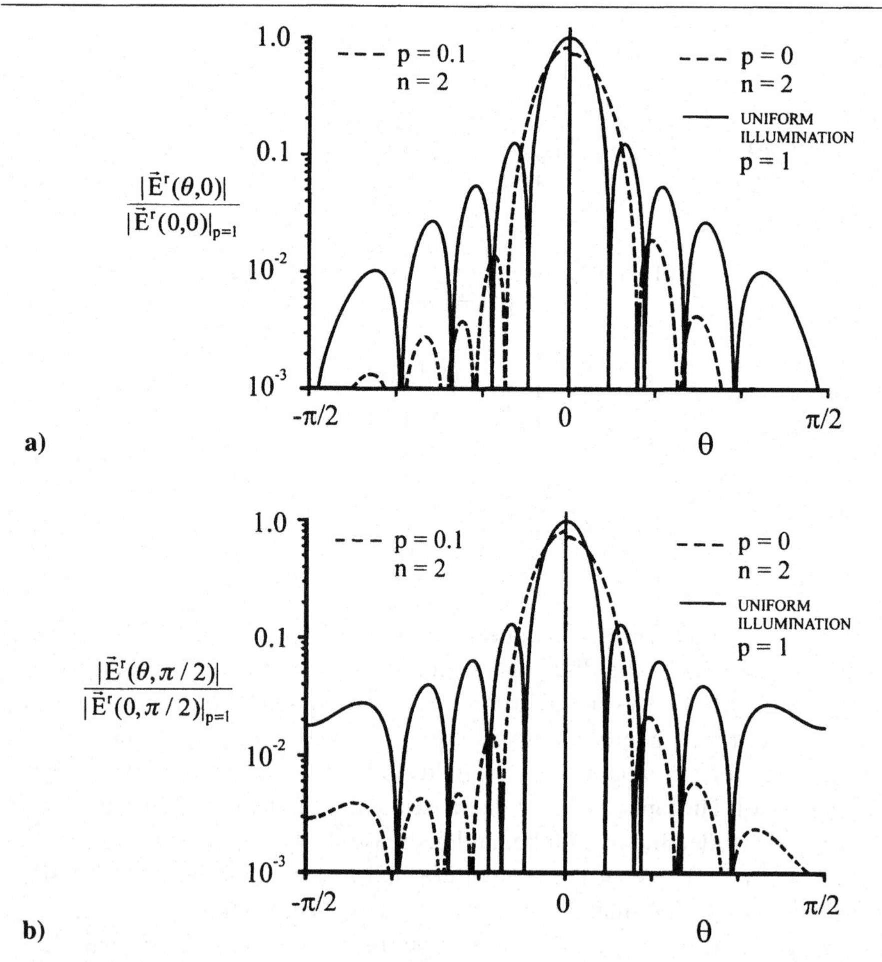

Fig. 3.32. Normalized electric field patterns for circular aperture with various illuminations, for $d/\lambda_o = 5.0$. a) Principal H-plane patterns, $\phi = 0\ (\pi)$. b) Principal E-plane patterns, $\phi = \pi/2\ (3\pi/2)$.

are simply related:

$$G(\theta,\phi) = \frac{\text{time-average power radiated per unit area in the direction } \theta,\phi}{(\text{time-average power supplied to the antenna})/4\pi r^2}$$
$$= \eta D_A(\theta,\phi). \tag{3.138}$$

Here the *efficiency factor* η accounts for the fraction of the power supplied to the antenna that is not radiated by the aperture:

$$\eta = \frac{\text{time-average power radiated by aperture}}{\text{time-average power supplied to antenna}} \leq 1. \tag{3.139}$$

For our simple model η is composed of two terms:

$$\eta = \eta_d \eta_s. \tag{3.140}$$

The factor η_d accounts for the power dissipated in the antenna (e.g., the heating of

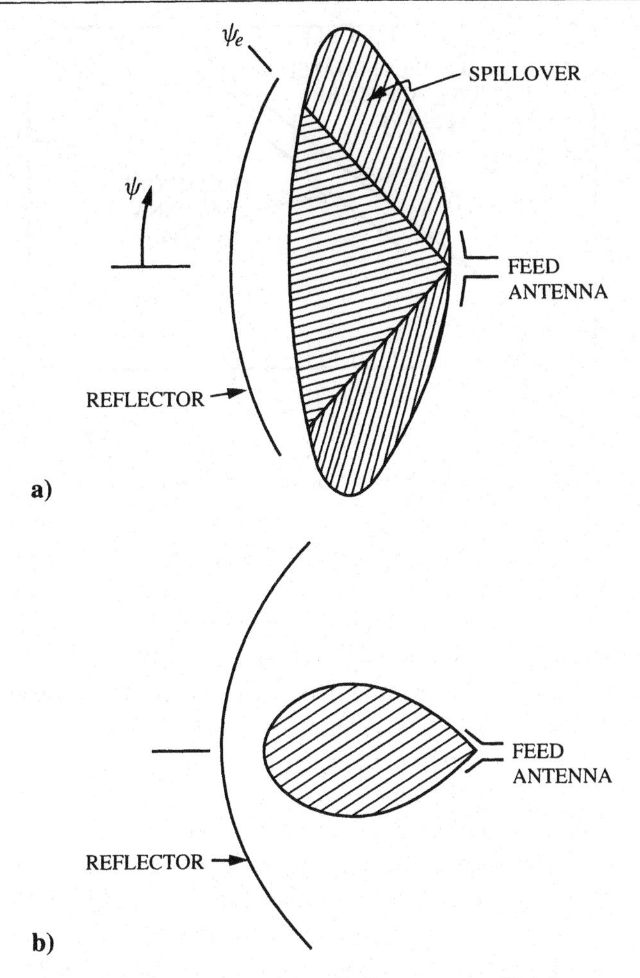

Fig. 3.33. Schematic representations for radiation pattern of feed antenna for a) maximum directivity of aperture and b) negligible spillover.

metal parts), and the factor η_s accounts for the "spillover" – power that is radiated by the feed and not intercepted by the reflector. The ray AE in Figure 3.30 represents spillover. For more complicated models additional factors may be included in η, such as the loss associated with power being radiated into undesired states of polarization.

Now let us consider the gain of our simple paraboloidal reflector antenna. The radiation pattern of the feed antenna determines the illumination of the aperture. We can obtain the maximum directivity D_A with a uniformly illuminated aperture. This requires a radiation pattern for the feed antenna that increases in amplitude toward the edges of the reflector to compensate for the aforementioned natural tapering of the aperture field. A pattern of this form is shown schematically in Figure 3.33a. Since the pattern cannot go abruptly to zero at the edges of the reflector ($\psi = \psi_e$), there will be significant spillover associated with this feed ($\eta_s < 1$). As an alternative we can make this spillover negligible ($\eta_s \approx 1$) by using

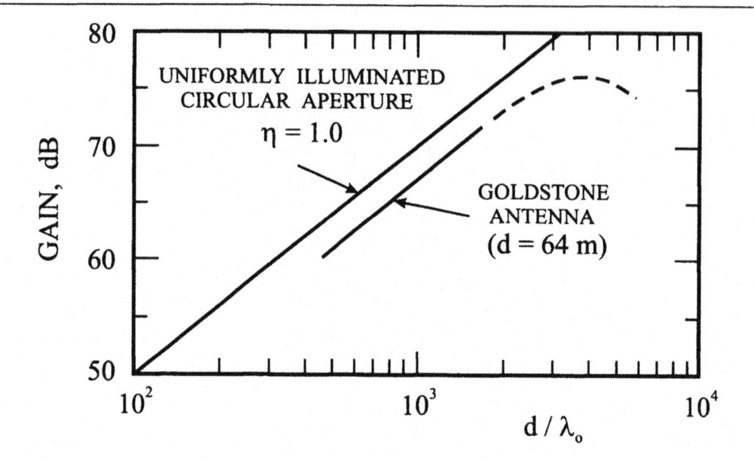

Fig. 3.34. Gain versus d/λ_o. (Data for Goldstone antenna from Love [23].)

a feed antenna that produces little radiation in the direction of the reflector edge: one with a radiation pattern like that in Figure 3.33b. However, this pattern will produce a strong taper to the aperture field and a directivity D_A substantially less than the maximum. Thus, we see that there is a practical difficulty associated with maximizing the gain (3.138) by maximizing both the directivity of the aperture D_A and the spillover efficiency η_s. A good design is one that includes some taper to the aperture illumination and for which neither D_A nor η_s is maximum.

The gain for the Goldstone antenna (Figure 3.29a) is graphed as a function of d/λ_o in Figure 3.34. The solid portion of the curve was obtained from measured data; the dashed portion is an extrapolation of these data [23]. For comparison the gain for an antenna with a uniformly illuminated aperture and 100% efficiency ($\eta = 1.0$) is also shown. For values of d/λ_o less than about 2×10^3, the two curves are parallel and separated by about 3 dB; this suggests that gain for the Goldstone antenna is proportional to $(d/\lambda_o)^2$ and is thus about 50% of the gain for the uniformly illuminated aperture. When d/λ_o is greater than 2×10^3, the increase in the gain with d/λ_o is less, and it eventually decreases with increasing d/λ_o. The decrease in the gain is caused by the roughness of the surface of the reflector. Surface distortions, which may result from thermal strains, gravity, wind, etc., limit the gain of all electrically large reflector antennas.

3.7.3 Gaussian beam – paraxial approximation

In our discussion of the radiated field, we found that at a sufficient distance from any source the field exhibits a definite behavior: It propagates outward in the radial direction with a fixed angular or transverse dependence. Thus, a field confined to a narrow range of solid angle, a beam, propagates in a predictable manner, as exemplified by the tube of rays in Figure 3.19. In certain applications, a knowledge of the radiated field is not sufficient, and a more detailed model for a beam that applies at any distance from the source is required. The *Gaussian beam* is a model frequently used in optics.

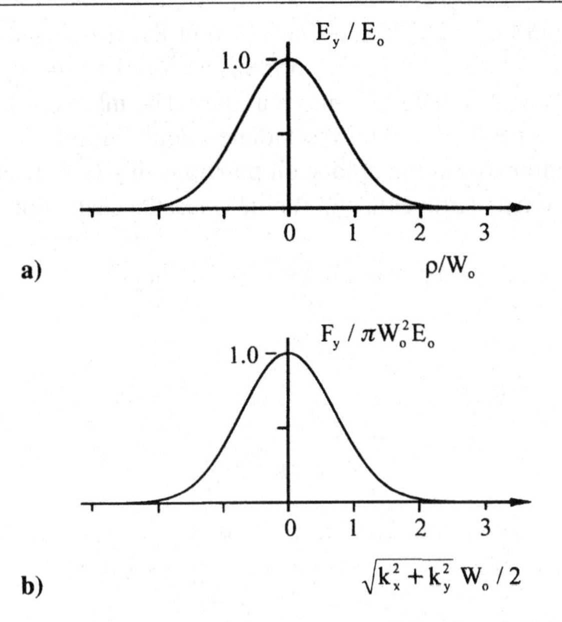

Fig. 3.35. Gaussian beam. a) Tangential component of the electric field on the plane $z = 0$. b) Spectral function for the field of a).

We will begin our discussion by assuming that the tangential component of the electric field on the reference plane $z = 0$ is described by the Gaussian function $\exp(-\rho^2/W_o^2)$:

$$\vec{E}(x, y, z = 0) = E_o e^{-\rho^2/W_o^2}\hat{y}, \qquad \rho = \sqrt{x^2 + y^2}, \tag{3.141}$$

which is graphed in Figure 3.35a. From Table 3.3, the spectral function corresponding to this field is

$$F_y(k_x, k_y) = E_o \int_{x=-\infty}^{\infty} e^{(-x^2/W_o^2 + jk_x x)} dx \int_{y=-\infty}^{\infty} e^{(-y^2/W_o^2 + jk_y y)} dy. \tag{3.142}$$

Each of the integrals in (3.142) can be evaluated [22] using

$$\int_{-\infty}^{\infty} e^{(-p^2u^2 \pm qu)} du = \frac{\sqrt{\pi}}{p} e^{q^2/4p^2}, \qquad p > 0 \tag{3.143}$$

to give

$$F_y(k_x, k_y) = \pi W_o^2 E_o e^{-(k_x^2 + k_y^2)(W_o/2)^2}, \tag{3.144}$$

which is graphed in Figure 3.35b. Notice that (3.144) shows an interesting property of the Gaussian function: Its Fourier transform is also a Gaussian function.

The electromagnetic field in the half space $z > 0$ is obtained from the expressions in Table 3.3; the electric field is

$$\vec{E}(\vec{r}) = \frac{1}{(2\pi)^2} \int_{-\infty}^{\infty} \int_{-\infty}^{\infty} \left(\hat{y} - \frac{k_y}{k_z}\hat{z}\right) F_y(k_x, k_y) e^{-j(k_x x + k_y y + k_z z)} dk_x dk_y. \tag{3.145}$$

If we inserted (3.144) into (3.145), we would find that for a general position $\vec{r}$ the resulting integrals could not be performed in closed form. This is the same situation we encountered in the previous examples. The integrals in (3.145) could be evaluated approximately to obtain the radiated field, but this would be of little help since we are interested in the field at all positions in space. There is, however, another approximation that proves useful for the Gaussian beam; this is the so-called *paraxial approximation.*

Recall that the vector wave number (3.77) associated with each point of the integration (3.145) is

$$\vec{k} = k_x\hat{x} + k_y\hat{y} + k_z\hat{z},$$

with

$$k_z = \sqrt{k_o^2 - (k_x^2 + k_y^2)}.$$

In the paraxial approximation we assume that the spectral function is of negligible magnitude except for vector wave numbers that are nearly parallel to the z axis; that is,

$$F_y(k_x, k_y) \approx 0, \tag{3.146}$$

except for

$$\sqrt{k_x^2 + k_y^2} \ll k_o. \tag{3.147}$$

This inequality is used to approximate k_z in the exponential term of (3.145) by

$$k_z = \sqrt{k_o^2 - (k_x^2 + k_y^2)} \approx k_o\left(1 - \frac{k_x^2 + k_y^2}{2k_o^2}\right) \tag{3.148}$$

and in the algebraic term by

$$k_z \approx k_o. \tag{3.149}$$

Thus, in the paraxial approximation, the electric field (3.145) becomes

$$\vec{E}(\vec{r}) = \frac{1}{(2\pi)^2} e^{-jk_oz} \int_{-\infty}^{\infty}\int_{-\infty}^{\infty} \left(\hat{y} - \frac{k_y}{k_o}\hat{z}\right) F_y(k_x, k_y)$$

$$\exp\left[-jk_xx + j(k_x^2/2k_o)z\right] \exp\left[-jk_yy + j(k_y^2/2k_o)z\right] dk_x dk_y, \tag{3.150}$$

which can be rewritten as

$$\vec{E}(\vec{r}) = \frac{1}{(2\pi)^2} e^{-jk_oz} \left(\hat{y} - \hat{z}\frac{j}{k_o}\frac{\partial}{\partial y}\right) \int_{-\infty}^{\infty}\int_{-\infty}^{\infty} F_y(k_x, k_y)$$

$$\exp\left[-jk_xx + j(k_x^2/2k_o)z\right] \exp\left[-jk_yy + j(k_y^2/2k_o)z\right] dk_x dk_y. \tag{3.151}$$

Note that this approximation has excluded all evanescent waves from the spectrum.

Now let us apply these results to the Gaussian beam. First we will consider the conditions for the paraxial approximation, (3.146) and (3.147). These conditions will be satisfied by the spectral function (3.144) provided

$$e^{-k_o^2(W_o/2)^2} = e^{-(k_o W_o/2)^2} \ll 1,$$

or

$$\left(\frac{k_o W_o}{2}\right)^2 \gg 1. \tag{3.152}$$

Thus, $k_o W_o = 2\pi W_o/\lambda_o$ must be large for the paraxial approximation to hold for the Gaussian beam. Of course, precisely how large $k_o W_o$ must be will depend on the accuracy required for the approximation.

The electric field of the Gaussian beam is obtained by substituting (3.144) into (3.151):

$$\vec{E}(\vec{r}) = \frac{(W_o/2)^2}{\pi} E_o e^{-jk_o z}\left(\hat{y} - \hat{z}\frac{j}{k_o}\frac{\partial}{\partial y}\right)$$
$$\int_{-\infty}^{\infty}\int_{-\infty}^{\infty}\Big(\exp\left\{-k_x^2[(W_o/2)^2 - jz/2k_o] - jk_x x\right\}$$
$$\exp\left\{-k_y^2[(W_o/2)^2 - jz/2k_o] - jk_y y\right\}\Big)dk_x dk_y. \tag{3.153}$$

The integrals in this expression can be evaluated using (3.143), and after some rearrangement, we obtain the final result:

$$\vec{E}(\vec{r}) = \frac{E_o}{W(z)/W_o} e^{j\psi(z)}\left[\hat{y} + j\frac{\gamma(y)}{W(z)/W_o} e^{j\psi(z)}\hat{z}\right] e^{-[\rho/W(z)]^2} e^{-j[k_o z + k_o \rho^2/2R(z)]}, \tag{3.154a}$$

where ρ is the radial distance on a plane of constant z (the circular cylindrical coordinates of Appendix B) and

$$\gamma(u) = \frac{2u/W_o}{k_o W_o}, \tag{3.154b}$$

$$\psi(u) = \tan^{-1}[\gamma(u)], \tag{3.154c}$$

$$W(u) = W_o[1 + \gamma^2(u)]^{\frac{1}{2}}, \tag{3.154d}$$

$$R(u) = u[1 + \gamma^{-2}(u)]. \tag{3.154e}$$

The structure of the Gaussian beam is shown in Figures 3.36 and 3.37; these results are for $k_o W_o = 5.0$. There are two components to the electric field, E_y and E_z. E_y is rotationally symmetric about the z axis, whereas E_z is maximum in the plane $x = 0$ for ρ fixed; notice that $|E_z|$ is plotted for this plane in Figure 3.36. The ratio of the maximum values for these two components is $|E_z|_{\max}/|E_y|_{\max} = \sqrt{2}e^{-1/2}/k_o W_o = 0.8577\ldots/k_o W_o$; hence, E_z is negligible when $k_o W_o$ is large, as is required for the paraxial approximation (3.152). The component E_y is approximately a Gaussian function on each plane z/W_o because the term $\exp\{-[\rho/W(z)]^2\}$ dominates the amplitude of the field (3.154a). The factor $W(z)$ fixes the width of the

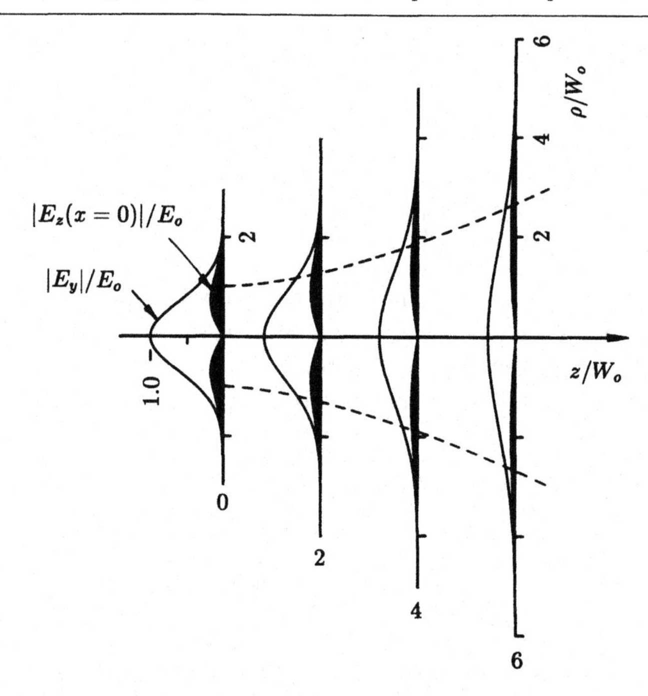

Fig. 3.36. Electric field of Gaussian beam in region near plane $z = 0$, for $k_o W_o = 5.0$.

beam. Since $W(z)$ increases with increasing z, the beam spreads out as it propagates away from the plane $z/W_o = 0$.

The radius or "spot size" of the beam is defined as the value of ρ at which the amplitude of the field has decreased by the factor $1/e$ from its value on the axis, that is,

$$[\rho/W(z)]^2 = 1.$$

After substituting (3.154d), we see that this is the equation for a hyperbola:

$$(\rho/W_o)^2 - \left(\frac{2}{k_o W_o}\right)^2 (z/W_o)^2 = 1. \tag{3.155}$$

The radius of the beam is graphed as a dashed line in Figures 3.36 and 3.37. The beam is narrowest at its "waist," $z/W_o = 0$, where its radius is W_o. At large distances from the plane $z/W_o = 0$, the radius of the beam subtends the angle 2α at the origin, where

$$\alpha = \tan^{-1}(\rho/z) = \tan^{-1}(2/k_o W_o). \tag{3.156}$$

The phase of the dominant component (y) of the field (3.154a) is

$$\psi(z) - k_o z - k_o\rho^2/2R(z), \tag{3.157}$$

making the equation for a surface of a constant phase that includes the point z_o on the axis ($\rho = 0$)

$$\psi(z_o) - \psi(z) - k_o(z_o - z) + k_o\rho^2/2R(z) = 0,$$

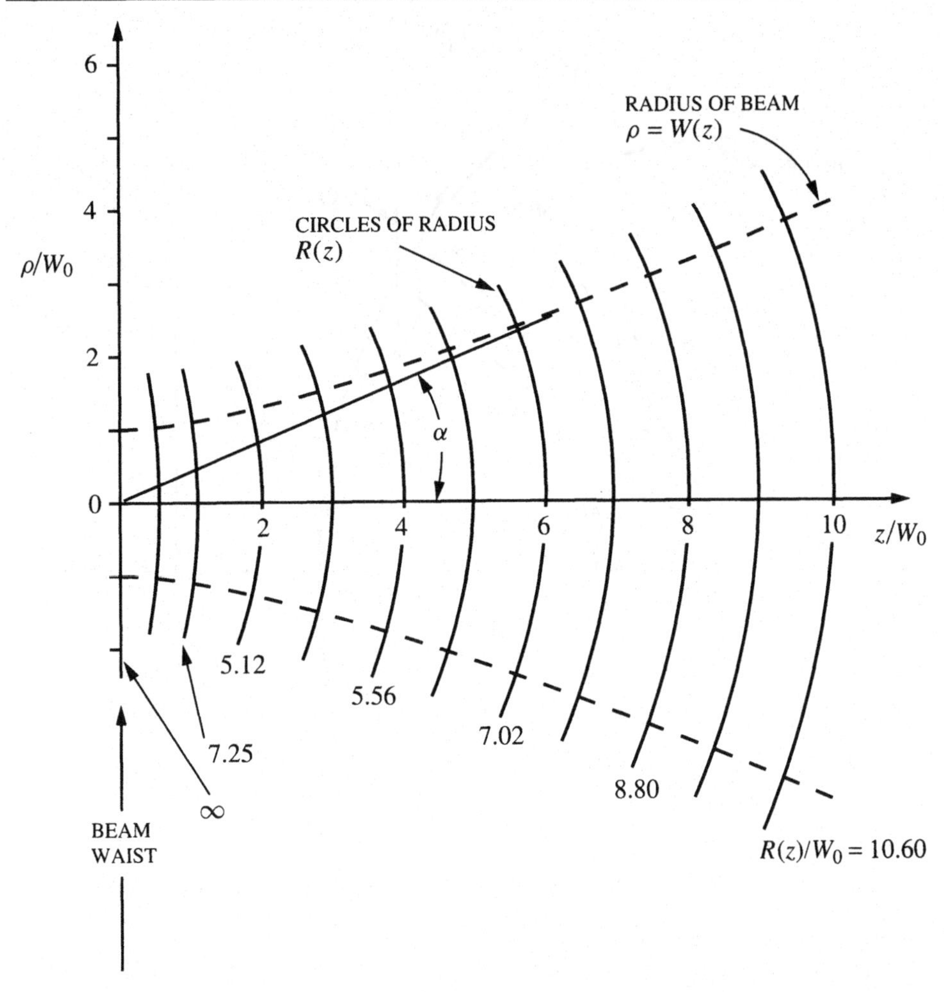

Fig. 3.37. Propagation of Gaussian beam, for $k_o W_o = 5.0$.

or

$$\rho(z) = \left(2R(z)\left\{(z_o - z) - \frac{1}{k_o}\left[\psi(z_o) - \psi(z)\right]\right\}\right)^{\frac{1}{2}}. \tag{3.158}$$

Since this surface is rotationally symmetric about the z axis, it can be characterized by the plane curve (solid line) shown in Figure 3.38. Now we wish to determine the radius of curvature of this surface on the axis of the beam, that is, the radius of the circle $R_c(z_o)$ (dashed line) that has the same curvature as our curve at the point $\rho = 0$, z_o.

Consider two points, 1 and 2, on the curve. The tangent to the curve makes the angle τ with the z axis. In moving from point 1 to point 2, this angle changes by $\Delta\tau$, and the arc length increases by Δs. The curvature is defined [24] as

$$\chi(z) = \lim_{\Delta\tau\to 0}\left(\frac{\Delta\tau}{\Delta s}\right) = \frac{d\tau}{ds} = \left(\frac{d\tau}{dz}\right)\Bigg/\left(\frac{ds}{dz}\right). \tag{3.159}$$

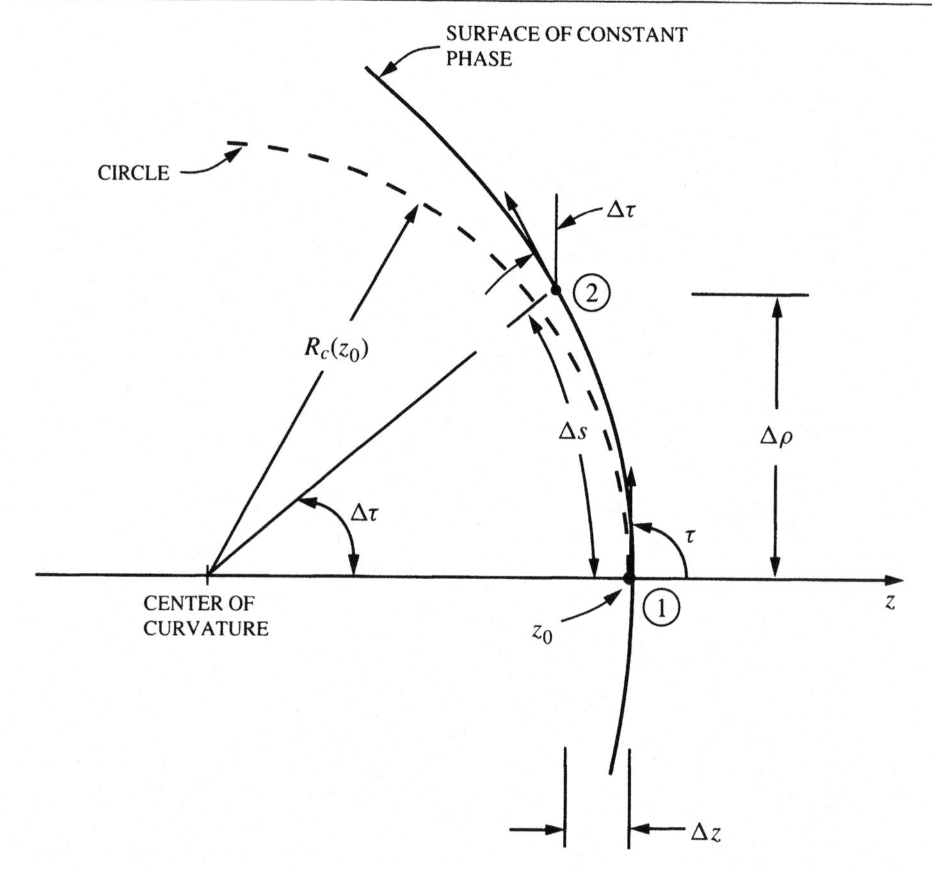

Fig. 3.38. Detail for determining the radius of curvature for the wavefronts.

Since the curvature for the circle is $\chi_c = 1/R_c(z)$, and because this must equal the curvature (3.159), the radius of curvature we seek is

$$R_c(z) = \frac{1}{\chi(z)} = \left(\frac{ds}{dz}\right) \Big/ \left(\frac{d\tau}{dz}\right). \tag{3.160}$$

Now we notice from Figure 3.38 that

$$\tan\tau = \lim_{\Delta z \to 0} \left(\frac{\Delta\rho}{\Delta z}\right) = \frac{d\rho}{dz},$$

so

$$\frac{d\tau}{dz} = \frac{d}{dz}\left[\tan^{-1}\left(\frac{d\rho}{dz}\right)\right] = \left[1 + \left(\frac{d\rho}{dz}\right)^2\right]^{-1} \frac{d^2\rho}{dz^2}$$

and

$$\frac{ds}{dz} = \lim_{\Delta z \to 0}\left(\frac{\Delta s}{\Delta z}\right) = \lim_{\Delta z \to 0}\left[\frac{\sqrt{(\Delta\rho)^2 + (\Delta z)^2}}{\Delta z}\right] = \sqrt{1 + \left(\frac{d\rho}{dz}\right)^2}.$$

The radius of curvature (3.160) then becomes

$$R_c(z) = \left\{ \left[1 + \left(\frac{d\rho}{dz} \right)^2 \right]^{\frac{3}{2}} \Big/ \frac{d^2\rho}{dz^2} \right\}. \tag{3.161}$$

After substituting (3.158) into (3.161) and evaluating the result on the axis ($z = z_o$), we find that

$$R_c(z_o) = R(z_o) \left\{ 1 - \frac{2}{(k_o W_o)^2 [1 + \gamma^2(z_o)]} \right\} \approx R(z_o), \tag{3.162}$$

where, on the right-hand side, we have simplified our result by noting that $(k_o W_o)^2$ must be large (3.152) for the paraxial approximation. We now see the significance of the parameter $R(z)$ in Equation (3.154a); it is the radius of curvature of the surfaces of constant phase at points along the axis of the beam (z axis). Circles of radius $R(z)$ are shown as solid lines in Figure 3.37. At the beam waist $z/W_o = 0$, the radius of curvature is infinite; the wavefront is planar. As z/W_o increases the radius of curvature decreases rapidly, becoming a minimum $R/W_o = k_o W_o$ when $z/W_o = k_o W_o/2$ (at $z/W_o = 2.5$ for the example in Figure 3.37). Past this point, the radius of curvature increases and approaches $R/W_o = z/W_o$ as z/W_o becomes large; the wavefronts then become spheres centered on the origin.

At large distances from the plane $z/W_o = 0$, the field of the Gaussian beam, shown in Figure 3.37, clearly has the asymptotic form we expect: spherical wavefronts with a fixed angular dependence for the field, as indicated by the radius of the beam occurring at the fixed angle α. The radiated electric field is obtained by substituting (3.144) into the formulas in Table 3.4:

$$\vec{E}^r(\vec{r}) = \left[F_\theta(\theta, \phi)\hat{\theta} + F_\phi(\theta, \phi)\hat{\phi} \right] e^{-jk_o r}/r, \tag{3.163a}$$

with

$$F_\theta(\theta, \phi) = \frac{2j}{k_o} \left(\frac{k_o W_o}{2} \right)^2 E_o e^{-[(k_o W_o/2)\sin\theta]^2} \sin\phi \tag{3.163b}$$

$$F_\phi(\theta, \phi) = \frac{2j}{k_o} \left(\frac{k_o W_o}{2} \right)^2 E_o e^{-[(k_o W_o/2)\sin\theta]^2} \cos\theta \cos\phi. \tag{3.163c}$$

It is interesting to compare the radiated field for the Gaussian beam with those for the uniformly illuminated circular aperture and the circular aperture with tapered illumination. The side lobes that are prominent for the uniformly illuminated aperture decrease with the addition of taper and completely disappear for the Gaussian beam. This phenomenon is easily understood when we recall that the radiated field is related to the Fourier transform of the field on the plane $z = 0$. It is a general property of the Fourier transform: The transform of a smooth function, such as the Gaussian, is again smooth, whereas the transform of a function with abrupt changes in amplitude, such as the pulse, is oscillatory.

The Gaussian beam plays an important role in the analysis of lasers, such as the gas laser shown in Figure 3.20c. The spherical mirrors for this laser form an open-cavity resonator. An electromagnetic field that satisfies Maxwell's equations within the cavity and satisfies the boundary conditions at the perfectly conducting mirrors is called a *mode*. Several modes can exist in the cavity; the simplest of these, the lowest order mode, is a Gaussian beam [25]. This mode is usually designated $\mathrm{TEM}_{0,0}$.

Before we show that the Gaussian beam is a solution for the cavity resonator, we will make a few general observations concerning the field (3.154a). Our first point is that we could perform our analysis for the electromagnetic field in the left half space (LHS) $z < 0$, rather than in the right half space (RHS) $z > 0$. We would start with the same electric field on the plane $z = 0$ (3.141). The spectral representation for the field would be the same as for the field in the right half space (3.145), the only difference being that the evanescent waves would decay with decreasing z, rather than with increasing z. In terms of the wave numbers, this means that

$$k_z = \begin{cases} \sqrt{k_o^2 - k_x^2 - k_y^2}, & 0 \le k_x^2 + k_y^2 \le k_o^2 \\ -j\sqrt{k_x^2 + k_y^2 - k_o^2}, & \text{RHS} \\ j\sqrt{k_x^2 + k_y^2 - k_o^2}, & \text{LHS} \end{cases} \Bigg\} \; k_x^2 + k_y^2 > k_o^2,$$

or simply $\vec{k}_{\mathrm{LHS}} = \vec{k}^*_{\mathrm{RHS}}$. In the paraxial approximation the evanescent waves are omitted, so conjugation leaves the wave number unchanged. Thus, the electric field (3.154a) represents the Gaussian beam in the left half space as well as in the right half space. We will rewrite this field as

$$\vec{E}^+(\vec{r}) = \frac{E_o^+}{W(z)/W_o} e^{j\psi(z)} e^{-[\rho/W(z)]^2} e^{-j[k_o z + k_o \rho^2/2R(z)]} \hat{y}. \tag{3.164}$$

Here the $+$ sign indicates that the beam is propagating to the right ($+z$), and we have omitted the z component of the field, which is usually negligible.

Our second point is that we could solve for a Gaussian beam that propagates to the left ($-z$), rather than to the right. The electric field would then be (3.164) with j replaced by $-j$:

$$\vec{E}^-(\vec{r}) = \frac{E_o^-}{W(z)/W_o} e^{-j\psi(z)} e^{-[\rho/W(z)]^2} e^{j[k_o z + k_o \rho^2/2R(z)]} \hat{y}. \tag{3.165}$$

Inside the laser cavity the electric field consists of a superposition of two Gaussian beams, one propagating to the right and the other propagating to the left. We will assume that these beams have equal amplitude; then, the electric field in the cavity is the superposition of (3.164) and (3.165) with $E_o^+ = E_o^- = E_o$:

$$\vec{E}(\vec{r}) = \frac{2E_o}{W(z)/W_o} e^{-[\rho/W(z)]^2} \cos\left[\psi(z) - k_o z - k_o \rho^2/2R(z)\right] \hat{y}. \tag{3.166}$$

This standing wave must satisfy the boundary conditions at the perfectly conducting mirrors (i.e., the tangential component of the electric field must be zero). If we confine our attention to the region in the vicinity of the axis, then the y component of the field must be zero at the mirrors. The spherical mirrors are of radius R_m and are located at $z = \pm d/2$, as in Figure 3.20c. The boundary conditions are satisfied by (3.166) when the radius of curvature of the beam matches the radius of curvature of the mirrors,

$$R(\pm d/2) = R_m, \tag{3.167}$$

so that the argument of the cosine term is constant on the mirrors, and

$$\cos\left[\psi(z) - k_o z - k_o \rho^2/2R(z)\right]\Big|_{\substack{\rho=0\\ z=\pm d/2}} = 0,$$

or, equivalently,

$$k_o d/2 - \psi(d/2) = (2n+1)\pi/2, \qquad n = 0, 1, 2, \ldots. \tag{3.168}$$

After introducing (3.154c) and (3.154d), the conditions (3.167) and (3.168) become

$$k_o W_o = \left[k_o d\left(\frac{2R_m}{d} - 1\right)^{\frac{1}{2}}\right]^{\frac{1}{2}} \tag{3.169}$$

and

$$k_o d/2 - \tan^{-1}\left(\frac{d/W_o}{k_o W_o}\right) = (2n+1)\pi/2. \tag{3.170}$$

If we had chosen $E_o^+ = -E_o^- = E_o$ for our superposition of Gaussian beams, we would have obtained (3.169) and, in place of (3.170),

$$k_o d/2 - \tan^{-1}\left(\frac{d/W_o}{k_o W_o}\right) = n\pi. \tag{3.171}$$

The two solutions represented by (3.170) and (3.171) can be combined in a single equation:

$$k_o d/2 - \tan^{-1}\left(\frac{d/W_o}{k_o W_o}\right) = n\pi/2. \tag{3.172}$$

When the argument of the arc tangent is large ($d/W_o \gg k_o W_o$), the following approximation applies:

$$\tan^{-1}(u) \approx \pi/2 - 1/u,$$

and (3.172) becomes

$$k_o d = (n+1)\pi - \frac{2k_o W_o}{d/W_o},$$

or

$$\lambda_o = \left[2d + \frac{4W_o}{(d/W_o)}\right] \Big/ (n+1), \qquad n = 0, 1, 2, \ldots. \tag{3.173}$$

Equations (3.169) and (3.173) determine the conditions under which the superposition of Gaussian beams is a solution to Maxwell's equations within the cavity. From (3.173) we see that the wavelength must be slightly greater than $2d/(n+1)$. The frequencies ($f = c/\lambda_o$) that correspond to these wavelengths are the resonant frequencies of the cavity. The radius or spot size of the beam at its waist is given by (3.169).

In this section we have used the plane-wave spectrum to study a few simple problems. The plane-wave spectrum has many other applications in electromagnetics; these include practical problems associated with antenna analysis and synthesis [26, 27] and antenna measurements [28, 29]. Its application to the scalar field of optics is the basis for "Fourier optics" [19, 30].

References

[1] H. G. Booker and P. C. Clemmow, "The Concept of an Angular Spectrum of Plane Waves, and Its Relation to that of Polar Diagram and Aperture Distribution," *Proc. IEE*, Vol. 97, Pt. III, pp. 11–17, January 1950.

[2] P. C. Clemmow, *The Plane Wave Spectrum Representation of Electromagnetic Fields*, Pergamon Press, New York, 1966.

[3] A. Papoulis, *The Fourier Integral and Its Applications*, McGraw-Hill, New York, 1962.

[4] G. Kirchhoff, *Vorlesungen über Mathematische Physik*, Vol. 2, Mathematische Optik, B. G. Teubner, Berlin, 1891.

[5] J. J. Bowman, T. B. A. Senior, and P. L. E. Uslenghi, *Electromagnetic and Acoustic Scattering by Simple Shapes*, Ch. 4, "The Strip," North-Holland, Amsterdam, 1969.

[6] G. F. Carrier, M. Krook, and C. E. Pearson, *Functions of a Complex Variable Theory and Technique*, Ch. 6, McGraw-Hill, New York, 1966.

[7] A. Erdélyi, *Asymptotic Expansions*, Dover Publications, New York, 1956.

[8] M. Abramowitz and I. A. Stegun, *Handbook of Mathematical Functions*, U.S. Government Printing Office, Washington, DC, 1964.

[9] G. G. Stokes, "On the Numerical Calculation of a Class of Definite Integrals and Infinite Series," *Trans. Cambridge. Philos. Soc.*, Vol. 9, pp. 166–87, 1850. Reprinted in *Mathematical and Physical Papers*, Vol. II, pp. 329–57, Johnson Reprint Corp., New York, 1966.

[10] Lord Kelvin (W. Thomson), "On the Waves Produced by a Single Impulse in Water of Any Depth, or in a Dispersive Medium," *Philos. Mag.*, Vol. 23, pp. 252–5, March 1887. Reprinted in *Mathematical and Physical Papers*, Vol. 4, pp. 303–6, Cambridge University Press, London, 1910.

[11] P. I. Ufimtsev, "Secondary Diffraction of Electromagnetic Waves by a Strip," *Soviet Phys.-Techn. Phys.*, Vol. 3, pp. 535–48, March 1958.

[12] M. Born and E. Wolf, *Principles of Optics*, 3rd Edition, Pergamon Press, New York, 1965.

[13] M. Kline and I. W. Kay, *Electromagnetic Theory and Geometric Optics*, Wiley, New York, 1965.

[14] W. R. Scott, Jr., "A General Program for Plotting Three-Dimensional Antenna Patterns," *IEEE Antennas and Propagation Newsletter*, Vol. 31, pp. 6–11, December 1989.

[15] S. Silver, *Microwave Antenna Theory and Design*, Massachusetts Institute of Technology Radiation Laboratory Series Vol. 12, McGraw-Hill, New York, 1949.
[16] C. L. Andrews, "Diffraction Pattern of a Circular Aperture at Short Distances," *Phys. Rev.*, Vol. 71, pp. 777–86, June 1, 1947.
[17] —, "Diffraction Pattern in a Circular Aperture Measured in the Microwave Region," *J. Appl. Phys.*, Vol. 21, pp. 761–7, August 1950.
[18] H. Severin, "Beugung elektromagnetischer Zentimeterwellen an metallischen Blenden," *Zeits. f. Naturforschung*, Vol. 1, pp. 487–95, 1946.
[19] J. W. Goodman, *Introduction to Fourier Optics*, McGraw-Hill, New York, 1968.
[20] G. B. Airy, "On the Diffraction of an Object-Glass with Circular Aperture," *Trans. Cambridge Philos. Soc.*, Vol. 5, pp. 283–91, 1835.
[21] A. W. Rudge, K. Milne, A. D. Olver, and P. Knight, *The Handbook of Antenna Design*, Vol. 1, Ch. 3, Peter Peregrinus, London, 1982.
[22] I. S. Gradshteyn and I. W. Ryzhik, *Tables of Integrals, Series, and Products*, Academic Press, New York, 1980.
[23] A. W. Love, "Some Highlights in Reflector Antenna Development," *Radio Sci.*, Vol. 11, pp. 671–84, August–September 1976.
[24] W. Gellert, H. Kustner, M. Hellwich, and H. Kastner, *The VNR Concise Encyclopedia of Mathematics*, Ch. 19, Van Nostrand Reinhold, New York, 1977.
[25] J. T. Verdeyen, *Laser Electronics*, Ch. 3, Prentice-Hall, Englewood Cliffs, New Jersey, 1981.
[26] R. H. Clarke and J. Brown, *Diffraction Theory and Antennas*, Wiley, New York, 1980.
[27] D. R. Rhodes, *Synthesis of Planar Aperture Antenna Sources*, Oxford University Press, London, 1974.
[28] D. M. Kerns, *Plane-Wave Scattering-Matrix Theory of Antenna – Antenna Interactions*, National Bureau of Standards, Monograph 162, U.S. Government Printing Office, Washington, DC, 1981.
[29] D. T. Paris, W. M. Leach, Jr., and E. B. Joy, "Basic Theory of Probe-Compensated Near-Field Measurements," *IEEE Trans. Antennas and Propagat.*, Vol. AP-26, pp. 373–9, May 1978.
[30] K. E. Oughstun, editor, *Selected Papers on Scalar Wave Diffraction*, SPIE Milestone Series, Volume MS51, Ch. 5, "Angular Spectrum of Plane Waves Representation," SPIE, Belingham, WA, 1992.
[31] A. Sommerfeld, *Optics*, Academic Press, New York, 1954.
[32] B. N. Harden, "Diffraction of Electromagnetic Waves by a Half-Plane," *Proc. IEE*, Vol. 99, Pt. III, pp. 229–35, September 1952.
[33] A. Erdélyi, editor, *Tables of Integral Transforms*, Volume I, McGraw-Hill, New York, 1954.

Problems

3.1 Show that the electric field (3.7) satisfies the Helmholtz equation (3.5) when $\vec{k} \cdot \vec{k} = k_o^2$ (3.8). In this exercise, it will be helpful to express the vectors $\vec{E}_o$ and $\vec{k}$ in terms of their components in the rectangular Cartesian coordinate system (x, y, z).

3.2 An inhomogeneous plane wave has the following electric field:

$$\vec{E} = A(-\hat{x} + \hat{y} + \hat{z})e^{-j\vec{k}\cdot\vec{r}},$$

with

$$\vec{k} = (\vec{k}_r + j\vec{k}_i)$$

and

$$\vec{k}_r = k_o(\hat{x} + \hat{z}).$$

Determine the vector $\vec{k}_i$ and the magnetic field $\vec{B}$.

3.3 Obtain Equations (3.26a–c) and (3.27a–c) from Maxwell's equations and the relations (3.24a–c) and (3.25a–c).

3.4 Show that the two-dimensional TM plane wave [Equations (3.31a,b)] becomes (3.32a,b) when expressed in terms of the electric field E_{ox} rather than the magnetic field B_{oy}.

3.5 In this problem, we will examine an alternate derivation for the spectral representation of the two-dimensional field. Consider the TE field $\vec{E}^{TE} = E^{TE}(x, z)\hat{y}$. This field must be a solution to the scalar Helmholtz equation (3.26c)

$$\nabla^2 E^{TE} + k_o^2 E^{TE} = 0,$$

where the operator ∇^2 involves only derivatives with respect to x and z. Take the Fourier transform of this equation with respect to x. Solve the resulting equation for the transformed field $E^{TE}(k_x, z)$. Now equate this field to the transform $F_y(k_x)$ of the field specified on the plane $z = 0$:

$$F_y(k_x) = \int_{-\infty}^{\infty} E^{TE}(x, 0)e^{jk_x x}dx.$$

Take the inverse transform of $E^{TE}(k_x, z)$; you should obtain the result in Table 3.1 for $\vec{E}^{TE}(\vec{r})$.

3.6 The antenna shown in Figure 3.29 is a *Cassegrain reflector*; it is based on its optical counterpart – the Cassegrain telescope. This is a dual-reflector antenna that uses a paraboloid as the primary reflector and a hyperboloid as the secondary reflector. As shown in Figure P3.1, one of the foci of the hyperboloid is coincident with the focus of the paraboloid E, and the other is at the feed antenna A. This system places the feed antenna and electronics near the primary reflector, where they are easily supported. This is to be compared with a prime focus feed (Figure 3.30), where the feed antenna must be supported at the focus of the paraboloid.

a) For a hyperboloid, the difference of the distances from the foci to a point on the surface is a constant (i.e., AB – BE = constant in Figure P3.1). Using the ray ABCD, show that all rays from the feed antenna travel the

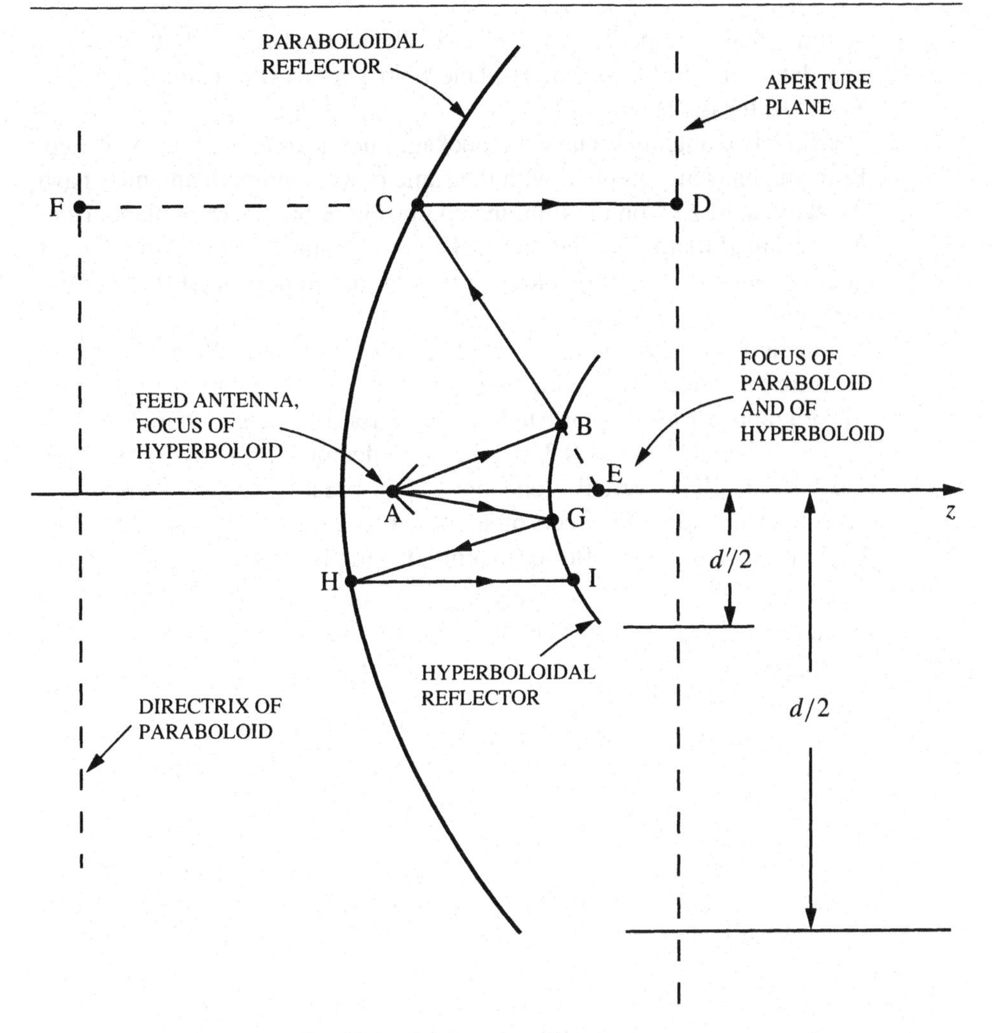

Fig. P3.1. Cassegrain reflector antenna.

same distance to arrive at the aperture plane. Thus, when all rays leave the feed with the same phase, the aperture field has uniform phase.

b) The secondary reflector blocks a portion of the radiation from the primary reflector from reaching the aperture plane (e.g., the ray AGHI in Figure P3.1). This blockage can be represented by setting the field equal to zero over a disc of radius $d'/2$ on the aperture plane. For simplicity assume that the field on the aperture plane is linearly polarized and has uniform amplitude and phase outside the area of blockage:

$$\vec{E}_y(x, y, z = 0) = \begin{cases} E_o \hat{y}, & d'/2 < \rho < d/2 \\ 0, & \text{elsewhere.} \end{cases}$$

Determine the spectral function $F_y(k_x, k_y)$ for this distribution.

c) Assume that the apertures are electrically large ($k_o d \gg 1$, $k_o d' \gg 1$) and determine the principal H-plane field pattern (the pattern function $F_\phi(\theta, 0)$) for the antenna.

d) Consider two antennas, one with blockage and the other without blockage. Both antennas are supplied with the same power, and both antennas have the same field $E_o\hat{y}$ on the illuminated portion of the aperture plane. Here it is assumed that the radiation blocked by the subreflector is lost (i.e., it does not contribute to the field over the unblocked portion of the aperture plane).
Compare the axial gains $G(\theta = 0)$ of the two antennas. What effect does aperture blockage have on the gain?

e) Make graphs of the H-plane field patterns for antennas with $k_o d = 400$, $d'/d = 0$, and $d'/d = 0.316$. The subreflector with $d'/d = 0.316$ is much larger than normally encountered in practice; it was chosen this large to exaggerate the effects of aperture blockage. The numerical values in Table 3.5 will be helpful for making the graphs.

What effect does the aperture blockage have on the level of the side lobes? Would you expect the results to be similar if the aperture field were tapered as in Equation (3.132)?

3.7 A helium-neon laser operates at the wavelength $\lambda_o = 0.6328\ \mu$m. The cavity for the laser (Figure 3.20c) is confocal ($R_m = d$) with the length $d = 1$ m.

a) What is the beam waist W_o?
b) What is the spot size for the laser beam at the mirror, $z = d/2$?
c) What is the angle α (3.156) that determines the spot size for the radiated field? What is the spot size for the laser beam at the distance $z = 100$ m?

3.8 A TE plane wave is obliquely incident on a perfectly conducting screen containing a slit of width d. The geometry is similar to that in Figure 3.7, with the angle between the incident wave vector and the z axis being θ_i. Assume that the tangential component of the electric field on the plane $z = 0$ is

$$\vec{E}_y(x, z = 0) = \begin{cases} E_o^i e^{-jk_o x \sin\theta_i}\hat{y}, & |x| < d/2 \\ 0, & |x| \geq d/2 \end{cases}.$$

This aperture distribution has a phase that varies linearly with the position x – a "linear phase taper."

a) Determine the spectral function $F_y(k_x)$ and the radiated electric field $\vec{E}^r(\vec{r})$.
b) Let the slit width be $d/\lambda_o = 5$ and $\theta_i = 30°$. Make a graph of the normalized radiated field, as in Figure 3.15. At what angle does the main lobe of the pattern occur?
c) Using the results you obtained above, explain how the main beam of a paraboloidal reflector antenna could be scanned (its angular position varied) by changing the transverse position of the prime focus feed.

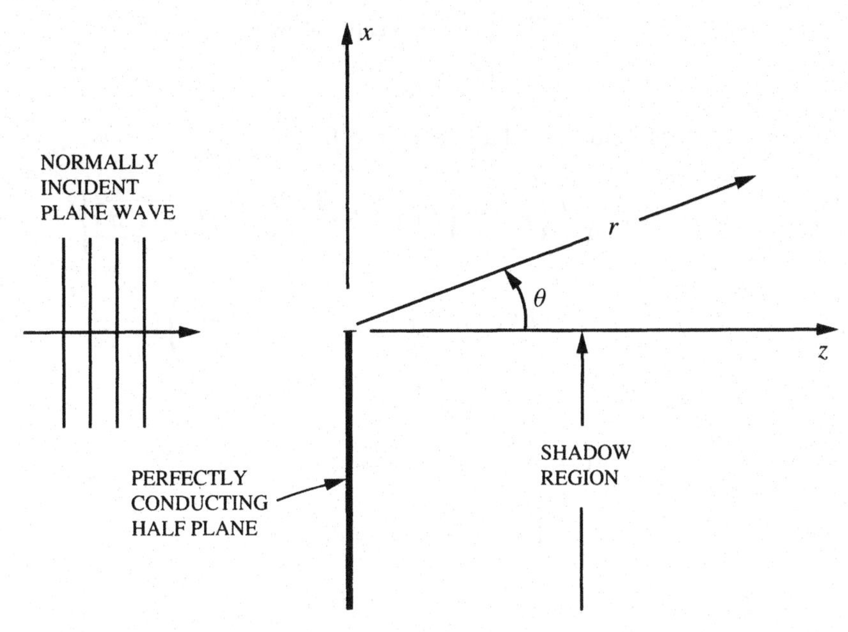

Fig. P3.2. Plane wave normally incident on an infinitesimally thin, perfectly conducting half plane.

3.9 a) Consider a Gaussian aperture distribution for which $k_o W_o \gg 1$ (an electrically large aperture). Determine the directivity $D_A(\theta = 0)$ for this aperture by using the same approximations that were used for the electrically large, uniformly illuminated circular aperture.

b) Compare your result from part a with the directivity of an electrically large, uniformly illuminated circular aperture (3.124). For the two aperture distributions to have the same directivity, what is the relationship between the radius $d/2$ of the circular aperture and the parameter W_o for the Gaussian distribution?

3.10 Show that the directivity of an electrically large aperture with tapered illumination is given by (3.137). The following definite integrals will be of help:

$$\int_0^\infty \frac{J_{n+1}^2(u)}{u^{2n+1}} du = \frac{1}{2^{2n+1}(2n+1)(n!)^2}$$

$$\int_0^\infty \frac{J_1(u)J_{n+1}(u)}{u^{n+1}} du = \frac{1}{2^{n+1}(n+1)n!},$$

where n is an integer.

3.11 Figure P3.2 illustrates a famous two-dimensional problem from the theory of diffraction: diffraction of an incident plane wave by an infinitessimally thin, perfectly conducting half plane. The *exact* solution to this problem was first obtained by Arnold Sommerfeld in 1896. A full account of the solution is given in Sommerfeld's book *Optics* [31] and in Reference [12].

For the case of a normally incident TE plane wave,

$$\vec{E}^i(\vec{r}) = E_o^i e^{-jk_oz}\hat{y},$$

the exact electric field in the half space $z \geq 0$ is

$$\vec{E}_y(\vec{r}) = E_o^i\sqrt{\frac{j}{2}}\left(e^{-jk_oz}\left\{F(\infty) + \frac{x}{|x|}F\left[\sqrt{2k_o(r-z)\pi}\right]\right\}\right.$$

$$\left. - e^{jk_oz}\left\{F(\infty) - F\left[\sqrt{2k_o(r+z)/\pi}\right]\right\}\right)\hat{y}.$$

Here

$$F(x) = \int_0^x e^{-j\frac{\pi}{2}u^2}du = C(x) - jS(x),$$

where $C(x)$ and $S(x)$ are the Fresnel integrals [8]:

$$C(x) = \int_0^x \cos\left(\frac{\pi}{2}u^2\right)du, \qquad S(x) = \int_0^x \sin\left(\frac{\pi}{2}u^2\right)du,$$

and

$$F(\infty) = \frac{1}{2}(1-j) = \sqrt{\frac{-j}{2}}.$$

The normalized magnitude of this field is graphed as a function of x/λ_o for a distance three wavelengths behind the screen, $z/\lambda_o = 3.0$, in Figure P3.3a. Figure P3.3b is a similar graph for $z/\lambda_o = 10.0$. The normalized field is seen to oscillate about the value 1.0 in the illuminated region and to monotonically decay in the shadow region. Measured results are also shown for the case $z/\lambda_o = 3.0$, and these are seen to be in excellent agreement with the theory. The measurements are for $\lambda_o = 3.18$ cm (a microwave frequency) and a half plane of thickness $0.025\lambda_o$ [32].

a) An approximate solution to this problem can be obtained by assuming the electric field on the plane $z = 0$ to be

$$\vec{E}_y(x, z=0) = E_o^i U(x)\hat{y},$$

where U is the Heaviside unit step function (Section 5.2)

$$U(x) = \begin{cases} 0, & x < 0 \\ 1, & x > 0 \end{cases}.$$

Starting with this field, use the spectral representation for two-dimensional fields and the paraxial approximation to show that in the half space $z \geq 0$

$$\vec{E}_y(\vec{r}) = E_o^i e^{-jk_oz}\left[\frac{1}{2} + \sqrt{\frac{j}{2}}F\left(\sqrt{\frac{k_o}{\pi z}}x\right)\right]\hat{y}.$$

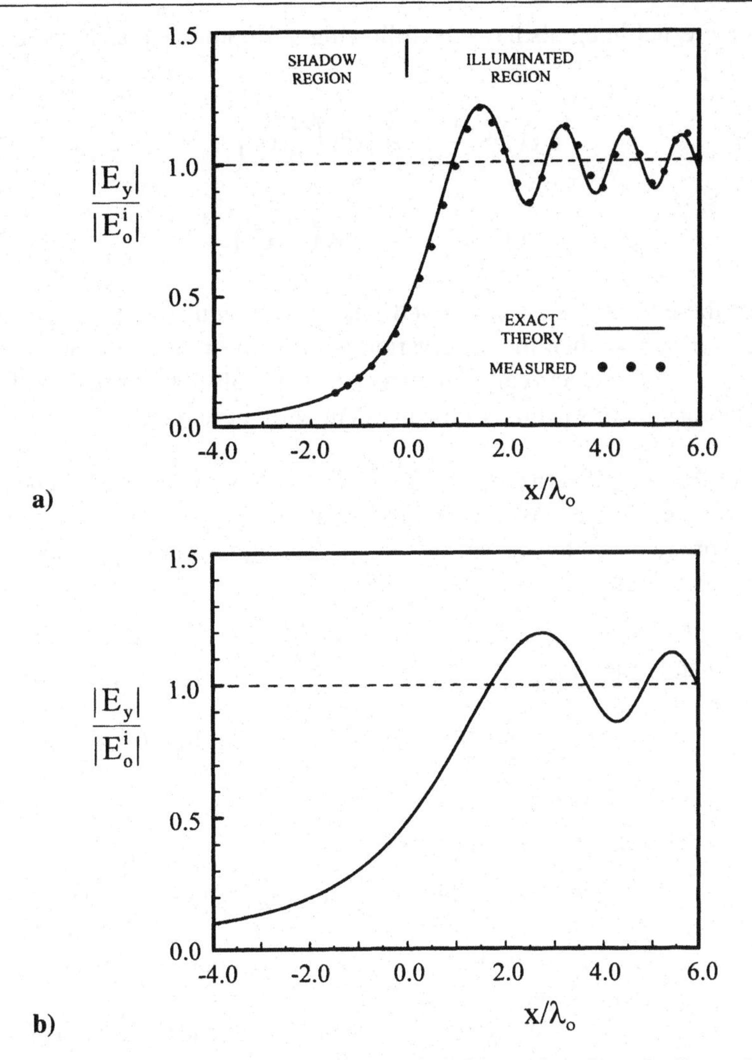

Fig. P3.3. Normalized magnitude of the electric field as a function of x/λ_o at the distances a) $z/\lambda_o = 3.0$ and b) $z/\lambda_o = 10.0$ behind the half plane. (Measured data from Harden [32].)

The Fourier transform for the unit step function [3] is

$$\int_{-\infty}^{\infty} U(x)e^{jk_x x}dx = \pi\delta(k_x) - \frac{1}{jk_x},$$

and the definite integral [33]

$$\int_0^{\infty} \frac{e^{jau^2}}{u}\sin(ux)du = \pi\sqrt{\frac{j}{2}}F\left(\sqrt{\frac{1}{2\pi a}}x\right)$$

may be of help.

b) Compare the approximate field from part a with the exact field by graphing the approximate solution on Figures 3.3a and b. Is the agreement better when $z/\lambda_o = 3.0$ or when $z/\lambda_0 = 10.0$?

c) The Fresnel integrals have the following asymptotic expansions for $x \to \infty$ [8]:

$$C(x) \sim \frac{1}{2} + \frac{1}{\pi x} \sin\left(\frac{\pi}{2}x^2\right) + \cdots$$

$$S(x) \sim \frac{1}{2} - \frac{1}{\pi x} \cos\left(\frac{\pi}{2}x^2\right) + \cdots.$$

Use these formulas to show that the exact field reduces to the approximate field of part a when the observation point is located both near the z axis ($x^2 \ll z^2$) and electrically far away from the half plane ($\sqrt{k_o z} \gg 1$). Are your conclusions from part b consistent with these results?

3.12 Starting from the electromagnetic field for an inhomogeneous plane wave, show that the complex Poynting vector and the time average of the real Poynting vector are given by (3.19) and (3.20). You may want to use the fact that $\vec{E} \times \vec{E}^*$ is an imaginary quantity in obtaining your results.

3.13 Consider the following integral representation for the Bessel function of the first kind and order zero:

$$J_o(x) = \frac{2}{\pi} \int_0^\infty \sin(x \cosh u) du = \frac{1}{\pi} \int_{-\infty}^\infty \sin(x \cosh u) du, \qquad x > 0.$$

Use the method of stationary phase to obtain an asymptotic approximation for $J_o(x)$ that applies in the limit $x \to \infty$.

3.14 *Frustrated total internal reflection.* In the arrangement shown in Figure P3.4, the parallel faces of two dielectric regions ($\varepsilon = \varepsilon_r \varepsilon_o$, $\mu = \mu_o$) are separated by a gap of free space of width d. A homogeneous plane wave is incident at the angle ψ_i in the left-hand dielectric.

First, consider the case where d is extremely large; the right-hand dielectric is essentially absent. When ψ_i is greater than the critical angle

$$\psi_c = \sin^{-1}\left(\sqrt{1/\varepsilon_r}\right),$$

total internal reflection occurs at the dielectric/free space interface. There is a homogeneous reflected wave in the left-hand dielectric, and the magnitude of the electric field for this wave is equal to that of the incident wave. There is also an inhomogeneous wave in free space. It decays with increasing z. As we have shown, this inhomogeneous wave carries no time-average power per unit area through a plane of constant z.

Now, consider the case where the distance d is very small. A thorough analysis shows that a homogeneous plane wave is produced in the right-hand dielectric. This homogeneous wave, of course, carries time-average power per unit area through a plane of constant z. The amount of power coupled into the right-hand dielectric is a function of the width of the gap d. This phenomenon is known as frustrated total internal reflection, and is

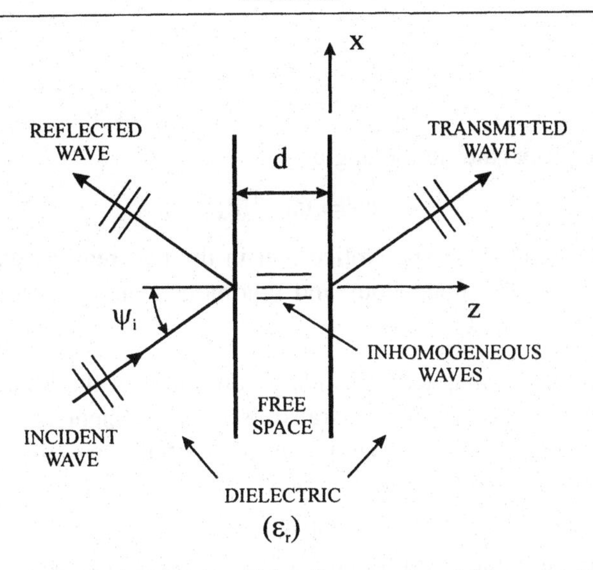

Fig. P3.4. Illustration for describing frustrated total internal reflection.

used to make optical devices such as beam splitters and interference filters [12].

Notice that there is an apparent contradiction in the above description: In the region of free space, there is no time-average power per unit area carried by the inhomogeneous wave through a plane of constant z, yet in the right-hand dielectric, there is time-average power per unit area carried by the homogeneous plane wave through a plane of constant z. This contradiction is resolved by noticing that there must be two inhomogeneous waves in the region of free space: the wave that decays with increasing z plus a second wave that increases with increasing z. The latter is caused by the reflection of the former at the free space/dielectric interface.

For our analysis, we will only consider the planar interface between free space and a dielectric (Figure P3.4 with the left-hand dielectric absent). An inhomogeneous TE plane wave with the electric field

$$\vec{E}^i = E_{oy} e^{-j(k_x x + k_z z)} \hat{y}$$

and

$$k_z = -j\sqrt{k_x^2 - k_o^2}, \qquad k_x^2 > k_o^2$$

is incident from free space onto the dielectric. The electric field for the reflected inhomogeneous plane wave in free space is

$$\vec{E}^r = \Gamma E_{oy} e^{-j(k_x x - k_z z)} \hat{y},$$

with k_z as given above. In the dielectric, there is a transmitted wave that is homogeneous and has an electric field given by

$$\vec{E}^t = \tau E_{oy} e^{-j(k_x x + k'_z z)} \hat{y},$$

where

$$k_z' = \sqrt{\varepsilon_r k_o^2 - k_x^2}, \qquad k_x^2 < \varepsilon_r k_o^2.$$

This wave propagates at the angle

$$\psi_t = \tan^{-1}(k_x / k_z')$$

with respect to the z axis. Notice that in the two regions we will have the types of waves (homogeneous and inhomogeneous) described above, provided $k_o^2 < k_x^2 < \varepsilon_r k_o^2$.

a) Use the electromagnetic boundary conditions at the interface, $z = 0$, to determine the reflection and transmission coefficients, Γ and τ.
b) Determine the time-average power per unit area passing through a plane of constant z for the following waves:
   i) the incident wave in free space,
   ii) the combination of the incident and reflected waves in free space,
   iii) the transmitted wave in the dielectric.

Discuss your results; in particular, show that the contradiction mentioned above is resolved.

3.15 We wish to produce a Gaussian beam in the half space $z > 0$. Propagation is to be in the $+z$ direction, with the waist of the beam at the position $z = z_o$. In this situation, the electromagnetic energy introduced at the plane $z = 0$ is brought to a focus at the plane $z = z_o$.

a) What should the tangential electric field on the plane $z = 0$ be to form this beam?
b) Simplify your results from part a by assuming that

$$\gamma(z_o) = \frac{2z_o}{k_o W_o^2} \gg 1.$$

What are the dependences on ρ for the magnitude and phase of the electric field on the plane $z = 0$?

3.16 The electric field on the plane $z = 0$ is the difference between two sinc functions:

$$\vec{E}(x, z = 0) = \frac{A_o}{\pi}\left[\alpha \operatorname{sinc}(\alpha x) - \beta \operatorname{sinc}(\beta x)\right]\hat{y},$$

with $\alpha > \beta$.

a) Using the following definite integral, determine the spectral function for this field:

$$\int_{-\infty}^{\infty} \left[\frac{\sin(\gamma x)}{\pi x}\right] e^{j\kappa x} dx = \begin{cases} 1, & |\kappa| < \gamma \\ 0, & |\kappa| > \gamma \end{cases}.$$

Make a graph of the spectral function versus k_x.

b) Determine the radiated field $\vec{E}^r$ and make a polar graph of the magnitude of this field versus the angle θ. Use the values $\alpha = \sqrt{3}k_o/2$, $\beta = k_o/2$ for the graph.

c) Describe the field in the half space $z > 0$ in terms of the homogeneous and inhomogeneous plane waves in its spectrum for the following cases:

i) $\alpha = k_o, \qquad \beta = 0$

ii) $\alpha = 2k_o, \qquad \beta = k_o.$

4

Electromagnetic analogues of some optical principles

In the introduction to Chapter 1, we mentioned that Maxwell's theory provided a unified explanation for electric, magnetic, and optical phenomena. Prior to Maxwell, classical optical phenomena were explained, more or less, by theories based on scalar fields. After Maxwell, these phenomena were shown to follow from Maxwell's theory for the vector electromagnetic field and to exist for general frequencies, not just for those in the optical range. We have seen graphic examples of this in our discussions of the experiments of Hertz (Section 2.7) and the diffraction by a circular aperture (Section 3.7).

Many of the older principles developed in classical optics have analogues in electromagnetic theory. Often the beauty and utility of the original optical principle is preserved in the electromagnetic counterpart. Of course, the former is usually for a scalar field, whereas the latter is for the vector electromagnetic field. Consequently, the statement of the principle, its derivation, and the requirements for its use may be quite different for the two cases.

In this chapter, we will obtain the electromagnetic analogues for a few of the more important optical principles and illustrate their use with examples. To simplify this task, whenever possible we will make use of the results already developed in Chapter 3 for the plane-wave spectrum.

4.1 Huygens' principle: An alternate representation

The representation we obtained in Chapter 3 considers the electromagnetic field in the half space $z > 0$ to be composed of a spectrum of plane waves; this is shown schematically in Figure 4.1a. Here the field on the reference plane $z = 0$ is assumed to extend over the aperture $|\vec{r}'| \leq d/2$. The entire spectrum of waves is required to represent the field at a point in the aperture. However, only a single plane wave propagating in the direction of observation determines the field at a large distance, the radiated field.

An alternate representation is shown schematically in Figure 4.1b. A set of discrete elements, small patches of area, fills the aperture. Each of these elements acts as a source radiating a spherically symmetric electromagnetic wave. The superposition of the waves from these elements forms the electromagnetic field

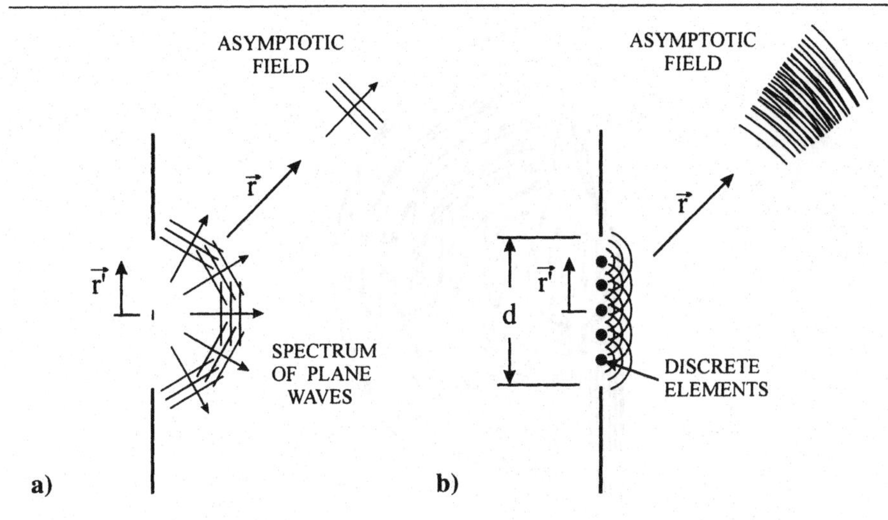

Fig. 4.1. Representations for the electromagnetic field in the half space $z > 0$. a) Spectral representation. b) Huygens' principle.

in the half space $z > 0$. At a point in the aperture, one element determines the field, whereas at a large distance, all of the elements collectively determine the radiated field.

It is interesting to contrast these two points of view. In the former the entire spectrum of plane waves collectively determines the field in the aperture, but a single wave determines the radiated field. In the latter a single element determines the field in the aperture, whereas all of the elements collectively determine the radiated field.

The alternate representation we have outlined above historically preceded the spectral representation. Its optical counterpart is attributed first to Christiaan Huygens (1629–1695). Later mathematical formulations that express this idea are often referred to as *Huygens' principle*. Figure 4.2 is a reproduction of a drawing from Huygens' treatise [1]. It shows his interpretation of the light produced by the flame of a candle: Each of the points in the flame, A, B, and C, acts as a source producing a spherical wave; these waves combine to determine the wavefront at a distance from the candle. Huygens' model, although novel for his time, lacked many of the features that are incorporated in electromagnetic theory. For instance, interference phenomena were not included. The radiation from individual elements may arrive at a point of observation with different relative phases and, therefore, add destructively as well as constructively. The principle of interference was first introduced into optics by Thomas Young (1773–1829) in 1801 [2].

The two approaches, the spectral representation and Huygens' principle, are totally equivalent; when correctly applied to a specific problem they produce the same numerical results for the electromagnetic field. However, the methodologies underlying the two approaches are quite different, and a knowledge of both increases our basic understanding of electromagnetic radiation.

Fig. 4.2. Illustration from C. Huygens' *Traité de la Lumière*, 1690 [1].

4.1.1 General formulation

We will now obtain a rigorous mathematical expression for the electromagnetic field that incorporates Huygens' idea. We will restrict our argument to the geometry of Figure 4.1; that is, we seek an expression for the field in the half space $z > 0$ due to a specified field on the plane $z = 0$. There are many approaches we could use to derive this expression; however, since the spectral representation is available, we will simply rearrange it to obtain the desired form.

From Table 3.3, the spectral representation for the electric field is

$$\vec{E}(\vec{r}) = \frac{1}{(2\pi)^2} \int_{-\infty}^{\infty} \int_{-\infty}^{\infty} \left\{ F_x(k_x, k_y)\hat{x} + F_y(k_x, k_y)\,\hat{y} \right.$$
$$\left. - \left(\frac{k_o}{k_z}\right)\left[\left(\frac{k_x}{k_o}\right)F_x(k_x, k_y) + \left(\frac{k_y}{k_o}\right)F_y(k_x, k_y)\right]\hat{z}\right\}$$
$$e^{-j(k_x x + k_y y + k_z z)} dk_x dk_y. \tag{4.1}$$

First, we will write this expression in a more compact form. Note that (4.1) is equivalent to

$$\vec{E}(\vec{r}) = \frac{j}{(2\pi)^2} \int_{-\infty}^{\infty} \int_{-\infty}^{\infty} \left[(F_x\hat{x} + F_y\hat{y})\frac{\partial}{\partial z} - \left(F_x\frac{\partial}{\partial x} + F_y\frac{\partial}{\partial y}\right)\hat{z}\right]$$
$$\frac{e^{-j(k_x x + k_y y + k_z z)}}{k_z} dk_x dk_y, \tag{4.2}$$

which, after introducing

$$\vec{U} = U_x\hat{x} + U_y\hat{y} = \hat{z} \times \vec{F}\,\frac{e^{-j(k_x x + k_y y + k_z z)}}{k_z}$$

$$= -F_y\,\frac{e^{-j(k_x x + k_y y + k_z z)}}{k_z}\,\hat{x} + F_x\,\frac{e^{-j(k_x x + k_y y + k_z z)}}{k_z}\,\hat{y},$$

becomes

$$\vec{E}(\vec{r}) = \frac{j}{(2\pi)^2}\int_{-\infty}^{\infty}\int_{-\infty}^{\infty}\left[\frac{\partial U_y}{\partial z}\hat{x} - \frac{\partial U_x}{\partial z}\hat{y} + \left(\frac{\partial U_x}{\partial y} - \frac{\partial U_y}{\partial x}\right)\hat{z}\right]dk_x dk_y. \quad (4.3)$$

Now, we recognize the integrand of (4.3) to be $-\nabla \times \vec{U}$; thus,

$$\vec{E}(\vec{r}) = \frac{-j}{(2\pi)^2}\int_{-\infty}^{\infty}\int_{-\infty}^{\infty}\nabla \times \vec{U} dk_x dk_y$$

$$= \frac{-j}{(2\pi)^2}\nabla \times \int_{-\infty}^{\infty}\int_{-\infty}^{\infty}\hat{z} \times \vec{F}(k_x, k_y)\,\frac{e^{-j(k_x x + k_y y + k_z z)}}{k_z}dk_x dk_y. \quad (4.4)$$

When the spectral function from Table 3.3, that is,

$$\hat{z} \times \vec{F}(k_x, k_y) = \int_{-\infty}^{\infty}\int_{-\infty}^{\infty}\hat{z} \times \vec{E}(x', y', z' = 0)e^{j(k_x x' + k_y y')}dx'dy',$$

is substituted, and the order of the integrations is interchanged, (4.4) becomes

$$\vec{E}(\vec{r}) = 2\nabla \times \left\{\int_{y'=-\infty}^{\infty}\int_{x'=-\infty}^{\infty}\hat{z} \times \vec{E}(x', y', z' = 0)\right.$$

$$\left.\left[\frac{-j}{8\pi^2}\int_{k_y=-\infty}^{\infty}\int_{k_x=-\infty}^{\infty}\frac{e^{-j[k_x(x-x') + k_y(y-y') + k_z z]}}{k_z}dk_x dk_y\right]dx'dy'\right\}, \quad (4.5)$$

with k_z given by (3.81a), (3.81b), and $z \geq 0$.

Surprisingly, the formidable looking integration within the brackets can be evaluated in closed form:

$$G_o(x, y, z; x', y', z' = 0) = \frac{-j}{8\pi^2}\int_{k_y=-\infty}^{\infty}\int_{k_x=-\infty}^{\infty}\frac{e^{-j[k_x(x-x') + k_y(y-y') + k_z z]}}{k_z}dk_x dk_y$$

$$= \frac{e^{-jk_o\sqrt{(x-x')^2 + (y-y')^2 + (z)^2}}}{4\pi\sqrt{(x-x')^2 + (y-y')^2 + (z)^2}}. \quad (4.6)$$

The function G_o is called the *free-space, scalar Green's function for harmonic time dependence* (George Green, 1793–1841); it can be written in a more compact form by introducing the vectors

$$\vec{r} = x\hat{x} + y\hat{y} + z\hat{z},$$

$$\vec{r}\,' = x'\hat{x} + y'\hat{y} + z'\hat{z},$$

and

$$\vec{R} = R\hat{R} = \vec{r} - \vec{r}';$$

then

$$G_o(\vec{r}, \vec{r}') = \frac{e^{-jk_o R}}{4\pi R} = \frac{e^{-jk_o|\vec{r}-\vec{r}'|}}{4\pi|\vec{r} - \vec{r}'|}, \tag{4.7}$$

where it is understood that $z' = 0$ for our case. We will have more to say about this function in Section 5.5.

For those interested in the mathematical details, we will briefly describe the evaluation of the definite integral in (4.6); others may proceed directly to Equation (4.8). First, we will write the exponent in terms of the vectors $\vec{R}$ and $\vec{k}$:

$$G_o = \frac{-j}{8\pi^2} \int_{k_y=-\infty}^{\infty} \int_{k_x=-\infty}^{\infty} \frac{e^{-j\vec{k}\cdot\vec{R}}}{k_z} dk_x dk_y.$$

Next, the position in space, $\vec{R}$, and the vector wave number, $\vec{k}$, are expressed in terms of spherical coordinates. This was done in Section 3.6: The spherical coordinates in Figure 3.17a are used for $\vec{R}$, with r replaced by R, and the spherical coordinates in Figure 3.17b are used for $\vec{k}$. The integral then becomes

$$G_o = \frac{-jk_o}{8\pi^2} \int_{C_{1/2}} \int_{\chi=-\pi}^{\pi} e^{-j\vec{k}\cdot\vec{R}} \sin\psi d\chi d\psi,$$

where $C_{1/2}$ is the right half of the contour shown in Figure 3.6.

Now we will use a clever procedure introduced by Hermann Weyl to quickly evaluate the integrals [3, 4]. The vector wave number $\vec{k}$ can be expressed in terms of the new coordinates k_x', k_y', k_z' (spherical coordinates k_o, α, β) shown in Figure 4.3. These new coordinates are oriented so that the polar axis, $\vec{k}_z'$, coincides with the direction of the position vector $\vec{R}$. The angle between $\vec{k}$ and $\vec{R}$ is then the new polar angle α, so the exponent is simply

$$-j\vec{k} \cdot \vec{R} = -jk_o R \cos\alpha.$$

Notice that $\sin\psi d\chi d\psi$ is the differential solid angle $d\Omega$, which becomes $\sin\alpha d\beta d\alpha$ in the new coordinates (Appendix B). Using these observations, the integral becomes

$$G_o = \frac{-jk_o}{8\pi^2} \iint e^{-jk_o R\cos\alpha} \sin\alpha d\beta d\alpha.$$

Weyl recognized from the theory of complex variables that the limits on the above integrals involving the variables α and β would be the same as they were on the original integrals involving the variables ψ and χ; that is, α goes from 0 to $\pi/2 + j\infty$ (the end points of the contour $C_{1/2}$ in Figure 3.6) and β goes from $-\pi$ to π. The evaluation of the integral is now straightforward:

$$G_o = \frac{-jk_o}{8\pi^2} \int_{\beta=-\pi}^{\pi} \int_{\alpha=0}^{\pi/2+j\infty} e^{-jk_o R\cos\alpha} \sin\alpha d\alpha d\beta$$

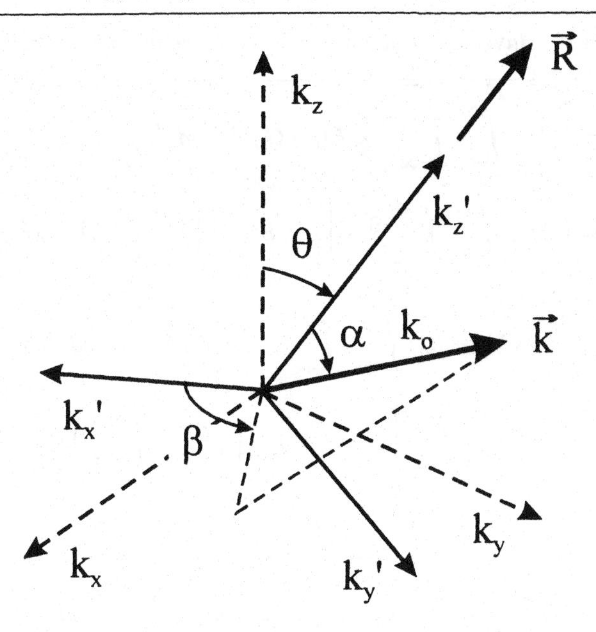

Fig. 4.3. The new coordinate system k_x', k_y', k_z' (k_o, α, β) for the vector wave number. The drawing is for a wave number with real components.

$$= \frac{-jk_o}{4\pi}\left\{\frac{1}{jk_oR}\left[e^{-jk_oR\cos\alpha}\right]_{\alpha=0}^{\pi/2+j\infty}\right\}$$

$$= \frac{1}{4\pi R}\left[e^{-jk_oR} - e^{-k_oR\sinh(\infty)}\right] = \frac{e^{-jk_oR}}{4\pi R}.$$

Our final result for the electric field is obtained by substituting (4.7) into (4.5):

$$\vec{E}(\vec{r}) = 2\nabla \times \int_{-\infty}^{\infty}\int_{-\infty}^{\infty} \hat{z} \times \vec{E}(\vec{r}\,')G_o(\vec{r}, \vec{r}\,')dx'dy'. \tag{4.8}$$

This is a mathematical expression stating Huygens' principle for our electromagnetic problem. At each point on the plane $z = 0$, the field $\vec{E}(\vec{r}\,')$ is the source of a spherically symmetric wave of the form $G_o(\vec{r}, \vec{r}\,')$. The superposition of these waves, after the operation $2\nabla \times (\hat{z}\times)$, comprises the electric field in the half space $z > 0$. The magnetic field that accompanies (4.8) is

$$\vec{B}(\vec{r}) = \frac{j2}{\omega}(\nabla\nabla\cdot + k_o^2)\int_{-\infty}^{\infty}\int_{-\infty}^{\infty} \hat{z} \times \vec{E}(\vec{r}\,')G_o(\vec{r}, \vec{r}\,')dx'dy'. \tag{4.9}$$

Table 4.1 summarizes these results together with those for the case where the tangential components of the magnetic field are specified on the plane $z = 0$. The second forms for the expressions in Table 4.1 are obtained by moving the differential operators under the integral signs (Problem 4.18).[1]

[1] Here we have simplified expressions by using notation such as

$$(1 - \hat{R}\hat{R}\cdot)\vec{A} = \vec{A} - \hat{R}(\hat{R}\cdot\vec{A}).$$

Table 4.1. *Huygens' principle for fields: harmonic time dependence*

$$\vec{E}(\vec{r}) = 2\,\nabla \times \int_{-\infty}^{\infty}\int_{-\infty}^{\infty} \hat{z} \times \vec{E}(\vec{r}')\, G_o(\vec{r},\vec{r}')dx'dy'$$

$$= -2jk_o \int_{-\infty}^{\infty}\int_{-\infty}^{\infty} \hat{R} \times \left[\hat{z} \times \vec{E}(\vec{r}')\right]\left(1 - \frac{j}{k_o R}\right) G_o(\vec{r},\vec{r}')dx'dy'$$

$$\vec{B}(\vec{r}) = \frac{2j}{\omega}\,(\nabla\nabla\cdot + k_o^2)\int_{-\infty}^{\infty}\int_{-\infty}^{\infty} \hat{z} \times \vec{E}(\vec{r}')\, G_o(\vec{r},\vec{r}')dx'dy'$$

$$= \frac{2jk_o}{c}\int_{-\infty}^{\infty}\int_{-\infty}^{\infty}\left[(1 - \hat{R}\hat{R}\cdot) - \left(\frac{j}{k_o R} + \frac{1}{k_o^2 R^2}\right)(1 - 3\hat{R}\hat{R}\cdot)\right]$$

$$[\hat{z} \times \vec{E}(\vec{r}')]G_o(\vec{r},\vec{r}')dx'dy'$$

$$\vec{E}(\vec{r}) = \frac{-2jc}{k_o}(\nabla\nabla\cdot + k_o^2)\int_{-\infty}^{\infty}\int_{-\infty}^{\infty} \hat{z} \times \vec{B}(\vec{r}')\, G_o(\vec{r},\vec{r}')dx'dy'$$

$$= -2jk_o c\int_{-\infty}^{\infty}\int_{-\infty}^{\infty}\left[(1 - \hat{R}\hat{R}\cdot) - \left(\frac{j}{k_o R} + \frac{1}{k_o^2 R^2}\right)(1 - 3\hat{R}\hat{R}\cdot)\right]$$

$$[\hat{z} \times \vec{B}(\vec{r}')]G_o(\vec{r},\vec{r}')dx'dy'$$

$$\vec{B}(\vec{r}) = 2\,\nabla \times \int_{-\infty}^{\infty}\int_{-\infty}^{\infty} \hat{z} \times \vec{B}(\vec{r}')\, G_o(\vec{r},\vec{r}')dx'dy'$$

$$= -2jk_o \int_{-\infty}^{\infty}\int_{-\infty}^{\infty} \hat{R} \times \left[\hat{z} \times \vec{B}(\vec{r}')\right]\left(1 - \frac{j}{k_o R}\right) G_o(\vec{r},\vec{r}')dx'dy'$$

$$G_o(\vec{r},\vec{r}') = \frac{e^{-jk_o R}}{4\pi R} = \frac{e^{-jk_o|\vec{r}-\vec{r}'|}}{4\pi|\vec{r}-\vec{r}'|}$$

4.1.2 Radiated field

At a large distance from the aperture, the expressions in Table 4.1 for the electromagnetic field simplify to become the radiated field. The radial distance r must satisfy

$$k_o r \gg 1 \tag{4.10}$$

and

$$r'/r \le d/2r \ll 1. \tag{4.11}$$

The first of these inequalities states that r must be large compared to the wavelength

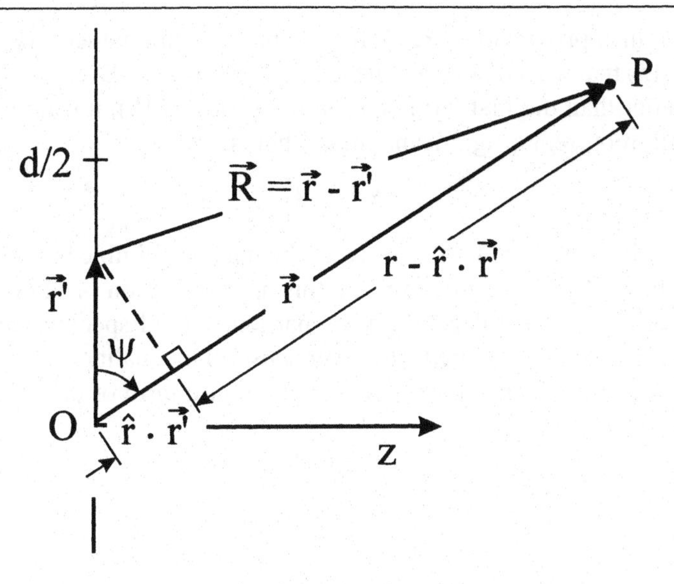

Fig. 4.4. Coordinates for evaluation of the radiated field.

$[k_o r = 2\pi(r/\lambda_o)]$; the second states that r must be large compared to the maximum dimension (d) of the aperture.

The Green's function $G_o(\vec{r}, \vec{r}')$ (4.7) depends upon the distance R, so first we will consider the approximation of this distance using (4.11). From Figure 4.4,

$$R = |\vec{r} - \vec{r}'| = \sqrt{r^2 + r'2 - 2rr' \cos\psi} = r\sqrt{1 + (r'/r)^2 - 2(r'/r)\cos\psi}, \tag{4.12}$$

where ψ is the angle between the vectors $\vec{r}$ and $\vec{r}'$. The binomial series for R is

$$R = r\left\{1 + \frac{1}{2}[(r'/r)^2 - 2(r'/r)\cos\psi] - \frac{1}{8}[(r'/r)^2 - 2(r'/r)\cos\psi]^2 + \cdots\right\}$$

$$= r - r'\cos\psi + \frac{1}{2}r'(r'/r)\sin^2\psi + \cdots. \tag{4.13}$$

In applying this result to the Green's function (4.7), we will consider the amplitude and the phase of the function separately. The *amplitude* of $G_o(\vec{r}, \vec{r}')$ is

$$\frac{1}{R} = \frac{1}{r}\left[1 - (r'/r)\cos\psi + \frac{1}{2}(r'/r)^2\sin^2\psi + \cdots\right]^{-1} \approx \frac{1}{r}, \tag{4.14}$$

where the approximation clearly follows from (4.11). The *phase* of $G_o(\vec{r}, \vec{r}')$ is

$$k_o R = k_o r - k_o r'\cos\psi + \frac{1}{2}k_o r'(r'/r)\sin^2\psi + \cdots$$

$$\approx k_o r - k_o r'\cos\psi = k_o(r - \hat{r}\cdot\vec{r}'), \tag{4.15}$$

where in the last line, we have recognized $r'\cos\psi$ to be the projection of $\vec{r}'$ onto $\hat{r}$ (see Figure 4.4).

Notice that in approximating R, we have kept an additional term ($k_o r' \cos\psi$) in the phase (4.15) that we did not keep in the amplitude (4.14). Even though this term is much smaller than the first term ($k_o r$) in the series (4.15), it cannot be ignored for it may still represent a significant phase; that is,

$$e^{jk_o r' \cos\psi} \neq 1.$$

The third term in the series (4.15) can always be made negligible because of (4.11).

Precisely how large r must be to make this approximation (4.15) for the phase acceptable, of course, will depend on the particular field specified on the plane $z = 0$ and on the accuracy desired. However, a useful inequality can be established by noting that $r' \leq d/2$ and $\sin\psi \leq 1$; thus, the third term of the series (4.15) is negligible, provided

$$\frac{1}{2}\left(\frac{k_o d}{2}\right)\left(\frac{d}{2r}\right) = \frac{1}{2}k_o r\left(\frac{d}{2r}\right)^2 \ll 1,$$

or

$$r \gg \frac{\pi}{8}\left(\frac{2d^2}{\lambda_o}\right). \tag{4.16}$$

In general antenna analysis, this inequality is often replaced by [5]

$$\frac{1}{2}k_o r\left(\frac{d}{2r}\right)^2 \leq \pi/8,$$

or

$$r \geq \frac{2d^2}{\lambda_o}. \tag{4.17}$$

In other words, an error in the phase as large as $\pi/8 = 22.5°$ is allowed. We should emphasize again that the adequacy of a criterion like (4.17) for using the radiated field depends on the particular problem being considered and the accuracy desired.

After substituting (4.14) and (4.15), the approximation for the Green's function is obtained:

$$G_o^r(\vec{r}, \vec{r}') = \frac{e^{-jk_o(r - \hat{r}\cdot\vec{r}')}}{4\pi r}. \tag{4.18}$$

The electric field (Table 4.1) is now

$$\vec{E}^r(\vec{r}) = -2jk_o \int_{-\infty}^{\infty}\int_{-\infty}^{\infty} \hat{r} \times \hat{z} \times \vec{E}(\vec{r}')\left(1 - \frac{j}{k_o r}\right)\frac{e^{-jk_o(r - \hat{r}\cdot\vec{r}')}}{4\pi r} dx'dy',$$

where we have used (4.14) and made the following approximation for the unit vector $\hat{R}$:

$$\hat{R} = \frac{\vec{r} - \vec{r}'}{R} \approx \hat{r} - \left(\frac{r'}{r}\right)\hat{r}' \approx \hat{r}. \tag{4.19}$$

Table 4.2. *Huygens' principle for radiated fields: harmonic time dependence*

$$\vec{E}^r(\vec{r}) = \frac{-jk_o}{2\pi}\frac{e^{-jk_or}}{r}\hat{r}\times\hat{z}\times\int_{-\infty}^{\infty}\int_{-\infty}^{\infty}\vec{E}(\vec{r}')e^{jk_o\hat{r}\cdot\vec{r}'}dx'dy'$$

$$\vec{B}^r(\vec{r}) = \frac{-jk_o}{2\pi c}\frac{e^{-jk_or}}{r}\hat{r}\times\hat{r}\times\hat{z}\times\int_{-\infty}^{\infty}\int_{-\infty}^{\infty}\vec{E}(\vec{r}')e^{jk_o\hat{r}\cdot\vec{r}'}dx'dy'$$

$$= \frac{1}{c}\hat{r}\times\vec{E}^r(\vec{r})$$

$$\vec{E}^r(\vec{r}) = \frac{jk_oc}{2\pi}\frac{e^{-jk_or}}{r}\hat{r}\times\hat{r}\times\hat{z}\times\int_{-\infty}^{\infty}\int_{-\infty}^{\infty}\vec{B}(\vec{r}')e^{jk_o\hat{r}\cdot\vec{r}'}dx'dy'$$

$$\vec{B}^r(\vec{r}) = \frac{-jk_o}{2\pi}\frac{e^{-jk_or}}{r}\hat{r}\times\hat{z}\times\int_{-\infty}^{\infty}\int_{-\infty}^{\infty}\vec{B}(\vec{r}')e^{jk_o\hat{r}\cdot\vec{r}'}dx'dy'$$

$$= \frac{1}{c}\hat{r}\times\vec{E}^r(\vec{r})$$

After using (4.10), we obtain our final result for the radiated electric field:

$$\vec{E}^r(\vec{r}) = \frac{-jk_o}{2\pi}\frac{e^{-jk_or}}{r}\hat{r}\times\hat{z}\times\int_{-\infty}^{\infty}\int_{-\infty}^{\infty}\vec{E}(\vec{r}')e^{jk_o\hat{r}\cdot\vec{r}'}dx'dy'. \tag{4.20}$$

The accompanying radiated magnetic field can be obtained in a similar manner:

$$\vec{B}^r(\vec{r}) = \frac{-jk_o}{2\pi c}\frac{e^{-jk_or}}{r}\hat{r}\times\hat{r}\times\hat{z}\times\int_{-\infty}^{\infty}\int_{-\infty}^{\infty}\vec{E}(\vec{r}')e^{jk_o\hat{r}\cdot\vec{r}'}dx'dy'$$
$$= \frac{1}{c}\hat{r}\times\vec{E}^r(\vec{r}). \tag{4.21}$$

Since neither of the vectors ($\vec{E}^r$, $\vec{B}^r$) has a radial component, it is evident that the radiated field is transverse to the direction of propagation ($\hat{r}$) – a feature expected from our previous analysis. The formulas for the radiated field are summarized in Table 4.2.

The inequalities we have used in obtaining the radiated field are

$$k_or \gg 1, \tag{4.22a}$$

$$d/2r \ll 1, \tag{4.22b}$$

and

$$\frac{1}{2}k_o r\left(\frac{d}{2r}\right)^2 \ll 1. \tag{4.22c}$$

The three are not independent; if the strengths of the inequalities are taken to be the same, then any two of the inequalities imply the third. For example, when (4.22a) and (4.22b) are satisfied such that $k_o r = 100 \gg 1$ and $d/2r = 1/100 \ll 1$, then (4.22c) is also satisfied, since $(1/2)k_o r(d/2r)^2 = 1/200 \ll 1$. It is gratifying to see that these inequalities for the radiated field are essentially the same as those obtained earlier from spectral analysis, $k_o r \gg 1$ and $(4/\pi)k_o r(d/2r)^2 \ll 1$ (3.74).

To illustrate the use of these formulas, we will again consider the uniformly illuminated circular aperture with the geometry shown in Figure 3.21 and with the electric field on the aperture plane given by (3.104). On substitution of (3.104) and a change of coordinates from rectangular (x', y') to circular cylindrical (ρ', ϕ'), the radiated electric field (4.20) becomes

$$\vec{E}^r(\vec{r}) = \frac{jk_o}{2\pi}E_o^i\frac{e^{-jk_o r}}{r}(\sin\phi\hat{\theta} + \cos\phi\cos\theta\hat{\phi})$$

$$\int_{\rho'=0}^{d/2}\int_{\phi'=0}^{2\pi} e^{jk_o\rho'\sin\theta\cos(\phi-\phi')}\rho' d\phi' d\rho' \tag{4.23}$$

(see Problem 4.1). We have considered the integrals appearing in this expression before; using (3.109) and (3.111) they can be evaluated to give

$$\vec{E}^r(\vec{r}) = \frac{j}{2k_o}\left(\frac{k_o d}{2}\right)^2 E_o^i\frac{e^{-jk_o r}}{r}\left[\frac{2J_1\left(\frac{k_o d}{2}\sin\theta\right)}{\frac{k_o d}{2}\sin\theta}\right](\sin\phi\,\hat{\theta} + \cos\phi\cos\theta\,\hat{\phi}).$$

Notice that, as expected, this is the same result obtained using the plane-wave spectrum (3.113).

4.1.3 Discussion

Formulas that can be used to continue a field from a surface into an adjoining region of space are prevalent in electromagnetics and optics. Their inception dates from the treatise of Huygens that we have already mentioned. These formulas are not necessarily restricted to plane surfaces as are the results presented in this section. For example, when all sources are exterior to the volume V (Figure 4.5), the electromagnetic field interior to the volume is given by the following formulas [6–9]:

$$\vec{E}(\vec{r}) = \nabla\times\oiint_S [\hat{n}\times\vec{E}(\vec{r}')]G_o(\vec{r},\vec{r}')dS'$$

$$-\frac{jc}{k_o}(\nabla\nabla\cdot + k_o^2)\oiint_S [\hat{n}\times\vec{B}(\vec{r}')]G_o(\vec{r},\vec{r}')dS' \tag{4.24}$$

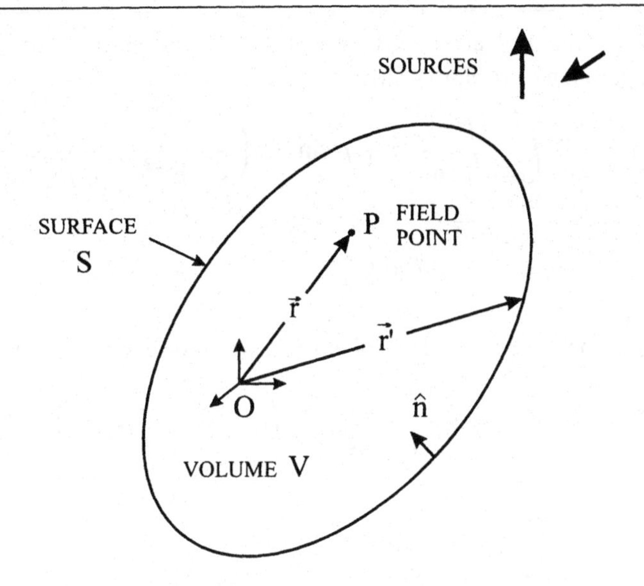

Fig. 4.5. Geometry for volume V with all sources exterior.

$$\vec{B}(\vec{r}) = \nabla \times \oiint_S \left[\hat{n} \times \vec{B}(\vec{r}')\right] G_o(\vec{r}, \vec{r}')dS'$$

$$+ \frac{j}{\omega}(\nabla\nabla \cdot + k_o^2) \oiint_S \left[\hat{n} \times \vec{E}(\vec{r}')\right] G_o(\vec{r}, \vec{r}')dS'. \tag{4.25}$$

Here the integrals involve only the tangential components of the field on the closed surface S with inward pointing normal $\hat{n}$. When the surface S is the plane $z = 0$ closed by a hemisphere of infinite radius (the volume V is the half space $z > 0$), these formulas can be related to the results in Table 4.1 (Problem 4.9).

4.2 Fresnel zones

Accurate evaluation of the integral in the electromagnetic Huygens' principle (4.8) is generally a formidable task, made practical only by the advent of electronic computers. Simple techniques, based on physical principles, can be used to approximate the integral. Although these techniques do not give precise values, they often can increase our understanding of the electromagnetic phenomena. We will examine a technique based on procedures first introduced into optics by Augustin Jean Fresnel (1788–1827) [10]. Recall that Huygens did not include interference phenomena in his original principle and that Young introduced the interference principle into optics about one hundred years later. Fresnel combined the ideas of Huygens and Young in a geometrical construction, which he used to explain, for the first time, many of the observations of optical diffraction.

We will confine our discussion to the evaluation of the electric field on the axis of a uniformly illuminated circular aperture. The geometry is the same as in Figure 3.21, with the field on the aperture plane specified by (3.104). The circular cylindrical coordinates for a point on the aperture plane are ρ, ϕ, $z = 0$; those for the point of observation on the axis are $x = 0$, $y = 0$, z. The integral for the electric field (Table 4.1) is then (see Problem 4.20)

$$\vec{E}(z) = \frac{E_o^i}{2\pi} \int_{\phi=0}^{2\pi} \int_{\rho=0}^{d/2} (z\hat{y} + \rho \sin\phi \hat{z}) \left(\frac{1}{\rho^2 + z^2} + \frac{jk_o}{\sqrt{\rho^2 + z^2}} \right)$$

$$\frac{e^{-jk_o\sqrt{\rho^2+z^2}}}{\sqrt{\rho^2 + z^2}} \rho d\rho d\phi. \tag{4.26}$$

After evaluating the integral with respect to ϕ, we find that $\vec{E}(z) = E_y(z)\hat{y}$ with

$$E_y(z) = -E_o^i e^{-jk_o z} \int_{\rho=0}^{d/2} \frac{z}{\sqrt{\rho^2 + z^2}} \left(1 - \frac{j}{k_o\sqrt{\rho^2 + z^2}} \right) (-jk_o\rho)$$

$$\frac{e^{-jk_o\left(\sqrt{\rho^2+z^2}-z\right)}}{\sqrt{\rho^2 + z^2}} d\rho. \tag{4.27}$$

Now we will concentrate on the all-important exponential term in the integrand of (4.27). The exponent is the difference in phase for two waves: a wave that propagates from the radial position ρ in the aperture to the observation point over the distance $r = \sqrt{\rho^2 + z^2}$ and a wave that propagates from the center of the aperture to the observation point over the distance z. With z held fixed, this phase difference will be a multiple of π when

$$k_o(r_n - z) = k_o\left(\sqrt{\rho_n^2 + z^2} - z \right) = n\pi, \qquad n = 1, 2, 3, \ldots \tag{4.28}$$

or

$$\rho_n = \sqrt{\left(\frac{n\lambda_o}{2} \right) \left(\frac{n\lambda_o}{2} + 2z \right)}. \tag{4.29}$$

The annular regions that these circles define on the aperture plane are called *Fresnel half-period zones*; they are illustrated in Figure 4.6. The first zone extends from $\rho = 0$ to $\rho = \rho_1$, the second from $\rho = \rho_1$ to $\rho = \rho_2$, etc.

An integer number m of these zones just fits within the aperture of radius $d/2$ when the observation point is at z_m. To determine z_m, we let $n = m$, $\rho_n = d/2$, and $z = z_m$ in (4.29):

$$\frac{d}{2} = \sqrt{\left(\frac{m\lambda_o}{2} \right) \left(\frac{m\lambda_o}{2} + 2z_m \right)},$$

which, when rearranged, becomes

$$z_m/\lambda_o = \frac{1}{4m}\left[(d/\lambda_o)^2 - m^2 \right], \qquad m = 1, 2, 3, \ldots, m_{\max}, \tag{4.30}$$

where $m_{\max}$ = Integer Part (d/λ_o) is the maximum number of zones that can fit within the aperture. For this case (the observation point is at one of the z_m), the

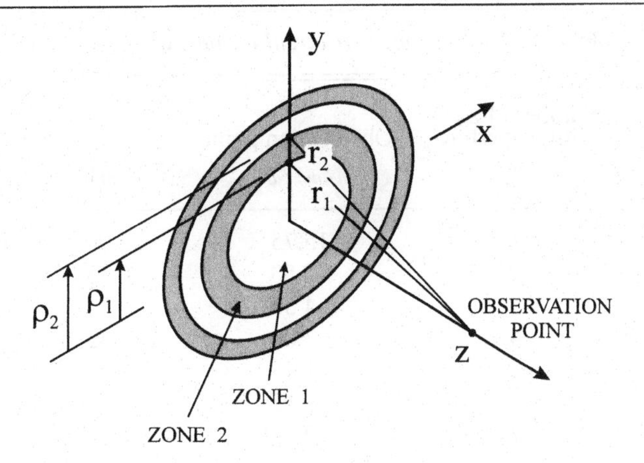

Fig. 4.6. Fresnel zones on the aperture plane. The first four zones are shown.

integral in (4.27) can be expressed as a sum of m integrals, each over one of the zones:

$$E_y(z_m) = -E_o^i e^{-jk_o z_m} \sum_{i=1}^{m} \left[\int_{\rho=\rho_{i-1}}^{\rho_i} \frac{z_m}{\sqrt{\rho^2 + z_m^2}} \left(1 - \frac{j}{k_o\sqrt{\rho^2 + z_m^2}}\right) \right.$$
$$\left. \times (-jk_o\rho) \frac{e^{-jk_o(\sqrt{\rho^2+z_m^2}-z_m)}}{\sqrt{\rho^2 + z_m^2}} d\rho \right]. \tag{4.31}$$

In this formula $\rho_0 = 0$.

To this point, we have not introduced any approximations into the evaluation of the integral. Let us now assume that $k_o\sqrt{\rho^2 + z_m^2} \gg 1$, so that we can drop the second term in the parentheses above. In addition, we will assume that the first factor in the integrand varies little over each Fresnel zone; hence, it can be replaced by its value at the inner radius of each zone; thus, in the i-th zone,

$$\frac{z_m}{\sqrt{\rho^2 + z_m^2}} \approx \frac{z_m}{\sqrt{\rho_{i-1}^2 + z_m^2}}.$$

With these approximations, (4.31) becomes

$$E_y(z_m) = -E_o^i e^{-jk_o z_m} \sum_{i=1}^{m} \frac{z_m}{\sqrt{\rho_{i-1}^2 + z_m^2}} \int_{\rho=\rho_{i-1}}^{\rho_i} (-jk_o\rho) \frac{e^{-jk_o(\sqrt{\rho^2+z_m^2}-z_m)}}{\sqrt{\rho^2 + z_m^2}} d\rho. \tag{4.32}$$

Since

$$\frac{d}{d\rho} e^{-jk_o(\sqrt{\rho^2+z_m^2}-z_m)} = (-jk_o\rho) \frac{e^{-jk_o(\sqrt{\rho^2+z_m^2}-z_m)}}{\sqrt{\rho^2 + z_m^2}},$$

Table 4.3. *Fresnel zones for circular aperture, $d/\lambda_o = 8.0$*

Number of zones filling aperture m	Observation point z_m/λ_o	$\lvert E_y(z_m)/E_o^i\rvert$
1	15.75	2.00
2	7.50	0.13
3	4.58	1.84
4	3.00	0.45
5	1.95	1.59
6	1.17	0.90
7	0.54	1.51
8	0.0	–

the integrals in (4.32) are easily evaluated to give

$$E_y(z_m) = -E_o^i e^{-jk_o z_m} \sum_{i=1}^{m} \frac{z_m}{\sqrt{\rho_{i-1}^2 + z_m^2}} \left[e^{-jk_o(\sqrt{\rho_i^2+z_m^2}-z_m)} - e^{-jk_o(\sqrt{\rho_{i-1}^2+z_m^2}-z_m)}\right],$$

or, on inserting (4.28) and (4.29),

$$E_y(z_m) = 2E_o^i e^{-jk_o z_m} \sum_{i=1}^{m} A_i,$$

where

$$A_i = \frac{(-1)^{i+1}}{1 + (i-1)\lambda_o/2z_m}. \tag{4.33}$$

The properties of the series in (4.33) are best described by a numerical example. We will let the electrical size of the aperture be $d/\lambda_o = 8.0$, since we have already considered this case in detail (Figure 3.26). The values of z_m/λ_o and $|E_y(z_m)/E_o^i|$ for $m = 1, 2, \ldots, 8$ are given in Table 4.3 and are graphed in Figure 4.7a. The curve was drawn freehand between the calculated points. A comparison of Figure 4.7a with Figure 3.26c shows that our estimate for the field based on (4.33) is in good qualitative agreement with our previous more accurate results.

The field is seen to oscillate with z/λ_o and to have a relative maximum whenever an odd number of zones fills the aperture, $m = 1, 3, \ldots$, and a relative minimum whenever an even number of zones fills the aperture, $m = 2, 4, \ldots$. We can easily explain these observations by recalling that the phase of the radiation received from a point in the aperture increases by π as the point moves radially outward through each Fresnel zone. As a result, radiation from a point in the zone n is nearly cancelled by the radiation from a corresponding point in the adjacent zone $n + 1$. Thus, whenever there is an even number of zones in the aperture the net radiation from the aperture is small.

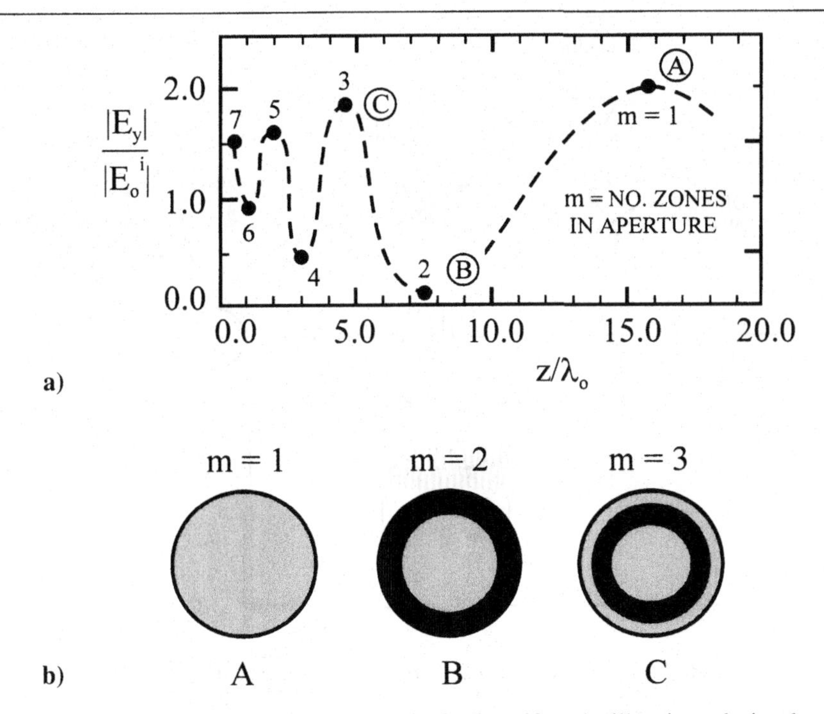

Fig. 4.7. a) Normalized field along z axis for a uniformly illuminated circular aperture with $d/\lambda_o = 8.0$. Points are calculated from Fresnel zones, and curve is sketched freehand. b) Fresnel zones in aperture when observation point is at various positions, z/λ_o (A, B, C).

Once z/λ_o is greater than about 16, less than one Fresnel zone fills the aperture. The field then ceases to oscillate with increasing z/λ_o and decays monotonically, eventually assuming the form for the radiated field: $E_y \propto \exp(-jk_o z)/z$.

The construction based on Fresnel zones, although approximate, provides a useful physical interpretation for the structure of the field near the aperture. The structure results from constructive and destructive interference of the radiation from different parts of the aperture. The Fresnel zones provide an approximate method of accounting for the phase difference that occurs when radiation propagates from different regions in the aperture to an observation point. They are used in the design of practical devices such as the zone plates discussed in Problem 4.2.

4.3 The scattered field

In Chapter 3 we considered problems like those shown in Figures 3.7 and 3.21, where an incident electromagnetic field (a plane wave) interacts with a perfectly conducting object (a plane screen with aperture). For these problems, the total electromagnetic field ($\vec{E}^t$, $\vec{B}^t$) throughout space can be thought of as being composed of two terms: the specified incident field ($\vec{E}^i$, $\vec{B}^i$) and the *scattered* or *reradiated field* ($\vec{E}^s$, $\vec{B}^s$), which is the field produced by the charge and current induced in the object by the incident field:

$$\vec{E}^t = \vec{E}^i + \vec{E}^s, \qquad \vec{B}^t = \vec{B}^i + \vec{B}^s. \tag{4.34}$$

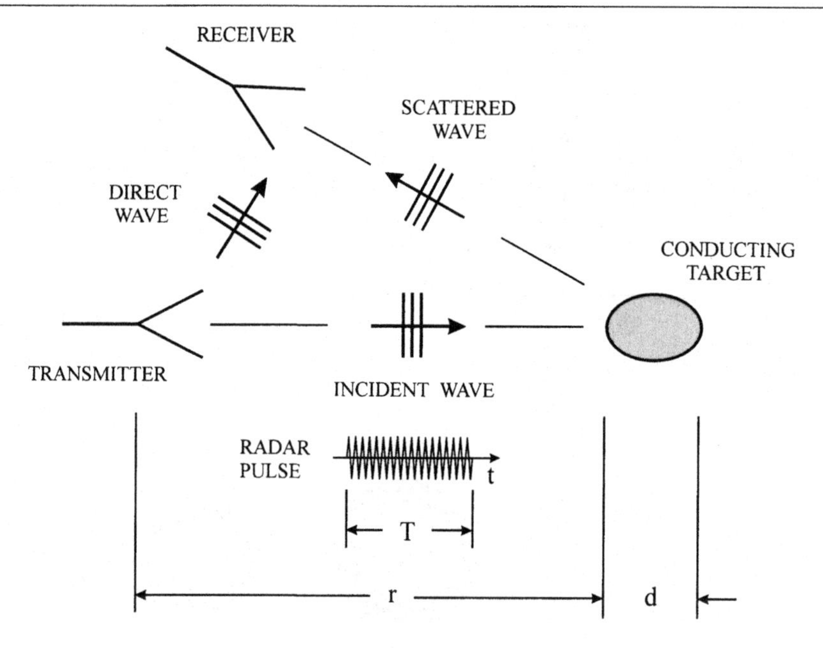

Fig. 4.8. Schematic drawing of simple radar system.

Inside the perfectly conducting object, the total field is zero; that is, the scattered field cancels the incident field.

The scattered field contains information about the object, such as its shape and orientation, and is important in radar. Figure 4.8 shows a schematic drawing for a simple radar system. The transmitting antenna radiates the incident field, which closely approximates a plane wave at the target. The current induced in the target by the incident wave produces the scattered wave, which is approximately a plane wave at the receiving antenna. The received signal is processed to obtain information about the target.

The radar signal is a pulse-modulated carrier of frequency f (see the inset in Figure 4.8 for a typical signal). The duration T of the pulse is chosen so that it contains many cycles of the carrier ($Tf \gg 1$). When the length of the pulse in space, cT, is large compared to the dimensions of the target ($cT \gg d$), the interaction with the target is approximately the same as if the incident field were monochromatic (purely time harmonic).

Notice that it is the pulsing of the signal that allows the separation of the incident and scattered fields in this radar system. The transmitter and receiver are often at the same location (monostatic radar) at the distance r from the target. The transmitter produces a pulsed incident field at the receiver, the direct wave in Figure 4.8, which is followed by the pulsed scattered wave at a time approximately $t = 2r/c$ later. The pulses for the incident and scattered fields are separated in time, provided $2r/c > T$.

For the cases we have considered, the object producing the scattered field is fairly simple: a plane, infinitesimally thin, perfectly conducting screen containing

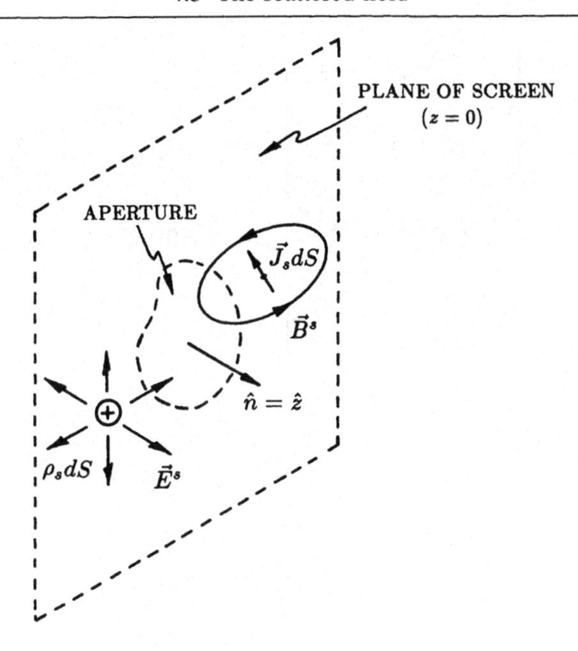

Fig. 4.9. Scattered field produced by charge and current in plane screen.

an aperture. The geometrical and electrical simplicity of this scatterer leads to certain symmetry for the scattered field, which we will examine next. Recall that the incident field induces charge and current in the perfectly conducting screen and that the charge and current, in turn, produce the scattered field. For calculating the scattered field, we can then remove the screen and replace it by the surface densities of charge and current. The perfectly conducting screen first can be thought of as having finite thickness; there are then surface densities of charge and current on both sides of the screen: ρ_s^+, $\vec{J}_s^+$; ρ_s^-, $\vec{J}_s^-$. As the thickness of the screen is made infinitesimal, the surface densities associated with the two sides combine into one, and the screen can be replaced by the surface densities $\rho_s = \rho_s^+ + \rho_s^-$, $\vec{J}_s = \vec{J}_s^+ + \vec{J}_s^-$.

The symmetry of the scattered field is easily deduced by considering the elements of charge and current, $\rho_s dS$ and $\vec{J}_s dS$, in the screen (see Figure 4.9). The tangential components of the electric field are the same on both sides ($z = 0+$, $z = 0-$) of the screen, whereas the normal components are equal in magnitude but point in opposite directions. Within the aperture the normal component of the scattered electric field is zero. The tangential components of the magnetic field are equal in magnitude but opposite in direction on the two sides of the screen, whereas the normal components are the same. Within the aperture the tangential component of the scattered magnetic field is zero.

The symmetry observed for the fields on the two sides of the screen extends throughout the half spaces. This can be seen from the plane-wave spectral representation, where the field in the half space $z > 0$ ($z < 0$) is determined from the field on $z = 0+$ ($z = 0-$). These results are summarized below and in

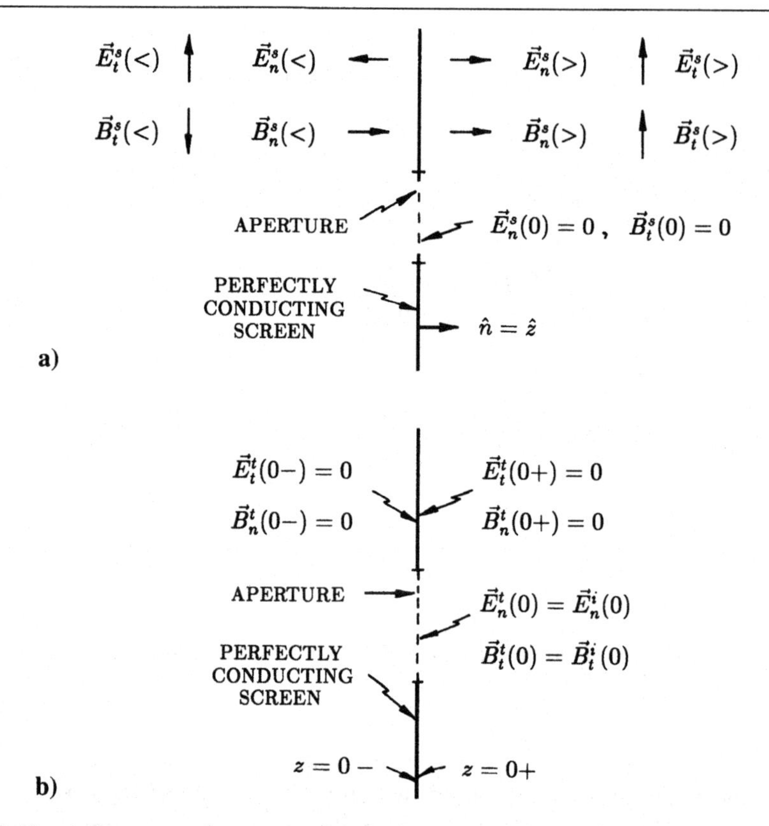

Fig. 4.10. a) Symmetry for scattered field on the two sides of the screen. b) Components of the total field on the screen and in the aperture.

Figure 4.10a:

$$\vec{E}^s_t(<) = \vec{E}^s_t(>), \qquad \hat{n} \times \vec{E}^s(<) = \hat{n} \times \vec{E}^s(>) \tag{4.35a}$$

$$\vec{E}^s_n(<) = -\vec{E}^s_n(>), \qquad \hat{n} \cdot \vec{E}^s(<) = -\hat{n} \cdot \vec{E}^s(>) \tag{4.35b}$$

$$\vec{E}^s_n(0) = 0, \qquad \hat{n} \cdot \vec{E}^s(0) = 0, \quad \text{in the aperture} \tag{4.35c}$$

$$\vec{B}^s_t(<) = -\vec{B}^s_t(>), \qquad \hat{n} \times \vec{B}^s(<) = -\hat{n} \times \vec{B}^s(>) \tag{4.35d}$$

$$\vec{B}^s_n(<) = \vec{B}^s_n(>), \qquad \hat{n} \cdot \vec{B}^s(<) = \hat{n} \cdot \vec{B}^s(>) \tag{4.35e}$$

$$\vec{B}^s_t(0) = 0, \qquad \hat{n} \times \vec{B}^s(0) = 0, \quad \text{in the aperture.} \tag{4.35f}$$

Here we have introduced a shorthand notation (symbols in the arguments of the field) to indicate the location in space. For a point in the right half space (x, y, z, with $z > 0$), we use the symbol $>$; for the corresponding point in the left half space ($x, y, -z$), we use the symbol $<$; and for a point on the plane $z = 0$, we use 0. The subscripts t and n refer to the components of the field that are tangential and normal to the plane $z = 0$, respectively.

These results, together with the electromagnetic boundary conditions for a perfect conductor (Table 1.6), determine the total field (4.34) on the surface of the screen and in the aperture. This is summarized in Figure 4.10b. Notice, in particular, that in the aperture the normal component of the total electric field and the tangential components of the total magnetic field are the same as these components for the incident field:

$$\vec{E}^t_n(0) = \vec{E}^i_n(0), \qquad \hat{n}\cdot\vec{E}^t(0) = \hat{n}\cdot\vec{E}^i(0), \quad \text{in the aperture} \qquad (4.36a)$$

$$\vec{B}^t_t(0) = \vec{B}^i_t(0), \qquad \hat{n}\times\vec{B}^t(0) = \hat{n}\times\vec{B}^i(0), \quad \text{in the aperture.} \qquad (4.36b)$$

The results we have obtained show that once the field (total or scattered) on one side of the screen ($z > 0$ or $z < 0$) is known, the field on the other side is also known. For example, in our discussion of the circular aperture in Section 3.7, we obtained an approximation for the total field $\vec{E}^t$, $\vec{B}^t$ on the "shadow side" of the screen, $z > 0$. The field on the "illuminated side" of the screen, $z < 0$, is easily obtained from these results using (4.34) and (4.35a–f).

As a simple, almost trivial, illustration of these results, consider the geometry in Figure 3.21 for the case when there is no aperture in the screen ($d = 0$); a normally incident plane wave then illuminates an infinitely large, perfectly conducting screen. Of course, the total field in the right half space ($z > 0$) is zero:

$$\vec{E}^t(>) = 0,$$

or

$$\vec{E}^s(>) = -\vec{E}^i(>) = -E^i_o e^{-jk_oz}\,\hat{y}.$$

After noting that the electric field has only a tangential component, we have from (4.35a)

$$\vec{E}^s(<) = \vec{E}^s(>) = -E^i_o e^{jk_oz}\,\hat{y}$$

and from (4.34)

$$\begin{aligned}\vec{E}^t(<) &= \vec{E}^i(<) + \vec{E}^s(<)\\ &= E^i_o e^{-jk_oz}\hat{y} - E^i_o e^{jk_oz}\hat{y}\\ &= -2jE^i_o\sin(k_oz)\,\hat{y}.\end{aligned}$$

As expected, there is a pure standing wave in the region $z < 0$. Recall that this is what Hertz observed in his famous experiment in which an electromagnetic wave was reflected from a large, metallic sheet (Section 2.7).

4.4 Babinet's principle

4.4.1 Optical formulation

In Figure 4.11a we show the optical counterpart of the problems we have been discussing: The light from a source falls on a plane, opaque screen containing an aperture. Notice that in optics the screen is considered to be opaque rather than perfectly conducting. We will call the arrangement in Figure 4.11a the *original problem*. There is a corresponding *complementary problem* shown in Figure 4.11b. Here the areas on the plane $z = 0$ occupied by the opaque screen and the aperture in

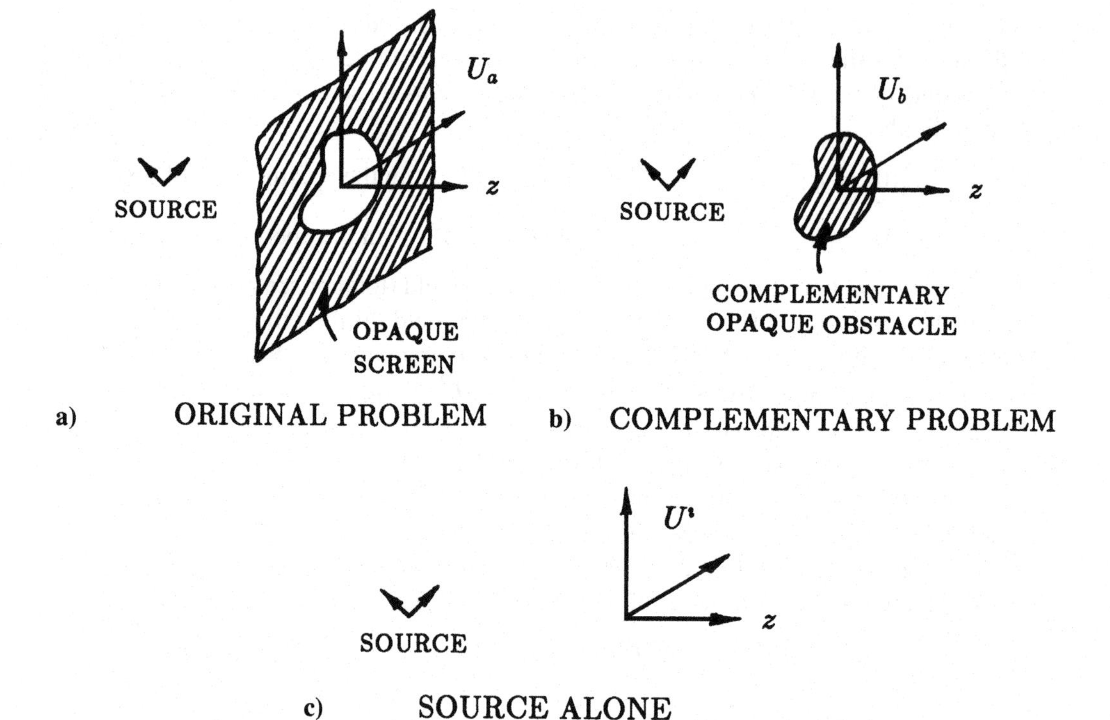

Fig. 4.11. Diagrams describing Babinet's principle in optics.

the original problem are interchanged; the region occupied by the aperture is now the obstacle.

In 1837 Jacques Babinet (1794–1872) briefly mentioned in a memoir on meteorological optics a connection between the optical effects produced in these two problems [11]. The mathematical statement of this relationship has since become known as *Babinet's principle* [7, 12]. Consider the light source shown in Figure 4.11c. In the absence of any screen or obstacle, the source produces the incident field U^i. (Here U is the scalar field of classical optics and is not to be confused with the electromagnetic field.) The irradiance or intensity is $I = |U|^2$.

When this source is used with the screen in Figure 4.11a, the field is U_a for $z > 0$; whereas, when it is used with the obstacle in Figure 4.11b, the field is U_b for $z > 0$. Babinet's principle is that

$$U_a + U_b = U^i, \tag{4.37a}$$

or

$$U_b = U^i - U_a. \tag{4.37b}$$

Hence, once the solution to the original problem is known, the solution to the complementary problem is automatically known.

The relationship between the irradiances or intensities is

$$I_b = |U_b|^2 = I^i + I_a - 2\mathrm{Re}(U^i U_a^*), \tag{4.38a}$$

or

$$I_a = |U_a|^2 = I^i + I_b - 2\mathrm{Re}(U^i U_b^*). \tag{4.38b}$$

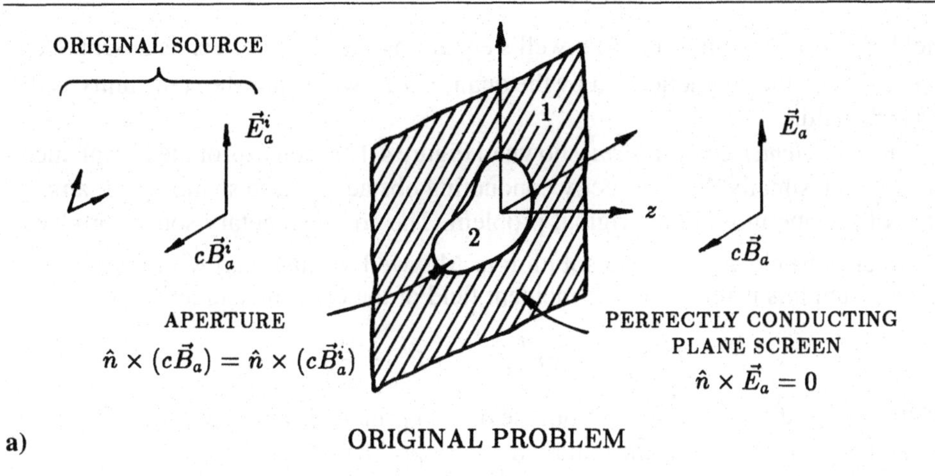

a) ORIGINAL PROBLEM

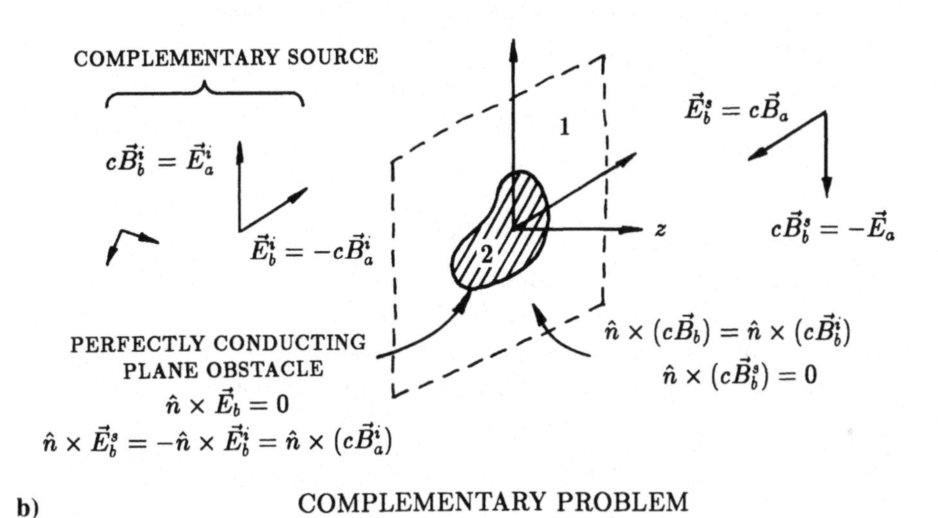

b) COMPLEMENTARY PROBLEM

Fig. 4.12. Diagrams describing Babinet's principle in electromagnetics.

An interesting consequence of the principle is that at any point where the irradiance is zero for one of the problems, the irradiance for the other problem equals that of the incident field. For example, if the irradiance in the aperture in Figure 4.11a is assumed to be that of the incident field, $I_a = I^i$, then the irradiance directly behind the obstacle in Figure 4.11b is zero, $I_b = 0$ (a perfect shadow).

4.4.2 Electromagnetic formulation

The electromagnetic counterpart of Babinet's optical principle is best explained by first stating the principle and then showing that it follows directly from Maxwell's equations and the symmetry for the field presented earlier in Section 4.3.

Consider the sketches in Figure 4.12. The *original problem*, shown in Figure 4.12a, consists of the original source and a plane, infinitesimally thin, perfectly conducting screen containing an aperture. The original source produces the incident field $\vec{E}^{i}_{a}$, $c\vec{B}^{i}_{a}$ (a solution to Maxwell's equations) when the screen is absent. Here we represent the magnetic field by the quantity $c\vec{B}$, which has the same units as the electric field.

The *complementary problem*, shown in Figure 4.12b, consists of the complementary infinitesimally thin, perfectly conducting obstacle whose shape is the same as that of the aperture in the original problem. The complementary source produces the incident field $\vec{E}^{i}_{b}$, $c\vec{B}^{i}_{b}$ (a solution to Maxwell's equations) with the obstacle absent, and this is related to the incident field of the original source:[2]

$$\vec{E}^{i}_{b} = -c\vec{B}^{i}_{a}, \qquad c\vec{B}^{i}_{b} = \vec{E}^{i}_{a}. \tag{4.39}$$

Notice a significant difference from the optical principle, where the sources are the same for the original and the complementary problems.

The electromagnetic form of Babinet's principle states that in the half space $z > 0$ the electromagnetic fields for the original and the complementary problems satisfy the relations [13, 14]

$$\vec{E}^{s}_{b} = (\vec{E}_{b} - \vec{E}^{i}_{b}) = c\vec{B}_{a} \tag{4.40a}$$

and

$$c\vec{B}^{s}_{b} = (c\vec{B}_{b} - c\vec{B}^{i}_{b}) = -\vec{E}_{a}, \tag{4.40b}$$

where we have used the notation for the scattered field introduced in (4.34).[3]

To prove this principle, we will consider the equations that determine the total field ($\vec{E}_{a}$, $c\vec{B}_{a}$) for the original problem and the scattered field ($\vec{E}^{s}_{b}$, $c\vec{B}^{s}_{b}$) for the complementary problem. For the original problem, the total field must satisfy Maxwell's equations in the half space $z > 0$,

$$\nabla \times \vec{E}_{a} = -jk_{o}(c\vec{B}_{a}), \qquad \nabla \times (c\vec{B}_{a}) = jk_{o}\vec{E}_{a}, \tag{4.41}$$

and on the plane $z = 0$,

$$\hat{n} \times \vec{E}_{a} = 0 \qquad \text{on 1 (screen)} \tag{4.42a}$$

$$\hat{n} \times (c\vec{B}_{a}) = \hat{n} \times (c\vec{B}^{i}_{a}) \qquad \text{on 2 (aperture)}, \tag{4.42b}$$

where the last result follows from (4.36b).

The scattered field for the complementary problem is due to the charge and current in the obstacle; it is also a solution to Maxwell's equations in the half space $z > 0$,

$$\nabla \times \vec{E}^{s}_{b} = -jk_{o}(c\vec{B}^{s}_{b}), \qquad \nabla \times (c\vec{B}^{s}_{b}) = jk_{o}\vec{E}^{s}_{b}, \tag{4.43}$$

[2] Some authors define the complementary source so that the minus sign is in the relation for $c\vec{B}^{i}_{b}$ and not in the one for $\vec{E}^{i}_{b}$. Actually, only one of the two relations in (4.39) need be specified; the other follows from the fact that both incident fields (a and b) satisfy Maxwell's equations (Problem 4.7).

[3] Here we have dropped the superscript t used previously in Section 4.3 to indicate the total field.

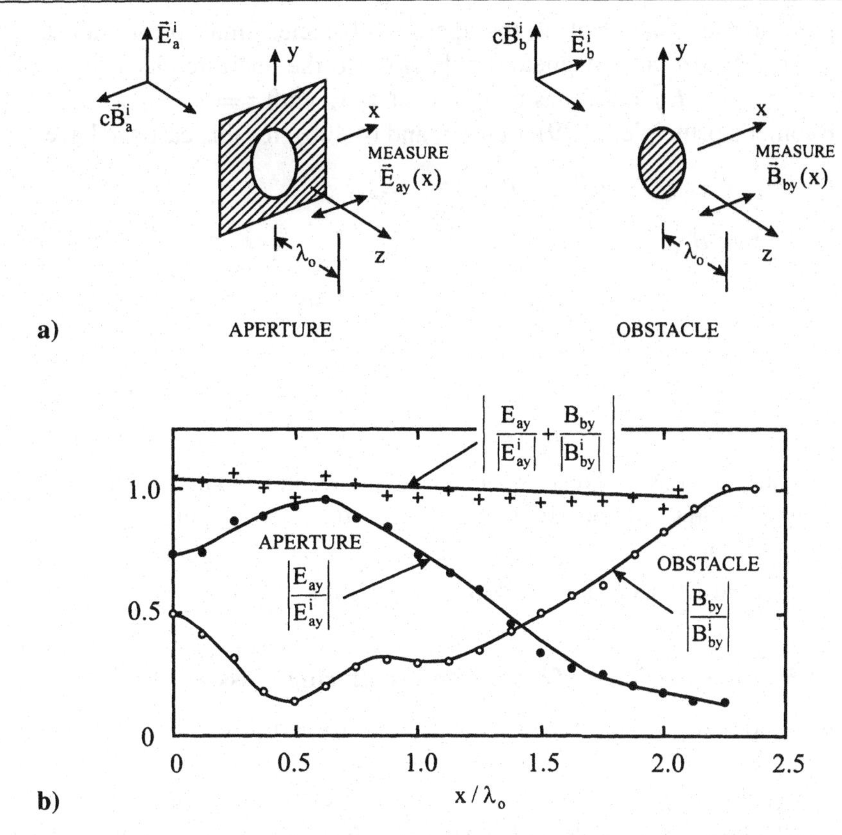

Fig. 4.13. Experimental confirmation for the electromagnetic Babinet's principle for a circular aperture with $d/\lambda_o = 3.0$. a) Experimental arrangement. b) Measured results. (Measured data from Ehrlich et al. [15].)

and on the plane $z = 0$,

$$\hat{n} \times (c\vec{B}_b^s) = 0 \qquad \text{on 1} \tag{4.44a}$$

$$\hat{n} \times \vec{E}_b^s = -\hat{n} \times \vec{E}_b^i = \hat{n} \times (c\vec{B}_a^i) \qquad \text{on 2 (obstacle).} \tag{4.44b}$$

In obtaining the last result, (4.39) was used.

Now we notice that Equations (4.43), (4.44a), and (4.44b) become Equations (4.41), (4.42a), and (4.42b) when we make the substitution $\vec{E}_b^s \to c\vec{B}_a$, $c\vec{B}_b^s \to -\vec{E}_a$. Hence, the solutions to Maxwell's equations for the fields of the original problem and complementary problem are related by (4.40a) and (4.40b); that is, they are related by Babinet's principle.

Convincing experimental verification for the electromagnetic Babinet's principle can be obtained from microwave measurements such as those described earlier in Section 3.7 and Figure 3.26a. The arrangement for one such measurement is sketched in Figure 4.13a [15]. Here the original problem is for a circular aperture in a plane metallic sheet, and the complementary problem is for a circular metallic disc (obstacle). The original and complementary sources produce approximately plane waves, normally incident on the aperture/obstacle. Small probes are used to measure the electric and magnetic fields close to the aperture/obstacle on the shadow side ($z > 0$): $\vec{E}_y$ and $\vec{B}_y$ as functions of x/λ_o with $z = \lambda_o$.

From Babinet's principle, (4.39), (4.40a), and (4.40b), for this case we have

$$cB_{by} - cB^i_{by} = -E_{ay},$$

or, on dividing by cB^i_{by},

$$\frac{B_{by}}{B^i_{by}} - 1 = -\frac{E_{ay}}{cB^i_{by}} = -\frac{E_{ay}}{E^i_{ay}},$$

$$\frac{E_{ay}}{E^i_{ay}} + \frac{B_{by}}{B^i_{by}} = 1. \tag{4.45}$$

Figure 4.13b shows measured results for a circular aperture and disc with diameter $d/\lambda_o = 3.0$ [15]. The solid dots show E_{ay}/E^i_{ay} for the aperture, and the open dots show B_{by}/B^i_{by} for the disc. The crosses show the sum of these two measurements, which should equal 1.0 according to Babinet's principle (4.45).

4.5 Transmission coefficients and scattering cross sections

It is customary to describe the transmission through an aperture and the scattering by an obstacle in terms of certain quantities: transmission coefficients and scattering cross sections. When the screen and obstacle are formed from plane, perfectly conducting sheets, these quantities for complementary apertures and obstacles are simply related through Babinet's principle.

The directional characteristics of the radiation from an aperture are described by the *directivity of the aperture*, D_A, which was introduced in Section 3.7. When the aperture is illuminated by a plane wave incident in the direction $\hat{k}_i(\theta_i, \phi_i)$, as in Figure 4.14, the directivity in the direction $\hat{k}(\theta, \phi)$ is

$$D_A(\hat{k}, \hat{k}_i) = \frac{\text{time-average power per unit area radiated in the direction } \theta, \phi}{\text{time-average power radiated per unit area averaged over all directions}}$$

$$= \frac{4\pi r^2 |\vec{E}^r(\theta, \phi)|^2}{\int_{\phi'=0}^{2\pi} \int_{\theta'=0}^{\pi/2} |\vec{E}^r(\theta', \phi')|^2 r^2 \sin\theta' d\theta' d\phi'}. \tag{4.46}$$

The *transmission coefficient* is a measure of the total power transmitted through the aperture of area A:

$$\tau(\hat{k}_i) = \frac{\text{total time-average power transmitted through the aperture}}{A \cdot (\text{time-average power per unit area of the incident plane wave})}$$

$$= \frac{\int_{\phi'=0}^{2\pi} \int_{\theta'=0}^{\pi/2} |\vec{E}^r(\theta', \phi')|^2 r^2 \sin\theta' d\theta' d\phi'}{A |\vec{E}^i_o|^2}. \tag{4.47}$$

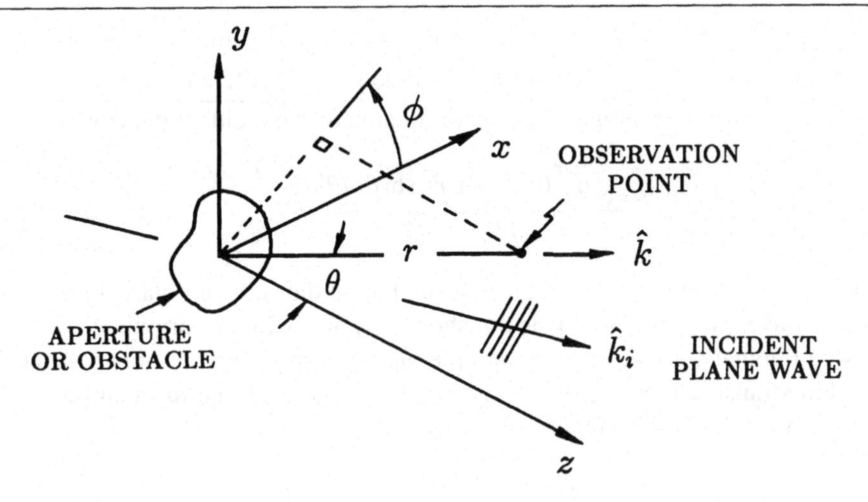

Fig. 4.14. Coordinates for describing transmission through an aperture and scattering by an obstacle.

The scattering cross sections describe the power scattered by an obstacle relative to the power per unit area of the incident plane wave; they have the dimension of area [16, 17]. For the coordinates in Figure 4.14, the *bistatic* or *differential cross section* is

$$\sigma(\hat{k}, \hat{k}_i) =$$

$$\lim_{r\to\infty}\left[\frac{4\pi r^2 \cdot (\text{time-average power per unit area scattered in the direction } \theta, \phi)}{\text{time-average power per unit area of the incident plane wave}}\right]$$

$$= \frac{4\pi r^2 |\vec{E}^{sr}(\theta, \phi)|^2}{|\vec{E}_o^i|^2}. \tag{4.48}$$

Notice that the numerator of this expression is the total power scattered by a fictitious obstacle that radiates the same field in all directions (isotropic scatterer) as the actual obstacle radiates in the direction $\hat{k}$. For the special case $\hat{k} = -\hat{k}_i$ $(\theta = \pi - \theta_i,\ \phi = \phi_i + \pi)$, the bistatic cross section is called the *monostatic* or *backscattering cross section*:

$$\sigma_B(\hat{k}_i) = \sigma(-\hat{k}_i, \hat{k}_i). \tag{4.49}$$

This cross section is particularly important for radar systems that have their transmitter and receiver at the same location.[4]

The *total scattering cross section* is a measure of the total power scattered by the

[4] When the radar uses the same antenna for transmission and reception, the backscattering cross section is sometimes referred to as the *radar cross section*, and it is defined so as to include the state of polarization of the antenna. The numerator of (4.48) is then the part of the power in the scattered wave that can be received by the antenna [18].

obstacle:

$$\sigma_T(\hat{k}_i) = \frac{\text{total time-average power scattered}}{\text{time-average power per unit area of the incident plane wave}}$$

$$= \frac{\int_{\phi'=0}^{2\pi}\int_{\theta'=0}^{\pi}|\vec{E}^{sr}(\theta', \phi')|^2 r^2 \sin\theta' d\theta' d\phi'}{|\vec{E}_o^i|^2}. \tag{4.50}$$

All of these quantities, (4.46)–(4.50), depend upon the state of polarization for the incident wave, even though this is not shown explicitly in the formulas.

Using Babinet's principle, it is a simple matter to show that the above quantities describing transmission and scattering for complementary apertures and obstacles are related (Problem 4.3) by the equations

$$\sigma(\hat{k}, \hat{k}_i) = \tau(\hat{k}_i)\, A\, D_A(\hat{k}, \hat{k}_i), \tag{4.51}$$

$$\sigma_B(\hat{k}_i) = \tau(\hat{k}_i)\, A\, D_A(\hat{k}_i, \hat{k}_i), \tag{4.52}$$

and

$$\sigma_T(\hat{k}_i) = 2A\, \tau(\hat{k}_i). \tag{4.53}$$

We will illustrate the use of these formulas by again considering the complementary problems of transmission through a circular aperture and scattering by a circular disc. Recall that in Section 3.7 we used Kirchhoff's approximation with the plane-wave spectrum to estimate the field transmitted through a circular aperture illuminated by a normally incident plane wave ($\hat{k}_i = \hat{z}$). We can now use these results to estimate the field scattered by the circular disc.

We will only consider the radiated field, for which

$$\vec{E}^r = -\hat{k} \times (c\vec{B}^r), \qquad c\vec{B}^r = \hat{k} \times \vec{E}^r,$$

making Babinet's principle (4.40a, b)

$$\vec{E}^{sr}_{\text{Disc}} = \hat{k} \times \vec{E}^r_{\text{Aperture}}, \tag{4.54a}$$

$$\vec{B}^{sr}_{\text{Disc}} = \hat{k} \times \vec{B}^r_{\text{Aperture}}, \tag{4.54b}$$

or

$$|\vec{E}^{sr}_{\text{Disc}}| = |\vec{E}^r_{\text{Aperture}}|, \tag{4.55a}$$

$$|\vec{B}^{sr}_{\text{Disc}}| = |\vec{B}^r_{\text{Aperture}}|. \tag{4.55b}$$

Thus, the field patterns for the circular aperture (Figures 3.23 and 3.24) apply to the scattered field for the circular disc. They are actually $\sigma^{1/2}(\hat{k}, \hat{z})$ multiplied by a constant. Notice, however, that the principal E-plane (H-plane) pattern for the aperture is the principal H-plane (E-plane) pattern for the disc. From these results, we see that a normally incident plane wave is scattered by an electrically large disc most strongly in the forward ($\hat{k} = \hat{k}_i = \hat{z}$) and backward ($\hat{k} = -\hat{k}_i = -\hat{z}$) directions.

The radiated field (3.113) and directivity (3.116), previously obtained for the aperture using Kirchhoff's approximation, can be used with Equations (4.47), (4.52), and (4.53) to determine the transmission coefficient of the aperture and the scattering cross sections for the disc at normal incidence ($\hat{z}_i = \hat{z}$) (Problem 4.4):

$$\tau(\hat{z}) \approx \frac{(k_o d)^2}{4D_A(\hat{z},\,\hat{z})} = \frac{\pi^2 d^2}{\lambda_o^2 D_A(\hat{z},\,\hat{z})}, \tag{4.56}$$

$$\sigma_B(\hat{z}) = \tau(\hat{z})\,A\,D_A(\hat{z},\,\hat{z}) \approx \frac{(k_o d)^2 A}{4} = \frac{\pi^3 d^4}{4\lambda_o^2}, \tag{4.57}$$

$$\sigma_T(\hat{z}) = 2A\,\tau(\hat{z}) \approx \frac{(k_o d)^2 A}{2D_A(\hat{z},\,\hat{z})} = \frac{\pi^3 d^4}{2\lambda_o^2 D_A(\hat{z},\,\hat{z})}, \tag{4.58}$$

where $A = \pi d^2/4$ is the area of the disc. These approximate results are compared with exact results and measurements in Figure 4.15 [19–22]. As expected from our earlier discussion in Chapter 3, the approximate results are in reasonable agreement with the exact results only for electrically large apertures and discs, $k_o d \gtrsim 10$ ($d/\lambda_o \gtrsim 1.6$), in which case $D_A(\hat{z},\,\hat{z}) \approx 4\pi A/\lambda_o^2$ (3.129a), making

$$\tau(\hat{z}) \approx 1 \tag{4.59}$$

and

$$\sigma_T(\hat{z}) \approx 2A = \pi d^2/2. \tag{4.60}$$

We will say more about the exact results in Figure 4.15 when we discuss scattering from electrically small discs in Section 7.6.

As you might expect, the power transmitted through the electrically large aperture is simply the physical area A times the power per unit area of the normally incident plane wave, making the transmission coefficient (4.59) one. For the electrically large disc and normal incidence, the total scattering cross section (4.60) is *twice* the physical area A. This may seem surprising, since at first glance it may seem that the disc should block, hence scatter, the amount of power contained in the area A of the incident wave, not in the area $2A$. There is a simple physical explanation for this result.

On a plane directly behind the electrically large disc (Figure 4.16a), the total field is approximately zero in the shadow and equal to the incident field on the remainder of the plane. Hence, the scattered field is approximately equal to the negative of the incident field in the shadow and zero on the remainder of the plane. This means that the total power scattered into the right half space ($z > 0$) is approximately AS_o^i, where S_o^i is the time-average power per unit area of the incident plane wave. Owing to the symmetry of the problem, the currents in the disc radiate the same power into the right and the left half spaces, so the total power scattered is $2AS_o^i$, making $\sigma_T \approx 2AS_o^i/S_o^i = 2A$. This result, $\sigma_T \approx 2A$, is true for an electrically large, planar obstacle of any shape illuminated at normal incidence.

Our knowledge about the scattering from the electrically large disc can be used to gain some physical insight into the scattering from other electrically large obstacles

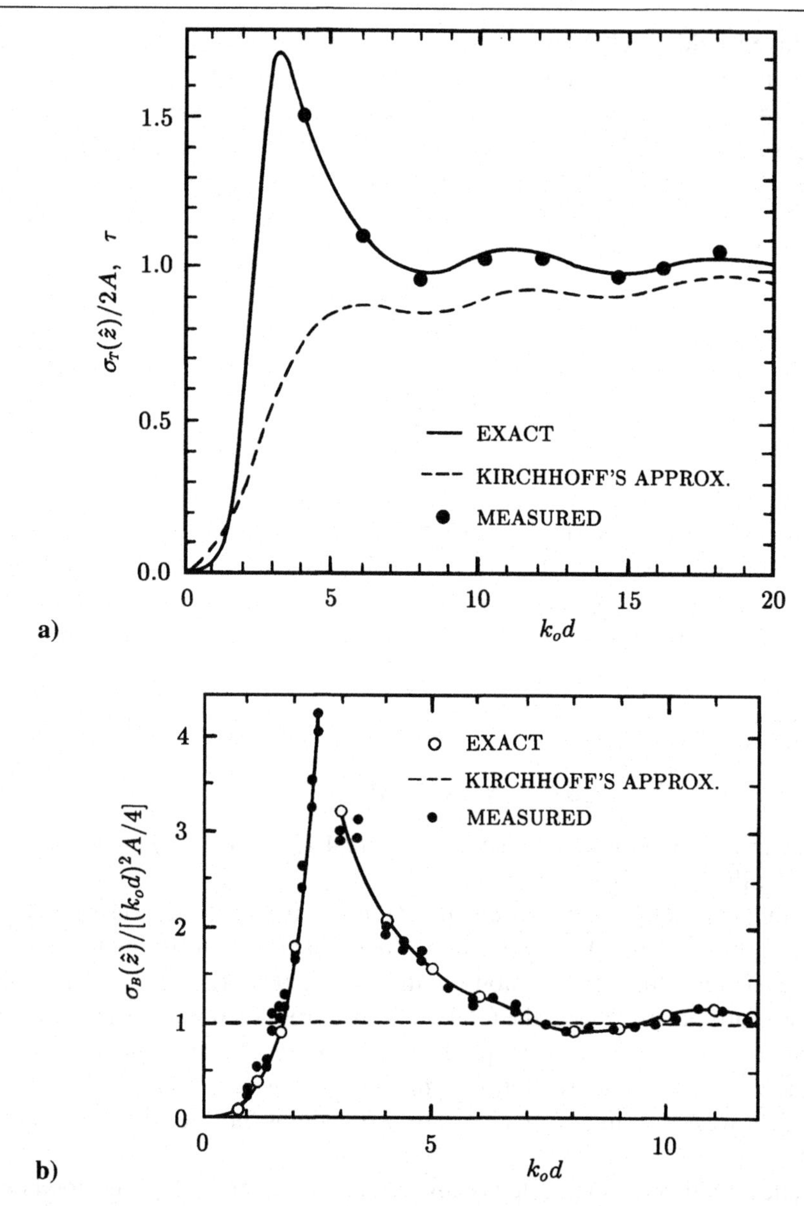

Fig. 4.15. Cross sections for a disc irradiated at normal incidence. a) Total scattering cross section versus $k_o d$. b) Backscattering cross section versus $k_o d$. (Theoretical results from Andrejewski [19] and Hodge [22]. Measured data from Huang et al. [20] and Hey et al. [21].)

that are not necessarily planar. Here we will only consider the case of scattering from a perfectly conducting sphere. Patterns for the bistatic cross sections of the disc (normal incidence) and the sphere are in Figure 4.17 [16, 19]. These are graphs of the exact results for σ versus the angle θ defined in Figure 4.14. The top (bottom) half of each pattern is for the E plane (H plane). The disc and sphere have the same diameter and are electrically large ($k_o d = 20.0$). The figures are enlarged in the regions where the scattered field is small to show the details of the patterns.

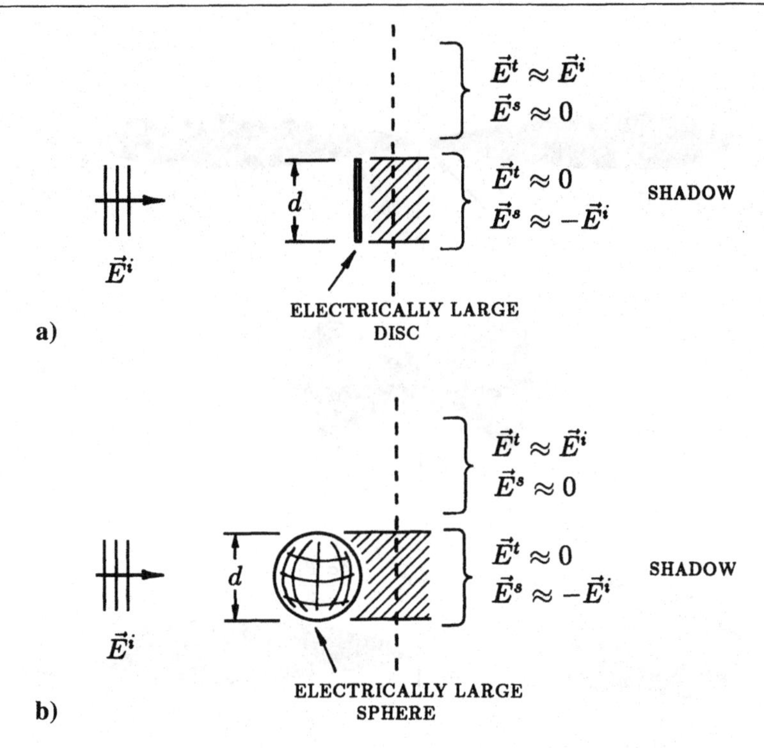

Fig. 4.16. Approximations for the field of a) an electrically large disc, and b) an electrically large sphere.

As we saw earlier, the disc scatters most strongly in the forward and backward directions ($\theta = 0°, 180°$). The sphere, however, scatters predominantly in the forward direction (right half space), where the pattern is almost the same as that for the disc. This similarity can be explained using Figure 4.16b. On the plane directly behind the electrically large sphere, the field is nearly the same as that for the disk: approximately zero in the shadow and equal to the incident field elsewhere. Hence, the scattered field to the right of the plane (calculated using the plane-wave spectrum or Huygens' principle) must also be nearly the same as that for the disc.

Detailed calculations show that the total scattering cross section, σ_T, for the electrically large sphere is the same as that for the disc (4.60) [17]. Thus, the two objects scatter the same total power. For the disc, half of the power is concentrated in each of the beams in the forward and backward directions. For the sphere, one half of the power is concentrated in the beam in the forward direction, as for the disc, and the other half of the power is very roughly distributed uniformly over all directions, as can be seen in the enlarged portion of Figure 4.17b [23] (Problem 4.6).

One might ask, what approximation for the planar obstacle (e.g. disc) is equivalent to Kirchhoff's approximation for the field in the aperture? Again, we can use Babinet's principle to answer this question. Kirchhoff's approximation for the field within the aperture is

$$\hat{z} \times \vec{E}_a(z = 0+) = \hat{z} \times \vec{E}^i_a(z = 0). \tag{4.61}$$

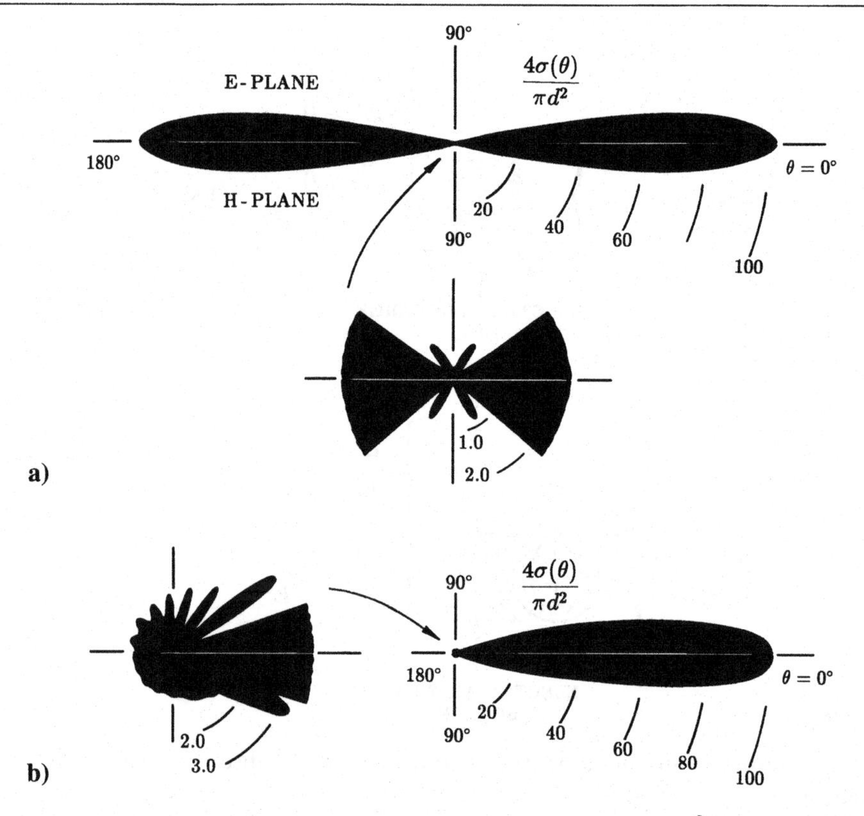

Fig. 4.17. Patterns for the bistatic scattering cross section, $4\sigma(\theta)/\pi d^2$, for a) an electrically large disc (normal incidence) and b) an electrically large sphere. Both results are for $k_o d = 20.0$. Notice that the top half of each pattern is for the E plane, while the bottom half is for the H plane. (Data for disc from Andrejewski [19]. Data for sphere from King and Wu [16].)

Now from equations (4.39) and (4.40b) and the symmetry (4.35d), the scattered and total magnetic fields at the front and back surfaces of the complementary obstacle are

$$\hat{z} \times \vec{B}^s_b(z = 0+) = -\hat{z} \times \vec{B}^i_b(z = 0) \tag{4.62a}$$

$$\hat{z} \times \vec{B}^s_b(z = 0-) = \hat{z} \times \vec{B}^i_b(z = 0) \tag{4.62b}$$

and

$$\hat{z} \times \vec{B}_b(z = 0+) = 0 \tag{4.63a}$$

$$\hat{z} \times \vec{B}_b(z = 0-) = 2\hat{z} \times \vec{B}^i_b(0). \tag{4.63b}$$

Using the boundary condition for a perfect conductor (Table 1.6) with (4.63a) and (4.63b), the surface current densities on the shadow ($z = 0+$) and illuminated

$(z = 0-)$ sides of the obstacle are

$$\vec{J}_s = \begin{cases} \hat{z} \times \vec{H}_b(z = 0+) = 0, & \text{shadow side} \\ -\hat{z} \times \vec{H}_b(z = 0-) = -2\hat{z} \times \vec{H}^i_b(z = 0-), & \text{illuminated side.} \end{cases} \tag{4.64}$$

After dropping the subscript b and introducing the unit, outward-pointing normal $\hat{n}$ to the surface of the obstacle, (4.64) becomes

$$\vec{J}_s = \begin{cases} 0, & \text{shadow side} \\ 2\hat{n} \times \vec{H}^i, & \text{illuminated side.} \end{cases} \tag{4.65}$$

This approximation for the surface current density on the obstacle, which is equivalent to Kirchhoff's approximation for the field in an aperture, is known as the *physical optics current*. There is a simple physical explanation that accompanies the approximation. When the object is an infinitely large, perfectly conducting sheet, the surface currents on the two sides of the sheet are given *exactly* by (4.65). So this approximation holds under the assumption that the current at each point on the surface of the planar obstacle of finite size is the same as it would be on an infinite sheet illuminated by the same incident field. The physical optics approximation (4.65) is used to estimate the current on electrically large objects of general shape, not just for the planar obstacles we have discussed here.

Before we leave this topic, we will mention an important theorem derived in Problem 4.5: the *optical* or *forward scattering theorem* [12, 24]. The transmission coefficient of an aperture and total scattering cross section of an obstacle are given by

$$\tau(\hat{k}_i) = \frac{\lambda_o}{A} \text{Im}\left[\frac{(\vec{E}^i_o)^* \cdot \vec{E}^r(\hat{k}_i) r e^{jk_i \cdot \vec{r}}}{|\vec{E}^i_o|^2}\right] \tag{4.66}$$

and

$$\sigma_T(\hat{k}_i) = \frac{-4\pi}{k_o} \text{Im}\left[\frac{(\vec{E}^i_o)^* \cdot \vec{E}^{sr}(\hat{k}_i) r e^{jk_i \cdot \vec{r}}}{|\vec{E}^i_o|^2}\right]. \tag{4.67}$$

Here $\vec{E}^r(\hat{k}_i)$ indicates the electric field radiated in the direction of the incident wave $\hat{k}_i$: the forward direction. Notice that the left-hand sides of these relations are for quantities that involve an integration of the radiated field over all directions, (4.47) and (4.50), whereas, the right-hand sides involve the field in a single direction, the direction of the incident wave $\hat{k}_i$. The value of these relations is now obvious; the quantities τ and σ_T can be determined from a calculated or measured value of the field in a single direction. Although (4.67) is derived in Problem 4.5 for a planar, perfectly conducting obstacle, it can be shown to apply to nonabsorbing obstacles of general shape.

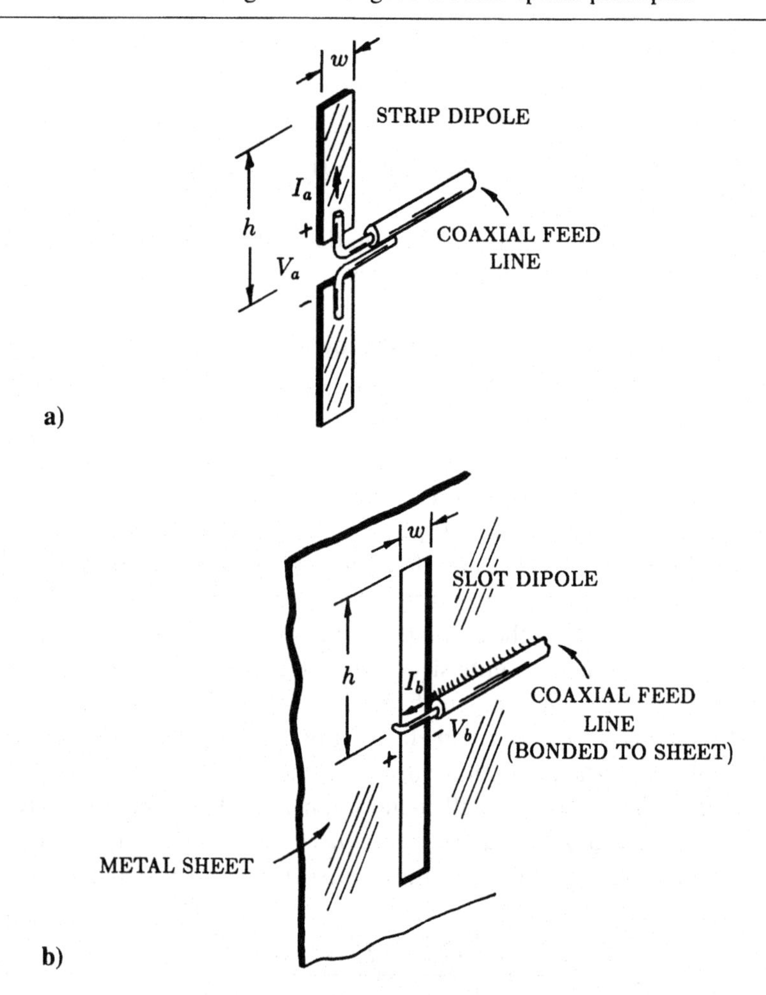

Fig. 4.18. Practical construction for a) a strip dipole and b) a complementary slot dipole.

4.6 Complementary antennas

Figure 4.18a shows a simple dipole antenna, each arm of which is a thin, metallic strip of width w and length h. A signal from the coaxial line connected at the midpoint excites the antenna. The voltage and current at the drive point, V_a and I_a, determine the input impedance of the strip dipole:

$$Z_{\text{Strip}} = V_a / I_a. \tag{4.68}$$

The complementary antenna to the strip dipole is the slot dipole shown in Figure 4.18b. It consists of a slot of width w and length $2h$ in a thin, metallic sheet, and it is also excited by a signal from a coaxial line connected at its midpoint. The input impedance of the slot dipole is

$$Z_{\text{Slot}} = V_b / I_b. \tag{4.69}$$

The complementary geometry for these two antennas suggests that their performance is related. We could determine this relationship using Babinet's principle

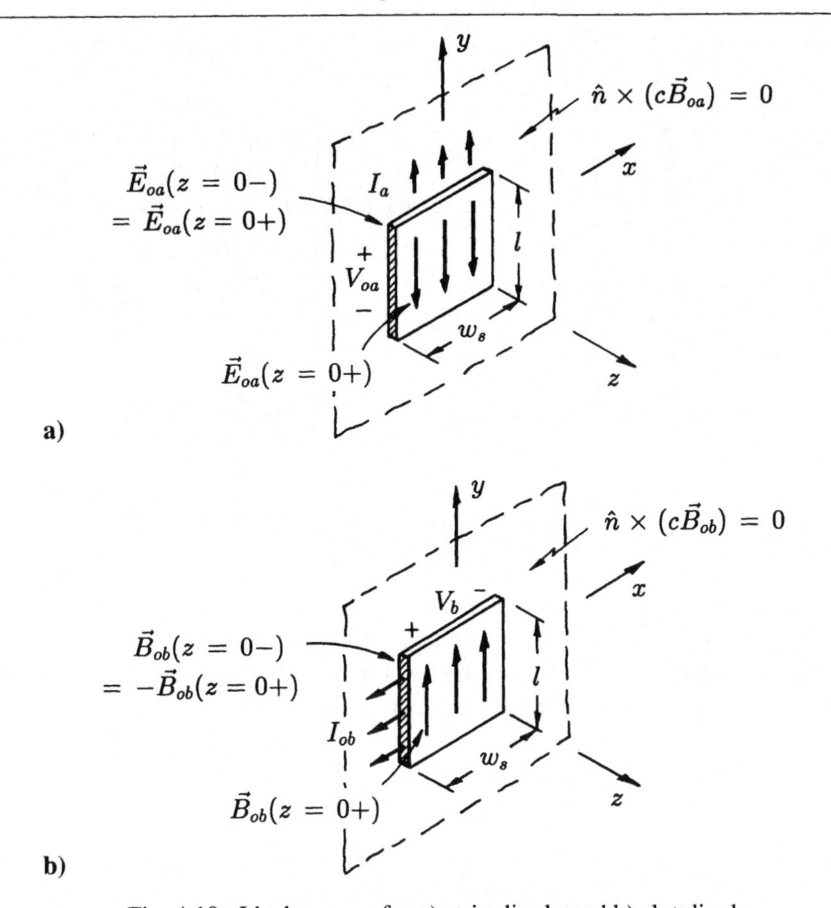

Fig. 4.19. Ideal sources for a) strip dipole and b) slot dipole.

directly; however, it is more instructive to use the basic symmetry of the electromagnetic field for this purpose, and we will take this approach. We begin by replacing the practical antennas in Figure 4.18 by ideal models in which the plane sheets are infinitesimally thin, perfect conductors and the sources, which excite the antennas, are the ideal ones shown in Figure 4.19.

The source for the strip dipole is an infinitesimally thin region of area $w_s \times l$ on the plane $z = 0$. It maintains the tangential electric field

$$\vec{E}_{oa} = E_{oa}\hat{y}, \qquad z = 0+,\, 0- \tag{4.70}$$

on its front and rear surfaces, and it produces no tangential magnetic field on the remainder of the plane $z = 0$ $[\hat{n} \times (c\vec{B}_{oa}) = 0]$. The source occupies an electrically small volume, so the voltage across its terminals is simply

$$V_{oa} = -\int_{y=-l/2}^{l/2} \vec{E}_{oa} \cdot \hat{y}dy = -E_{oa}l,$$

or

$$E_{oa} = -V_{oa}/l. \tag{4.71}$$

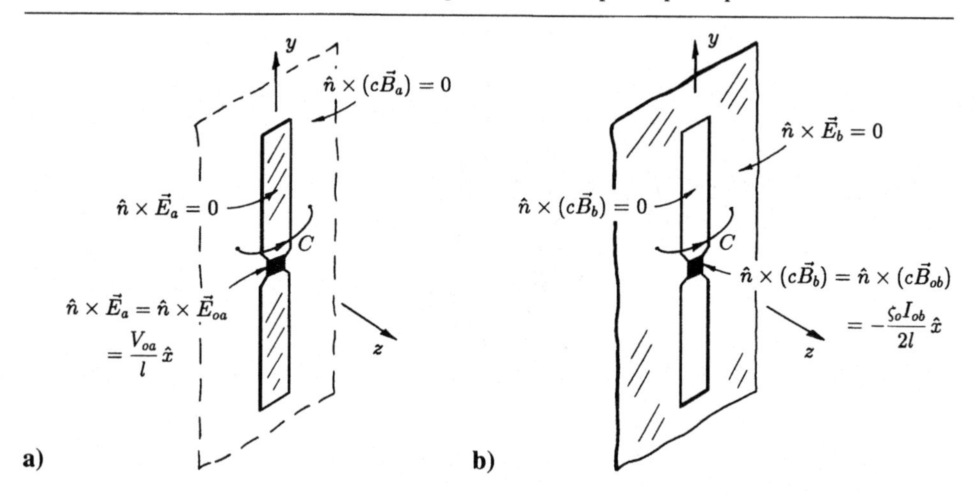

Fig. 4.20. Tangential components of the electromagnetic field on the plane $z = 0+$ for ideal models: a) strip dipole, b) slot dipole.

This source is equivalent to the ideal voltage generator of circuit theory; it maintains the voltage V_{oa} independent of the current I_a through the source.

The source for the slot dipole is of the same physical size as the source for the strip dipole. It produces the tangential magnetic field

$$c\vec{B}_{ob} = \begin{cases} (cB_{ob})\hat{y}, & z = 0+ \\ -(cB_{ob})\hat{y}, & z = 0- \end{cases} \tag{4.72}$$

on its front and rear surfaces and no tangential magnetic field on the remainder of the plane $z = 0$ $[\hat{n} \times (c\vec{B}_{ob}) = 0]$. The current through the electrically small source is simply

$$I_{ob} = \int_{y=-l/2}^{l/2} \vec{H}_{ob}(z = 0+) \cdot \hat{y}dy + \int_{y=l/2}^{-l/2} \vec{H}_{ob}(z = 0-) \cdot \hat{y}dy = \frac{2}{\zeta_o}(cB_{ob})l,$$

or

$$cB_{ob} = \zeta_o I_{ob}/2l. \tag{4.73}$$

This source is equivalent to the ideal current generator of circuit theory; it maintains the current I_{ob} independent of the voltage V_b across the source. Here we will not examine the physical construction for these ideal sources; this is discussed in some detail in Problem 4.10.

The ideal sources are shown with the corresponding antennas in Figure 4.20. The electromagnetic fields for both antennas ($\vec{E}_a$, $c\vec{B}_a$ and $\vec{E}_b$, $c\vec{B}_b$) must satisfy Maxwell's equations in the right half space ($z > 0$) and have the specified values on the plane $z = 0+$.[5] The latter are given in Figure 4.20.

[5] Here we are dealing with the total field as defined in Equation (4.34). Therefore, for both antennas, in the open areas of the plane $z = 0$

$$\hat{n} \times (c\vec{B}) = \hat{n} \times (c\vec{B}_o) + \hat{n} \times (c\vec{B}^s) = 0,$$

because $\hat{n} \times (c\vec{B}_o) = 0$ for either source and because $\hat{n} \times (c\vec{B}^s) = 0$ for the charge and current in the sheet.

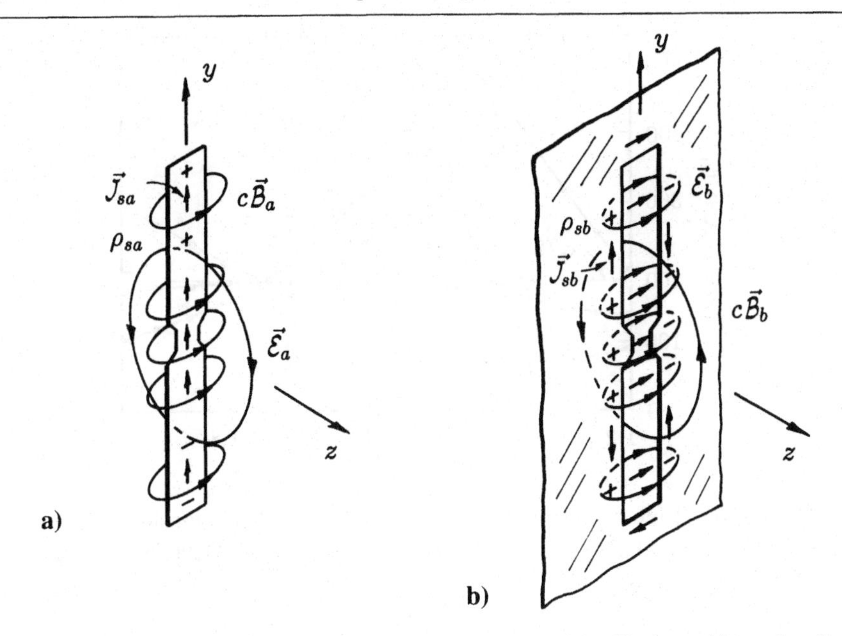

Fig. 4.21. Sketches of the electromagnetic field near a) a strip dipole and b) a slot dipole.

Now we notice that Maxwell's equations and the specified values for the field of problem (a) become those of problem (b) on the interchange

$$\vec{E}_b = c\vec{B}_a \tag{4.74a}$$

$$c\vec{B}_b = -\vec{E}_a. \tag{4.74b}$$

For the ideal sources this means that

$$c\vec{B}_{ob} = -\vec{E}_{oa}, \tag{4.75}$$

which on substituting (4.70)–(4.73) becomes

$$I_{ob} = 2V_{oa}/\zeta_o. \tag{4.76}$$

Thus, we see that the electromagnetic field of the slot dipole (b) can be obtained from that of the strip dipole (a) on the simple interchange of field components (4.74a, b). Of course, the two antennas must be excited by sources that satisfy (4.76).

The electromagnetic field in the vicinity of the strip and the slot are sketched in Figure 4.21. These sketches show the field at a single instant of time and were constructed using calculated and measured data [25, 26]. The magnetic field lines, $c\vec{\mathcal{B}}_a$, encircle the axial current, $\vec{\mathcal{J}}_{sa}$, in the strip, while the electric field lines, $\vec{\mathcal{E}}_a$, run from the positive charge, ρ_{sa}, on the upper arm to the negative charge on the lower arm. The sketch for the field lines of the slot is similar, only the electric and magnetic fields are interchanged from those for the strip. The electric field lines, $\vec{\mathcal{E}}_b$, run from the positive charge, ρ_{sb}, on the left-hand side of the slot to the negative charge on the right-hand side, while the magnetic field lines, $c\vec{\mathcal{B}}_b$, encircle the transverse electric field in the slot (actually, the displacement current $\partial\vec{\mathcal{D}}_b/\partial t = \varepsilon_o\partial\vec{\mathcal{E}}_b/\partial t$ in the slot).

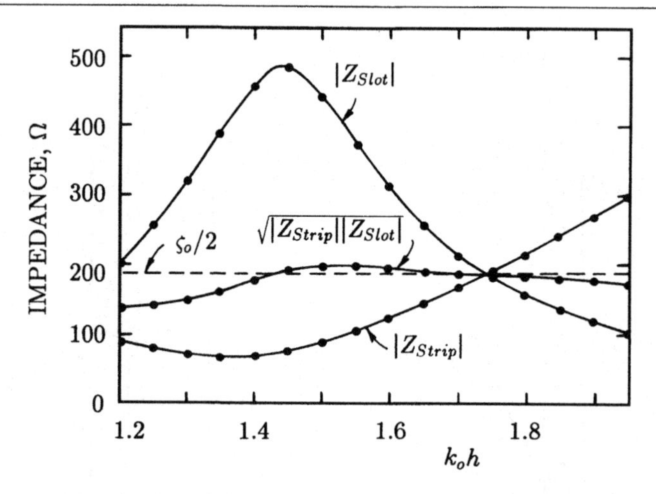

Fig. 4.22. Measured input impedances illustrating Booker's relationship. (Data for strip dipole (actually, a circular, cylindrical dipole) from Scott [28]. Data for slot dipole from Long [29].)

The input impedances of these two antennas obey a simple relationship, which is easily obtained from the results at hand. The input impedances of the strip dipole (4.68) and the slot dipole (4.69) are

$$Z_{\text{Strip}} = \frac{V_a}{I_a} = \frac{V_{oa}}{I_a} = \frac{V_{oa}}{2\int_C \vec{H}_a \cdot d\vec{l}} = \frac{\zeta_o V_{oa}}{2\int_C c\vec{B}_a \cdot d\vec{l}} \tag{4.77}$$

and

$$Z_{\text{Slot}} = \frac{V_b}{I_b} = \frac{V_b}{I_{ob}} = \frac{\int_C \vec{E}_b \cdot d\vec{l}}{I_{ob}}. \tag{4.78}$$

Here the current I_a and the voltage V_b are expressed as integrals over the contour C in the drive-point region (see Figure 4.20). On using (4.74a) and (4.76) with (4.78), we find that

$$Z_{\text{Slot}} = \frac{\zeta_o}{2V_{oa}} \int_C c\vec{B}_a \cdot d\vec{l} = \frac{\zeta_o^2}{4} \frac{1}{Z_{\text{Strip}}}$$

or

$$Z_{\text{Strip}} Z_{\text{Slot}} = \zeta_o^2/4. \tag{4.79}$$

This result, known as *Booker's relation*, shows that the input impedance of the slot dipole is simply related to that of the strip dipole [13, 27].

The validity of this relationship is illustrated by the measurements presented in Figure 4.22. These results are for antennas that are only approximately complementary: The dipole is formed from round wire, not a strip (diameter d, $d/h = 6.1 \times 10^{-2}$); and the slot is in a sheet that is not infinitesimally thin (thickness t, $t/w = 0.64$, $w/h = 8.0 \times 10^{-2}$) [28, 29]. The magnitudes of the impedances, $|Z_{\text{Strip}}|$ and $|Z_{\text{Slot}}|$, are shown as functions of the electrical length of the strip/slot, $k_o h = 2\pi h/\lambda_o$, for frequencies near the resonance of the antennas ($k_o h \approx \pi/2$).

The square root of their product, $\sqrt{|Z_{\text{Strip}}||Z_{\text{Slot}}|}$, is seen to be roughly equal to $\zeta_o/2$, in agreement with Booker's relation (4.79).

4.7 Images

The very word image brings to mind the familiar experience of observing one's reflection in a pool of water or in a mirror, and, no doubt, the origin of this word is in such observations. Our physical description of the optical image will make use of the sketch in Figure 4.23a. Here an object is placed in front of a planar mirror: a polished, metallic surface assumed to be a perfect reflector (perfect conductor). The light from the object in front of the mirror is reflected from the surface of the mirror and enters the eye of the observer. As the rays in the figure show, to the observer the light appears to come from a virtual image behind the mirror.

The apparent location of each point in the image is simply related to the corresponding point in the object; for example, the position vector $\vec{r}$ to point A in the object is

$$\vec{r} = x\hat{x} + y\hat{y} + z\hat{z}, \tag{4.80}$$

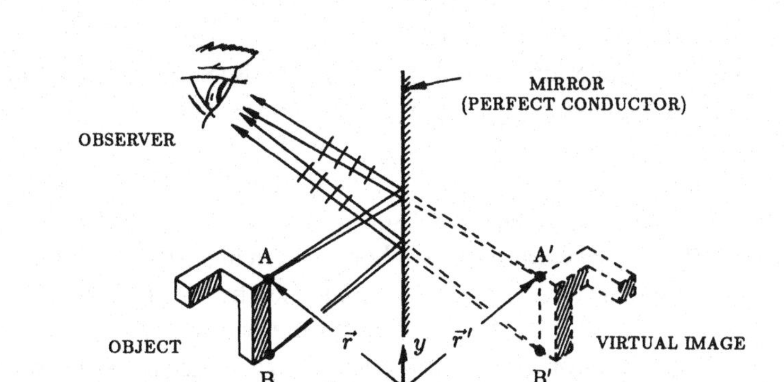

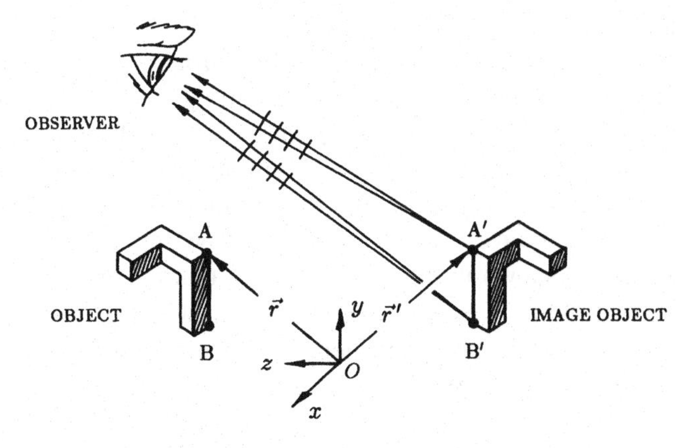

Fig. 4.23. a) Virtual image of object in planar mirror. b) Replacement of mirror by physical image object.

while the position vector $\vec{r}'$ to the corresponding point A′ in the virtual image is

$$\vec{r}' = x'\hat{x} + y'\hat{y} + z'\hat{z} = x\hat{x} + y\hat{y} - z\hat{z}. \tag{4.81}$$

This relationship between $\vec{r}$ and $\vec{r}'$ is the mathematical statement of a reflection in a plane, the plane $z = 0$ in this case. It is important to recognize that the image in Figure 4.23a cannot be obtained by simply rotating the object to the other side of the mirror (for example, by a rotation through the angle π about the y axis).

Our experience suggests that we could replace the mirror in Figure 4.23a with a real image object, one that is identical to the original object except that the positions are interchanged according to (4.81) as in Figure 4.23b, and the observer would not be able to tell the difference. The two arrangements would be equivalent as far as an observer located in the left half space ($z > 0$) is concerned. Since light is an electromagnetic field, we might expect this equivalence to apply, in some sense, to general electromagnetic fields, not just to light. Recall, however, that observations with light, like those discussed above, involve the intensity, which is the magnitude of the time average of the Poynting vector (3.130). Thus, they tell us little about the relationships between the electric and magnetic fields, charges and currents, etc. for the original object and the image object in Figure 4.23b.

From the uniqueness theorem for time-harmonic fields (Section 1.7) we know the conditions that must be satisfied for the two arrangements depicted in Figure 4.23 to have the same electromagnetic field in the left half space ($z > 0$): The impressed currents in the half space must be the same, and the tangential components of either $\vec{E}$ or $\vec{B}$ must be the same on the boundary surface of the half space.[6] The boundary surface for the half space is composed of a hemispherical surface of infinite radius and the plane $z = 0$. On the former, we will assume that the field is zero, based on the same argument as in Section 3.3. On the latter, for the arrangement in Figure 4.23a, the tangential components of the electric field are zero, because the mirror is a perfect conductor. For the arrangement in Figure 4.23b, the image object must be chosen to make the tangential components of the electric field zero on the plane $z = 0$; that is, the original object and the image object must produce electric fields whose tangential components cancel on the plane $z = 0$.

We can obtain the necessary electromagnetic properties of the image object using a simple heuristic argument based on the sketch in Figure 4.24. Here the original object and the image object are in free space and carry the impressed charges and currents ρ_i, $\vec{J}_i$ and ρ_i', $\vec{J}_i'$, respectively. From our discussion in Chapter 3, we know that the electromagnetic field of the original (image) object can be expressed as a spectrum of plane waves all propagating to the right (left). A few of the propagating waves are shown in the sketch. Notice that the vector wave numbers $\vec{k}$ and $\vec{k}'$ for corresponding waves from the original object and the image object are drawn as reflections of each other; their components satisfy (4.81). This is in keeping with the directions for the rays drawn in Figure 4.23.

[6] We have now assumed that the field has harmonic time dependence. For a situation like that in Figure 4.23, the source illuminating the object would be monochromatic.

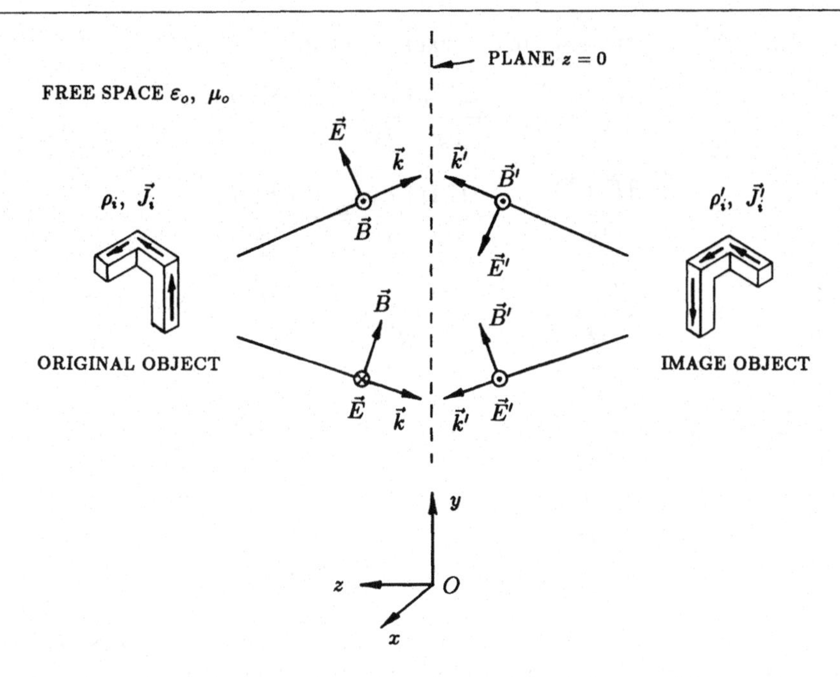

Fig. 4.24. Sketch showing plane waves from original object and image object.

Now we see that the tangential components of the electric field will be zero on the plane $z = 0$ if the electric fields for the plane waves satisfy the relationships

$$E_x'(\vec{r}') = -E_x(\vec{r}), \tag{4.82a}$$

$$E_y'(\vec{r}') = -E_y(\vec{r}), \tag{4.82b}$$

$$E_z'(\vec{r}') = E_z(\vec{r}). \tag{4.82c}$$

Because we are dealing with plane waves, this implies that the magnetic fields must satisfy the relationships

$$B_x'(\vec{r}') = B_x(\vec{r}), \tag{4.83a}$$

$$B_y'(\vec{r}') = B_y(\vec{r}), \tag{4.83b}$$

$$B_z'(\vec{r}') = -B_z(\vec{r}). \tag{4.83c}$$

If we assume that the above symmetry for the electromagnetic field extends into the regions of the sources, these results can be used with Maxwell's equations to determine the relationships for the impressed charge and current in the original and image objects. From Gauss' electric law, we have for the original object

$$\nabla \cdot \vec{E}(\vec{r}) = \frac{\partial}{\partial x}E_x(\vec{r}) + \frac{\partial}{\partial y}E_y(\vec{r}) + \frac{\partial}{\partial z}E_z(\vec{r}) = \frac{1}{\varepsilon_o}\rho_i(\vec{r}) \tag{4.84}$$

and for the image object

$$\nabla' \cdot \vec{E}'(\vec{r}') = \frac{\partial}{\partial x'}E_x'(\vec{r}') + \frac{\partial}{\partial y'}E_y'(\vec{r}') + \frac{\partial}{\partial z'}E_z'(\vec{r}') = \frac{1}{\varepsilon_o}\rho_i'(\vec{r}').$$

From (4.82a–c), the derivatives in this expression are

$$\frac{\partial E'_x(\vec{r}')}{\partial x'} = -\frac{\partial E_x(\vec{r})}{\partial x}\frac{\partial x}{\partial x'} = -\frac{\partial E_x(\vec{r})}{\partial x}$$

$$\frac{\partial E'_y(\vec{r}')}{\partial y'} = -\frac{\partial E_y(\vec{r})}{\partial y}\frac{\partial y}{\partial y'} = -\frac{\partial E_y(\vec{r})}{\partial y}$$

$$\frac{\partial E'_z(\vec{r}')}{\partial z'} = \frac{\partial E_z(\vec{r})}{\partial z}\frac{\partial z}{\partial z'} = -\frac{\partial E_z(\vec{r})}{\partial z},$$

so

$$\nabla' \cdot \vec{E}'(\vec{r}') = -\frac{\partial}{\partial x}E_x(\vec{r}) - \frac{\partial}{\partial y}E_y(\vec{r}) - \frac{\partial}{\partial z}E_z(\vec{r}) = \frac{1}{\varepsilon_o}\rho'_i(\vec{r}'). \tag{4.85}$$

A comparison of (4.85) with (4.84) shows that the volume densities of impressed charge must satisfy the relationship

$$\rho'_i(\vec{r}') = -\rho_i(\vec{r}). \tag{4.86}$$

In a similar manner, the relationships for the volume densities of impressed current can be obtained from the Ampère-Maxwell law (Problem 4.14):

$$J'_{ix}(\vec{r}') = -J_{ix}(\vec{r}) \tag{4.87a}$$

$$J'_{iy}(\vec{r}') = -J_{iy}(\vec{r}) \tag{4.87b}$$

$$J'_{iz}(\vec{r}') = J_{iz}(\vec{r}). \tag{4.87c}$$

For future reference, the results obtained above, which are often referred to collectively as the *method of images*, are summarized in Figure 4.25. Here we have generalized the argument to include perfect conductors, and regions of simple materials. The *original problem* consists of the impressed sources ρ_i, $\vec{J}_i$, perfect conductors and various regions of simple materials (ε, μ, σ) located in front of an infinite, perfectly conducting, plane surface. In the *image equivalent problem*, the perfectly conducting plane is replaced by the impressed sources ρ'_i, $\vec{J}'_i$ and perfect conductors and material regions that are the images of those in the original problem. All positions $\vec{r}'$ in the image objects are related to the corresponding positions in the original objects $\vec{r}$ by

$$\vec{r} = \vec{r}_t + \vec{r}_n, \tag{4.88a}$$

$$\vec{r}' = \vec{r}_t - \vec{r}_n, \tag{4.88b}$$

where the subscripts n and t refer to components normal to and tangential to the plane, respectively. The relationships between the impressed sources are

$$\rho'_i(\vec{r}') = -\rho_i(\vec{r}) \tag{4.89a}$$

and

$$\vec{J}'_i(\vec{r}') = -\vec{J}_{it}(\vec{r}) + \vec{J}_{in}(\vec{r}), \tag{4.89b}$$

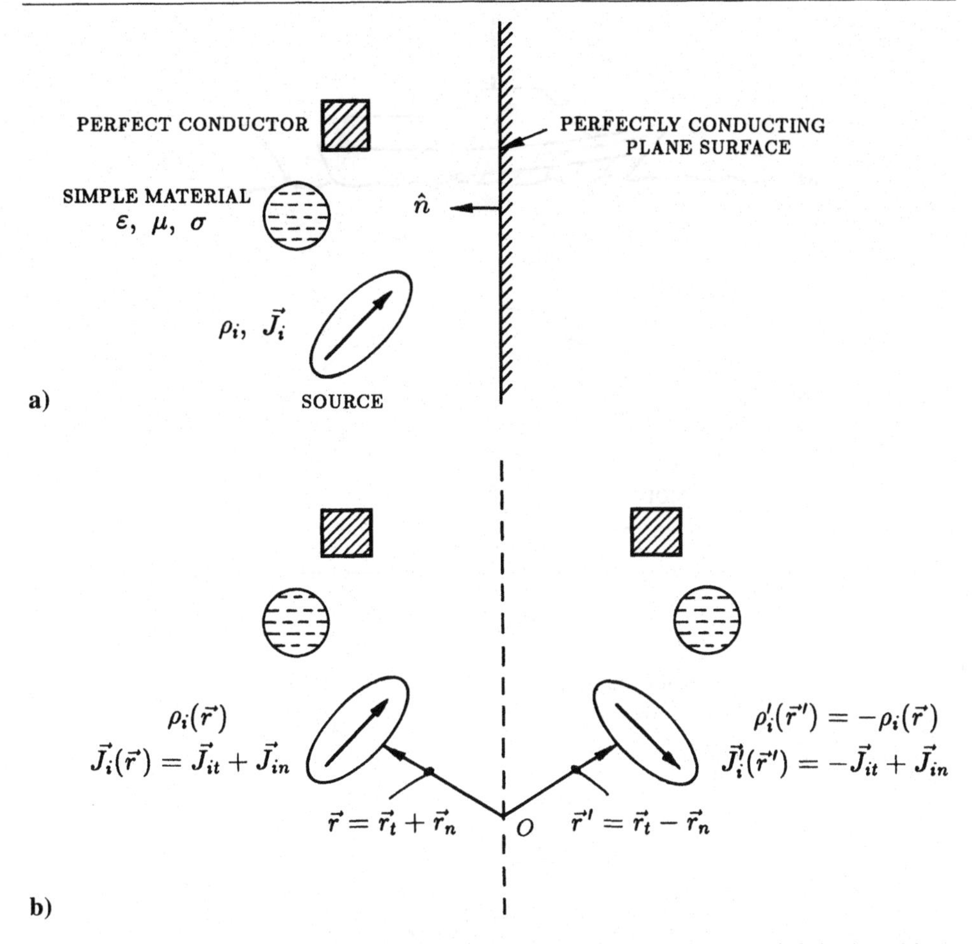

Fig. 4.25. Schematic drawing summarizing the method of images. a) Original problem. b) Image equivalent problem.

where

$$\vec{J}_i(\vec{r}) = \vec{J}_{it}(\vec{r}) + \vec{J}_{in}(\vec{r}). \tag{4.89c}$$

The solution to Maxwell's equations in the left half space ($z > 0$) is the same for these two problems. Sometimes the solution to the image equivalent problem is easier to obtain, hence, the utility of the method of images.

Figure 4.26 illustrates a particularly practical use of the method of images. Here the original problem is to determine the field radiated by an antenna on a ship on the sea. At the frequency of interest, the ship and the sea can both be approximated by perfect conductors. One way to determine a pattern for the radiated field would be to move a probe antenna over a large, spherical surface surrounding the ship, sampling the field at a number of points. Although this measurement is impractical for a real ship, it can be easily made using a scale model for the ship, as shown in Figure 4.26b. Notice that the sea has been replaced by a large, metallic plate in the model. The dimensions of the model ship are decreased by the scale factor $k_l < 1$

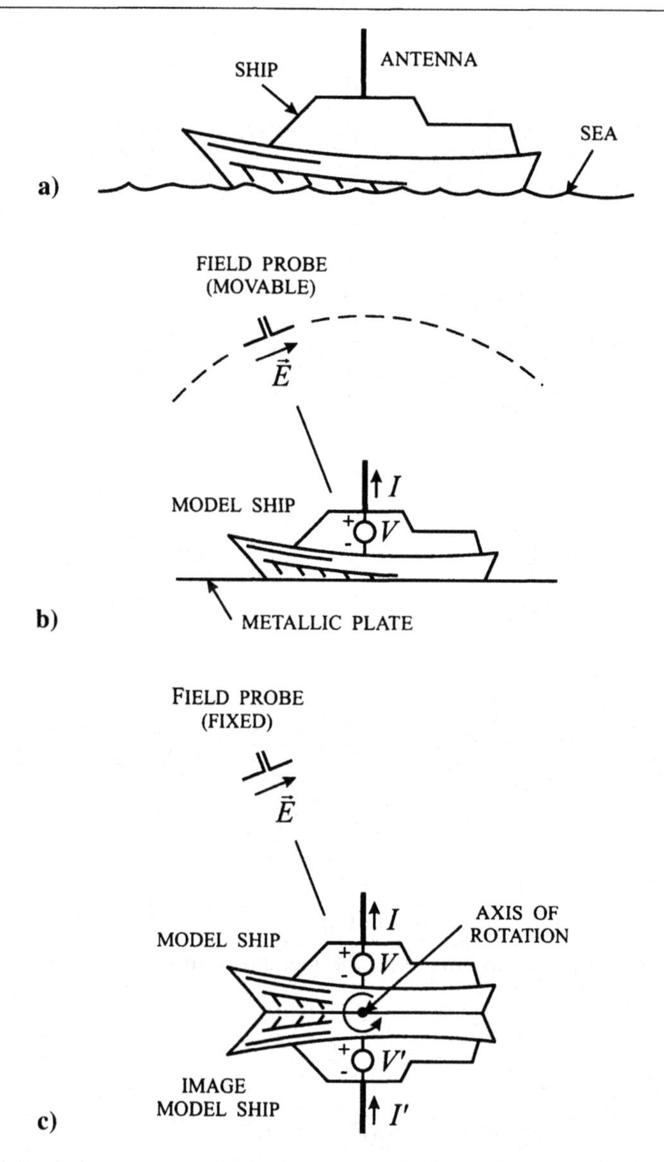

Fig. 4.26. a) Ship on the sea. b) Scale model of ship. c) Image equivalent scale model.

from those of the full-size ship, and the frequency used with the model is increased by the factor $1/k_l > 1$ over that used with the full-size system [30, 31]. The electric field measured with the model is simply related to the electric field for the full-size system (Problem 4.17). For a typical case, $k_l = 1/50$, so a 100-m-long ship with a communications system operating at 30 MHz would be modeled by a 2-m-long miniature ship operating at 1.5 GHz.

The measurement can be made even simpler by using the method of images (Figure 4.26c). The metallic image plane can be replaced by an image object for the model ship. Notice that the antennas on the model ship and image model ship

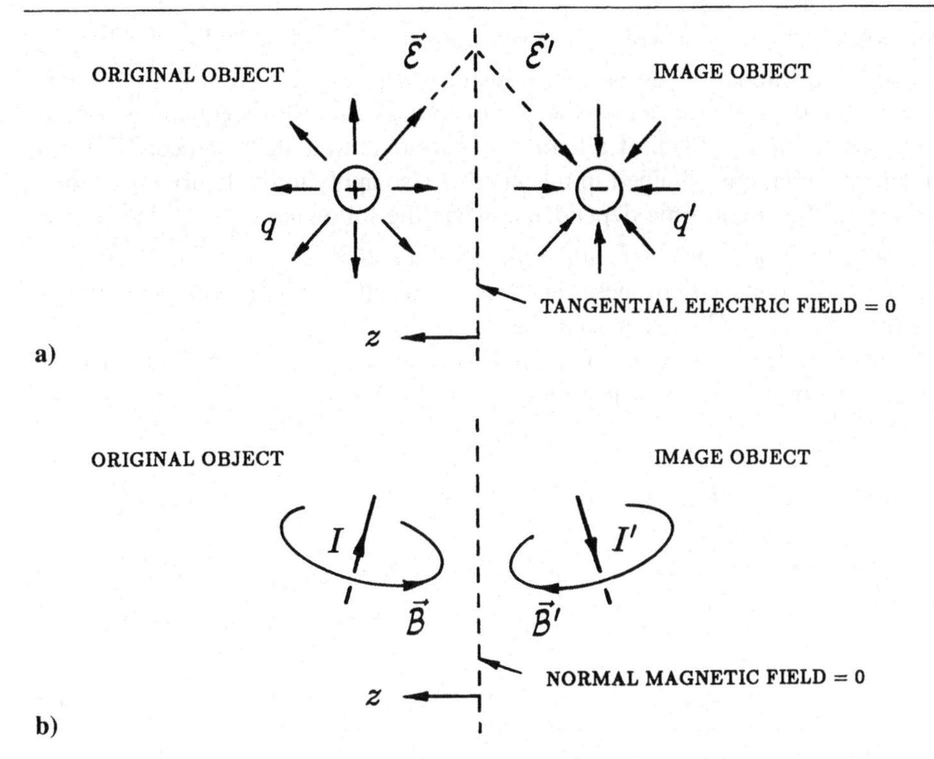

Fig. 4.27. Illustrations for the method of images. a) Electrostatic charges. b) Magnetostatic currents.

are fed so that the impressed currents I and I' satisfy (4.89b). The pattern for the radiation is determined by rotating the model ship plus image, with the position of the probe antenna fixed.

The method of images was apparently first used by William Thomson (Lord Kelvin) in 1847 [32, 33], and it was later presented in detail by Maxwell in his famous treatise [34]. Thomson and Maxwell were primarily concerned with static problems. For static problems, it is particularly easy to see from a few simple sketches (Figure 4.27) that the fields of the charges (electrostatic) and currents (magnetostatic) in the object and the image object add to produce the correct fields on the plane $z = 0$: The tangential components of the electric field are zero, and the normal component of the magnetic field is zero. Such sketches provide a convenient means for remembering the method of images.

4.8 General time dependence

To this point, the discussion in this chapter has been for the special case of fields with harmonic time dependence, and all of the results have been given in terms of the vector phasors $\vec{E}(\vec{r})$, $\vec{B}(\vec{r})$ for the field. However, the fundamental relations we have obtained, such as Huygens' principle, Babinet's principle, and the method of images, are applicable to fields with general time dependence. This is most easily seen through the use of the Fourier transform. Recall from Section 1.7 that Maxwell's equations for the Fourier transform, $\tilde{\vec{E}}(\vec{r},\omega)$, $\tilde{\vec{B}}(\vec{r},\omega)$, of a field with general time dependence are of the same form as Maxwell's equations for the vector phasors, $\vec{E}(\vec{r})$, $\vec{B}(\vec{r})$, of a field with harmonic time dependence. Thus, the fundamental relations obtained in this chapter also apply to the Fourier transform of a field with general time dependence when the following changes in notation are made: $\vec{E}(\vec{r})$ becomes $\tilde{\vec{E}}(\vec{r},\omega)$, $\vec{B}(\vec{r})$ becomes $\tilde{\vec{B}}(\vec{r},\omega)$, etc. After taking the inverse Fourier transform of these relations, the results for fields with general time dependence, $\vec{\mathcal{E}}(\vec{r},t)$, $\vec{\mathcal{B}}(\vec{r},t)$, are obtained.

We will illustrate this procedure for Huygens' principle. From Table 4.1 the Fourier transform of the electric field is

$$\begin{aligned}\tilde{\vec{E}}(\vec{r},\omega) &= -2jk_o\int_{-\infty}^{\infty}\int_{-\infty}^{\infty}\hat{R}\times\left[\hat{z}\times\tilde{\vec{E}}(\vec{r}\,',\omega)\right]\left(1-\frac{j}{k_oR}\right)G_o(\vec{r},\vec{r}\,')dx'dy'\\ &= \frac{-1}{2\pi}\int_{-\infty}^{\infty}\int_{-\infty}^{\infty}\frac{1}{R}\hat{R}\times\left\{\frac{j\omega}{c}\left[\hat{z}\times\tilde{\vec{E}}(\vec{r}\,',\omega)\right]e^{-j\omega R/c}\right.\\ &\quad\left.+\frac{1}{R}\left[\hat{z}\times\tilde{\vec{E}}(\vec{r}\,',\omega)\right]e^{-j\omega R/c}\right\}dx'dy'.\end{aligned}\tag{4.90}$$

If we take the inverse Fourier transform (1.200), we can obtain the result for general time dependence:

$$\begin{aligned}\vec{\mathcal{E}}(\vec{r},t) &= \frac{-1}{2\pi}\int_{-\infty}^{\infty}\int_{-\infty}^{\infty}\frac{1}{R}\hat{R}\\ &\quad\times\left(\frac{1}{c}\left\{\frac{1}{2\pi}\int_{\omega=-\infty}^{\infty}j\omega\left[\hat{z}\times\tilde{\vec{E}}(\vec{r}\,',\omega)\right]e^{-j\omega R/c}\,e^{j\omega t}d\omega\right\}\right.\\ &\quad\left.+\frac{1}{R}\left\{\frac{1}{2\pi}\int_{\omega=-\infty}^{\infty}\left[\hat{z}\times\tilde{\vec{E}}(\vec{r}\,',\omega)\right]e^{-j\omega R/c}\,e^{j\omega t}d\omega\right\}\right)dx'dy',\end{aligned}$$

or

$$\begin{aligned}\vec{\mathcal{E}}(\vec{r},t) &= \frac{-1}{2\pi}\int_{-\infty}^{\infty}\int_{-\infty}^{\infty}\left(\frac{1}{R^2}\hat{R}\times\left[\hat{z}\times\vec{\mathcal{E}}(\vec{r}\,',t')\right]_{t'=t-R/c}\right.\\ &\quad\left.+\frac{1}{cR}\hat{R}\times\left\{\frac{\partial}{\partial t'}\left[\hat{z}\times\vec{\mathcal{E}}(\vec{r}\,',t')\right]\right\}_{t'=t-R/c}\right)dx'dy'.\end{aligned}\tag{4.91}$$

Table 4.4. *Huygens' principle for fields: general time dependence*

$$\vec{\mathcal{E}}(\vec{r},t) = \frac{-1}{2\pi}\int_{-\infty}^{\infty}\int_{-\infty}^{\infty}\left(\frac{1}{R^2}\hat{R}[\hat{z}\times\vec{\mathcal{E}}(\vec{r}\,',t')]_{t'=t_r}\right.$$
$$\left. + \frac{1}{cR}\hat{R}\times\left\{\frac{\partial}{\partial t'}[\hat{z}\times\vec{\mathcal{E}}(\vec{r}\,',t')]\right\}_{t'=t_r}\right)dx'dy'$$

$$\vec{\mathcal{B}}(\vec{r},t) = \frac{1}{2\pi c}\int_{-\infty}^{\infty}\int_{-\infty}^{\infty}\left(\frac{c}{R^3}(1-3\hat{R}\hat{R}\cdot)\int_{t'=-\infty}^{t_r}\left[\hat{z}\times\vec{\mathcal{E}}(\vec{r}\,',t')\right]dt' + \frac{1}{R^2}(1-3\hat{R}\hat{R}\cdot)\right.$$
$$\left.[\hat{z}\times\vec{\mathcal{E}}(\vec{r}\,',t')]_{t'=t_r} + \frac{1}{cR}(1-\hat{R}\hat{R}\cdot)\left\{\frac{\partial}{\partial t'}[\hat{z}\times\vec{\mathcal{E}}(\vec{r}\,',t')]\right\}_{t'=t_r}\right)dx'dy'$$

$$\vec{\mathcal{E}}(\vec{r},t) = \frac{-c}{2\pi}\int_{-\infty}^{\infty}\int_{-\infty}^{\infty}\left(\frac{c}{R^3}(1-3\hat{R}\hat{R}\cdot)\int_{t'=-\infty}^{t_r}[\hat{z}\times\vec{\mathcal{B}}(\vec{r}\,',t')]dt' + \frac{1}{R^2}(1-3\hat{R}\hat{R}\cdot)\right.$$
$$\left.[\hat{z}\times\vec{\mathcal{B}}(\vec{r}\,',t')]_{t'=t_r} + \frac{1}{cR}(1-\hat{R}\hat{R}\cdot)\left\{\frac{\partial}{\partial t'}[\hat{z}\times\vec{\mathcal{B}}(\vec{r}\,',t')]\right\}_{t'=t_r}\right)dx'dy'$$

$$\vec{\mathcal{B}}(\vec{r},t) = \frac{-1}{2\pi}\int_{-\infty}^{\infty}\int_{-\infty}^{\infty}\left(\frac{1}{R^2}\hat{R}\times[\hat{z}\times\vec{\mathcal{B}}(\vec{r}\,',t')]_{t'=t_r} + \frac{1}{cR}\hat{R}\right.$$
$$\left.\times\left\{\frac{\partial}{\partial t'}[\hat{z}\times\vec{\mathcal{B}}(\vec{r}\,',t')]\right\}_{t'=t_r}\right)dx'dy'$$

$$t_r = t - R/c$$

The same operations applied to the magnetic field give (see Problem 4.19)

$$\vec{\mathcal{B}}(\vec{r},t) = \frac{1}{2\pi c}\int_{-\infty}^{\infty}\int_{-\infty}^{\infty}\left(\frac{c}{R^3}(1-3\hat{R}\hat{R}\cdot)\int_{t'=-\infty}^{t-R/c}[\hat{z}\times\vec{\mathcal{E}}(\vec{r}\,',t')]dt'\right.$$
$$+ \frac{1}{R^2}(1-3\hat{R}\hat{R}\cdot)[\hat{z}\times\vec{\mathcal{E}}(\vec{r}\,',t')]_{t'=t-R/c}$$
$$\left. + \frac{1}{cR}(1-\hat{R}\hat{R}\cdot)\left\{\frac{\partial}{\partial t'}[\hat{z}\times\vec{\mathcal{E}}(\vec{r}\,',t')]\right\}_{t'=t-R/c}\right)dx'dy'. \quad (4.92)$$

In Table 4.4 we summarize these results together with those for the case in which the tangential components of the magnetic field are specified on the plane $z = 0$. Notice that the terms in the integrands of these expressions are evaluated at the earlier time $t' = t - R/c$, which we will call the retarded time t_r. We will have much more to say about the retarded time in Chapters 5 and 6.

As an example, we will apply these results (4.91) to the case of a uniformly illuminated circular aperture. The geometry is the same one we have used before (Figure 3.21), only now the field in the aperture has general time dependence:

$$\vec{\mathcal{E}}_y(x, y, z = 0) = \begin{cases} \mathcal{E}_o^i(t)\hat{y}, & \rho = \sqrt{x^2 + y^2} < d/2 \\ 0, & \rho \geq d/2. \end{cases} \tag{4.93}$$

After substituting (4.93) into (4.91) and changing to circular cylindrical coordinates (ρ, ϕ, z) on the aperture plane, we have

$$\vec{\mathcal{E}}(\vec{r}, t) = \frac{-1}{2\pi} \int_{\phi=0}^{2\pi} \int_{\rho=0}^{d/2} \left\{ \frac{1}{R^2} \mathcal{E}_o^i(t - R/c) + \frac{1}{cR} \left[\frac{\partial \mathcal{E}_o^i(t')}{\partial t'} \right]_{t'=t-R/c} \right\} \hat{x} \times \hat{R} \; \rho d\rho d\phi. \tag{4.94}$$

We will confine our attention to points along the z axis; then (4.94) simplifies to become

$$\vec{\mathcal{E}}(z, t) = \hat{y}\, z \int_{\rho=0}^{d/2} \left\{ \frac{1}{(\rho^2 + z^2)^{3/2}} \mathcal{E}_o^i\left[t - (\rho^2 + z^2)^{1/2}/c\right] + \frac{1}{c(\rho^2 + z^2)} \left[\frac{\partial \mathcal{E}_o^i(t')}{\partial t'} \right]_{t'=t-(\rho^2+z^2)^{1/2}/c} \right\} \rho d\rho. \tag{4.95}$$

Notice that the integrand of (4.95) can be rewritten as

$$-\frac{\partial}{\partial \rho} \left\{ \frac{1}{(\rho^2 + z^2)^{1/2}} \mathcal{E}_o^i\left[t - (\rho^2 + z^2)^{1/2}/c\right] \right\};$$

therefore, the electric field on the z axis is

$$\vec{\mathcal{E}}(z, t) = \hat{y} \left\{ \frac{-z}{(\rho^2 + z^2)^{1/2}} \mathcal{E}_o^i\left[t - (\rho^2 + z^2)^{1/2}\right] \right\}_{\rho=0}^{d/2},$$

or

$$\vec{\mathcal{E}}(z, t) = \left(\mathcal{E}_o^i(t - z/c) - \frac{z}{[(d/2)^2 + z^2]^{1/2}} \mathcal{E}_o^i\left\{t - \left[(d/2)^2 + z^2\right]^{1/2}/c\right\} \right) \hat{y}. \tag{4.96}$$

At large distances from the aperture, (4.96) becomes the radiated field, $\vec{\mathcal{E}}^r(z, t)$. Taking the limit as $z/d \to \infty$, we have

$$\vec{\mathcal{E}}^r(z, t) \approx \mathcal{E}_o^i(t - z/c) - \frac{1}{1 + (d/z)^2/8} \mathcal{E}_o^i\left[t - z/c - (d/8c)(d/z)\right]\hat{y}$$

$$\approx \left[\mathcal{E}_o^i(t - z/c) - \mathcal{E}_o^i(t - z/c) + \frac{\partial \mathcal{E}_o^i(t - z/c)}{\partial t} (d/8c)(d/z) \right] \hat{y},$$

or

$$\vec{\mathcal{E}}^r(z,t) = \frac{d^2}{8cz}\left[\frac{\partial \mathcal{E}_o^i(t - z/c)}{\partial t}\right]\hat{y}. \tag{4.97}$$

To graphically illustrate these results, we will let the electric field within the aperture be a Gaussian pulse in time; that is, in (4.93) we will let

$$\mathcal{E}_o^i(t) = E_o e^{-t^2/\tau^2}. \tag{4.98}$$

A picture of this waveform is in Figure 1.36. Figure 4.28 shows the electric field as a function of time for three points along the z axis; the pulse width is chosen so that $d/c\tau = 32$. Close to the aperture ($z = d/2$ in Figure 4.28a), two distinct pulses, marked A and B, are present; these correspond to the two terms in (4.96) for the electric field. The first pulse, A, arrives at time $t = z/c$, the time it takes light to travel from the center of the aperture to the observation point. For a diffraction problem like that shown in Figure 3.21, this pulse represents the incident field unperturbed by the presence of the screen. The second pulse, B, arrives at time $t = \sqrt{(d/2)^2 + z^2}/c$ (at $(t - z/c)/\tau = 6.63$ in Figure 4.28a). This is the time it takes for light to travel from the rim of the aperture to the observation point. This pulse accounts for the presence of the screen. As the observation point is moved farther out along the z axis, the two pulses, A and B, approach each other, as in Figure 4.28b, which is for $z = d$. At large distances from the aperture, as in Figure 4.28c where $z = 10d$, the two pulses merge to form a waveform that is the derivative of the field in the aperture (derivative of the Gaussian pulse), as predicted by (4.97).

The explanation offered in this example gives us yet another way of viewing the field for a diffraction problem: as the superposition of two terms, the undisturbed incident field (the geometrical optics field) and a field originating at the edge of the screen (the diffracted field). This explanation was first used by Thomas Young in 1802 to describe optical diffraction [35, 36]. More recently it was incorporated into the geometrical theory of diffraction for electromagnetic waves [37].

4.9 Discussion

In Chapters 3 and 4, we have introduced a number of concepts, all of which are associated with obtaining and interpreting solutions to Maxwell's equations. In some cases the results are exact and in other cases they are approximate, and it is important to keep in mind which results belong in these two categories.

The plane-wave spectrum and Huygens' principle are methods for obtaining a solution to Maxwell's equations in the half space $z > 0$, given the tangential components of the electric or magnetic field on the plane $z = 0$. They provide a rigorous solution in the half space, and they are self-consistent in that they reproduce the prescribed field on the plane $z = 0$. However, these formulas do not directly provide a solution to diffraction problems, such as those shown in Figures 3.7 and 3.21. That is, they do not tell us what exact field is produced in the aperture and on the perfectly conducting screen by the incident plane wave. If the exact

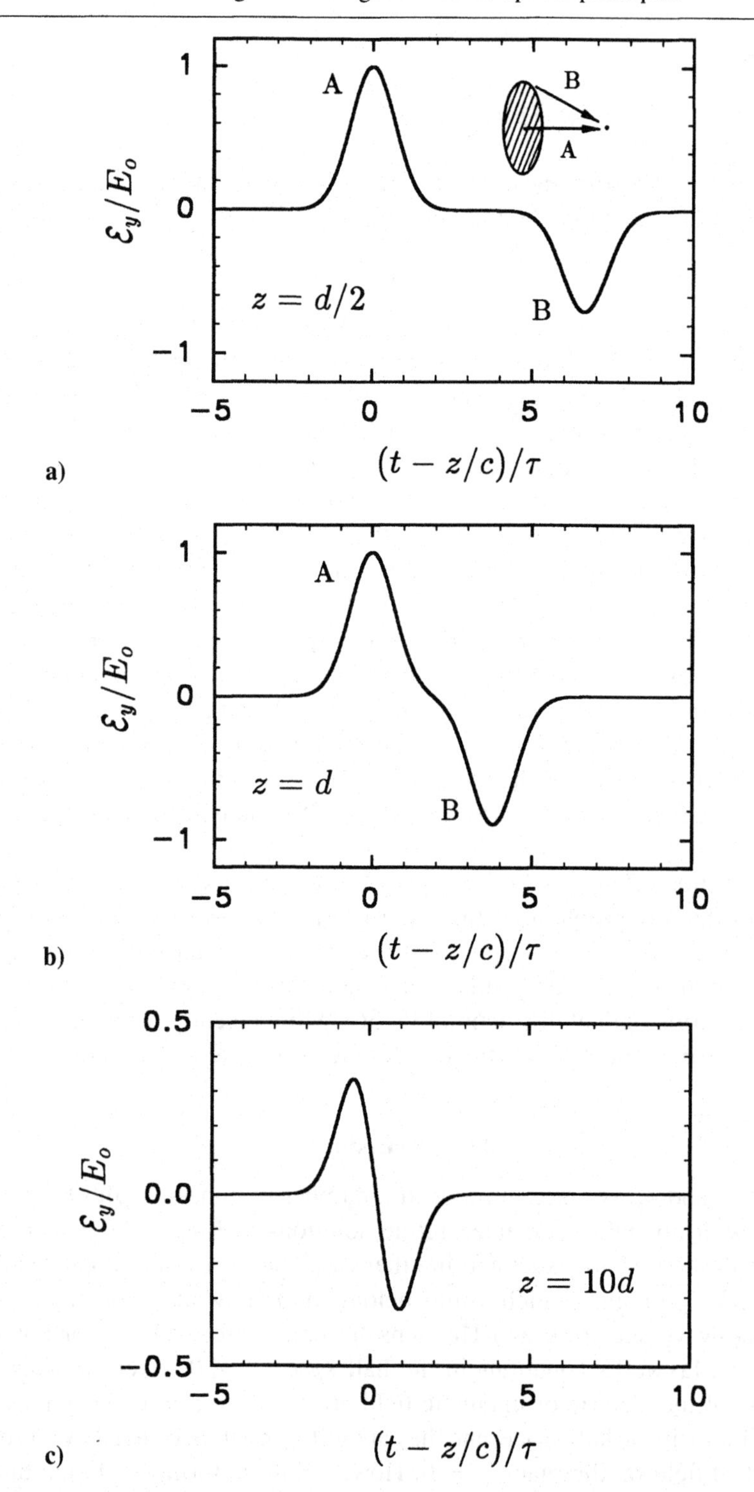

Fig. 4.28. Electric field as a function of time for various points along z axis. Uniformly illuminated circular aperture excited by a Gaussian pulse in time, $d/c\tau = 32.0$.

field is known in the aperture and on the screen, then these formulas can be used to determine the exact field in the half space. However, if an approximate field is specified in the aperture and on the screen, such as with Kirchhoff's approximation, then these formulas can provide only an approximate field in the half space.

Similar comments to those above apply to our other results: Babinet's principle, the transmission coefficients and scattering cross sections, the optical theorem, and Booker's relation. These are exact relationships; however, when they are applied to approximate results, they can only produce approximate answers.

References

[1] C. Huygens, *Traité de la Lumière*, Van der Aa, Leiden, 1690. English translation, S. P. Thompson, *Treatise on Light*, Dover, New York, 1962.

[2] T. Young, "On the Theory of Light and Colour," and "An Account of Some Cases of the Production of Colour, Not Hitherto Described," *Philos. Trans. Roy. Soc. London*, Vol. 92, pp. 12–48, 387–97, 1802.

[3] H. Weyl, "Ausbreitung elektromagnetischer Wellen über einem ebenen Leiter," *Ann. Physik*, Vol. 60, pp. 481–500, 1919.

[4] A. Baños, Jr., *Dipole Radiation in the Presence of a Conducting Half-Space*, Ch. 2, Pergamon Press, New York, 1966.

[5] *IEEE Standard Test Procedures for Antennas*, The Institute of Electrical and Electronic Engineers, Inc., Std. 149–1979, 1979.

[6] W. Franz, "Zur Formulierung des Huygensschen Prinzips," *Zeits. f. Naturforschung*, Vol. 3A, pp. 500–6, 1948.

[7] A. Sommerfeld, *Optics*, Academic Press, New York, 1954.

[8] C.-T. Tai, "Kirchhoff Theory: Scalar, Vector, or Dyadic," *IEEE Trans. Antennas Propagat.*, Vol. AP-20, pp. 114–5, January 1972.

[9] D. S. Jones, *The Theory of Electromagnetism*, Ch. 5, Pergamon Press, Oxford, 1964.

[10] A. Fresnel, "Sur la Diffraction de la lumière, oú l'on examine particulièrement le phénomène des franges colorées que présentent les ombres des corps éclairés par un point lumineux," *Annales de Chimie et de Physique*, Vol. 1, pp. 239–81, March 1830.

[11] J. Babinet, "Mémoires d'optique météorologique," *Comptes Rendus,* Vol. 4, pp. 638–48, May 3, 1837.

[12] M. Born and E. Wolf, *Principles of Optics,* 3rd Edition, Pergamon Press, New York, 1965.

[13] H. G. Booker, "Slot Aerials and their Relation to Complementary Wire Aerials (Babinet's Principle)," *J. IEE*, Vol. 93, Pt. III-A, pp. 620–6, 1946.

[14] B. B. Baker and E. T. Copson, *The Mathematical Theory of Huygens' Principle*, 2nd Edition, Oxford University Press, Oxford, 1950.

[15] M. J. Ehrlich, S. Silver, and G. Held, "Studies of the Diffraction of Electromagnetic Waves by Circular Apertures and Complementary Obstacles: The Near Field," *J. Appl. Phys.*, Vol. 26, pp. 336–45, March 1955.

[16] R. W. P. King and T. T. Wu, *The Scattering and Diffraction of Waves*, Harvard University Press, Cambridge, MA, 1959.

[17] J. J. Bowman, T. B. A. Senior, and P. L. E. Uslenghi, *Electromagnetic and Acoustic Scattering by Simple Shapes*, North-Holland, Amsterdam, 1969.

[18] P. Blacksmith, Jr., R. E. Hyatt, and R. B. Mack, "Introduction to Radar Cross-Section Measurements," *Proc. IEEE*, Vol. 53, pp. 901–20, August 1965.

[19] W. Andrejewski, "Die Beugung elektromagnetischer Wellen an der leitenden Kreisscheibe und an der kreisförmigen Öffnung im leitenden ebenen Schirm," *Z. Angew. Phys.* Vol. 5, pp. 178–86, May 1953.
[20] C. Huang, R. D. Kodis, and H. Levine, "Diffraction by Apertures," *J. Appl. Phys.*, Vol. 26, pp. 151–65, February 1955.
[21] J. S. Hey, G. S. Stewart, J. T. Pinson, and P. E. V. Prince, "The Scattering of Electromagnetic Waves by Conducting Spheres and Discs," *Proc. Phys. Soc.* (London), Vol. 69B, pp. 1,038–49, October 1956.
[22] D. B. Hodge, "Scattering by Circular Metallic Disks," *IEEE Trans. Antennas Propagat.*, Vol AP-28, pp. 707–11, September 1980.
[23] L. Brillouin, "The Scattering Cross Section of Spheres for Electromagnetic Waves," *J. Appl. Phys.*, Vol. 20, pp. 1,110–25, November, 1949.
[24] H. C. Van de Hulst, "On the Attenuation of Plane Waves by Obstacles of Arbitrary Size and Form," *Physica*, Vol. 15, pp. 740–6, September 1949.
[25] J. L. Putman, B. Russel, and W. Walkinshaw, "Field Distributions Near a Center-Fed Half-Wave Radiating Slot," *J. IEE*, Vol. 95, Pt. III, pp. 282–9, July 1948.
[26] R. King and G. H. Owyang, "An Experimental Study of the Slot Aerial and the Three-Element Collinear Array of Slot Aerials," *Proc. IEE*, Vol. 107, Pt. C, Monograph No. 365E, pp. 216–27, March 1960.
[27] V. H. Rumsey, *Frequency Independent Antennas*, Academic Press, New York, 1966.
[28] W. R. Scott, Jr., "Dielectric Spectroscopy Using Open-Circuited Coaxial Lines and Monopole Antennas of General Length," Ph.D. Dissertation, School of Elec. Engr., Georgia Inst. Technol., Atlanta, GA, 1985.
[29] S. A. Long, "Experimental Study of the Impedance of Cavity-Backed Slot Antennas," *IEEE Trans. Antennas Propagat.*, Vol. AP-23, pp. 1–7, January 1975.
[30] R. W. P. King and G.S. Smith, *Antennas in Matter*, Ch. 12, MIT Press, Cambridge, MA, 1981.
[31] K. F. Woodman and B. E. Stemp, "Shipborne Antenna Modeling,"*Microwave Journal*, Vol. 22, pp. 73–81, April 1979.
[32] W. Thomson, "On Electrical Images," *Notices and Abstracts of Communications Brit. Assoc. Adv. Sci.*, pp. 6–7, June 1847.
[33] —, *Reprints of Papers on Electrostatics and Magnetism*, Ch. 5, Macmillan, London, 1872.
[34] J. C. Maxwell, *A Treatise on Electricity and Magnetism*, 3rd Edition, Clarendon Press, Oxford, 1891. Republication, Dover Publications, New York, 1954.
[35] T. Young, "On the Theory of Light and Colors," *Philos. Trans. Roy. Soc. London*, Vol. 92, pp. 12–48, 1802.
[36] A. Rubinowicz, "Thomas Young and the Theory of Diffraction," *Nature*, Vol. 180, pp. 160–2, July 27, 1957.
[37] J. B. Keller, "Geometrical Theory of Diffraction," *J. Opt. Soc. Am.*, Vol. 52, pp. 116–30, February 1962.
[38] M. Sussman, "Elementary Diffraction Theory of Zone Plates," *Am. J. Phys.*, Vol. 28, pp. 394–8, April 1960.
[39] L. F. Van Buskirk and C. E. Hendrix, "The Zone Plate as a Radio-Frequency Focusing Element," *IRE Trans. Antennas Propagat.*, Vol. AP-9, pp. 319–20, May 1961.

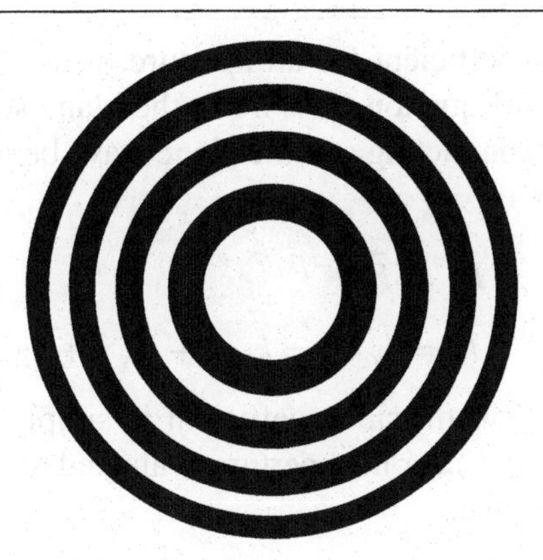

Fig. P4.1. Zone plate. Black rings block radiation, while white rings pass radiation.

Problems

4.1 Apply Huygens' principle to the uniformly illuminated, circular aperture and obtain the integral (4.23) for the radiated electric field.

4.2 *Zone plate* We used the concept of Fresnel zones to explain the behavior of the field on the axis of a uniformly illuminated circular aperture. Alternate zones were found to add destructively to the field. Thus, if the contribution from alternate zones could be blocked, the field on the axis could be increased. A zone plate is a device that is placed in the aperture to block the radiation from alternate zones. At optical frequencies, it is formed by simply placing an opaque material over the annular regions for alternate zones (Figure P4.1) [38]. At radio and microwave frequencies a thin metal conductor is used to cover alternate zones [39].

a) Consider a circular aperture of diameter $d/\lambda_o = 8.0$ containing a zone plate. Calculate the electric field $|E_y(z_m)/E_o^i|$ at the points z_m given in Table 4.3. Assume that at each position the zone plate blocks the radiation from the alternate zones $m = 2, 4, \ldots$.
b) A zone plate can also be constructed that blocks the radiation from the zones $m = 1, 3, \ldots$. The plate would be as in Figure P4.1, except the light and the dark regions would be interchanged. Calculate $|E_y(z_m)/E_o^i|$ for this plate at the same points as used in part a.
c) The zone plate produces a focusing effect similar to that of a lens. For this example, what is the maximum increase in the field that can be obtained using a zone plate [maximum value of $|E_y(z_m)/E_o^i|$]?

4.3 Use Babinet's principle to obtain Equations (4.51)–(4.53), which relate the scattering cross sections for a plane, perfectly conducting obstacle to the transmission coefficient and directivity for the complementary aperture in a plane, perfectly conducting screen.

4.4 Obtain the expressions in (4.56)–(4.58) for the transmission coefficient and scattering cross sections of the electrically large, circular aperture and disc.

4.5 The transmission coefficient for an aperture of area A in an opaque plane screen is defined in Equation (4.47). Let the plane screen be infinitesimally thin and perfectly conducting and the plane wave be incident in the direction $\hat{k}_i$, as in Figure 4.14:

$$\vec{E}_i = \vec{E}_o^{\,i} e^{-j\vec{k}_i\cdot\vec{r}}$$

$$\vec{k}_i = k_o\hat{k}_i = k_{ix}\hat{x} + k_{iy}\hat{y} + k_{iz}\hat{z}.$$

a) Show that the transmission coefficient is simply related to the spectral functions F_x and F_y for the aperture evaluated at the wave number of the incident plane wave:

$$\tau(\hat{k}_i) = \frac{-\mathrm{Re}[(\vec{E}_o^{\,i})^* \cdot (\hat{k}_i \times \{\hat{z} \times [F_x(k_{ix}, k_{iy})\hat{x} + F_y(k_{ix}, k_{iy})\hat{y}]\})]}{A|\vec{E}_o^{\,i}|^2}.$$

Hint: Make use of the fact that within the aperture, the component of the magnetic field tangential to the screen is exactly equal to that of the incident magnetic field.

b) Show that the result from part a, when expressed in terms of the pattern functions F_θ and F_ϕ, becomes

$$\tau(\hat{k}_i) = \frac{2\pi\,\mathrm{Im}\{(\vec{E}_o^{\,i})^* \cdot [F_\theta(\theta_i, \phi_i)\hat{\theta}_i + F_\phi(\theta_i, \phi_i)\hat{\phi}_i]\}}{k_o A|\vec{E}_o^{\,i}|^2},$$

or, when expressed in terms of the radiated electric field $\vec{E}^r$ in the direction $\hat{k}_i$,

$$\tau(\hat{k}_i) = \frac{\lambda_o}{A}\mathrm{Im}\left[\frac{(\vec{E}_o^{\,i})^*}{|\vec{E}_o^{\,i}|} \cdot \frac{\vec{E}^r(\hat{k}_i) r e^{j\vec{k}_i\cdot\vec{r}}}{|\vec{E}_o^{\,i}|}\right].$$

c) Using Babinet's principle and the results from part b, show that the total scattering cross section for a planar obstacle is given by (4.67).

4.6 Consider the results for the electrically large sphere ($k_o d = 20$) shown in Figure 4.17b. Assume that half of the scattered power is in the forward beam, and that the remainder is roughly distributed uniformly over all directions. Also assume that the total scattering cross sections for the electrically large sphere and disc are approximately the same.

a) What is the value of $4\sigma/\pi d^2$ associated with the uniform distribution? Compare your answer with the numerical results in Figure 4.17b for angles in the range $30° \lesssim \theta \leq 180°$.

b) What is the backscattering cross section when these approximations are assumed?

4.7 Show that only one of the two relations for the complementary incident fields (4.39) need be specified. The second relation follows from the fact that both incident fields (a and b) satisfy Maxwell's equations.

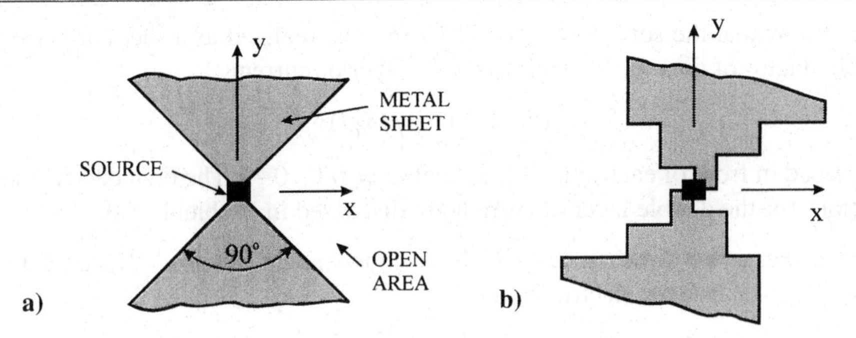

Fig. P4.2. Two self-complementary antennas.

4.8 *Self-complementary antennas* Antennas like those shown in Figure P4.2 have a special property: If the metal portions are rotated about the z axis through an angle of 90°, they coincide with the areas that were originally open. Such antennas are called self complementary [27].

a) Show that the input impedance of a self-complementary antenna is $Z = \zeta_o/2$.

b) In Figure P4.2 the metal portions of the antenna are assumed to be infinitely large (i.e., they extend out to an infinite radial distance in the x-y plane). What effect do you think there would be on the performance of the antennas if the metal plates were truncated at some finite radius, as must be done for practical antennas?

4.9 a) Apply Huygens' principle as given in (4.24) and (4.25), which is for the general volume V with the surface S, to the special case where V is the half space $z > 0$ and S is the plane $z = 0$. Compare these results with those derived directly for the half space (Table 4.1).

Show that the equation for the electric field $\vec{E}$ (4.24) is equal to the mean of the results for the electric field given in Table 4.1 (i.e., the mean of the results when $\vec{E}$ is specified on the plane $z = 0$ and the results when $\vec{B}$ is specified on the plane $z = 0$). Similarly, show that the equation for the magnetic field $\vec{B}$ (4.25) is the mean of the results for the magnetic field in Table 4.1.

b) To use the equations obtained in part a, one must know the tangential components of both $\vec{E}$ and $\vec{B}$ on the plane $z = 0$. Does this agree with the uniqueness theorem? What will be the consequences of using these equations as compared to directly using those in Table 4.1?

4.10 In this problem we will examine simple physical arrangements of current that produce the ideal sources shown in Figure 4.19. The source in Figure 4.19b can be realized as a uniform sheet of surface current

$$\vec{J}_{sb} = -(I_{ob}/l)\hat{x}$$

over the area $w_s \times l$. This produces the desired change in $\vec{B}_{ob}$ (4.72) on crossing the plane $z = 0$.

Show that the source in Figure 4.19a can be realized as a sheet of perfect conductor of area $w_s \times l$ with a double layer of current

$$\vec{C}_J = -(V_{oa}/j\omega_o l)\hat{y}$$

placed in front of each side of the sheet ($z = 0+,\ 0-$). The boundary conditions for the double layer of current are discussed in Problem 1.20.

4.11 Consider a two-dimensional field that is independent of the coordinate y. On the plane $z = 0$ the electric field is

$$\vec{E}(x, z = 0) = E_x(x, z = 0)\hat{x} + E_y(x, z = 0)\hat{y},$$

where the first term is associated with the TM field and the second term is associated with the TE field in the half space $z > 0$. Starting from Equation (4.8), show that Huygens' principle for the two-dimensional electric field is

$$\vec{E}(\vec{r}) = 2\nabla \times \int_{-\infty}^{\infty} \hat{z} \times \vec{E}(\vec{r}\,')\overline{G}_o(\vec{r}, \vec{r}\,')dx'.$$

Here it is understood that the vector $\vec{r}\,'$ has the components $y' = 0$ and $z' = 0$ and that the vector $\vec{r}$ has the component $y = 0$. The function $\overline{G}_o(\vec{r}, \vec{r}\,')$ is the *two-dimensional free-space, scalar Green's function for harmonic time dependence*, which can be expressed in terms of a Hankel function (Bessel function of the third kind) of order zero,

$$\overline{G}_o(\vec{r}, \vec{r}\,') = \frac{-j}{4} H_o^{(2)}(k_o|\vec{r} - \vec{r}\,'|).$$

The Hankel function has the following integral representation:

$$H_o^{(2)}(x) = \frac{j}{\pi} \int_{-\infty}^{\infty} \frac{\exp(-j\sqrt{x^2 + u^2})}{\sqrt{x^2 + u^2}} du, \qquad x > 0.$$

4.12 The radiated electromagnetic field for a strip dipole of resonant length ($k_o h = \pi/2$) is approximately (Section 8.5)

$$E^r_{\theta a} = \frac{j\zeta_o I_a}{2\pi} \frac{e^{-jk_o r}}{r} \frac{\cos[(\pi/2)\cos\theta]}{\sin\theta}$$

$$B^r_{\phi a} = \frac{1}{c} E^r_{\theta a}$$

when the dipole has the orientation shown in Figure P4.3a, where I_a is the total current at the drive point.

a) What is the radiated electromagnetic field of the complementary, resonant, slot dipole antenna that has the same orientation as the strip dipole in Figure P4.3a?
b) Consider the orthogonal arrangement of a strip monopole and a slot dipole antenna shown in Figure P4.3b. The current at the drive point of the strip monopole is I_a, and the voltage at the drive point of the slot dipole is $V_b = -\zeta_o I_a/2$. Make sketches of the electric field pattern in the upper

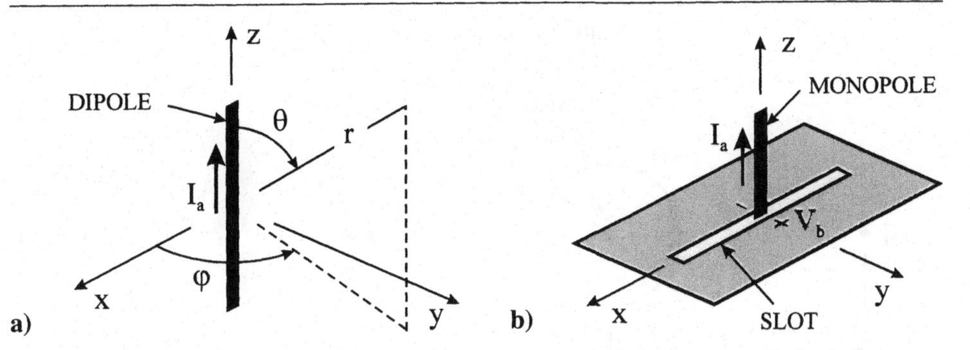

Fig. P4.3. a) Strip dipole. b) Strip monopole with orthogonal slot dipole.

half space ($z > 0$) for this combination of antennas (i.e., a graph of $|\vec{E}^r|$ versus angle for the x-y plane).

4.13 The zone plate complementary to the one shown in Figure P4.1 (Problem 4.2) has the open and opaque regions interchanged. If the number of rings is large enough so that the outer rings contribute little to the field at the focus, then it is not necessary to replace the last open region in the original plate, which extends to infinity in the x-y plane, by an opaque region in the complementary plate.

Let the plane waves normally incident on both plates (the original plate and the complementary plate) have the orientation shown in Figure 3.21. Using Babinet's principle, compare the electromagnetic fields for the zone plate and its complement for points near the focus. Assume that near the focus the magnitude of the total field is much greater than the magnitude of the incident field.

Calculate the time-average power per unit area on a plane of constant z near the focus, and compare the results for the two plates.

4.14 a) Obtain the relationships (4.87a–c) for the volume densities of impressed current in the original and image objects by starting from the Ampère-Maxwell law.

b) Show that the volume densities of impressed charge and current for the image object, (4.86) and (4.87a–c), satisfy the equation of continuity if the volume densities of impressed charge and current in the original object satisfy the equation of continuity.

4.15 In a particular direction, the radiation from an antenna is right-hand circularly polarized. This radiation is normally incident on a perfectly conducting, plane surface. Assume that the radiation can be approximated by a plane wave in the following calculations.

a) Use the method of images to determine the state of polarization for the wave after reflection from the plane surface.

b) Will the antenna that radiated the incident wave receive the reflected wave? (You may need to review the material in Section 2.6 before doing this part).

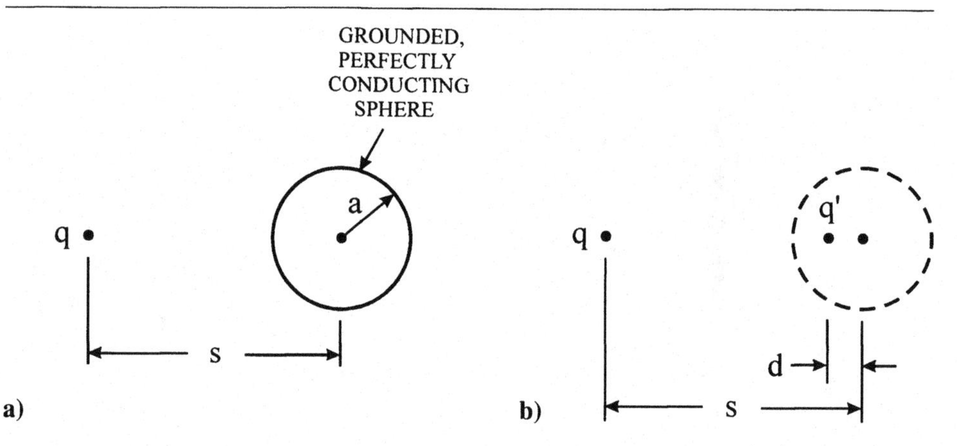

Fig. P4.4. The electrostatic problem of a point charge and a perfectly conducting sphere. a) Original problem. b) Equivalent problem.

4.16 In electrostatics the method of images can be applied to a few geometries in addition to the infinite, plane conductor. Figure P4.4 shows one such case. Here the original problem is a point charge q located a distance s from the center of a grounded, perfectly conducting sphere of radius a. The equivalent problem replaces the sphere by an image point charge q' located a distance d from the center of the sphere on the line through the center and q. The values of q' and d are chosen to make the tangential component of the electric field zero everywhere on the spherical surface of radius a (dashed line in Figure P4.4b). The electric fields for the two problems are then the same everywhere outside the spherical surface. Determine q' and d in terms of q, a, and s.

4.17 *Scale model* Consider a full-size system and scale model such as those shown in Figures 4.26a and b. The essential elements in both can be considered to be perfect conductors surrounded by free space. Let the electromagnetic field, angular frequency, and position vector in the full-size system be

$$\vec{E}_f, \quad \vec{H}_f, \quad \omega_f, \quad \vec{r}_f,$$

and let the corresponding quantities in the model be

$$\vec{E}_m, \quad \vec{H}_m, \quad \omega_m, \quad \vec{r}_m.$$

The quantities in the full-size system and the model are related by the scale factors k_E, k_H, k_ω, and k_l:

$$\vec{E}_m(\vec{r}_m) = k_E \vec{E}_f(\vec{r}_f), \qquad \vec{H}_m(\vec{r}_m) = k_H \vec{H}_f(\vec{r}_f),$$
$$\omega_m = k_\omega \omega_f, \qquad \vec{r}_m = k_l \vec{r}_f.$$

Assume that the field $\vec{E}_f$, $\vec{H}_f$ satisfies Maxwell's equations for free space. Show that the field for the model $\vec{E}_m$, $\vec{H}_m$ will also satisfy Maxwell's equations for free space if k_l and k_ω are chosen so that $k_\omega = 1/k_l$. Also show that this choice makes $k_E = k_H$. An electromagnetic measurement with the scale model can directly yield information about the full-size system.

4.18 Obtain the alternate expressions in Table 4.1 for $\vec{E}(\vec{r})$ and $\vec{B}(\vec{r})$ when $\vec{E}(\vec{r}')$ is given. Do this by moving the differential operators in the original expressions under the integral signs.

4.19 With $\vec{\mathcal{E}}(\vec{r}, t)$ specified on the plane $z = 0$, obtain the expression for $\vec{\mathcal{B}}(\vec{r}, t)$ given in Table 4.4. Use the Fourier transform with the results in Table 4.1.

4.20 Obtain the integral in (4.26) for the electric field on the axis of a uniformly illuminated circular aperture.

5

Radiation from distributions of charge and current: General formulation

In Chapters 3 and 4 we studied the problem sketched in Figure 5.1a, in which the electromagnetic field was determined in the half space $z > 0$ from the tangential components of the field specified on the plane surface $z = 0$. All of the sources for the field, the charge and the current, were assumed to be in the half space $z < 0$. Now we will examine this problem from a different point of view (Figure 5.1b): The situation in which the electromagnetic field is determined directly from the specified distribution of charge and current.

A summation procedure, which follows directly from the linearity of Maxwell's equations, will be used for this calculation. The electromagnetic field at the observation point P will be determined for an infinitesimal element of volume dV within the distribution (Figure 5.1b). The total field at the observation point will then be obtained by summing the contributions from all of the infinitesimal elements that form the distribution. The results obtained will have general application; the source may be as simple as a moving charged particle or as complex as the charge and current on a large antenna, such as the one in Figure 3.29a.

Recall that the radiated field from a time-harmonic source was found to fall off as $1/r$, where r is the radial distance from the source. This is to be compared with the field of an electrostatic source, which falls off at least as rapidly as $1/r^2$. The slower rate of decay for the time-harmonic field, and for time-varying fields in general, makes possible such important practical applications of electromagnetism as radio communication. In the remainder of the book, we will examine in detail the special characteristics of time-varying sources that give rise to radiated electromagnetic fields.

We will restrict our discussion to sources in free space with two exceptions: the description of Cherenkov radiation in Section 6.4 and the discussion of the related problem of radiation from the insulated linear antenna in Section 8.4.

5.1 Electromagnetic potentials

The determination of the electromagnetic field from Maxwell's equations can often be aided by the introduction of auxiliary functions – electromagnetic potentials. The use of potentials is usually first encountered in electrostatics and magnetostatics, and we will review these special cases first [1–4].

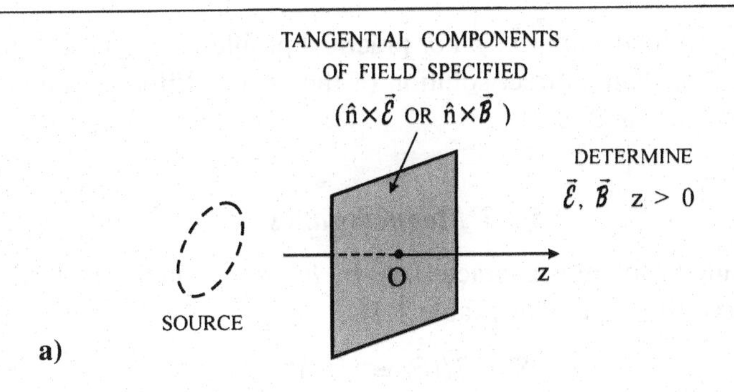

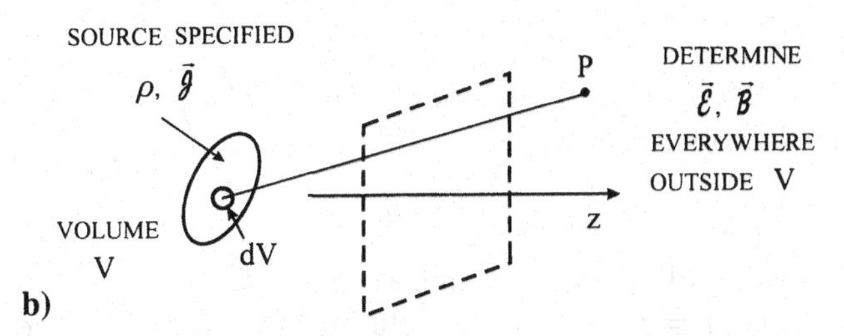

Fig. 5.1. a) Electromagnetic field in half space ($z > 0$) determined from tangential components of field on plane $z = 0$. b) Electromagnetic field determined directly from source (charge and current).

5.1.1 Electrostatics

For the electrostatic problem we have $\partial/\partial t = 0$ and $\vec{\mathcal{J}} = 0$; thus, Maxwell's equations (Table 1.3) reduce to

$$\nabla \times \vec{\mathcal{E}}(\vec{r}) = 0 \tag{5.1}$$

and

$$\nabla \cdot \vec{\mathcal{D}}(\vec{r}) = \rho(\vec{r}). \tag{5.2}$$

Since $\nabla \times (\nabla\Phi) = 0$ for any scalar function Φ (Appendix B), Equation (5.1) will be satisfied automatically if we let

$$\vec{\mathcal{E}} = -\nabla\Phi. \tag{5.3}$$

The function Φ is called the *electrostatic scalar potential*. If we assume the medium is free space, $\vec{\mathcal{D}} = \varepsilon_o\vec{\mathcal{E}}$, and substitute (5.3) into (5.2), we obtain Poisson's partial differential equation for Φ (Siméon Denis Poisson, 1781–1840):

$$-\varepsilon_o\nabla \cdot (\nabla\Phi) = \rho,$$

or

$$\nabla^2\Phi = -\rho/\varepsilon_o. \tag{5.4}$$

When the charge density ρ is specified, this scalar differential equation, together with appropriate boundary conditions, can be solved for the scalar potential and the electric field determined from (5.3). For practical problems this procedure is often easier to implement than a direct solution of the vector differential equations for the electric field, (5.1) and (5.2).

5.1.2 *Magnetostatics*

The magnetostatic problem is characterized by $\partial/\partial t = 0$ and $\rho = 0$, which make the relevant Maxwell's equations (Table 1.3)

$$\nabla \times \vec{\mathcal{H}}(\vec{r}) = \vec{\mathcal{J}}(\vec{r}) \tag{5.5}$$

and

$$\nabla \cdot \vec{\mathcal{B}}(\vec{r}) = 0. \tag{5.6}$$

Now $\nabla \cdot (\nabla \times \vec{\mathcal{A}}) = 0$ for any vector function $\vec{\mathcal{A}}$ (Appendix B); thus, Equation (5.6) will be satisfied automatically if we let

$$\vec{\mathcal{B}} = \nabla \times \vec{\mathcal{A}}. \tag{5.7}$$

The function $\vec{\mathcal{A}}$ is called the *magnetostatic vector potential.* After assuming the medium is free space, $\vec{\mathcal{H}} = \vec{\mathcal{B}}/\mu_o$, and substituting (5.7), (5.5) becomes

$$\nabla \times \nabla \times \vec{\mathcal{A}} = \mu_o \vec{\mathcal{J}},$$

or

$$\nabla^2 \vec{\mathcal{A}} - \nabla\nabla \cdot \vec{\mathcal{A}} = -\mu_o \vec{\mathcal{J}}, \tag{5.8}$$

where we used the vector relation $\nabla^2 \vec{\mathcal{A}} \equiv \nabla(\nabla \cdot \vec{\mathcal{A}}) - \nabla \times (\nabla \times \vec{\mathcal{A}})$ (Appendix B).

To this point, we have gained nothing by introducing the vector potential $\vec{\mathcal{A}}$; Equation (5.8) is not simpler than Equation (5.5); each is a set of three *coupled* partial differential equations. For example, in the rectangular Cartesian coordinate system, (5.8) is equivalent to the three equations

$$\frac{\partial^2 \mathcal{A}_u}{\partial v^2} + \frac{\partial^2 \mathcal{A}_u}{\partial w^2} - \frac{\partial}{\partial u}\left(\frac{\partial \mathcal{A}_v}{\partial v} + \frac{\partial \mathcal{A}_w}{\partial w}\right) = -\mu_o \mathcal{J}_u, \tag{5.9}$$

where u, v, w stands for a cyclic permutation of x, y, z.

A further simplification of Equation (5.8) can be accomplished once we recognize that the following transformation

$$\vec{\mathcal{A}} \rightarrow \vec{\mathcal{A}}' = \vec{\mathcal{A}} + \nabla\chi \tag{5.10}$$

leaves the magnetic field unchanged:

$$\begin{aligned}\vec{\mathcal{B}}' &= \nabla \times \vec{\mathcal{A}}' = \nabla \times \vec{\mathcal{A}} + \nabla \times (\nabla\chi) \\ &= \nabla \times \vec{\mathcal{A}} = \vec{\mathcal{B}}.\end{aligned} \tag{5.11}$$

Here χ is an arbitrary scalar function. A direct substitution of (5.10) into (5.8) shows that the differential equation for $\vec{\mathcal{A}}'$,

$$\nabla^2 \vec{\mathcal{A}}' - \nabla\nabla \cdot \vec{\mathcal{A}}' = -\mu_o \vec{\mathcal{J}}, \tag{5.12}$$

is the same as the equation for $\vec{\mathcal{A}}$. This is not surprising, because (5.8) is really an equation for $\vec{\mathcal{B}}$, and $\vec{\mathcal{A}}$ and $\vec{\mathcal{A}}'$ have the same relationship to $\vec{\mathcal{B}}$. Equation (5.8) can also be written as

$$\nabla^2\vec{\mathcal{A}}' - \nabla(\nabla^2\chi + \nabla\cdot\vec{\mathcal{A}}) = -\mu_o\vec{\mathcal{J}}, \tag{5.13}$$

where we have substituted (5.10) into only the term $\nabla^2\vec{\mathcal{A}}$ in (5.8). Now we are free to choose the arbitrary function χ. We will let χ be a solution to the equation

$$\nabla^2\chi = -\nabla\cdot\vec{\mathcal{A}}, \tag{5.14}$$

which makes (5.13)

$$\nabla^2\vec{\mathcal{A}}' = -\mu_o\vec{\mathcal{J}}. \tag{5.15}$$

From (5.10) we have

$$\nabla\cdot\vec{\mathcal{A}}' = \nabla\cdot\vec{\mathcal{A}} + \nabla^2\chi,$$

which on substituting (5.14) becomes

$$\nabla\cdot\vec{\mathcal{A}}' = 0. \tag{5.16}$$

Our new equation for $\vec{\mathcal{A}}'$ (5.15) is simpler than our original equation for $\vec{\mathcal{A}}$ (5.8). In the rectangular Cartesian coordinate system, it is equivalent to three *uncoupled* partial differential equations:

$$\frac{\partial^2\mathcal{A}'_u}{\partial u^2} + \frac{\partial^2\mathcal{A}'_u}{\partial v^2} + \frac{\partial^2\mathcal{A}'_u}{\partial w^2} = -\mu_o\mathcal{J}_u, \tag{5.17}$$

where u, v, w again stands for a cyclic permutation of x, y, z. Each rectangular component of the current density, $\mathcal{J}_u$, now produces only a component of the vector potential in the same direction, $\mathcal{A}'_u$.

Even after our restriction (5.16), there is still some flexibility in the choice for the vector potential. To see this, consider a second transformation from $\vec{\mathcal{A}}'$ to $\vec{\mathcal{A}}''$:

$$\vec{\mathcal{A}}' \to \vec{\mathcal{A}}'' = \vec{\mathcal{A}}' + \nabla\chi', \tag{5.18}$$

which again leaves $\vec{\mathcal{B}}$ unchanged. After the substitution of (5.18), (5.15) becomes

$$\nabla^2\vec{\mathcal{A}}'' - \nabla(\nabla^2\chi') = -\mu_o\vec{\mathcal{J}}. \tag{5.19}$$

If we let χ' be a solution to the differential equation

$$\nabla^2\chi' = 0, \tag{5.20}$$

then (5.19) becomes

$$\nabla^2\vec{\mathcal{A}}'' = -\mu_o\vec{\mathcal{J}}, \tag{5.21}$$

and from (5.18)

$$\nabla\cdot\vec{\mathcal{A}}'' = \nabla\cdot\vec{\mathcal{A}}' + \nabla^2\chi',$$

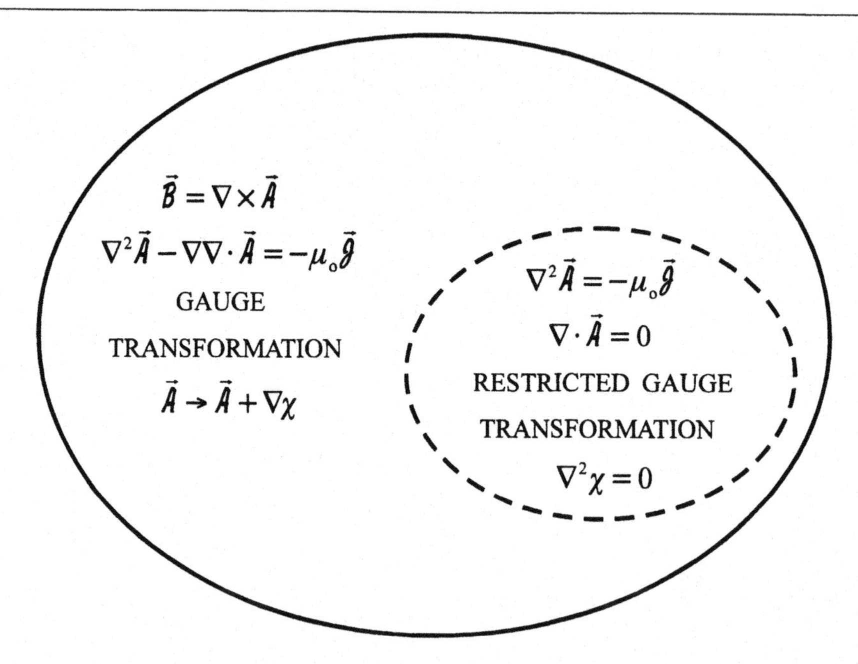

Fig. 5.2. Schematic drawing showing relationship between magnetostatic vector potentials ($\vec{\mathcal{A}}$) that determine the same magnetic field.

or

$$\nabla \cdot \vec{\mathcal{A}}'' = 0. \tag{5.22}$$

The new potential $\vec{\mathcal{A}}''$ satisfies the same equations as $\vec{\mathcal{A}}'$: (5.15) and (5.16).

Now we will summarize our findings with the help of Figure 5.2. We will simplify the notation by omitting the primes and double primes on $\vec{\mathcal{A}}$. With the current density $\vec{\mathcal{J}}$ specified, there exists a set of functions $\vec{\mathcal{A}}$, all of which satisfy the differential equation

$$\nabla^2 \vec{\mathcal{A}} - \nabla\nabla \cdot \vec{\mathcal{A}} = -\mu_o \vec{\mathcal{J}} \tag{5.23}$$

and give the same magnetic field

$$\vec{\mathcal{B}} = \nabla \times \vec{\mathcal{A}}. \tag{5.24}$$

Two different functions in this set are related by the *gauge transformation*

$$\vec{\mathcal{A}} \to \vec{\mathcal{A}} + \nabla\chi, \tag{5.25}$$

where χ is an arbitrary scalar function. A subset of this set has $\vec{\mathcal{A}}$ as a solution to the differential equation

$$\nabla^2 \vec{\mathcal{A}} = -\mu_o \vec{\mathcal{J}} \tag{5.26}$$

and satisfies

$$\nabla \cdot \vec{\mathcal{A}} = 0. \tag{5.27}$$

The functions within this subset are related by a *restricted gauge transformation*: (5.25) with χ restricted to be a solution to the differential equation[1]

$$\nabla^2 \chi = 0. \tag{5.28}$$

5.1.3 Electrodynamics

We will now return to the case of primary interest – the electromagnetic field with general time dependence. In Chapter 1 we argued that this field is determined by the three independent Maxwell's equations

$$\nabla \times \vec{\mathcal{E}}(\vec{r}, t) = -\frac{\partial \vec{\mathcal{B}}(\vec{r}, t)}{\partial t} \tag{5.29}$$

$$\frac{1}{\mu_o} \nabla \times \vec{\mathcal{B}}(\vec{r}, t) = \vec{\mathcal{J}}(\vec{r}, t) + \varepsilon_o \frac{\partial \vec{\mathcal{E}}(\vec{r}, t)}{\partial t} \tag{5.30}$$

$$\nabla \cdot \vec{\mathcal{J}}(\vec{r}, t) = -\frac{\partial \rho(\vec{r}, t)}{\partial t}, \tag{5.31}$$

which contain the two dependent Maxwell's equations

$$\varepsilon_o \nabla \cdot \vec{\mathcal{E}}(\vec{r}, t) = \rho(\vec{r}, t) \tag{5.32}$$

$$\nabla \cdot \vec{\mathcal{B}}(\vec{r}, t) = 0. \tag{5.33}$$

Note that the medium is assumed to be free space in these equations. Recall that we are interested in determining the electromagnetic field $\vec{\mathcal{E}}$, $\vec{\mathcal{B}}$ for a specified distribution of charge and current. Hence, we will assume that the specified densities ρ and $\vec{\mathcal{J}}$ satisfy the equation of continuity for electric charge (5.31) and determine the potential functions so that the two remaining independent equations, (5.29) and (5.30), are satisfied. Equation (5.33) and our discussion of the magnetostatic case suggest that we let

$$\vec{\mathcal{B}} = \nabla \times \vec{\mathcal{A}}, \tag{5.34}$$

where $\vec{\mathcal{A}}$ will now be called simply the *vector potential*. After inserting (5.34), the first of the independent equations (5.29) becomes

$$\nabla \times \vec{\mathcal{E}} = -\frac{\partial}{\partial t} \nabla \times \vec{\mathcal{A}},$$

or

$$\nabla \times \left(\vec{\mathcal{E}} + \frac{\partial \vec{\mathcal{A}}}{\partial t} \right) = 0. \tag{5.35}$$

[1] The condition $\nabla \cdot \vec{\mathcal{A}}$ can be motivated by another argument. From vector analysis we know that a vector function $\vec{\mathcal{A}}$ is uniquely determined within a region by $\nabla \times \vec{\mathcal{A}}$, $\nabla \cdot \vec{\mathcal{A}}$, and some additional information about the behavior of $\vec{\mathcal{A}}$ on the surface of the region [5]. The magnetic field (5.24) only fixes $\nabla \times \vec{\mathcal{A}}$; we are free to choose $\nabla \cdot \vec{\mathcal{A}}$. The choice $\nabla \cdot \vec{\mathcal{A}} = 0$ (5.27) eliminates the term $\nabla\nabla \cdot \vec{\mathcal{A}}$ from the differential equation (5.23).

Our discussion of the electrostatic case suggests that we let

$$\vec{\mathcal{E}} + \frac{\partial \vec{\mathcal{A}}}{\partial t} = -\nabla \Phi,$$

or

$$\vec{\mathcal{E}} = -\nabla \Phi - \frac{\partial \vec{\mathcal{A}}}{\partial t}, \tag{5.36}$$

where Φ will now be called simply the *scalar potential.*

With the potentials defined as in (5.34) and (5.36), the second of the independent equations (5.30) becomes

$$\nabla \times \nabla \times \vec{\mathcal{A}} = \mu_o \vec{\mathcal{J}} - \varepsilon_o \mu_o \frac{\partial}{\partial t}\left(\nabla \Phi + \frac{\partial \vec{\mathcal{A}}}{\partial t}\right),$$

or, on rearranging terms,

$$\nabla^2 \vec{\mathcal{A}} - \frac{1}{c^2}\frac{\partial^2 \vec{\mathcal{A}}}{\partial t^2} = -\mu_o \vec{\mathcal{J}} + \nabla\left(\nabla \cdot \vec{\mathcal{A}} + \frac{1}{c^2}\frac{\partial \Phi}{\partial t}\right). \tag{5.37}$$

An auxiliary equation is obtained by inserting (5.36) into (5.32):

$$\nabla^2 \Phi = -\frac{1}{\varepsilon_o}\rho - \frac{\partial}{\partial t}(\nabla \cdot \vec{\mathcal{A}}).$$

On adding $-(\partial^2 \Phi/\partial t^2)/c^2$ to both sides, this becomes

$$\nabla^2 \Phi - \frac{1}{c^2}\frac{\partial^2 \Phi}{\partial t^2} = -\frac{1}{\varepsilon_o}\rho - \frac{\partial}{\partial t}\left(\nabla \cdot \vec{\mathcal{A}} + \frac{1}{c^2}\frac{\partial \Phi}{\partial t}\right). \tag{5.38}$$

Before we proceed, let's pause and summarize what we have done to this point. A distribution of charge and current was specified such that ρ and $\vec{\mathcal{J}}$ satisfy the equation of continuity (5.31). The potentials $\vec{\mathcal{A}}$ and Φ were defined as in (5.34) and (5.36) so that the first of the independent Maxwell's equations (5.29) is automatically satisfied. For the remaining independent equation (5.30) to be satisfied, the potentials must be a solution to Equation (5.37). As in the magnetostatic case, this equation can be simplified by a suitable gauge transformation.

Consider the *gauge transformation*

$$\vec{\mathcal{A}} \to \vec{\mathcal{A}}' = \vec{\mathcal{A}} + \nabla \psi, \qquad \Phi \to \Phi' = \Phi - \frac{\partial \psi}{\partial t}, \tag{5.39}$$

which leaves the electromagnetic field unchanged:

$$\vec{\mathcal{E}}' = -\nabla \Phi' - \frac{\partial \vec{\mathcal{A}}'}{\partial t} = -\nabla \Phi + \nabla \frac{\partial \psi}{\partial t} - \frac{\partial \vec{\mathcal{A}}}{\partial t} - \frac{\partial}{\partial t}\nabla \psi$$

$$= -\nabla \Phi - \frac{\partial \vec{\mathcal{A}}}{\partial t} = \vec{\mathcal{E}},$$

$$\vec{\mathcal{B}}' = \nabla \times \vec{\mathcal{A}}' = \nabla \times \vec{\mathcal{A}} + \nabla \times \nabla \psi$$

$$= \nabla \times \vec{\mathcal{A}} = \vec{\mathcal{B}}.$$

Direct substitution of (5.39) into (5.37) and (5.38) shows that the new pair of potentials $\vec{\mathcal{A}}'$, Φ' satisfies the same differential equations as the original pair $\vec{\mathcal{A}}$, Φ:

$$\nabla^2\vec{\mathcal{A}}' - \frac{1}{c^2}\frac{\partial^2\vec{\mathcal{A}}'}{\partial t^2} = -\mu_o\vec{\mathcal{J}} + \nabla\left(\nabla\cdot\vec{\mathcal{A}}' + \frac{1}{c^2}\frac{\partial\Phi'}{\partial t}\right), \tag{5.40}$$

$$\nabla^2\Phi' - \frac{1}{c^2}\frac{\partial^2\Phi'}{\partial t^2} = -\frac{1}{\varepsilon_o}\rho - \frac{\partial}{\partial t}\left(\nabla\cdot\vec{\mathcal{A}}' + \frac{1}{c^2}\frac{\partial\Phi'}{\partial t}\right). \tag{5.41}$$

Equations (5.37) and (5.38) can also be written as

$$\nabla^2\vec{\mathcal{A}}' - \frac{1}{c^2}\frac{\partial^2\vec{\mathcal{A}}'}{\partial t^2} = -\mu_o\vec{\mathcal{J}} + \nabla\left(\nabla^2\psi - \frac{1}{c^2}\frac{\partial^2\psi}{\partial t^2} + \nabla\cdot\vec{\mathcal{A}} + \frac{1}{c^2}\frac{\partial\Phi}{\partial t}\right), \tag{5.42}$$

$$\nabla^2\Phi' - \frac{1}{c^2}\frac{\partial^2\Phi'}{\partial t^2} = -\frac{1}{\varepsilon_o}\rho - \frac{\partial}{\partial t}\left(\nabla^2\psi - \frac{1}{c^2}\frac{\partial^2\psi}{\partial t^2} + \nabla\cdot\vec{\mathcal{A}} + \frac{1}{c^2}\frac{\partial\Phi}{\partial t}\right). \tag{5.43}$$

Here we have substituted (5.39) into only the left-hand sides of (5.37) and (5.38).

Since the scalar function ψ is arbitrary, we can choose it so as to simplify Equations (5.42) and (5.43). If we choose ψ to be a solution to the differential equation

$$\nabla^2\psi - \frac{1}{c^2}\frac{\partial^2\psi}{\partial t^2} = -\nabla\cdot\vec{\mathcal{A}} - \frac{1}{c^2}\frac{\partial\Phi}{\partial t}, \tag{5.44}$$

then Equations (5.42) and (5.43) decouple to become wave equations for $\vec{\mathcal{A}}'$ and Φ':

$$\nabla^2\vec{\mathcal{A}}' - \frac{1}{c^2}\frac{\partial^2\vec{\mathcal{A}}'}{\partial t^2} = -\mu_o\vec{\mathcal{J}}, \tag{5.45}$$

$$\nabla^2\Phi' - \frac{1}{c^2}\frac{\partial^2\Phi'}{\partial t^2} = -\frac{1}{\varepsilon_o}\rho. \tag{5.46}$$

Substitution of (5.39) into (5.44) shows that the restriction on ψ is equivalent to choosing

$$\nabla\cdot\vec{\mathcal{A}}' + \frac{1}{c^2}\frac{\partial\Phi'}{\partial t} = 0, \tag{5.47}$$

which is known as the *Lorentz condition.*

The simplification gained by introducing the potential functions is now evident. The coupled partial differential equations for $\vec{\mathcal{E}}$ and $\vec{\mathcal{B}}$, the two Maxwell's equations (5.29) and (5.30), have been transformed into the two uncoupled partial differential equations for $\vec{\mathcal{A}}'$ and Φ', (5.45) and (5.46). In the rectangular Cartesian coordinate system, each component of the current density, $\mathcal{J}_u$, produces only a component of the vector potential in the same direction, $\mathcal{A}'_u$:

$$\frac{\partial^2\mathcal{A}'_u}{\partial u^2} + \frac{\partial^2\mathcal{A}'_u}{\partial v^2} + \frac{\partial^2\mathcal{A}'_u}{\partial w^2} - \frac{1}{c^2}\frac{\partial^2\mathcal{A}'_u}{\partial t^2} = -\mu_o\mathcal{J}_u, \tag{5.48}$$

where u, v, w stands for a cyclic permutation of x, y, z. The scalar potential Φ' and the three components of the vector potential $\vec{\mathcal{A}}'$ satisfy the same scalar partial differential equation – the scalar wave equation.

Further restrictions can still be placed on the potentials without affecting the electromagnetic field. To see this, consider a second gauge transformation from $\vec{\mathcal{A}}'$, Φ' to $\vec{\mathcal{A}}''$, Φ'':

$$\vec{\mathcal{A}}' \to \vec{\mathcal{A}}'' = \vec{\mathcal{A}}' + \nabla\psi', \qquad \Phi' \to \Phi'' = \Phi' - \frac{\partial \psi'}{\partial t}, \tag{5.49}$$

which again leaves the electromagnetic field unchanged. On substitution of (5.49), (5.45) and (5.46) become

$$\nabla^2 \vec{\mathcal{A}}'' - \frac{1}{c^2}\frac{\partial^2 \vec{\mathcal{A}}''}{\partial t^2} - \nabla\left(\nabla^2\psi' - \frac{1}{c^2}\frac{\partial^2 \psi'}{\partial t^2}\right) = -\mu_o \vec{\mathcal{J}}, \tag{5.50}$$

$$\nabla^2 \Phi'' - \frac{1}{c^2}\frac{\partial^2 \Phi''}{\partial t^2} + \frac{\partial}{\partial t}\left(\nabla^2\psi' - \frac{1}{c^2}\frac{\partial^2 \psi'}{\partial t^2}\right) = -\frac{1}{\varepsilon_o}\rho. \tag{5.51}$$

If we choose ψ' to be a solution to the differential equation

$$\nabla^2\psi' - \frac{1}{c^2}\frac{\partial^2 \psi'}{\partial t^2} = 0, \tag{5.52}$$

(5.50) and (5.51) reduce to

$$\nabla^2 \vec{\mathcal{A}}'' - \frac{1}{c^2}\frac{\partial^2 \vec{\mathcal{A}}''}{\partial t^2} = -\mu_o \vec{\mathcal{J}}, \tag{5.53}$$

$$\nabla^2 \Phi'' - \frac{1}{c^2}\frac{\partial^2 \Phi''}{\partial t^2} = -\frac{1}{\varepsilon_o}\rho, \tag{5.54}$$

and from (5.49)

$$\nabla \cdot \vec{\mathcal{A}}'' + \frac{1}{c^2}\frac{\partial \Phi''}{\partial t} = \nabla \cdot \vec{\mathcal{A}}' + \frac{1}{c^2}\frac{\partial \Phi'}{\partial t} + \nabla^2\psi' - \frac{1}{c^2}\frac{\partial^2 \psi'}{\partial t^2},$$

which on substitution of (5.47) and (5.52) becomes

$$\nabla \cdot \vec{\mathcal{A}}'' + \frac{1}{c^2}\frac{\partial \Phi''}{\partial t} = 0. \tag{5.55}$$

The new potential pair $\vec{\mathcal{A}}''$, Φ'' satisfies the same equations as $\vec{\mathcal{A}}'$, Φ': (5.45)–(5.47).

Let's summarize our findings. As for the magnetostatic case, we will use a schematic drawing (Figure 5.3) to help with this process, and we will omit the prime and double prime superscripts for clarity. A distribution of charge and current is specified such that ρ and $\vec{\mathcal{J}}$ satisfy the equation of continuity (5.31). There exists a set of potential function pairs $\vec{\mathcal{A}}$, Φ, all of which satisfy the differential equations

$$\nabla^2 \vec{\mathcal{A}} - \frac{1}{c^2}\frac{\partial^2 \vec{\mathcal{A}}}{\partial t^2} = -\mu_o \vec{\mathcal{J}} + \nabla\left(\nabla \cdot \vec{\mathcal{A}} + \frac{1}{c^2}\frac{\partial \Phi}{\partial t}\right) \tag{5.56}$$

and

$$\nabla^2 \Phi - \frac{1}{c^2}\frac{\partial^2 \Phi}{\partial t^2} = -\frac{1}{\varepsilon_o}\rho - \frac{\partial}{\partial t}\left(\nabla \cdot \vec{\mathcal{A}} + \frac{1}{c^2}\frac{\partial \Phi}{\partial t}\right) \tag{5.57}$$

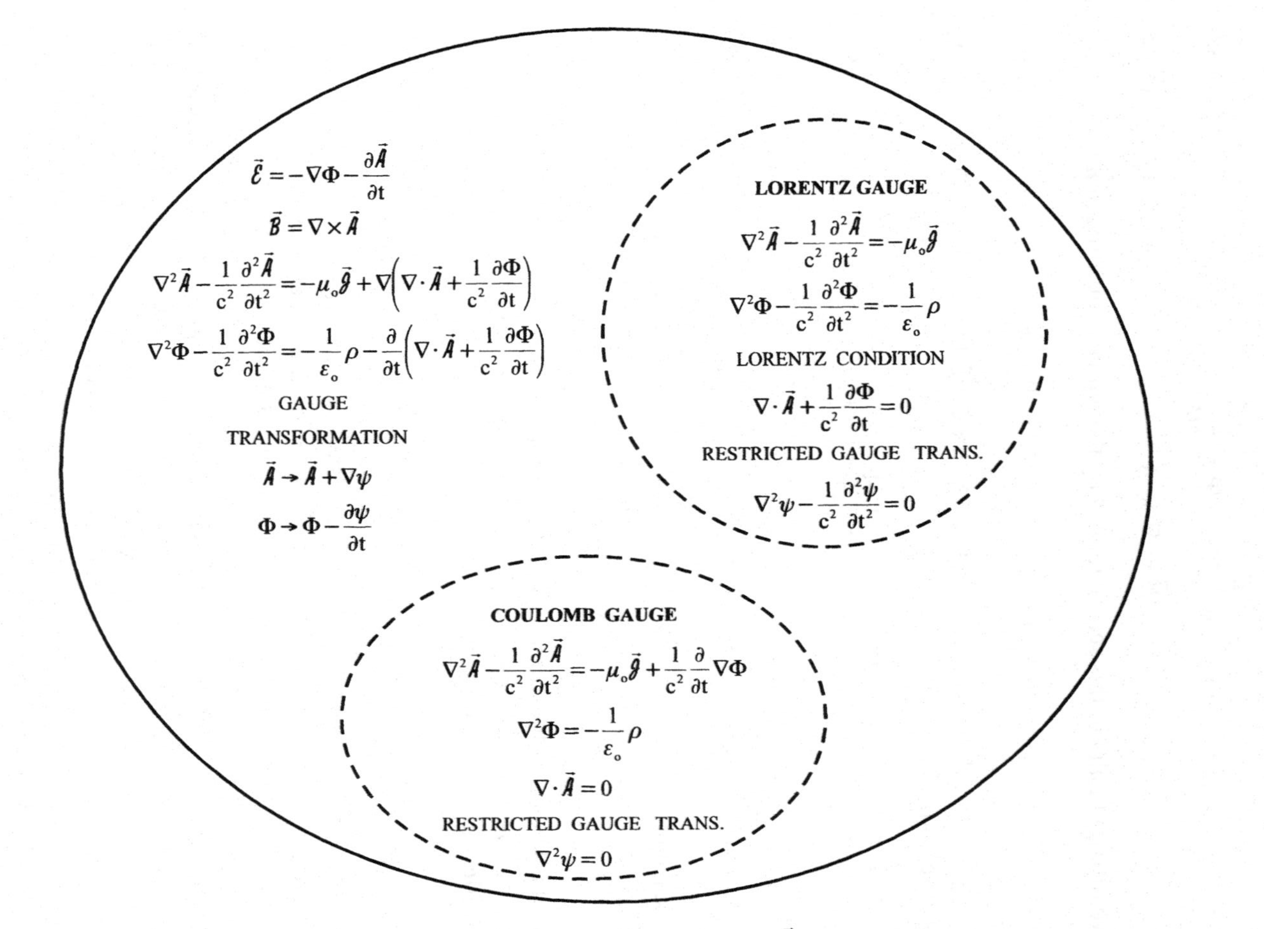

Fig. 5.3. Schematic drawing showing relationship between potential pairs $(\Phi, \vec{A})$ that determine the same electromagnetic field.

and give the same electromagnetic field

$$\vec{\mathcal{E}} = -\nabla\Phi - \frac{\partial \vec{\mathcal{A}}}{\partial t}, \tag{5.58}$$

$$\vec{\mathcal{B}} = \nabla \times \vec{\mathcal{A}}. \tag{5.59}$$

Two different pairs in this set are related by the *gauge transformation*

$$\vec{\mathcal{A}} \to \vec{\mathcal{A}} + \nabla\psi, \qquad \Phi \to \Phi - \frac{\partial \psi}{\partial t}, \tag{5.60}$$

where ψ is an arbitrary scalar function. A subset of this set has $\vec{\mathcal{A}}$, Φ solutions to the differential equations

$$\nabla^2\vec{\mathcal{A}} - \frac{1}{c^2}\frac{\partial^2\vec{\mathcal{A}}}{\partial t^2} = -\mu_o\vec{\mathcal{J}} \tag{5.61}$$

and

$$\nabla^2\Phi - \frac{1}{c^2}\frac{\partial^2\Phi}{\partial t^2} = -\frac{1}{\varepsilon_o}\rho \tag{5.62}$$

and satisfies

$$\nabla\cdot\vec{\mathcal{A}} + \frac{1}{c^2}\frac{\partial\Phi}{\partial t} = 0, \tag{5.63}$$

which is known as the *Lorentz condition*. Potential pairs within this subset are related by a *restricted gauge transformation* – Equation (5.60) with ψ a solution to the differential equation

$$\nabla^2\psi - \frac{1}{c^2}\frac{\partial^2\psi}{\partial t^2} = 0. \tag{5.64}$$

These potential pairs are said to belong to the *Lorentz gauge*.

We now have two options when we wish to determine the electromagnetic field of a specified distribution of charge and current that satisfies the equation of continuity (5.31): to solve Maxwell's equations (5.29) and (5.30) directly for $\vec{\mathcal{E}}$ and $\vec{\mathcal{B}}$ or, equivalently, to solve the differential equations (5.61) and (5.62) for $\vec{\mathcal{A}}$ and Φ and then determine $\vec{\mathcal{E}}$ and $\vec{\mathcal{B}}$ from (5.58) and (5.59). It is often easier to perform the latter.

In the discussion that follows, we will only make use of potentials within the Lorentz gauge. However, we should mention that other gauges can be introduced and that the potentials within these gauges may be of use in solving certain problems. As an example we mention the potential pairs $\vec{\mathcal{A}}, \phi$ within the *Coulomb gauge*, which satisfy the differential equations [4, 6]

$$\nabla^2\vec{\mathcal{A}} - \frac{1}{c^2}\frac{\partial^2\vec{\mathcal{A}}}{\partial t^2} = -\mu_o\vec{\mathcal{J}} + \frac{1}{c^2}\frac{\partial}{\partial t}\nabla\Phi, \tag{5.65}$$

$$\nabla^2\Phi = -\frac{1}{\varepsilon_o}\rho, \tag{5.66}$$

$$\nabla\cdot\vec{\mathcal{A}} = 0. \tag{5.67}$$

The restricted gauge transformation for the Coulomb gauge is (5.60) with ψ a solution to the differential equation (Problem 5.3)

$$\nabla^2 \psi = 0. \tag{5.68}$$

Note that (5.67) now plays the same role as the Lorentz condition (5.63). The potentials for the Coulomb gauge are also shown in the schematic drawing of Figure 5.3.

We have introduced the potential functions $\vec{\mathcal{A}}$, Φ with one purpose in mind: to simplify the solution of Maxwell's equations for the electromagnetic field $\vec{\mathcal{E}}$, $\vec{\mathcal{B}}$. We can attach physical significance to $\vec{\mathcal{E}}$ and $\vec{\mathcal{B}}$ because they are directly connected to the observable force on charge and current through the Lorentz force expression (1.12). The following question is often asked: Is there any physical significance to the potential functions?[2] For classical electrodynamics the answer is probably no; however, for quantum mechanics the answer is probably yes.

In quantum mechanics a particle such as an electron is described by a wave function. When an electron passes through a region of space where there is a vector potential $\vec{\mathcal{A}}$ but no electromagnetic field ($\vec{\mathcal{E}} = 0, \vec{\mathcal{B}} = 0$), the wave function changes. This change has been experimentally observed using the interference pattern produced by two overlapping electron beams. A similar effect occurs when there is a scalar potential Φ in the field-free region. Thus, in quantum mechanics the vector and scalar potentials are said to have a physical significance that they lack in classical electrodynamics. This interesting result is known as the *Aharonov-Bohm effect* [8–10].

5.2 Dirac delta function – concept of a point charge

In physical situations we sometimes encounter phenomena that are very localized in one or more spatial dimensions or in time. They extend over distances or times that are very small compared to the resolving power of our measuring instrumentation. For practical purposes then, these phenomena exist only at a point, on a line, or on a plane in three-dimensional space, or at a point on the time axis. The *Dirac delta function* is a device that mathematically represents such localized phenomena.

We have already encountered an example. In Section 1.3 we recognized that the volume density of charge ρ may be localized at a surface boundary, extending a very small distance from the boundary into either material. In this case, the charge essentially exists only on the surface boundary and can be represented by a surface density of charge ρ_s. When the surface boundary is the plane $z = 0$, this localization of the charge is described by the one-dimensional delta function $\delta(z)$ (1.56):

$$\rho(x, y, z) = \rho_s(x, y)\delta(z). \tag{5.69}$$

In this chapter, we will make frequent use of the delta function in one or more dimensions; thus, at the outset it is worth summarizing the properties of this important

[2] Heaviside's opinion of the potential functions, which is often shared by students, was stated in the preface to his book *Electromagnetic Theory*, "... the potential functions ... are such powerful aids to obscuring and complicating the subject, and hiding from view useful and sometimes important relations" [7].

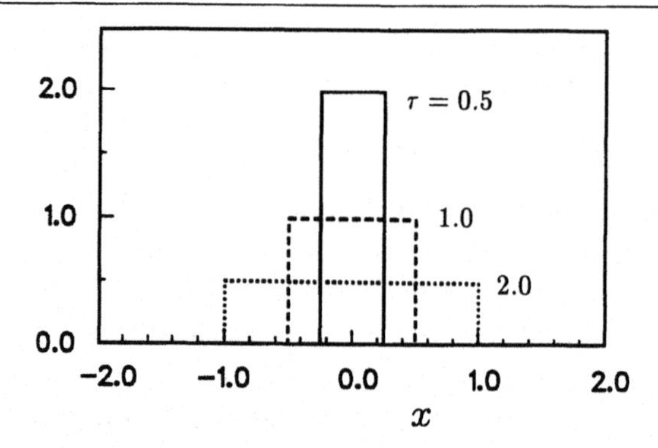

Fig. 5.4. Members of the sequence of rectangular functions representing the delta function, $\delta(x)$.

device. Our approach will be simple, emphasizing graphical representations; a more precise mathematical approach is contained in the theory of generalized functions or distributions [11–14].

We will begin with a review of the properties for the one-dimensional delta function.

5.2.1 *One-dimensional delta function*

The one-dimensional delta function or impulse function $\delta(x)$ has the following properties:

$$\delta(x) = 0, \qquad x \neq 0 \tag{5.70}$$

$$\int_{-\infty}^{\infty} \delta(x)dx = 1. \tag{5.71}$$

The delta function can be viewed as a limit of a sequence of ordinary functions; for example, we can take

$$\delta(x) = \lim_{\tau \to 0} \left[\frac{1}{\tau} \Pi(x/\tau) \right], \tag{5.72}$$

where Π is the *rectangular function*

$$\Pi(\xi) = \begin{cases} 1, & |\xi| < \frac{1}{2} \\ 0, & |\xi| > \frac{1}{2}. \end{cases} \tag{5.73}$$

Each member of this sequence is a rectangular pulse of width τ and height $1/\tau$, as in Figure 5.4.[3] As $\tau \to 0$, the height of the pulse becomes infinite, and the width of the pulse goes to zero, so $\delta(x) = 0$, $x \neq 0$ in agreement with (5.70). The area

[3] Notice that the one-dimensional delta function has the same dimensions as $1/\tau$. When the dimensions for x and τ are length L, the dimensions for the delta function are reciprocal length L^{-1}. The units in an equation such as (5.69) are then consistent: ρ with the SI units C/m^3, ρ_s with the units C/m^2, and δ with the units 1/m.

under the pulse remains constant in the limit,[4]

$$\int_{-\infty}^{\infty} \delta(x)dx = \lim_{\tau\to 0}\left(\frac{1}{\tau}\int_{-\tau/2}^{\tau/2} dx\right) = 1, \tag{5.74}$$

in agreement with (5.71). The delta function is clearly not an ordinary function; it is infinite at $x = 0$.

The sequence (5.72) for the delta function is not unique; many other sequences exist that have the properties (5.70) and (5.71). For example, in some instances it is desirable to have a sequence of smooth functions representing $\delta(x)$, such as the sequence of Gaussian functions

$$\delta(x) = \lim_{\tau\to 0}\left(\frac{1}{\sqrt{\pi}\tau}e^{-x^2/\tau^2}\right), \tag{5.75}$$

with

$$\int_{-\infty}^{\infty} \delta(x)dx = \lim_{\tau\to 0}\left(\frac{1}{\sqrt{\pi}}\int_{-\infty}^{\infty} e^{-\xi^2}d\xi\right) = 1, \tag{5.76}$$

where we have used the definite integral [15]

$$\int_{0}^{\infty} e^{-\eta^2}d\eta = \sqrt{\pi}/2. \tag{5.77}$$

Members of this sequence are shown in Figure 5.5a.

Other properties of $\delta(x)$ can be deduced from these sequences. For a well-behaved function $f(x)$, we have from (5.75)

$$\begin{aligned}\int_{-\infty}^{\infty} f(x)\delta(x)dx &= \lim_{\tau\to 0}\left[\frac{1}{\sqrt{\pi}\tau}\int_{-\infty}^{\infty} f(x)e^{-x^2/\tau^2}dx\right]\\ &= \lim_{\tau\to 0}\left\{\frac{1}{\sqrt{\pi}\tau}\int_{-\infty}^{\infty}\left[f(0) + f'(0)x\right.\right.\\ &\qquad \left.\left. + \frac{1}{2}f''(0)x^2 + \cdots\right]e^{-x^2/\tau^2}dx\right\}\\ &= \lim_{\tau\to 0}\left[f(0) + \frac{\tau^2}{4}f''(0) + \cdots\right] = f(0),\end{aligned} \tag{5.78}$$

which is known as the "sifting property" of the delta function. Also from (5.75) we see that the integral of the delta function is

$$\begin{aligned}\int_{-\infty}^{x} \delta(\xi)d\xi &= \lim_{\tau\to 0}\left(\frac{1}{\sqrt{\pi}\tau}\int_{-\infty}^{x} e^{-\xi^2/\tau^2}d\xi\right)\\ &= \frac{1}{2}\left\{1 + \lim_{\tau\to 0}\left[\mathrm{erf}(x/\tau)\right]\right\},\end{aligned} \tag{5.79}$$

[4] We will assume that the order of integration and taking the limit can be interchanged here and in obtaining future results.

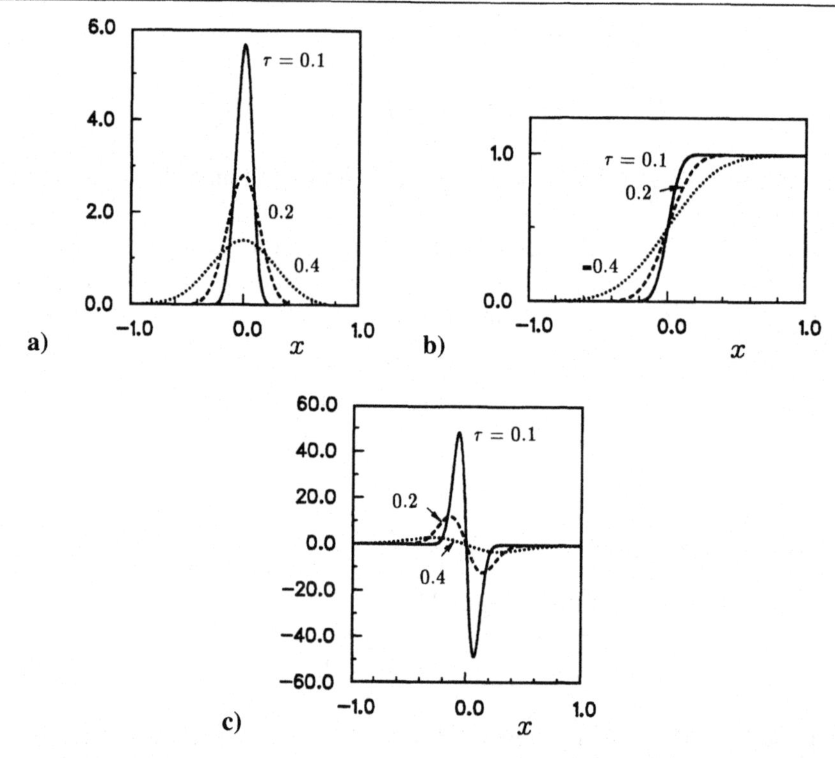

Fig. 5.5. Members of the sequence of functions representing a) the delta function, b) the integral of the delta function, and c) the derivative of the delta function.

where we have introduced the error function [15]

$$\operatorname{erf}(z) = \frac{2}{\sqrt{\pi}} \int_0^z e^{-\eta^2} d\eta \tag{5.80}$$

and used the definite integral (5.77). Members of this sequence are shown in Figure 5.5b. In the limit, we have from (5.77),

$$\lim_{\tau \to 0} \left[\operatorname{erf}(x/\tau)\right] = \begin{cases} -1, & x < 0 \\ 1, & x > 0, \end{cases}$$

so that (5.79) becomes

$$\int_{-\infty}^{x} \delta(\xi)\, d\xi = U(x), \tag{5.81}$$

where $U(x)$ is the *Heaviside unit-step function* $U(x)$:

$$U(x) = \begin{cases} 0, & x < 0 \\ 1, & x > 0. \end{cases} \tag{5.82}$$

The sequence of smooth functions (5.75) can be used to obtain the derivative of

Table 5.1. *Relationships for the one-dimensional delta function*

i. $\int_{-\infty}^{\infty} \delta(x)dx = 1$	*ii.* $\int_{-\infty}^{x} \delta(\xi)d\xi = U(x)$
iii. $\int_{-\infty}^{\infty} f(x)\delta(x)dx = f(0)$	*iv.* $\int_{a}^{b} f(x)\delta(x)dx = f(0), \quad a < 0 < b$
v. $\int_{-\infty}^{\infty} f(x)\delta(x-a)dx = f(a)$	*vi.* $\int_{-\infty}^{\infty} f(x)\delta(ax)dx = \frac{1}{\lvert a\rvert} f(0)$
vii. $\int_{-\infty}^{\infty} f(x)\delta'(x)dx = -f'(0)$	*viii.* $\int_{-\infty}^{\infty} f(x)\delta^{(n)}(x)dx = (-1)^n f^{(n)}(0)$
ix. $\int_{-\infty}^{\infty} f(x)\delta[g(x)]dx = \sum_{i=1}^{n} f(x_i)/\lvert g'(x_i)\rvert$, $g(x_i) = 0, \; g'(x_i) \neq 0, \; i = 1, 2, \ldots, n$	*x.* $\delta(x) = \frac{1}{2\pi}\int_{-\infty}^{\infty} e^{j\eta x} d\eta$

the delta function:

$$\delta'(x) = \lim_{\tau \to 0} \left(\frac{-2x}{\sqrt{\pi}\tau^3} e^{-x^2/\tau^2} \right); \tag{5.83}$$

members of this sequence are shown in Figure 5.5c. The integral of the product of a well-behaved function $f(x)$ with $\delta'(x)$ is then

$$\int_{-\infty}^{\infty} f(x)\delta'(x)dx = \lim_{\tau \to 0} \left\{ \frac{-2}{\sqrt{\pi}\tau^3} \int_{-\infty}^{\infty} \left[f(0) + f'(0)x + \frac{1}{2} f''(0)x^2 + \frac{1}{6} f'''(0)x^3 + \cdots \right] x\, e^{-x^2/\tau^2} dx \right\}$$

$$= \lim_{\tau \to 0} \left[-f'(0) - \frac{\tau^2}{4} f'''(0) + \cdots \right] = -f'(0). \tag{5.84}$$

The above relationships for the one-dimensional delta function and others that can be obtained using the sequences (5.72) and (5.75) are summarized in Table 5.1 (Problem 5.4).

5.2.2 Three-dimensional delta function

It is easy to extend the concept of the delta function to higher dimensions. When a phenomenon is localized at the origin in three-dimensional space, the three-dimensional delta function, $\delta(\vec{r})$, is the appropriate representation; in the *rectangular*

Cartesian system (x, y, z)

$$\delta(\vec{r}) = \delta(x)\delta(y)\delta(z), \tag{5.85}$$

where $\vec{r} = x\hat{x} + y\hat{y} + z\hat{z}$ is the position vector. Clearly, we have

$$\delta(\vec{r}) = 0, \qquad \vec{r} \neq \vec{0} = 0\hat{x} + 0\hat{y} + 0\hat{z} \tag{5.86a}$$

and

$$\iiint_{\text{all space}} \delta(\vec{r})dV = \int_{-\infty}^{\infty}\int_{-\infty}^{\infty}\int_{-\infty}^{\infty} \delta(x)\delta(y)\delta(z)dxdydz = 1. \tag{5.86b}$$

A sequence of ordinary functions that corresponds to (5.85) is

$$\delta(\vec{r}) = \lim_{\tau \to 0}\left[\frac{1}{\tau^3}\Pi(x/\tau)\Pi(y/\tau)\Pi(z/\tau)\right]. \tag{5.87}$$

When the delta function is shifted from the origin to the point (x_o, y_o, z_o) it assumes the form

$$\delta(\vec{r} - \vec{r}_o) = \delta(x - x_o)\delta(y - y_o)\delta(z - z_o), \tag{5.88}$$

with

$$\delta(\vec{r} - \vec{r}_o) = 0, \qquad \vec{r} \neq \vec{r}_o, \tag{5.89a}$$

and

$$\iiint_V \delta(\vec{r} - \vec{r}_o)dV = \begin{cases} 1, & \vec{r}_o \text{ in } V \\ 0, & \vec{r}_o \text{ outside } V. \end{cases} \tag{5.89b}$$

In the *spherical coordinate system* (r, θ, ϕ), when $r_o \neq 0, \theta_o \neq 0, \pi$

$$\delta(\vec{r} - \vec{r}_o) = \frac{\delta(r - r_o)\delta(\theta - \theta_o)\delta(\phi - \phi_o)}{r^2 \sin\theta}, \tag{5.90a}$$

or

$$\delta(\vec{r} - \vec{r}_o) = \frac{\delta(r - r_o)\delta(\cos\theta - \cos\theta_o)\delta(\phi - \phi_o)}{r^2}. \tag{5.90b}$$

Notice that (5.90b) contains a delta function whose argument is a function of θ, so on integration with respect to θ, we must use relationship ix from Table 5.1.

When the delta function is at the origin ($r_o = 0$), a singular point of the spherical coordinate system, the following special representation is used:

$$\delta(\vec{r}) = \delta(r)/4\pi r^2. \tag{5.91}$$

Here we take

$$\int_0^{\infty} \delta(r)dr = 1, \tag{5.92}$$

so that

$$\iiint_{\text{all space}} \delta(\vec{r})dV = \int_{\phi=0}^{2\pi}\int_{\theta=0}^{\pi}\int_{r=0}^{\infty} \frac{\delta(r)}{4\pi r^2} r^2 \sin\theta dr d\theta d\phi = 1. \tag{5.93}$$

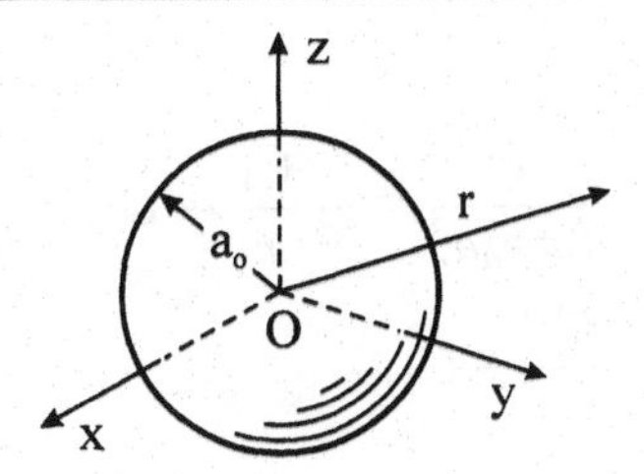

Fig. 5.6. Uniformly charged spherical particle.

Sequences of ordinary functions that correspond to $\delta(r)$ and $\delta(\vec{r})$ and satisfy (5.92) and (5.93) are

$$\delta(r) = \lim_{\tau \to 0} \left[\frac{2}{\tau} \Pi(r/\tau) \right] \tag{5.94}$$

and

$$\delta(\vec{r}) = \lim_{\tau \to 0} \left[\frac{6}{\pi \tau^3} \Pi(r/\tau) \right]. \tag{5.95}$$

Notice that there is a 2 in the argument of (5.94) that does not appear in the argument (5.72); this factor is necessary because of the difference in the limits of integration for (5.92) and (5.71).

5.2.3 Electrostatic point charge

We will illustrate the use of the delta function and its physical interpretation with a simple electrostatic example. A small sphere of charge, as in Figure 5.6, is a model for a charged particle. If the radius of the sphere is allowed to shrink to zero while the total charge q within the sphere remains fixed, we obtain a theoretical idealization, the point charge, which can be represented mathematically by the three-dimensional delta function.

We will assume that the charge q is uniformly distributed over the small spherical volume of radius a_o, located at the origin of the spherical coordinate system (r, θ, ϕ). The volume density of charge is then

$$\rho(\vec{r}) = \frac{3q}{4\pi a_o^3} \Pi(r/2a_o) = \begin{cases} \dfrac{3q}{4\pi a_o^3}, & r < a_o \\ 0, & r > a_o. \end{cases} \tag{5.96}$$

The electrostatic scalar potential Φ is a solution to Poisson's equation (5.4)

$$\nabla^2 \Phi = -\rho/\varepsilon_o,$$

which simplifies to

$$\frac{1}{r^2} \frac{\partial}{\partial r} \left[r^2 \frac{\partial \Phi(\vec{r})}{\partial r} \right] = -\rho(\vec{r})/\varepsilon_o, \tag{5.97}$$

because of the spherical symmetry of ρ (Appendix B). On substitution of (5.96), this equation can be integrated to obtain the potential both inside and outside the sphere:

$$\Phi(\vec{r}) = \begin{cases} -\dfrac{q}{8\pi\varepsilon_o a_o^3} r^2 - \dfrac{C_1}{r} + C_2, & r < a_o \\ -\dfrac{C_3}{r} + C_4, & r > a_o. \end{cases} \tag{5.98}$$

We will assume that $\Phi \to 0$ as $r \to \infty$, that Φ is finite at $r = 0$, and that Φ and its radial derivative (the electric field) are continuous at the surface of the sphere, $r = a_o$. The constants in (5.98) are then

$$C_1 = 0, \qquad C_2 = \frac{3q}{8\pi\varepsilon_o a_o},$$
$$C_3 = \frac{-q}{4\pi\varepsilon_o}, \qquad C_4 = 0,$$

making the scalar potential

$$\Phi(\vec{r}) = \begin{cases} \dfrac{q}{8\pi\varepsilon_o a_o}\left[3 - \left(\dfrac{r}{a_o}\right)^2\right], & r < a_o \\ \dfrac{q}{4\pi\varepsilon_o r}, & r > a_o \end{cases} \tag{5.99}$$

and the electric field

$$\vec{\mathcal{E}}(\vec{r}) = -\nabla\Phi = \begin{cases} \dfrac{q}{4\pi\varepsilon_o a_o^2}\left(\dfrac{r}{a_o}\right)\hat{r}, & r < a_o \\ \dfrac{q}{4\pi\varepsilon_o r^2}\hat{r}, & r > a_o. \end{cases} \tag{5.100}$$

These results and the volume density of charge (5.96) are graphed as a function of r/a_o in Figure 5.7 (the solid lines, which are for the case $a/a_o = 1$). The

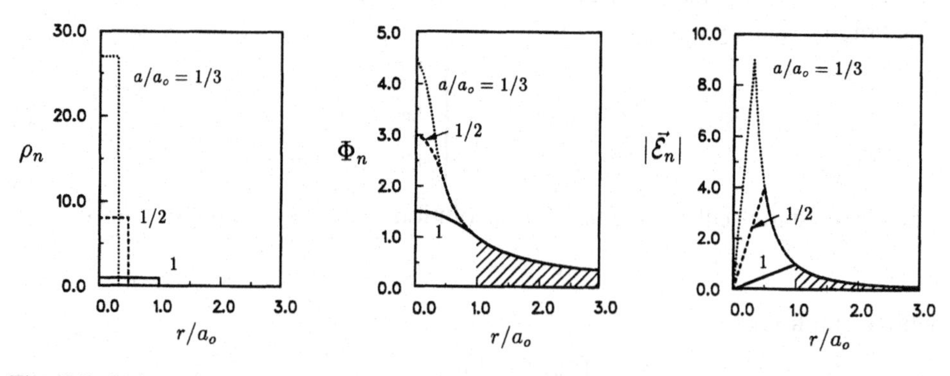

Fig. 5.7. Volume density of charge, potential, and electric field for uniformly charged spherical particle. Radius of particle a/a_o is shrunk, while total charge within particle is held fixed.

normalizations used in the figures are

$$\rho_n = \rho/\rho_o, \qquad \rho_o = 3q/4\pi a_o^3$$
$$\Phi_n = \Phi/\Phi_o, \qquad \Phi_o = q/4\pi\varepsilon_o a_o$$
$$|\vec{\mathcal{E}}_n| = |\vec{\mathcal{E}}|/\mathcal{E}_o, \qquad \mathcal{E}_o = q/4\pi\varepsilon_o a_o^2.$$

Now we ask the question: What happens when the sphere is shrunk from radius a_o to radius a ($a < a_o$), while the total charge within the sphere remains fixed and uniformly distributed? The charge density, potential, and electric field are now given by (5.96), (5.99), and (5.100), respectively, with a_o replaced everywhere by a. Results for the radii $a/a_o = 1, 1/2$, and $1/3$ are shown in Figure 5.7. Notice that the potential and the electric field for $r/a_o > 1$ (shaded region) are the same for all cases; the potential and the electric field outside our initial volume of radius a_o do not change as we shrink the sphere. In the limit as the radius a goes to zero ($a/a_o \to 0$), we obtain a point charge with the volume density of charge [(5.96) with (5.95)]

$$\rho(\vec{r}) = q\left\{\lim_{a\to 0}\left[\frac{6}{\pi(2a)^3}\Pi(r/2a)\right]\right\} = q\delta(\vec{r}), \tag{5.101}$$

the potential (5.99)

$$\Phi(\vec{r}) = \frac{q}{4\pi\varepsilon_o r}, \qquad r > 0, \tag{5.102}$$

and the electric field (5.100)

$$\vec{\mathcal{E}}(\vec{r}) = \frac{q}{4\pi\varepsilon_o r^2}\hat{r}, \qquad r > 0. \tag{5.103}$$

The point charge is a theoretical model, an idealization, for a charged particle of small but finite size. It is represented mathematically by the three-dimensional delta function. This model is of value because, in certain situations, the electromagnetic behavior of the point charge and the charged particle are the same and because the point charge is generally easier to treat analytically. We have already discussed an instance where the point charge and the actual charged particle are equivalent: The potential and the electric field outside the particle ($r/a_o > 1$) are exactly the same as those for the point charge [compare (5.99) and (5.100) with (5.102) and (5.103)].

The point charge is also a useful model in situations where the size of the particle is much smaller than the size of an instrument (theoretical or experimental) used to measure a related quantity, such as the electrostatic potential. To illustrate this point, we consider a spherical probe of radius r_o that measures the average of the potential throughout its volume. When the probe is centered on the particle of radius a_o ($a_o < r_o$), it measures the average potential

$$\langle\Phi\rangle_{\text{particle}} = \frac{3q}{8\pi\varepsilon_o r_o}\left[1 - \frac{1}{5}\left(\frac{a_o}{r_o}\right)^2\right].$$

For the equivalent point charge, the probe measures the average potential

$$\langle\Phi\rangle_{\text{point charge}} = \frac{3q}{8\pi\varepsilon_o r_o}.$$

Clearly, the two results are equivalent whenever the radius of the probe is much larger than the radius of the particle ($r_o \gg a_o$).

The point charge is not equivalent to the charged particle for all calculations. For example, the total energy stored in the electrostatic field (1.112),

$$\mathcal{U}_e = \frac{1}{2}\iiint_V \varepsilon_o|\vec{\mathcal{E}}|^2 dV,$$

is finite for the charged particle,

$$\mathcal{U}_e = \frac{3q^2}{20\pi\varepsilon_o a_o},$$

but infinite for the point charge.

In the approach taken above, we have emphasized the physical connection between the point charge and the charged particle of small but finite radius. The potential of the point charge is simply the limit of the potential of the particle as its radius goes to zero. Another approach is to obtain the potential for the point charge directly by solving Poisson's equation (5.97) with the volume density of charge (5.101):

$$\nabla^2\Phi(\vec{r}) = \frac{1}{r^2}\frac{\partial}{\partial r}\left[r^2\frac{\partial\Phi(\vec{r})}{\partial r}\right] = -q\delta(\vec{r})/\varepsilon_o. \tag{5.104}$$

An examination of this approach is a helpful preliminary to our discussion of the electrodynamic case in the next section.

Away from the origin ($r \neq 0$), the right-hand side of this equation is zero, and its solution is (5.98)

$$\Phi(\vec{r}) = -\frac{C}{r}, \tag{5.105}$$

where we have assumed that $\Phi \to 0$ as $r \to \infty$. The unknown constant C is determined by integrating (5.104) over a small spherical volume V of radius r_o centered at the origin:

$$\iiint_V \nabla\cdot\nabla\Phi dV = \oiint_S \nabla\Phi\cdot d\vec{S} = \frac{-q}{\varepsilon_o}\iiint_V \delta(\vec{r})dV, \tag{5.106}$$

where the divergence theorem (Appendix B) was used to convert the volume integral on the left to a surface integral.[5] After substituting

$$\nabla\Phi = \frac{\partial}{\partial r}\left(\frac{-C}{r}\right)\hat{r} = \frac{C}{r^2}\hat{r},$$

[5] The use of the divergence theorem here may seem questionable, because the theorem requires continuity of the function and its first partial derivatives within V, but from (5.105) we see that $\nabla\Phi$ is singular at the origin. Notice, however, that (5.106) must be correct, for on substituting $\nabla\Phi = -\vec{\mathcal{E}} = -\vec{\mathcal{D}}/\varepsilon_o$, this equation is seen to be Gauss' electric law (1.3).

the integral on the left-hand side of (5.106) is readily evaluated:

$$\oiint_S \nabla\Phi \cdot d\vec{S} = \int_{\phi=0}^{2\pi}\int_{\theta=0}^{\pi} \frac{C}{r_o^2}\hat{r}\cdot\hat{r}r_o^2 \sin\theta d\theta d\phi = 4\pi C,$$

and the integral on the right-hand side is just (5.86b). Thus,

$$4\pi C = -q/\varepsilon_o,$$

or

$$C = \frac{-q}{4\pi\varepsilon_o},$$

making the potential of the point charge (5.105)

$$\Phi(\vec{r}) = \frac{q}{4\pi\varepsilon_o r}, \tag{5.107}$$

which is the same as our previous result (5.102).

5.3 Retarded potentials and electromagnetic field

With the preliminary results obtained in the last two sections, we are now in a position to complete the calculation we set as our goal at the beginning of this chapter – determination of the electromagnetic field for a specified distribution of charge and current (Figure 5.1b). We will first solve the partial differential equations (5.61) and (5.62) for the potentials of time-varying point sources. These results will then be used to obtain the potentials for a general distribution of charge and current, and finally the electromagnetic field will be determined from the potentials using (5.58) and (5.59).

A time-varying charge $q(t)$ is located at the origin of the spherical coordinate system (r, θ, ϕ). The volume density for this point charge is

$$\rho(\vec{r}, t) = q(t)\delta(\vec{r}), \tag{5.108}$$

and the accompanying scalar potential is a solution to the partial differential equation (5.62)

$$\nabla^2\Phi(\vec{r}, t) - \frac{1}{c^2}\frac{\partial^2\Phi(\vec{r}, t)}{\partial t^2} = -q(t)\delta(\vec{r})/\varepsilon_o, \tag{5.109}$$

or

$$\frac{1}{r^2}\frac{\partial}{\partial r}\left[r^2\frac{\partial\Phi(\vec{r}, t)}{\partial r}\right] - \frac{1}{c^2}\frac{\partial^2\Phi(\vec{r}, t)}{\partial t^2} = -q(t)\delta(\vec{r})/\varepsilon_o, \tag{5.110}$$

where the spherical symmetry of the charge density ρ in space has been used to simplify the equation.

We will examine this equation first for points away from the origin ($r \neq 0$); the right-hand side is then zero, so

$$\frac{1}{r^2}\frac{\partial}{\partial r}\left[r^2\frac{\partial\Phi(\vec{r}, t)}{\partial r}\right] - \frac{1}{c^2}\frac{\partial^2\Phi(\vec{r}, t)}{\partial t^2} = 0. \tag{5.111}$$

After substituting

$$\Phi(\vec{r}, t) = \frac{\chi(\vec{r}, t)}{r}, \tag{5.112}$$

(5.111) becomes

$$\left(\frac{\partial^2}{\partial r^2} - \frac{1}{c^2}\frac{\partial^2}{\partial t^2}\right)\chi(\vec{r}, t) = 0. \tag{5.113}$$

This is the one-dimensional wave equation that we encountered earlier in our discussion of plane waves (2.7). The two solutions to the equation are (2.11a,b):

$$\chi_+(\vec{r}, t) = \mathcal{F}_+(t - r/c)$$

and

$$\chi_-(\vec{r}, t) = \mathcal{F}_-(t + r/c),$$

making the corresponding potentials (5.112)

$$\Phi_+(\vec{r}, t) = \frac{\mathcal{F}_+(t - r/c)}{r} \tag{5.114}$$

and

$$\Phi_-(\vec{r}, t) = \frac{\mathcal{F}_-(t + r/c)}{r}. \tag{5.115}$$

The functions $\mathcal{F}_+$ and $\mathcal{F}_-$ in (5.114) and (5.115) are still unknown. They will be determined by a procedure similar to that used in the electrostatic case (5.106). Equation (5.109) is integrated over a small spherical volume of radius r_o centered at the origin:

$$\iiint_V \nabla \cdot \nabla\Phi dV - \frac{1}{c^2}\iiint_V \frac{\partial^2\Phi}{\partial t^2} dV = \frac{-q(t)}{\varepsilon_o}\iiint_V \delta(\vec{r}) dV. \tag{5.116}$$

After using the divergence theorem (Appendix B), the first integral on the left-hand side becomes

$$\oiint_S \nabla\Phi \cdot d\vec{S} = \int_{\phi=0}^{2\pi}\int_{\theta=0}^{\pi}\left[\frac{1}{r_o}\frac{\partial}{\partial r}\mathcal{F}_\pm(t \mp r_o/c) - \frac{1}{r_o^2}\mathcal{F}_\pm(t \mp r_o/c)\right] r_o^2 \sin\theta d\theta d\phi$$

$$= -4\pi\left[\mathcal{F}_\pm(t \mp r_o/c) - r_o\frac{\partial}{\partial r}\mathcal{F}_\pm(t \mp r_o/c)\right],$$

and the second integral is evaluated using integration by parts:

$$\iiint_V \frac{\partial^2\Phi}{\partial t^2} dV = \int_{\phi=0}^{2\pi}\int_{\theta=0}^{\pi}\int_{r=0}^{r_o} \frac{1}{r}\frac{\partial^2}{\partial t^2}\mathcal{F}_\pm(t \mp r/c) r^2 \sin\theta dr d\theta d\phi$$

$$= 4\pi c^2 \int_{r=0}^{r_o} \frac{\partial^2}{\partial r^2}\mathcal{F}_\pm(t \mp r/c) r dr$$

$$= 4\pi c^2\left[r_o\frac{\partial}{\partial r}\mathcal{F}_\pm(t \mp r_o/c) - \int_{r=0}^{r_o}\frac{\partial}{\partial r}\mathcal{F}_\pm(t \mp r/c)dr\right]$$

$$= 4\pi c^2\left[r_o\frac{\partial}{\partial r}\mathcal{F}_\pm(t \mp r_o/c) - \mathcal{F}_\pm(t \mp r_o/c) + \mathcal{F}_\pm(t)\right].$$

When these results and (5.86b) are inserted into (5.116), we find that

$$\mathcal{F}_\pm(t) = q(t)/4\pi\varepsilon_o;$$

thus,

$$\mathcal{F}_\pm(t \mp r/c) = q(t \mp r/c)/4\pi\varepsilon_o,$$

and the two solutions for the scalar potential are simply

$$\Phi_+(\vec{r}, t) = \frac{q(t - r/c)}{4\pi\varepsilon_o r} \tag{5.117}$$

and

$$\Phi_-(\vec{r}, t) = \frac{q(t + r/c)}{4\pi\varepsilon_o r}. \tag{5.118}$$

Notice that these solutions reduce to the electrostatic potential for a point charge (5.107) when q is constant in time.

From our discussion of plane waves in Section 2.1, we know that the first of these solutions (5.117) is a signal traveling radially outward from the origin, whereas the second (5.118) is a signal traveling radially inward toward the origin. For the former (5.117), the signal received at radius r at time t was at the charge (origin) at the earlier (retarded) time $t - r/c$; hence, this solution is called the *retarded scalar potential.* For the latter (5.118), the signal received at radius r at time t will be at the charge (origin) at the later (advanced) time $t + r/c$; hence, this solution is called the *advanced scalar potential.* If we believe that the charge is the source of the signal, then the retarded potential is the causal solution – the signal must reach the radius r sometime after it was at the charge (origin $r = 0$). For the rest of our discussion, we will consider only the causal solution (5.117).

The vector potential for a point current can be determined using the same procedure used to determine the scalar potential of a point charge. The volume density of current for the time-varying point current is

$$\vec{\mathcal{J}}(\vec{r}, t) = \vec{j}(t)\delta(\vec{r}). \tag{5.119}$$

Notice that the dimensions for $\vec{j}(t)$ are current times length, IL (SI units A · m); the vector $\vec{j}(t)$ can be thought of as the moment for a filament of current of infinitesimal length pointing in the direction $\vec{j}(t)/|\vec{j}(t)|$ (Problem 5.5). The vector potential is a solution to the partial differential equation (5.61)

$$\nabla^2\vec{A}(\vec{r}, t) - \frac{1}{c^2}\frac{\partial^2\vec{A}(\vec{r}, t)}{\partial t^2} = -\mu_o\vec{j}(t)\delta(\vec{r}), \tag{5.120}$$

which can be written in component form as

$$\frac{1}{r^2}\frac{\partial}{\partial r}\left[r^2\frac{\partial \mathcal{A}_i(\vec{r},t)}{\partial r}\right] - \frac{1}{c^2}\frac{\partial^2 \mathcal{A}_i(\vec{r},t)}{\partial t^2} = -\mu_o j_i(t)\delta(\vec{r}), \tag{5.121}$$

where $i = x, y, z$. Notice that this is the same differential equation we studied for Φ (5.110); thus, the solutions are readily obtained by substituting $\mu_o j_i$ for q/ε_o in (5.117) and (5.118):

$$\mathcal{A}_{+i}(\vec{r},t) = \frac{\mu_o j_i(t - r/c)}{4\pi r}, \qquad i = x, y, z$$

$$\mathcal{A}_{-i}(\vec{r},t) = \frac{\mu_o j_i(t + r/c)}{4\pi r}, \qquad i = x, y, z,$$

or on combining components

$$\vec{\mathcal{A}}_{+}(\vec{r},t) = \frac{\mu_o \vec{j}(t - r/c)}{4\pi r}, \tag{5.122}$$

$$\vec{\mathcal{A}}_{-}(\vec{r},t) = \frac{\mu_o \vec{j}(t + r/c)}{4\pi r}. \tag{5.123}$$

As for the scalar potential, these two solutions are referred to as the *retarded vector potential* and the *advanced vector potential*, respectively.

The retarded potentials, (5.117) and (5.122), are for point sources located at the origin. The spatial coordinates enter these expressions only through the distance $r = |\vec{r}|$ between the *source point* (origin) and the *observation point* at $\vec{r}$. When the sources are moved from the origin to the point at $\vec{r}'$, the distance between the source point and the observation point becomes $|\vec{r} - \vec{r}'|$, and the retarded potentials are[6]

$$\Phi(\vec{r},\vec{r}';t) = \frac{q(t - |\vec{r} - \vec{r}'|/c)}{4\pi\varepsilon_o|\vec{r} - \vec{r}'|}, \tag{5.124}$$

$$\vec{\mathcal{A}}(\vec{r},\vec{r}';t) = \frac{\mu_o \vec{j}(t - |\vec{r} - \vec{r}'|/c)}{4\pi|\vec{r} - \vec{r}'|}. \tag{5.125}$$

The argument of q and $\vec{j}$ is still a retarded time.

With the retarded potentials known for the point sources, it is a simple matter to obtain these potentials for a general, continuous distribution described by the volume densities $\rho(\vec{r},t)$ and $\vec{\mathcal{J}}(\vec{r},t)$ (Figure 5.8). The charge and current are confined to the volume V and we will only determine the potentials at points outside this volume. We will consider the scalar potential first.

The infinitesimal element of volume dV' located at the source point $\vec{r}'$ contains the charge

$$dq(t) = \rho(\vec{r}',t)dV'.$$

[6] Since we will only be dealing with the retarded potentials, the subscript "+" will be omitted hereafter.

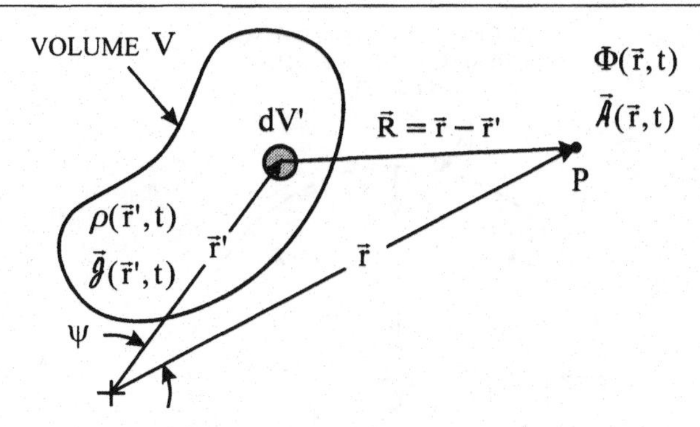

Fig. 5.8. Coordinates for determining the potentials (Φ, $\vec{A}$) of a distribution of charge and current (ρ, $\vec{\mathcal{J}}$).

This point charge produces a potential $d\Phi$ at the observation point $\vec{r}$ (5.124):

$$d\Phi(\vec{r},\vec{r}\,';t) = \frac{dq(t-|\vec{r}-\vec{r}\,'|/c)}{4\pi\varepsilon_o|\vec{r}-\vec{r}\,'|} = \frac{\rho(\vec{r}\,',t-|\vec{r}-\vec{r}\,'|/c)}{4\pi\varepsilon_o|\vec{r}-\vec{r}\,'\,|}dV'.$$

Owing to the linearity of Maxwell's equations and (5.62), we can sum the contributions from all the volume elements within V to obtain the total retarded scalar potential at $\vec{r}$:

$$\Phi(\vec{r},t) = \iiint_V d\Phi(\vec{r},\vec{r}\,';t),$$

or

$$\Phi(\vec{r},t) = \frac{1}{\varepsilon_o}\iiint_V \frac{\rho(\vec{r}\,',t-|\vec{r}-\vec{r}\,'|/c)}{4\pi|\vec{r}-\vec{r}\,'|}dV'$$

$$= \frac{1}{\varepsilon_o}\iiint_V \frac{\rho(\vec{r}\,',t-R/c)}{4\pi R}dV'. \tag{5.126}$$

A similar argument shows that the retarded vector potential is

$$\vec{A}(\vec{r},t) = \mu_o\iiint_V \frac{\vec{\mathcal{J}}(\vec{r}\,',t-|\vec{r}-\vec{r}\,'|/c)}{4\pi|\vec{r}-\vec{r}\,'|}dV'$$

$$= \mu_o\iiint_V \frac{\vec{\mathcal{J}}(\vec{r}\,',t-R/c)}{4\pi R}dV'. \tag{5.127}$$

On the right-hand sides of (5.126) and (5.127), we have introduced the notation

$$\vec{R} = R\hat{R} = \vec{r}-\vec{r}\,',$$

with

$$R = |\vec{r}-\vec{r}\,'|.$$

The role of the retarded time in these expressions is nicely illustrated by an example whose geometry is sketched in Figure 5.9 [16]. An observer located at the point P is measuring the electromagnetic field (retarded potentials) at time t. The

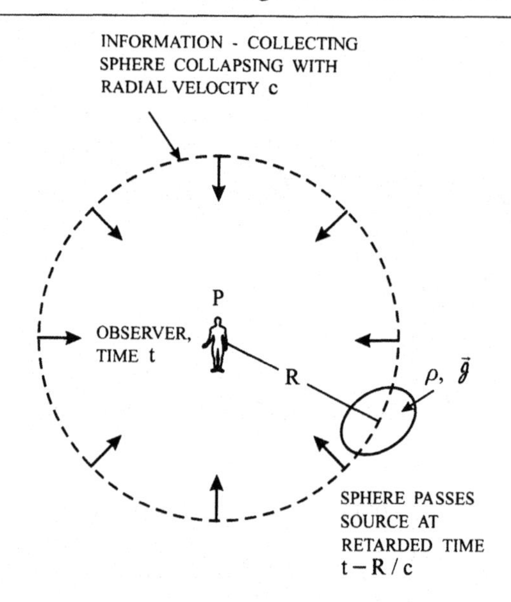

Fig. 5.9. Information-collecting sphere collapsing on observer at time t.

signals the observer receives can be thought of as being gathered by an information-collecting sphere that is collapsing on this point with a radial velocity equal to the speed of light c. The sphere is similar to a balloon with the air being let out at a rate that makes the surface contract with a constant radial velocity. The sphere enclosed all space at an infinitely long time ago and collapses to the point P at time t. Any signal originating at a source $(\rho, \vec{\mathcal{J}})$ as the sphere passes that source reaches P at time t. All sources at the radial distance R are passed at the same retarded time $t - R/c$. In this manner, the observer receives at time t signals that originated throughout all space at the appropriate earlier times.

The constraint appearing in (5.126) and (5.127), which essentially tells us that we must evaluate the sources at the earlier time $t - R/c$, can be difficult to implement in analytical calculations. It can be eliminated at a price – an additional integration. We can use the sifting property of the delta function (v in Table 5.1), to show that

$$\begin{aligned}\rho(\vec{r}', t - R/c) &= \int_{t'=-\infty}^{\infty} \rho(\vec{r}', t')\delta\big[t' - (t - R/c)\big]dt' \\ &= \int_{t'=-\infty}^{\infty} \rho(\vec{r}', t')\delta(t - t' - R/c)dt'.\end{aligned}$$

In obtaining the last line, we changed the sign of the argument of the delta function; this does not affect the value of the integral as can be seen from vi in Table 5.1. Now we can write the retarded potentials (5.126) and (5.127) as

$$\Phi(\vec{r}, t) = \frac{1}{\varepsilon_0}\int_{t'=-\infty}^{\infty}\iiint_V \rho(\vec{r}', t')\frac{\delta(t - t' - R/c)}{4\pi R}dV'dt' \tag{5.128}$$

and

$$\vec{A}(\vec{r},t) = \mu_o \int_{t'=-\infty}^{\infty} \iiint_V \vec{\mathcal{J}}(\vec{r}',t') \frac{\delta(t-t'-R/c)}{4\pi R} dV'dt'. \tag{5.129}$$

It is the convention to call the common factor in the integrands of these expressions the *free-space, scalar Green's function*:[7]

$$\mathcal{G}_o(\vec{r},\vec{r}';t,t') = \frac{\delta(t-t'-|\vec{r}-\vec{r}'|/c)}{4\pi|\vec{r}-\vec{r}'|} = \frac{\delta(t-t'-R/c)}{4\pi R}; \tag{5.130}$$

then the retarded potentials become

$$\Phi(\vec{r},t) = \frac{1}{\varepsilon_o} \int_{t'=-\infty}^{\infty} \iiint_V \rho(\vec{r}',t')\mathcal{G}_o(\vec{r},\vec{r}';t,t')dV'dt', \tag{5.131}$$

$$\vec{A}(\vec{r},t) = \mu_o \int_{t'=-\infty}^{\infty} \iiint_V \vec{\mathcal{J}}(\vec{r}',t')\mathcal{G}_o(\vec{r},\vec{r}';t,t')dV'dt'. \tag{5.132}$$

Once the retarded potentials are evaluated for a specified distribution of charge and current, the electromagnetic field is obtained by inserting them into the relations (5.58) and (5.59),

$$\vec{\mathcal{E}}(\vec{r},t) = -\nabla\Phi(\vec{r},t) - \frac{\partial \vec{A}(\vec{r},t)}{\partial t} \tag{5.133}$$

$$\vec{\mathcal{B}}(\vec{r},t) = \nabla \times \vec{A}(\vec{r},t), \tag{5.134}$$

and performing the indicated differentiation. This is the result we have been seeking: the electromagnetic field that is a solution to Maxwell's equations for a specified distribution of charge and current that satisfies the equation of continuity. The last part of this statement is particularly important, for we have assumed that ρ and $\vec{\mathcal{J}}$ satisfy the equation of continuity (5.31) throughout our derivation. Our findings are summarized in Table 5.2.

5.4 Radiated field

At large distances from the charge and current, the retarded potentials and electromagnetic field simplify to become the radiated potentials and field. The approximations involved in the simplification are similar to those we used earlier in Section 4.1 for obtaining Huygens' principle for the radiated field. A major difference, however,

[7] Notice that the scalar potential (5.131) is simply

$$\Phi(\vec{r},t) = \mathcal{G}_o(\vec{r},\vec{r}_o;t,t_o),$$

when the charge density is

$$\rho(\vec{r}',t') = \varepsilon_o\delta(\vec{r}'-\vec{r}_o)\delta(t'-t_o).$$

Thus, $\mathcal{G}_o(\vec{r},\vec{r}_o;t,t_o)$ is a solution to the partial differential equation (5.62):

$$\nabla^2\mathcal{G}_o(\vec{r},\vec{r}_o;t,t_o) - \frac{1}{c^2}\frac{\partial^2\mathcal{G}_o(\vec{r},\vec{r}_o;t,t_o)}{\partial t^2} = -\delta(\vec{r}-\vec{r}_o)\delta(t-t_o).$$

The Green's function is the causal solution to the scalar wave equation for a point source in space and in time, that is, a source located at a point in space and impulsive in time.

Table 5.2. *Potential functions and electromagnetic field: general time dependence*

General field

$$\Phi(\vec{r}, t) = \frac{1}{\varepsilon_o} \int_{t'=-\infty}^{\infty} \iiint_V \rho(\vec{r}', t') \mathcal{G}_o(\vec{r}, \vec{r}'; t, t') dV' dt'$$

$$\vec{A}(\vec{r}, t) = \mu_o \int_{t'=-\infty}^{\infty} \iiint_V \vec{\mathcal{J}}(\vec{r}', t') \mathcal{G}_o(\vec{r}, \vec{r}'; t, t') dV' dt'$$

$$\mathcal{G}_o(\vec{r}, \vec{r}'; t, t') = \frac{\delta(t - t' - R/c)}{4\pi R} = \frac{\delta(t - t' - |\vec{r} - \vec{r}'|/c)}{4\pi |\vec{r} - \vec{r}'|}$$

$$\vec{\mathcal{E}}(\vec{r}, t) = -\nabla \Phi(\vec{r}, t) - \frac{\partial \vec{A}(\vec{r}, t)}{\partial t}$$

$$\vec{\mathcal{B}}(\vec{r}, t) = \nabla \times \vec{A}(\vec{r}, t)$$

Radiated field

$$\vec{\mathcal{E}}^r(\vec{r}, t) = \frac{-\mu_o}{4\pi r}(1 - \hat{r}\hat{r}\cdot) \iiint_V \left[\frac{\partial}{\partial t'} \vec{\mathcal{J}}(\vec{r}', t')\right]_{t'=t_r} dV'$$

$$= \frac{\mu_o}{4\pi r} \hat{r} \times \hat{r} \times \iiint_V \left[\frac{\partial}{\partial t'} \vec{\mathcal{J}}(\vec{r}', t')\right]_{t'=t_r} dV'$$

$$\vec{\mathcal{B}}^r(\vec{r}, t) = \frac{-\mu_o}{4\pi r c} \hat{r} \times \iiint_V \left[\frac{\partial}{\partial t'} \vec{\mathcal{J}}(\vec{r}', t')\right]_{t'=t_r} dV'$$

$$= \frac{1}{c} \hat{r} \times \vec{\mathcal{E}}^r(\vec{r}, t)$$

$$t_r = t - (r - \hat{r} \cdot \vec{r}')/c$$

is that the earlier discussion is for harmonic time dependence, whereas, the current discussion is for general time dependence.

We will assume that the maximum dimension of the volume V containing the charge and current is d and that the origin O for the coordinates in Figure 5.8 is chosen so that $r' \leq d/2$. Figure 4.4 then applies to the present discussion. When the observation point P is a large distance from V,

$$r'/r \leq d/2r \ll 1, \tag{5.135}$$

according to (4.13), the distance R is approximately

$$R \approx r - r' \cos \psi + \frac{1}{2} r'(r'/r) \sin^2 \psi, \tag{5.136}$$

where ψ is the angle between $\vec{r}$ and $\vec{r}'$ in Figure 5.8.

Now we will apply this approximation to the integrand of the expression for the scalar potential Φ (5.126). The factor $1/R$, as shown earlier (4.14), is simply

$$\frac{1}{R} \approx \frac{1}{r}. \tag{5.137}$$

The charge density ρ in (5.126) depends upon R through the retarded time, which from (5.136) is approximately

$$t - R/c \approx t - r/c + (r'/c)\cos\psi - \frac{1}{2}(r'/c)(r'/r)\sin^2\psi.$$

Clearly, we can drop the fourth term on the right-hand side of this expression, for it can be made as small as desired because of (5.135); thus,

$$t - R/c = t - (r - \hat{r}\cdot\vec{r}')/c \approx t - r/c + (r'/c)\cos\psi. \tag{5.138}$$

We cannot, however, drop the third term, for we would then be ignoring variations in the charge density ρ that occur over times of the order $r'/c \leq d/2c$, which may not be small.

To show more precisely what we are neglecting by making the approximation (5.138) for the retarded time, we will examine the Taylor series for the charge density ρ at the time $t' = t - r/c + (r'/c)\cos\psi$:

$$\rho(\vec{r}', t - R/c) = \left\{\rho(\vec{r}', t') + \left[\frac{\partial\rho(\vec{r}', t')}{\partial t'}\right] \times \left[-\frac{1}{2}(r'/c)(r'/r)\sin^2\psi\right] + \cdots\right\}_{t'=t-r/c+(r'/c)\cos\psi}. \tag{5.139}$$

From this result, we see that the approximation is equivalent to assuming

$$\left|\frac{1}{2}(r'/c)(r'/r)\sin^2\psi\frac{\partial\rho}{\partial t}\right| \ll |\rho|,$$

or, on using the maximum values $r' = d/2$ and $\sin\psi = 1$ $(\psi = \pi/2)$,[8]

$$\frac{d^2}{8cr}\left|\frac{\partial\rho}{\partial t}\right| \ll |\rho|. \tag{5.140}$$

To help with the physical interpretation of this requirement (5.140), in Figure 5.10 we have redrawn the geometry for the special case $r' = d/2$, $\psi = \pi/2$. The evaluation of the potential Φ at point P at the time t requires knowledge of the charge density ρ at point A at the earlier, retarded time $t - R/c$. For this special case, our approximation (5.138) for the retarded time is $t - R/c \approx t - r/c$. Consequently, the error made in the retarded time by using this approximation is

$$\Delta t = (R - r)/c = \delta/c.$$

[8] This inequality is to be satisfied in an average sense over the time of observation. Clearly, it will not be satisfied at an instant when $|\rho| = 0$, $|\partial\rho/\partial t| \neq 0$.

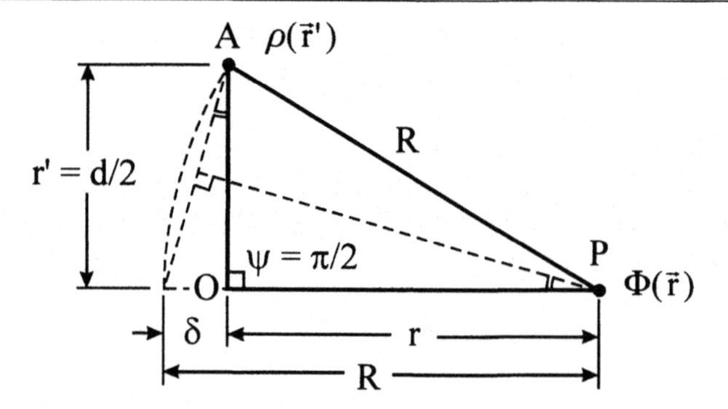

Fig. 5.10. Coordinates for evaluating the error in the retarded time, $\Delta t = \delta/c$, for the special case $r' = d/2$, $\psi = \pi/2$.

From Figure 5.10 and the rules for similar triangles, we can determine the distance δ in the above equation:

$$\frac{\delta}{\sqrt{(d/2)^2 + \delta^2}} = \frac{\sqrt{(d/2)^2 + \delta^2}}{2(r + \delta)},$$

or

$$\delta/r = \sqrt{1 + (d/2r)^2} - 1 \approx \frac{1}{8}\left(\frac{d}{r}\right)^2,$$

where we have assumed that $(d/2r)^2 \ll 1$. Thus, the error in the retarded time is

$$\Delta t \approx \frac{d^2}{8cr}.$$

Notice that this is the coefficient on the left-hand side of (5.140).

We can now offer a physical interpretation for the requirement (5.140): The error, Δt, in the retarded time made by using the approximation (5.138) (by using r instead of R for the distance between the points A and P in Figure 5.10) must be so small that the change in the charge density, $\Delta t|\partial\rho/\partial t|$, during that time is negligible. Stated differently, making the error Δt in the retarded time has no consequence if the charge density does not change substantially during that time.

With (5.137) and (5.138), the radiated scalar potential becomes

$$\Phi^r(\vec{r}, t) = \frac{1}{4\pi\varepsilon_o r}\iiint_V \rho\big[\vec{r}', t - (r - \hat{r}\cdot\vec{r}')/c\big] dV', \tag{5.141}$$

and a similar argument shows that the radiated vector potential is

$$\vec{A}^r(\vec{r}, t) = \frac{\mu_o}{4\pi r}\iiint_V \vec{J}\big[\vec{r}', t - (r - \hat{r}\cdot\vec{r}')/c\big] dV'. \tag{5.142}$$

The radiated electromagnetic field is obtained by substituting the radiated potentials, (5.141) and (5.142), into (5.133) and (5.134). After interchanging the order of differentiation and integration, the expression for the radiated electric field becomes[9]

$$\vec{\mathcal{E}}^r(\vec{r}, t) = \frac{-1}{4\pi\varepsilon_o} \iiint_V \left(\nabla \left\{ \frac{\rho[\vec{r}', t - (r - \hat{r}\cdot\vec{r}')/c]}{r} \right\} \right.$$
$$\left. + \frac{1}{c^2 r}\frac{\partial}{\partial t}\left\{ \vec{\mathcal{J}}\left[\vec{r}', t - (r - \hat{r}\cdot\vec{r}')/c\right]\right\} \right) dV'. \tag{5.143}$$

The first term in the integrand of (5.143) is

$$\begin{aligned}
\nabla\left[\frac{\rho(\vec{r}', t_r)}{r}\right] &= \nabla\left(\frac{1}{r}\right)\rho(\vec{r}\,', t_r) + \frac{1}{r}\nabla\rho(\vec{r}\,', t_r) \\
&= -\frac{1}{r^2}\rho(\vec{r}', t_r)\hat{r} + \frac{1}{r}\left[\frac{\partial\rho(\vec{r}', t')}{\partial t'}\right]_{t'=t_r} \nabla t_r \\
&= -\frac{1}{r^2}\rho(\vec{r}', t_r)\hat{r} - \frac{1}{cr}\left[\frac{\partial\rho(\vec{r}', t')}{\partial t'}\right]_{t'=t_r} \nabla(r - \hat{r}\cdot\vec{r}') \\
&= -\frac{1}{r^2}\rho(\vec{r}', t_r)\hat{r} - \frac{1}{cr}\left[\frac{\partial\rho(\vec{r}', t')}{\partial t'}\right]_{t'=t_r} \left\{\hat{r} - \frac{1}{r}[\vec{r}' - \hat{r}(\hat{r}\cdot\vec{r}')]\right\} \\
&\approx -\frac{1}{cr}\left[\frac{\partial\rho(\vec{r}\,', t')}{\partial t'}\right]_{t'=t_r} \hat{r}.
\end{aligned} \tag{5.144}$$

Here we have used the following notation for the retarded time

$$t_r = t - (r - \hat{r}\cdot\vec{r}')/c, \tag{5.145}$$

and we have dropped terms proportional to $1/r^2$ in the last line. The equation of continuity (5.31) can now be used to express the right-hand side of (5.144) in terms of the volume current density $\vec{\mathcal{J}}$:

$$\begin{aligned}
\nabla\left[\frac{\rho(\vec{r}', t_r)}{r}\right] &\approx -\frac{1}{cr}\left[\frac{\partial\rho(\vec{r}\,', t')}{\partial t'}\right]_{t'=t_r} \hat{r} = \frac{1}{cr}\left[\nabla'\cdot\vec{\mathcal{J}}(\vec{r}', t')\right]_{t'=t_r}\hat{r} \\
&= \hat{r}\frac{1}{cr}\left\{\nabla'\cdot\vec{\mathcal{J}}(\vec{r}', t_r) - \left[\frac{\partial}{\partial t'}\vec{\mathcal{J}}(\vec{r}', t')\right]_{t'=t_r}\cdot\nabla' t_r\right\}
\end{aligned}$$

[9] The integrands for the radiated potentials are well behaved, so the interchange of the order of differentiation and integration causes no problem. If we had substituted the general expressions for the potentials, (5.126) and (5.127), into (5.133) and (5.134), we would have had to be more careful about the interchange of the order of differentiation and integration. The integrands of the general expressions for the potentials are singular when $R = 0$ ($\vec{r}' = \vec{r}$). Moving the differentiation under the integral increases the strength of the singularity; it goes from R^{-1} to R^{-3} for the electric field when expressed in terms of only $\vec{\mathcal{J}}$. Methods for handling the singularities for this case are discussed in Reference [17].

$$= \hat{r}\frac{1}{cr}\left\{\nabla'\cdot\vec{\mathcal{J}}(\vec{r}\,',t_r) - \left[\frac{\partial}{\partial t'}\vec{\mathcal{J}}(\vec{r}\,',t')\right]_{t'=t_r}\cdot\frac{\hat{r}}{c}\right\}$$

$$= \hat{r}\frac{1}{c^2 r}\left\{c\nabla'\cdot\vec{\mathcal{J}}(\vec{r}\,',t_r) - \hat{r}\cdot\left[\frac{\partial}{\partial t'}\vec{\mathcal{J}}(\vec{r}\,',t')\right]_{t'=t_r}\right\}. \tag{5.146}$$

After substituting (5.146) into (5.143), we have

$$\vec{\mathcal{E}}^r(\vec{r},t) = \frac{-\mu_o}{4\pi r}\iiint_V (1-\hat{r}\hat{r}\cdot)\left[\frac{\partial}{\partial t'}\vec{\mathcal{J}}(\vec{r}\,',t')\right]_{t'=t_r} dV'$$

$$-\hat{r}\frac{\mu_o c}{4\pi r}\iiint_V \nabla'\cdot\vec{\mathcal{J}}(\vec{r}\,',t_r)dV'$$

$$= \frac{-\mu_o}{4\pi r}\iiint_V (1-\hat{r}\hat{r}\cdot)\left[\frac{\partial}{\partial t'}\vec{\mathcal{J}}(\vec{r}\,',t')\right]_{t'=t_r} dV' - \hat{r}\frac{\mu_o c}{4\pi r}\oiint_S \vec{\mathcal{J}}(\vec{r}\,',t_r)\cdot d\vec{S}\,',$$

where the divergence theorem (Appendix B) was used to convert the second volume integral to a surface integral. The only requirement for the volume V is that it contains all of the charge and current. Therefore, we are free to choose V so that its surface S lies entirely outside the region containing all current. This makes the surface integral zero, and the radiated electric field becomes simply

$$\vec{\mathcal{E}}^r(\vec{r},t) = \frac{-\mu_o}{4\pi r}(1-\hat{r}\hat{r}\cdot)\iiint_V \left[\frac{\partial}{\partial t'}\vec{\mathcal{J}}(\vec{r}\,',t')\right]_{t'=t_r} dV'$$

$$= \frac{\mu_o}{4\pi r}\hat{r}\times\hat{r}\times\iiint_V \left[\frac{\partial}{\partial t'}\vec{\mathcal{J}}(\vec{r}\,',t')\right]_{t'=t_r} dV'. \tag{5.147}$$

A similar argument (Problem 5.7) shows that the radiated magnetic field is

$$\vec{\mathcal{B}}^r(\vec{r},t) = \frac{-\mu_o}{4\pi rc}\hat{r}\times\iiint_V \left[\frac{\partial}{\partial t'}\vec{\mathcal{J}}(\vec{r}\,',t')\right]_{t'=t_r} dV'$$

$$= \frac{1}{c}\hat{r}\times\vec{\mathcal{E}}^r(\vec{r},t). \tag{5.148}$$

These results for the radiated field are also summarized in Table 5.2. As we observed in our earlier discussions of the radiated field (Chapters 3 and 4), the electric and magnetic fields are proportional to $1/r$ and are normal to each other and to $\hat{r}$.

5.5 Harmonic time dependence

For the special case of harmonic time dependence, further simplification of the formulas in Table 5.2 is possible. Recall, from the discussion in Section 1.7, that for harmonic time dependence, the electric and magnetic fields, volume densities of charge and current, and potential functions all can be expressed in terms of complex phasors. For example, the volume density of charge and scalar potential are

$$\rho(\vec{r}, t) = \text{Re}\left[\rho(\vec{r})e^{j\omega t}\right]$$

and

$$\Phi(\vec{r}, t) = \text{Re}\left[\Phi(\vec{r})e^{j\omega t}\right],$$

where $\rho(\vec{r})$ and $\Phi(\vec{r})$ are the complex phasors.

The relationship between the scalar potential and volume charge density (5.128) then becomes

$$\text{Re}\left[\Phi(\vec{r})e^{j\omega t}\right] = \frac{1}{\varepsilon_o}\int_{t'=-\infty}^{\infty}\iiint_V \text{Re}\left[\rho(\vec{r}')e^{j\omega t'}\right]\mathcal{G}_o(\vec{r}, \vec{r}'; t, t')dV'dt'$$

$$= \text{Re}\left(\left\{\frac{1}{\varepsilon_o}\iiint_V \rho(\vec{r}')\left[\int_{t'=-\infty}^{\infty}\mathcal{G}_o(\vec{r}, \vec{r}'; t, t')e^{j\omega(t'-t)}dt'\right]dV'\right\}e^{j\omega t}\right),$$

or, according to the procedure in Section 1.7,

$$\Phi(\vec{r}) = \frac{1}{\varepsilon_o}\iiint_V \rho(\vec{r}')\left[\int_{t'=-\infty}^{\infty}\mathcal{G}_o(\vec{r}, \vec{r}'; t, t')e^{j\omega(t'-t)}dt'\right]dV'.$$

The term in brackets is readily evaluated on substitution of (5.130),

$$\int_{t'=-\infty}^{\infty}\mathcal{G}_o(\vec{r}, \vec{r}\,'; t, t')e^{j\omega(t'-t)}dt' = \int_{t'=-\infty}^{\infty}\frac{\delta(t - t' - R/c)}{4\pi R}e^{-j\omega(t-t')}dt'$$

$$= \frac{e^{-jk_oR}}{4\pi R} = G_o(\vec{r}, \vec{r}'), \tag{5.149}$$

and it is recognized to be the *free-space, scalar Green's function for harmonic time dependence* discussed in Section 4.1. Thus, we can write the phasor for the scalar potential as

$$\Phi(\vec{r}) = \frac{1}{\varepsilon_o}\iiint_V \rho(\vec{r}')G_o(\vec{r}, \vec{r}')dV', \tag{5.150}$$

and by a similar argument we can express the phasor for the vector potential as

$$\vec{A}(\vec{r}) = \mu_o\iiint_V \vec{J}(\vec{r}')G_o(\vec{r}, \vec{r}')dV'. \tag{5.151}$$

The phasors for the electromagnetic field follow directly from (5.133) and (5.134):

$$\vec{E}(\vec{r}) = -\nabla\Phi(\vec{r}) - j\omega\vec{A}(\vec{r}) \tag{5.152}$$

$$\vec{B}(\vec{r}) = \nabla \times \vec{A}(\vec{r}). \tag{5.153}$$

Equation (5.149) establishes the relationship between the free-space, scalar Green's functions for general time dependence, $\mathcal{G}_o$, and that for harmonic time dependence G_o. This relationship can be simplified further when we recognize that $\mathcal{G}_o$ is a function of the differences $R = \left|\vec{r} - \vec{r}'\right|$ and $T = t - t'$, whereas G_o is a

Table 5.3. *Potential functions and electromagnetic field: harmonic time dependence*

General field
$\Phi(\vec{r}) = \frac{1}{\varepsilon_o} \iiint_V \rho(\vec{r}\,')G_o(\vec{r}, \vec{r}\,')dV'$
$\vec{A}(\vec{r}) = \mu_o \iiint_V \vec{J}(\vec{r}\,')G_o(\vec{r}, \vec{r}\,')dV'$
$G_o(\vec{r}, \vec{r}\,') = \frac{e^{-jk_o R}}{4\pi R} = \frac{e^{-jk_o\lvert\vec{r}-\vec{r}\,'\rvert}}{4\pi\lvert\vec{r}-\vec{r}\,'\rvert}$
$\vec{E}(\vec{r}) = -\nabla\Phi(\vec{r}) - j\omega\vec{A}(\vec{r})$
$\vec{B}(\vec{r}) = \nabla \times \vec{A}(\vec{r})$
Radiated field
$\vec{E}^r(\vec{r}) = -j\omega\mu_o \frac{e^{-jk_o r}}{4\pi r}(1 - \hat{r}\hat{r}\cdot) \iiint_V \vec{J}(\vec{r}\,')e^{jk_o\hat{r}\cdot\vec{r}\,'}dV'$
$= j\omega\mu_o \frac{e^{-jk_o r}}{4\pi r}\hat{r} \times \hat{r} \times \iiint_V \vec{J}(\vec{r}\,')e^{jk_o\hat{r}\cdot\vec{r}\,'}dV'$
$\vec{B}^r(\vec{r}) = -j\mu_o k_o \frac{e^{-jk_o r}}{4\pi r}\hat{r} \times \iiint_V \vec{J}(\vec{r}\,')e^{jk_o\hat{r}\cdot\vec{r}\,'}dV'$
$= \frac{1}{c}\hat{r} \times \vec{E}^r(\vec{r})$

function of only R. On expressing (5.149) in terms of these variables, we have

$$G_o(R) = \int_{T=-\infty}^{\infty} \mathcal{G}_o(R, T)e^{-j\omega T}dT, \tag{5.154}$$

which shows that G_o is the Fourier transform (1.201) of $\mathcal{G}_o$.

The radiated field is obtained by assuming harmonic time dependence in (5.147) and (5.148). The result for the electric field is

$$\begin{aligned}
\mathrm{Re}\big[\vec{E}^r(\vec{r})e^{j\omega t}\big] &= \frac{\mu_o}{4\pi r}\hat{r} \times \hat{r} \times \iiint_V \left\{\frac{\partial}{\partial t'}\mathrm{Re}\big[\vec{J}(\vec{r}\,')e^{j\omega t'}\big]\right\}_{t'=t_r} dV' \\
&= \frac{\mu_o}{4\pi r}\hat{r} \times \hat{r} \times \iiint_V \mathrm{Re}\big\{j\omega\vec{J}(\vec{r}\,')e^{j\omega[t-(r-\hat{r}\cdot\vec{r}\,')/c]}\big\}dV' \\
&= \mathrm{Re}\left\{\left[j\omega\mu_o \frac{e^{-jk_o r}}{4\pi r}\hat{r} \times \hat{r} \times \iiint_V \vec{J}(\vec{r}\,')e^{jk_o\hat{r}\cdot\vec{r}\,'}dV'\right]e^{j\omega t}\right\},
\end{aligned}$$

or

$$\vec{E}^r(\vec{r}) = j\omega\mu_o \frac{e^{-jk_o r}}{4\pi r} \hat{r} \times \hat{r} \times \iiint_V \vec{J}(\vec{r}\,')e^{jk_o \hat{r}\cdot\vec{r}\,'} dV'. \qquad (5.155)$$

For the magnetic field, similar steps give

$$\vec{B}^r(\vec{r}) = -j\mu_o k_o \frac{e^{-jk_o r}}{4\pi r} \hat{r} \times \iiint_V \vec{J}(\vec{r}\,')e^{jk_o \hat{r}\cdot\vec{r}\,'} dV'$$

$$= \frac{1}{c} \hat{r} \times \vec{E}^r(\vec{r}). \qquad (5.156)$$

The requirements for use of the radiated field are (5.135) ($d/2r \ll 1$) and (5.140). The latter, when evaluated in a root-mean-square sense, becomes

$$\frac{d^2}{8cr} \left\{ \frac{1}{T} \int_0^T \left[\frac{\partial \rho(\vec{r}, t)}{\partial t} \right]^2 dt \right\}^{1/2} \ll \left[\frac{1}{T} \int_0^T \rho^2(\vec{r}, t) dt \right]^{1/2},$$

or, for harmonic time dependence,

$$\frac{d^2}{8cr}\omega = \frac{1}{2} k_o r \left(\frac{d}{2r} \right)^2 \ll 1. \qquad (5.157)$$

These are the same requirements we found in Chapter 4 for the use of the radiated field from an aperture of maximum dimension d (4.22b, c).

The results for harmonic time dependence are summarized in Table 5.3.

References

[1] J. A. Stratton, *Electromagnetic Theory*, McGraw-Hill, New York, 1941.

[2] A. Sommerfeld, *Electrodynamics*, Lectures on Theoretical Physics, Vol. II, Academic Press, New York, 1952.

[3] D. S. Jones, *The Theory of Electromagnetism*, Pergamon Press, Oxford, 1964.

[4] J. D. Jackson, *Classical Electrodynamics*, 2nd Edition, Wiley, New York, 1975.

[5] M. Mason and W. Weaver, *The Electromagnetic Field*, Mathematical Appendix, University of Chicago Press, Chicago, 1929.

[6] O. L. Brill and B. Goodman, "Causality in the Coulomb Gauge," *Am. J. Phys.*, Vol. 35, pp. 832–7, September 1967.

[7] O. Heaviside, *Electromagnetic Theory*, Volume I, Preface, The Electrician Printing and Publishing Co., London, 1893. Republication, Chelsea, New York, 1971.

[8] Y. Aharonov and D. Bohm, "Significance of Electromagnetic Potentials in Quantum Mechanics," *Phys. Rev.*, Vol. 115, pp. 485–91, August 1959.

[9] H. Erlichson, "Aharonov-Bohm Effect – Quantum Effects on Charged Particles in Field-Free Regions," *Am. J. Phys.*, Vol. 38, pp. 162–73, February 1970.

[10] M. Peshkin and A. Tonomura, *The Aharonov-Bohm Effect*, Springer-Verlag, Berlin, 1989.

[11] R. N. Bracewell, *The Fourier Transform and Its Applications*, 2nd Edition, Ch. 5, McGraw-Hill, New York, 1978.

[12] A. Papoulis, *The Fourier Integral and Its Applications*, Appendix I, McGraw-Hill, New York, 1962.

[13] M. J. Lighthill, *Introduction to Fourier Analysis and Generalized Functions*, Cambridge University Press, Cambridge 1958.
[14] I. Stakgold, *Boundary Value Problems of Mathematical Physics*, McGraw-Hill, New York, Vol. 1, Ch. 1, 1967; Vol. 2, Ch. 5, 1968.
[15] M. Abramowitz and I. A. Stegun, *Handbook of Mathematical Functions*, U.S. Government Printing Office, Washington, DC, 1964.
[16] W. K. H. Panofsky and M. Phillips, *Classical Electricity and Magnetism*, 2nd Edition, Addison-Wesley, Reading, MA, 1962.
[17] J. Van Bladel, *Singular Electromagnetic Fields and Sources*, Oxford University Press, Oxford, 1991.

Problems

5.1 Consider a region of free space containing no sources: $\rho = 0$, $\vec{\mathcal{J}} = 0$. Show that the electromagnetic field within this region can be expressed entirely in terms of the vector potential $\vec{\mathcal{A}}$. Hint: Assume that $\vec{\mathcal{A}}$, Φ satisfy (5.61)–(5.63), then perform a gauge transformation, (5.60) with (5.64), selecting ψ so as to make the transformed scalar potential zero.

5.2 a) One of the differential equations (5.65) for the potentials in the Coulomb gauge involves both $\vec{\mathcal{A}}$ and Φ. Obtain an equation involving only $\vec{\mathcal{A}}$ by eliminating Φ from this equation. To accomplish this, use the fact that a vector field, like the current density $\vec{\mathcal{J}}$, can be expressed as the sum of the longitudinal or irrotational part $\vec{\mathcal{J}}_\ell$, with $\nabla \times \vec{\mathcal{J}}_\ell = 0$, and the transverse or solenoidal part $\vec{\mathcal{J}}_t$, with $\nabla \cdot \vec{\mathcal{J}}_t = 0$ [5]:

$$\vec{\mathcal{J}} = \vec{\mathcal{J}}_\ell + \vec{\mathcal{J}}_t.$$

You will also have to use the fact that a vector is uniquely determined by its divergence and curl, so that two vectors that have the same divergence and curl are equal [5]. Your final differential equation for $\vec{\mathcal{A}}$ should only contain $\vec{\mathcal{J}}_t$.

b) Starting with the differential equation from part a and Equation (5.66), obtain the potentials $\vec{\mathcal{A}}$ and Φ in the Coulomb gauge for a point current and a point charge, respectively. The procedures used in obtaining the retarded potentials, (5.124) and (5.125), for the Lorentz gauge can serve as a guide.

c) In your answer to part b, the scalar potential should not involve the retarded time (i.e., Φ should depend on t not $t - |\vec{r} - \vec{r}\,'|/c$). Thus, the scalar potential at any position $\vec{r}$ varies instantaneously as the point charge $q(t)$ varies. How can this be consistent with causality? For a detailed discussion see Reference [6].

5.3 In the Coulomb gauge, show that (5.65)–(5.67) are invariant under the gauge transformation (5.60) when ψ is a solution to (5.68).

5.4 Derive relationships *v*, *viii*, and *ix* in Table 5.1 for the one-dimensional delta function.

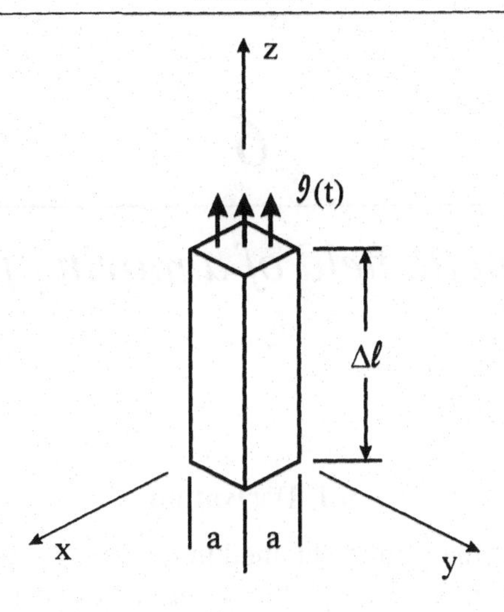

Fig. P5.1. Filament of current.

5.5 The volume density for the point current (5.119) can be thought of as belonging to a filament of current of infinitesimal length. To show this, consider the rectangular volume in Figure P5.1 with square cross section of side a and length $\Delta\ell$ ($\Delta\ell \gg a$). Within this volume a current $\mathcal{I}(t)$ is in the z direction. Obtain an expression for the volume density of current $\vec{\mathcal{J}}$; use the rectangular function (5.73) to confine the current to the volume.

In the limits as $a \to 0$ and $\Delta\ell \to 0$, show that the current density becomes

$$\vec{\mathcal{J}} = \lim_{\Delta\ell \to 0} [\mathcal{I}(t)\Delta\ell\hat{z}]\delta(\vec{r}) = \vec{j}(t)\delta(\vec{r}),$$

where the moment $\vec{j}(t)$ for the infinitesimal current element has been defined.

5.6 Verify that the retarded potentials (5.128) and (5.129) satisfy the Lorentz condition (5.63). Hint: Convert the derivatives with respect to the unprimed quantities (x, y, z, t) to derivatives with respect to the primed quantities (x', y', z', t'), then use integration by parts. What role does the equation of continuity for electric charge play in the argument?

5.7 Obtain the expression for the radiated magnetic field $\vec{\mathcal{B}}^r$ (5.148) by inserting the expression for the radiated vector potential (5.142) into (5.134) and performing the indicated operations.

5.8 a) A charge distribution varies in time as a Gaussian function:

$$\rho = \rho_o e^{-t^2/\tau^2}.$$

Assume that the inequality (5.140) is to hold when the time average is used on each side. How far must one be from the distribution to observe the radiated field? The answer should be in terms of r, d, and τ.

b) Assume that the width of the Gaussian pulse is approximately 4τ in time or $4c\tau$ in space. What must the quantity $r/4c\tau$ be in terms of the quantity $d/4c\tau$ to observe the radiated field?

6

Electromagnetic field of a moving point charge

6.1 Derivation

Our first application of the results obtained in Chapter 5 will be to the calculation of the electromagnetic field of a moving point charge. The point charge may be thought of as a classical model for a charged particle such as the electron. We will be particularly interested in the effect the motion of the charge has on the electromagnetic field, that is, how the motion alters the field from that of a stationary point charge, the electrostatic field given in (5.103).

We will assume that the point charge moves on a prescribed trajectory with its position $\vec{r}_q(t)$ a known function of time. Its velocity and acceleration are

$$\vec{v}(t) = \frac{d\vec{r}_q(t)}{dt}, \qquad \vec{a}(t) = \frac{d^2\vec{r}_q(t)}{dt^2}.$$

The volume densities of charge and current are then

$$\rho(\vec{r}, t) = q\delta\big[\vec{r} - \vec{r}_q(t)\big] \tag{6.1}$$

and

$$\vec{\mathcal{J}}(\vec{r}, t) = q\vec{v}(t)\delta\big[\vec{r} - \vec{r}_q(t)\big], \tag{6.2}$$

which make the retarded potentials (5.131) and (5.132)

$$\Phi(\vec{r}, t) = \frac{q}{\varepsilon_o}\int_{t'=-\infty}^{\infty}\iiint_{\text{all space}} \delta\big[\vec{r}' - \vec{r}_q(t')\big]\mathcal{G}_o(\vec{r}, \vec{r}'; t, t')dV'dt', \tag{6.3}$$

$$\vec{\mathcal{A}}(\vec{r}, t) = \mu_o q\int_{t'=-\infty}^{\infty}\iiint_{\text{all space}} \vec{v}(t')\delta\big[\vec{r}' - \vec{r}_q(t')\big]\mathcal{G}_o(\vec{r}, \vec{r}'; t, t')dV'dt'. \tag{6.4}$$

A portion of the particle's trajectory is shown in Figure 6.1 along with the coordinates used in expressions (6.3) and (6.4). This figure requires some explanation. The retarded potentials are to be evaluated at point P, position $\vec{r}$ at time t. The spatial integral is over all space with the volume element dV' at position $\vec{r}'$. The temporal integral extends over all time and is with respect to the variable t'. At time t' the charge is at position $\vec{r}_q(t')$, and the vector $\vec{R}_q(t')$ is from this position to the

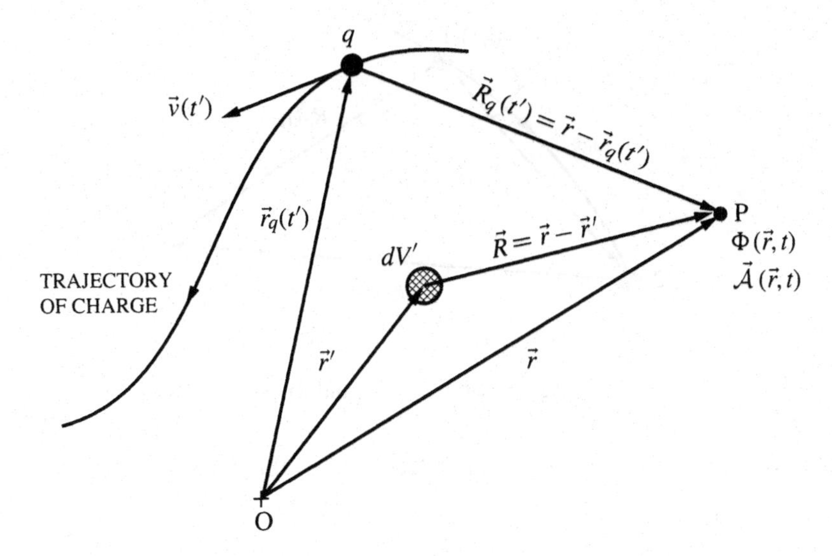

Fig. 6.1. Trajectory of moving point charge and coordinates to be used in evaluating the retarded potentials.

observation point P:

$$\vec{R}_q(t') = R_q(t')\hat{R}_q(t') = \vec{r} - \vec{r}_q(t'). \tag{6.5}$$

6.1.1 Liénard-Wiechert potentials

Because the form of Equations (6.3) and (6.4) is similar, we only need to consider the evaluation of the integral for the scalar potential and then infer the results for the vector potential. After substituting the Green's function (5.130), (6.3) becomes

$$\Phi(\vec{r}, t) = \frac{q}{4\pi\varepsilon_o}\int_{t'=-\infty}^{\infty}\iiint_{\text{all space}} \frac{\delta[\vec{r}' - \vec{r}_q(t')]\delta(t - t' - R/c)}{R} dV'dt'. \tag{6.6}$$

We will consider the spatial integral first. The volume density dV' ranges over all space during this integration. However, due to the "sifting property" of the first delta function in the integrand, this integral has a value only when dV' includes the location of the point charge at time t'; that is, $\vec{r}' = \vec{r}_q(t')$ and $\vec{R} = \vec{R}_q(t')$ in Figure 6.1. Hence, the scalar potential becomes

$$\Phi(\vec{r}, t) = \frac{q}{4\pi\varepsilon_o}\int_{t'=-\infty}^{\infty} \frac{\delta[t - t' - R_q(t')/c]}{R_q(t')} dt', \tag{6.7}$$

and there is a similar expression for the vector potential:

$$\vec{A}(\vec{r}, t) = \frac{\mu_o q}{4\pi}\int_{t'=-\infty}^{\infty} \frac{\vec{v}(t')\delta[t - t' - R_q(t')/c]}{R_q(t')} dt'. \tag{6.8}$$

The argument of the remaining delta function in (6.7) is a function of the variable

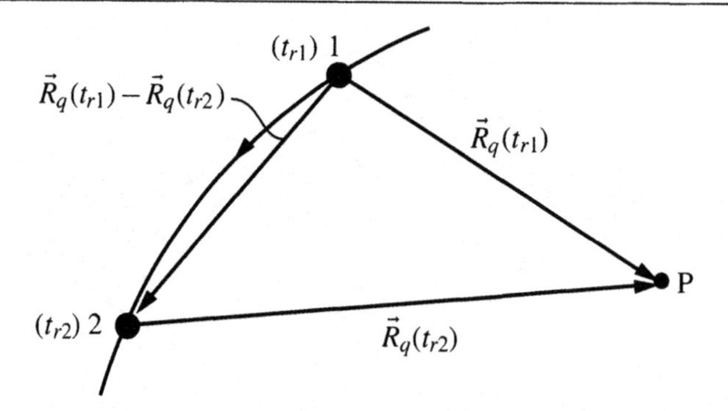

Fig. 6.2. Charge at two points (1 and 2) on trajectory.

of integration t', so we must use *ix* in Table 5.1 to evaluate the integral. With

$$g(t') = t - t' - R_q(t')/c = t - t' - \left|\vec{r} - \vec{r}_q(t')\right|/c, \tag{6.9}$$

we have (Problem 6.2)

$$\frac{dg(t')}{dt'} = -1 - \frac{1}{c}\frac{dR_q(t')}{dt'} = -\left[1 - \hat{R}_q(t') \cdot \vec{v}(t')/c\right], \tag{6.10}$$

or

$$\left|\frac{dg(t')}{dt'}\right| = \left|1 - \hat{R}_q(t') \cdot \vec{v}(t')/c\right|,$$

and $g(t') = 0$ implies that t' is a solution to

$$t' = t - R_q(t')/c;$$

this is the retarded time

$$t_r = t - R_q(t_r)/c. \tag{6.11}$$

With these results, (6.7) and (6.8) become

$$\Phi(\vec{r}, t) = \frac{q}{4\pi\varepsilon_o}\left\{\frac{1}{R_q(t')|1 - \hat{R}_q(t') \cdot \vec{v}(t')/c|}\right\}_{t'=t_r}, \tag{6.12}$$

$$\vec{A}(\vec{r}, t) = \frac{\mu_o q}{4\pi}\left\{\frac{\vec{v}(t')}{R_q(t')|1 - \hat{R}_q(t') \cdot \vec{v}(t')/c|}\right\}_{t'=t_r}. \tag{6.13}$$

These are the *Liénard-Wiechert potentials*; they were first obtained by Alfred Liénard and Emil Wiechert in the period 1898–1900 [1, 2].

For a point charge traveling with a velocity less than the speed of light in free space ($|\vec{v}| < c$), there is only one retarded time t_r [one position $\vec{r}_q(t_r)$] that satisfies Equation (6.11) for each observation time t. To see this, assume that there are two points on the trajectory of the charge that satisfy (6.11), 1 and 2 (t_{r1} and t_{r2}) in Figure 6.2. Then we have

$$t_{r1} = t - R_q(t_{r1})/c, \qquad t_{r2} = t - R_q(t_{r2})/c,$$

or

$$c|t_{r2} - t_{r1}| = \left|R_q(t_{r1}) - R_q(t_{r2})\right| = \left|\left|\vec{R}_q(t_{r2})\right| - \left|\vec{R}_q(t_{r1})\right|\right|.$$

It is easy to show that

$$\left|\left|\vec{R}_q(t_{r2})\right| - \left|\vec{R}_q(t_{r1})\right|\right| \leq \left|\vec{R}_q(t_{r2}) - \vec{R}_q(t_{r1})\right|,$$

so

$$c\left|t_{r2} - t_{r1}\right| \leq \left|\vec{R}_q(t_{r2}) - \vec{R}_q(t_{r1})\right| \leq \begin{cases} \text{distance charge travels} \\ \text{in time interval } |t_{r2} - t_{r1}|. \end{cases} \tag{6.14}$$

In words, the distance light travels in the time interval $|t_{r2} - t_{r1}|$ must be less than or equal to the distance the charge travels in the same time interval. Clearly, this inequality cannot be satisfied unless the charge travels at a speed greater than or equal to the speed of light in free space – a possibility excluded by the special theory of relativity.[1] Thus, there can be only one solution to (6.11), that is, one retarded time t_r or position $\vec{r}_q(t_r)$ corresponding to each observation time t. Notice that the condition $|\vec{v}| < c$ also means that $1 - \hat{R}_q \cdot \vec{v}/c > 0$, which allows the removal of the absolute value sign in (6.12), (6.13), and subsequent formulas.

It is instructive to compare our expression for the scalar potential of the moving point charge (6.12) with the results for a stationary point charge (5.102) [4]. The motion of the charge has introduced two factors into the potential, the retarded time t_r and the term $|1 - \hat{R}_q \cdot \vec{v}/c|$ in the denominator. The retarded time is a result of the finite speed of propagation for electromagnetic signals. It is not the position $\vec{r}_q(t)$ of the charge at the observation time t that enters (6.12), but the position $\vec{r}_q(t_r)$ of the charge at the earlier time t_r. The time interval $t - t_r$ just equals the time it takes light to travel from the earlier position of the charge $\vec{r}_q(t_r)$ to the point of observation $\vec{r}$.

To explain the additional term in the denominator of (6.12), we will graphically examine the evaluation of the integral for the scalar potential (5.126):

$$\Phi(\vec{r}, t) = \frac{1}{\varepsilon_o} \iiint_V \frac{\rho(\vec{r}\,', t - R/c)}{4\pi R} dV'.$$

The volume of integration V can be divided into a large number n of spherical shells centered on the observation point P, where the thickness of each shell is ΔR, as shown in Figure 6.3a. The potential at P is then given approximately by the finite sum

$$\Phi(\vec{r}, t) \approx \frac{1}{\varepsilon_o} \sum_{i=1}^{N} \frac{\Delta q_i(t_{ri})}{4\pi R_i}, \tag{6.15}$$

where R_i is the mean radius of the i-th shell and $\Delta q_i(t_{ri})$ is the charge in the i-th shell at the retarded time $t_{ri} = t - R_i/c$.

[1] Special relativity actually excludes the possibility of a particle whose speed is less than c being accelerated to a speed greater than c, for this would require an infinite amount of energy. Particles whose speed is always greater than c are not excluded. Such particles have been proposed, and they are called tachyons, after the Greek word *tachys*, which means swift [3].

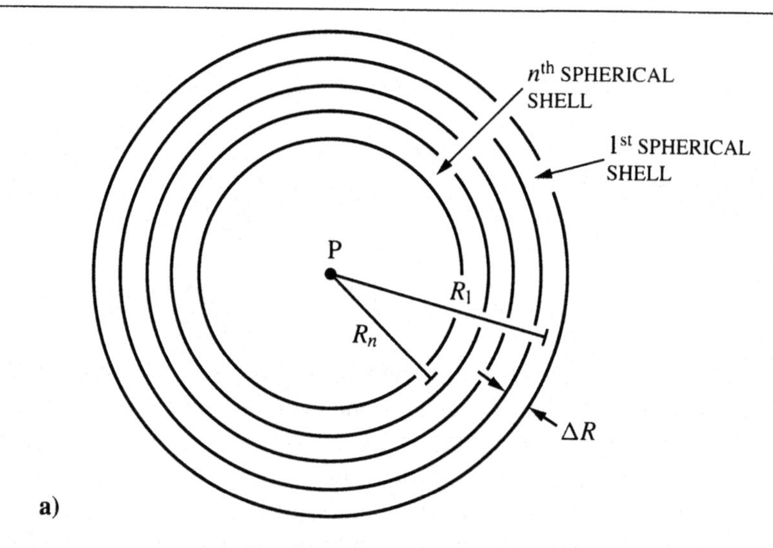

Fig. 6.3. (*continued on next page*)

For our argument, we will consider a charge of finite size: a sphere with uniform charge density and radius a_o in the rest frame of the observer. The charge is moving away from the observation point P with the radial velocity $\vec{v}$. In Figure 6.3b, the n spherical shells of the volume V are arranged so that they just enclose the charge. If we ignore retardation and evaluate all of the elements of charge Δq_i at the same time, (6.15) gives

$$\Phi(\vec{r}, t) \approx \frac{1}{4\pi\varepsilon_o R_q} \sum_{i=1}^{n} \Delta q_i \approx \frac{q}{4\pi\varepsilon_o R_q}; \tag{6.16}$$

here we have assumed that n is large ($\Delta R \ll a_o$) and $R_i \gg a_o$, and we have taken R_q to be the distance from the center of the sphere to the observation point.[2]

Now we will consider the effects of retardation on the evaluation of (6.15). In performing the summation, the charge in each spherical shell is sampled at a different retarded time; for the i-th shell, the retarded time is t_{ri}. The right-hand edge of the charge is in the first shell ($i = 1$) at the time this shell is sampled t_{r1} (Figure 6.3c). Since the charge is moving to the right, the last shell occupied by the left-hand edge of the charge at the retarded time for which the shell is sampled is the m-th shell ($m < n$) (Figure 6.3d). In other words, the charge is never in the shells $m+1, m+2, \ldots, n$ at the retarded times at which these shells are sampled $t_{rm+1}, t_{rm+2}, \ldots, t_{rn}$. For n large ($\Delta R \ll a_o$) and $R_i \gg a_o$, the potential (6.15) is now approximately

$$\Phi(\vec{r}, t) \approx \frac{1}{4\pi\varepsilon_o R_q} \sum_{i=1}^{m} \Delta q_i(t_{ri}) \approx \frac{q}{4\pi\varepsilon_o R_q}\left(\frac{m}{n}\right), \tag{6.17}$$

where $R_q \approx R_{q1}(t_{r1}) \approx R_{qm}(t_{rm})$.

[2] In Figure 6.3, n is chosen to be small ($n = 5$) to clarify the presentation; this exaggerates the size of ΔR relative to a_o.

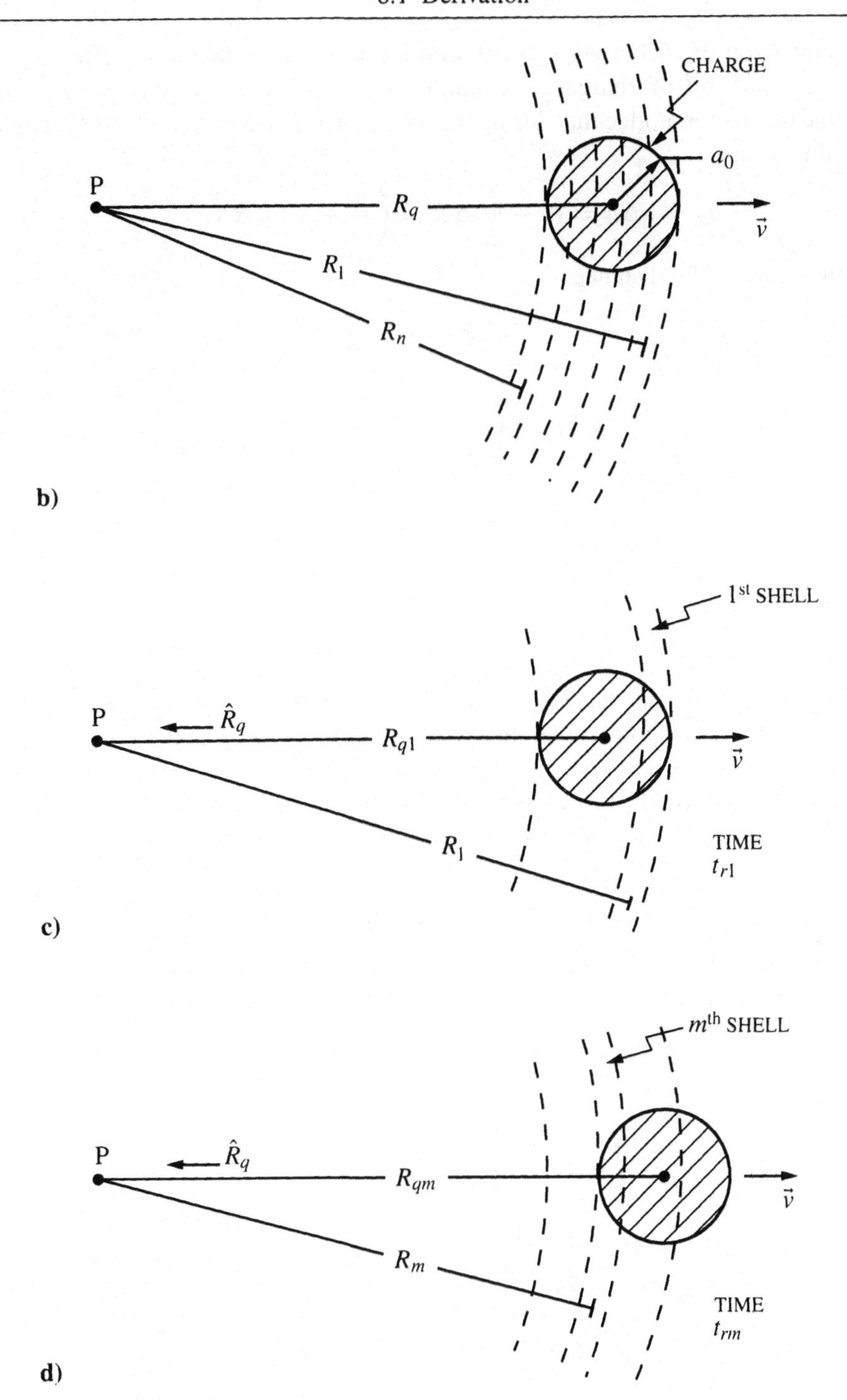

Fig. 6.3. a) Observation point P surrounded by spherical shells. b)–d) Sampling of charge in the various shells.

A comparison of (6.17) with (6.16) shows that the retardation has effectively reduced the amount of charge determining the potential by the factor m/n. To determine this factor, notice that during the time interval $\Delta t_r = t_{rm} - t_{r1}$ the charge travels the distance

$$d_q = v\Delta t_r = (n - m)\Delta R = \left(\frac{n}{m} - 1\right) m\Delta R,$$

while light travels the distance

$$d_c = c\Delta t_r = m\Delta R;$$

thus,

$$v\Delta t_r = \left(\frac{n}{m} - 1\right) c\Delta t_r,$$

or

$$m/n = 1/(1 + v/c).$$

Therefore, the potential (6.17) becomes[3]

$$\Phi(\vec{r}, t) \approx \frac{q}{4\pi\varepsilon_o R_q(1 + v/c)}, \tag{6.18}$$

which is just (6.12) for our special case: $\hat{R}_q \cdot \vec{v} = -v$. If the charge had been moving toward the observation point rather than away from it, the factor in the denominator of (6.18) would be $(1 - v/c)$ rather than $(1 + v/c)$. In general it is the velocity of the charge in the direction of the observation point that affects the integration, which explains the factor $|1 - \hat{R}_q \cdot \vec{v}/c|$ in (6.12).

6.1.2 Electromagnetic field

The electromagnetic field of the moving point charge is obtained by inserting the Liénard-Wiechert potentials into (5.58) and (5.59):

$$\begin{aligned}\vec{\mathcal{E}}(\vec{r}, t) &= -\nabla\Phi(\vec{r}, t) - \frac{\partial\vec{\mathcal{A}}(\vec{r}, t)}{\partial t}\\ &= \frac{-q}{4\pi\varepsilon_o}\int_{t'=-\infty}^{\infty}\left(\nabla\left\{\frac{\delta[t - t' - R_q(t')/c]}{R_q(t')}\right\}\right.\\ &\quad\left. + \frac{\vec{v}(t')}{c^2 R_q(t')}\left\{\frac{\partial}{\partial t}\delta[t - t' - R_q(t')/c]\right\}\right)dt',\end{aligned} \tag{6.19}$$

$$\begin{aligned}\vec{\mathcal{B}}(\vec{r}, t) &= \nabla \times \vec{\mathcal{A}}(\vec{r}, t)\\ &= \frac{\mu_o q}{4\pi}\int_{t'=-\infty}^{\infty}\nabla \times \left(\vec{v}(t')\left\{\frac{\delta[t - t' - R_q(t')/c]}{R_q(t')}\right\}\right)dt'.\end{aligned} \tag{6.20}$$

[3] Notice that the assumed finite radius of the charge a_o does not enter our final expression (6.18) for the potential; hence, we expect this result to apply to the point charge.

Here we have used the Liénard-Weichert potentials in the form (6.7), (6.8) rather than in the form (6.12), (6.13). This avoids difficulties encountered when differentiating terms with the constraint $t' = t_r = t - R_q(t_r)/c$ [5].

The two terms in the integrand of (6.19) are

$$\nabla\left\{\frac{\delta[t - t' - R_q(t')/c]}{R_q(t')}\right\} = \nabla\left(\frac{1}{R_q}\right)\delta[\,] - \frac{1}{cR_q}\delta'[\,]\nabla R_q$$

$$= \frac{-\hat{R}_q}{R_q}\left\{\frac{1}{R_q}\delta[\,] + \frac{1}{c}\delta'[\,]\right\}$$

and

$$\frac{\partial}{\partial t}\delta\big[t - t' - R_q(t')/c\big] = \delta'[\,],$$

where

$$\nabla R_q = \hat{R}_q \tag{6.21}$$

was used (Appendix B) and the prime indicates differentiation with respect to the argument. With these results, $\vec{\mathcal{E}}$ becomes

$$\vec{\mathcal{E}} = \frac{q}{4\pi\varepsilon_o}\int_{t'=-\infty}^{\infty}\left\{\frac{\hat{R}_q}{R_q^2}\delta[\,] + \frac{1}{cR_q}(\hat{R}_q - \vec{v}/c)\delta'[\,]\right\}dt'. \tag{6.22}$$

To evaluate the remaining integral, we must change the derivative of the delta function to one with respect to t': Using $g(t')$ as in (6.9), we get

$$\frac{d}{dt'}\delta\big[g(t')\big] = \delta'\big[g(t')\big]\frac{dg(t')}{dt'},$$

or

$$\delta'\big[g(t')\big] = -\left\{\frac{d}{dt'}\delta\big[g(t')\big]\right\}\Big/\big[1 - \hat{R}_q(t')\cdot\vec{v}(t')/c\big],$$

where (6.10) was used in the last step. The electric field (6.22) is then

$$\vec{\mathcal{E}} = \frac{q}{4\pi\varepsilon_o}\int_{t'=-\infty}^{\infty}\left\{\frac{\hat{R}_q}{R_q^2}\delta[\,] - \frac{(\hat{R}_q - \vec{v}/c)}{cR_q(1 - \hat{R}_q\cdot\vec{v}/c)}\frac{d}{dt'}\delta[\,]\right\}dt',$$

which, on integration of the second term by parts, becomes

$$\vec{\mathcal{E}} = \frac{q}{4\pi\varepsilon_o}\int_{t'=-\infty}^{\infty}\left\{\frac{\hat{R}_q}{R_q^2} + \frac{d}{dt'}\left[\frac{(\hat{R}_q - \vec{v}/c)}{cR_q(1 - \hat{R}_q\cdot\vec{v}/c)}\right]\right\}\delta[\,]dt'.$$

The remaining integral can be evaluated by the same procedure we used earlier with the potentials:

$$\vec{\mathcal{E}}(\vec{r},t) = \frac{q}{4\pi\varepsilon_o}\left[\frac{1}{|1-\hat{R}_q(t')\cdot\vec{v}(t')/c|}\right.$$

$$\left.\left(\frac{\hat{R}_q(t')}{R_q^2(t')}+\frac{d}{dt'}\left\{\frac{[\hat{R}_q(t')-\vec{v}(t')/c]}{cR_q(t')[1-\hat{R}_q(t')\cdot\vec{v}(t')/c]}\right\}\right)\right]_{t'=t_r}. \tag{6.23}$$

The expression for the magnetic field (6.20) is evaluated in the same manner as that for the electric field (Problem 6.3):

$$\vec{\mathcal{B}}(\vec{r},t) = \frac{-\mu_o q}{4\pi}\left[\frac{1}{|1-\hat{R}_q(t')\cdot\vec{v}(t')/c|}\right.$$

$$\left.\left(\frac{\hat{R}_q(t')\times\vec{v}(t')}{R_q^2(t')}+\frac{1}{c}\frac{d}{dt'}\left\{\frac{\hat{R}_q(t')\times\vec{v}(t')}{R_q(t')[1-\hat{R}_q(t')\cdot\vec{v}(t')/c]}\right\}\right)\right]_{t'=t_r}. \tag{6.24}$$

The differentiations that remain in (6.23) and (6.24) are straightforward but tedious; making use of

$$\frac{d\vec{R}_q}{dt'} = -\vec{v}, \qquad \frac{dR_q}{dt'} = -\hat{R}_q\cdot\vec{v}, \qquad \frac{d\hat{R}_q}{dt'} = -\frac{1}{R_q}\left[\vec{v}-(\hat{R}_q\cdot\vec{v})\hat{R}_q\right], \tag{6.25}$$

we can combine terms to obtain our final results:

$$\vec{\mathcal{E}}(\vec{r},t) = \frac{q}{4\pi\varepsilon_o}\left(\frac{1}{|1-\hat{R}_q\cdot\vec{v}/c|^3}\right.$$

$$\left.\left\{\frac{(1-v^2/c^2)(\hat{R}_q-\vec{v}/c)}{R_q^2}+\frac{\hat{R}_q\times[(\hat{R}_q-\vec{v}/c)\times\vec{a}]}{c^2R_q}\right\}\right)_{t_r}, \tag{6.26}$$

$$\vec{\mathcal{B}}(\vec{r},t) = \frac{-\mu_o cq}{4\pi}\left[\frac{1}{|1-\hat{R}_q\cdot\vec{v}/c|^3}\right.$$

$$\left.\left(\frac{(1-v^2/c^2)[\hat{R}_q\times(\vec{v}/c)]}{R_q^2}-\frac{\hat{R}_q\times\{\hat{R}_q\times[(\hat{R}_q-\vec{v}/c)\times\vec{a}]\}}{c^2R_q}\right)\right]_{t_r}. \tag{6.27}$$

Notice that the electric and magnetic fields are simply related by

$$\vec{\mathcal{B}}(\vec{r},t) = \frac{1}{c}\hat{R}_q(t_r)\times\vec{\mathcal{E}}(\vec{r},t); \tag{6.28}$$

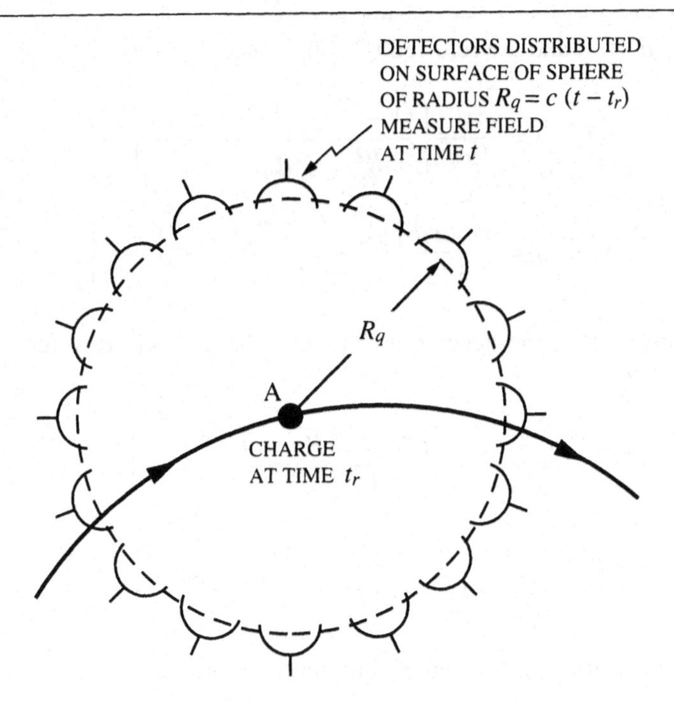

Fig. 6.4. Diagram for visualizing time retardation for the field of the moving charge.

the magnetic field is always orthogonal to the electric field and orthogonal to the vector $\hat{R}_q$.

These expressions completely describe the electromagnetic field of the moving point charge. They are complex and require careful examination. For now we will just look at some of the general characteristics of the field. In the next section, we will examine the field in detail for a few important trajectories of the charge [specified $\vec{r}_q(t)$].

The electromagnetic field specified by Equations (6.26) and (6.27) is often difficult to visualize because of its dependence on the retarded time; recall that the quantities $\vec{R}_q$, $\vec{v}$, and $\vec{a}$ are all functions of the retarded time. One way of viewing this dependence is to consider the electromagnetic field produced by the moving charge at a particular retarded time t_r (position A in Figure 6.4). Equations (6.26) and (6.27) then describe this field at the later time t everywhere on the surface of a sphere of radius $R_q = c(t - t_r)$ centered on A. We can think of the sphere as being covered with an array of detectors (antennas). All of the detectors measure the field at the same time t, and their readings tell us the distribution in space of the field produced by the charge at the earlier time t_r. In this way we can speak of a pattern for the field, that is, a graph showing the amplitude of the field versus the angular position on the surface of the sphere.

The electromagnetic field clearly contains terms proportional to R_q^{-2} as well as terms proportional to R_q^{-1}. The former are the only terms present when there is no acceleration ($\vec{a} = 0$) and are referred to as the *velocity field*:

$$\vec{\mathcal{E}}_v(\vec{r}, t) = \frac{q}{4\pi\varepsilon_o}\left[\frac{(1 - v^2/c^2)(\hat{R}_q - \vec{v}/c)}{R_q^2|1 - \hat{R}_q \cdot \vec{v}/c|^3}\right]_{t_r}, \tag{6.29}$$

$$\vec{\mathcal{B}}_v(\vec{r}, t) = \frac{-\mu_o c q}{4\pi}\left\{\frac{(1 - v^2/c^2)[\hat{R}_q \times (\vec{v}/c)]}{R_q^2|1 - \hat{R}_q \cdot \vec{v}/c|^3}\right\}_{t_r}. \tag{6.30}$$

The latter result from the acceleration of the charge and are referred to as the *acceleration field*:

$$\vec{\mathcal{E}}_a(\vec{r}, t) = \frac{q}{4\pi\varepsilon_o c^2}\left\{\frac{\hat{R}_q \times [(\hat{R}_q - \vec{v}/c) \times \vec{a}]}{R_q|1 - \hat{R}_q \cdot \vec{v}/c|^3}\right\}_{t_r}, \tag{6.31}$$

$$\vec{\mathcal{B}}_a(\vec{r}, t) = \frac{\mu_o q}{4\pi c}\left(\frac{\hat{R}_q \times \{\hat{R}_q \times [(\hat{R}_q - \vec{v}/c) \times \vec{a}\,]\}}{R_q|1 - \hat{R}_q \cdot \vec{v}/c|^3}\right)_{t_r}. \tag{6.32}$$

Notice that the vectors $\vec{\mathcal{E}}_a$, $\vec{\mathcal{B}}_a$, and $\hat{R}_q$ are mutually orthogonal for the acceleration field:

$$(\hat{R}_q)_{t_r} \cdot \vec{\mathcal{E}}_a = 0, \qquad (\hat{R}_q)_{t_r} \cdot \vec{\mathcal{B}}_a = 0, \qquad \vec{\mathcal{B}}_a = \frac{1}{c}(\hat{R}_q)_{t_r} \times \vec{\mathcal{E}}_a. \tag{6.33}$$

The ratio of the acceleration field to the velocity field is

$$\left|\frac{\vec{\mathcal{E}}_a}{\vec{\mathcal{E}}_v}\right| = \left[\frac{R_q a|\hat{R}_q \times (\hat{F} \times \hat{a})|}{c^2(1 - v^2/c^2)}\right]_{t_r}, \tag{6.34}$$

with the unit vector

$$\hat{F} = \frac{(\hat{R}_q - \vec{v}/c)}{|\hat{R}_q - \vec{v}/c|}.$$

Assuming the triple product of unit vectors in (6.34) is not zero and $a \neq 0$, the distance R_q can always be made so large that the velocity field is negligible. Thus, the acceleration field predominates at large distances from the charge; it falls off as R_q^{-1}, and it obeys the relations in (6.33). Recall that these are the characteristics we found earlier for the radiated field. We conclude that at a large distance, the acceleration field is the radiated field of the moving charge. A charge in *free space* must undergo acceleration to produce electromagnetic radiation.

We can check our formulas, (6.26) and (6.27), for a few simple limiting cases. When the charge is stationary ($\vec{v} = 0, \vec{a} = 0$), we have the electric field

$$\vec{\mathcal{E}}(\vec{r}) = \frac{q}{4\pi\varepsilon_o R_q^2}\hat{R}_q.$$

This is just the electrostatic field of a point charge (5.103). When the charge is moving at constant velocity ($\vec{a} = 0$), such that $v/c \ll 1$, and there is negligible time retardation ($t_r \approx t$), we have the magnetic field

$$\vec{\mathcal{B}}(\vec{r}) = \frac{\mu_o}{4\pi R_q^2}(q\vec{v} \times \hat{R}_q).$$

We recognize this result to be of the same form as the Biot-Savart formula for the magnetic field of the current element $\mathcal{I}d\vec{\ell}$ (Problem 1.3):

$$\vec{\mathcal{B}}(\vec{r}) = \frac{\mu_o}{4\pi R_q^2}(\mathcal{I}d\vec{\ell} \times \hat{R}_q).$$

The electromagnetic field of the moving point charge can be expressed in an alternative form, which was first given by Oliver Heaviside in about 1904 and then rediscovered and popularized by Richard Philips Feynman (1918–1988) about sixty years later [4, 6–8]. The Heaviside-Feynman form for the field can be obtained directly from (6.23) and (6.24). First we notice that

$$\begin{aligned} \frac{d}{dt}[f(t')]_{t_r} &= \left[\frac{df(t')}{dt'}\right]_{t_r} \frac{d}{dt}[t']_{t_r} \\ &= \left[\frac{df(t')}{dt'}\right]_{t_r} \left\{1 - \frac{1}{c}\frac{d}{dt}[R_q(t')]_{t_r}\right\}, \end{aligned} \tag{6.35}$$

where f is a general function. Letting $f = R_q$ in (6.35) and making use of (6.25), we find that

$$\frac{d}{dt}[R_q]_{t_r} = -[\hat{R}_q \cdot \vec{v}]_{t_r}\left\{1 - \frac{1}{c}\frac{d}{dt}[R_q]_{t_r}\right\},$$

which can be rearranged to give

$$\frac{d}{dt}[R_q]_{t_r} = \left[\frac{-\hat{R}_q \cdot \vec{v}}{1 - \hat{R}_q \cdot \vec{v}/c}\right]_{t_r}, \tag{6.36}$$

or

$$1 - \frac{1}{c}\frac{d}{dt}[R_q]_{t_r} = \left[\frac{1}{1 - \hat{R}_q \cdot \vec{v}/c}\right]_{t_r}. \tag{6.37}$$

Substituting (6.37) into (6.35) yields

$$\frac{d}{dt}[f]_{t_r} = \left[\frac{1}{(1 - \hat{R}_q \cdot \vec{v}/c)}\frac{df}{dt'}\right]_{t_r}. \tag{6.38}$$

The electric field (6.23) is

$$\vec{\mathcal{E}}(\vec{r},t) = \frac{q}{4\pi\varepsilon_o}\left(\left[\frac{\hat{R}_q}{R_q^2}\right]_{t_r}\left[\frac{1}{1-\hat{R}_q\cdot\vec{v}/c}\right]_{t_r}\right.$$

$$+\frac{1}{c}\left\{\frac{1}{(1-\hat{R}_q\cdot\vec{v}/c)}\frac{d}{dt'}\left[\frac{1}{(1-\hat{R}_q\cdot\vec{v}/c)}\frac{\hat{R}_q}{R_q}\right.\right.$$

$$\left.\left.\left.+\frac{1}{cR_q}\frac{1}{(1-\hat{R}_q\cdot\vec{v}/c)}\frac{d\vec{R}_q}{dt'}\right]\right\}_{t_r}\right),$$

which on using (6.37) and (6.38) becomes

$$\vec{\mathcal{E}}(\vec{r},t) = \frac{q}{4\pi\varepsilon_o}\left[\left[\frac{\hat{R}_q}{R_q^2}\right]_{t_r}\left\{1-\frac{1}{c}\frac{d}{dt}[R_q]_{t_r}\right\}\right.$$

$$\left.+\frac{1}{c}\frac{d}{dt}\left(\left[\frac{\hat{R}_q}{R_q}\right]_{t_r}\left\{1-\frac{1}{c}\frac{d}{dt}[R_q]_{t_r}\right\}+\frac{1}{c}\left[\frac{1}{R_q}\right]_{t_r}\frac{d}{dt}[\vec{R}_q]_{t_r}\right)\right].$$

After combining terms in this expression, we obtain the Heaviside-Feynman form for the electric field:

$$\vec{\mathcal{E}}(\vec{r},t) = \frac{q}{4\pi\varepsilon_o}\left\{\left[\frac{\hat{R}_q}{R_q^2}\right]_{t_r}+\frac{1}{c}[R_q]_{t_r}\frac{d}{dt}\left[\frac{\hat{R}_q}{R_q^2}\right]_{t_r}+\frac{1}{c^2}\frac{d^2}{dt^2}[\hat{R}_q]_{t_r}\right\}. \quad (6.39)$$

Similar operations applied to (6.24) give the magnetic field (Problem 6.3):

$$\vec{\mathcal{B}}(\vec{r},t) = \frac{\mu_o q}{4\pi}\left\{\left[\frac{\hat{R}_q}{R_q}\right]_{t_r}\times\frac{d}{dt}[\hat{R}_q]_{t_r}+\frac{1}{c}[\hat{R}_q]_{t_r}\times\frac{d^2}{dt^2}[\hat{R}_q]_{t_r}\right\}. \quad (6.40)$$

Though these expressions appear to be quite different from our earlier results, (6.26) and (6.27), they are completely equivalent.

Both Heaviside and Feynman give interesting interpretations of these formulas that we will not repeat here [4, 6]. We will only mention that the radiated electric field is given by the third term in (6.39), which is proportional to $d^2[\hat{R}_q]_{t_r}/dt^2$. Recall that $[\hat{R}_q]_{t_r}$ is the unit vector pointing from the position of the charge at the retarded time t_r to the observation point (Figure 6.1). For this representation, we "watch" the motion of this vector and take the second derivative at the time of observation to determine the radiated field.

This result is to be compared with our original expression for the radiated electric field (6.31), which is proportional to the acceleration of the charge at the retarded time, $[\vec{a}]_{t_r} = [d^2\vec{r}_q/dt^2]_{t_r}$. For this representation, we "watch" the motion of the charge, the vector $\vec{r}_q$, and take the second derivative at the retarded time to determine the radiated field.

Perhaps it is the difference in these two representations that prompted

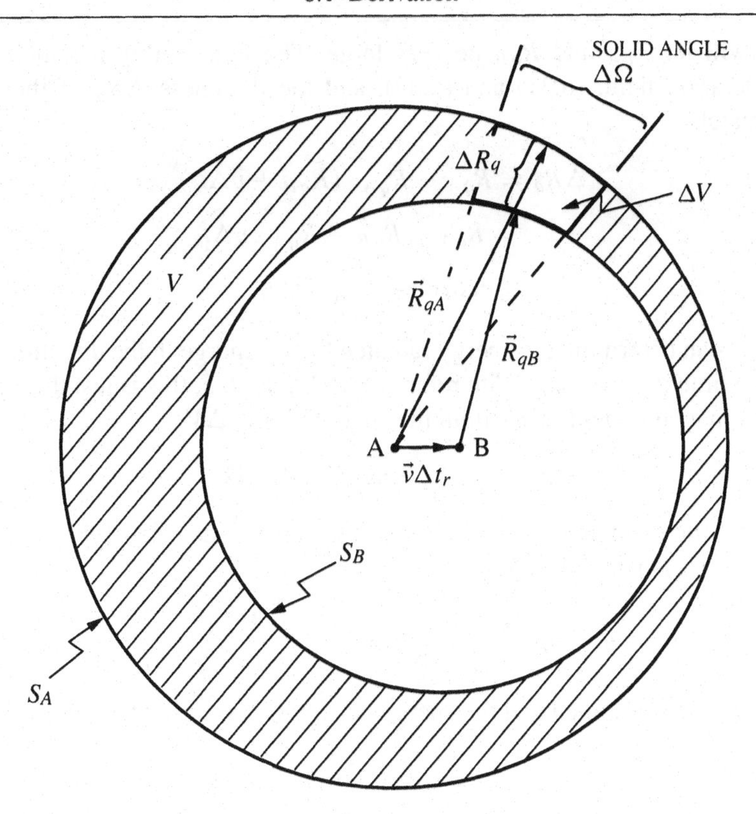

Fig. 6.5. Region containing energy radiated by the charge during the time interval $\Delta t_r = t_{rB} - t_{rA}$.

Heaviside [6] to refer to the former (6.39) as "a very curious way of representing $\vec{\mathcal{E}}$, and physically very unnatural."

6.1.3 Radiated power

As the point charge accelerates, it radiates energy into the surrounding space [9–11]. To determine the rate at which energy leaves the charge, we will consider the motion of the charge over a very short interval of the retarded time, $\Delta t_r = t_{rB} - t_{rA}$. During this interval the velocity and acceleration are approximately constant, and the charge moves the distance $v\Delta t_r$ from point A to point B, as shown in Figure 6.5.

At the observation time t, the field radiated by the charge at the earlier time t_{rA} is on the spherical surface S_A of radius

$$R_{qA} = c(t - t_{rA});$$

the field radiated at the earlier time t_{rB} is on the spherical surface S_B of radius

$$R_{qB} = c(t - t_{rB}).$$

The field within the shell of volume V bounded by these two spheres (the shaded region in Figure 6.5) is entirely due to radiation from the charge during the time interval Δt_r. The field outside S_A was radiated before t_{rA} and that inside S_B after

t_{rB}. We will let R_{qA} and R_{qB} be very large. The field within the shell is then the acceleration field, (6.31) and (6.32), and the thickness ΔR_q of the shell is approximately

$$\begin{aligned}\Delta R_q &= R_{qA} - \hat{R}_{qA} \cdot (\vec{R}_{qB} + \vec{v}\Delta t_r) \\ &\approx R_{qA} - R_{qB} - \hat{R}_{qA} \cdot \vec{v}\Delta t_r \\ &\approx c\Delta t_r(1 - \hat{R}_{qA} \cdot \vec{v}/c).\end{aligned} \tag{6.41}$$

Now we can determine the energy radiated by the charge in the direction of $\hat{R}_{qA}$ during the time interval Δt_r. With reference to Figure 6.5, this is just the energy of the electromagnetic field $\Delta\mathcal{U}$ within the small volume ΔV:

$$\Delta\mathcal{U} = u\Delta V = uR_{qA}^2\Delta R_q\Delta\Omega. \tag{6.42}$$

Here $\Delta\Omega$ is the element of solid angle; u is the energy density of the field, which is given by (1.114) and (1.115):

$$u = u_e + u_m = \frac{1}{2}\varepsilon_o|\vec{\mathcal{E}}_a|^2 + \frac{1}{2}\frac{1}{\mu_o}|\vec{\mathcal{B}}_a|^2. \tag{6.43}$$

In (6.42) we have assumed that the energy density is approximately uniform within ΔV.

For the acceleration field, the electric and magnetic energy densities are equal (Problem 6.4):

$$\begin{aligned}u_m &= \frac{1}{2}\frac{1}{\mu_o}|\vec{\mathcal{B}}_a|^2 = \frac{1}{2}\frac{1}{\mu_o c^2}|\hat{R}_q \times \vec{\mathcal{E}}_a|^2 \\ &= \frac{1}{2}\varepsilon_o|\vec{\mathcal{E}}_a|^2 = u_e.\end{aligned} \tag{6.44}$$

So, after inserting (6.41), (6.43), and (6.44), (6.42) becomes

$$\Delta\mathcal{U} = c\varepsilon_o|\vec{\mathcal{E}}_a|^2 R_{qA}^2(1 - \hat{R}_{qA} \cdot \vec{v}/c)\Delta t_r\Delta\Omega,$$

or

$$\frac{\Delta\mathcal{U}}{\Delta t_r\Delta\Omega} = c\varepsilon_o|\vec{\mathcal{E}}_a|^2\, R_{qA}^2(1 - \hat{R}_{qA} \cdot \vec{v}/c).$$

In the limit as $\Delta t_r \to 0$, this is just the energy per unit time, the power $\mathcal{P}$, radiated per unit solid angle in the direction of $\hat{R}_{qA}$:

$$\frac{d\mathcal{P}_q(t_r)}{d\Omega} = c\varepsilon_o|\vec{\mathcal{E}}_a|^2 R_q^2(1 - \hat{R}_q \cdot \vec{v}/c). \tag{6.45}$$

After inserting (6.31) we have

$$\frac{d\mathcal{P}_q(t_r)}{d\Omega} = \frac{q^2}{16\pi^2\varepsilon_o c^3}\left\{\frac{|\hat{R}_q \times [(\hat{R}_q - \vec{v}/c) \times \vec{a}]|^2}{(1 - \hat{R}_q \cdot \vec{v}/c)^5}\right\}_{t_r}, \tag{6.46}$$

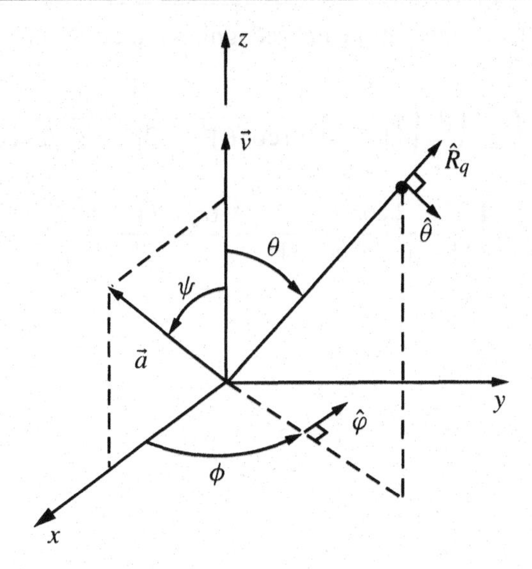

Fig. 6.6. Orientation of the vectors $\vec{v}$, $\vec{a}$, and $\hat{R}_q$.

or, on performing the vector operations and rearranging terms,

$$\frac{d\mathcal{P}_q(t_r)}{d\Omega} = \frac{q^2}{16\pi^2\varepsilon_o c^3}$$

$$\left\{\frac{(1-\hat{R}_q\cdot\vec{v}/c)^2a^2 + 2(1-\hat{R}_q\cdot\vec{v}/c)(\vec{a}\cdot\vec{v}/c)(\hat{R}_q\cdot\vec{a}) - (1-v^2/c^2)(\hat{R}_q\cdot\vec{a})^2}{(1-\hat{R}_q\cdot\vec{v}/c)^5}\right\}_{t_r}. \tag{6.47}$$

Equation (6.47) expresses the power radiated per unit solid angle in the direction of $\hat{R}_q$; to get the total power radiated by the charge we must integrate (6.47) over all solid angles. We will use the coordinates shown in Figure 6.6: θ is the angle between $\vec{v}$ and $\hat{R}_q$, $\vec{v}$ and $\vec{a}$ are in the plane $\phi = 0$, and the angle between $\vec{v}$ and $\vec{a}$ is ψ. Then the total power radiated becomes

$$\begin{aligned}
\mathcal{P}_q(t_r) &= \int_{\phi=0}^{2\pi}\int_{\theta=0}^{\pi} \frac{d\mathcal{P}(t_r)}{d\Omega}\sin\theta d\theta d\phi \\
&= \frac{q^2}{16\pi^2\varepsilon_o c^3}\int_{\phi=0}^{2\pi}\int_{\theta=0}^{\pi}\Bigg(a^2\Bigg\{\frac{1}{[1-(v/c)\cos\theta]^3} \\
&\quad + \frac{2(v/c)\cos\psi}{[1-(v/c)\cos\theta]^4}[\sin\psi\sin\theta\cos\phi + \cos\psi\cos\theta] \\
&\quad - \frac{(1-v^2/c^2)}{[1-(v/c)\cos\theta]^5}[\sin\psi\sin\theta\cos\phi + \cos\psi\cos\theta]^2\Bigg\}\Bigg)_{t_r}\sin\theta d\theta d\phi,
\end{aligned}$$

or, on performing the ϕ integration and recognizing that $va\cos\psi = \vec{v}\cdot\vec{a}$,

$$\mathcal{P}_q(t_r) = \frac{q^2}{8\pi\varepsilon_o c^3}\int_{\theta=0}^{\pi}\left(a^2\left\{\frac{1}{[1-(v/c)\cos\theta]^3} - \frac{(1-v^2/c^2)\sin^2\theta}{2[1-(v/c)\cos\theta]^5}\right\}\right.$$

$$\left. + \frac{1}{cv}(\vec{v}\cdot\vec{a})^2\left\{\frac{2\cos\theta}{[1-(v/c)\cos\theta]^4} - \frac{(c/v)(1-v^2/c^2)(3\cos^2\theta - 1)}{2[1-(v/c)\cos\theta]^5}\right\}\right)_{t_r}$$

$$\sin\theta d\theta. \tag{6.48}$$

With the change of variable

$$w = 1 - (v/c)\cos\theta,$$

(6.48) becomes

$$\mathcal{P}_q(t_r) = \frac{q^2(c/v)}{8\pi\varepsilon_o c^3}\int_{w=1-v/c}^{1+v/c}\left(a^2\left\{\frac{1}{w^3} - \frac{1}{2w^5}(1-v^2/c^2)\left[1-(c/v)^2(1-w)^2\right]\right\}\right.$$

$$+ \frac{(\vec{v}\cdot\vec{a})^2}{cv}\left\{\frac{2}{w^4}(c/v)(1-w) - \frac{1}{2w^5}(c/v)(1-v^2/c^2)\right.$$

$$\left.\left.\left[3(c/v)^2(1-w)^2 - 1\right]\right\}\right)_{t_r} dw. \tag{6.49}$$

The integrand now only involves reciprocal powers of w, and the integral is easily evaluated. After rearranging terms, we have our final result for the total power radiated by the charge (Problem 6.5):

$$\mathcal{P}_q(t_r) = \frac{q^2}{6\pi\varepsilon_o c^3}\left(\frac{1}{(1-v^2/c^2)^2}\left\{a^2 + \frac{[(\vec{v}/c)\cdot\vec{a}]^2}{(1-v^2/c^2)}\right\}\right)_{t_r}. \tag{6.50}$$

The power radiated is seen to be proportional to the square of the acceleration and to increase drastically as the velocity of the charge approaches the speed of light due to the powers of $(1-v^2/c^2)$ in the denominator. The relative directions of the velocity and acceleration affect the power radiated through the factor $\vec{v}\cdot\vec{a}$. We will examine this dependence in more detail in the next section.

The calculation we just performed could have been approached from a different point of view. We will outline this alternative, since it provides additional insight into the radiation process. The Poynting vector for the acceleration field is obtained from (6.31) and (6.32):

$$\vec{\mathcal{S}}_a(\vec{r},t) = \frac{1}{\mu_o}\vec{\mathcal{E}}_a(\vec{r},t)\times\vec{\mathcal{B}}_a(\vec{r},t)$$

$$= \frac{q^2}{16\pi^2\varepsilon_o c^3}\left\{\frac{|\hat{R}_q\times[(\hat{R}_q - \vec{v}/c)\times\vec{a}]|^2}{R_q^2(1-\hat{R}_q\cdot\vec{v}/c)^6}\hat{R}_q\right\}_{t_r}. \tag{6.51}$$

This is the energy per unit time passing through a unit area of a spherical surface surrounding the charge, such as the sphere shown in Figure 6.4 or the sphere of radius R_{qA} in Figure 6.5. It is the power density that an observer on the sphere would measure at time t. To obtain the power per unit solid angle in the direction of $\hat{R}_q$ we must take $R_q^2 \hat{R}_q \cdot \vec{S}_a$:

$$\frac{d\mathcal{P}_o(t)}{d\Omega} = \frac{q^2}{16\pi^2\varepsilon_o c^3}\left\{\frac{|\hat{R}_q \times [(\hat{R}_q - \vec{v}/c) \times \vec{a}]|^2}{(1 - \hat{R}_q \cdot \vec{v}/c)^6}\right\}_{t_r}. \tag{6.52}$$

Notice that (6.52) is not the same as (6.46); the former is the energy per unit time (t), per unit solid angle as measured by the observer on the sphere,

$$\frac{d\mathcal{P}_o(t)}{d\Omega} = \lim_{\Delta t \to 0}\left[\frac{\Delta \mathcal{U}}{\Delta t \Delta\Omega}\right],$$

whereas the latter is the energy per unit time (t_r), per unit solid angle radiated by the charge,

$$\frac{d\mathcal{P}_q(t_r)}{d\Omega} = \lim_{\Delta t_r \to 0}\left[\frac{\Delta \mathcal{U}}{\Delta t_r \Delta\Omega}\right].$$

The time intervals Δt and Δt_r are not equal. This can be clearly seen from Figure 6.5. The time interval Δt_r, during which the radiation is produced, is fixed. However, the time interval Δt, during which the radiation is measured on the sphere of radius R_{qA}, depends on the direction of observation; it is equal to the thickness of the shell ΔR_q divided by the speed of light:

$$\Delta t = \Delta R_q/c = (1 - \hat{R}_q \cdot \vec{v}/c)\Delta t_r,$$

or[4]

$$\frac{dt}{dt_r} = (1 - \hat{R}_q \cdot \vec{v}/c)_{t_r}. \tag{6.53}$$

The power radiated by the charge per unit solid angle in the direction of $\hat{R}_q$ is now easily obtained from (6.52):

$$\begin{aligned}\frac{d\mathcal{P}_q(t_r)}{d\Omega} &= \frac{d\mathcal{P}_o(t)}{d\Omega}\frac{dt}{dt_r} \\ &= \frac{q^2}{16\pi^2\varepsilon_o c^3}\left\{\frac{|\hat{R}_q \times [(\hat{R}_q - \vec{v}/c) \times \vec{a}]|^2}{(1 - \hat{R}_q \cdot \vec{v}/c)^5}\right\}_{t_r},\end{aligned}$$

which agrees with our previous result (6.46).

6.2 Special cases

The formulas obtained in the last section completely describe the classical electromagnetic radiation from a moving point charge. To acquire physical understanding

[4] This result can also be obtained by letting f equal t' in (6.38).

Table 6.1. *Moving point charge in free space*

	General: $v/c < 1$	$v/c \ll 1$
	Velocity field	
$\vec{\mathcal{E}}_v(\vec{r}, t)$	$\dfrac{q}{4\pi\varepsilon_o}\left[\dfrac{\hat{R}_q - \vec{\beta}}{\gamma^2 R_q^2 (1 - \hat{R}_q \cdot \vec{\beta})^3}\right]_{t_r}$	$\dfrac{q}{4\pi\varepsilon_o}\left[\dfrac{\hat{R}_q}{R_q^2}\right]_{t_r}$
$\vec{\mathcal{B}}_v(\vec{r}, t)$	$\dfrac{-\mu_o c q}{4\pi}\left[\dfrac{\hat{R}_q \times \vec{\beta}}{\gamma^2 R_q^2 (1 - \hat{R}_q \cdot \vec{\beta})^3}\right]_{t_r}$	$\dfrac{-\mu_o c q}{4\pi}\left[\dfrac{\hat{R}_q \times \vec{\beta}}{R_q^2}\right]_{t_r}$
	Acceleration field	
$\vec{\mathcal{E}}_a(\vec{r}, t)$	$\dfrac{q}{4\pi\varepsilon_o c^2}\left\{\dfrac{\hat{R}_q \times [(\hat{R}_q - \vec{\beta}) \times \vec{a}]}{R_q (1 - \hat{R}_q \cdot \vec{\beta})^3}\right\}_{t_r}$	$\dfrac{q}{4\pi\varepsilon_o c^2}\left[\dfrac{\hat{R}_q \times (\hat{R}_q \times \vec{a})}{R_q}\right]_{t_r}$
$\vec{\mathcal{B}}_a(\vec{r}, t)$	$\dfrac{\mu_o q}{4\pi c}\left(\dfrac{\hat{R}_q \times \{\hat{R}_q \times [(\hat{R}_q - \vec{\beta}) \times \vec{a}]\}}{R_q (1 - \hat{R}_q \cdot \vec{\beta})^3}\right)_{t_r}$	$\dfrac{-\mu_o q}{4\pi c}\left[\dfrac{\hat{R}_q \times \vec{a}}{R_q}\right]_{t_r}$
	Power radiated per unit solid angle	
$\dfrac{d\mathcal{P}_q(t_r)}{d\Omega}$	$\dfrac{q^2}{16\pi^2\varepsilon_o c^3}\left\{\dfrac{\lvert\hat{R}_q \times [(\hat{R}_q - \vec{\beta}) \times \vec{a}]\rvert^2}{(1 - \hat{R}_q \cdot \vec{\beta})^5}\right\}_{t_r}$	$\dfrac{q^2}{16\pi^2\varepsilon_o c^3}[\lvert\hat{R}_q \times \vec{a}\rvert^2]_{t_r}$
	Power radiated per unit solid angle measured by observer on surrounding sphere	
$\dfrac{d\mathcal{P}_o(t)}{d\Omega}$	$\dfrac{q^2}{16\pi^2\varepsilon_o c^3}\left\{\dfrac{\lvert\hat{R}_q \times [(\hat{R}_q - \vec{\beta}) \times \vec{a}]\rvert^2}{(1 - \hat{R}_q \cdot \vec{\beta})^6}\right\}_{t_r}$	$\dfrac{q^2}{16\pi^2\varepsilon_o c^3}[\lvert\hat{R}_q \times \vec{a}\rvert^2]_{t_r}$
	Total power radiated	
$\mathcal{P}_q(t_r)$	$\dfrac{q^2}{6\pi\varepsilon_o c^3}\left\{\gamma^4[a^2 + \gamma^2(\vec{\beta} \cdot \vec{a})^2]\right\}_{t_r}$	$\dfrac{q^2}{6\pi\varepsilon_o c^3}[a^2]_{t_r}$

$t_r = t - R_q(t_r)/c, \qquad \vec{\beta} = \vec{v}/c, \qquad \beta = \lvert\vec{\beta}\rvert, \qquad \gamma = (1 - \beta^2)^{-1/2}$

of these rather complicated looking results, we will examine in detail a few particular cases, that is, a few particular trajectories for the charge. Before we begin this examination, we will summarize the results of the previous section in Table 6.1. Here we have included the expressions for the electromagnetic field, both the velocity and acceleration terms, the power radiated per unit solid angle, and the total power radiated by the charge. The notation

$$\vec{\beta} \equiv \vec{v}/c, \qquad \beta \equiv |\vec{\beta}|,$$

and

$$\gamma \equiv (1 - v^2/c^2)^{-1/2} = (1 - \beta^2)^{-1/2} \tag{6.54}$$

has been introduced to simplify the formulas.

6.2.1 *Low velocity: v/c ≪ 1*

When the velocity of the charge is much less than the speed of light, $v/c \ll 1$ ($\beta \ll 1$, $\gamma \approx 1$), the expressions in Table 6.1 simplify to those given on the right-hand side. Notice for this case that the relative orientation of the velocity with respect to the acceleration does not enter these expressions. So when we introduce the angular coordinates of Figure 6.6, we can choose the velocity to be parallel to the acceleration ($\psi = 0$) without loss of generality. Then the vector products in these expressions become

$$\hat{R}_q \times \vec{a} = -a \sin\theta\hat{\phi},$$

$$\hat{R}_q \times (\hat{R}_q \times \vec{a}) = a \sin\theta\hat{\theta}, \tag{6.55}$$

and the radiated electromagnetic field (acceleration field) is

$$\vec{\mathcal{E}}_a = \frac{q}{4\pi\varepsilon_o c^2}\left[\frac{a\sin\theta}{R_q}\hat{\theta}\right]_{t_r}, \tag{6.56}$$

$$\vec{\mathcal{B}}_a = \frac{\mu_o q}{4\pi c}\left[\frac{a\sin\theta}{R_q}\hat{\phi}\right]_{t_r}. \tag{6.57}$$

For this simple case ($\beta \ll 1$), the power radiated by the charge per unit solid angle $d\mathcal{P}_q/d\Omega$ (6.46) is the same as the power per unit solid angle measured by an observer $d\mathcal{P}_o/d\Omega$ (6.52):

$$\frac{d\mathcal{P}_q(t_r)}{d\Omega} = \frac{d\mathcal{P}_o(t)}{d\Omega} = \frac{q^2}{16\pi^2\varepsilon_o c^3}[a^2\sin^2\theta]_{t_r}. \tag{6.58}$$

In Figure 6.7a, the directional characteristics of the radiation are shown in a three-dimensional representation [12]. In this figure the radial distance from the origin is proportional to the power per unit solid angle radiated in the direction determined by the angles θ and ϕ. The pattern has the shape of a fat doughnut or fat tire, with a null in the direction of the acceleration and a maximum in the plane normal to the acceleration. In Figure 6.7b a quadrant of the figure is removed to show the cross section of the pattern.

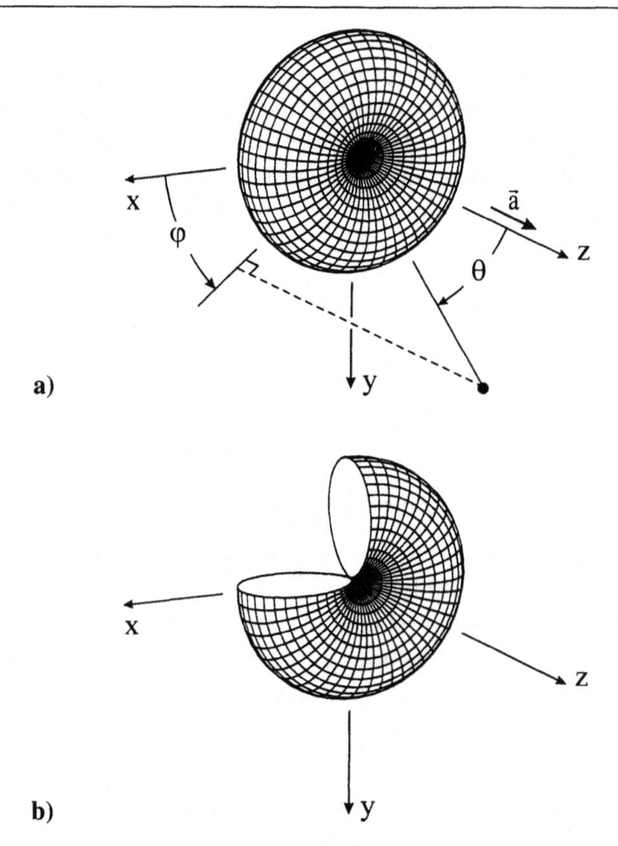

Fig. 6.7. Directional characteristics of the radiation, $d\mathcal{P}_q/d\Omega$, from a charge moving at low velocity ($v/c \ll 1$).

When the pattern is for $d\mathcal{P}_o/d\Omega$, we have the following physical explanation. With reference to Figure 6.4, the radial distance in Figure 6.7a is proportional to the power per unit solid angle measured at time t by an array of detectors distributed on a spherical surface that was centered on the charge at the earlier retarded time t_r. Of course, the radius of the sphere R_q must be large enough that only the radiated field is significant on the surface.

The total power radiated by the charge is given by *Larmor's formula* (Sir Joseph Larmor, 1857–1942) [13, 14],

$$\mathcal{P}_q(t_r) = \frac{q^2}{6\pi\varepsilon_o c^3}[a^2]_{t_r}. \tag{6.59}$$

Notice that the power is proportional to the square of the acceleration.

6.2.2 Relativistic velocity, $\vec{a}$ parallel to $\vec{v}$

The general expressions for the radiation from a moving point charge (left-hand side of Table 6.1) contain the factors β and γ, which cause a marked departure from the low-velocity results when the velocity of the charge is close to the speed of light, $v/c \approx 1$ ($\beta \approx 1,\ 1/\gamma \approx 0$). For this case, the relative orientation of the velocity

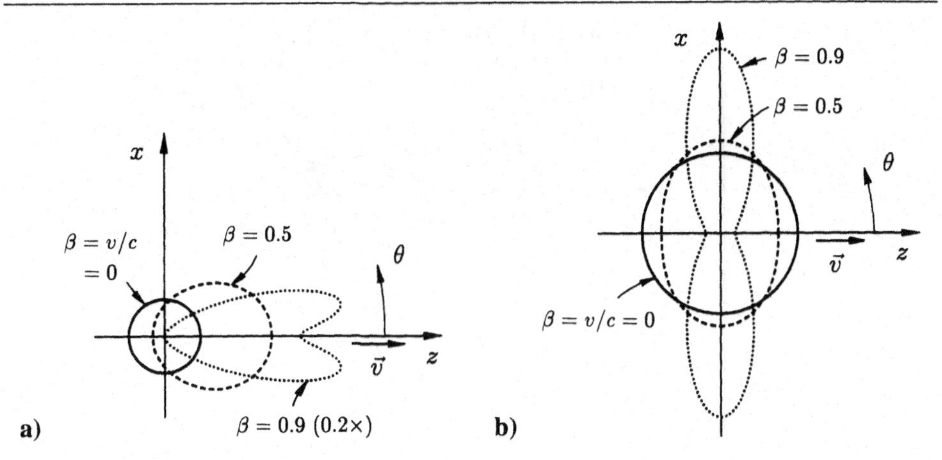

Fig. 6.8. Magnitude of electric field versus angle θ for a charge moving with constant velocity $\vec{v}$. a) Field on spherical surface centered on charge at retarded time t_r. b) Field on spherical surface centered on charge at current time t.

with respect to the acceleration is important. Two orientations of practical interest are when the acceleration is parallel to the velocity and when the acceleration is perpendicular to the velocity. We will consider the former first.

Before we introduce an acceleration, we will consider a charge moving with a constant velocity along the z axis ($\vec{a} = 0$). We will then only have the velocity field, and it is easily expressed in terms of the angular coordinates in Figure 6.6:

$$\vec{\mathcal{E}}_v = \frac{q}{4\pi\varepsilon_o}\left[\frac{1}{\gamma^2 R_q^2(1-\beta\cos\theta)^3}(\hat{R}_q - \beta\hat{z})\right]_{t_r}, \tag{6.60}$$

$$\vec{\mathcal{B}}_v = \frac{\mu_o c q}{4\pi}\left[\frac{\beta\sin\theta}{\gamma^2 R_q^2(1-\beta\cos\theta)^3}\hat{\phi}\right]_{t_r}. \tag{6.61}$$

The electric field is rotationally symmetric about the z axis; Figure 6.8a shows a graph of the field in the x-z plane ($\phi = 0$). Here we plotted the magnitude of the electric field as a function of the angle θ for different values of β. This is the field that would be measured at time t by an array of detectors on a spherical surface of radius $R_q(t_r)$, centered on the charge at the *retarded time* t_r (Figure 6.4). When $\beta = 0$ we have the familiar electrostatic field, and the pattern is a circle. As β increases, the pattern becomes elongated in the direction of the velocity ($\theta = 0$) due to the factor $(1-\beta\cos\theta)^3$ in the denominator of (6.60). Notice that there is a scale change for $\beta = 0.9$; the results are multiplied by 0.2.

Because the charge is moving with a constant velocity, we know its position at any time. Hence, it is possible to determine the field that would be measured on a spherical surface of radius $R_q(t)$, centered on the charge at the *current time* t [15]. To do this, we must express the electric field (6.60) in terms of the radius $R_q(t)$ and the angle $\theta(t)$ shown in Figure 6.9. Notice that the electric field at each point P on the circle $R_q(t) = $ constant is associated with a different retarded time:

$$R_q^2(t_r) = \left[c(t-t_r)\right]^2 = \left[v(t-t_r)\right]^2 + R_q^2(t) - 2v(t-t_r)R_q(t)\cos\left[\pi - \theta(t)\right].$$

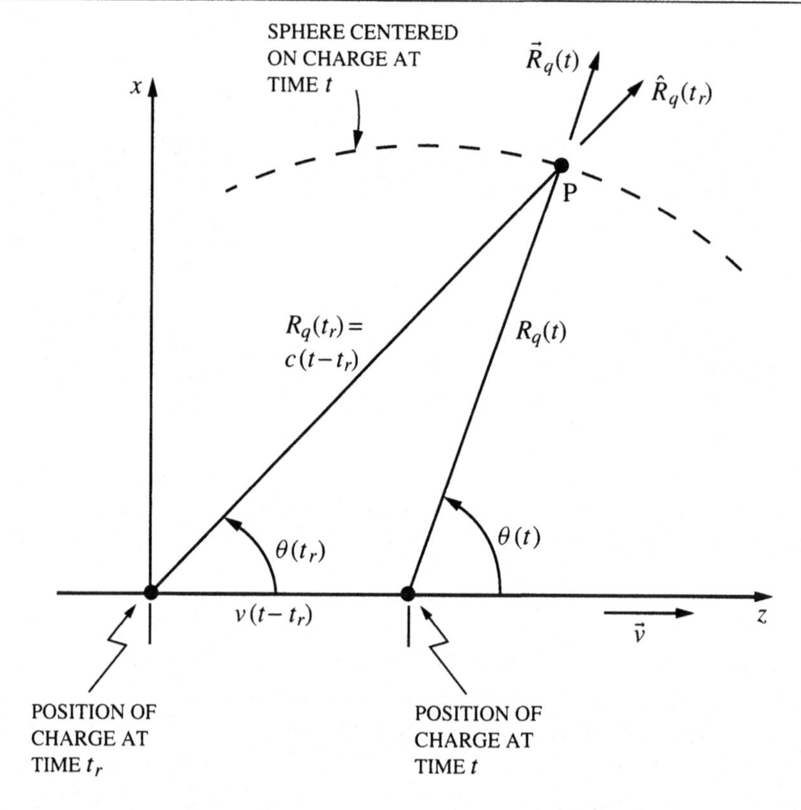

Fig. 6.9. Coordinates describing field for charge moving with constant velocity.

Solving this quadratic equation for t_r yields

$$t_r\big[R_q(t), \theta(t)\big] = t - \big[R_q(t)/c\big]\gamma^2\big[\beta \cos\theta(t) + \sqrt{1 - \beta^2 \sin^2\theta(t)}\,\big]. \quad (6.62)$$

The corresponding electromagnetic field is (see Problem 6.7)

$$\vec{\mathcal{E}}_v = \frac{q}{4\pi\varepsilon_o} \frac{1}{\gamma^2 R_q^2(t)[1 - \beta^2 \sin^2\theta(t)]^{3/2}} \hat{R}_q(t), \quad (6.63)$$

$$\vec{\mathcal{B}}_v = \frac{\mu_o c q}{4\pi} \frac{\beta \sin\theta(t)}{\gamma^2 R_q^2(t)[1 - \beta^2 \sin^2\theta(t)]^{3/2}} \hat{\phi}. \quad (6.64)$$

The electric field is now in the radial direction, $\hat{R}_q(t)$, whereas in the previous representation (6.60) it contained both radial, $\hat{R}_q(t_r)$, and axial, $\hat{z}$, components.

Figure 6.8b is a graph of the magnitude of the electric field for the new coordinates. The pattern is now symmetric about the x and z axes. As β increases, the pattern becomes elongated in the direction normal to the velocity, $\theta(t) = \pi/2$. These results suggest that the field for a charge moving with a constant velocity is most simply expressed in terms of the coordinates centered on the position of the charge at the current time.

The graphs in Figure 6.8 show the relative magnitude of the electric field; they

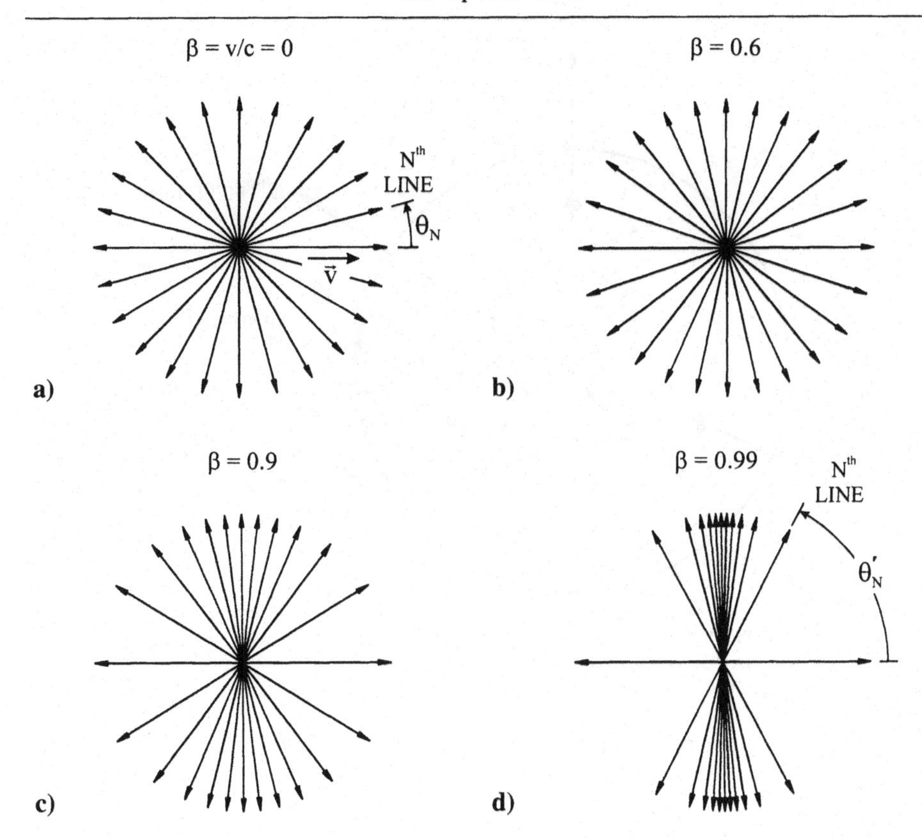

Fig. 6.10. Electric field lines for a charge moving with constant velocity $\vec{v}$.

say little about the direction of the field. A different representation can be used to convey information about the direction. For a stationary charge, $\beta = 0$, we use the familiar radial vectors shown in Figure 6.10a to indicate the direction of the electrostatic field. The vectors are equally spaced to indicate that the field is spherically symmetric; on a sphere of fixed radius it has the same strength in every direction θ, ϕ.

For the uniform field in Figure 6.10a, we have made the lines equally spaced in two dimensions, on the plane piece of paper. The arc length separating any two adjacent lines is a constant. What we would really like to have for the uniform field is a three-dimensional model with the radial field lines uniformly spaced over a spherical surface. Since the area of the sphere is proportional to the radius squared, the number of field lines per unit area would decrease as the reciprocal of the radius squared. This is the same dependence as for the electric field in (6.63); it is proportional to $1/R_q^2(t)$. Thus, the number of radial lines per unit area of the spherical surface would correctly indicate the strength of the electric field.[5] For a nonuniform radial field, we could adjust the number of lines per unit area on the spherical surface to be proportional to the field strength. This model can

[5] Our picture in Figure 6.10a is actually correct for the electric field of an infinitely long line charge. The number of lines per unit area decreases as the reciprocal of the radius, which is the same as the dependence of the electric field for the line charge.

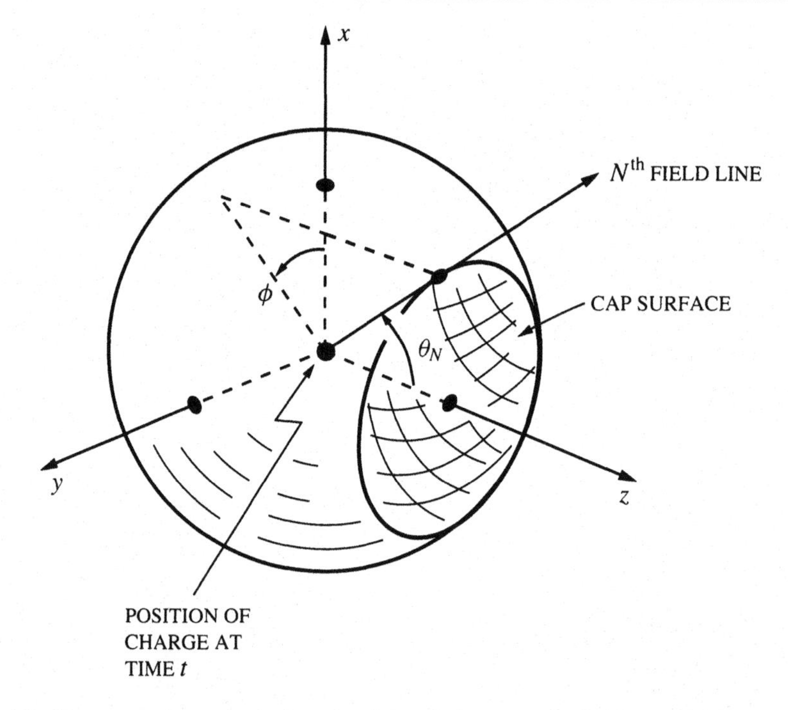

Fig. 6.11. Geometry for calculating the flux of the electric field through a cap surface bounded by the angle θ_N.

be visualized as a spherical pin cushion with pins distributed over the surface to represent the radial field lines; the density of the pins would represent the field strength. Because we are dealing with plain paper, we must settle for the two-dimensional representation in Figure 6.10.

For the uniform field in Figure 6.10a, the flux of the electric field through a spherical surface centered on the charge is easily calculated. With reference to Figure 6.11, the flux through the cap surface bounded by the field line at angle θ_N is

$$\frac{q}{4\pi\varepsilon_o}\int_{\phi=0}^{2\pi}\int_{\theta=0}^{\theta_N}\frac{1}{R_q^2}R_q^2\sin\theta d\theta d\phi = \frac{q}{2\varepsilon_o}(1-\cos\theta_N), \tag{6.65}$$

and the flux between two adjacent field lines, the N-th and the $(N+1)$-th, is

$$\frac{-q}{2\varepsilon_o}(\cos\theta_{N+1} - \cos\theta_N).$$

The flux is clearly not the same between any two adjacent field lines, as Figure 6.10a might imply.

Now let us consider the field of a charge moving with constant velocity (6.63). We will use the same number of field lines as for the stationary charge and require the flux of the field through the cap surfaces bounded by the N-th field line to be the same for the stationary and moving charges (cap surfaces with angle θ_N in

Figure 6.10a and angle θ'_N in Figure 6.10d):

$$\frac{q}{4\pi\varepsilon_o}\int_{\phi=0}^{2\pi}\int_{\theta=0}^{\theta'_N}\frac{1}{\gamma^2 R_q^2(1-\beta^2\sin^2\theta)^{3/2}}R_q^2\sin\theta d\theta d\phi$$

$$=\frac{q}{2\varepsilon_o}\left[1-\frac{\cos\theta'_N}{(1-\beta^2\sin^2\theta'_N)^{1/2}}\right]$$

$$=\frac{q}{2\varepsilon_o}(1-\cos\theta_N),$$

or

$$\frac{\cos\theta'_N}{(1-\beta^2\sin^2\theta'_N)^{1/2}}=\cos\theta_N.$$

After inverting and squaring this result, subtracting 1 from each side, and rearranging, we have

$$(1-\beta^2)\tan^2\theta'_N=\tan^2\theta_N,$$

or

$$\theta'_N=\tan^{-1}(\gamma\tan\theta_N). \tag{6.66}$$

Notice that the *total* flux through the spherical surface surrounding the charge must be the same for all cases (all values of $\beta=v/c$) because the total charge contained in the sphere is always the same. This is easily verified: $\theta'_N=\theta_N=\pi/2$ satisfies (6.66), which means the flux through the hemisphere is a constant, independent of β.

This relationship (6.66) was used to draw the field lines for the cases $\beta=0.6, 0.9$, and 0.99 in Figures 6.10b, c, and d. As the velocity approaches the speed of light, the radial field lines are seen to concentrate in the direction normal to the velocity, indicating an increase in the field strength in this direction.

So far we have only considered representations for the electric field of a charge moving with constant velocity. Now we will introduce an acceleration that is parallel to the velocity. We will assume that the charge initially moves with constant velocity $\vec{v}_i=v_i\hat{z}$, and that at time $t=0$ it is located at the origin $z_A=0$. The acceleration $\vec{a}=a\hat{z}$ begins at time $t=0$ and continues uniformly until time $t=T$, when the position is

$$z_B=v_iT+\frac{1}{2}aT^2=v_iT+\frac{1}{2}\Delta vT=\bar{v}T. \tag{6.67}$$

The charge then continues to move with velocity $\vec{v}_f=v_f\hat{z}$. At the observation time $t=t_o$ ($t_o>T$), the charge is located at the position

$$z_D=v_iT+\frac{1}{2}aT^2+v_f(t_o-T)=\bar{v}T+v_f(t_o-T)=v_ft_o-\frac{1}{2}\Delta vT. \tag{6.68}$$

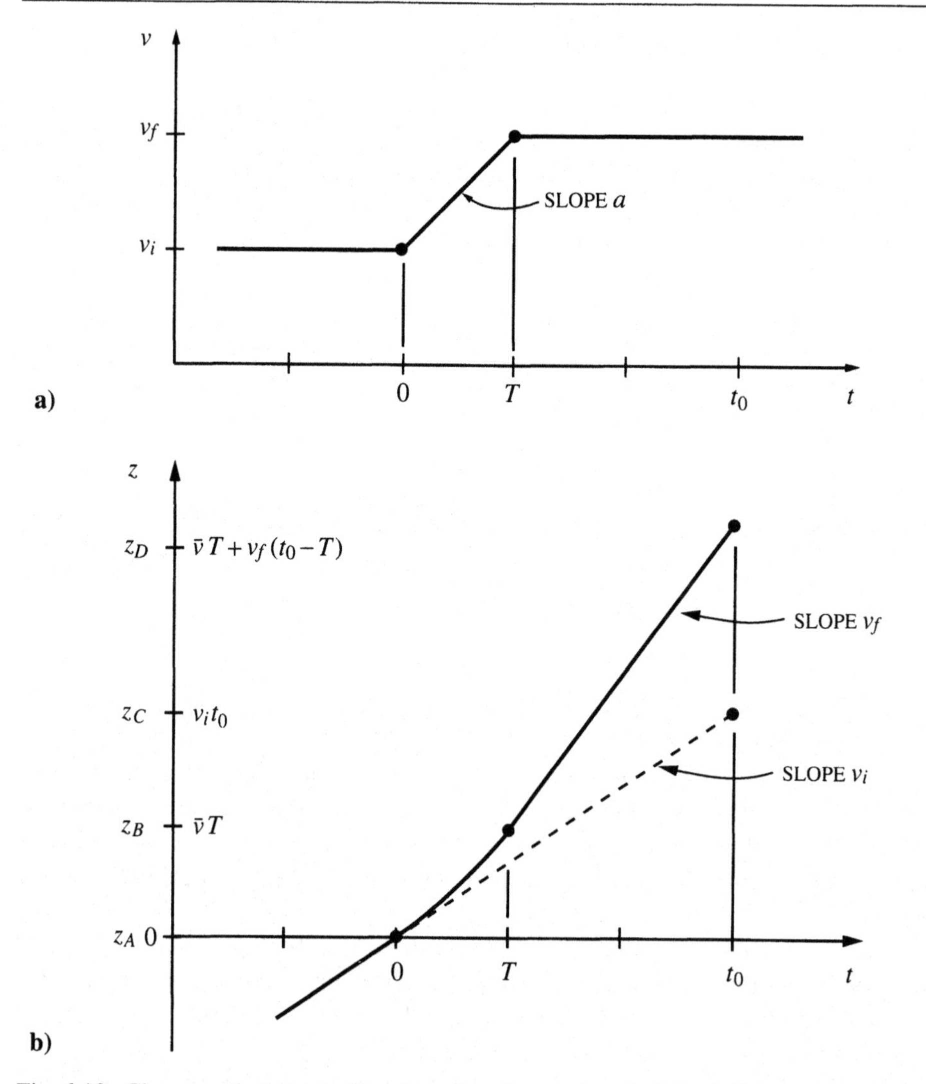

Fig. 6.12. Charge undergoing uniform acceleration over period T. a) Velocity versus time. b) Position versus time.

In the above expressions, the velocity difference and the average velocity on the interval T are

$$\Delta v = aT = v_f - v_i \tag{6.69a}$$

and

$$\bar{v} = v_i + \frac{1}{2}aT = (v_f + v_i)/2. \tag{6.69b}$$

Figure 6.12 shows the velocity and the position of the charge as a function of time. The dashed line in this figure shows the positions the charge would have had if it had continued with the initial velocity v_i. At time t_o it would have been at the

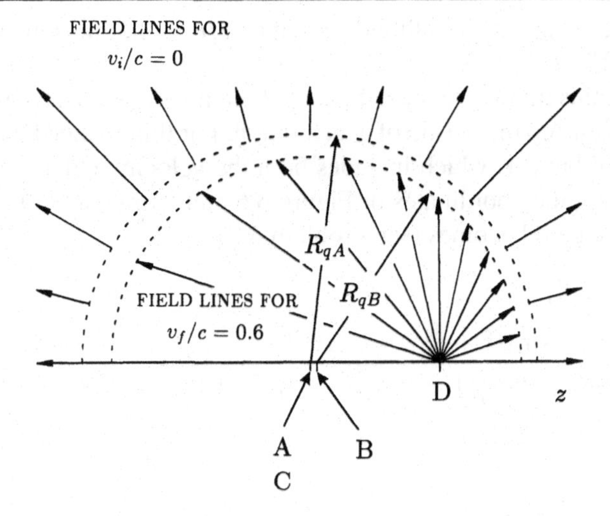

Fig. 6.13. Construction for the electric field lines of the velocity fields: $v_i/c = 0$, $v_f/c = 0.6$. Time of observation $t_o/T = 10.0$.

position

$$z_C = v_i t_o. \tag{6.70}$$

Now we will consider a specific case; we will let

$$v_i/c = 0.0, \qquad v_f/c = 0.6, \qquad t_o/T = 10.0.$$

In Figure 6.13 we have marked with the letters A, B, C, and D the points corresponding to z_A, z_B, z_C, and z_D for this case. Also shown in the drawing are circles centered at A and B with the radii

$$R_{qA} = ct_o \tag{6.71}$$

and

$$R_{qB} = c(t_o - T), \tag{6.72}$$

where R_{qA} is the distance an electromagnetic signal from point A (the start of the acceleration) will travel in the time interval t_o, and R_{qB} is the distance a signal from point B (the end of the acceleration) will travel in the time interval $t_o - T$. The electric field lines produced during the period of acceleration must be within the space bounded by these two circles.

For points outside the circle of radius R_{qA}, the field lines must be those produced before time $t = 0$. They are graphed for the observation time $t = t_o$, so they must be centered at point C – the position z_C the charge would occupy if it had continued with the initial velocity v_i. Since the charge was initially at rest, point C coincides with point A. These field lines are for a stationary charge ($v_i/c = 0$); thus, they are the same as the lines in Figure 6.10a.

For points inside the circle of radius R_{qB}, the field lines must be those produced after time $t = T$. They are graphed for the observation time $t = t_o$, so they must be centered at the point D – the position z_D of the charge at the observation time.

These lines are for a charge with the velocity $v_f/c = 0.6$; thus, they are the same as the lines in Figure 6.10b.

To complete this figure, we must draw the field lines between the two circles, the lines produced during the period of acceleration. For this we need the general form of the field in Table 6.1, which includes both the velocity and acceleration terms. After introducing the coordinates of Figure 6.6 and letting $\psi = 0$ and $\phi = 0$, the electromagnetic field becomes (see Problem 6.8)

$$\vec{\mathcal{E}}(\vec{r},t) = \vec{\mathcal{E}}_v + \vec{\mathcal{E}}_a$$
$$= \frac{q}{4\pi\varepsilon_o}\left[\frac{\sin\theta\hat{x} + (\cos\theta - \beta)\hat{z}}{\gamma^2 R_q^2(1-\beta\cos\theta)^3}\right]_{t_r} + \frac{q}{4\pi\varepsilon_o c^2}\left[\frac{a\sin\theta}{R_q(1-\beta\cos\theta)^3}\hat{\theta}\right]_{t_r}, \tag{6.73}$$

$$\vec{\mathcal{B}}(\vec{r},t) = \vec{\mathcal{B}}_v + \vec{\mathcal{B}}_a$$
$$= \frac{\mu_o c q}{4\pi}\left[\frac{\beta\sin\theta}{\gamma^2 R_q^2(1-\beta\cos\theta)^3}\hat{\phi}\right]_{t_r} + \frac{\mu_o q}{4\pi c}\left[\frac{a\sin\theta}{R_q(1-\beta\cos\theta)^3}\hat{\phi}\right]_{t_r}. \tag{6.74}$$

For each point (x, z) between the two circles in Figure 6.13, we must determine the retarded time t_r, the position $z_q(t_r)$ occupied by the charge during the period of acceleration (on the line AB), the radius $R_q(t_r)$, and the angle $\theta(t_r)$. From Figure 6.14a,

$$z_q(t_r) = v_i t_r + \frac{1}{2}at_r^2, \tag{6.75}$$

$$R_q(t_r) = c(t_o - t_r) = \sqrt{x^2 + \left[z - z_q(t_r)\right]^2}, \tag{6.76}$$

$$\theta(t_r) = \sin^{-1}\left[x/R_q(t_r)\right], \tag{6.77}$$

and, on combining (6.75) and (6.76), a quartic equation is obtained for t_r:

$$\left(\frac{t_r}{T}\right)^4 + \frac{4v_i}{\Delta v}\left(\frac{t_r}{T}\right)^3 + \frac{4(v_i^2 - c^2 - z\Delta v/T)}{\Delta v^2}\left(\frac{t_r}{T}\right)^2$$
$$+ \frac{8[c^2(t_o/T) - zv_i/T]}{\Delta v^2}\left(\frac{t_r}{T}\right) + \frac{4[(x/T)^2 + (z/T)^2 - c^2(t_o/T)^2]}{\Delta v^2} = 0. \tag{6.78}$$

After solving this equation for t_r and evaluating $z_q(t_r)$, $R_q(t_r)$, and $\theta(t_r)$, the electric field at the point x, z is determined from (6.73).

Now the field lines can be completed. First, as shown in Figure 6.14b, a small vector of length Δs is drawn in the direction of the field, $\vec{\mathcal{E}}/|\vec{\mathcal{E}}|$, at the point x_1, z_1, which is at the end of the field line for v_f/c. Then a second small vector of length Δs is drawn in the direction of the field at the point x_2, z_2, which is at the end of

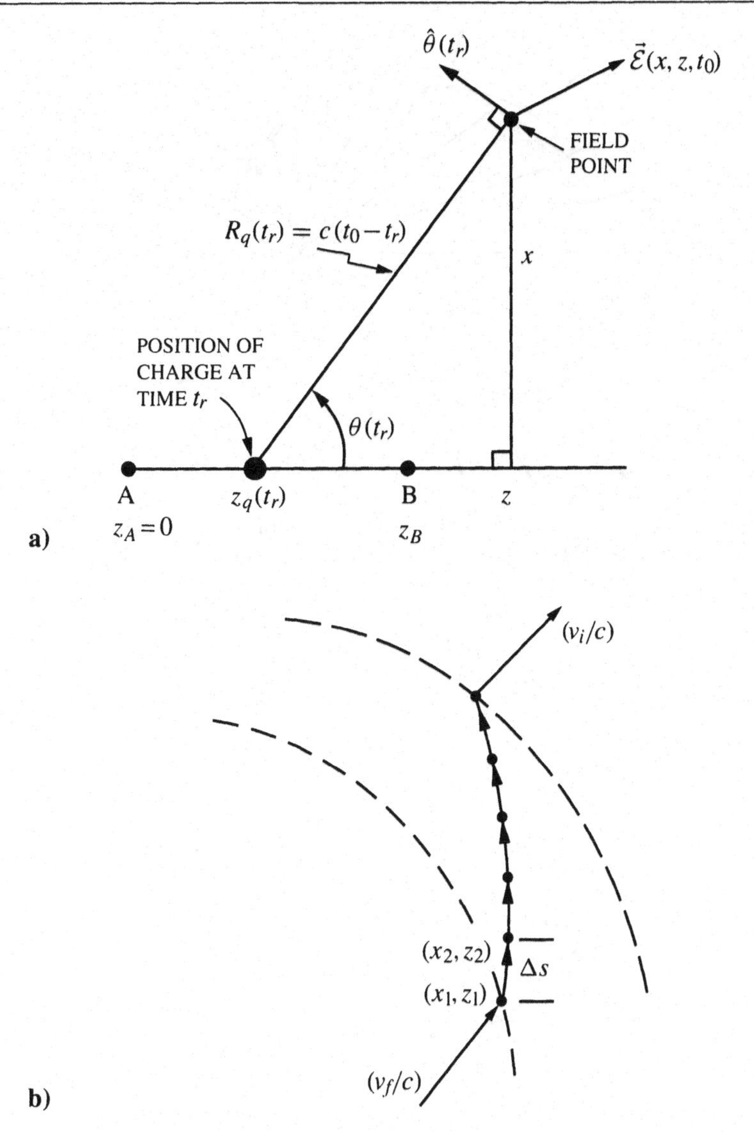

Fig. 6.14. Details for drawing electric field lines of the total field (velocity field plus acceleration field).

the first vector. This procedure is repeated until the curve intersects the field line for v_i/c.

Figure 6.15 is the complete graph for the electric field lines. The acceleration field is dominant in the ring between the two circles, so the field lines are nearly transverse to the direction of propagation in this region. With advancing time the ring moves outward, and the field lines become roughly tangent to an expanding circle. This is illustrated in Figure 6.16, where the field within the ring is shown for the times $t_o/T = 4$, 18, and 32.

Notice that a distant observer on the positive z axis in Figure 6.15 will receive a pulse of the electromagnetic field whose duration is shorter than the period of

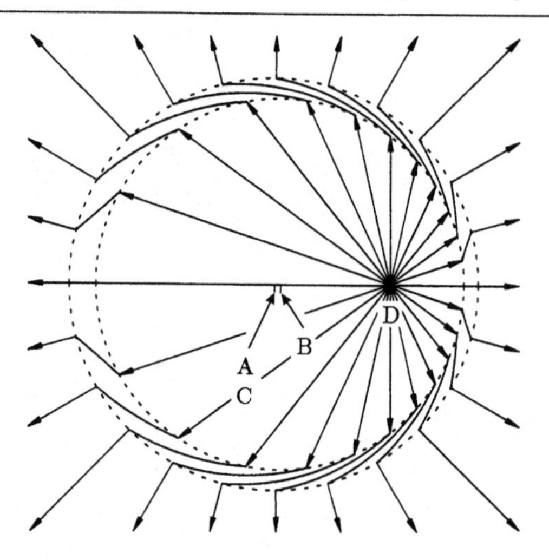

Fig. 6.15. Electric field lines for charge uniformly accelerated from rest, $v_i/c = 0$, to the velocity $v_f/c = 0.6$ during the period T. Time of observation $t_o/T = 10.0$.

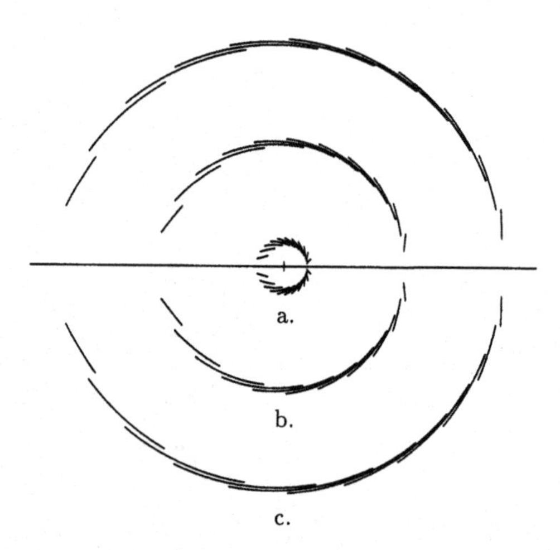

Fig. 6.16. Electric field lines produced during period T of uniform acceleration for different observation times: a) $t_o/T = 4.0$, b) $t_o/T = 18.0$, c) $t_o/T = 32.0$. $v_i/c = 0$, $v_f/c = 0.6$.

acceleration T:

$$T_+ = \left[R_{qA} - (R_{qB} + \bar{v}T)\right]/c = (1 - \bar{v}/c)T,$$

whereas a distant observer on the negative z axis will receive a pulse whose duration is longer than T:

$$T_- = \left[R_{qA} - (R_{qB} - \bar{v}T)\right]/c = (1 + \bar{v}/c)T.$$

After combining these results, we have

$$T_\pm/T = (1 \mp \bar{v}/c). \tag{6.79}$$

This is the *Doppler effect* (Christian Johann Doppler, 1803–1853). If the period T

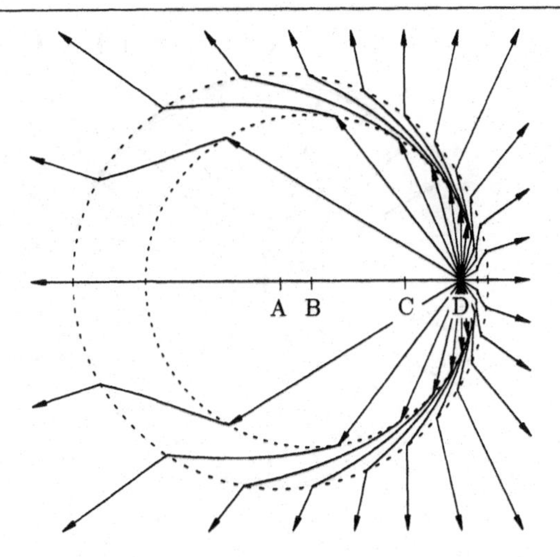

Fig. 6.17. Electric field lines for charge uniformly accelerated from the velocity $v_i/c = 0.6$ to the velocity $v_f/c = 0.9$ during the period T. Time of observation $t_o/T = 5.0$.

we have been discussing is one cycle of an oscillating signal of frequency $f = 1/T$, the observers would measure the frequencies

$$\frac{f_\pm}{f} = \frac{1}{(1 \mp \bar{v}/c)}. \tag{6.80}$$

When the source is approaching, the observer measures a frequency higher than f, and when it is receding, the observer measures a frequency lower than f.

It is important to bear in mind that all three periods (T, T_+, and T_-) in the above discussion are measured by stationary observers in the *same* inertial reference frame. We are *not* comparing periods in different reference frames, such as the instantaneous rest frame of the moving charge and the rest frame of the observers.

Figures 6.17 and 6.18 show the field lines for additional cases. In Figure 6.17 a charge with initial velocity $v_i/c = 0.6$ is uniformly accelerated to the final velocity $v_f/c = 0.9$. Because the charge is not initially at rest, points A and C are not coincident, as they are in Figure 6.15. In Figure 6.18 a charge is uniformly decelerated from the initial velocity $v_i/c = 0.9$ to rest, $v_f/c = 0.0$. After the period of deceleration, the field lines become those of a static charge, and since the charge is no longer moving, points B and D coincide. Diagrams like those we have presented in Figures 6.15–6.18 were apparently first discussed by Sir Joseph John Thomson (1856–1940), the discoverer of the electron [16]. Since then they have been described by several authors [17–20].

The power per unit solid angle radiated by the charge is given in Table 6.1; recall that this is a result of the acceleration field. When we introduce the coordinates of Figure 6.6 with $\psi = 0$ and use the fact that the velocity and acceleration are parallel, we get

$$\frac{d\mathcal{P}_q(t_r)}{d\Omega} = \frac{q^2}{16\pi^2\varepsilon_o c^3}\left[\frac{a^2 \sin^2\theta}{(1-\beta\cos\theta)^5}\right]_{t_r}. \tag{6.81}$$

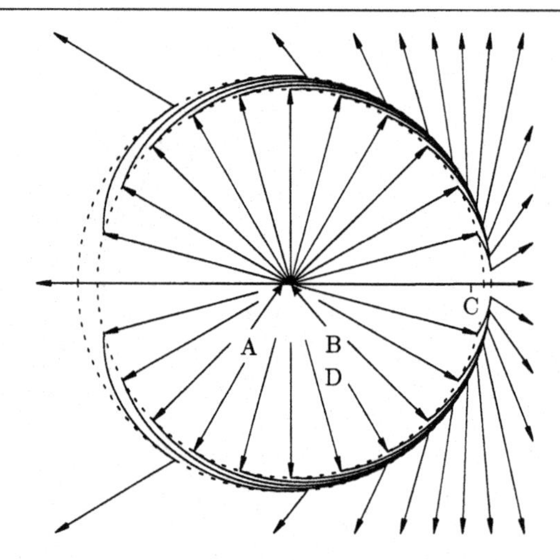

Fig. 6.18. Electric field lines for charge uniformly decelerated from the velocity $v_i/c = 0.9$ to rest, $v_f/c = 0$, during the period T. Time of observation $t_o/T = 15.0$.

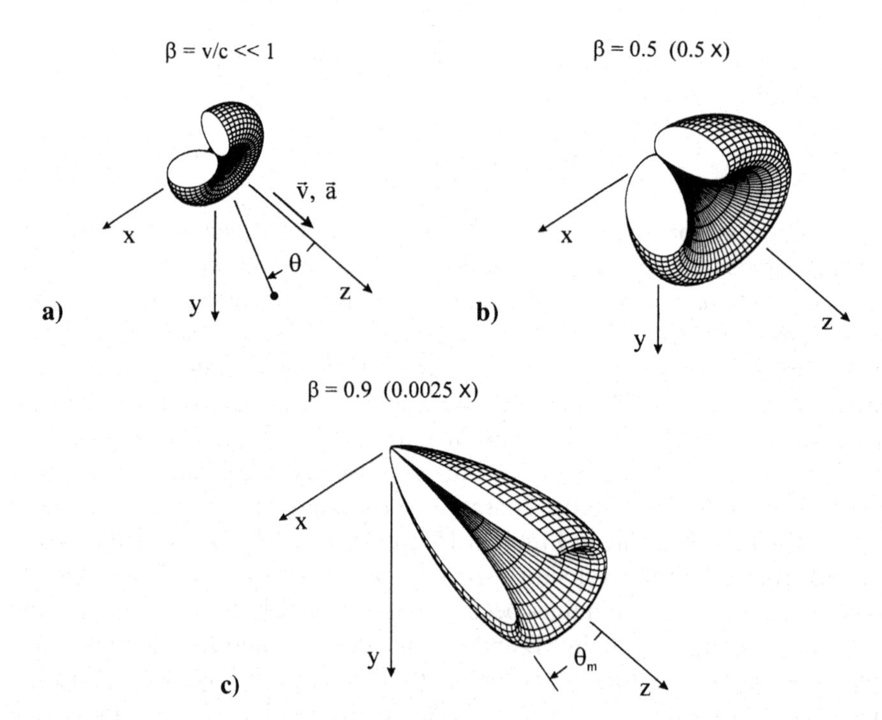

Fig. 6.19. Directional characteristics of the radiation, $d\mathcal{P}_q/d\Omega$, when the acceleration is parallel to the velocity. Note scale factors in parentheses.

A comparison shows that this is the same as the low-velocity results ($\beta \ll 1$) (6.58), except for the factor $(1 - \beta\cos\theta)^5$ in the denominator, which has a marked effect on the distribution of the radiation when $\beta \approx 1$.

Figure 6.19 shows patterns for the radiation for different values of β. Here the radial distance from the origin is proportional to $d\mathcal{P}_q/d\Omega$ in the direction determined by the angles θ and ϕ. Notice that the patterns are rotationally symmetric about the direction of the velocity/acceleration (z axis) as expected from (6.81). As the velocity increases from $\beta \ll 1$ to $\beta = 0.9$, the pattern becomes elongated in the direction of the velocity/acceleration; however, there is always a null on the axis, at $\theta = 0$. The maximum of the pattern occurs at the angle

$$\theta_m = \cos^{-1}\left[\frac{1}{3\beta}\left(\sqrt{1+15\beta^2} - 1\right)\right], \tag{6.82}$$

which for $\beta \approx 1$ becomes simply

$$\theta_m \approx \frac{1}{2\gamma}. \tag{6.83}$$

The scale changes in Figure 6.19 indicate that the maximum for the pattern increases drastically as β approaches 1. Notice that the concentration of radiation in the forward direction for β near 1 is also shown qualitatively in Figures 6.17 and 6.18: The density of the electric field lines within the ring is greatest at a small angle θ. The patterns in Figure 6.19 apply to a charge undergoing acceleration or deceleration in the direction of the velocity, since a^2 appears in (6.81).

The total power radiated by the charge is

$$\mathcal{P}_q(t_r) = \frac{q^2}{6\pi\varepsilon_o c^3}[a^2\gamma^6]_{t_r}. \tag{6.84}$$

This is Larmor's formula for $\beta \ll 1$ (6.59) multiplied by the factor γ^6, which rapidly increases as β approaches one.

The radiation produced during the deceleration of a charged particle is called *bremsstrahlung*, which is German for "braking radiation." This term was coined during the early history of X rays. X-ray radiation was experimentally discovered by Wilhelm Conrad Röntgen (1845–1923) in 1895 [21]. Additional experiments were immediately carried out by many investigators, and several explanations were put forth for the physical mechanisms that produce X rays. These efforts culminated with the development of quantum mechanics in the 1920s [22].

A schematic drawing for a "Coolidge" X-ray tube is shown in Figure 6.20a [23]. Electrons leave the heated filament, are accelerated to high velocity by the potential difference V_o, then impact the metallic target where they produce X-ray radiation. Two types of radiation are produced in the X-ray tube: the characteristic X-ray spectrum and the continuous X-ray spectrum [24, 25]. Both types of radiation are evident in the measured results presented in Figure 6.20b [26]. Here the intensity of the radiation emitted when a beam of electrons strikes a molybdenum target is graphed as a function of the wavelength, $\lambda_o = c/f$.

The *characteristic X-ray spectrum* comprises radiation in narrow frequency ranges, a series of lines called a line spectrum. In the quantum explanation it is caused by an incident electron knocking out one of the inner electrons from an atom in the target; an outer electron then makes the transition to the vacated state, producing electromagnetic radiation. The *continuous X-ray spectrum* was named bremsstrahlung by A. Sommerfeld [22]. The classical electromagnetic explanation

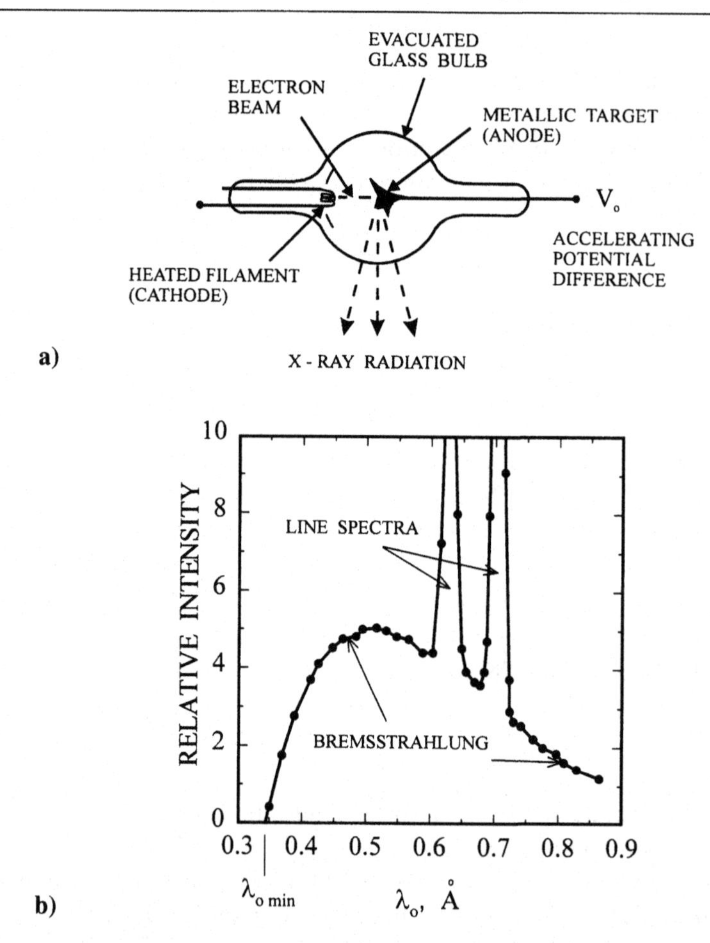

Fig. 6.20. a) Schematic drawing of "Coolidge" X-ray tube. b) Spectrum of radiation for X-ray tube with molybdenum target, $V_o = 35,000$ V. (Measured data from Ulrey [26].)

is that this radiation is caused by the deceleration (braking) of the electrons on impact with the target. The deceleration of an electron produces a temporal pulse of radiation. The nature of the collision of the electron with the atoms of the target determines the shape and duration of the pulse and hence the characteristics of the continuous spectrum (Fourier transform of the pulse).

Notice in Figure 6.20b that there is no radiation produced for wavelengths shorter than $\lambda_{o\,\text{min}}$. This is a quantum effect known as the "Duane-Hunt limit." The kinetic energy an incident electron gains in moving through the accelerating potential difference V_o of the electrostatic field in the X-ray tube is

$$\mathcal{K} = eV_o, \tag{6.85}$$

where the electronic charge is $e = 1.60 \times 10^{-19}$ C. Because the energy of an X-ray photon emitted from an atom in the target can be no greater than the kinetic energy of the incident electron, we have

$$hf_{\text{max}} = hc/\lambda_{o\,\text{min}} = \mathcal{K} = eV_o,$$

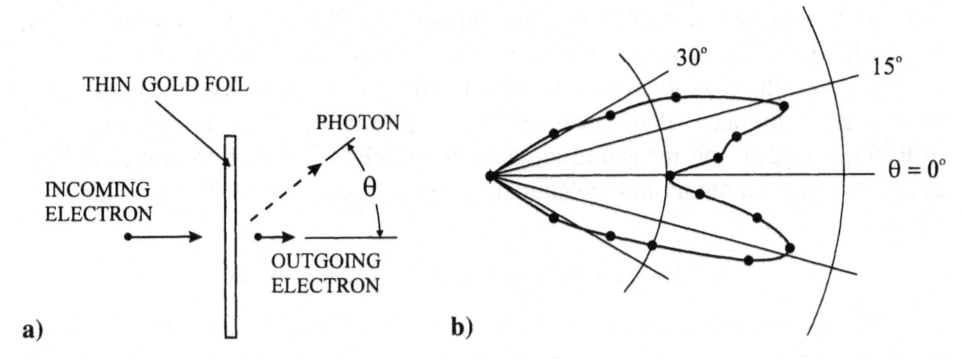

Fig. 6.21. a) Schematic drawing for experiment in which electrons impact a thin, gold foil with a kinetic energy of 300 keV. b) Angular distribution of photons with an energy of 130 keV. (Measured data from Nakel [28].)

or

$$\lambda_{o\,\min} = hc/eV_o, \tag{6.86}$$

where Planck's constant $h = 6.63 \times 10^{-34}$ J·s (Max Karl Ernst Ludwig Planck, 1858–1947). For the results in Figure 6.20b, the potential difference is $V_o = 35,000$ V, making $\lambda_{o\,\min} \approx 3.55 \times 10^{-11}$ m = 0.355 Å.

For this example, we can estimate the velocity of the electrons when they struck the target. From the special theory of relativity, the kinetic energy of an electron moving with velocity v is [27]

$$\mathcal{K} = (\gamma - 1)mc^2, \tag{6.87}$$

where the mass of the electron is $m = 9.11 \times 10^{-31}$ kg. After substituting (6.87) into (6.85) and solving for β, we have

$$\beta = \sqrt{1 - 1/(1 + eV_o/mc^2)^2}, \tag{6.88}$$

which, on inserting numerical values, gives $\beta = v/c \approx 0.35$. The velocity of the electrons was about 35% of the speed of light when they struck the target.

When electrons collide with the atoms of the metallic target their motion is altered. For example, an electron may penetrate an atom and be deflected by the force of the Coulomb field of the nucleus. The electron does not necessarily experience a deceleration that is parallel to its velocity. Only under special circumstances would we expect the radiation from the electrons to have the directional characteristics shown in Figure 6.19.

Figure 6.21 presents results from an experiment in which an electron beam impacts a thin gold foil (thickness = 250 Å) [28]. The kinetic energy of the electrons in the beam is $\mathcal{K} = eV_o = 300$ keV (1 eV = 1.60×10^{-19} J); thus, from (6.88) $\beta = v/c \approx 0.78$; the velocity of the electrons is about 78% of the speed of light. In this experiment, only the electrons emerging from the foil with a velocity nearly parallel to the direction of the incident beam are considered (Figure 6.21a). Hence, the deceleration of the electrons is nearly parallel to their velocity. The radiation (photons) produced by these electrons is measured as a function of the angle θ. The measured angular distribution in Figure 6.21b is for photons with the energy 130 keV. The directional characteristics are seen to be qualitatively the same as those from the classical electromagnetic analysis (Figure 6.19c): a symmetric pattern about the axis of the beam with a large maximum at a small angle θ_m. In Reference [29], the measured angular distribution is shown to agree with the predictions from a quantum mechanical analysis.

6.2.3 Relativistic velocity, $\vec{a}$ perpendicular to $\vec{v}$

When the acceleration is perpendicular to the velocity ($\psi = \pi/2$ for the coordinates in Figure 6.6), the radiated electromagnetic field from Table 6.1 is (see Problem 6.8)

$$\vec{\mathcal{E}}_a(r,t) = \frac{q}{4\pi\varepsilon_o c^2}\left\{\frac{a}{R_q}\left[\frac{\sin\phi}{(1-\beta\cos\theta)^2}\hat{\phi} - \frac{\cos\phi(\cos\theta-\beta)}{(1-\beta\cos\theta)^3}\hat{\theta}\right]\right\}_{t_r}, \tag{6.89}$$

$$\vec{\mathcal{B}}_a(r,t) = \frac{-\mu_o q}{4\pi c}\left\{\frac{a}{R_q}\left[\frac{\sin\phi}{(1-\beta\cos\theta)^2}\hat{\theta} + \frac{\cos\phi(\cos\theta-\beta)}{(1-\beta\cos\theta)^3}\hat{\phi}\right]\right\}_{t_r}, \tag{6.90}$$

and the power radiated per unit solid angle becomes

$$\frac{d\mathcal{P}_q(t_r)}{d\Omega} = \frac{q^2}{16\pi^2\varepsilon_o c^3}\left\{\frac{a^2}{(1-\beta\cos\theta)^3}\left[1 - \frac{\sin^2\theta\cos^2\phi}{\gamma^2(1-\beta\cos\theta)^2}\right]\right\}_{t_r}. \tag{6.91}$$

Patterns for the radiation field (6.91) are shown in Figure 6.22. For $\beta \ll 1$ (Figure 6.22a), we have the same pattern as for the case when the acceleration and velocity are parallel (Figure 6.19a). Recall that the relative orientation of the velocity with respect to the acceleration is not important for the patterns when $\beta \ll 1$. Although the patterns in Figures 6.19a and 6.22a appear to be different, they are the same – a fat doughnut or fat tire; the apparent difference is caused by the difference in the orientation of the contours of the two figures.

As the velocity is increased, the patterns become elongated in the direction of the velocity (z axis), eventually becoming a beam in this direction (Figure 6.22d). The half width of the beam is roughly the angle between the z axis and the null in the pattern, which occurs in the plane containing the velocity and the acceleration (x-z plane, $\phi = 0$):

$$\theta_n = \cos^{-1}(\beta), \tag{6.92}$$

which for $\beta \approx 1$ becomes simply

$$\theta_n \approx \sqrt{2(1-\beta)} = \sqrt{\frac{2}{1+\beta}\frac{1}{\gamma}} \approx \frac{1}{\gamma}. \tag{6.93}$$

We have scale changes in Figure 6.22, and again they indicate that the maximum for the pattern increases drastically as β approaches one.

As β approaches one, the electric field of the beam also becomes nearly linearly polarized in the direction $-\hat{x}$, along the line of the acceleration. To show this, in Figure 6.23 we have constructed patterns for the rectangular components (x, y, z)

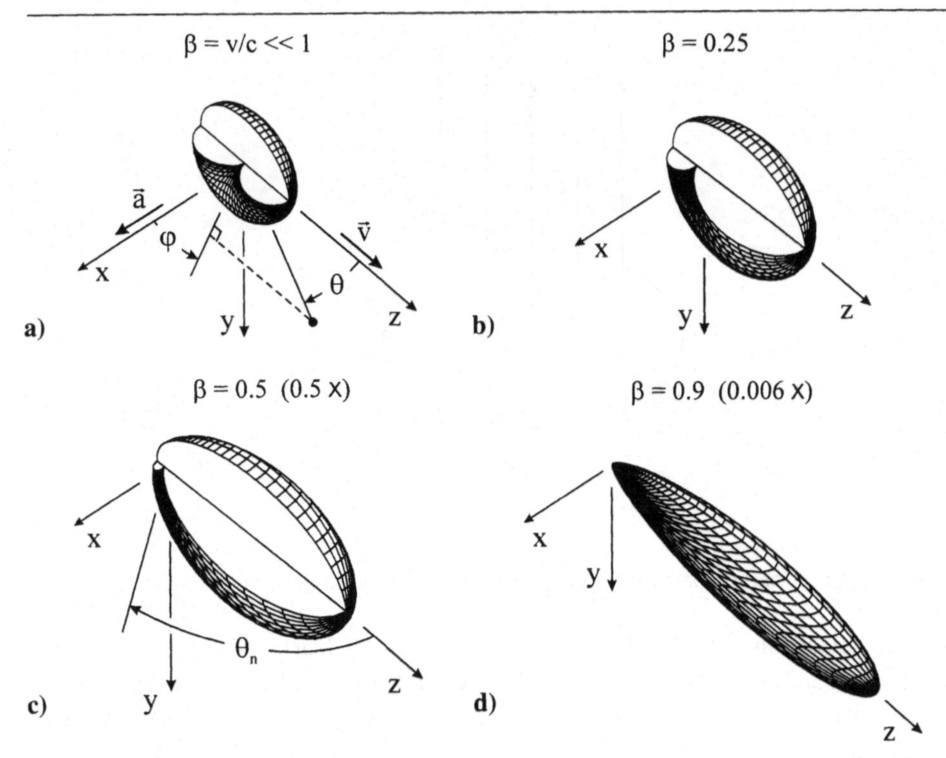

Fig. 6.22. Directional characteristics of the radiation, $d\mathcal{P}_q/d\Omega$, when the acceleration is perpendicular to the velocity. Note scale factors in parentheses.

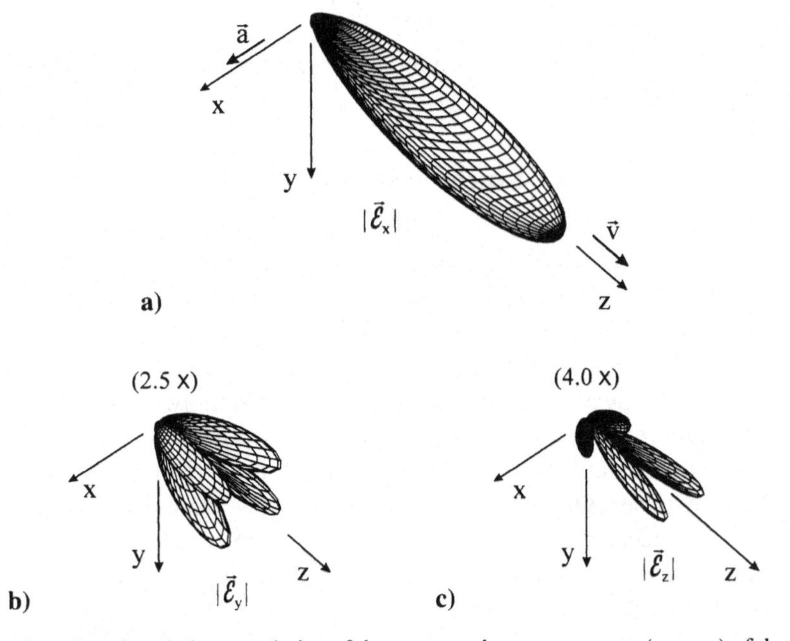

Fig. 6.23. Directional characteristics of the rectangular components (x, y, z) of the radiated electric field when the acceleration is perpendicular to the velocity; $\beta = v/c = 0.9$. Note scale factors in parentheses.

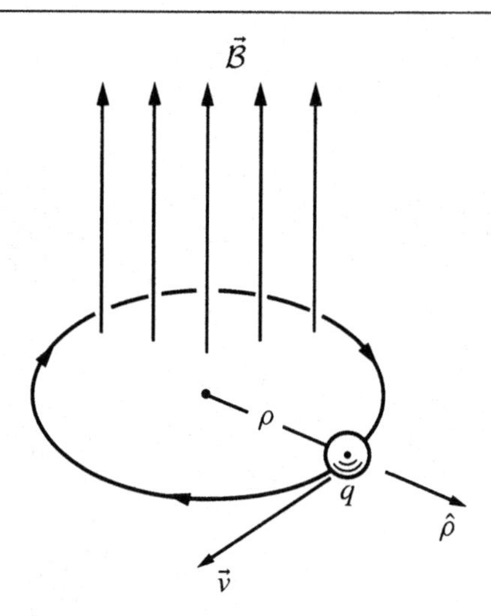

Fig. 6.24. Charged particle moving in uniform magnetic field, such that $\vec{v} \perp \vec{B}$.

of the acceleration field (6.89) for the case $\beta = 0.9$. Clearly, near the axis of the beam, the y and z components of the electric field (Figures 6.23b and c) are small compared to the x component (Figure 6.23a). Note the scale changes for these figures.

The total power radiated by the charge is

$$\mathcal{P}_q(t_r) = \frac{q^2}{6\pi\varepsilon_o c^3}[\gamma^4 a^2]_{t_r}, \tag{6.94}$$

which is Larmor's formula for $\beta \ll 1$ (6.59) with the added factor γ^4, which rapidly increases as β approaches one.

6.3 Synchrotron radiation

Particle accelerators are a basic tool used in experimental high-energy physics. These machines are generally of two types: linear accelerators in which a charged particle travels in a straight line, being accelerated by an electric field applied at a large number of points along the path; and circular or cyclic accelerators in which the charged particle traverses a circular orbit a large number of times, being accelerated each time by an electric field applied at one or more points on the orbit. Most modern high-energy particle accelerators are of the circular type.

The first practical circular machine for accelerating charged particles was the cyclotron built by Ernest Orlando Lawrence (1901–1958) around 1930 [30–32]. A number of circular accelerators were proposed after the cyclotron: the betatron, the microtron, etc., and in 1945 the synchrotron [33, 34]. The basic operating principle of the synchrotron is still used in modern high-energy particle accelerators [35, 36].

To understand the operation of the synchrotron, we will first examine the motion of a charged particle in a uniform magnetic field. We will assume that the velocity of the particle $\vec{v}$ is in a direction normal to the magnetic field $\vec{B}$ (Figure 6.24). The particle will then travel in a circular orbit of radius ρ, such that the centripetal force equals the Lorentz force (1.16) of the magnetic field:

$$m\vec{a} = \frac{mv^2}{\rho}(-\hat{\rho}) = q\vec{v} \times \vec{\mathcal{B}},$$

and so

$$\rho = \frac{mv}{q\mathcal{B}}. \tag{6.95}$$

The angular frequency at which the particle circulates is the *Larmor frequency*,

$$\omega = \frac{v}{\rho} = \frac{q\mathcal{B}}{m}, \tag{6.96}$$

and the kinetic energy of the particle is

$$\mathcal{K} = \frac{1}{2}mv^2. \tag{6.97}$$

These results are for a particle moving at a velocity much less than the speed of light ($\beta \ll 1$). Particles in modern accelerators travel at velocities close to the speed of light, so we must use the relativistic generalizations of these equations in our discussion [27]:

$$\rho = \gamma\left(\frac{mv}{q\mathcal{B}}\right) = \frac{1}{qc\mathcal{B}}\sqrt{\mathcal{E}^2 - \mathcal{E}_o^2}, \tag{6.98}$$

$$\omega = \frac{1}{\gamma}\left(\frac{q\mathcal{B}}{m}\right) = \frac{qc^2\mathcal{B}}{\mathcal{E}}, \tag{6.99}$$

$$\mathcal{K} = (\gamma - 1)mc^2 = \mathcal{E} - \mathcal{E}_o. \tag{6.100}$$

Here the total energy of the particle is[6]

$$\mathcal{E} = \gamma mc^2 = \gamma\mathcal{E}_o, \tag{6.101}$$

m is the mass of the particle, and $\mathcal{E}_o$ is the rest energy of the particle,

$$\mathcal{E}_o = mc^2. \tag{6.102}$$

The schematic drawing in Figure 6.25 shows the essential elements of a synchrotron. The operation of the machine is fairly simple. A bunch of charged particles, which usually comes from a lower energy accelerator, is injected into the synchrotron (upper right-hand corner of Figure 6.25). The particles travel in a vacuum chamber that is approximately a circular ring of radius ρ. The magnetic field $\vec{\mathcal{B}}$ that forces the trajectory of the particles to be along the center line of the ring is provided by a series of "bending magnets." "Focusing magnets" are distributed along the ring to maintain a small cross section for the bunch.

The charged particles traveling in the ring pass once per cycle through a radio-frequency cavity (bottom of Figure 6.25), where they are accelerated by an electric field of angular frequency ω_o, which increases their energy. For this process to

[6] The total energy $\mathcal{E}$ is not to be confused with the electric field $\vec{\mathcal{E}}$.

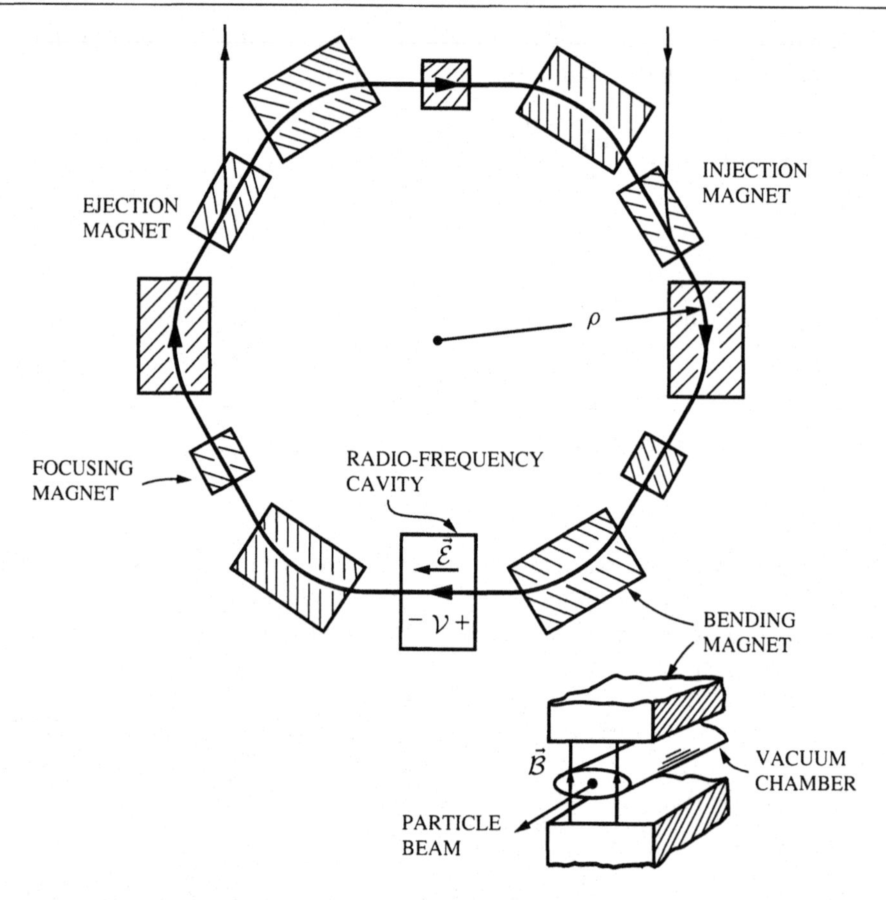

Fig. 6.25. Schematic drawing showing the essential elements of a synchrotron.

be repeated on each cycle and for the energy of the particles to be continuously increased, the angular frequency at which the particles move around the ring (6.99) must equal the fixed angular frequency of the electric field,

$$\omega = \frac{qc^2\mathcal{B}(t)}{\mathcal{E}(t)} = \omega_o, \tag{6.103}$$

a condition known as resonance. In the synchrotron this relationship is maintained by smoothly increasing the field $\mathcal{B}(t)$ of the bending magnets as the energy $\mathcal{E}(t)$ of the particles increases. During this process the particles stay roughly in the center of the ring, since the radius of their orbit (6.98) is approximately independent of $\mathcal{E}(t)$ for high-energy particles (for ultra-relativistic particles with $\beta \approx 1$, $\mathcal{E}(t) \gg \mathcal{E}_o$):[7]

$$\begin{aligned}\rho(t) &= \frac{\mathcal{E}(t)}{qc\mathcal{B}(t)}\sqrt{1 - \mathcal{E}_o^2/\mathcal{E}^2(t)} = \frac{c}{\omega_o}\sqrt{1 - \mathcal{E}_o^2/\mathcal{E}^2(t)} \\ &\approx (c/\omega_o).\end{aligned} \tag{6.104}$$

[7] Our discussion basically holds for an electron synchrotron. In a proton synchrotron the frequency ω_o of the electric field and the magnetic field $\vec{\mathcal{B}}$ are both varied with time to satisfy (6.103) and keep the particles in the center of ring (Problem 6.11).

When the charged particles have reached their final energy, they are deflected from the ring by an ejection magnet (upper left-hand corner of Figure 6.25) and used in an experiment.

The synchrotron possesses a special property known as "phase stability," which we will describe next [37, 38].[8] The electric field in the radio-frequency cavity of the synchrotron produces the accelerating potential difference

$$\mathcal{V}(t) = V_o \cos(\omega_o t), \tag{6.105}$$

which is sketched in Figure 6.26. We will assume that the energy of the particle and the magnetic field initially satisfy the resonance condition (6.103):

$$\omega = \frac{qc^2\mathcal{B}}{\mathcal{E}} = \omega_o. \tag{6.106}$$

During the next cycle, the particle passes through the cavity at time t_e gaining the energy increment

$$\Delta\mathcal{E}_e = qV_e = qV_o \cos(\omega_o t_e), \tag{6.107}$$

and the magnetic field is increased by the increment $\Delta\mathcal{B}_e$. To maintain resonance, the new energy $(\mathcal{E} + \Delta\mathcal{E}_e)$ and the new magnetic field $(\mathcal{B} + \Delta\mathcal{B}_e)$ must also satisfy (6.103):

$$\omega = \frac{qc^2(\mathcal{B} + \Delta\mathcal{B}_e)}{(\mathcal{E} + \Delta\mathcal{E}_e)} = \omega_o. \tag{6.108}$$

On substitution of (6.106) and (6.107), this requirement becomes

$$\Delta\mathcal{E}_e = \frac{\mathcal{E}}{\mathcal{B}}\Delta\mathcal{B}_e = \frac{qc^2}{\omega_o}\Delta\mathcal{B}_e,$$

or

$$\cos(\psi_e) = \cos(\omega_o t_e) = \frac{c^2\Delta\mathcal{B}_e}{\omega_o V_o}. \tag{6.109}$$

The angle ψ is called the "phase of the particle," and ψ_e is the "equilibrium phase." On consecutive cycles the particle must arrive at the cavity when the potential (6.105) has the phase $\psi_e + 2n\pi$ $(n = 0, 1, 2, \ldots)$ to be continually accelerated. A particle with the equilibrium phase arrives at the cavity at the times marked A, A′, A″, etc., in Figure 6.26.

Now we ask what happens if the particle does not pass through the cavity with equilibrium phase? We will consider the case where the particle arrives at the cavity too early, point B in Figure 6.26 $(\psi_B < \psi_e)$. The accelerating potential difference is higher than the equilibrium value V_e, so the particle gains an amount of energy greater than the equilibrium increment, $\Delta\mathcal{E} > \Delta\mathcal{E}_e$. From (6.108) we see that this extra energy will cause the angular frequency of the particle to be less than the equilibrium value $(\omega < \omega_o)$. Thus, on the next pass, the particle will arrive at the cavity at a later time in the cycle of the potential, point B′ $(\psi_{B'} > \psi_B + 2\pi)$, and

[8] The principle of phase stability for the synchrotron was discovered independently around 1945 by Edwin Mattison McMillan (1907–) in the United States and Vladimir Ioifovich Veksler (1907–1966) in the former Soviet Union [33, 34].

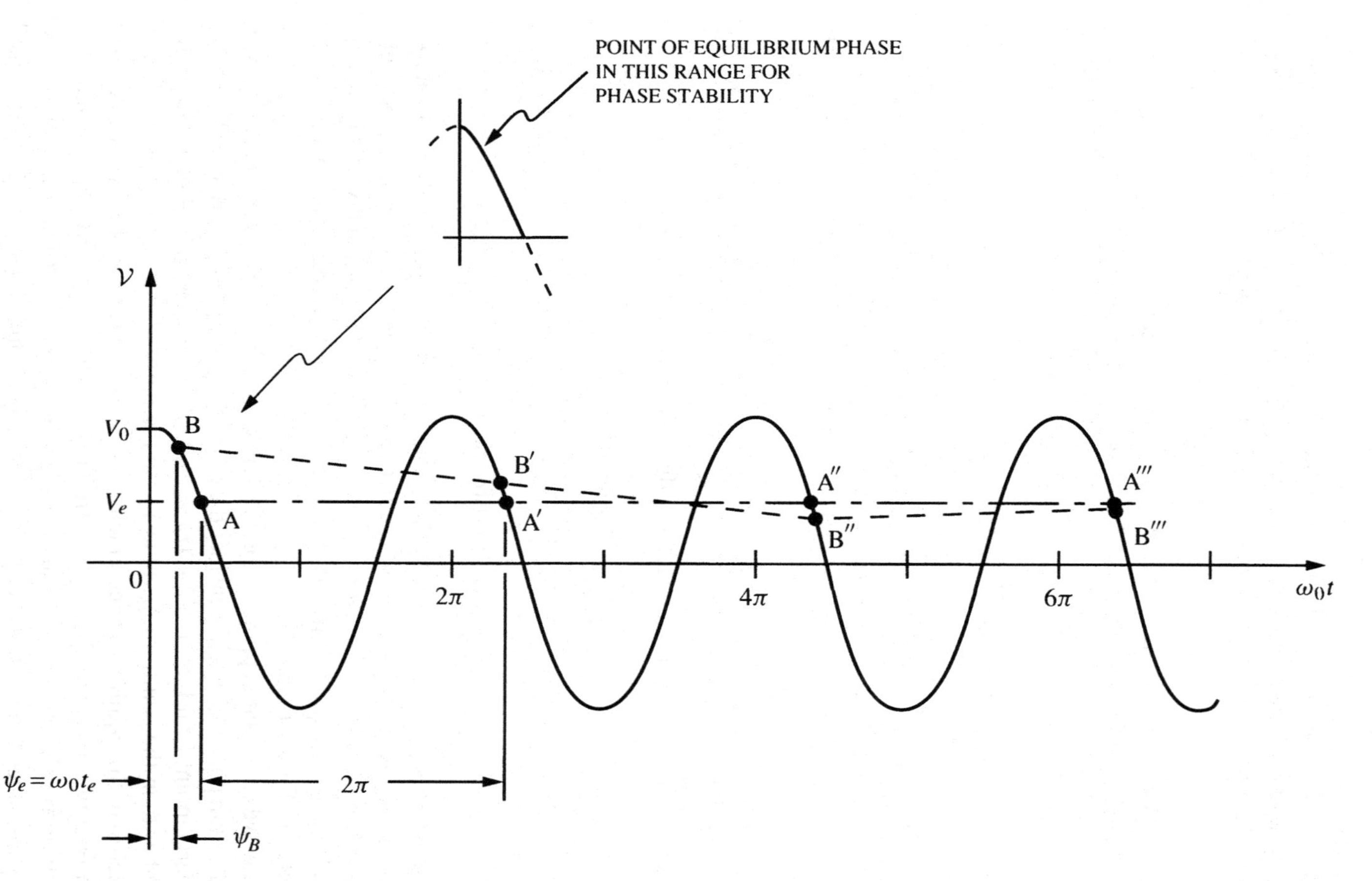

Fig. 6.26. Graph of the accelerating potential difference in the radio-frequency cavity, illustrating the principle of phase stability.

have a phase closer to the equilibrium value. This process is repeated over many cycles, with the phase and the energy of the particle oscillating about the equilibrium values, points B″, B‴, etc.[9] The analogy of this behavior to the hunting in phase of the armature in a synchronous electric motor was the reason for choosing the name synchrotron [32, 33].

Notice that the potential difference must be positive at the point of equilibrium phase to accelerate the particle as it passes through the cavity. In addition, for phase stability the slope of the curve for the potential difference must be negative at the point of equilibrium phase. Thus, only points on the right half of the positive lobe of the potential are suitable as points of equilibrium phase (see the inset in Figure 6.26).

Phase stability causes a bunch of particles injected into the synchrotron to group together in a stable orbit and obtain nearly equal energies during the process of acceleration. The maximum energy reached by the particles is determined by the maximum magnetic field $\mathcal{B}_{\text{max}}$, from (6.103),

$$\mathcal{E}_{\text{max}} = \frac{qc^2\mathcal{B}_{\text{max}}}{\omega_o} = \frac{qc\rho\mathcal{B}_{\text{max}}}{\beta}, \tag{6.110}$$

which for ultra-relativistic particles, $\beta \approx 1$, is simply

$$\mathcal{E}_{\text{max}} \approx qc\rho\mathcal{B}_{\text{max}},$$

or

$$\mathcal{E}_{\text{max}}(\text{eV}) = 3.00 \times 10^8 \rho(\text{m})\mathcal{B}_{\text{max}}(\text{T})N, \tag{6.111}$$

where the energy is expressed in electron volts and N is the ratio of the charge on the particle to the electronic charge, $q = Ne$.

Clearly, a high-energy synchrotron must have a large radius and powerful bending magnets. As an example we mention the ill-fated Superconducting Supercollider (SSC). This proton synchrotron was to be the highest energy machine in existence. It was under construction in 1993 when the U.S. Congress cancelled funding for the project. The SSC had a radius $\rho \approx 13.3$ km and superconducting bending magnets with $\mathcal{B}_{\text{max}} \approx 6.6$ T; thus, the maximum energy (6.111) would have been about 2.6×10^{13} eV $= 26$ TeV , and the velocity of a proton in this machine would have been about $v/c \approx 0.99999999935$ [39].

A charged particle in a synchrotron continually undergoes centripetal acceleration; therefore, it continually radiates electromagnetic energy. Our results from the previous section for the case with the acceleration perpendicular to the velocity are directly applicable to the synchrotron. The radiated electric field (acceleration field) is (6.89)

$$\vec{\mathcal{E}}_a(\vec{r}, t) = \frac{q}{4\pi\varepsilon_o c^2}\left\{\frac{a}{R_q}\left[\frac{\sin\phi}{(1-\beta\cos\theta)^2}\hat{\phi} - \frac{\cos\phi(\cos\theta-\beta)}{(1-\beta\cos\theta)^3}\hat{\theta}\right]\right\}_{t_r}. \tag{6.112}$$

[9] In Figure 6.26 we have exaggerated the changes in phase per cycle to compress the presentation. Actually, much smaller phase changes occur per cycle.

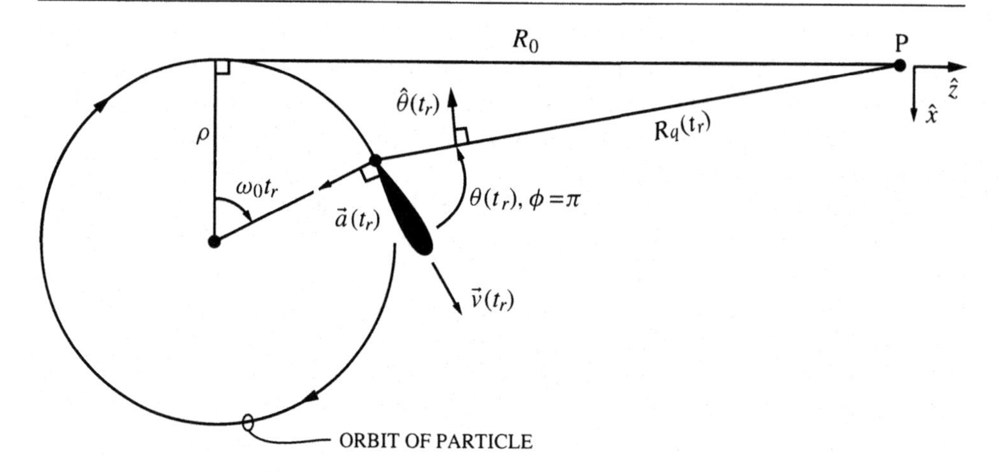

Fig. 6.27. Coordinates for describing the radiated electric field in the plane of the particle's orbit $\phi = 0,\ \pi$.

To simplify the examination of this field, we will only consider the geometry shown in Figure 6.27. The observation point P is in the plane of the particle's orbit ($\phi = 0, \pi$) and at a large distance from the orbit ($R_o \gg \rho$). The distance R_q, the angle θ, and the unit vector $\hat{\theta}$ are then approximately

$$R_q(t_r) \approx R_o\left[1 - \left(\frac{\rho}{R_o}\right)\sin(\omega_o t_r)\right], \tag{6.113}$$

$$\theta(t_r) \approx \omega_o t_r, \tag{6.114}$$

and

$$\cos\phi\,\hat{\theta}(t_r) \approx -\hat{x}, \tag{6.115}$$

where $\hat{x}$ is a unit vector at the observation point (see Figure 6.27). The retarded time is a solution to the equation

$$t_r = t - R_q(t_r)/c \approx t - \frac{R_o}{c}\left[1 - \left(\frac{\rho}{R_o}\right)\sin(\omega_o t_r)\right],$$

or, since $\omega_o = v/\rho$,

$$\omega_o(t - R_o/c) = \omega_o t_r - \beta\sin(\omega_o t_r). \tag{6.116}$$

When these results are inserted into (6.112), the electric field in the plane of the particle's orbit becomes

$$\vec{\mathcal{E}}_a(\vec{r}, t) = \mathcal{E}_{ax}(R_o, t)\hat{x},$$

with

$$\mathcal{E}_{ax}(R_o, t) \approx \frac{-q}{4\pi\varepsilon_o c^2 R_o}\left\{\frac{a[\cos(\omega_o t_r) - \beta]}{[1 - \beta\cos(\omega_o t_r)]^3}\right\}_{\omega_o(t-R_o/c)=\omega_o t_r - \beta\sin(\omega_o t_r)}. \tag{6.117}$$

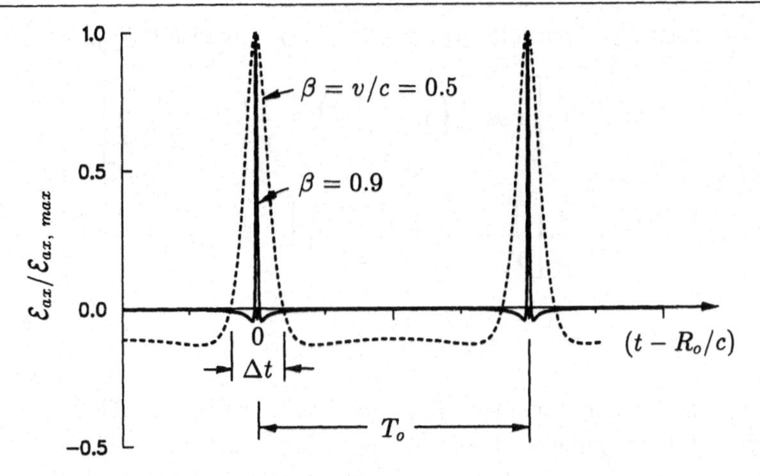

Fig. 6.28. Radiated electric field as a function of time at a point in the plane of the particle's orbit.

This field, in normalized form, is graphed as a function of the time $t - R_o/c$ in Figure 6.28 for two values of the velocity, $\beta = 0.5$ and 0.9. The field is a periodic waveform, with consecutive positive pulses separated by the time $T_o = 2\pi/\omega_o$. The pulses become narrower as β increases. These pulses are caused by the beam of radiation from the charge (Figure 6.23a) periodically sweeping over the observer. An analogous situation would occur if a small laser were fastened to a wheel with the laser beam tangential to the rim. As the wheel spins, an observer in the plane of the wheel sees a narrow pulse of light each time the beam points in his direction.

The width of the positive pulses in Figure 6.28 is roughly equal to the time Δt between the zero crossings of the waveform. From (6.117) the maximum of the first pulse occurs when $t_r = 0$ and $t - R_o/c = 0$, and a zero crossing occurs when

$$t_r = \left[\cos^{-1}(\beta)\right]/\omega_o$$

and

$$t - R_o/c = \Delta t/2.$$

On substituting these values into (6.116), we can solve for the pulse width Δt:

$$\Delta t = \frac{2}{\omega_o}\left\{\cos^{-1}(\beta) - \beta\sin\left[\cos^{-1}(\beta)\right]\right\}$$
$$= \frac{2}{\omega_o}\left[\cos^{-1}(\beta) - \beta\sqrt{1-\beta^2}\right],$$

or, since $\beta = \sqrt{1 - 1/\gamma^2}$,

$$\Delta t = \frac{2}{\omega_o}\left[\cos^{-1}\left(\sqrt{1 - 1/\gamma^2}\right) - (1/\gamma)\sqrt{1 - 1/\gamma^2}\right]. \tag{6.118}$$

For an ultra-relativistic particle ($\beta \approx 1,\ 1/\gamma \approx 0$) this result simplifies to become

$$\begin{aligned}\Delta t &\approx \frac{2}{\omega_o}\left[\cos^{-1}\left(1-\frac{1}{2\gamma^2}\right)-\frac{1}{\gamma}\left(1-\frac{1}{2\gamma^2}\right)\right]\\ &\approx \frac{2}{\omega_o}\left[\frac{1}{\gamma}-\frac{1}{\gamma}\left(1-\frac{1}{2\gamma^2}\right)\right]\\ &\approx \frac{1}{\omega_o\gamma^3}=\frac{T_o}{2\pi\gamma^3}.\end{aligned} \tag{6.119}$$

We could have easily obtained this result from our previous discussions. Recall that for $\beta \approx 1$, the angular width of the beam of radiation from (6.93) is $2\theta_n \approx 2/\gamma$. In terms of the retarded time, t_r, the charge rotates through this angle in the interval of time $\Delta t_r \approx 2\theta_n/\omega_o = T_o/\pi\gamma$. The charge is moving toward the observer when the pulse is emitted, so there is a Doppler effect (6.79), and the observer measures the time interval (width of the pulse)

$$\Delta t \approx (1-\beta)\Delta t_r = \frac{T_o}{\pi}\frac{(1-\beta)}{\gamma} = \frac{T_o}{\pi}\left[\frac{1-\sqrt{1-1/\gamma^2}}{\gamma}\right] \approx \frac{T_o}{2\pi\gamma^3}$$

as the beam sweeps over him. This is just Equation (6.119).

The periodic electric field (6.117) can be expressed as the Fourier series

$$\mathcal{E}_{ax}(R_o,t) = E_o + \sum_{n=1}^{\infty} E_n \cos\left[n\omega_o(t-R_o/c)\right], \tag{6.120}$$

where the coefficients are (from Problem 6.9)

$$\begin{aligned}E_o &= \frac{1}{T_o}\int_{t-R_o/c=-T_o/2}^{t-R_o/c=T_o/2} \mathcal{E}_{ax}(R_o,t)\,dt\\ &= \frac{-qa}{8\pi^2\varepsilon_o c^2 R_o}\int_{\xi=-\pi}^{\pi}\frac{\cos\xi-\beta}{(1-\beta\cos\xi)^2}d\xi = 0\end{aligned} \tag{6.121}$$

and

$$\begin{aligned}E_n &= \frac{2}{T_o}\int_{t-R_o/c=-T_o/2}^{t-R_o/c=T_o/2} \mathcal{E}_{ax}(R_o,t)\cos\left[n\omega_o(t-R_o/c)\right]dt\\ &= \frac{-qa}{4\pi^2\varepsilon_o c^2 R_o}\int_{\xi=-\pi}^{\pi}\frac{\cos\xi-\beta}{(1-\beta\cos\xi)^2}\cos\left[n(\xi-\beta\sin\xi)\right]d\xi.\end{aligned} \tag{6.122}$$

The narrowness of the radiated pulses for $\beta \approx 1$, that is, $\Delta t/T_o = 1/2\pi\gamma^3 = (1-\beta^2)^{3/2}/2\pi \ll 1$, means that the Fourier series coefficients (6.122) will be significant for frequencies much greater than the fundamental frequency ω_o, that is, for harmonic numbers n much greater than one. We can estimate the harmonic number n_m for which the absolute value of the coefficient $|E_n|$ is maximum. This will occur roughly when the half-period of the cosinusoid in (6.122) ($T_n/2 =$

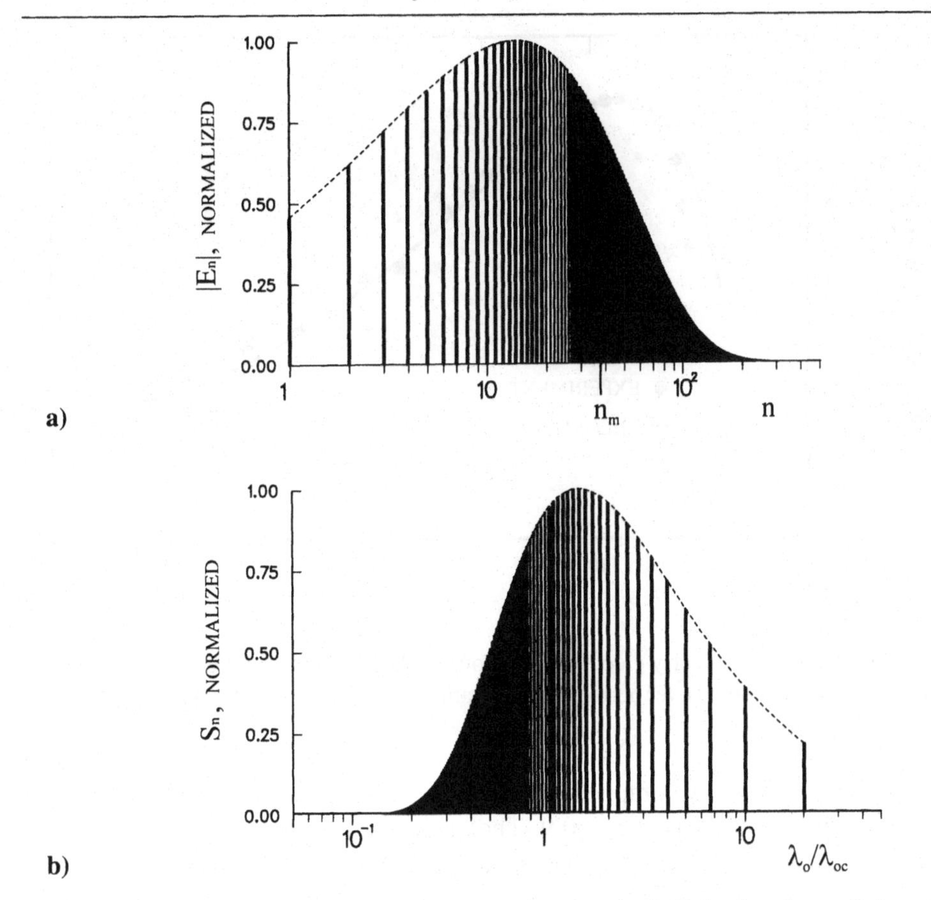

Fig. 6.29. Fourier series coefficients. a) $|E_n|$, for electric field in the plane of the orbit versus the harmonic number n: $\beta = 0.9$. b) S_n, for the time-average power per unit area in the plane of the orbit versus the normalized wavelength λ_o/λ_{oc}: $\beta = 0.9$.

$\pi/n_m\omega_o$) equals the pulse width ($\Delta t \approx 1/\omega_o\gamma^3$), or

$$n_m = [\pi\gamma^3]. \tag{6.123}$$

Here $[x]$ indicates the largest integer $\leq x$. When $\beta \approx 1$, n_m can be quite large; for example, with $\beta = 0.9$, $n_m = 37$, and with $\beta = 0.9999$, $n_m = 10^6$.

In Figure 6.29a the Fourier series coefficients $|E_n|$ are graphed as a function of the harmonic number n for the case $\beta = 0.9$. The first point is for $n = 1$; the coefficient for $n = 0$ is zero: The mean value of the radiated field is always zero (6.121). The coefficients are seen to gradually increase with increasing n, reach a maximum when n is near n_m, as expected, then rapidly decrease with increasing n.

For the n-th harmonic, the time-average power radiated per unit area in the plane of the orbit is

$$S_n = \langle \hat{z} \cdot \vec{S}_n(R_o, t) \rangle = \frac{1}{T_o} \int_{t-R_o/c=-T_o/2}^{t-R_o/c=T_o/2} \frac{1}{\zeta_o} \Big\{ E_n \cos\big[n\omega_o(t - R_o/c)\big] \Big\}^2 dt$$

$$= \frac{E_n^2}{2\zeta_o}. \tag{6.124}$$

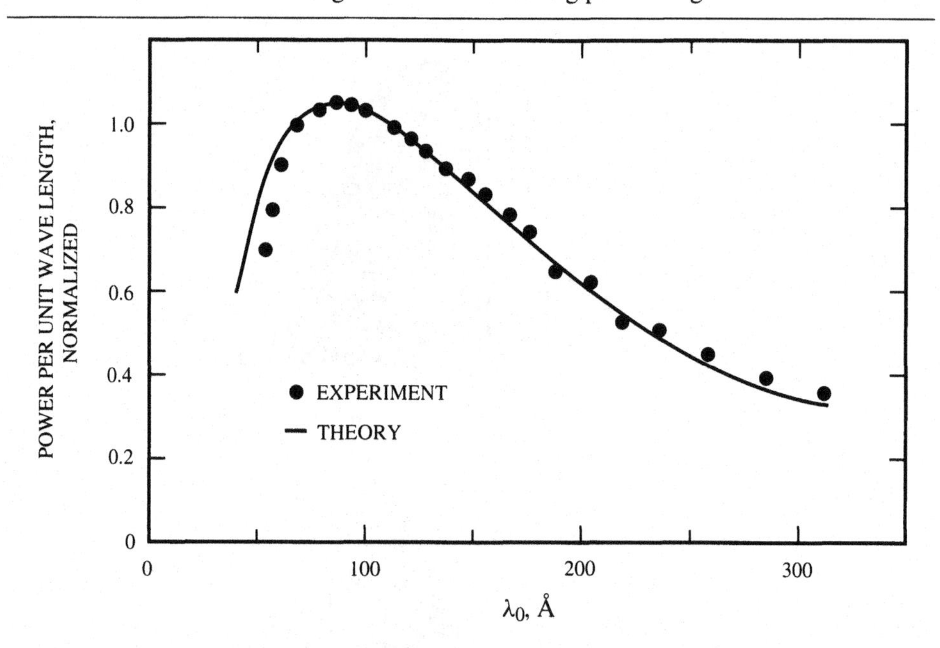

Fig. 6.30. Results from early electron synchrotron. Plotted is the total power radiated per unit wavelength averaged over period of acceleration versus wavelength. (After Tomboulian and Hartman [40].)

In Figure 6.29b we have graphed S_n as a function of the normalized wavelength λ_o/λ_{oc}, again for the case $\beta = 0.9$. The critical wavelength λ_{oc} is a parameter customarily used to describe the spectrum of synchrotron radiation:

$$\lambda_{oc} \equiv \frac{4\pi}{3}\rho\,(\mathcal{E}_o/\mathcal{E})^3\,; \tag{6.125}$$

it is roughly equal to twice the wavelength λ_{om} of the harmonic n_m:

$$\lambda_{oc} = \frac{4\pi}{3}\frac{c\beta}{\omega_o}\frac{1}{\gamma^3} \approx \frac{4\pi^2}{3}\beta\frac{c}{n_m\omega_o} = \frac{2\pi}{3}\beta\lambda_{om}.$$

In an actual synchrotron, a bunch of particles travels in the orbit. The positions of the particles within a bunch vary by a random fraction of a wavelength for wavelengths as small as λ_{oc}. Hence, the particles radiate incoherently, and the total power radiated is proportional to the sum of the powers radiated by the individual particles. Small fluctuations in the particles' energies cause the spectrum in this range ($\lambda_o \approx \lambda_{oc}$) to be continuous, rather than discrete as in Figure 6.29b. The measured spectrum of the radiation from an actual synchrotron, an early electron synchrotron with $\mathcal{E}_{\text{max}} \approx 320$ MeV, is shown in Figure 6.30 [40]. This is a graph of the total power radiated (power radiated in all directions) per unit wavelength interval averaged over the entire time of acceleration. A theoretical curve, with parameters adjusted for this machine, is also shown. The similarity of these results to those from our much simpler calculation (Figure 6.29b) is evident.

The radiation of electromagnetic energy is a parasitic effect in high-energy particle accelerators. It is energy that must be supplied by the radio-frequency source that otherwise could be used to accelerate the particles. The energy radiated by a particle during one cycle in a synchrotron can be determined from (6.94) (Problem 6.10):

$$\Delta\mathcal{E} = T_o\mathcal{P}_q = \left(\frac{2\pi}{\omega_o}\right)\left(\frac{q^2\gamma^4a^2}{6\pi\varepsilon_o c^3}\right)$$

$$= \frac{q^2\beta^3}{3\varepsilon_o\rho}\left(\frac{\mathcal{E}}{\mathcal{E}_o}\right)^4. \tag{6.126}$$

The fraction of the particle's total energy lost to radiation per cycle is then

$$\frac{\Delta\mathcal{E}}{\mathcal{E}} = \frac{q^2\beta^3}{3\varepsilon_o\rho}\frac{\mathcal{E}^3}{(mc^2)^4} \approx \frac{q^2}{3\varepsilon_o\rho}\frac{\mathcal{E}^3}{(mc^2)^4}, \tag{6.127}$$

where the result on the right-hand side is for an ultra-relativistic particle ($\beta \approx 1$). Notice that this factor is inversely proportional to the fourth power of the mass of the particle. The mass of a proton is 1.84×10^3 times that of an electron, so when these particles have equal total energy $\mathcal{E}$, the electron loses 1.15×10^{13} times more energy to radiation than the proton! This loss imposes a practical constraint on the highest energies obtainable with electron synchrotrons. The highest energy particle accelerators are generally proton synchrotrons.

The radiation that has been the bane of the electron synchrotrons used for high-energy particle physics has been a boon for other areas of physics. The radiation from an electron synchrotron has several useful properties. It is a high intensity beam with a broad, predictable, continuous spectrum and an electric field that is nearly linearly polarized. Electron synchrotrons produce radiation at wavelengths ranging from the infrared to the γ-ray regions of the spectrum (see Figure 6.31). These wavelengths are of the same order as the dimensions of atoms and molecules and, therefore, are useful as probes for determining the structure of matter. Other applications of synchrotron radiation are in medical imaging and X-ray lithography for microelectronic circuits [41–45].

The history of the calculations we have been discussing is interesting in itself. The electromagnetic radiation from charged particles moving in circular orbits was first examined near the turn of the century in connection with studies of models for the atom [17, 46, 47]. Discrete charges were thought to travel in orbits inside the atom; one model had negatively charged electrons traveling in circular orbits around a positively charged nucleus (Problem 6.13). The objective of these studies was to give theoretical explanations for observed phenomena, such as the emission of spectral lines. Of course, these problems were solved with the advent of quantum mechanics, and interest in the classical calculations diminished.

Charged particles moving in circular orbits again became of interest when circular, high-energy particle accelerators were being built. The earlier calculations were refined, and radiation, with the characteristic we have just described, was predicted from these machines [48–53]. This radiation was first observed in a 70 MeV ($\lambda_{oc} \approx 4{,}800$ Å) synchrotron by researchers at the General Electric Co. in 1947 [54–56]. The radiation was seen as a small spot of brilliant white light. This observation

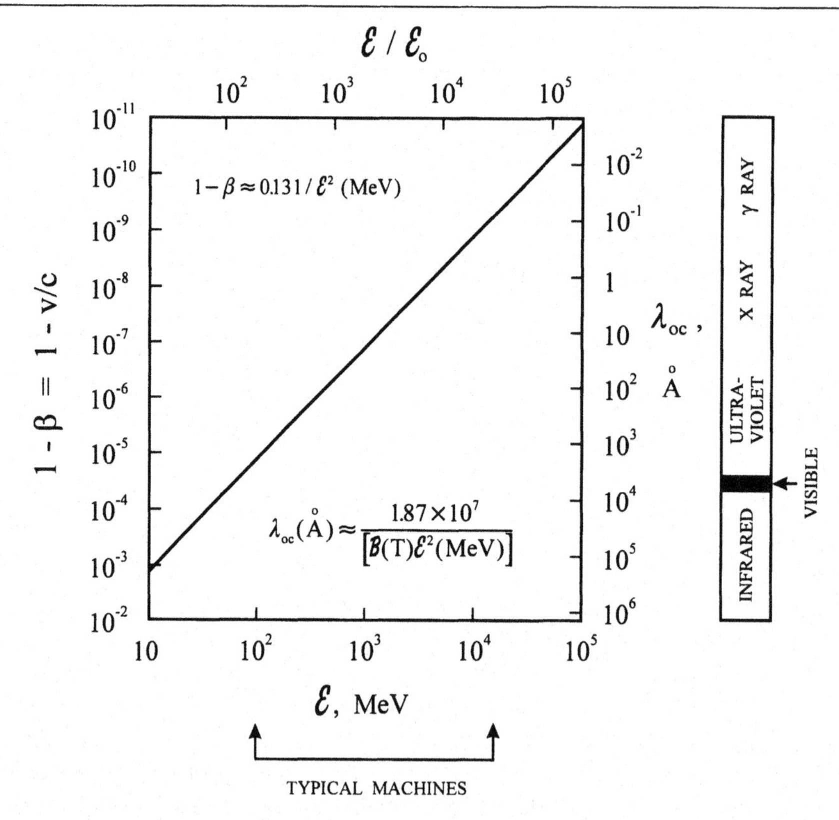

Fig. 6.31. Critical wavelength, λ_{oc}, of synchrotron radiation and velocity, $1 - v/c$, as functions of the total energy, $\mathcal{E}$, for ultra-relativistic electrons ($\beta = v/c \approx 1$). For the calculation of λ_{oc}, the magnetic field is taken to be 1.0 T.

inspired the name "synchrotron radiation" for the radiation produced whenever a charged particle moves in a magnetic field.

The occurrence of synchrotron radiation is not limited to high-energy particle accelerators; it exists throughout the universe. This last statement is best illustrated by the story of the Crab Nebula [57–59]. When a star dies, its center contracts. If the mass of the star is sufficient, the contraction is followed by a violent explosion, a supernova. The outer layers of the star are ejected, leaving behind a neutron star at the center. The Crab Nebula is a supernova remnant, the remains of a supernova observed by Chinese astronomers in 1054.

Optical observations of the Crab Nebula show two distinct features: an amorphous structure more or less oval in shape (Figure 6.32a) and a network of filaments (Figure 6.32b). The radiation from the former has a continuous spectrum, whereas the radiation from the latter consists of a system of spectral lines. The continuous spectrum accounts for about 90% of the intensity of the optical radiation. For some time astronomers had difficulty explaining the mechanism responsible for the continuous spectrum. Then in 1953 the astrophysicist Iosif Samuilovich Shklovsky (1916–1985) proposed that the continuous spectrum was synchrotron radiation produced by relativistic electrons moving in a magnetic field [60, 61]. Relativistic electrons are now thought to be ejected by the neutron star and produce synchrotron radiation when they pass through the star's magnetic field.

Since, as we have seen, synchrotron radiation is polarized, the proof for Shklovsky's hypothesis came with the experimental observations that showed the optical radiation from the Crab Nebula to be polarized [62–65]. In Figure 6.32a the measured polarization pattern for optical radiation (the small, white lines) is shown superimposed on an optical photograph of the amorphous structure of the Crab Nebula [57, 64]. The radiation from the nebula is only partially polarized; it consists of a polarized component plus an unpolarized component. In this figure the length of a line is proportional to the degree of linear polarization d_l, where[10]

$$d_l = \frac{\text{irradiance of linearly polarized component}}{\text{irradiance of total}}, \tag{6.128}$$

and the direction of a line is that of the electric field for the linearly polarized component. An estimate for $d_l = 1$ is in the lower right-hand corner of the figure. The magnetic field in the nebula is roughly in a direction normal to these lines. Recall that synchrotron radiation from relativistic particles is linearly polarized with the electric field in the plane of the orbit, the plane normal to the magnetic field (Figure 6.33).

[10] The degree of linear polarization is discussed in more detail in Section 7.7.

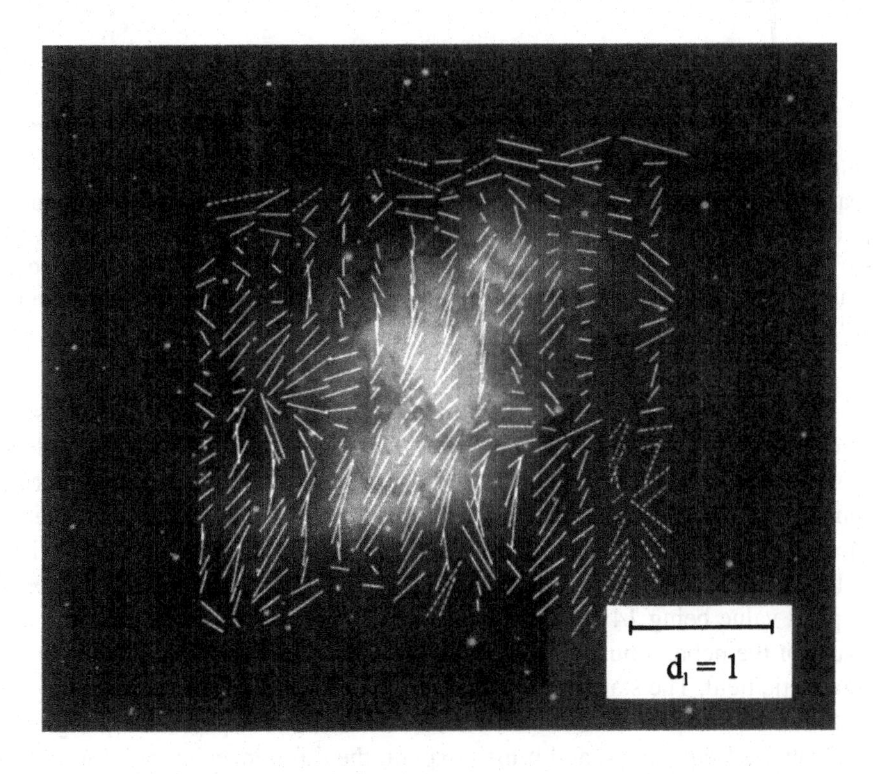

a)

Fig. 6.32. (*continued on the next page*)

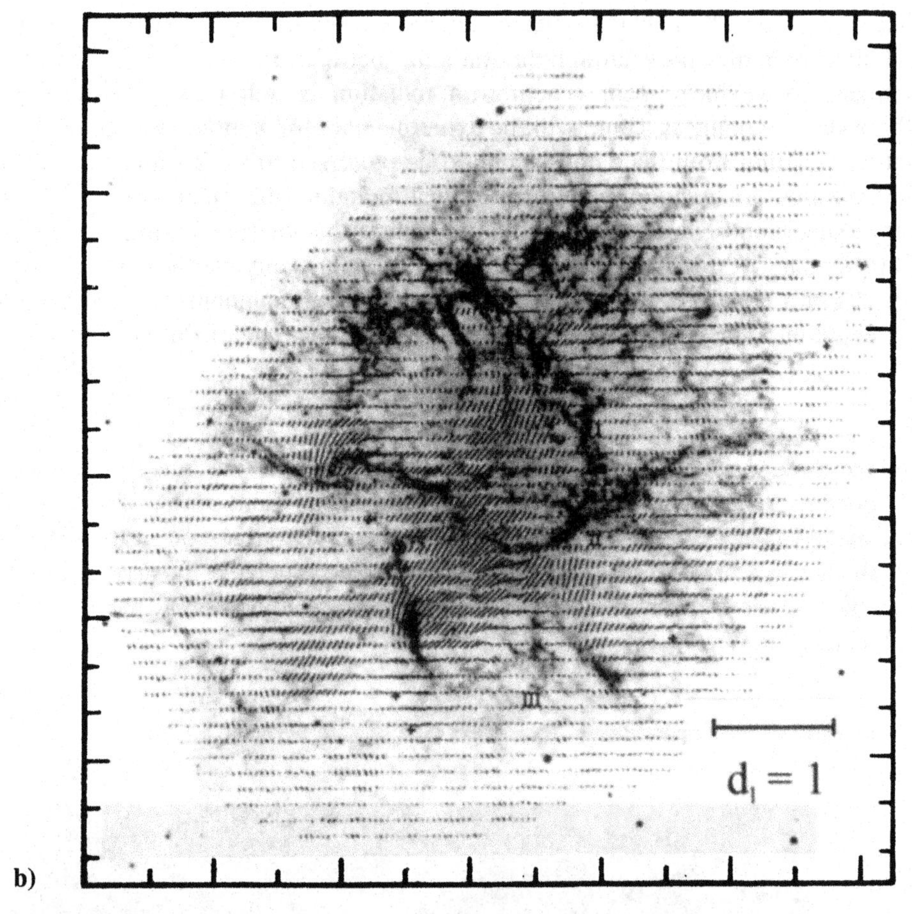

Fig. 6.32. Polarization patterns for the Crab Nebula. a) Optical radiation, with polarization pattern superimposed on optical photograph of amorphous structure. (Reproduced from Oort [57], by permission of *The Observatories of the Carnegie Institution of Washington.*) b) Microwave radiation, $f = 5$ GHz, with polarization pattern superimposed on optical photograph of filamentary structure. (Reproduced from Wilson [67], by permission of Blackwell Science, Ltd.)

Observations of the polarization made at microwave frequencies are similar to those made at optical frequencies [66, 67]. Figure 6.32b shows the polarization pattern measured at the frequency 5.0 GHz; the lines (black) are now superimposed on an optical photograph of the filamentary structure [67]. At optical frequencies the degree of polarization d_l commonly reaches 35%, with a typical value being 25%, whereas at the microwave frequency 5.0 GHz, d_l commonly reaches 20%, with a typical value being 14% [67]. To reach the observer, the radiation from different parts of the nebula must pass through electron clouds of nonuniform density in a magnetic field. The state of the polarization of the radiation is changed by Faraday rotation in this process (Problem 2.12). This effect is particularly significant at microwave frequencies, and it may explain the differences in the observed optical and microwave polarization patterns. In addition, nonuniform Faraday rotation may

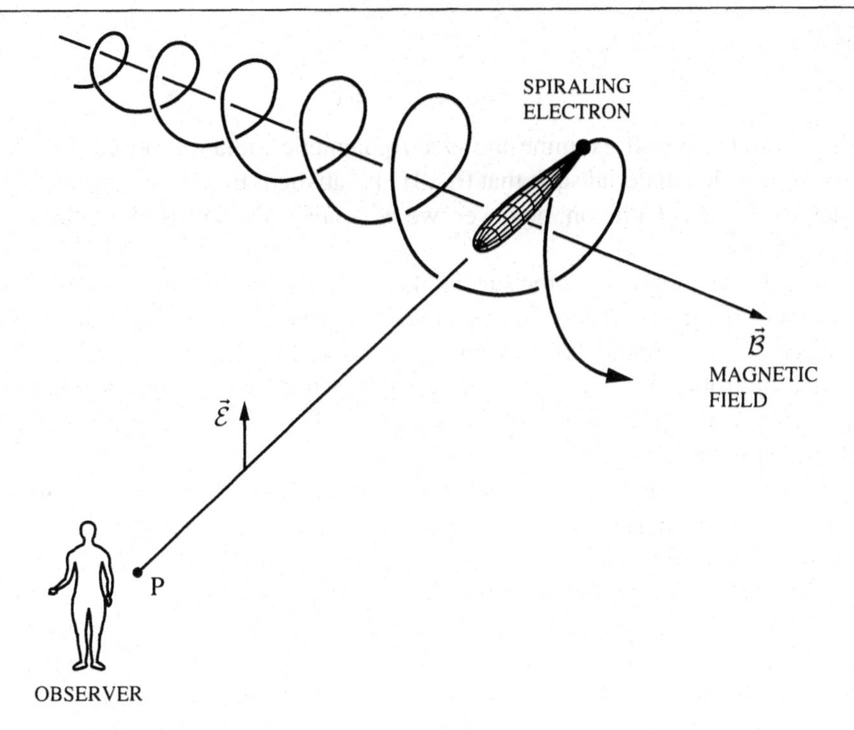

Fig. 6.33. Schematic drawing showing polarization of synchrotron radiation from an electron spiraling in a magnetic field.

cause depolarization of the microwave radiation, producing lower observed values for d_l than at optical frequencies.

6.4 Cherenkov radiation

Up to this point, our discussion of the radiation from a moving point charge has been restricted to charges in free space. We have seen several interesting effects develop as we allowed the speed of the charge to approach the speed of light in free space, such as the concentrating of the radiation in a particular direction (Figures 6.19 and 6.22). However, we have not considered what happens when the speed of the charge exceeds the speed of light in free space, since this possibility is excluded by the special theory of relativity.

In a simple dielectric material, described by the parameters $\varepsilon = \varepsilon_r \varepsilon_o$, $\mu = \mu_o$, and $\sigma = 0$, the speed of light is

$$c_n = 1/\sqrt{\varepsilon \mu_o} = c/\sqrt{\varepsilon_r} = c/n. \tag{6.129}$$

It is less than the speed of light in free space by the factor $1/n$, where

$$n = \sqrt{\varepsilon_r} \tag{6.130}$$

is the *index of refraction* for the material. Thus, we see that it is possible for a charge moving in a material to have a speed v that is greater than the speed of light in the material and still not violate the restriction that its speed be less than the speed of light in free space, that is,

$$c_n = c/n < v < c. \tag{6.131}$$

In this section we will examine the electromagnetic radiation produced when a charge moves in a material such that (6.131) is satisfied. Before we get involved in the details of the calculation, however, we will say a few words about the history of this subject.

Before the development of the special theory of relativity, it was not uncommon to consider charged particles moving at speeds greater than the speed of light in free space. In fact, around 1904 Sommerfeld worked out the theoretical details for the radiation from a charged particle moving with uniform velocity such that $v > c$ [68, 69]. The acceptance of the special theory of relativity with its restriction $v < c$ made such calculations of little interest.

In the 1930s, Pavel Alekseevich Cherenkov (1904–1990) and his colleagues made an extensive experimental investigation of the faint bluish light emitted by pure liquids when exposed to γ rays from a radium source [70–80].[11] They soon realized that the light was not directly caused by the γ rays but was caused by fast moving electrons ejected from the atoms of the liquid by the γ rays. Additional experiments performed by Cherenkov using β rays (fast moving electrons) and by other investigators using electron beams and proton beams from particle accelerators confirmed that high-velocity, charged particles moving through the material were the source of the emitted radiation (light) [82–84].

After recognizing that the key to understanding Cherenkov's observations was that the charged particles were moving with a speed greater than the speed of light in the material, $v > c_n$, Il'ia Mikhailovich Frank (1908–) and Igor' Evgen'evich Tamm (1895–1971) produced a theory, based on classical electromagnetism, that completely explained the experimental results [85, 86]. Their theory assumed a point charge moving with uniform velocity, and some of their results were comparable to those from the earlier work of Sommerfeld. A novel feature of their theory was it showed that a charge moving in a material did not have to accelerate to radiate electromagnetic energy. This is to be contrasted with a charge in free space, which, as we have seen, must accelerate to radiate.

In honor of Cherenkov's discoveries, the radiation produced when a charged particle moves through a material at a speed greater than the speed of light in the material is now called *Cherenkov radiation.*[12] Cherenkov identified several distinctive characteristics of this radiation; we will only describe the experimental determination of three of the more important characteristics. We will not do this in the context of Cherenkov's original experiments. Instead, we will describe simple arrangements that combine elements of Cherenkov's experiments with those of later investigators.

[11] You have probably seen a picture of a nuclear reactor showing the reactor core immersed in water. The bluish glow emanating from the water surrounding the core is due to the same effect studied by Cherenkov; it is Cherenkov radiation [81].

[12] In the former Soviet Union it is sometimes called Cherenkov-Vavilov or Vavilov-Cherenkov radiation to recognize the role Sergei Ivanovich Vavilov (1891–1951) played in the discoveries [80].

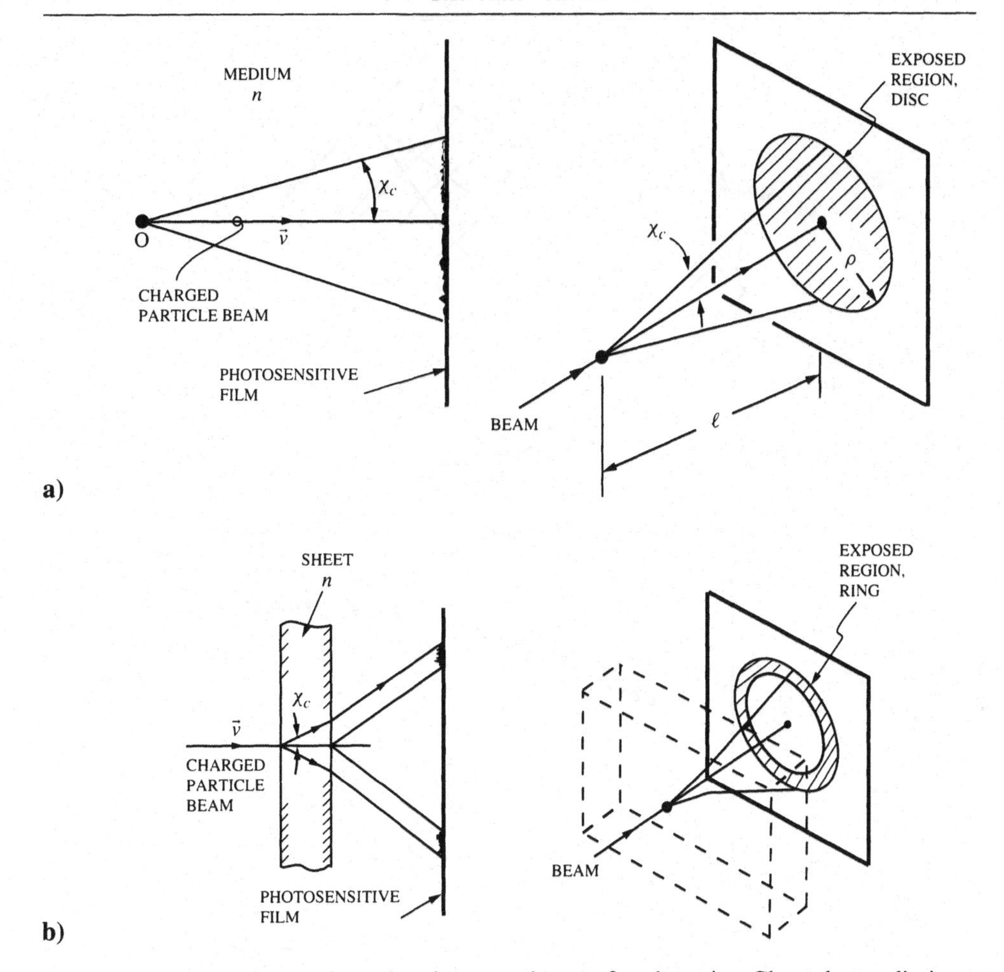

Fig. 6.34. Schematic drawings showing experiments for observing Cherenkov radiation.

i) Cherenkov radiation is asymmetric (directive), being larger in the direction of motion of the charged particle. This is illustrated in Figure 6.34a. Here a beam of particles is traveling through a material over a straight path of length ℓ, and the visible radiation emitted is observed with a photosensitive film placed normal to the path. The beam could be composed of electrons coming from a particle accelerator at point O and traveling through air (which has an index of refraction n that is slightly greater than one) with a velocity such that $v > c_n$. The exposed region on the film is a disc of radius ρ, so radiation is emitted in the forward direction within a cone of angle $\chi_c = \tan^{-1}(\rho/\ell)$.

Figure 6.34b shows a similar experimental arrangement, only now the beam of particles passes through a sheet of material with the index of refraction n. The exposed region on the film is a ring. After tracing an emitted ray back into the material, taking into account the refraction at the surface of the sheet, the direction of the Cherenkov radiation is found to make an angle χ_c with the path of the particles. The sheet may be rotated as in Figure 6.35a, so that its surface is normal to the direction of the emitted ray. This eliminates changes in

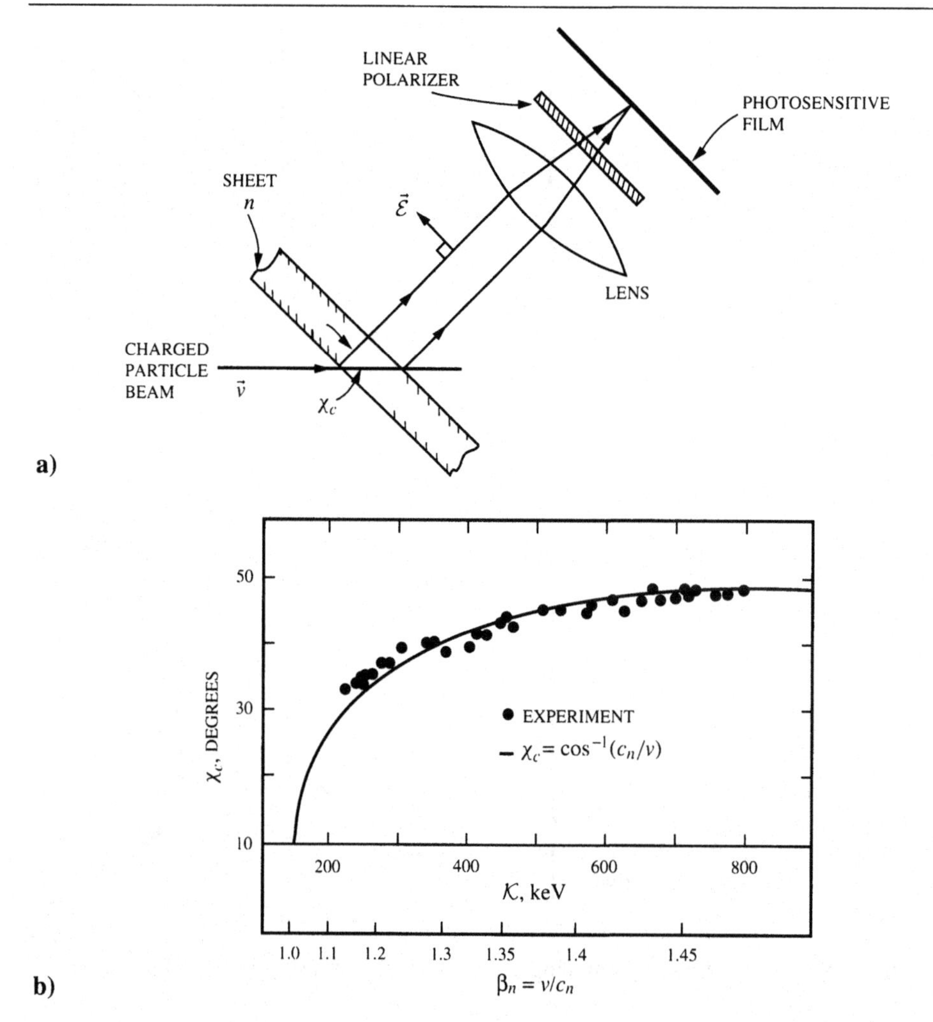

Fig. 6.35. a) Experiment for observing the angular dependence and polarization of Cherenkov radiation. b) Angle of emission χ_c versus particle velocity for mica ($n = 1.59$). (Measured data from Wyckoff and Henderson [83].)

the angle associated with refraction and also prevents total internal reflection, which would trap the radiation within the sheet.

When this experiment is performed using particles with different velocities (energies) and materials with different indices of refraction, the following relationship is observed for the angle χ_c:

$$\cos \chi_c = c_n/v, \qquad v > c_n. \tag{6.132}$$

Figure 6.35b shows results for an electron beam of variable energy (variable velocity) passing through a sheet of mica ($n = 1.59$) [83]. The measured points are seen to agree with the above relationship (6.132).

ii) Cherenkov radiation is linearly polarized, with a component of the electric field parallel to the path of the particles. Rotation of a linear polarizer, placed in the radiation as in Figure 6.35a, shows that the electric field has the orientation shown in the figure [84].

iii) The visible spectrum of Cherenkov radiation is continuous, with an energy per unit wavelength varying as $1/\lambda_o^3$ (where λ_o is the free-space wavelength). Figure 6.36a shows a simple experiment for measuring the spectrum [87]. Either a γ-ray source or a standard tungsten lamp is placed in a large tank filled with deionized water. As mentioned earlier, the γ rays eject fast electrons from the atoms in the water, and the electrons produce the Cherenkov radiation. A spectrometer is used to measure the spectra for both sources, and the spectrum for the Cherenkov radiation is determined by comparison with the known spectrum for the standard tungsten lamp.

The measured spectrum, energy per unit wavelength, for optical wavelengths (3,200 Å to 6,000 Å) shown in Figure 6.36b is seen to fit the $1/\lambda_o^3$ behavior [87]. Notice that these results confirm the bluish color observed for the radiation; the radiation is more intense at the shorter wavelengths (blue) than at the longer wavelengths (red).

We can make use of our previous analysis for a moving point charge in free space to study Cherenkov radiation; we just need to determine the modifications necessary when we remove the restriction that the speed of the charge must be less than the speed of light in the medium. We will only consider a point charge moving with constant velocity along the z axis, $\vec{v} = v\hat{z}$. The coordinates are those in Figure 6.9.

Recall that, for a point charge moving *in free space*, the field at point P and time t is caused by the charge at an earlier time, the retarded time t_r. This is illustrated in Figure 6.37a for a charge moving with constant velocity ($\beta = v/c = 0.5$). The field at time t on the spherical surface S_i ($i = 1, 2, 3, \ldots$) was produced by the charge at the earlier time t_{ri}. For any observation point P in space, a *single* retarded time t_r can always be found. When the position of the point P is expressed in terms of the spherical coordinates $R_q(t)$, $\theta(t)$ centered on the charge at the current time t, the relationship between t and t_r, obtained earlier (6.62), is

$$t - t_r = \left[\frac{R_q(t)}{c}\right]\left(\frac{1}{1-\beta^2}\right)\left[\beta\cos\theta(t) + \sqrt{1-\beta^2\sin^2\theta(t)}\,\right],$$

$$\beta < 1,\ 0 \le \theta(t) \le \pi. \quad (6.133)$$

Notice that, as expected from causality, $t - t_r$ is always positive. This is easily seen from (6.133) and the relation

$$1 - \beta^2\sin^2\theta > (1 - \beta^2\sin^2\theta) - (1 - \beta^2) = \beta^2\cos^2\theta.$$

Also, notice that it would make no sense to choose the other possibility for the square root in (6.133) (change the + sign in front of the square root to a − sign), for $t - t_r$ would then always be negative and violate causality.

When a charge is moving *in a material* with constant velocity such that its speed v is greater than the speed of light in the material c_n, we have the situation shown

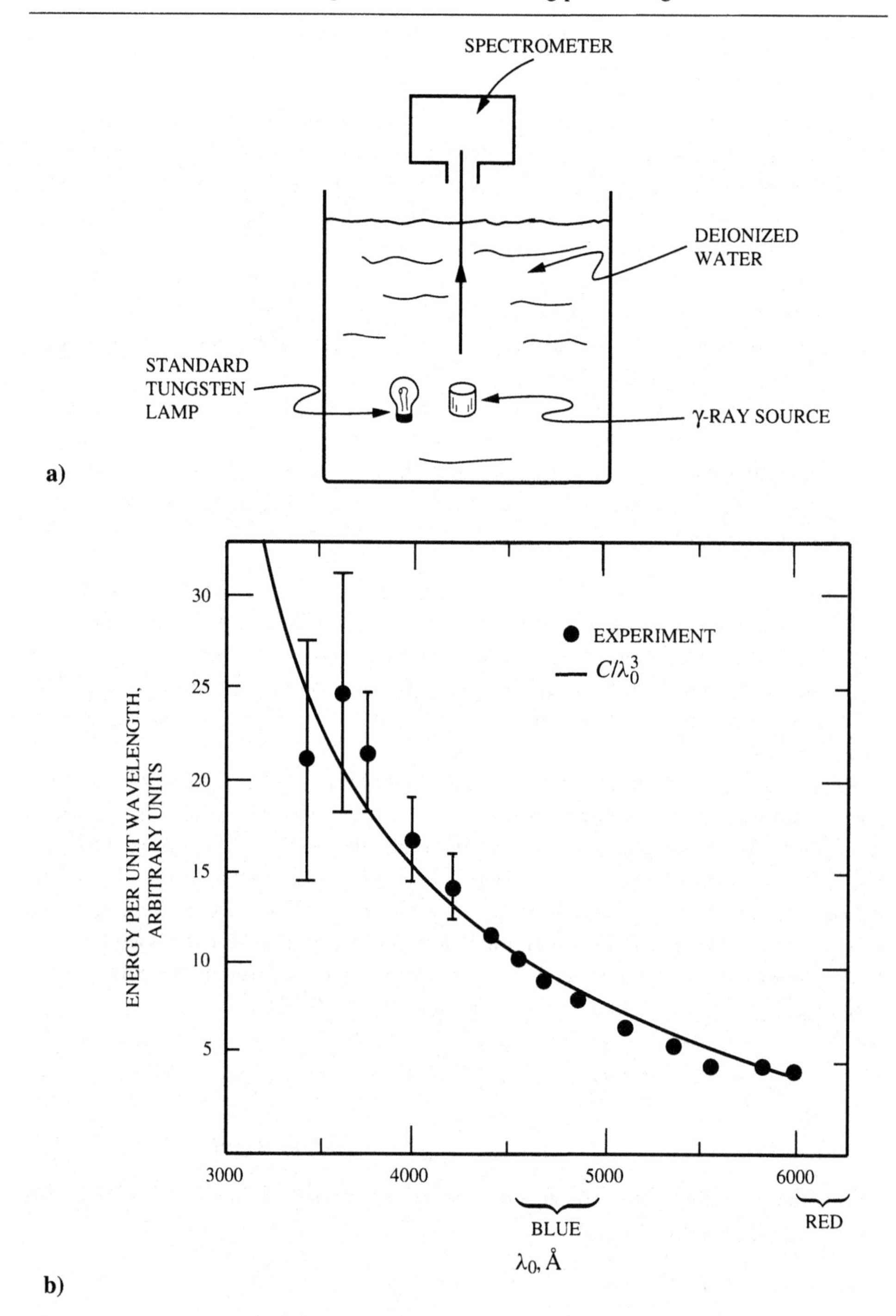

Fig. 6.36. a) Experiment for observing the spectrum of Cherenkov radiation. b) Energy per unit wavelength versus wavelength for optical spectrum. (Measured data from Rich et al. [87].)

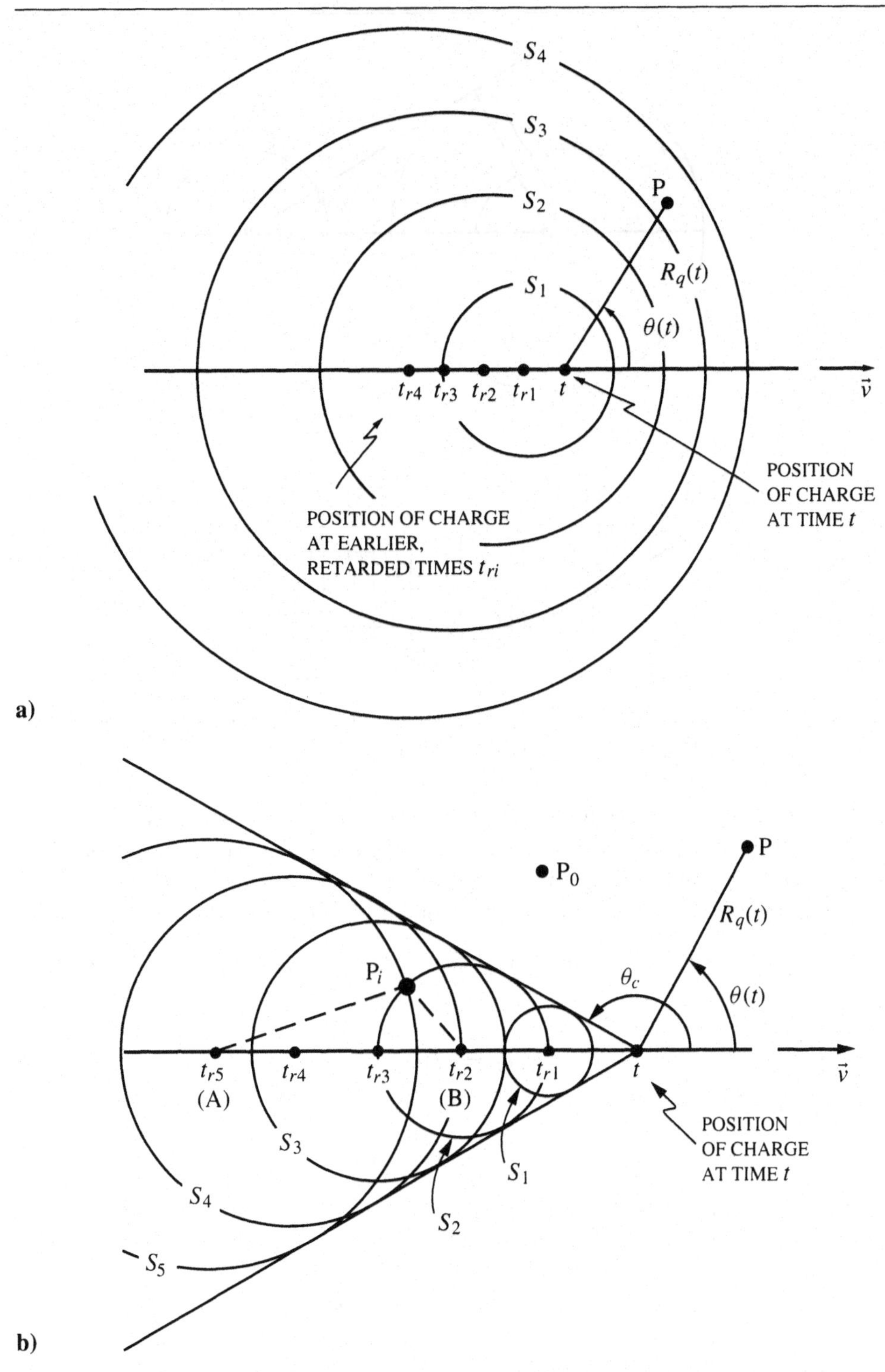

Fig. 6.37. a) Spherical surfaces $S_1, S_2, \ldots$ associated with retarded times $t_{r1}, t_{r2}, \ldots$ for a charge moving with constant velocity in free space, such that $\beta = v/c = 0.5$. b) Spherical surfaces $S_1, S_2, \ldots$ associated with retarded times $t_{r1}, t_{r2}, \ldots$ for a charge moving with constant velocity in a medium, such that $\beta_n = v/c_n = 2.0$.

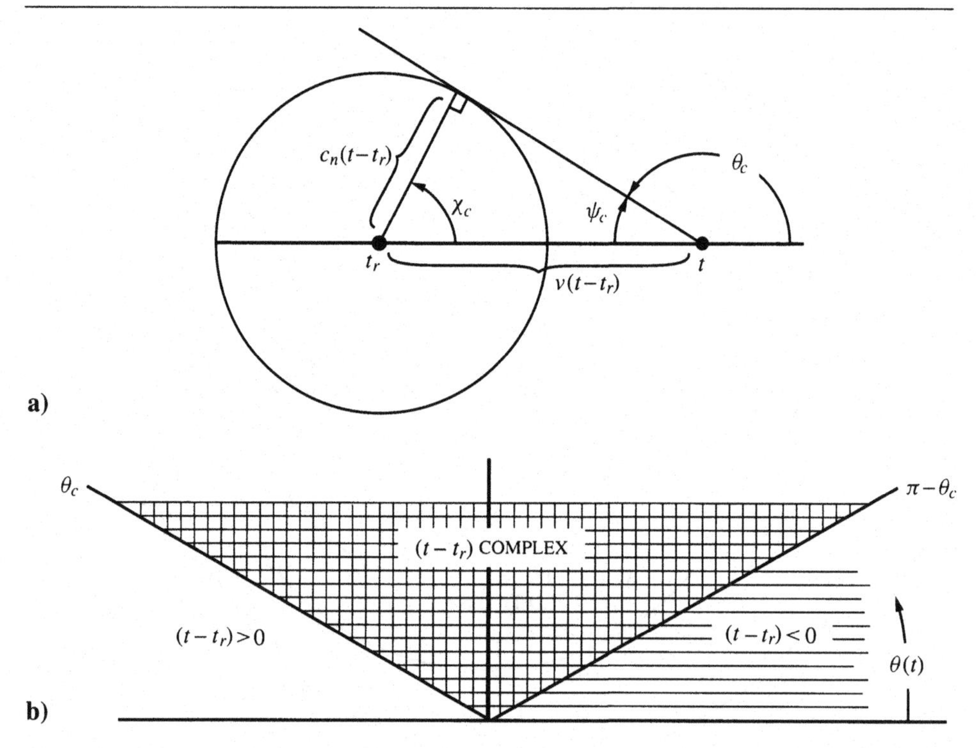

Fig. 6.38. a) Geometry for obtaining angles θ_c, ψ_c, and χ_c. b) Form of solutions for the retarded time for different angles of observation $\theta(t)$; $\beta_n > 1$.

in Figure 6.37b. Again, the field at time t on the spherical surface S_i was produced by the charge at the earlier time t_{ri}. Only now we see that the space surrounding the charge is divided into two distinct regions by a conical surface defined by the exterior, obtuse half angle θ_c. From the geometry in Figure 6.38a,

$$\theta_c = \pi - \sin^{-1}(1/\beta_n), \qquad \pi/2 < \theta_c < \pi, \tag{6.134}$$

where the notation

$$\beta_n = v/c_n = n(v/c) = n\beta \tag{6.135}$$

is used. The auxiliary angles

$$\psi_c = \sin^{-1}(1/\beta_n), \tag{6.136a}$$

which is the supplement of θ_c, and

$$\chi_c = \cos^{-1}(1/\beta_n) = \theta_c - \pi/2 = \pi/2 - \psi_c, \tag{6.136b}$$

which is the compliment of ψ_c, are also shown. In these expressions, the principal values of the functions $\sin^{-1}$ and $\cos^{-1}$ are assumed (i.e., $-\pi/2 \leq \sin^{-1}() \leq \pi/2$, $0 \leq \cos^{-1}() \leq \pi$). In Figure 6.37b, $\beta_n = 2.0$, making $\theta_c = 150^\circ$ ($\psi_c = 30^\circ$ and $\chi_c = 60^\circ$).

For points that are *outside the cone* ($0 \leq \theta(t) < \theta_c$), such as P_o, no retarded time exists, and the field must be zero. All of the radiation produced by the charge at times earlier than t has not had time to reach this region. For points that are *inside the cone* ($\theta_c \leq \theta(t) \leq \pi$), such as P_i, there are *two* retarded times, indicated by A (t_{rA}) and B (t_{rB}) in Figure 6.37b. The charge radiates a signal at position A and then quickly moves to position B where it radiates a second signal. Both signals arrive at point P_i at the same time t.

The conclusions we have just reached by examining the graphical construction in Figure 6.37b can also be obtained from the expression for the retarded time (6.133). After recognizing that $\beta = \beta_n > 1$ for this case and including the + and − signs in front of the square root, Equation (6.133) becomes

$$t - t_r = \left[\frac{R_q(t)}{c_n}\right]\left(\frac{1}{\beta_n^2 - 1}\right)\left[-\beta_n \cos\theta(t) \pm \sqrt{1 - \beta_n^2 \sin^2\theta(t)}\right]. \tag{6.137}$$

Clearly, the argument of the square root must be positive, otherwise $t - t_r$ would be a complex number and meaningless; thus,

$$\sin\theta(t) \leq 1/\beta_n = \sin\theta_c,$$

which means $\theta(t)$ must be in one of the following ranges:

$$0 \leq \theta(t) \leq \pi - \theta_c \tag{6.138a}$$

or

$$\theta_c \leq \theta(t) \leq \pi. \tag{6.138b}$$

The range of angles excluded by (6.138a) and (6.138b) is $\pi - \theta_c < \theta(t) < \theta_c$; this range is marked by a crosshatched region in Figure 6.38b. Now we have

$$\beta_n^2 \cos^2\theta = (1 - \beta_n^2 \sin^2\theta) + \beta_n^2 - 1 > 1 - \beta_n^2 \sin^2\theta, \tag{6.139}$$

for $\beta_n > 1$. For the first range of angles (6.138a), $\cos\theta$ is positive, which with (6.137) and (6.139) means that $t - t_r$ is negative, independent of the choice for the sign in front of the square root in (6.137). Invoking causality, the range of angles (6.138a) must also be excluded and is marked by parallel-line shading in Figure 6.38b. For the remaining range of angles (6.138b), $\cos\theta$ is negative, which with (6.137) and (6.139) means that $t - t_r$ is positive, independent of the choice for the sign in front of the square root in (6.137). Hence, there are two solutions for the retarded time in this range:

$$\left.\begin{array}{l} t - t_{rA} \\ t - t_{rB} \end{array}\right\} = \left[\frac{R_q(t)}{c_n}\right]\left(\frac{1}{\beta_n^2 - 1}\right)\left[\beta_n|\cos\theta(t)| \pm \sqrt{1 - \beta_n^2 \sin^2\theta(t)}\right],$$

$$\beta_n > 1,\ \theta_c \leq \theta(t) \leq \pi. \tag{6.140}$$

These results confirm those we obtained earlier from the graphical construction: For an observation point outside the cone, there is no solution (positive real number) for the retarded time, so the field is zero, whereas for an observation point inside the cone, there are two solutions (6.140) for the retarded time.

Before we can determine the electromagnetic field of the point charge moving with constant velocity such that $\beta_n > 1$, we must examine how the restrictions we have discussed for the retarded time alter the potential functions (6.12) and (6.13). The fact that the potentials must be zero for angles $\theta(t) < \theta_c$ can be incorporated using the Heaviside unit-step function (5.82):

$$\Phi(\vec{r},t) = \frac{q}{4\pi\varepsilon}\left[\frac{1}{R_q(t')|1-\hat{R}_q(t')\cdot\vec{v}/c_n|}\right]_{t'=t_{rA}} U\big[\cos\theta_c - \cos\theta(t)\big]$$

$$+\frac{q}{4\pi\varepsilon}\left[\frac{1}{R_q(t')|1-\hat{R}_q(t')\cdot\vec{v}/c_n|}\right]_{t'=t_{rB}} U\big[\cos\theta_c - \cos\theta(t)\big], \quad (6.141)$$

$$\vec{A}(\vec{r},t) = \frac{\mu_o q}{4\pi}\left[\frac{\vec{v}}{R_q(t')|1-\hat{R}_q(t')\cdot\vec{v}/c_n|}\right]_{t'=t_{rA}} U\big[\cos\theta_c - \cos\theta(t)\big]$$

$$+\frac{\mu_o q}{4\pi}\left[\frac{\vec{v}}{R_q(t')|1-\hat{R}_q(t')\cdot\vec{v}/c_n|}\right]_{t'=t_{rB}} U\big[\cos\theta_c - \cos\theta(t)\big]. \quad (6.142)$$

Notice that $\cos\theta_c$ is negative since $\pi/2 < \theta_c < \pi$, which makes U zero for $0 \le \theta < \theta_c$.

The denominator of these expressions must be evaluated at the two retarded times, t_{rA} and t_{rB}, given by (6.140). Utilizing the diagram in Figure 6.9, the denominator can be expressed as

$$\big|R_q(t_r) - \beta_n\hat{z}\cdot\vec{R}_q(t_r)\big| = \Big|R_q(t_r) - \beta_n\hat{z}\cdot\big[(t-t_r)\vec{v} + \vec{R}_q(t)\big]\Big|$$

$$= \Big|c_n(t-t_r) - \beta_n\big[v(t-t_r) + R_q(t)\cos\theta(t)\big]\Big|$$

$$= \big|c_n(\beta_n^2 - 1)(t-t_r) + \beta_n R_q(t)\cos\theta(t)\big|,$$

which on substituting the retarded times (6.140) becomes

$$\big|R_q(t_r) - \beta_n\hat{z}\cdot\vec{R}_q(t_r)\big|_{t_{rA},t_{rB}}$$

$$= \Big|\beta_n R_q(t)\big|\cos\theta(t)\big| \pm R_q(t)\sqrt{1-\beta_n^2\sin^2\theta(t)} - \beta_n R_q(t)\big|\cos\theta(t)\big|\Big|$$

$$= R_q(t)\sqrt{1-\beta_n^2\sin^2\theta(t)}, \quad (6.143)$$

where we have used the fact that $\pi/2 < \theta_c < \theta < \pi$ to write $\cos\theta = -|\cos\theta|$. Notice that (6.143) has the same value for the two retarded times, t_{rA} and t_{rB}: The two contributions to the potentials at P_i are the same, even though they arise from different earlier times and positions of the charge. With (6.143), the potentials (6.141) and (6.142) become

$$\Phi(\vec{r},t) = \left\{\frac{q}{2\pi\varepsilon R_q(t)[1-\beta_n^2\sin^2\theta(t)]^{1/2}}\right\}_{\Phi} U\big[\cos\theta_c - \cos\theta(t)\big] \quad (6.144)$$

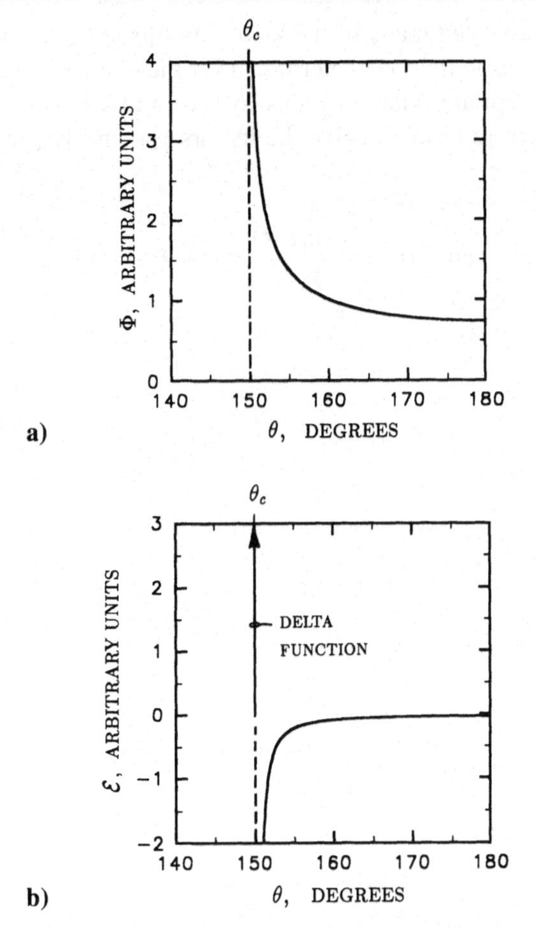

Fig. 6.39. a) Scalar potential and b) electric field for point charge as a functions of the angle θ; $\beta_n = v/c_n = 2.0$, $\theta_c = 150°$.

and

$$\vec{\mathcal{A}}(\vec{r}, t) = \left\{ \frac{\mu_o q v \hat{z}}{2\pi R_q(t)[1 - \beta_n^2 \sin^2 \theta(t)]^{1/2}} \right\}_{\mathcal{A}} U\big[\cos\theta_c - \cos\theta(t)\big]. \quad (6.145)$$

Here the subscripts Φ and $\mathcal{A}$ are placed on the brackets so that these terms can be identified for use in future calculations. The scalar potential is graphed as a function of the angle $\theta(t)$ in Figure 6.39a for the case $\beta_n = 2.0$. The discontinuity in the potential at the angle $\theta = \theta_c = 150°$ is evident.

The electromagnetic field is obtained by differentiating the potential functions; symbolically, the electric field is (5.58)

$$\vec{\mathcal{E}}(\vec{r}, t) = -\nabla\Phi - \frac{\partial \vec{\mathcal{A}}}{\partial t}$$

$$= \left(-\nabla \Big\{\Big\}_{\Phi} - \frac{\partial}{\partial t}\Big\{\Big\}_{\mathcal{A}} \right) U - \Big\{\Big\}_{\Phi} \nabla U - \Big\{\Big\}_{\mathcal{A}} \frac{\partial}{\partial t} U. \quad (6.146)$$

We have already determined many of the factors in this expression. The first term is just U times the electric field we determined for the charge in free space (6.63), evaluated at the appropriate retarded times (6.140) and with $\varepsilon_o \to \varepsilon$, $\beta \to \beta_n$, etc. The second and third terms involve the factors $\{\}_\Phi$ and $\{\}_{\mathcal{A}}$ from (6.144) and (6.145) and the terms

$$\nabla U\big[\cos\theta_c - \cos\theta(t)\big] = \frac{\sin\theta(t)}{R_q(t)}\delta\big[\cos\theta_c - \cos\theta(t)\big]\hat{\theta}(t)$$

and

$$\frac{\partial}{\partial t}U\big[\cos\theta_c - \cos\theta(t)\big] = \frac{v\sin^2\theta(t)}{R_q(t)}\delta\big[\cos\theta_c - \cos\theta(t)\big].$$

After combining these results, the electric field becomes

$$\begin{aligned}\vec{\mathcal{E}}(\vec{r},t) = {} & \frac{q}{2\pi\varepsilon}\frac{\left(\sqrt{\beta_n^2 - 1}\,\right)/\beta_n}{R_q^2(t)[1-\beta_n^2\sin^2\theta(t)]^{1/2}}\delta\big[\cos\theta_c - \cos\theta(t)\big]\hat{R}_q(t) \\ & - \frac{q}{2\pi\varepsilon}\frac{\beta_n^2 - 1}{R_q^2(t)[1-\beta_n^2\sin^2\theta(t)]^{3/2}}U\big[\cos\theta_c - \cos\theta(t)\big]\hat{R}_q(t).\end{aligned} \tag{6.147}$$

A similar calculation for the magnetic field gives

$$\begin{aligned}\vec{\mathcal{B}}(\vec{r},t) &= \frac{\mu_o c_n q}{2\pi}\frac{\sqrt{\beta_n^2 - 1}\sin\theta(t)}{R_q^2(t)[1-\beta_n^2\sin^2\theta(t)]^{1/2}}\delta\big[\cos\theta_c - \cos\theta(t)\big]\hat{\phi} \\ &= \frac{\mu_o c_n q}{2\pi}\frac{\beta_n(\beta_n^2 - 1)\sin\theta(t)}{R_q^2(t)[1-\beta_n^2\sin^2\theta(t)]^{3/2}}U\big[\cos\theta_c - \cos\theta(t)\big]\hat{\phi} \\ &= \frac{1}{c_n}\beta_n\sin\theta(t)\big[-\hat{\theta}(t)\times\vec{\mathcal{E}}(\vec{r},t)\big].\end{aligned} \tag{6.148}$$

Figure 6.39b is a graph of the electric field as a function of the angle $\theta(t)$. There are two distinct contributions to the field (6.147). The first contribution, the term that contains the delta function, exists only on the surface of the cone, $\theta(t) = \theta_c$, where it is an infinite radial field pointing away from the charge. The second contribution, the term that contains the step function, is a radial field pointing toward the charge; within the cone it increases as $\theta(t)$ approaches θ_c, becoming infinite at θ_c. It is zero outside the cone ($\theta(t) < \theta_c$). The directions for these two contributions to the electric field are shown clearly in the schematic drawing in Figure 6.40a.

We will now show that this electric field is consistent with Gauss' electric law

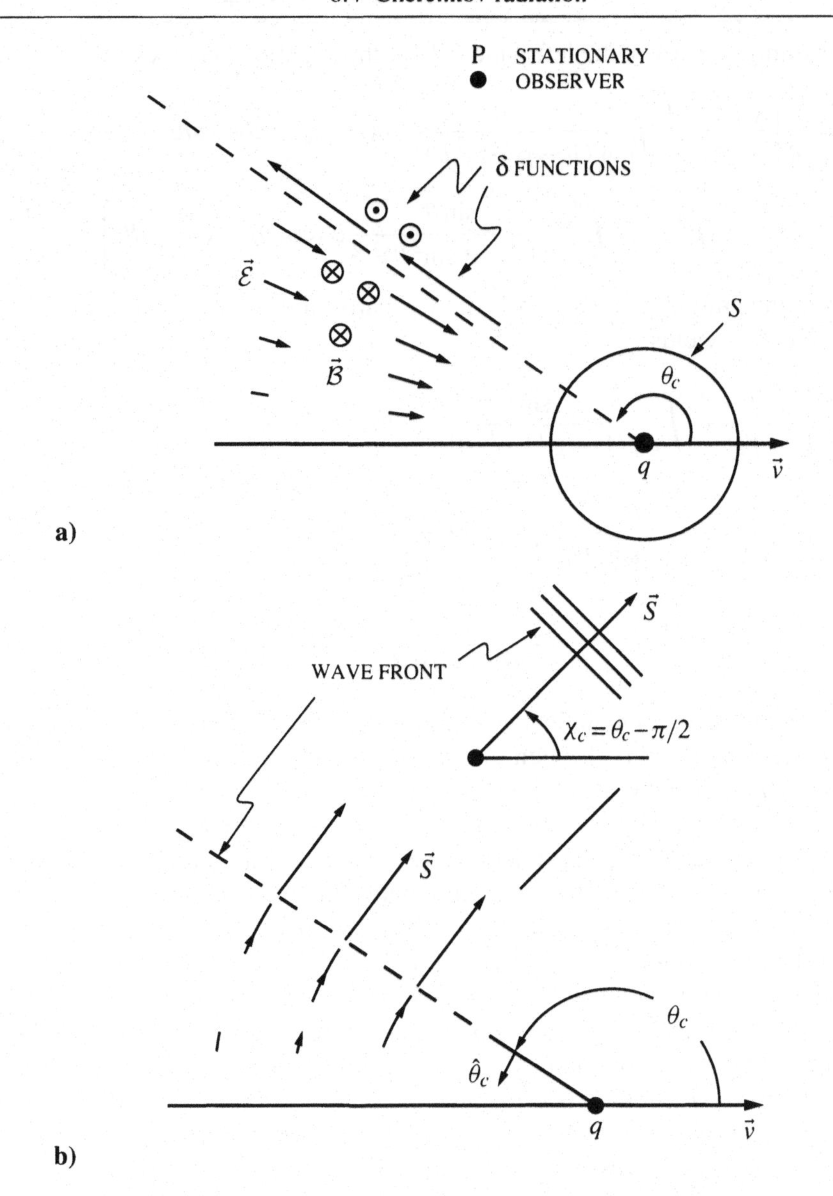

Fig. 6.40. Schematic drawings showing directions for a) electromagnetic field and b) Poynting vector.

(1.3), that is,

$$\oiint_S \vec{\mathcal{E}}(\vec{r}, t) \cdot d\vec{S} = \frac{1}{\varepsilon} \iiint_V \rho(\vec{r}, t) dV.$$

When the surface S is a sphere of radius $R_q(t)$ centered on the charge as in Figure 6.40a, this equation simplifies to become

$$2\pi R_q^2 \int_{\theta=0}^{\pi} \hat{R}_q \cdot \vec{\mathcal{E}} \sin\theta d\theta = q/\varepsilon.$$

After substituting (6.147), the left-hand side of the equation becomes

$$\frac{q}{\varepsilon}\left[\frac{\sqrt{\beta_n^2-1}}{\beta_n}\int_{\theta=0}^{\pi}\frac{\sin\theta}{(1-\beta_n^2\sin^2\theta)^{1/2}}\delta(\cos\theta_c-\cos\theta)d\theta\right.$$

$$\left.-(\beta_n^2-1)\int_{\theta=0}^{\pi}\frac{\sin\theta}{(1-\beta_n^2\sin^2\theta)^{3/2}}U(\cos\theta_c-\cos\theta)d\theta\right],$$

which, on integrating the second term by parts and noticing that $\cos\theta_c = -(\sqrt{\beta_n^2-1})/\beta_n$, yields

$$\frac{q}{\varepsilon}\left\{\frac{\sqrt{\beta_n^2-1}}{\beta_n}\int_{\theta=0}^{\pi}\frac{\sin\theta}{(1-\beta_n^2\sin^2\theta)^{1/2}}\delta(\cos\theta_c-\cos\theta)d\theta\right.$$

$$-\left[\frac{\cos\theta}{(1-\beta_n^2\sin^2\theta)^{1/2}}U(\cos\theta_c-\cos\theta)\right]_{\theta=0}^{\pi}$$

$$\left.\frac{-\sqrt{\beta_n^2-1}}{\beta_n}\int_{\theta=0}^{\pi}\frac{\sin\theta}{(1-\beta_n^2\sin^2\theta)^{1/2}}\delta(\cos\theta_c-\cos\theta)d\theta\right\} = q/\varepsilon.$$

Thus, we see that even though the two contributions to the electric field in (6.147) separately give infinite flux through the spherical surface, their sum gives a finite flux equal to q/ε, in agreement with Gauss' electric law.

It is interesting that the first term in the expression for the electric field (6.147), the term that contains the delta function, was missing in the original analysis of Tamm [86]. Thus, the net flux for the electric field pointed toward the charge, rather than away from the charge as it should for a positive charge. This discrepancy was pointed out and corrected by later authors [88, 89].

The direction of the Poynting vector, $\vec{\mathcal{S}}(t) = [\vec{\mathcal{E}}(t)\times\vec{\mathcal{B}}(t)]/\mu_o$, is shown in Figure 6.40b. Notice that $\vec{\mathcal{S}}$ always points in the direction $-\hat{\theta}(t) = \hat{\psi}(t)$. Thus, $\vec{\mathcal{S}}$ is normal to the wavefront defined by the conical surface of angle θ_c and points into the region where the field is zero.

As time advances the pictures shown in Figure 6.40 move to the right with velocity $\vec{v}$. A stationary observer at the point P sees zero field until the conical wavefront reaches him; then he sees the field within the cone as it passes by. We can easily estimate the temporal variation of the field seen by a distant, stationary observer. With reference to Figure 6.41, we will choose $t = 0$ to be the time at which the conical wavefront reaches the observer. At a later time the observer sees the field at the point $R_q(t)$, $\theta(t) = \theta_c + \Delta\theta(t)$, where

$$R_q(t) = \sqrt{R_q^2(0) + (vt)^2 - 2R_q(0)vt\cos\theta_c},$$

$$\Delta\theta(t) = \sin^{-1}\left[vt\sin\theta_c/R_q(t)\right].$$

Now we will assume that $R_q(0)$ is very large, that is,

$$R_q(0) \gg vt \tag{6.149}$$

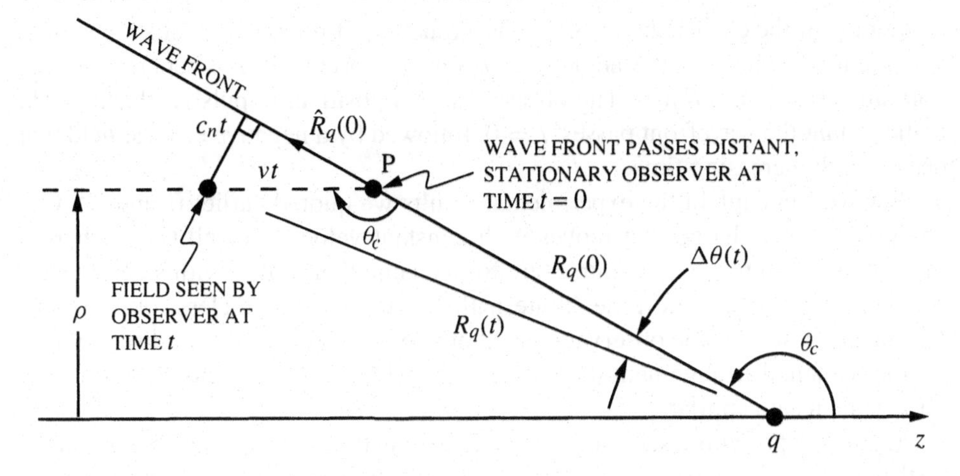

Fig. 6.41. Geometry for determining the temporal variation of the field at a distant, stationary observer.

for all times of interest. Then $R_q(t)$ and $\Delta\theta(t)$ are approximately

$$R_q(t) \approx R_q(0)$$

and

$$\Delta\theta(t) \approx vt \sin\theta_c / R_q(0) = c_n t / R_q(0) \ll 1,$$

making

$$\begin{aligned}\cos\theta_c - \cos\theta(t) &= \cos\theta_c - \cos\left[\theta_c + \Delta\theta(t)\right] \\ &\approx \sin\theta_c \Delta\theta(t) = c_n t / \beta_n R_q(0)\end{aligned}$$

and

$$\begin{aligned}1 - \beta_n^2 \sin^2\theta(t) &= 1 - \beta_n^2 \sin^2\left[\cos\theta_c + \Delta\theta(t)\right] \\ &\approx -2\beta_n^2 \sin\theta_c \cos\theta_c \Delta\theta(t) = 2\sqrt{\beta_n^2 - 1}\, c_n t / R_q(0).\end{aligned}$$

After substituting these results into (6.147), the electric field seen by a distant observer becomes[13]

$$\vec{\mathcal{E}}(\vec{r}, t) \approx \frac{q(\beta_n^2 - 1)^{1/4}}{2^{3/2}\pi\varepsilon c_n^{3/2} R_q^{1/2}(0)} \left[t^{-1/2}\delta(t) - \frac{1}{2} t^{-3/2} U(t) \right] \hat{R}_q(0), \tag{6.150}$$

and the accompanying magnetic field from (6.148) is

$$\begin{aligned}\vec{\mathcal{B}}(\vec{r}, t) &\approx \frac{1}{c_n} \beta_n \sin\theta_c \left[-\hat{\theta}_c \times \vec{\mathcal{E}}(\vec{r}, t) \right] \\ &= \frac{1}{c_n} \left[-\hat{\theta}_c \times \vec{\mathcal{E}}(\vec{r}, t) \right].\end{aligned} \tag{6.151}$$

[13] Here we have used relationship vi from Table 5.1 to write $\delta(ax)$ as $\delta(x)/|a|$.

A graph of the electric field (6.150) as a function of time would look something like Figure 6.39b, with the angle $\theta - \theta_c$ on the horizontal axis replaced by the normalized time $c_n t/R_q(0)$. The observer sees a positive, impulsive electric field at the instant the wavefront passes, $t = 0$, followed by a negative electric field that decays with increasing time.

Now we can explain the experimental results we quoted earlier (i and ii). Our analysis is for a charge that moves with constant velocity for all time, whereas the experimental results, of course, are for a charge that moves with a nearly constant velocity for a finite time (finite path length).[14] Nevertheless, our analysis adequately describes the observations. Although we will confine our discussion to the experimental arrangement shown in Figure 6.34a, the explanations for the other arrangements are similar.

Figure 6.42 presents a series of sketches showing the spherical surfaces produced by the charge at a number of equally spaced points (times) along the path from point O to the photosensitive film. Only a portion of each spherical surface is shown to simplify the presentation. In the first sketch, Figure 6.42a, the charge has not reached the film; hence, no region on the film has been exposed. For our simple model, the electric field is infinite along the wavefront AB, and it falls off rapidly away from this line. In the second sketch, Figure 6.42b, the charge has passed the film, and the wavefront AB has intercepted the film and exposed a disc of radius CD. In the last sketch, Figure 6.42c, the entire wavefront AB has passed the film leaving the disc of radius CE exposed. Notice that the film is illuminated beyond point E in Figure 6.42c; however, the illumination is a field that is much weaker than that on the wavefront AB and can be ignored. The angle χ_c over which the radiation is emitted is seen to be

$$\chi_c = \tan^{-1}(\rho/\ell) = \cos^{-1}(c_n/v),$$

in agreement with the experimental observations (6.132).

The electric field predicted by the theory is given by (6.150) and shown in Figure 6.40a. It points in the direction $\hat{R}_q(0)$ along the wavefront, in agreement with the experimental observations for the polarization (Figure 6.35a).

The frequency spectrum for the electric field seen by a distant observer is obtained by taking the Fourier transform of (6.150):

$$\vec{E}(\vec{r}, \omega) = \int_{-\infty}^{\infty} \vec{\mathcal{E}}(\vec{r}, t) e^{-j\omega t} dt$$

$$= \frac{q(\beta_n^2 - 1)^{1/4}}{2^{3/2}\pi\varepsilon c_n^{3/2} R_q^{1/2}(0)} \hat{R}_q(0) \int_{-\infty}^{\infty} \left[t^{-1/2}\delta(t) - \frac{1}{2} t^{-3/2} U(t) \right] e^{-j\omega t} dt. \tag{6.152}$$

To evaluate this integral, we integrate the first term in the integrand by parts and

[14] A related problem, that of radiation from an insulated linear antenna, is discussed in Section 8.4. The analysis presented there can be used to study Cherenkov radiation from a charge moving on a path of finite length.

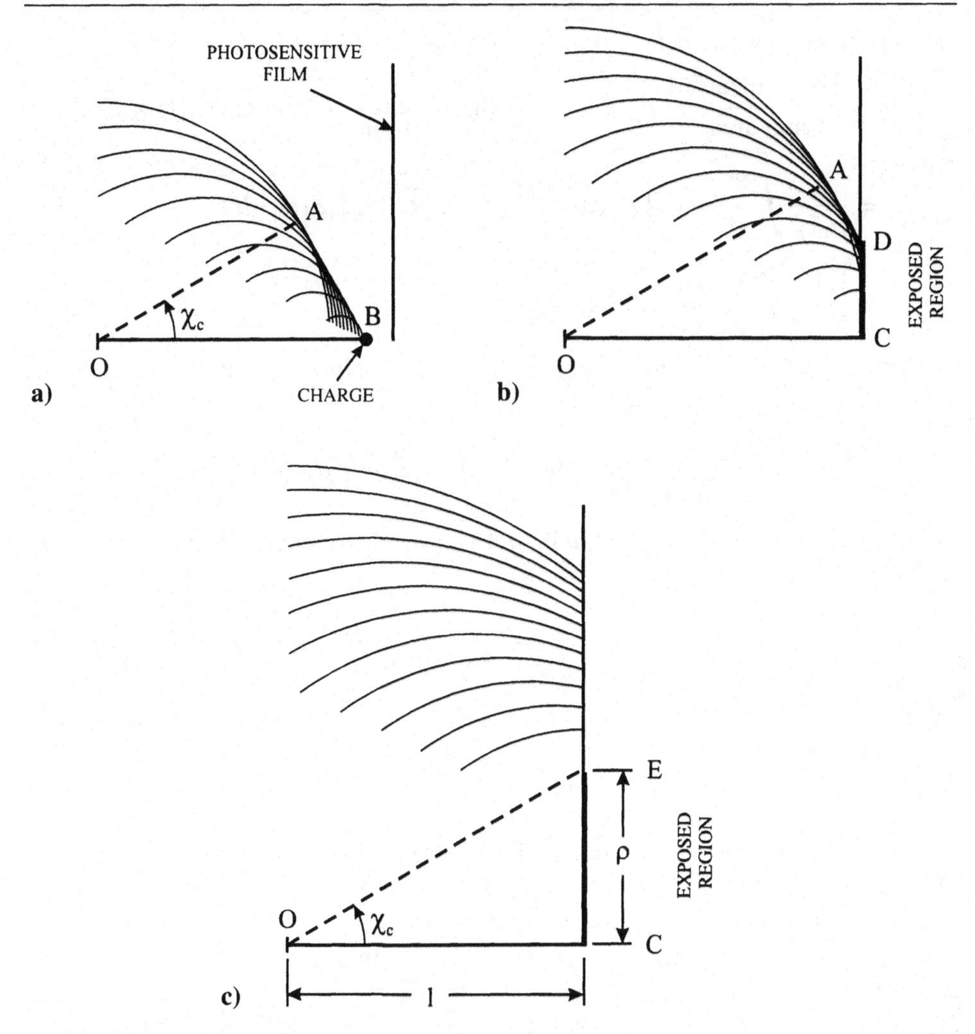

Fig. 6.42. Sketches showing wavefront AB impinging on the photosensitive film for the experiment described in Figure 6.34a.

combine terms to obtain

$$\vec{E}(\vec{r},\omega)=\frac{j\omega q(\beta_n^2-1)^{1/4}}{2^{3/2}\pi\varepsilon c_n^{3/2}R_q^{1/2}(0)}\hat{R}_q(0)\int_0^{\infty}t^{-1/2}e^{-j\omega t}dt,$$

which is a tabulated integral [90]:

$$\vec{E}(\vec{r},\omega)=\frac{q(\beta_n^2-1)^{1/4}\omega^{1/2}}{4\sqrt{\pi}\varepsilon c_n^{3/2}R_q^{1/2}(0)}(1+j)\hat{R}_q(0). \qquad (6.153)$$

Let's assume that the observer has a detector that can measure the energy received per unit area. We can think of the detector as being a square aperture, oriented so that its normal is in the direction $-\hat{\theta}_c$. The total energy (for all time) entering a unit area of the aperture is then

$$\frac{dW}{dA} = \frac{\text{energy received}}{\text{unit area}} = \int_{-\infty}^{\infty} -\hat{\theta}_c \cdot \vec{\mathcal{S}}(t)dt = \frac{1}{\mu_o}\int_{-\infty}^{\infty} -\hat{\theta}_c \cdot \left[\vec{\mathcal{E}}(t) \times \vec{\mathcal{B}}(t)\right]dt$$

$$= \frac{1}{\mu_o c_n}\int_{-\infty}^{\infty} -\hat{\theta}_c \cdot \left\{\hat{R}_q(0) \times \left[-\hat{\theta}_c \times \hat{R}_q(0)\right]\right\}\left|\vec{\mathcal{E}}(t)\right|^2 dt$$

$$= \frac{1}{\mu_o c_n}\int_{-\infty}^{\infty} \left|\vec{\mathcal{E}}(t)\right|^2 dt, \tag{6.154}$$

where (6.151) was used in obtaining the last line. Now from Parseval's formula (1.214) we know that

$$\int_{-\infty}^{\infty} \left|\vec{\mathcal{E}}(t)\right|^2 dt = \frac{1}{\pi}\int_0^{\infty} \left|\vec{E}(\omega)\right|^2 d\omega,$$

when $\vec{\mathcal{E}}(t) \leftrightarrow \vec{E}(\omega)$ [91]. Thus, dW/dA can be easily expressed in terms of the spectrum for the electric field (6.153):

$$\frac{dW}{dA} = \frac{1}{\pi\mu_o c_n}\int_0^{\infty} \left|\vec{E}(\omega)\right|^2 d\omega = \frac{\mu_o q^2}{8\pi^2 R_q(0)}\int_0^{\infty} (\beta_n^2 - 1)^{1/2}\omega d\omega. \tag{6.155}$$

With the change in variable $\omega = 2\pi c/\lambda_o$ $[d\omega = (-2\pi c/\lambda_o^2)d\lambda_o]$, we have

$$\frac{dW}{dA} = \frac{q^2}{2\varepsilon_o R_q(0)}\int_0^{\infty} (\beta_n^2 - 1)^{1/2}\frac{1}{\lambda_o^3}d\lambda_o. \tag{6.156}$$

The integrand of this expression is the total energy received per unit area, per unit wavelength; it is proportional to $1/\lambda_o^3$, in agreement with the experimental observations we quoted earlier (*iii* and Figure 6.36b).

It is customary when discussing Cherenkov radiation to give the total energy radiated per unit path length. This is the total energy passing through a unit length of a cylindrical surface of radius ρ whose axis coincides with the path of the charge; it is easily obtained from (6.155):

$$\frac{dW}{d\ell} = \frac{\text{energy radiated}}{\text{unit path length}} = 2\pi\rho\left[\hat{\rho} \cdot (-\hat{\theta}_c)\right]\frac{dW}{dA}$$

$$= -2\pi R_q(0)\sin\theta_c \cos\theta_c \frac{dW}{dA}$$

$$= \int_0^{\infty} \frac{\mu_o q^2}{4\pi}\left(1 - \frac{1}{\beta_n^2}\right)\omega d\omega. \tag{6.157}$$

The integrand of (6.157) is the energy radiated per unit path length, per unit frequency:

$$\frac{dW}{d\ell d\omega} = \frac{\mu_o q^2}{4\pi}\left(1 - \frac{1}{\beta_n^2}\right)\omega. \tag{6.158}$$

After a change of units, this becomes the result originally obtained by Frank and Tamm [85].

The model we have used in our analysis includes several approximations; probably the most severe is that we are dealing with a simple material whose permittivity ε_r, or, equivalently, index of refraction $n = \sqrt{\varepsilon_r}$, is a constant. The consequences of this assumption are that the electric field, given by (6.147) and shown in Figure 6.39b, is singular at $\theta = \theta_c$, and the energy, given by (6.155) or (6.156), becomes infinite as $\omega \to \infty$ or $\lambda_o \to 0$. To understand how this behavior should be modified for a real material, we must examine what happens on the microscopic level.

Atoms in a material have electric dipole moments that are either permanent or induced by an applied electric field. These microscopic dipole moments give the material its dielectric properties such as ε_r. In a *simple material*, the relationship between the dipole moment per unit volume or polarization $\vec{\mathcal{P}}$ and the electric field is

$$\vec{\mathcal{P}}(\vec{r}, t) = \varepsilon_o(\varepsilon_r - 1)\vec{\mathcal{E}}(\vec{r}, t).$$

So in a simple material, $\vec{\mathcal{P}}$ must respond instantaneously to any temporal variation in $\vec{\mathcal{E}}$; if $\vec{\mathcal{E}}$ varies infinitely fast then $\vec{\mathcal{P}}$ must do likewise. This is clearly impossible in a real material, because $\vec{\mathcal{P}}$ is associated with physical processes – the orientation and induction of microscopic dipole moments. Thus, in a real material, we expect extremely rapid temporal variations in electromagnetic quantities, such as the singular behavior for the electric field near $\theta = \theta_c$, to be absent. These features would be smeared out; for example, the delta function in $\vec{\mathcal{E}}$ would be a pulse of finite width and amplitude.

Since the rapidity of temporal variations in the field is limited in a real material, high frequencies must be absent from the spectrum for the field. Thus, results like (6.155) can only apply when the frequency is not too high. The good agreement with experimental results (Figure 6.36b) shows that this is the case at optical frequencies where Cherenkov radiation is normally observed.

Another approximation in our model is that of a point charge. For a single, elementary particle, like an electron, this is a good assumption, since the effects due to finite size are overshadowed by those that we have just described for the polarization. The latter generally become significant in the ultraviolet region, $\lambda_o \approx 10^3$ Å, whereas the former, for an electron, would not occur until $\lambda_o \approx 3 \times 10^{-4}$ Å [79].

When a bunch of elementary particles radiates coherently, they behave essentially like one large charge. The finite size of the bunch can then be an important factor in determining the characteristics of the Cherenkov radiation. Such bunches can be produced in the beams of particle accelerators [92]. A simple model for the finite-size charge or bunch has the volume densities of charge and current given by [93]

$$\rho(\vec{r}, t) = \frac{q}{\sqrt{\pi}\alpha}\delta(x)\delta(y)e^{-(z-vt)^2/\alpha^2} \tag{6.159}$$

and

$$\vec{\mathcal{J}}(\vec{r}, t) = \frac{qv}{\sqrt{\pi}\alpha}\delta(x)\delta(y)e^{-(z-vt)^2/\alpha^2}\hat{z}. \tag{6.160}$$

These equations describe a charge moving with constant velocity $\vec{v} = v\hat{z}$. It is infinitesimal in cross section (x, y) but has finite size in the direction of motion (z). In this direction, the charge density is a Gaussian function described by the parameter α; the length 2α can roughly be equated to the size of the charge. The total charge is q. Notice from (5.75) that in the limit $\alpha \to 0$, (6.159) and (6.160) reduce to the volume densities for a point charge moving with constant velocity, (6.1) and (6.2).

The volume density of charge (6.159) can also be expressed as

$$\rho(\vec{r}, t) = \int_{-\infty}^{\infty} [q\delta(x)\delta(y)\delta(z - v\tau)]\left\{\frac{v}{\sqrt{\pi}\alpha} e^{-[v(t-\tau)/\alpha]^2}\right\} d\tau, \tag{6.161}$$

which is the convolution of the volume density for the point charge (brackets) with the Gaussian function (braces). The other electromagnetic quantities, such as the volume density of current, potential functions, electric field, etc., can be expressed as similar convolution integrals. For the electric field seen by a distant observer, we have from (6.150)

$$\vec{\mathcal{E}}(\vec{r}, t) = \frac{q(\beta_n^2 - 1)^{1/4}}{2^{3/2}\pi\varepsilon c_n^{3/2} R_q^{1/2}(0)} \hat{R}_q(0)$$

$$\int_{-\infty}^{\infty} \left[\tau^{-1/2}\delta(\tau) - \frac{1}{2}\tau^{-3/2}U(\tau)\right]\left\{\frac{v}{\sqrt{\pi}\alpha} e^{-[v(t-\tau)/\alpha]^2}\right\} d\tau. \tag{6.162}$$

After integrating the first term in the integrand by parts, combining terms, and introducing the notation

$$\mathcal{E}_o = \frac{q\beta_n^{3/2}(\beta_n^2 - 1)^{1/4}}{4\pi^{3/2}\varepsilon\alpha^{3/2} R_q^{1/2}(0)},$$

$$T_o = \frac{\alpha}{v}, \tag{6.163}$$

Equation (6.162) becomes

$$\vec{\mathcal{E}}(\vec{r}, t) = -\left(\frac{2}{T_o}\right)^{3/2} \mathcal{E}_o \hat{R}_q(0) \int_0^{\infty} \tau^{-1/2}(t - \tau) e^{-(t-\tau)^2/T_o^2} d\tau.$$

The remaining integral can be expressed in terms of the modified Bessel functions I_ν and K_ν and the gamma function, Γ [93]:

$$\vec{\mathcal{E}}(\vec{r}, t) = \mathcal{E}_o \left|\frac{t}{T_o}\right|^{3/2} e^{-|t/T_o|^2/2} G(t) \hat{R}_q(0),$$

where

$$G(t) = \begin{cases} K_{1/4}(|t/T_o|^2/2) + K_{3/4}(|t/T_o|^2/2)\} & t < 0 \\ \left.\begin{array}{l} -K_{1/4}(|t/T_o|^2/2) + K_{3/4}(|t/T_o|^2/2) \\ +\sqrt{2}\,\pi\,[-I_{1/4}(|t/T_o|^2/2) + I_{3/4}(|t/T_o|^2/2)] \end{array}\right\} & t > 0 \\ \sqrt{(2)}\Gamma(3/4)\} & t = 0. \end{cases} \tag{6.164}$$

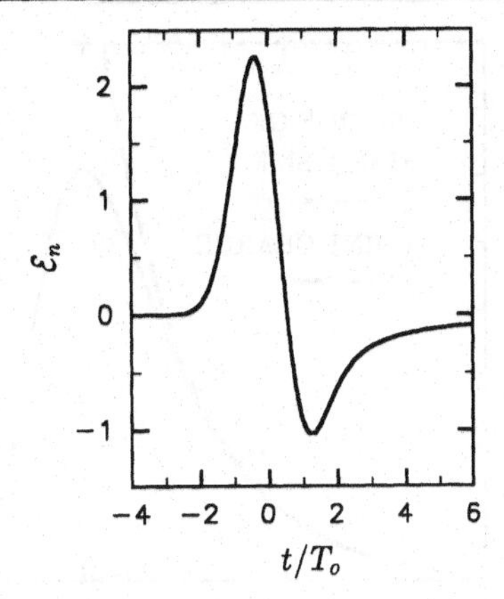

Fig. 6.43. Electric field as seen by a distant, stationary observer as a charge of finite size passes. (After Smith [93].)

Figure 6.43 shows the normalized electric field

$$\mathcal{E}_n(\vec{r}, t) = \mathcal{E}(\vec{r}, t)/\mathcal{E}_o$$

as a function of the normalized time t/T_o. This result should be compared with that for the point charge in Figure 6.39b. Giving the charge finite size is seen to eliminate the singularities in the field; the delta function becomes a pulse of finite width and amplitude. The duration of the pulse is a few T_o, which from (6.163) is roughly the time for the charge to travel a distance equal to its size.

The frequency spectrum for the electric field seen by a distant observer can be readily obtained from results at hand using the time convolution theorem [91]:

$$f_1(t) \leftrightarrow F_1(\omega), \qquad f_2(t) \leftrightarrow F_2(\omega)$$

$$\int_{-\infty}^{\infty} f_1(\tau) f_2(t-\tau)d\tau \leftrightarrow F_1(\omega)F_2(\omega). \tag{6.165}$$

Let's compare (6.162) with (6.165). If we let f_1 represent the electric field of the point charge, whose Fourier transform is given by (6.153), and let f_2 represent the Gaussian function, whose Fourier transform is

$$\frac{1}{\sqrt{\pi}T_o} e^{-t^2/T_o^2} \leftrightarrow e^{-\omega^2 T_o^2/4},$$

then the Fourier transform of the electric field for the charge of finite size is simply

$$\vec{E}(\vec{r}, \omega) = \left[\frac{q(\beta_n^2 - 1)^{1/4}\omega^{1/2}}{4\sqrt{\pi}\varepsilon c_n^{3/2} R_q^{1/2}(0)}(1 + j)\hat{R}_q(0)\right] e^{-\omega^2 T_o^2/4}.$$

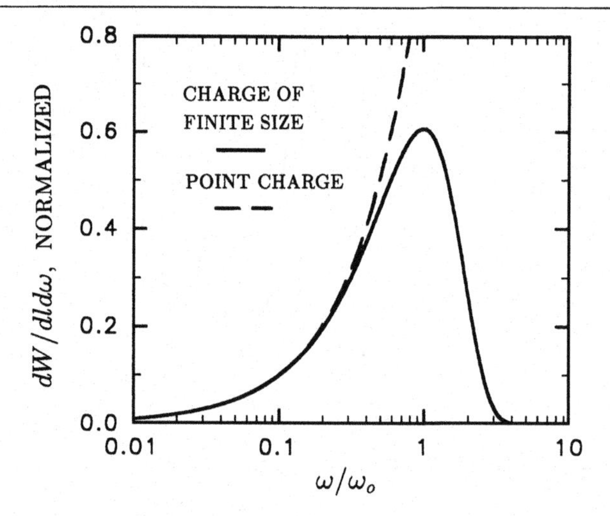

Fig. 6.44. Energy per unit path length per unit frequency for Cherenkov radiation seen by a distant, stationary observer as a charge passes. (After Smith [93].)

The energy radiated per unit path length, per unit frequency, for this field is[15]

$$\frac{dW}{d\ell d\omega} = \frac{2}{\pi}\left\{2\pi\rho\hat{\rho}\cdot\mathrm{Re}\left[\vec{S}_c(\vec{r},\omega)\right]\right\} = -2\sqrt{\frac{\varepsilon}{\mu_o}}\sin\theta_c\cos\theta_c R_q(0)\left|\vec{E}(\vec{r},\omega)\right|^2$$

$$= \frac{\mu_o q^2}{4\pi}\left(1-\frac{1}{\beta_n^2}\right)\omega e^{-(\omega T_o)^2/2}. \tag{6.166}$$

Figure 6.44 shows the magnitude of the normalized spectrum,

$$\frac{4\pi T_o}{\mu_o q^2(1-\beta_n^{-2})}\frac{dW}{d\ell d\omega}, \tag{6.167}$$

as a function of the normalized frequency ω/ω_o, where

$$\omega_o = 1/T_o = v/\alpha.$$

For low frequencies, $\omega/\omega_o \ll 1$, the spectrum is the same as that for the point charge (6.158) (dashed line). However, the two results differ significantly once $\omega/\omega_o \gtrsim 1$; the spectrum for the point charge continues to increase, whereas that for the charge of finite size rapidly decreases. The wavelength in free space corresponding to the frequency ω_o is

$$\lambda_o = 2\pi c/\omega_o = 2\pi\alpha/\beta.$$

We thus see that the spectrum decreases rapidly once the wavelength becomes smaller than the size of the charge.

We have limited the analysis and discussion above to the electric field seen by a

[15] The factor $2/\pi$ in (6.166) is due to the relationship between the energy per unit frequency and the Poynting vector expressed in terms of the Fourier transform of the field (1.216).

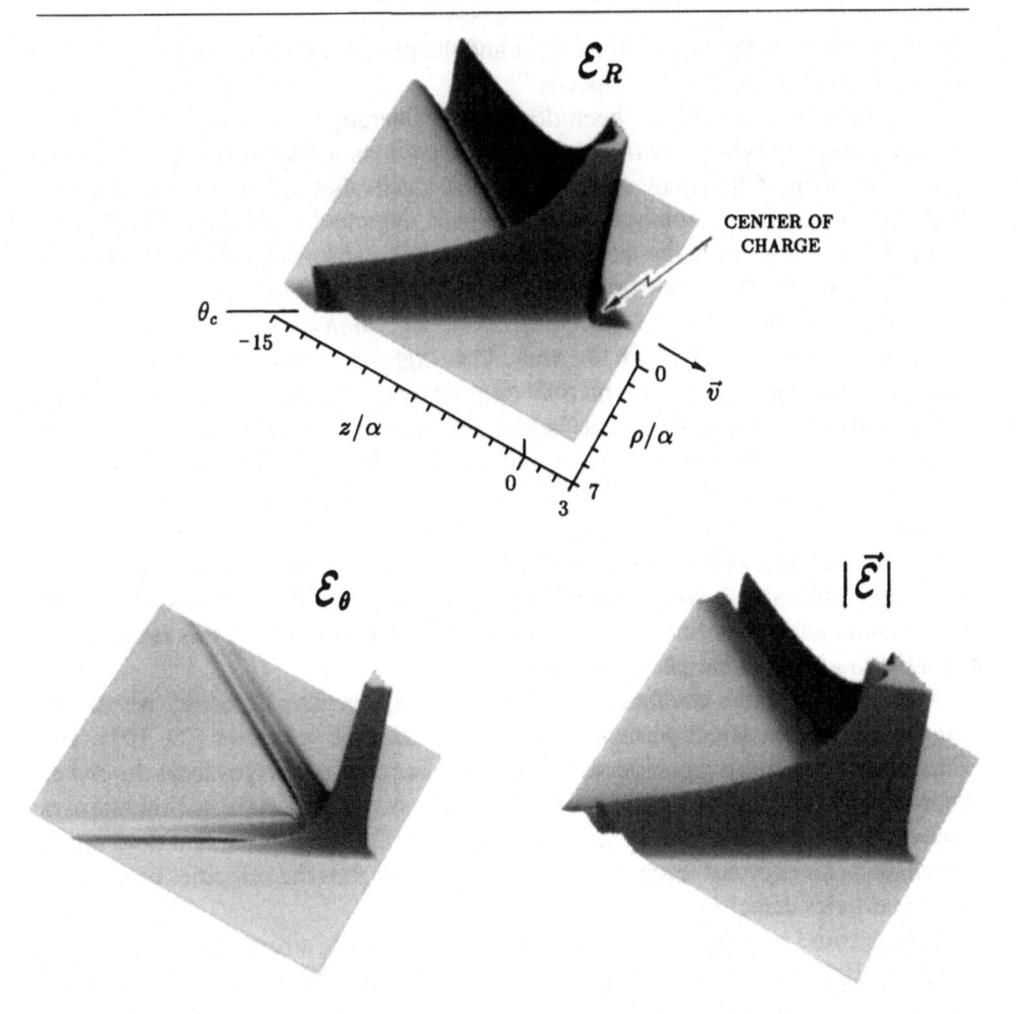

Fig. 6.45. Cherenkov radiation from charge of finite size. Components of electric field are shown as functions of the normalized position (ρ/α, z/α) for $\beta_n = 2.0$ ($\theta_c = 150°$). Singularity in the field on z axis (within charge) has been suppressed. (After Smith [93].)

distant, stationary observer; this is essentially the field near the conical wavefront away from the charge. The analysis for the charge of finite size, however, is easily extended to determine the field in all space [93]. Figure 6.45 shows the complete distribution for the electric field: The components $\mathcal{E}_R$, $\mathcal{E}_\theta$, and the magnitude $|\vec{\mathcal{E}}|$ are graphed in relief as functions of the normalized position (ρ/α, z/α) for the case $\beta_n = 2.0$. The conical wavefront attached to the charge is evident. Similar pictures are associated with other, more familiar, physical phenomena: the wake for a boat moving in shallow water (Figure 6.46a), and the disturbance (Mach cone) for a projectile moving in air (Figure 6.46b). In the former case, the boat is traveling at a speed greater than the speed for waves in shallow water, and in the latter case, the projectile is traveling at a speed greater than the speed of sound in air. In both cases, the angle between the path of the object and the disturbance is given by (6.134) on substitution of the appropriate speeds.[16]

The phenomenon we have been describing, Cherenkov radiation, is applied in a device for observing subatomic particles known as a *Cherenkov detector*. The operation of the Cherenkov detector is most easily described using one of the earliest configurations proposed for this device, depicted in Figure 6.47a [79, 99, 100]. The particle enters the detector from the left and travels along the axis of a dielectric rod with index of refraction n. For a particle with velocity $v > c_n = c/n$, Cherenkov radiation (light), is produced in the direction of the ray that makes the angle $\chi_c = \cos^{-1}(c_n/v)$ with the axis. This light is repeatedly reflected at the surface of the rod, by total internal reflection or by specular reflection from a metal coating, until it reaches the end of the rod, whereupon it enters the dielectric cone. The half angle of the cone is $\chi_c/2$, so all of the light rays reflected at the conical surface emerge parallel at the base. The lens brings these rays to a focus at the photomultiplier after passing through a diaphragm. This device is only useful for detecting particles with a narrow range of velocities (narrow range of angles about χ_c). This equates to a narrow range of energies if the particle's mass is known. Cherenkov radiation produced by particles with velocities outside this range is not brought to a focus at the photomultiplier.

Figure 6.47b shows another early design for a Cherenkov detector; this device can be adjusted to detect particles with a wide range of velocities [79, 101]. The Cherenkov radiation is again produced in a rod, but now it emerges from the end of the rod and is reflected by a cylindrical mirror. The stop and diaphragm limit the rays that can reach the photomultiplier. The separation D is adjustable and determines the angle χ_c for rays that reach the photomultiplier and thus the velocities or energies of the particles detected.

It may come as a surprise that the human eye can act directly as a Cherenkov detector [102–106]. After dark adaptation, an observer with closed eyes sometimes sees a luminous image, which may appear as a point flash, streak, or cloud. These images are called "phosphenes." The source of the phosphene may be mechanical (pressing on the eyeball through a closed lid), electrical, radiation that is normally not visible, such as X rays, or atomic particles.

The index of refraction of the vitreous humour of the eye is roughly that of water; $n \approx 1.33$ for visible light. Thus, when a fast, charged particle ($\beta \approx 1$) in air enters the eye, it will travel at a speed greater than the speed of light ($\beta > 1$) in the vitreous humour. The particle emits Cherenkov radiation (light), which is detected by the retina. The observer sees a "particle-induced visual sensation."

A noteworthy example of this phenomenon occurred during the Apollo space

[16] The results for water waves and sound waves can be much more complicated than our simple picture for Cherenkov radiation. Only for waves in shallow water (depth of water much less than the wavelength) is the velocity independent of the wavelength. For waves in deep water, the velocity depends upon the wavelength, and the wake for a boat is much more complicated [94, 95].
A projectile traveling at supersonic speed in air produces a shock wave, a result of the velocity being a function of the amplitude of the wave [96]. This changes the shape of the disturbance from that of the simple Mach cone [97, 98].

a)

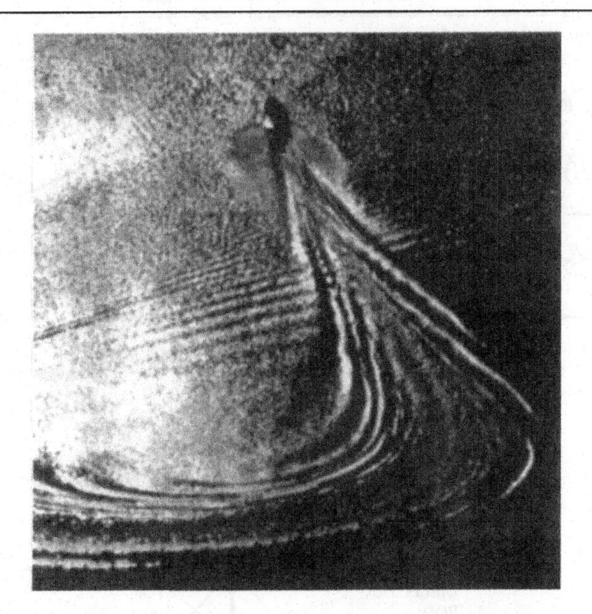

b)

Fig. 6.46. a) Wake of boat in shallow water. (Reproduced with permission from Stoker's *Water Waves* [94], © 1992, John Wiley & Sons, Inc.) b) Projectile traveling at supersonic speed. (Photograph courtesy of Aerodynamics Branch, Propulsion and Flight Division, U.S. Army Research Laboratory, Aberdeen Proving Ground, MD 21005.)

missions [102]. During translunar flight, astronauts, whose eyes were dark adapted and closed, observed white flashes of light. Some of these flashes are thought to be due to Cherenkov radiation produced when cosmic rays (fast particles) entered the astronauts' eyes.

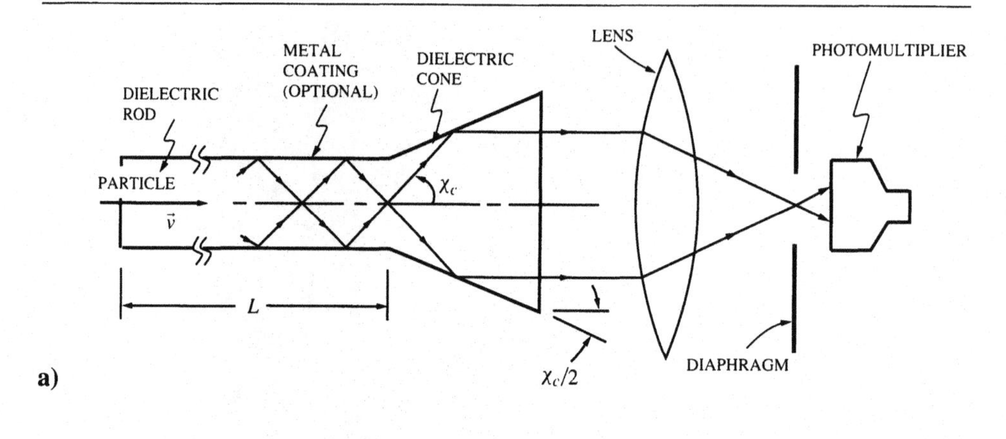

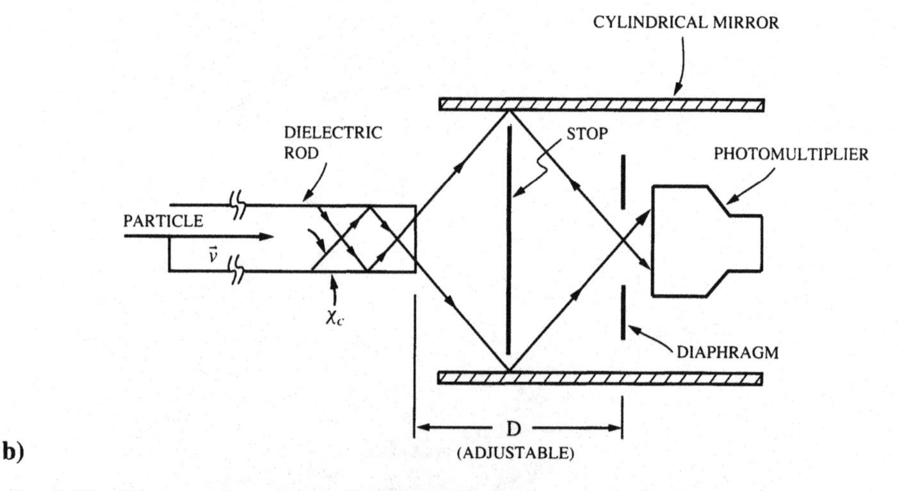

Fig. 6.47. Elements of two basic Cherenkov detectors: a) fixed focus, which detects particles with a narrow range of velocities, and b) adjustable focus, which can be adjusted to detect particles with different velocities.

6.5 Self force

We have seen that an accelerated point charge radiates energy into the surrounding space, and we expect the same behavior for a charged particle of small but finite size. If we assume that an external force is causing the acceleration of the particle, by conservation of energy, we expect this force to do work in the process. The particle must offer some resistance to the motion when it is being accelerated and radiating. This resistance is contained in the self force of the charged particle. Recall that we briefly discussed the self force in Section 1.1 when we introduced the Lorentz force expression; it is the net force on the charged particle that results from the particle's own electromagnetic field. Each element of charge within the particle produces an electromagnetic field and, therefore, a force on every other element of charge within the particle; the self force is the resultant of these forces.

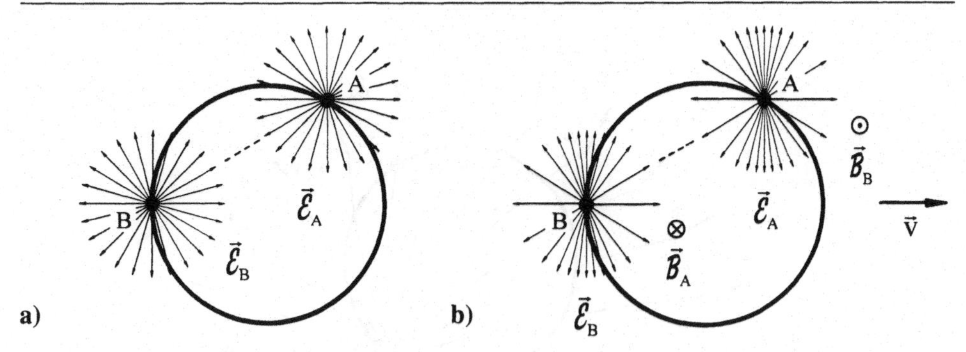

Fig. 6.48. Electric field lines for two elements of charge (A and B) on a charged particle. a) Stationary particle, $\beta = 0$. b) Particle with constant velocity, $\beta = 0.9$.

We can get a qualitative understanding of the origin of the self force using the pictures we have already developed for the electromagnetic field of moving point charges. For our simple illustrations, we will assume that in the rest frame of the observer at the instant of observation the particle is a sphere with the charge q uniformly distributed on its surface, that is, a spherical shell of charge in free space. We will not be concerned with what this assumption implies, from relativistic considerations, about the shape of the particle at other times or in other reference frames.

When the particle is stationary ($\beta = 0$), there is only an electrostatic field; this is shown in Figure 6.48a for two arbitrary elements of charge on the particle, A and B. When drawing these field lines, each element was considered a point charge, and the results from Figure 6.10a were used. The force on A due to the electric field of B is clearly equal in magnitude and opposite in direction to the force on B due to the electric field of A; thus, there is no net self force on the stationary charged particle.

In Figure 6.48b the particle is moving with a constant velocity, $\beta = 0.9$, in the rest frame of the observer, and the results from Figure 6.10c were used to draw the field lines. There are now forces on the particle due to both the electric and the magnetic fields. The force on A due to the electric field of B is still clearly equal in magnitude and opposite in direction to the force on B due to the electric field of A. The magnetic field (6.64) of current element B at A is equal in magnitude to that of current element A at B. The former points out of the page; the latter points into the page. Thus, the magnetic forces on A and B are along a line normal to the velocity and are also equal in magnitude and opposite in direction. Consequently, there is no net self force on the charged particle even when it is moving with constant velocity. As we shall see shortly, the situation changes once the particle is accelerated.

In Figure 6.49a we examine the electric field of a particle undergoing uniform acceleration. The particle is initially at rest and has attained the velocity $\beta = 0.6$ in the rest frame of the observer at the time of observation. To simplify the following

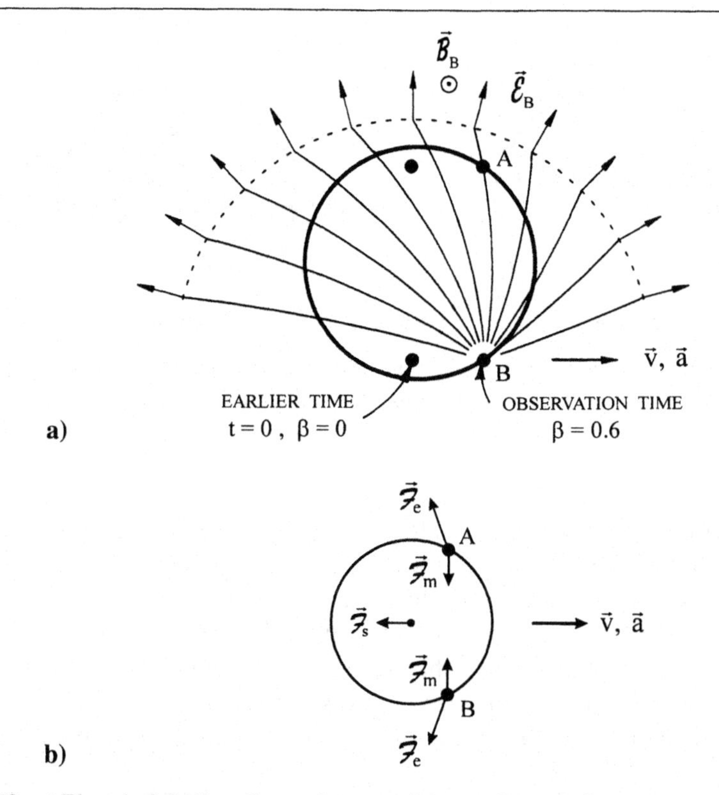

Fig. 6.49. a) Electric field lines for an element of charge (B) on a charged particle undergoing uniform acceleration. b) Forces on element A (B) due to field of B (A).

argument, the elements of charge, A and B, are now assumed to lie along a line normal to the velocity.

The electric field lines of element B are shown in Figure 6.49a; those of element A are easily obtained using symmetry. These lines were drawn using the procedure presented earlier (Figure 6.14). The magnetic field of B at A is equal in magnitude to that of A at B, with the former pointing out of the page and the latter pointing into the page ($\vec{\mathcal{B}} = \vec{R}_q \times \vec{\mathcal{E}}/c$). The forces on element A (B) due to the electromagnetic field of element B (A) are shown in Figure 6.49b. The magnetic forces $\vec{\mathcal{F}}_m$ on the two elements clearly cancel, whereas only the vertical components of the electric forces $\vec{\mathcal{F}}_e$ cancel. Hence, there is a net self force $\vec{\mathcal{F}}_s$ of electromagnetic origin for these two elements. The total self force for the particle is obtained by summing the forces on all elements of the particle.

Notice that our simple picture shows that electromagnetic forces generally do not satisfy Newton's third law: The force on element A due to element B is not equal and opposite to the force on element B due to element A.

Derivations of the self force are quite involved, and we will not reproduce one of them here. Instead, we will simply state the result for a small, uniformly charged, spherical shell of radius a_o moving at low velocity ($\beta \ll 1$) and discuss the implications of this result:[17]

$$\vec{\mathcal{F}}_s = -\left(\frac{q^2}{6\pi\varepsilon_o a_o c^2}\right)\vec{a} + \left(\frac{q^2}{6\pi\varepsilon_o c^3}\right)\frac{d\vec{a}}{dt}. \tag{6.168}$$

The term in parentheses that multiplies the acceleration is called the *electromagnetic mass* m_{em},

$$m_{\text{em}} = \frac{q^2}{6\pi\varepsilon_o a_o c^2}; \tag{6.169}$$

the second term, which involves the temporal derivative of the acceleration, is called the *radiation-reaction force*,

$$\vec{\mathcal{F}}_{\text{rad}} = \left(\frac{q^2}{6\pi\varepsilon_o c^3}\right)\frac{d\vec{a}}{dt}. \tag{6.170}$$

The equation of motion for the charged particle is

$$m'\vec{a} = \vec{\mathcal{F}}_s + \vec{\mathcal{F}}_{\text{ext}},$$

or

$$m\vec{a} = (m' + m_{\text{em}})\vec{a} = \vec{\mathcal{F}}_{\text{rad}} + \vec{\mathcal{F}}_{\text{ext}}, \tag{6.171}$$

where the *observed mass* m includes the electromagnetic mass m_{em} and all of the other contributions to the mass m'. Notice that because the particle is charged, when accelerated it resists the motion by appearing to have increased mass. The external force in (6.171) may include the force due to an applied electromagnetic field, as we discussed in Section 1.1.

A simple numerical calculation shows that the self force is negligible for a macroscopic particle. Consider a small plastic sphere with diameter $2a_o = 1$ mm, mass $m = 6.28 \times 10^{-7}$ kg, and charge $q = 6.95 \times 10^{-11}$ C. For this amount of charge, the electric field at the surface of the particle is 2.5×10^6 V/m; this is just about large enough to break down the surrounding air. The electromagnetic mass for the particle is very small compared to its observed mass:

$$\frac{m_{\text{em}}}{m} = 1.0 \times 10^{-18}.$$

Thus, the leading term in the expression for the self force, $-m_{\text{em}}\vec{a}$, is negligible. The self force can be significant for particles of subatomic dimensions, such as the "classical electron."

[17] The two terms in (6.168) are the first in a series for $\vec{\mathcal{F}}_s$. The higher-order terms are neglected when

$$\frac{2a_o}{c}\left|\frac{d^n\vec{a}}{dt^n}\right| \Big/ \left|\frac{d^{n-1}\vec{a}}{dt^{n-1}}\right| \ll 1, \qquad n = 2, 3, 4, \dots ;$$

that is, the temporal variation of $d^{n-1}\vec{a}/dt^{n-1}$ is negligible during the time for light to cross the particle ($2a_o/c$). The radius of the particle is assumed to be small; however, the limit $a_o \to 0$ cannot be strictly taken, since the first term in the expression (the electromagnetic mass) will become infinite.

Derivations of the self force and detailed discussions of the role it played in the history of classical models for the electron are presented in References [9, 10], and [107–111].

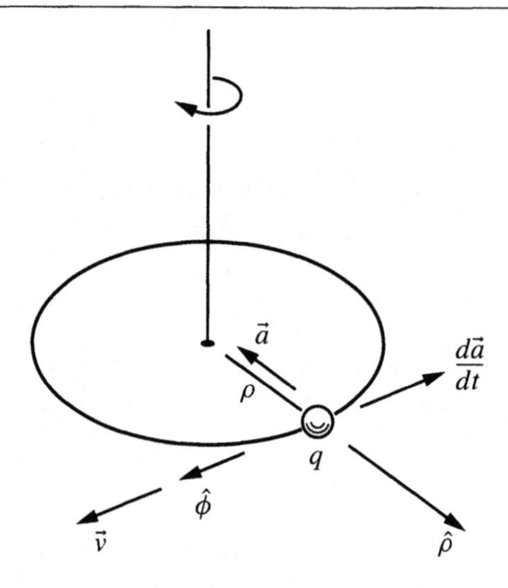

Fig. 6.50. Coordinates for charged particle moving in a circular orbit.

The concept of a self force is consistent with our earlier discussion of radiation from an accelerated charge. To illustrate this point, we will consider a simple example: a charge moving at low velocity ($\beta \ll 1$) in a fixed circular orbit. This could be a charged particle moving at low velocity in a synchrotron. With reference to the coordinates in Figure 6.50, the acceleration is

$$\vec{a} = -\rho\omega^2\hat{\rho},$$

and

$$\frac{d\vec{a}}{dt} = -\rho\omega^3\hat{\phi}.$$

The external force $\vec{\mathcal{F}}_{\text{ext}}$ required to maintain the particle in the circular orbit is

$$\begin{aligned}
\vec{\mathcal{F}}_{\text{ext}} &= m\vec{a} - \vec{\mathcal{F}}_{\text{rad}} \\
&= -m\rho\omega^2\hat{\rho} + \frac{q^2}{6\pi\varepsilon_o c^3}\rho\omega^3\hat{\phi},
\end{aligned}$$

and the work done by this force during a single orbit is

$$\begin{aligned}
\mathcal{W} &= \oint \vec{\mathcal{F}}_{\text{ext}} \cdot d\vec{\ell} = \int_{\phi=0}^{2\pi} \vec{\mathcal{F}}_{\text{ext}} \cdot \hat{\phi}\rho d\phi \\
&= \int_{\phi=0}^{2\pi} |\vec{\mathcal{F}}_{\text{rad}}|\rho d\phi = \frac{q^2\rho^2\omega^3}{3\varepsilon_o c^3}.
\end{aligned}$$

Thus, the time-average power supplied is

$$\mathcal{P} = \frac{\omega\mathcal{W}}{2\pi} = \frac{q^2(\rho\omega^2)^2}{6\pi\varepsilon_o c^3}$$
$$= \frac{q^2 a^2}{6\pi\varepsilon_o c^3}.$$

This is just Larmor's formula (6.59) for the time-average power radiated by the charge. Thus, we see that the work done against the self force (just against the radiation-reaction force for this example) is equal to the energy radiated by the charge.

The success of the self force in predicting the radiated power for the simple example presented above should not be interpreted as proof that the equation of motion (6.171) is correct for all situations. Indeed, in some cases, it can lead to absurd results. For example, in the absence of an external force, Equation (6.171) becomes

$$\left(\frac{q^2}{6\pi\varepsilon_o c^3}\right)\frac{d\vec{a}}{dt} - m\vec{a} = 0.$$

Assuming motion in a straight line ($\vec{a} = a\hat{z}$, $d\vec{a}/dt = da/dt\,\hat{z}$), a solution to this equation is

$$a(t) = a(t=0)e^{t/\tau},$$

where

$$\tau = q^2/(6\pi\varepsilon_o c^3 m).$$

This is the so-called runaway solution; the acceleration increases exponentially with time! For a discussion of such nonphysical solutions and attempts to remove them and other controversial issues associated with the self force, the interested reader should consult References [111–113].

References

[1] A. Liénard, "Champ Électrique et Magnétique," *L'Éclairage Électrique*, Vol. 16, pp. 5–14, 53–59, 106–112, July 1898.

[2] E. Wiechert, "Elektrodynamische Elementargesetze," *Archives Néerlandaises*, Series 2, Vol. 5, pp. 549–73, 1900.

[3] G. Feinberg, "Particles that go Faster than Light," *Sci. Am.*, Vol. 222, pp. 68–77, February 1970.

[4] R. P. Feynman, R. B. Leighton, and M. Sands, *The Feynman Lectures on Physics*, Vol. I, Ch. 28 and Ch. 34; Vol. II, Ch. 21, Addison-Wesley, Reading, MA, 1963, 1964.

[5] R. Becker and F. Sauter, *Electromagnetic Fields and Interactions*, Vol. 1, Ch. DIII, Blaisdell, New York, 1964.

[6] O. Heaviside, "The Radiation from an Electron Moving in an Elliptic, or any other Orbit," *Nature*, Vol. 69, pp. 342–3, February, 1904. Also in *Electromagnetic*

Theory, Volume III, pp. 173–4, "The Electrician" Printing and Publishing Co., London, 1912. Republication, Chelsea, New York, 1971.
[7] J. J. Monaghan, "The Heaviside-Feynman Expressions for the Fields of an Accelerated Dipole," *J. Phys. A*, Vol. 1, pp. 112–7, 1968.
[8] A. R. Janah, T. Padmanabhan, and T. P. Singh, "On Feynman's Formula for the Electromagnetic Field of an Arbitrarily Moving Charge," *Am. J. Phys.*, Vol. 56, pp. 1,036–8, November 1988.
[9] W. K. H. Panofsky and M. Phillips, *Classical Electricity and Magnetism*, 2nd Edition, Addison-Wesley, Reading, MA, 1962.
[10] P. C. Clemmow and J. P. Dougherty, *Electrodynamics of Particles and Plasmas*, Addison-Wesley, Reading, MA, 1969.
[11] E. J. Konopinksi, *Electromagnetic Fields and Relativistic Particles*, McGraw-Hill, New York, 1981.
[12] W. R. Scott, Jr., "A General Program for Plotting Three-Dimensional Antenna Patterns," *IEEE Antennas and Propagation Society Newsletter*, Vol. 31, pp. 6–11, December 1989.
[13] J. Larmor, "On the Theory of the Magnetic Influence on Spectra; and on the Radiation from Moving Ions," *Philos. Mag.*, Vol. 44, pp. 503–13, December 1887.
[14] J. Larmor, *Aether and Matter*, Cambridge University Press, Cambridge, 1900.
[15] A. Sommerfeld, *Electrodynamics*, Lectures on Theoretical Physics, Vol. II, Academic Press, New York, 1952.
[16] J. J. Thomson, *Electricity and Matter*, Charles Scribner's Sons, New York, 1904. Based on the 1903 Silliman Lectures at Yale University.
[17] G. A. Schott, *Electromagnetic Radiation*, Cambridge University Press, Cambridge, 1912.
[18] J. R. Tessman and J.T. Finnell, Jr., "Electric Field of an Accelerating Charge," *Am. J. Phys.*, Vol. 35, pp. 523–7, June 1967.
[19] R. Y. Tsien, "Pictures of Dynamic Electric Fields," *Am. J. Phys.*, Vol. 40, pp. 46–56, January 1972.
[20] E. M. Purcell, *Electricity and Magnetism*, Berkeley Physics Course, Vol. 2, 2nd Edition, Ch. 5, McGraw-Hill, New York, 1985.
[21] W. C. Röntgen, "On a New Kind of Rays," *Nature*, Vol. 53, pp. 274–7, January 23, 1896. An English language translation of an article that appeared in *Sitzungsberichte der Würzburger Physikalisch-medizinische Gesellschaft*, 1895.
[22] B. R. Wheaton, *The Tiger and the Shark: Empirical Roots of Wave-Particle Dualism*, Cambridge University Press, Cambridge, 1983.
[23] W. D. Coolidge, "A Powerful Röntgen Ray Tube with a Pure Electron Discharge," *Phys. Rev.*, Vol. 2, pp. 409–30, December 1913.
[24] A. H. Compton and S. K. Allison, *X-Rays in Theory and Experiment*, 2nd Edition, D. Van Nostrand, New York, 1935.
[25] E. H. Wichmann, *Quantum Physics*, Berkeley Physics Course, Vol. 4, Ch. 4, McGraw-Hill, New York, 1967.
[26] C. T. Ulrey, "An Experimental Investigation of the Energy in the Continuous X-Ray Spectra of Certain Elements," *Phys. Rev.* Vol. 11, pp. 401–10, 1918.
[27] D. Bohm, *The Special Theory of Relativity*, W. A. Benjamin, New York, 1965.
[28] W. Nakel, "Zum Elementarprozess der Bremsstrahlungserzeugung," *Phys. Lett.*, Vol. 25A, pp. 569–70, October 23, 1967.
[29] G. Elwert and E. Haug, "Calculation of Bremsstrahlung Cross Sections with Sommerfeld-Maue Eigenfunctions," *Phys. Rev.*, Vol. 183, pp. 90–105, July 5, 1969.

[30] E. O. Lawrence and N. E. Edelfsen, "On the Production of High Speed Protons," *Science*, Vol. 72, pp. 376–7, October 10, 1930.
[31] E. O. Lawrence and M. S. Livingston, "A Method for Producing High Speed Hydrogen Ions Without the Use of High Voltages," *Phys. Rev.*, Vol. 37, p. 1, 707, June 15, 1931.
[32] M. S. Livingston, *Particle Accelerators: A Brief History*, Harvard University Press, Cambridge, MA, 1969.
[33] E. M. McMillan, "The Synchrotron – A Proposed High Energy Particle Accelerator," *Phys. Rev.*, Vol. 68, pp. 143–4, September 1 and 15, 1945.
[34] V. Veksler, "A New Method of Acceleration of Relativistic Particles," *Journal of Physics*, Academy of Sciences USSR, Vol. 9, No. 3, pp. 153–8, 1945.
[35] R. R. Wilson, "Particle Accelerators," *Sci. Am.*, Vol. 198, pp. 64–76, March, 1958.
[36] —, "The Next Generation of Particle Accelerators," *Sci. Am.*, Vol. 242, pp. 42–57, January 1980.
[37] R. Kollath, *Particle Accelerators*, 2nd Edition, Isaac Pitman, London, 1967.
[38] L. A. Artsimovich and S. Yu. Lukyanov, *Motion of Charged Particles in Electric and Magnetic Fields*, Mir Publishers, Moscow, 1980.
[39] J. D. Jackson, M. Tigner, and S. Wojcicki, "The Superconducting Supercollider," *Sci. Am.*, Vol. 254, pp. 66–76, March 1986.
[40] D. H. Tomboulian and P. L. Hartman, "Spectral and Angular Distribution of Ultraviolet Radiation from the 300-MeV Cornell Synchrotron," *Phys. Rev.*, Vol. 102, pp. 1,423–47, June 15, 1956.
[41] N. G. Basov, editor, *Synchrotron Radiation*, Plenum, New York, 1976. An English translation of the Proceeding of the P. N. Lebedev Physics Institute, Vol. 80, Moscow, 1975.
[42] E. M. Rowe and J. H. Weaver, "The Uses of Synchrotron Radiation," *Sci. Am.*, Vol. 236, pp. 32–41, June 1977.
[43] H. Winick and S. Doniach, editors, *Synchrotron Radiation Research*, Plenum, New York, 1980.
[44] G. Margaritondo and J.H. Weaver, *Synchrotron Radiation Selected Reprints*, American Association of Physics Teachers, College Park, MD, 1986.
[45] H. Winick, "Synchrotron Radiation," *Sci. Am.*, Vol. 257, pp. 88–99, November 1987.
[46] J. J. Thomson, "The Magnetic Properties of Systems of Corpuscles Describing Circular Orbits," *Philos. Mag.*, Vol. 6, pp. 673–93, December 1903.
[47] G. A. Schott, "On the Electron Theory of Matter and on Radiation," *Philos. Mag.*, Vol. 13, pp. 189–213, February 1907.
[48] D. Iwanenko and I. Pomeranchuk, "On the Maximal Energy Attainable in a Betatron," *Phys. Rev.*, Vol. 65, p. 343, June 1 and 15, 1944.
[49] L. Arzimovich and I. Pomeranchuk, "The Radiation of Fast Electrons in the Magnetic Field," *Journal of Physics*, Academy of Science USSR, Vol. 9, No. 4, pp. 267–76, 1945.
[50] E. M. McMillan, "Radiation from a Group of Electrons Moving in a Circular Orbit," *Phys. Rev.*, Vol. 68, pp. 144–5, September 1 and 15, 1945.
[51] J. P. Blewett, "Radiation Loss in the Induction Accelerator," *Phys. Rev.*, Vol. 69, pp. 87–92, February 1 and 15, 1946.
[52] J. Schwinger, "Electron Radiation in High Energy Accelerators," *Phys. Rev.*, Vol. 70, pp. 798–9, November 1 and 15, 1946.
[53] —, "On the Classical Radiation of Accelerated Electrons," *Phys. Rev.*, Vol. 75,

pp. 1,912–25, June 15, 1949.

[54] F. R. Elder, A. M. Gurewitsch, R. V. Langmuir, and H. C. Pollock, "Radiation from Electrons in a Synchrotron," *Phys. Rev.,* Vol. 71, pp. 829–30, June 1, 1947.

[55] F. R. Elder, R. V. Langmuir, and H. C. Pollock, "Radiation from Electrons Accelerated in a Synchrotron," *Phys. Rev.*, Vol. 74, pp. 52–6, July 1, 1948.

[56] H. C. Pollock, "The Discovery of Synchrotron Radiation," *Am J. Phys.*, Vol. 51, pp. 278–80, March 1983.

[57] J. H. Oort, "The Crab Nebula," *Sci. Am.*, Vol. 196, pp. 52–60, March 1957.

[58] P. Gorenstein and W. Tucker, "Supernova Remnants," *Sci. Am.*, Vol. 225, pp. 74–85, July 1971.

[59] S. Mitton, *The Crab Nebula*, Charles Scribner's Sons, New York, 1978.

[60] I. S. Shklovsky, "On the Nature of the Radiation from the Crab Nebula," *Dokl. Akad. Nauk SSSR*, Vol. 90, No. 6, pp. 983–6, 1953.

[61] —, *Cosmic Radio Waves*, Harvard University Press, Cambridge, MA, 1960.

[62] J. H. Oort and Th. Walraven, "Polarization and Composition of the Crab Nebula," *Bull. Astron. Inst. Neth.*, Vol. 12, pp. 285–308, May 5, 1956.

[63] W. Baade, "The Polarization of the Crab Nebula on Plates Taken with the 200-Inch Telescope," *Bull. Astron. Inst. Neth.*, Vol. 12, p. 312, May 5, 1956.

[64] Th. Walraven, "Photo-Electric Observations of the Polarization and Surface Brightness of the Crab Nebula Made at the Observatoire de Haute Provence," *Bull. Astron. Inst. Neth.*, Vol. 13, pp. 293–301, September 9, 1957.

[65] L. Woltjer, "The Polarization and Intensity Distribution in the Crab Nebula Derived from Plates Taken with the 200-Inch Telescope by Dr. W. Baade," *Bull. Astron. Inst. Neth.*, Vol. 13, pp. 301–11, September 9, 1957.

[66] C. H. Mayer, T. P. McCullough, and R. M. Sloanaker, "Linear Polarization of the Centimeter Radiation of Discrete Sources," *Astrophys. J.*, Vol. 139, pp. 248–68, January–May 15, 1964.

[67] A. S. Wilson, "The Structure of the Crab Nebula at 2.7 and 5 GHz, " Parts I through IV, *Mon. Not. Roy. Astron. Soc.*, Vol. 157, No. 3, pp. 229–53, 1972; Vol. 160, No. 4, pp. 355–71, pp. 373–9, 1972; Vol. 166, No. 3, pp. 617–31, 1974.

[68] A. Sommerfeld, "Zur Elektronentheorie," Parts I–III, *Götting. Nachricht.*, pp. 99–130, 363–439, 1904,
pp. 201–35, 1905.

[69] —, *Optics*, Lectures on Theoretical Physics, Vol. IV, Academic Press, New York, 1954.

[70] P. Čerenkov, "Sichtbares Leuchten von reinen Flüssigkeiten unter der Einwirkung von γ–Strahlen," *Dokl. Akad. Nauk SSSR*, Vol. 2, pp. 451–7, 1934.

[71] S. Wawilow, "Über die möglichen Ursachen des blauen γ-Leuchtens von Flüssigkeiten," *Dokl. Akad. Nauk SSSR*, Vol. 2, pp. 457–61, 1934.

[72] P. A. Čerenkov, "Die Wirkung eines Magnetfeldes auf das durch Gamma-Strahlen hervorgerufene sichtbare Leuchten der Flüssigkeiten," *Dokl. Akad. Nauk SSSR*, Vol. 3, pp. 413–6, 1936.

[73] —, "Sichtbares Leuchten der reinen Flüssigkeiten unter der Einwirkung von harten β-Strahlen," *Dokl. Akad. Nauk SSSR*, Vol. 14, pp. 101–5, 1937.

[74] —, "Winkelverteilung der Intensität des Leuchtens, das durch γ-Strahlen in reinen Flüssigkeiten hervorgerufen wird," *Dokl. Akad. Nauk SSSR*, Vol. 14, pp. 105–8, 1937.

[75] —, "Visible Radiation Produced by Electrons Moving in a Medium with Velocities Exceeding that of Light," *Phys. Rev.*, Vol. 52, pp. 378–9, August 15, 1937.

[76] —, "The Spectrum of Visible Radiation Produced by Fast Electrons," *Dokl. Akad. Nauk SSSR*, Vol. 20, pp. 651–5, 1938.
[77] —, "Absolute Output of Radiation Caused by Electrons Moving Within a Medium with Super-Light Velocity," *Dokl. Akad. Nauk SSSR*, Vol. 21, pp. 116–21, 1938.
[78] —, "Spatial Distribution of Visible Radiation Produced by Fast Electrons," *Dokl. Akad. Nauk SSSR*, Vol. 21, pp. 319–21, 1938.
[79] J. V. Jelley, *Čerenkov Radiation and its Applications*, Pergamon Press, New York, 1958.
[80] V. P. Zrelov, *Čerenkov Radiation in High-Energy Physics*, Part I, Israel Program for Scientific Translations, Jerusalem, 1970. Translation of a Russian document dated 1968.
[81] W. H. Jordan, "Radiation from a Reactor," *Sci. Am.*, Vol. 185, pp. 54–55, October 1951.
[82] G. B. Collins and V.G. Reiling, "Čerenkov Radiation," *Phys. Rev.*, Vol. 54, pp. 499–503, October 1, 1938.
[83] H. O. Wyckoff and J. E. Henderson, "The Spatial Asymmetry of Cerenkov Radiation as a Function of Electron Energy," *Phys. Rev.*, Vol. 64, pp. 1–6, July 1 and 15, 1943.
[84] R. L. Mather, "Čerenkov Radiation from Protons and the Measurement of Proton Velocity and Kinetic Energy," *Phys. Rev.*, Vol. 84, pp. 181–90, October 15, 1951.
[85] I. Frank and I. Tamm, "Coherent Visible Radiation of Fast Electrons Passing Through Matter," *Dokl. Akad. Nauk SSSR*, Vol. 14, pp. 109–14, 1937.
[86] I. Tamm, "Radiation Emitted by Uniformly Moving Electrons," *Journal of Physics*, Academy of Sciences USSR, Vol. 1, pp. 439–54, 1939.
[87] J. A. Rich, R. E. Slovacek, and F. J. Studer, "Čerenkov Radiation from Co^{60} Source in Water," *J. Opt. Soc. Am.*, Vol. 43, pp. 750–2, September 1953.
[88] G. Zin, "Teoria generale della radiazione di Čerenkov," *Nuovo Cimento*, Vol. 22, pp. 706–78, November 16, 1961.
[89] G. M. Volkoff, "Electric Field of a Charge Moving in a Medium," *Am. J. Phys.*, Vol. 31, pp. 601–5, August 1963.
[90] F. Oberhettinger, *Tables of Fourier Transforms and Fourier Transforms of Distributions*, Springer-Verlag, Berlin, 1990.
[91] A. Papoulis, *The Fourier Integral and Its Applications*, Appendix I, McGraw-Hill, New York, 1962.
[92] F. R. Buskirk and J. R. Neighbours, "Čerenkov Radiation from Periodic Electron Bunches," *Phys. Rev. A*, Vol. 28, pp. 1,531–8, September 1983.
[93] G. S. Smith, "Cherenkov Radiation for a Charge of Finite Size or a Bunch of Charges," *Am. J. Phys.*, Vol. 61, pp. 147–55, February 1993.
[94] J. J. Stoker, *Water Waves*, Interscience, New York, 1957.
[95] R. A. R. Tricker, *Bores, Breakers, Waves, and Wakes*, American Elsevier, New York, 1965.
[96] R. P. Feynman, R. B. Leighton, and M. Sands, *The Feynman Lectures on Physics*, Vol. I, Ch. 51, Addison-Wesley, Reading, MA, 1963.
[97] A. H. Shapiro, *The Dynamics and Thermodynamics of Compressible Fluid Flow*, Vol. I, Ronald Press, New York, 1953.
[98] F. M. White, *Fluid Mechanics*, McGraw Hill, New York, 1979.
[99] I. A. Getting, "A Proposed Detector for High Energy Electrons and Mesons," *Phys. Rev.*, Vol. 71, pp. 123–4, January 15, 1947.
[100] R. H. Dicke, "Čerenkov Radiation Counter," *Phys. Rev.*, Vol. 71, p. 737, May 15, 1947.

[101] J. Marshall, "Particle Counting by Čerenkov Radiation," *Phys. Rev.*, Vol. 86, pp. 685–93, June 1, 1952.
[102] G. G. Fazio, J. V. Jelley, and W. N. Charman, "Generation of Cherenkov Light Flashes by Cosmic Radiation within the Eyes of the Apollo Astronauts," *Nature*, Vol. 228, pp. 260–4, October 17, 1970.
[103] P. J. McNulty, "Light Flashes Produced in the Human Eye by Extremely Relativistic Muons," *Nature*, Vol. 234, p. 110, November 12, 1971.
[104] P. J. McNulty, V. P. Pease, and V. P. Bond, "Visual Phenomena Induced by Relativistic Carbon Ions with and without Cerenkov Radiation," *Science*, Vol. 201, pp. 341–3, July 28, 1978.
[105] K. D. Steidley, R. M. Eastman, and R. J. Stabile, "Observations of Visual Sensations Produced by Cerenkov Radiation from High-Energy Electrons," *Int. J. Radiation Oncology Biol. Phys.*, Vol. 17, pp. 685–90, September 1989.
[106] K. D. Steidley, "The Radiation Phosphene," *Vision Res.*, Vol. 30, No. 8, pp. 1,139–43, 1990.
[107] J. D. Jackson, *Classical Electrodynamics*, 2nd Edition, Wiley, New York, 1975.
[108] T. Erber, "The Classical Theories of Radiation Reaction," *Fortschritte der Physik*, Vol. 9, pp. 343–92, 1961.
[109] F. Rohrlich, *Classical Charged Particles*, Addison-Wesley, Reading, MA, 1965.
[110] P. Pearle, "Classical Electron Models," Ch. 7 in D. Teplitz, editor, *Electromagnetism Paths to Research*, pp. 211–85, Plenum, New York, 1982.
[111] A. D. Yaghjian, *Relativistic Dynamics of a Charged Sphere*, Springer-Verlag, Berlin, 1992.
[112] T. H. Boyer, "Classical Model of the Electron and the Definition of Electromagnetic Field Momentum," and F. Rohrlich, "Comment on Preceding Paper by T.H. Boyer," *Phys. Rev. D*, Vol. 25, pp. 3,246–50, and 3,251–55, June 15, 1982.
[113] I. Campos and J. L. Jiménez, "Comment on the 4/3 Problem in the Electromagnetic Mass and the Boyer-Rohrlich Controversy," *Phys. Rev. D*, Vol. 33, pp. 607–10, January 15, 1986.

Problems

6.1 Show that the volume densities of charge and current for the point charge, (6.1) and (6.2), satisfy the equation of continuity for electric charge.

6.2 Take the derivative with respect to the time t' of (6.9) and show that the result can be expressed as (6.10).

6.3 The potentials for the moving point charge must be differentiated to obtain the electromagnetic field. This is a tedious process, but one that everyone studying this subject should try once. In the text the steps used in obtaining the electric field are outlined.

a) Obtain the magnetic field in the Liénard-Weichert form (6.27) starting with (6.20).
b) Obtain the magnetic field in the Heaviside-Feynman form (6.40) starting with (6.24).

6.4 Show that the electric and magnetic energy densities are equal for the acceleration field of the point charge (6.44).

6.5 Perform the integration indicated in Equation (6.49) to obtain the total power radiated by the point charge (6.50).

6.6 a) Rearrange the expression for the total power radiated by the point charge (6.50) to obtain

$$\mathcal{P}_q(t_r) = \frac{q^2}{6\pi\varepsilon_o c^3}\left\{\frac{1}{(1-v^2/c^2)^3}\left[a^2 - \left|(\vec{v}/c)\times\vec{a}\right|^2\right]\right\}_{t_r}.$$

b) Assume that the magnitudes of the velocity and acceleration, $|\vec{v}|$ and $|\vec{a}|$, are fixed. For what relative orientations of the velocity and acceleration will the total radiated power be maximum and minimum? What is the ratio of the powers for these two orientations?

6.7 Equations (6.60) and (6.61) are for the electromagnetic field of a point charge moving with constant velocity. They describe the field in terms of coordinates centered on the charge at the retarded time t_r. Starting with these equations, obtain Equations (6.63) and (6.64), which describe the field in terms of the coordinates centered on the charge at the current time t.

6.8 The formulas in Table 6.1 are for the electromagnetic field of a point charge with arbitrary relative orientation of the velocity and the acceleration. Write the formulas for the acceleration field in terms of the spherical coordinates (r, θ, ϕ) in Figure 6.6 for the following special cases:

a) When the acceleration is parallel to the velocity ($\psi = 0$), compare the formulas with (6.73) and (6.74).
b) When the acceleration is perpendicular to the velocity ($\psi = \pi/2$), compare the formulas with (6.89) and (6.90).

6.9 Starting with Equation (6.117) for the electric field $\vec{\mathcal{E}}_a$ in the plane of the particle's orbit, show that the Fourier series coefficients for the electric field of the synchrotron are given by (6.121) and (6.122).

6.10 Show that the energy radiated by a particle during one cycle in a synchrotron is given by (6.126) and (6.127).

6.11 In an electron synchrotron, the electrons are often injected into the machine with velocities very close to the speed of light ($\beta \approx 1$). Hence, the radius of their orbit ρ remains practically constant as their energy is increased, as shown by Equation (6.104).

In a proton synchrotron, the more massive protons are often injected with velocities significantly less than the speed of light. Hence, the radius of their orbit will change as their energy is increased. To keep the radius constant, the frequency ω_o of the electric field accelerating the particles (6.103) is also varied with time. Show that the frequency and magnetic field in the proton synchrotron must vary in accordance with the relationship

$$\omega_o(t) = \frac{c\mathcal{B}(t)}{\rho\sqrt{\mathcal{B}^2(t) + (\mathcal{E}_o/\rho q c)^2}}$$

to keep the radius of the orbit ρ constant.

6.12 This problem concerns the motion of a charged particle at relativistic velocities in a synchrotron.

a) Starting with the relativistic equation for the momentum of the particle

$$\vec{p} = m\gamma\vec{v},$$

show that the kinetic energy of the particle is (6.100)

$$\mathcal{K} = (\gamma - 1)mc^2 = \mathcal{E} - \mathcal{E}_o.$$

Hint: Equate the kinetic energy to the work done on the particle as it mores from position $\vec{r}_1$ where the velocity is zero to position $\vec{r}_2$ where the velocity is $\vec{v}$:

$$\mathcal{K} = \mathcal{W} = \int_{\vec{r}_1}^{\vec{r}_2} \vec{F}\cdot d\vec{r} = \int_{\vec{r}_1}^{\vec{r}_2} \frac{d\vec{p}}{dt}\cdot d\vec{r},$$

then write the result in the form

$$\mathcal{K} = \int_0^v f(v)dv,$$

and integrate this equation to obtain the final expression.

b) Show that the centripetal force on a particle traveling in a circular orbit of radius ρ with constant speed v is

$$\vec{\mathcal{F}} = \frac{d\vec{p}}{dt} = \frac{-m\gamma v^2}{\rho}\hat{\rho}.$$

c) Using the results from parts a and b, obtain Equations (6.98) and (6.99).

d) Show that (6.98)–(6.100) become (6.95)–(6.97) in the low-velocity limit ($\beta \ll 1$).

6.13 The failure of classical theories to explain observed atomic phenomena led to the development of quantum mechanics. One early, unsuccessful, classical model for the atom had discrete electrons traveling in circular orbits around a nucleus [47].

a) Consider this model for a hydrogen atom; a single electron orbits a single proton. The emission spectrum of hydrogen contains a line at the ultraviolet wavelength $\lambda_o = 911.76\ldots$ Å (the shortest wavelength in the Lyman series). Assume that the electron orbits the nucleus with the frequency corresponding to this wavelength. What is the radius of the orbit for the atom to be mechanically stable? In your calculation, assume the velocity of the electron is small compared to the velocity of light ($\beta \ll 1$) and verify this assumption after obtaining your result for the radius.

b) A major problem with this classical model is that the electron continuously experiences centripetal acceleration and, therefore, radiates energy. As a result the orbit is not stable. What is the energy radiated per cycle (time to complete one orbit) by the electron? Assuming the electron stays in this orbit, how long would it take for the electron to radiate an amount of energy equivalent to its kinetic energy?

6.14 Make a graph showing the normalized retarded time $(t-t_r)/[R_q(t)/c_n]$ versus β_n for a point charge moving with constant velocity in a material with index of refraction n. Let $\theta(t) = 150°$ and make the graph for the range $0 \le \beta_n \le 3.0$. Discuss the behavior of the curve as β_n increases.

6.15 A charge moving with constant velocity in a material such that $v > c/n$ radiates electromagnetic energy – Cherenkov radiation. Therefore, we would expect a self force on the charge. Consider the charge to be a spherical shell in the rest frame of the observer and use diagrams like those in Figures 6.48 and 6.49 to qualitatively describe this force. Represent the field due to different points (point charges) on the shell as in Figure 6.37b.

6.16 Consider a particle moving with a speed close to the speed of light ($v \approx c$, $v > c_n$) through a Cherenkov detector like the one shown in Figure 6.47a. Assume that all of the light emitted by the particle in the wavelength range $\lambda_{o1} \le \lambda_o \le \lambda_{o2}$ is collected. Only consider the light produced in the dielectric rod of length L. Ignore any light produced in the dielectric cone.

a) Obtain an expression for the number of photons produced per unit path length in the rod.

b) Evaluate your expression from part a for an electron, a dielectric rod with relative permittivity $\varepsilon_r = 2.25$, and the wavelengths $\lambda_{o1} = 3,900$ Å and $\lambda_{o2} = 7,800$ Å.

6.17 A particle must have a velocity greater than the threshold value $v_t = c_n = c/n$ before it will produce Cherenkov radiation in a detector. For a particular particle, there is a threshold energy $\mathcal{E}_t$ that corresponds to this velocity.

a) Show that

$$\mathcal{E}_t/\mathcal{E}_o = \frac{n}{\sqrt{n^2-1}},$$

where $\mathcal{E}_o$ is the rest energy of the particle.

b) Evaluate the expression in part a for a detector made of plastic ($n = 1.5$) and a detector filled with nitrogen gas at atmospheric temperature and pressure ($n = 1 + 3.5 \times 10^{-5}$).

c) Explain how a group of detectors, each made from a material with different index of refraction n, could be used to bracket the energy of a particle.

6.18 Consider the equation of motion (6.171) for a charged particle when the external force is constant over the time internal $0 \leq t \leq T$:

$$ma(t) - \left(\frac{q^2}{6\pi\varepsilon_o c^3}\right)\frac{da(t)}{dt} = F_o\big[U(t) - U(t-T)\big].$$

a) Obtain a solution to this equation, $a(t)$, that is a continuous function of time.
b) Adjust the constants in your solution from part a to make the solution causal: $a(t) = 0$, for $t < 0$; that is, ensure that the charge does not accelerate until the force is applied. Show that your solution is a "runaway solution."
c) Adjust the constants in your solution from part a to eliminate the "runaway solution" and discuss the problems that result. Is your solution causal?

For further discussion of this problem see Reference [111].

7

Dipole radiation

In Chapter 6, we considered in detail the radiation from a single, moving point charge. If the moving charge belongs to an electrically neutral system, a charge of equal magnitude but opposite sign must also be present. For the cases we have considered so far, this compensating charge can be thought of as being at an infinite distance from the moving charge.

In many practical situations, charge neutrality is often maintained locally, with equal amounts of positive and negative charge being confined to the same finite volume. The simplest model for this situation is the *electric dipole* shown in Figure 7.1a. Here, two time-varying, point charges of equal magnitude, $q(t)$, and opposite sign are separated by the fixed distance Δl. A time-varying current, $\mathcal{I}(t)$, in the direction $\hat{i}$ is between the charges. For conservation of charge, we must have (see Problem 7.1)

$$q(t) = \int_{-\infty}^{t} \mathcal{I}(t')dt', \tag{7.1}$$

or

$$\mathcal{I}(t) = \frac{dq(t)}{dt}. \tag{7.2}$$

The *current moment* $\vec{j}$ and *electric dipole moment* $\vec{p}$ for this distribution are

$$\vec{j}(t) = j(t)\hat{i} = \mathcal{I}(t)\Delta l\hat{i} \tag{7.3}$$

and

$$\vec{p}(t) = p(t)\hat{i} = q(t)\Delta l\hat{i}. \tag{7.4}$$

From (7.1) and (7.2), we see that these moments are simply related by

$$\vec{j}(t) = \frac{d\vec{p}(t)}{dt},$$

or

$$\vec{p}(t) = \int_{-\infty}^{t} \vec{j}(t')dt'.$$

The electric dipole and its magnetic counterpart, which we will discuss in Section 7.3, can be used to model a number of physical phenomena; these include the

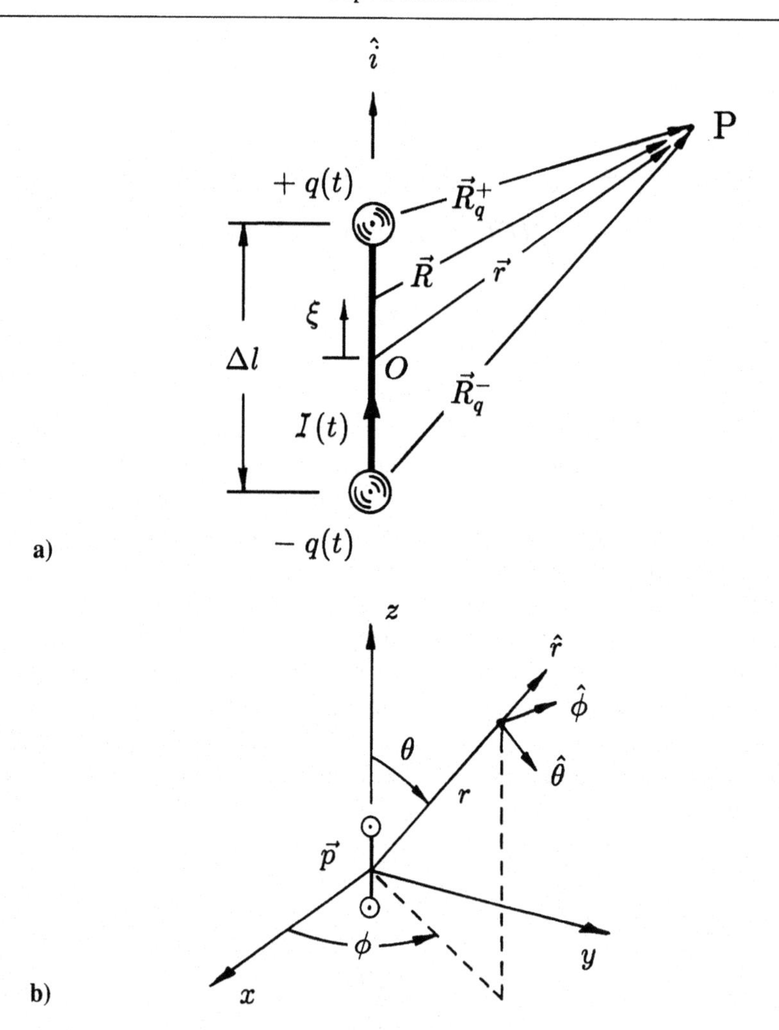

Fig. 7.1. a) Electric dipole or current element. b) Dipole at center of spherical coordinate system.

classical radiation from atoms and molecules, the radiation from electrically small antennas, and electromagnetic scattering by electrically small objects.

7.1 Infinitesimal electric dipole or current element

7.1.1 General time dependence

The *infinitesimal electric dipole* is the configuration of Figure 7.1a in the limit as the length of the dipole goes to zero ($\Delta l \to 0$) and the charge becomes infinite ($|q| \to \infty$), such that the dipole moment (7.4) remains finite. We will evaluate the expressions for the potentials and the electromagnetic field of the dipole in this limit. In doing this, we will be careful to identify the terms that become small and are therefore dropped in the limit. Knowledge of these terms will be essential when we use the infinitesimal dipole to model a practical dipole (one of finite physical size that is electrically small) because they allow us to determine when the results for the infinitesimal dipole are good approximations to those for the practical dipole.

The scalar potential for the two point charges is (5.124)

$$\Phi(\vec{r},t) = \frac{q(t - R_q^+/c)}{4\pi\varepsilon_o R_q^+} - \frac{q(t - R_q^-/c)}{4\pi\varepsilon_o R_q^-}, \tag{7.5}$$

and the vector potential for the current is (5.127)

$$\vec{A}(\vec{r},t) = \frac{\mu_o}{4\pi}\hat{i}\int_{\xi=-\Delta l/2}^{\Delta l/2} \frac{\mathcal{I}(t - R/c)}{R}\,d\xi, \tag{7.6}$$

where ξ is the distance along the current element, as shown in Figure 7.1a, and

$$R_q^{\pm} = \sqrt{r^2 + (\Delta l/2)^2 \mp \Delta l r(\hat{i}\cdot\hat{r})}, \tag{7.7}$$

$$R = \sqrt{r^2 + \xi^2 - 2\xi r(\hat{i}\cdot\hat{r})}. \tag{7.8}$$

In the limit as $\Delta l \to 0$, for $|\vec{r}| \neq 0$, we have

$$r \gg \Delta l. \tag{7.9}$$

We can use this inequality to obtain approximations for the potentials that apply in the limit. We will consider the vector potential first and express the integrand of (7.6) as a Taylor series about the origin, $\xi = 0$:

$$\frac{\mathcal{I}(t-R/c)}{R} = \frac{\mathcal{I}}{R}\bigg|_{\xi=0} + \frac{d}{d\xi}\left(\frac{\mathcal{I}}{R}\right)\bigg|_{\xi=0}\xi + \frac{1}{2!}\frac{d^2}{d\xi^2}\left(\frac{\mathcal{I}}{R}\right)\bigg|_{\xi=0}\xi^2 + \frac{1}{3!}\frac{d^3}{d\xi^3}\left(\frac{\mathcal{I}}{R}\right)\bigg|_{\xi=0}\xi^3 + \cdots, \tag{7.10}$$

or

$$\begin{aligned}\frac{\mathcal{I}(t-R/c)}{R} &= \frac{\mathcal{I}}{r} - \left(\frac{\dot{\mathcal{I}}}{cr} + \frac{\mathcal{I}}{r^2}\right)\frac{dR}{d\xi}\bigg|_{\xi=0}\xi \\ &+ \frac{1}{2}\left[\left(\frac{\ddot{\mathcal{I}}}{c^2r} + 2\frac{\dot{\mathcal{I}}}{cr^2} + 2\frac{\mathcal{I}}{r^3}\right)\left(\frac{dR}{d\xi}\bigg|_{\xi=0}\right)^2 - \left(\frac{\dot{\mathcal{I}}}{cr} + \frac{\mathcal{I}}{r^2}\right)\frac{d^2R}{d\xi^2}\bigg|_{\xi=0}\right]\xi^2 \\ &+ \frac{1}{6}\left[-\left(\frac{\dddot{\mathcal{I}}}{c^3r} + 3\frac{\ddot{\mathcal{I}}}{c^2r^2} + 6\frac{\dot{\mathcal{I}}}{cr^3} + 6\frac{\mathcal{I}}{r^4}\right)\left(\frac{dR}{d\xi}\bigg|_{\xi=0}\right)^3\right. \\ &+ \left(3\frac{\ddot{\mathcal{I}}}{c^2r} + 6\frac{\dot{\mathcal{I}}}{cr^2} + 6\frac{\mathcal{I}}{r^3}\right)\frac{dR}{d\xi}\frac{d^2R}{d\xi^2}\bigg|_{\xi=0} \\ &\left. - \left(\frac{\dot{\mathcal{I}}}{cr} + \frac{\mathcal{I}}{r^2}\right)\frac{d^3R}{d\xi^3}\bigg|_{\xi=0}\right]\xi^3 + \cdots,\end{aligned} \tag{7.11}$$

where the following shorthand notation is used:

$$\mathcal{I} = \mathcal{I}(t - R/c)\big|_{\xi=0},$$

$$\dot{\mathcal{I}} = \left.\frac{d\mathcal{I}(t - R/c)}{dt}\right|_{\xi=0}.$$

The various derivatives of R are

$$\left.\frac{dR}{d\xi}\right|_{\xi=0} = -(\hat{\imath}\cdot\hat{r}),$$

$$\left.\frac{d^2R}{d\xi^2}\right|_{\xi=0} = \frac{1}{r}[1 - (\hat{\imath}\cdot\hat{r})^2],$$

$$\left.\frac{d^3R}{d\xi^3}\right|_{\xi=0} = \frac{3}{r^2}(\hat{\imath}\cdot\hat{r})[1 - (\hat{\imath}\cdot\hat{r})^2],$$

which, on substitution into (7.11) and a rearrangement of terms, gives

$$\begin{aligned}\frac{\mathcal{I}(t - R/c)}{R} &= \frac{\mathcal{I}}{r} + \left(\frac{\dot{\mathcal{I}}}{cr} + \frac{\mathcal{I}}{r^2}\right)(\hat{\imath}\cdot\hat{r})\xi \\ &+ \frac{1}{2}\left\{\left(\frac{\ddot{\mathcal{I}}}{c^2r} + 2\frac{\dot{\mathcal{I}}}{cr^2} + 2\frac{\mathcal{I}}{r^3}\right)(\hat{\imath}\cdot\hat{r})^2 - \left(\frac{\dot{\mathcal{I}}}{cr^2} + \frac{\mathcal{I}}{r^3}\right)[1 - (\hat{\imath}\cdot\hat{r})^2]\right\}\xi^2 \\ &+ \frac{1}{6}\left\{3\left(\frac{\dot{\mathcal{I}}}{cr^3} + \frac{\mathcal{I}}{r^4}\right)(\hat{\imath}\cdot\hat{r})[5(\hat{\imath}\cdot\hat{r})^2 - 3] + \frac{\dddot{\mathcal{I}}}{c^3r}(\hat{\imath}\cdot\hat{r})^3 \right. \\ &\left. + 3\frac{\ddot{\mathcal{I}}}{c^2r^2}(\hat{\imath}\cdot\hat{r})[2(\hat{\imath}\cdot\hat{r})^2 - 1]\right\}\xi^3 + \cdots. \end{aligned} \tag{7.12}$$

When (7.12) is inserted into the expression for the vector potential (7.6), the resulting integrals are easily evaluated (those for odd powers of ξ are zero) (Problem 7.2):

$$\begin{aligned}\vec{A}(\vec{r}, t) &= \frac{\mu_o}{4\pi}\hat{\imath}\left(\frac{\mathcal{I}}{r}\Delta l + \frac{1}{24}\left\{\left(\frac{\ddot{\mathcal{I}}}{c^2r} + 2\frac{\dot{\mathcal{I}}}{cr^2} + 2\frac{\mathcal{I}}{r^3}\right)(\hat{\imath}\cdot\hat{r})^2 \right.\right. \\ &\left.\left. - \left(\frac{\dot{\mathcal{I}}}{cr^2} + \frac{\mathcal{I}}{r^3}\right)[1 - (\hat{\imath}\cdot\hat{r})^2]\right\}(\Delta l)^3 + \cdots\right) \\ &= \frac{\mu_o}{4\pi r}\left(\vec{j}(t - r/c)\left\{1 - \frac{1}{24}\left(\frac{\Delta l}{r}\right)^2[1 - 3(\hat{\imath}\cdot\hat{r})^2]\right\}\right.\end{aligned}$$

$$- \frac{1}{24}\left(\frac{\Delta l}{r}\right)\Delta t\, \dot{\vec{j}}(t - r/c)[1 - 3(\hat{\imath} \cdot \hat{r})^2]$$

$$+ \frac{1}{24}(\Delta t)^2\, \ddot{\vec{j}}(t - r/c)(\hat{\imath} \cdot \hat{r})^2 + \cdots \Bigg). \tag{7.13}$$

Here Δt is the time for light to travel the length of the dipole,

$$\Delta t = \frac{\Delta l}{c}. \tag{7.14}$$

Subject to (7.9) and the validity of the inequalities

$$\left|\left(\frac{\Delta l}{r}\right)\Delta t\, \dot{\vec{j}}\right| \ll |\vec{j}| \tag{7.15}$$

and

$$|(\Delta t)^2\, \ddot{\vec{j}}| \ll |\vec{j}|, \tag{7.16}$$

we can approximate $\vec{\mathcal{A}}$ as simply

$$\begin{aligned}\vec{\mathcal{A}}(\vec{r}, t) &= \frac{\mu_o}{4\pi r}\, \vec{j}(t - r/c) \\ &= \frac{\mu_o}{4\pi r}\, \dot{\vec{p}}\,(t - r/c).\end{aligned} \tag{7.17}$$

Now we will consider the scalar potential. The terms $q/R_q^{\pm}$ in (7.5) can be expressed as Taylor series about the origin. The steps are the same as those we just used with the vector potential to obtain the series for $\mathcal{I}/R$, so we can obtain our results by simply replacing $\mathcal{I}$ by q and ξ by $\pm\Delta l/2$ in (7.12):

$$\begin{aligned}\frac{q(t - R_q^{\pm}/c)}{R_q^{\pm}} &= \frac{q}{r} \pm \frac{1}{2}\left(\frac{\dot{q}}{cr} + \frac{q}{r^2}\right)(\hat{\imath} \cdot \hat{r})\Delta l \\ &+ \frac{1}{8}\left\{\left(\frac{\ddot{q}}{c^2 r} + 2\frac{\dot{q}}{cr^2} + 2\frac{q}{r^3}\right)(\hat{\imath} \cdot \hat{r})^2\right. \\ &\left. - \left(\frac{\dot{q}}{cr^2} + \frac{q}{r^3}\right)[1 - (\hat{\imath} \cdot \hat{r})^2]\right\}(\Delta l)^2 \\ &\pm \frac{1}{48}\left\{3\left(\frac{\dot{q}}{cr^3} + \frac{q}{r^4}\right)(\hat{\imath} \cdot \hat{r})[5(\hat{\imath} \cdot \hat{r})^2 - 3]\right. \\ &\left. + \frac{\dddot{q}}{c^3 r}(\hat{\imath} \cdot \hat{r})^3 + 3\frac{\ddot{q}}{c^2 r^2}(\hat{\imath} \cdot \hat{r})[2(\hat{\imath} \cdot \hat{r})^2 - 1]\right\}(\Delta l)^3 + \cdots.\end{aligned} \tag{7.18}$$

After substituting (7.18) into (7.5) and rearranging terms, the scalar potential becomes

$$\Phi(\vec{r},t) = \frac{1}{4\pi\varepsilon_o}\Bigg(\left(\frac{\dot{q}}{cr}+\frac{q}{r^2}\right)(\hat{\imath}\cdot\hat{r})\Delta l + \frac{1}{24}\Bigg\{3\left(\frac{\dot{q}}{cr^3}+\frac{q}{r^4}\right)(\hat{\imath}\cdot\hat{r})[5(\hat{\imath}\cdot\hat{r})^2-3]$$

$$+\frac{\dddot{q}}{c^3 r}(\hat{\imath}\cdot\hat{r})^3 + 3\frac{\ddot{q}}{c^2r^2}(\hat{\imath}\cdot\hat{r})[2(\hat{\imath}\cdot\hat{r})^2-1]\Bigg\}(\Delta l)^3+\cdots\Bigg)$$

$$= \frac{1}{4\pi\varepsilon_o r}\hat{r}\cdot\Bigg(\frac{1}{r}\int_{-\infty}^{t-r/c}\vec{j}(t')dt'\left\{1+\frac{1}{8}\left(\frac{\Delta l}{r}\right)^2[5(\hat{\imath}\cdot\hat{r})^2-3]\right\}$$

$$+\frac{1}{c}\vec{j}(t-r/c)\left\{1+\frac{1}{8}\left(\frac{\Delta l}{r}\right)^2[5(\hat{\imath}\cdot\hat{r})^2-3]\right\}$$

$$+\frac{1}{8}\left(\frac{1}{c}\right)\left(\frac{\Delta l}{r}\right)\Delta t\,\dot{\vec{j}}(t-r/c)[2(\hat{\imath}\cdot\hat{r})^2-1]$$

$$+\frac{1}{24}\left(\frac{1}{c}\right)(\Delta t)^2\,\ddot{\vec{j}}(t-r/c)(\hat{\imath}\cdot\hat{r})^2+\cdots\Bigg). \tag{7.19}$$

Making use of the inequalities we introduced earlier, (7.9), (7.15), and (7.16), we can approximate Φ as

$$\Phi(\vec{r},t) = \frac{1}{4\pi\varepsilon_o r}\left[\frac{1}{r}\int_{-\infty}^{t-r/c}\hat{r}\cdot\vec{j}(t')dt' + \frac{1}{c}\hat{r}\cdot\vec{j}(t-r/c)\right]$$

$$= \frac{1}{4\pi\varepsilon_o r}\left[\frac{1}{r}\hat{r}\cdot\vec{p}\,(t-r/c) + \frac{1}{c}\hat{r}\cdot\dot{\vec{p}}\,(t-r/c)\right]. \tag{7.20}$$

The electromagnetic field of the infinitesimal electric dipole is obtained by substituting the potential functions, (7.17) and (7.20), into (5.133) and (5.134). The electric field is

$$\vec{\mathcal{E}}(\vec{r},t) = -\nabla\Phi - \frac{\partial\vec{\mathcal{A}}}{\partial t}$$

$$= \frac{-1}{4\pi\varepsilon_o}\left[\nabla\left(\frac{p(t-r/c)}{r^3}+\frac{\dot{p}(t-r/c)}{cr^2}\right)(\hat{\imath}\cdot\vec{r})\right.$$

$$\left.+\left(\frac{p(t-r/c)}{r^3}+\frac{\dot{p}(t-r/c)}{cr^2}\right)\nabla(\hat{\imath}\cdot\vec{r})\right] - \frac{\mu_o}{4\pi r}\ddot{\vec{p}}\,(t-r/c). \tag{7.21}$$

The term $\nabla(\hat{\imath}\cdot\vec{r})$ is evaluated by temporarily assuming rectangular Cartesian coordinates (x, y, z) with $\hat{\imath} = \hat{z}$:

$$\nabla(\hat{\imath} \cdot \vec{r}) = \nabla\big[\hat{z} \cdot (x\hat{x} + y\hat{y} + z\hat{z})\big] = \nabla(z) = \hat{z} = \hat{\imath}.$$

After substituting this result into Equation (7.21), evaluating the remaining gradient in spherical coordinates, and rearranging terms, the final expression for the electric field is obtained:

$$\vec{\mathcal{E}}(\vec{r}, t) = \frac{-1}{4\pi\varepsilon_o}\left\{(1 - 3\hat{r}\hat{r}\cdot)\left[\frac{1}{r^3}\,\vec{p}\,(t - r/c) + \frac{1}{cr^2}\,\dot{\vec{p}}\,(t - r/c)\right]\right.$$

$$\left. - \frac{1}{c^2 r}\,\hat{r} \times \left[\hat{r} \times \ddot{\vec{p}}\,(t - r/c)\right]\right\}. \tag{7.22}$$

The magnetic field, using (7.17), is

$$\vec{\mathcal{B}}(\vec{r}, t) = \nabla \times \vec{\mathcal{A}} = \frac{\mu_o}{4\pi}\left\{\nabla\left[\frac{\dot{p}(t - r/c)}{r}\right] \times \hat{\imath}\right\}, \tag{7.23}$$

which, on evaluating the gradient in spherical coordinates and rearranging terms, becomes

$$\vec{\mathcal{B}}(\vec{r}, t) = \frac{-\mu_o}{4\pi}\left[\frac{1}{r^2}\,\hat{r} \times \dot{\vec{p}}\,(t - r/c) + \frac{1}{cr}\,\hat{r} \times \ddot{\vec{p}}\,(t - r/c)\right]. \tag{7.24}$$

For future reference, these results for the electromagnetic field of the infinitesimal electric dipole or current element are summarized in Table 7.1. In addition to the general vector formulas, (7.22) and (7.24), formulas are given for the special case where the dipole is at the origin of the spherical coordinate system (r, θ, ϕ) and points in the z direction ($\hat{\imath} = \hat{z}$), as in Figure 7.1b:

$$\vec{\mathcal{E}}(\vec{r}, t) = \frac{1}{4\pi\varepsilon_o}\left\{\left[\frac{1}{r^3}\,p(t - r/c) + \frac{1}{cr^2}\,\dot{p}(t - r/c)\right][2\cos\theta\,\hat{r} + \sin\theta\,\hat{\theta}]\right.$$

$$\left. + \frac{1}{c^2 r}\,\ddot{p}(t - r/c)\sin\theta\,\hat{\theta}\right\}, \tag{7.25}$$

$$\vec{\mathcal{B}}(\vec{r}, t) = \frac{\mu_o}{4\pi}\left[\frac{1}{r^2}\,\dot{p}(t - r/c) + \frac{1}{cr}\,\ddot{p}(t - r/c)\right]\sin\theta\,\hat{\phi}. \tag{7.26}$$

The electric field (7.25) is seen to consist of three terms with the radial dependences $1/r^3$, $1/r^2$, and $1/r$. The first term $(1/r^3)$ is proportional to the dipole moment, p; this is the familiar field of the electrostatic dipole with the static moment replaced by a time-varying moment. It predominates close to the dipole. The second term $(1/r^2)$ is proportional to the first temporal derivative of the dipole moment, $\dot{p}$, and the third term $(1/r)$ is proportional to the second temporal derivative of the dipole moment, $\ddot{p}$. The latter is the radiated field, and it predominates at large distances from the dipole ($\theta \neq 0$). The magnetic field (7.26) only contains terms with the radial dependences $1/r^2$ and $1/r$, and these are proportional to $\dot{p}$ and $\ddot{p}$,

Table 7.1. *Electromagnetic field of infinitesimal electric dipole or current element*

General time dependence

$$\vec{\mathcal{E}}(\vec{r},t) = \frac{-1}{4\pi\varepsilon_o}\left[(1-3\hat{r}\hat{r}\cdot)\left(\frac{1}{r^3}\vec{p}+\frac{1}{cr^2}\dot{\vec{p}}\right)-\frac{1}{c^2 r}\hat{r}\times(\hat{r}\times\ddot{\vec{p}})\right]_{t_r=t-r/c}$$

$$= \frac{1}{4\pi\varepsilon_o}\left[\left(\frac{1}{r^3}p+\frac{1}{cr^2}\dot{p}\right)(2\cos\theta\,\hat{r}+\sin\theta\,\hat{\theta})+\frac{1}{c^2 r}\ddot{p}\sin\theta\,\hat{\theta}\right]_{t_r}$$

$$\vec{\mathcal{B}}(\vec{r},t) = \frac{-\mu_o}{4\pi}\left(\frac{1}{r^2}\hat{r}\times\dot{\vec{p}}+\frac{1}{cr}\hat{r}\times\ddot{\vec{p}}\right)_{t_r=t-r/c}$$

$$= \frac{\mu_o}{4\pi}\left[\left(\frac{1}{r^2}\dot{p}+\frac{1}{cr}\ddot{p}\right)\sin\theta\,\hat{\phi}\right]_{t_r}$$

$$\vec{p}\,(t) = \int_{-\infty}^{t}\vec{j}(t')dt'$$

$$\vec{p}\,(t) = \lim_{\Delta l\to 0}\left[q(t)\Delta l\,\hat{i}\,\right] \qquad \vec{j}(t) = \lim_{\Delta l\to 0}\left[\mathcal{I}(t)\Delta l\,\hat{i}\right]$$

Harmonic time dependence

$$\vec{E}(\vec{r}) = \frac{-k_o^3}{4\pi\varepsilon_o}\left\{\left[\frac{1}{(k_o r)^3}+\frac{j}{(k_o r)^2}\right](1-3\hat{r}\hat{r}\cdot)\vec{p}+\left(\frac{1}{k_o r}\right)\hat{r}\times(\hat{r}\times\vec{p})\right\}e^{-jk_o r}$$

$$= \frac{k_o^3 p}{4\pi\varepsilon_o}\left\{2\cos\theta\left[\frac{1}{(k_o r)^3}+\frac{j}{(k_o r)^2}\right]\hat{r}+\sin\theta\left[\frac{1}{(k_o r)^3}+\frac{j}{(k_o r)^2}-\frac{1}{k_o r}\right]\hat{\theta}\right\}e^{-jk_o r}$$

$$\vec{B}(\vec{r}) = \frac{-\zeta_o k_o^3}{4\pi}\left[\frac{j}{(k_o r)^2}-\frac{1}{k_o r}\right]\hat{r}\times\vec{p}e^{-jk_o r}$$

$$= \frac{\zeta_o k_o^3 p}{4\pi}\sin\theta\left[\frac{j}{(k_o r)^2}-\frac{1}{k_o r}\right]\hat{\phi}\,e^{-jk_o r}$$

$$\vec{p} = \frac{-j}{\omega}\vec{j}$$

respectively, with the latter being the radiated field. Notice that, as expected, the radiated electric and magnetic fields are orthogonal to each other and to $\hat{r}$:

$$\vec{\mathcal{E}}^r(\vec{r},t) = \frac{\mu_o}{4\pi r}\sin\theta\,\ddot{p}(t-r/c)\hat{\theta}, \tag{7.27}$$

$$\vec{\mathcal{B}}^r(\vec{r},t) = \frac{1}{c}\hat{r}\times\vec{\mathcal{E}}^r(\vec{r},t) = \frac{\mu_o}{4\pi cr}\sin\theta\,\ddot{p}(t-r/c)\hat{\phi}. \tag{7.28}$$

The Poynting vector for the dipole field is (Problem 7.3)

$$\vec{\mathcal{S}}(\vec{r},t) = \frac{1}{\mu_o}\vec{\mathcal{E}}(\vec{r},t) \times \vec{\mathcal{B}}(\vec{r},t)$$

$$= \frac{1}{(4\pi)^2\varepsilon_o}\left\{\left[\frac{1}{r^5}\,p\dot{p} + \frac{1}{cr^4}(p\ddot{p} + \dot{p}^2) + \frac{2}{c^2r^3}\,\dot{p}\ddot{p} + \frac{1}{c^3r^2}\,\ddot{p}^2\right]\sin^2\theta\,\hat{r}\right.$$

$$\left. - 2\left[\frac{1}{r^5}\,p\dot{p} + \frac{1}{cr^4}(p\ddot{p} + \dot{p}^2) + \frac{1}{c^2r^3}\,\dot{p}\ddot{p}\right]\sin\theta\cos\theta\,\hat{\theta}\right\}_{t_r = t - r/c}, \quad (7.29)$$

from which the power instantaneously passing outward through a spherical surface of radius r is readily determined:

$$\oint\!\!\oint_S \hat{r}\cdot\vec{\mathcal{S}}(\vec{r},t)d\Omega = 2\pi\int_{\theta=0}^{\pi}\hat{r}\cdot\vec{\mathcal{S}}(\vec{r},t)r^2\sin\theta d\theta$$

$$= \frac{1}{6\pi\varepsilon_o}\left\{\frac{1}{r^3}\,p\dot{p} + \frac{1}{cr^2}(p\ddot{p} + \dot{p}^2) + \frac{2}{c^2r}\,\dot{p}\ddot{p} + \frac{1}{c^3}\,\ddot{p}^2\right\}_{t_r = t - r/c}. \quad (7.30)$$

At a large radial distance from the dipole ($\lim r \to \infty$), these expressions simplify to become the Poynting vector for the radiated field,

$$\vec{\mathcal{S}}^r(\vec{r},t) = \frac{1}{16\pi^2\varepsilon_o c^3}\frac{\sin^2\theta}{r^2}\left[\ddot{p}(t - r/c)\right]^2\hat{r} \quad (7.31)$$

and

$$\oint\!\!\oint_S \hat{r}\cdot\vec{\mathcal{S}}^r(\vec{r},t)d\Omega = \frac{1}{6\pi\varepsilon_o c^3}\left[\ddot{p}(t - r/c)\right]^2. \quad (7.32)$$

Close to the dipole, the terms proportional to $p\dot{p}$ are dominant in (7.29) and (7.30). The Poynting vector has components in both the $\hat{r}$ and $\hat{\theta}$ directions, and these may be positive or negative depending upon the sign of $p\dot{p}$. Instantaneously, power may be passing outward or inward through a sphere of constant radius, again depending upon the sign of $p\dot{p}$. At large distances from the dipole, however, where the radiated field is predominant, the Poynting vector (7.31) has only a radial component, and this always points outward; $[\ddot{p}]^2$ is always positive. Instantaneously, power is always passing outward through a sphere of constant radius (7.32).

From (7.31) and (7.32) we conclude that, at the retarded time $t_r = t - r/c$, the power radiated by the dipole per unit solid angle in the direction θ is

$$\frac{d\mathcal{P}_{\text{rad}}(t_r)}{d\Omega} = r^2\hat{r}\cdot\vec{\mathcal{S}}^r = \frac{1}{16\pi^2\varepsilon_o c^3}\sin^2\theta\left[\ddot{p}(t_r)\right]^2, \quad (7.33)$$

and the total power radiated in all directions is

$$\mathcal{P}_{\text{rad}}(t_r) = \frac{1}{6\pi\varepsilon_o c^3}\left[\ddot{p}(t_r)\right]^2. \quad (7.34)$$

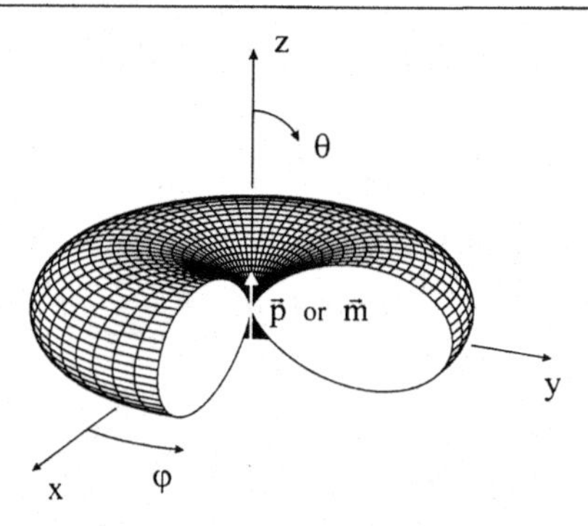

Fig. 7.2. Directional characteristics of the radiation from an infinitesimal electric or magnetic dipole. For general time dependence, the radial distance in the direction θ, ϕ, measured from the origin, is proportional to the power per unit solid angle, $d\mathcal{P}_{\text{rad}}(t_r)/d\Omega$. For harmonic time dependence, it is proportional to $d\langle\mathcal{P}_{\text{rad}}\rangle/d\Omega$.

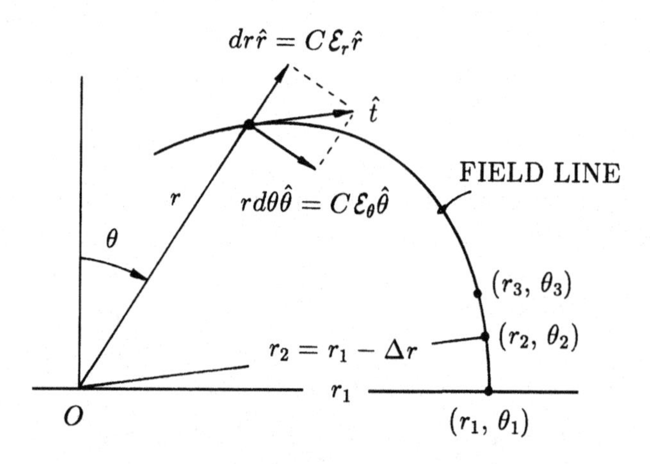

Fig. 7.3. Detail for the construction of electric field lines.

Figure 7.2 shows the directional characteristics of the radiation. Here the radial distance in the direction θ, ϕ, measured from the origin, is proportional to $d\mathcal{P}_{\text{rad}}/d\Omega$. A quadrant of the figure has been removed to show the cross section on a plane of constant ϕ. The pattern has the shape of a fat doughnut or fat tire; this is the same pattern we observed earlier in Figure 6.7 for the slowly moving point charge ($\beta \ll 1$).

The characteristics of the field are nicely illustrated by plots of the electric field lines surrounding the dipole, a procedure that dates back to the work of Hertz [1, 2]. Recall that a field line is a curve whose tangent at each point is in the direction of the electric field. With reference to Figure 7.3, the vector $\vec{t}$ that is tangent to the field line is

$$\vec{t} = dr\hat{r} + rd\theta\hat{\theta} = C(\mathcal{E}_r\hat{r} + \mathcal{E}_\theta\hat{\theta}), \tag{7.35}$$

where C is an arbitrary constant. After using (7.25) with (7.35), we obtain the following differential equation for the field lines of the dipole:

$$\frac{1}{r}\frac{dr}{d\theta} = \frac{\mathcal{E}_r}{\mathcal{E}_\theta} = \left\{ \left[2\cos\theta \left(\frac{p}{r^3} + \frac{\dot{p}}{cr^2} \right) \right] \Big/ \left[\sin\theta \left(\frac{p}{r^3} + \frac{\dot{p}}{cr^2} + \frac{\ddot{p}}{c^2 r} \right) \right] \right\}_{t_r},$$

or

$$\left(\frac{p}{r^2} + \frac{\dot{p}}{cr} + \frac{\ddot{p}}{c^2} \right)_{t_r} \frac{dr}{d\theta} = 2\cot\theta \left(\frac{p}{r} + \frac{\dot{p}}{c} \right)_{t_r}. \tag{7.36}$$

Next, we notice that

$$\frac{d}{d\theta}\left(\frac{p}{r} + \frac{\dot{p}}{c} \right)_{t_r} = \left\{ \left[\frac{d}{dr}\left(\frac{p}{r} + \frac{\dot{p}}{c} \right) \right]_{t_r} + \left[\frac{d}{dt}\left(\frac{p}{r} + \frac{\dot{p}}{c} \right) \right]_{t_r} \frac{dt_r}{dr} \right\} \frac{dr}{d\theta}$$

$$= -\left(\frac{p}{r^2} + \frac{\dot{p}}{cr} + \frac{\ddot{p}}{c^2} \right)_{t_r} \frac{dr}{d\theta},$$

which, on substitution into (7.36), gives

$$-\frac{d}{d\theta}\left(\frac{p}{r} + \frac{\dot{p}}{c} \right)_{t_r} = 2\cot\theta \left(\frac{p}{r} + \frac{\dot{p}}{c} \right)_{t_r}.$$

Using logarithmic differentiation, this becomes

$$\frac{d}{d\theta}\left[\ln\left(\frac{cp}{r} + \dot{p} \right)_{t_r} \right] = -2\cot\theta.$$

Now we will integrate this equation to obtain our final expression for the field lines:

$$\ln\left[\left(\frac{cp}{r} + \dot{p} \right)_{t_r} \right] = -2\int \cot\theta \, d\theta = \ln\left(\frac{K}{\sin^2\theta} \right),$$

or

$$\left\{ \frac{p[t - r(\theta)/c]}{r(\theta)/c} + \dot{p}\big[t - r(\theta)/c\big] \right\} \sin^2\theta = K, \tag{7.37}$$

where K is a constant of integration.

The following procedure is used with (7.37) to draw the field lines surrounding the dipole at a particular time t. A value of K is selected, with a different value being used for each field line. A starting point, r_1, θ_1, is determined, such that r_1 is the maximum value of r for the line and r_1, θ_1 satisfy Equation (7.37). Next, a new position, r_2, θ_2, is determined by decreasing the radius by a small amount, $r_2 = r_1 - \Delta r$, and solving (7.37) for θ_2. A segment of the field line is drawn between the points r_1, θ_1 and r_2, θ_2, as in Figure 7.3. Additional segments are drawn in the same manner until the field line is complete.

To illustrate these results, we will let the electric dipole moment vary in time as a Gaussian function with characteristic time τ:

$$\vec{p}\,(t) = p(t)\hat{z} = p_o e^{-(t/\tau)^2}\hat{z}. \tag{7.38}$$

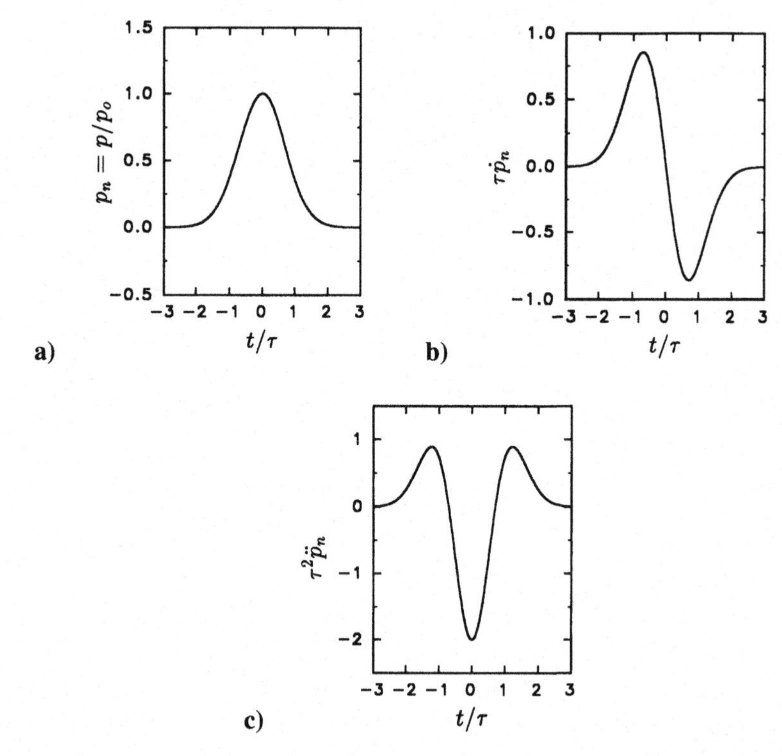

Fig. 7.4. a) Gaussian function for the electric dipole moment. b) First temporal derivative and c) second temporal derivative of the dipole moment.

Figure 7.4a is a graph of $p(t)$; the first and second temporal derivatives of $p(t)$, shown in Figures 7.4b and 7.4c, are

$$\dot{\vec{p}}\,(t) = \dot{p}(t)\hat{z} = -\frac{2}{\tau}\left(\frac{t}{\tau}\right)p_o e^{-(t/\tau)^2}\hat{z} \tag{7.39}$$

and

$$\ddot{\vec{p}}\,(t) = \ddot{p}(t)\hat{z} = -\frac{2}{\tau^2}\left[1 - 2(t/\tau)^2\right]p_o e^{-(t/\tau)^2}\hat{z}. \tag{7.40}$$

The electric field lines for this dipole moment are shown in Figures 7.5 and 7.6. Each of the six plots is for a different normalized time t/τ. Below each plot are graphs of the normalized electric and magnetic fields ($\mathcal{E}_{\theta n}$: solid line, $\mathcal{B}_{\phi n}$: dashed line) broadside to the dipole ($\theta = \pi/2$) versus the normalized radial position $r/c\tau$.

At the time $t/\tau = 0$, the dipole moment has just reached its maximum values ($p = p_o$). The field lines are very similar to those for an electrostatic dipole, going from the top of the dipole (positive charge) to the bottom of the dipole (negative charge). Near the dipole, the electric field, $\mathcal{E}_\theta$, falls off rapidly with increasing radial position due to the $1/r^3$ term in (7.25), and the magnetic field, $\mathcal{B}_\phi$, is very small, because the dominant term ($1/r^2$) in (7.26) has the coefficient $\dot{p}$, which is zero at time $t = 0$ (Figure 7.4b).

As time advances, kinks develop in the field lines, as seen in Figure 7.5 for $t/\tau = 1.10$. Later, at time $t/\tau = 1.35$, the kinks in the upper and lower half spaces

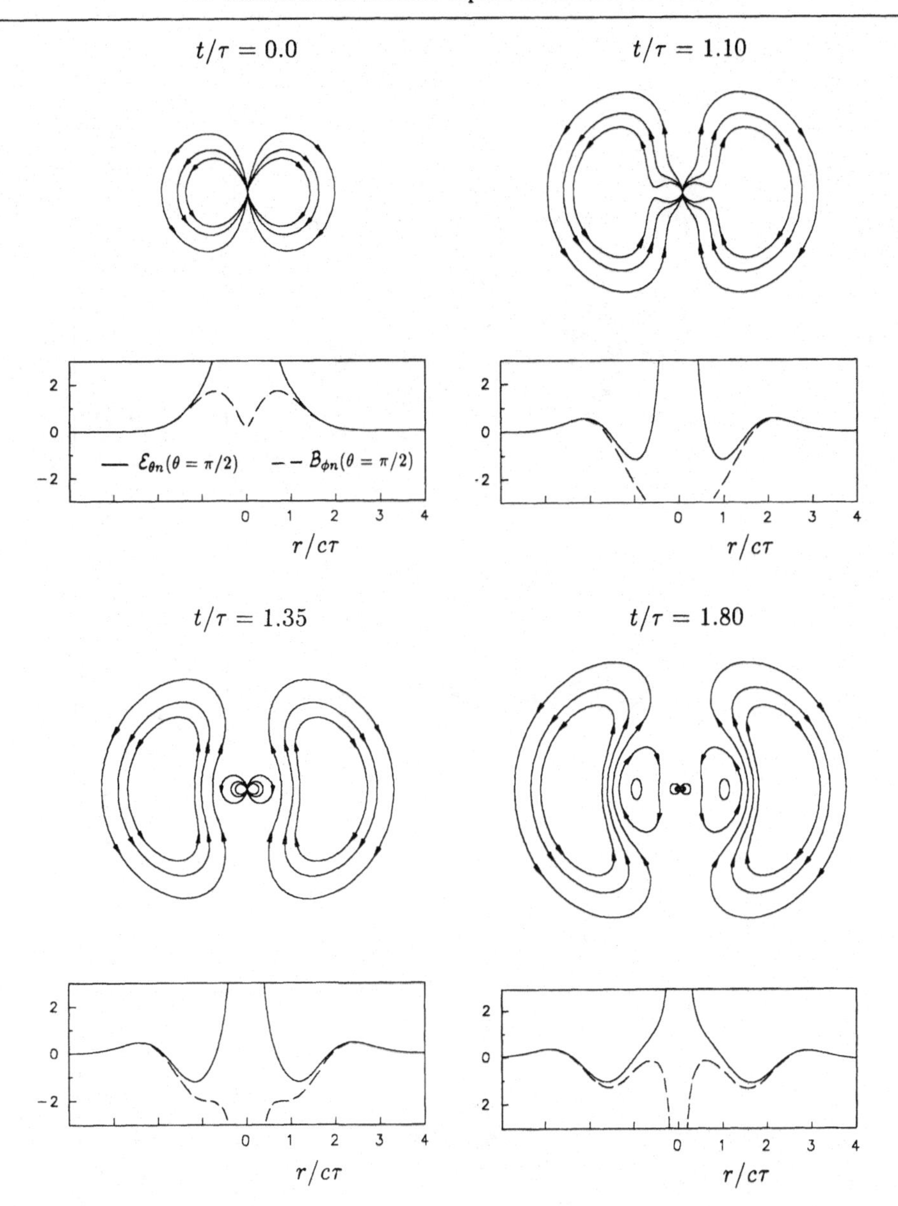

Fig. 7.5. Electric field lines for an electric dipole moment that is a Gaussian function of time. Each plot is for a different time t/τ, and the graph below each plot shows the field $(\mathcal{E}_\theta, \mathcal{B}_\phi)$ as a function of $r/c\tau$ for $\theta = \pi/2$.

have merged to form a continuous electric field line or loop away from the dipole. Near the dipole, the electric field lines still resemble those of the electrostatic dipole. The magnetic field, $\mathcal{B}_{\phi n}$, is now large near the dipole, because $\dot{p}$ is no longer zero. By the time $t/\tau = 1.80$, a second loop of opposite sense has formed, and as time advances, these two loops propagate away from the dipole, as seen in Figure 7.6.

A close examination shows that in the region near $\theta = \pi/2$, the field in Figure 7.6 behaves as the radiated field given by Equations (7.27) and (7.28). The electric and

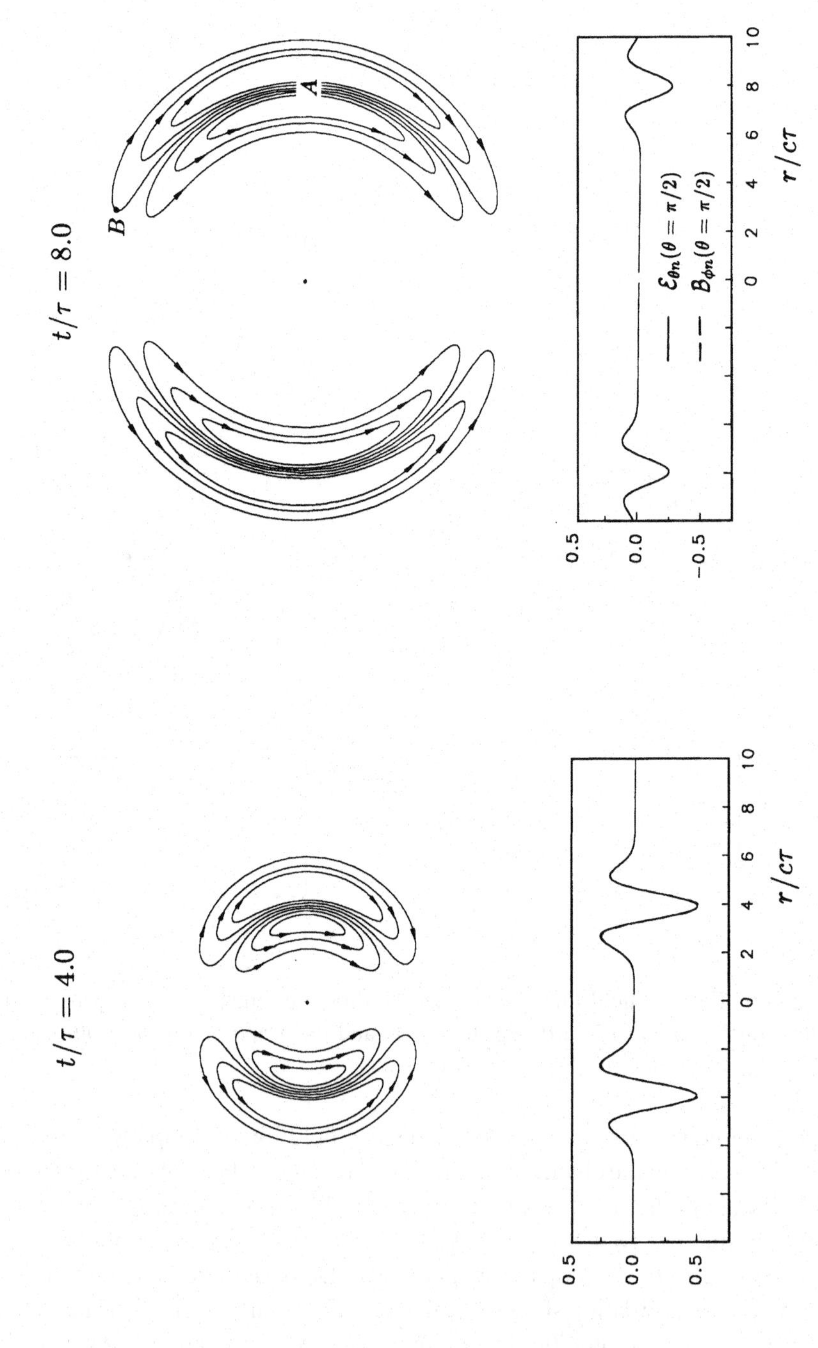

Fig. 7.6. Electric field lines for an electric dipole moment that is a Gaussian function of time. The field very close to the dipole has not been plotted.

magnetic fields, $\mathcal{E}_{\theta n}$ and $\mathcal{B}_{\phi n}$ (solid and dashed lines), are nearly identical, and they clearly resemble the second temporal derivative of the electric dipole moment, shown in Figure 7.4c. The field propagates outward with the speed of light and falls off as $1/r$: The negative peak moves the distance $\Delta r/c\tau \approx 4.0$ in the time $\Delta t/\tau = 4.0$, and it decreases in amplitude by a factor of about 2 in moving from $r/c\tau \approx 4$ to $r/c\tau \approx 8$.

It is important to keep in mind that a field line only shows the direction of the electric field; it says nothing directly about the amplitude of the field. To illustrate this point, we will consider the right-hand plot in Figure 7.6. The field at point A ($\theta = \pi/2$), as we have shown above, behaves as the radiated field. However, the field at the point B ($\theta \approx \pi/9$), which is on the same field line as A and at about the same radial distance from the dipole, clearly does not behave as the radiated field; it has a radial component.[1] The magnitude of the electric field at point A is about 12 times greater than that at point B. This is shown roughly by the density of the field lines near the two points.

When the radiated field is of primary interest, the representation shown in Figure 7.7 proves useful. Here we have plotted the radiated electric field, $\mathcal{E}_\theta^r$, of the dipole as a function of the normalized time, t/τ, for several values of the angle θ within the range $0 \le \theta \le \pi$.[2] The field is rotationally symmetric about the z axis, so the figure applies for any angle ϕ. The radiated field is clearly proportional to the second temporal derivative of the dipole moment ($\ddot{p}(t)$ in Figure 7.3c) and varies as $\sin\theta$ (7.27). It reaches a maximum broadside to the dipole ($\theta = \pi/2$) and falls off to zero at the ends of the dipole ($\theta = 0, \pi$).

7.1.2 Harmonic time dependence

For harmonic time dependence at the angular frequency ω, the electric dipole moment (7.4) is

$$\vec{p}\,(t) = p_o \cos(\omega t + \phi_o)\hat{z} = \mathrm{Re}(\vec{p}\, e^{j\omega t}),$$

where we have introduced the complex vector phasor

$$\vec{p} = p\hat{z} = p_o e^{j\phi_o}\hat{z}. \tag{7.41}$$

The electromagnetic field, (7.22) and (7.24), expressed in terms of vector phasors, is

$$\vec{E}(\vec{r}) = \frac{-k_o^3}{4\pi\varepsilon_o}\left\{\left[\frac{1}{(k_or)^3} + \frac{j}{(k_or)^2}\right](1 - 3\hat{r}\hat{r}\cdot)\vec{p} + \left(\frac{1}{k_or}\right)\hat{r}\times(\hat{r}\times\vec{p}\,)\right\}e^{-jk_or}, \tag{7.42}$$

$$\vec{B}(\vec{r}) = \frac{-\zeta_o k_o^3}{4\pi}\left[\frac{j}{(k_or)^2} - \frac{1}{k_or}\right]\hat{r}\times\vec{p}\,e^{-jk_or}. \tag{7.43}$$

[1] The radiated electric field is proportional to $(1/r)\sin\theta$, whereas the radial component of the electric field is proportional to $(1/r^3)\cos\theta$. Thus, at any finite radius, there will be a range of angles θ about $\theta = 0$ where the radial component of the field is larger than the radiated field.

[2] Note that the electric field is positive on the side of the axis indicated by the arrow at $\theta = 45°$.

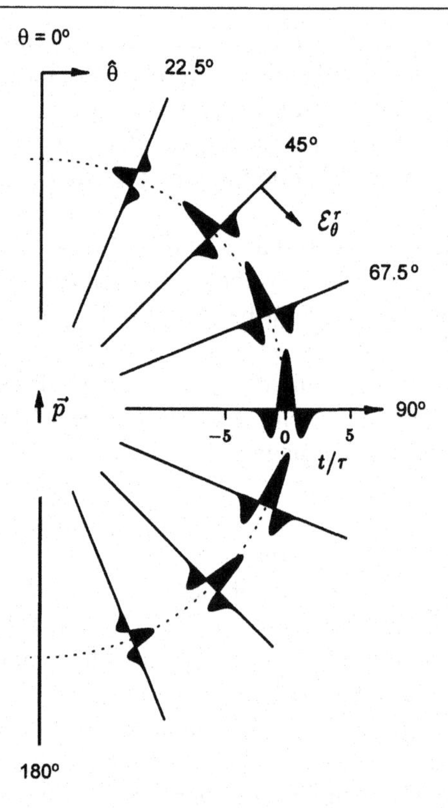

Fig. 7.7. Radiated electric field as a function of normalized time for several values of the angle θ. Electric dipole moment is a Gaussian function of time.

Using the spherical coordinates (r, θ, ϕ) shown in Figure 7.1b, the field can be expressed as

$$\vec{E}(\vec{r}) = \frac{k_o^3 p}{4\pi\varepsilon_o}\left\{2\cos\theta\left[\frac{1}{(k_o r)^3} + \frac{j}{(k_o r)^2}\right]\hat{r}\right.$$

$$\left. + \sin\theta\left[\frac{1}{(k_o r)^3} + \frac{j}{(k_o r)^2} - \frac{1}{k_o r}\right]\hat{\theta}\right\} e^{-jk_o r}, \tag{7.44}$$

$$\vec{B}(\vec{r}) = \frac{\zeta_o k_o^3 p}{4\pi}\sin\theta\left[\frac{j}{(k_o r)^2} - \frac{1}{k_o r}\right]\hat{\phi}\, e^{-jk_o r}. \tag{7.45}$$

These results are also summarized in Table 7.1.

The radiated electromagnetic field $(\vec{E}^r, \vec{B}^r)$ consists of the terms proportional to $1/k_o r$ in (7.44) and (7.45). Thus, the complex Poynting vector for the radiated field is

$$\vec{S}_c^r(\vec{r}) = \frac{1}{2\mu_o}\vec{E}^r(\vec{r}) \times \left[\vec{B}^r(\vec{r})\right]^*$$

$$= \frac{\zeta_o k_o^4 c^2 |p|^2}{32\pi^2 r^2}\sin^2\theta\,\hat{r} = \frac{1}{2}\pi^2\left(\frac{c}{\varepsilon_o}\right)\frac{|p|^2}{\lambda_o^4}\frac{1}{r^2}\sin^2\theta\,\hat{r},$$

making the time-average power per unit solid angle radiated by the dipole in the direction θ

$$\frac{d\langle\mathcal{P}_{\text{rad}}\rangle}{d\Omega} = \text{Re}[r^2\hat{r}\cdot\vec{S}_c^r]$$

$$= \frac{\zeta_o k_o^4 c^2 |p|^2}{32\pi^2}\sin^2\theta = \frac{1}{2}\pi^2\left(\frac{c}{\varepsilon_o}\right)\frac{|p|^2}{\lambda_o^4}\sin^2\theta \tag{7.46}$$

and the total time-average power radiated

$$\langle\mathcal{P}_{\text{rad}}\rangle = \int_{\phi=0}^{2\pi}\int_{\theta=0}^{\pi}\frac{d\langle\mathcal{P}_{\text{rad}}\rangle}{d\Omega}\sin\theta d\theta d\phi$$

$$= \frac{\mu_o\omega^4|p|^2}{12\pi c} = \frac{4}{3}\pi^3\left(\frac{c}{\varepsilon_o}\right)\frac{|p|^2}{\lambda_o^4}. \tag{7.47}$$

Notice that $\langle\mathcal{P}_{\text{rad}}\rangle$ increases as the fourth power of the frequency or as the inverse fourth power of the wavelength.

The directional characteristics of the radiation are again shown by Figure 7.2. The directivity (3.115) of the dipole in the horizontal plane ($\theta = \pi/2$) is easily obtained from (7.46) and (7.47):

$$D(\theta = \pi/2, \phi) = \frac{4\pi}{\langle\mathcal{P}_{\text{rad}}\rangle}\left[\frac{d\langle\mathcal{P}_{\text{rad}}(\theta = \pi/2)\rangle}{d\Omega}\right]$$

$$= 3/2 = 1.76\text{ dB}. \tag{7.48}$$

The electric field lines for the dipole with harmonic time dependence are shown in Figure 7.8 for the times $t/T = 0.0, 0.25, 0.50$, and 0.75, where $T = 2\pi/\omega$ is the period of the oscillation. Again, the graphs below each plot show the normalized electric and magnetic fields, $\mathcal{E}_{\theta n}$ and $\mathcal{B}_{\phi n}$, broadside to the dipole ($\theta = \pi/2$) versus the normalized radial position r/λ_o. The field lines are seen to resemble those of the dipole with the Gaussian excitation, consisting of a succession of pairs of loops (one with clockwise sense and one with counterclockwise sense), each pair resembling the single pair seen for the Gaussian excitation in Figure 7.6. A pair of loops is produced during each cycle (time T), and as time advances, they move away from the dipole. The comparison with the Gaussian excitation is developed further in Figure 7.9. The time-harmonic moment, shown in Figure 7.9a, can be viewed as the superposition in Figure 7.9b: a constant moment, which is negative, plus a positive moment that pulses in time with period T. The latter clearly resembles the

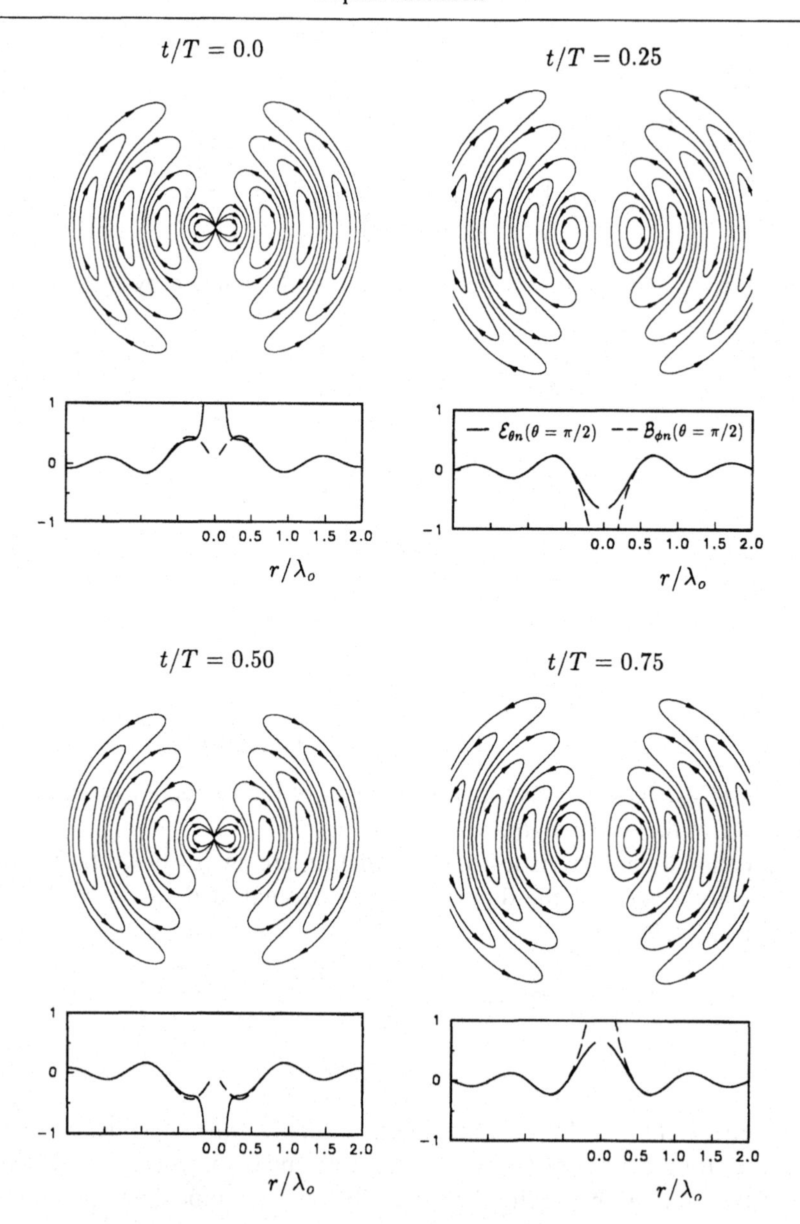

Fig. 7.8. Electric field lines for an electric dipole moment that is time harmonic. T is the period of oscillation. Each plot is for a different time t/T, and the graph below each plot shows the field $(\mathcal{E}_\theta, \mathcal{B}_\phi)$ as a function of r/λ_o for $\theta = \pi/2$.

train of Gaussian functions with characteristic time τ and spacing $T = 4\tau$ shown in Figure 7.9c. We thus see that the radiation of the time-harmonic signal is similar to that for the train of Gaussian pulses.

As we mentioned earlier, plots of the electric field lines surrounding a dipole (such as those in Figures 7.5, 7.6, and 7.8) were introduced by Heinrich Hertz around 1889 [1, 2], and since then they have been popular for instruction. For comparison, we

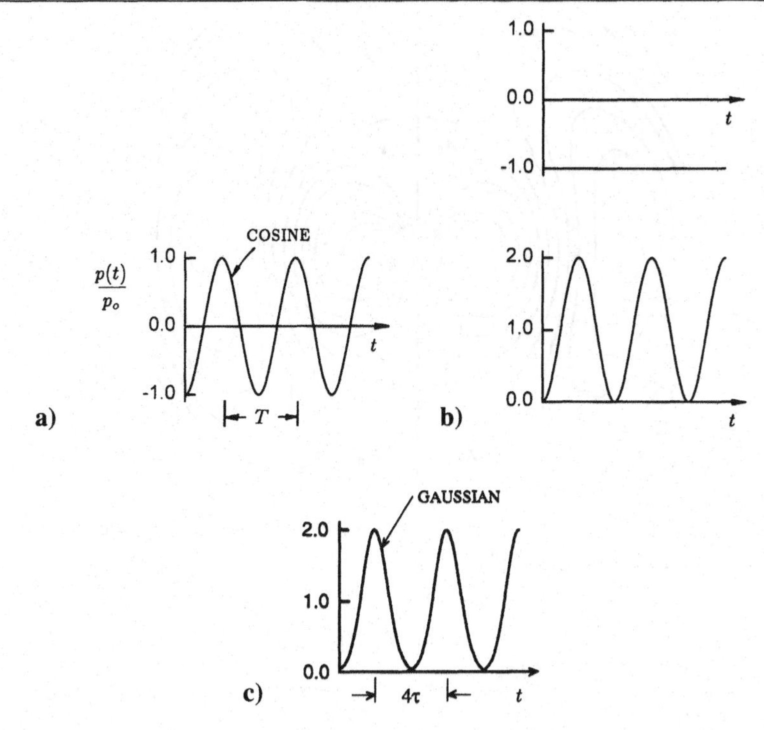

Fig. 7.9. a) Time-harmonic dipole moment with period T. b) Decomposition of time harmonic moment. c) Moment composed of a train of Gaussian functions with characteristic time τ and spacing $T = 4\tau$.

present one of Hertz's plots in Figure 7.10; it corresponds to the plot for $t/T = 0.0$ in Figure 7.8.[3] With these plots, Hertz explained the radiation from the dipole antennas he used in his classic experiments, which we discussed in Section 2.7. Hertz only considered a dipole moment with harmonic time dependence. Later, other researchers included damping to better represent the decaying oscillation of Hertz's spark-gap transmitter shown in Figure 2.20d. In fact, one paper published in 1900 contained 56 of these plots for the decaying oscillation; all of the plots were drawn by hand [3]!

7.2 Electrically short linear antennas

The linear antenna is simply a straight piece of wire driven at its midpoint by a voltage source. The geometry is shown in Figure 7.11a; each half of the antenna is a thin, metallic rod of radius a and length h; $a \ll h$. Figures 7.11b–d show variations of this basic structure. Figure 7.11b is the antenna used by Hertz; it is the linear antenna with square metal plates added at the ends. In Figure 7.11c, the plates are circular and positioned normal to the axis of the wire, and in Figure 7.11d they

[3] In Hertz's notation, the period of oscillation is $T = \pi/\omega$, which is one half of our value $T = 2\pi/\omega$; this must be kept in mind when comparing his plots with ours.

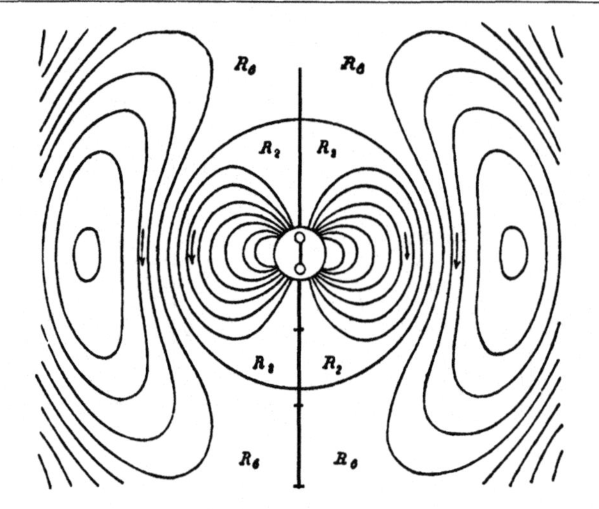

Fig. 7.10. One of Hertz's plots for the electric field lines surrounding an oscillating dipole (from Hertz's, *Electric Waves*, 1893 [2]).

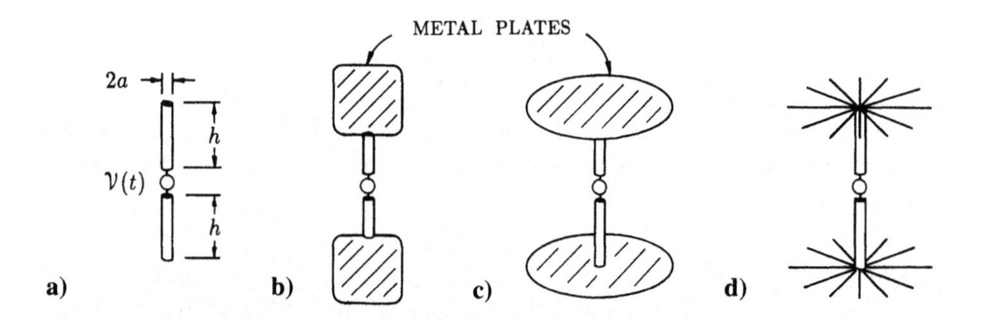

Fig. 7.11. a) Linear antenna. b)–d) Top-loaded linear antennas.

are replaced by a group of radial wires. The arrangements in Figures 7.11b–d are referred to as "top-loaded" antennas.[4]

To relate these antennas to the electric dipole, we will examine the current and charge (Figures 7.12 and 7.13). We will use the following notation. Since the axial current and the charge per unit length may vary with the position z along the wire, they are described by the distributions $\mathcal{I}(z, t)$ and $\mathcal{Q}(z, t)$, respectively. At the terminals, $z = 0$, the current and charge are $\mathcal{I}_o(t)$ and $\mathcal{Q}_o(t)$, respectively. The total charge on the top half of the antenna is $q(t)$.

These antennas are electrically short when the temporal variation of the source, the voltage $\mathcal{V}(t)$, is negligible during the time for light to travel the length of the antenna. We can then assume, for practical calculations, that the charge throughout the antenna changes almost instantaneously in response to a change in the applied

[4] The plates and wires used for top loading are generally much larger than those shown in the schematic drawings in Figures 7.11c and 7.11d. The radius of the plates or the length of the wires can be several times the height h; for an example see Figure 7.15a.

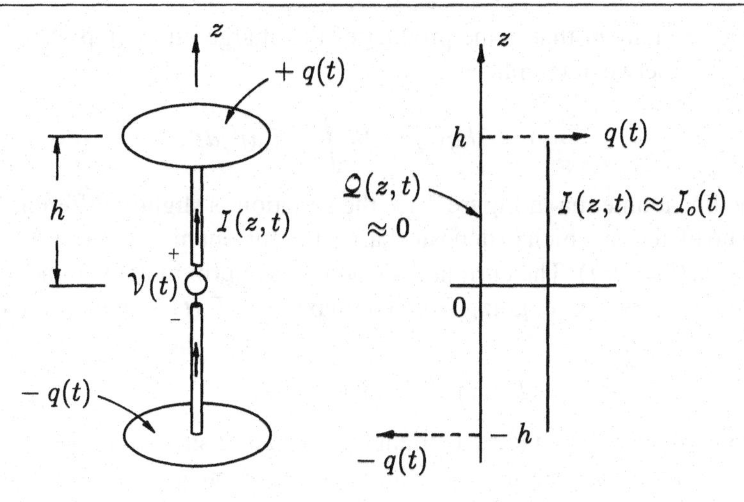

Fig. 7.12. Approximate distributions of charge and current for an electrically short, top-loaded linear antenna.

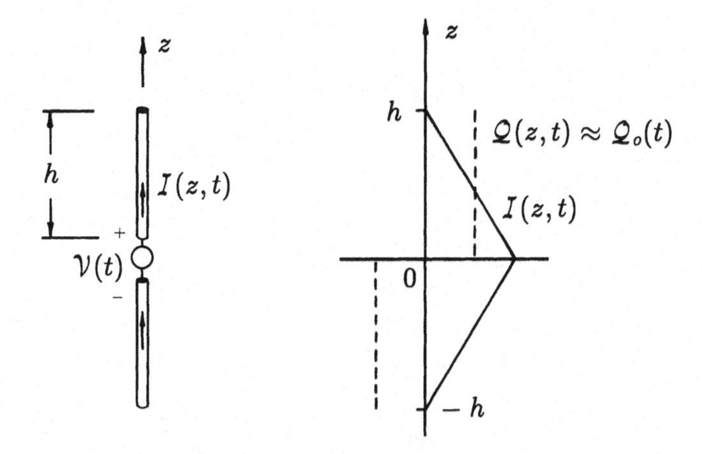

Fig. 7.13. Approximate distributions of charge and current for an electrically short linear antenna.

voltage; the antenna then behaves approximately as a capacitor C_a:

$$q(t) \approx C_a \mathcal{V}(t). \tag{7.49}$$

For the top-loaded antenna in Figure 7.12, the voltage $\mathcal{V}(t)$ produces a current distribution that is approximately uniform in the vertical wire, $\mathcal{I}(z, t) \approx \mathcal{I}_o(t)$. This current charges/discharges the plates. Negligible charge is on the wire $\mathcal{Q}(z, t) \approx 0$, so the total charge on the top half of the antenna, $q(t)$, is that on the top plate.[5] A comparison of Figure 7.1a and 7.12 shows that the antenna is simply an electric

[5] Radially directed currents also exist on the plates. These currents distribute the charge over the surfaces of the plates. The currents at diametrically opposite points on a plate are in opposite directions; thus, the fields of these currents approximately cancel at large distances from the antenna, and they contribute little to the radiation from the antenna.

dipole with a moment that is the product of the charge on the top plate $q(t)$ with the separation between the plates $2h$:

$$p(t) = 2hq(t) = 2h \int_{-\infty}^{t} \mathcal{I}_o(t')dt'.$$

For the linear antenna in Figure 7.13, the situation is slightly different. Because there are no plates on which to deposit charge, the current must be zero at the ends of the thin wire ($z = \pm h$). The current distribution is approximately a linear function of position along the wire, going from the maximum $\mathcal{I}_o(t)$ at the terminals to zero at the ends:

$$\mathcal{I}(z,t) \approx \mathcal{I}_o(t)(1 - |z|/h).$$

This is sometimes referred to as a "triangular current distribution." From the equation of continuity, we see that the charge distribution $\mathcal{Q}(z,t)$ is approximately uniform on the wires:

$$\frac{\partial \mathcal{Q}(z,t)}{\partial t} = -\frac{\partial \mathcal{I}(z,t)}{\partial z} \approx \pm \mathcal{I}_o(t)/h;$$

thus,

$$\mathcal{Q}(z,t) \approx \pm \mathcal{Q}_o(t) = \pm \frac{1}{h} \int_{-\infty}^{t} \mathcal{I}_o(t')dt',$$

where the $+$ sign applies for the top wire ($z > 0$) and the $-$ sign applies for the bottom wire ($z < 0$). The effective dipole moment is now

$$p(t) = \int_{-h}^{h} z\mathcal{Q}(z,t)dz = h^2 \mathcal{Q}_o(t) = h \int_{-\infty}^{t} \mathcal{I}_o(t')dt'.$$

For future use, it will be convenient to write the equivalent dipole moments for these two antennas as a single result:

$$p(t) = 2\gamma h \int_{-\infty}^{t} \mathcal{I}_o(t')dt', \tag{7.50}$$

or

$$\frac{dp}{dt} = 2\gamma h \mathcal{I}_o(t), \tag{7.51}$$

where[6]

$$\gamma = \begin{cases} 1 & \text{uniform current distribution (top-loaded, linear antenna)} \\ 1/2 & \text{triangular current distribution (linear antenna).} \end{cases}$$

When these practical dipoles are electrically short, we can model them as infinitesimal electric dipoles and determine their electromagnetic field by substituting (7.50) into the expressions in Table 7.1. Of course, the inequalities we used in obtaining the results for the infinitesimal dipole, (7.9), (7.15), and (7.16), must be

[6] In antenna engineering, the quantity $h_e = 2\gamma h$ is called the "effective length" for radiation in the direction $\theta = \pi/2$. For the antenna with uniform current, $h_e = 2h$, whereas for the antenna with triangular current, $h_e = h$.

satisfied. In terms of the quantities for these antennas, the inequalities are

$$r \gg 2h, \tag{7.52}$$

$$\left| \left(\frac{2h}{r} \right) \Delta t \dot{\mathcal{I}}_o \right| \ll |\mathcal{I}_o|, \tag{7.53}$$

and

$$\left| (\Delta t)^2 \ddot{\mathcal{I}}_o \right| \ll |\mathcal{I}_o|. \tag{7.54}$$

The first inequality says that the observation point must be at a large distance from the antenna in terms of the length $2h$. The second and third inequalities essentially say that the temporal variation in the current must be very small during the time for light to travel the length of the antenna, $\Delta t = 2h/c$. The inequalities (7.53) and (7.54) are to be satisfied in an average sense over the period of observation (Problem 7.4).[7]

The radiated field, (7.27) and (7.28), is proportional to $\ddot{p}$, which in terms of the current and voltage, is

$$\ddot{p} = 2\gamma h \dot{\mathcal{I}}_o \approx 2\gamma h C_a \ddot{\mathcal{V}}. \tag{7.55}$$

Thus, the radiated field is proportional to the first temporal derivative of the current or the second temporal derivative of the applied voltage. Notice that the radiated field of the uniform current distribution (top-loaded linear antenna) is twice that of the triangular current distribution (linear antenna) when the current at the terminals, $\mathcal{I}_o(t)$, is the same for both. The power radiated is proportional to $|\ddot{p}|^2$, so it is four times larger for the uniform distribution. The benefits of top loading are evident. For both antennas, the directional characteristics and the directivity are the same as for the infinitesimal electric dipole.

For the practical case of *harmonic time dependence*, the voltage and current are represented by the phasors V and I_o, respectively, and the phasor for the electric dipole moment is

$$p = -j \frac{2\gamma h}{\omega} I_o. \tag{7.56}$$

From (7.47) the time-average power radiated by the antenna is then

$$\langle \mathcal{P}_{\text{rad}} \rangle = \frac{\zeta_o}{12\pi} (k_o 2\gamma h)^2 |I_o|^2 = \frac{\pi \zeta_o}{3} \left(\frac{2\gamma h}{\lambda_o} \right)^2 |I_o|^2. \tag{7.57}$$

The simple series circuit of Figure 7.14 applies to the electrically short linear antenna [4]. In addition to the capacitance C_a, this circuit contains two resistors: the

[7] There are really two distinct assumptions involved in modeling the electrically short linear antenna as an infinitesimal electric dipole. The first is that the current distribution on the antenna is uniform (or triangular), and the second is that the inequalities (7.52)–(7.54) are satisfied, so that the electromagnetic field of the antenna is well approximated by that of the infinitesimal dipole.

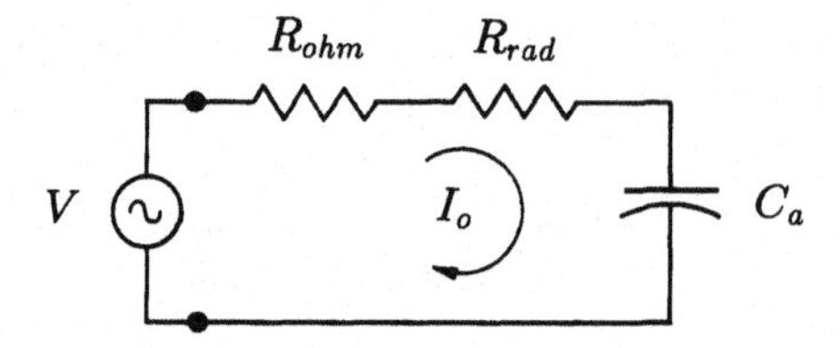

Fig. 7.14. Equivalent circuit for the electrically short linear antenna.

ohmic resistance, R_{ohm}, due to the losses in the antenna, and the *radiation resistance*, R_{rad}, that accounts for the energy radiated by the antenna.[8] The time-average power supplied to the antenna by the source is

$$\langle \mathcal{P}_{in} \rangle = \langle \mathcal{P}_{\text{ohm}} \rangle + \langle \mathcal{P}_{\text{rad}} \rangle = \frac{1}{2}|I_o|^2 R_{\text{ohm}} + \frac{1}{2}|I_o|^2 R_{\text{rad}}. \tag{7.58}$$

By equating (7.57) to the second term in (7.58), we can determine the radiation resistance,

$$R_{\text{rad}} = \frac{\zeta_o}{6\pi}(k_o 2\gamma h)^2 = \frac{2\pi\zeta_o}{3}\left(\frac{2\gamma h}{\lambda_o}\right)^2. \tag{7.59}$$

The advantage gained by making the antenna electrically longer is clear: The radiation resistance and radiated power increase as the square of the ratio of the height to the wavelength. Of course, we must remember that these results are for electrically short antennas; to use them, we must always have the length of the antenna much smaller than a wavelength, $2k_o h = 4\pi h/\lambda_o \ll 1$ (Problem 7.4).

To illustrate these results for the electrically short linear antenna, we will consider an example of historical interest. At the turn of this century, radio communication was in its infancy. It was believed that radio waves, like light, would be limited to line-of-sight paths; thus, the curvature of the Earth would prevent any long distance radio communication. To the contrary, in 1901 Guglielmo Marconi (1874–1937) experimentally demonstrated radio communication across the Atlantic [5–7]. Different mechanisms were proposed by which radio waves could propagate over such long distances; however, none was satisfactory. For example, diffraction around the curved surface of the Earth was proposed, but calculations showed that it was insufficient for the long distances experimentally observed.

At that time, the erroneous belief was held that long distance communications required low frequencies: The lower the frequency, the farther one would be able to communicate. Thus, the trend for a time was toward lower frequencies (longer wavelengths λ_o) and higher powers. As the wavelength was increased, the physical size of the antennas (height h) was also increased to maintain the power radiated, which from (7.57) is proportional to $(h/\lambda_o)^2$. In this process, some physically very large antennas were constructed.

[8] The resistances, R_{ohm} and R_{rad}, are usually much less than the capacitive reactance, $1/\omega C_a$. This is why we could ignore the resistances in our discussion for general time dependence and assume the approximation in (7.49), or equivalently $\mathcal{I}_o \approx C_a \dot{\mathcal{V}}$.

Figure 7.15a shows a top-loaded antenna built by Marconi around 1905 at Glace Bay, Nova Scotia [5–7]. It was to be used for trans-Atlantic communication at a frequency of about 82 kHz ($\lambda_o \approx 3.66$ km). The central, nearly vertical, portion of the antenna was 67 m (220 ft) high. The top loading was formed from 200 radial wires, each 305 m (1,000 ft) long. The structure was supported by wooden towers.

Marconi's antenna was a top-loaded monopole. The source was connected between the base of the vertical element and the ground. Wires were often placed over or in the ground below these antennas to increase the effective conductivity of the ground [8]. If we assume that the ground is a very good conductor, we can model it as a perfect conductor and replace the monopole by an equivalent dipole using the method of images (Section 4.7). This equivalence is sketched in Figure 7.15b. The current in the top half of the image equivalent is the same as that in the original antenna. The input impedance of the monopole is $Z_{\text{mono}} = V/I_o$, whereas that for the dipole is $Z_{\text{dip}} = 2V/I_o$; thus, the input impedance of the monopole is one half that of the dipole:[9]

$$Z_{\text{mono}} = \frac{1}{2} Z_{\text{dip}}.$$

Even though Marconi's antenna was physically very large, it was an electrically short antenna ($h/\lambda_o = 1.8 \times 10^{-2}$), so we can estimate its radiation resistance from (7.59):

$$R_{\text{rad mono}} = \frac{1}{2} R_{\text{rad dip}} = \frac{4\pi \zeta_o}{3}\left(\frac{h}{\lambda_o}\right)^2 = 0.53\ \Omega.$$

A very crude estimate for the capacitance of the antenna is obtained by replacing the array of radial wires by a solid plate and using the formula for an ideal, parallel-plate capacitor:[10]

$$C_a \approx \frac{\varepsilon_o A}{h} = 0.039\ \mu\text{F}.$$

[9] Another way to understand this relationship is to consider the power radiated by the two antennas. Let the power radiated into the upper half space by the monopole be

$$\langle P_{\text{rad}} \rangle_{\text{mono}} = \frac{1}{2} |I_o|^2 R_{\text{rad mono}}.$$

The dipole must radiate twice the power of the monopole; it radiates symmetrically into the upper and lower half spaces; thus,

$$\langle P_{\text{rad}} \rangle_{\text{dip}} = \frac{1}{2} |I_o|^2 R_{\text{rad dip}} = 2\langle P_{\text{rad}} \rangle_{\text{mono}} = |I_o|^2 R_{\text{rad mono}},$$

from which we see that

$$R_{\text{rad mono}} = \frac{1}{2} R_{\text{rad dip}}.$$

[10] The above values for R_{rad} and C_a are very small. For comparison, we mention that 0.53 Ω is the DC resistance of about 100 m of No. 12 copper wire (diameter 2.05 mm), the gauge often used for residential wiring. A commercial 0.04 μF capacitor made with Mylar film insulation is about 5 mm × 5 mm × 2 mm.

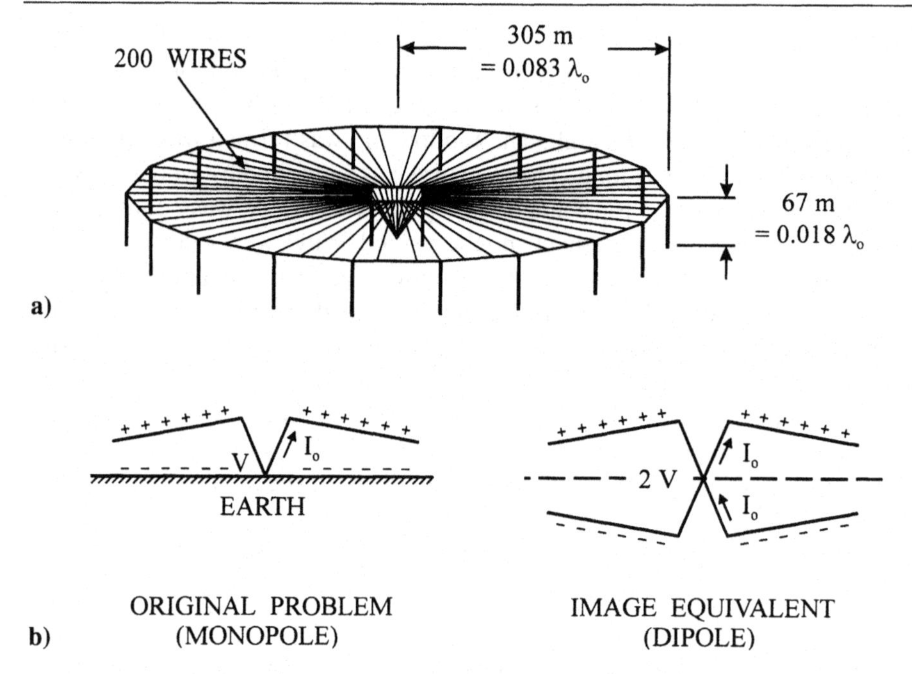

Fig. 7.15. a) Sketch of Marconi's top-loaded antenna at Glace Bay, Nova Scotia. b) Image equivalent for monopole antenna.

The capacitive reactance of the antenna is $1/\omega C_a \approx 50\ \Omega$, which is much greater than the radiation resistance. A tuning network (inductance) was often used at the base of these antennas to "tune out" the capacitance producing resonance.

The ohmic resistance, R_{ohm}, of the antenna is due to the losses in the wires and the ground. It is difficult to estimate because we do not know the conditions of the ground (electrical conductivity) under Marconi's antenna. However, for antennas like this, the ohmic resistance was often much greater than the radiation resistance; a typical value would be a few ohms [8]. Thus, the radiating efficiency for these antennas,

$$\eta = \frac{\text{time-average power radiated by antenna}}{\text{time-average power supplied to antenna}}$$

$$= \frac{R_{\text{rad}}}{R_{\text{ohm}} + R_{\text{rad}}}, \tag{7.60}$$

could be fairly small.

The radial wires forming the top loading for these antennas often drooped in going from the center to the supports at the ends; for Marconi's antenna, they went from a height of 67 m at the center to a height of 55 m at the ends. This makes the wires of the top loading appear as the ribs of an umbrella; hence, these antennas were called "umbrella antennas." Marconi's antenna at Glace Bay was not the highest umbrella antenna. In about 1911, the Telefunken Co. built an umbrella antenna at the Nauen station outside Berlin, Germany [9, 10]. The height for its central tower was 200 m (656 ft). For comparison, this is about one half of the height of the famous Empire State Building in New York City, which measures 381 m (1,250 ft).

In 1902, A. E. Kennelly and O. Heaviside independently hypothesized that the reflection of radio waves from a layer in the Earth's upper atmosphere (the Kennelly-Heaviside layer) might be what made long distance radio communication possible [11, 12]. If this layer were caused by the radiation from the sun, it would help to explain the observed differences in communication during the day and the night. Evidence gradually mounted for such a layer, with conclusive experimental proof for it coming in the 1920s [13–15]. The layer is now known as the ionosphere: A region of ionized gas, a plasma, that surrounds the Earth at altitudes from about 60 km to 600 km.

With the success of these experiments came the acceptance of the idea that long distance radio communication was made possible by the reflection of electromagnetic waves from the ionosphere, which could occur for frequencies up to about 30 MHz. The trend in long distance radio communication then reversed, going to higher frequencies (shorter wavelengths), lower powers, and physically smaller antennas [7].

7.3 Duality and the infinitesimal magnetic dipole or current loop

Maxwell's equations for free space (Table 1.3) are

$$\nabla \times \vec{\mathcal{E}} = -\frac{1}{c}\frac{\partial(c\vec{\mathcal{B}})}{\partial t} \tag{7.61}$$

$$\nabla \times (c\vec{\mathcal{B}}) = \zeta_o \vec{\mathcal{J}}_e + \frac{1}{c}\frac{\partial \vec{\mathcal{E}}}{\partial t} \tag{7.62}$$

$$\nabla \cdot \vec{\mathcal{E}} = c\zeta_o \rho_e \tag{7.63}$$

$$\nabla \cdot (c\vec{\mathcal{B}}) = 0 \tag{7.64}$$

$$\nabla \cdot \vec{\mathcal{J}}_e = -\frac{\partial \rho_e}{\partial t}. \tag{7.65}$$

Here we have included c with $\vec{\mathcal{B}}$ to form a quantity with the same units as the electric field $\vec{\mathcal{E}}$, and we have temporarily included the subscript e to indicate electric charge and current. In these equations, the volume densities of electric charge and current, ρ_e and $\vec{\mathcal{J}}_e$, can be viewed as the sources producing the electromagnetic field.

As we discussed in Problem 1.6, if magnetic charge were discovered, Maxwell's equations could be easily modified to include their effect. With the volume densities of magnetic charge and current, ρ_m and $\vec{\mathcal{J}}_m$, the only sources, Maxwell's equations in free space become

$$-\nabla \times \vec{\mathcal{E}} = \vec{\mathcal{J}}_m + \frac{1}{c}\frac{\partial(c\vec{\mathcal{B}})}{\partial t} \tag{7.66}$$

$$\nabla \times (c\vec{\mathcal{B}}) = \frac{1}{c}\frac{\partial \vec{\mathcal{E}}}{\partial t} \tag{7.67}$$

$$\nabla \cdot \vec{\mathcal{E}} = 0 \tag{7.68}$$

$$\nabla \cdot (c\vec{\mathcal{B}}) = c\rho_m \tag{7.69}$$

$$\nabla \cdot \vec{\mathcal{J}}_m = -\frac{\partial \rho_m}{\partial t}. \tag{7.70}$$

A comparison of these two sets of equations, (7.61)–(7.65) and (7.66)–(7.70), shows that the former becomes the latter on the interchange (Problem 7.5):

$$\zeta_o \rho_e \rightarrow \rho_m \tag{7.71}$$

$$\zeta_o \vec{\mathcal{J}}_e \rightarrow \vec{\mathcal{J}}_m \tag{7.72}$$

$$\vec{\mathcal{E}} \rightarrow c\vec{\mathcal{B}} \tag{7.73}$$

$$c\vec{\mathcal{B}} \rightarrow -\vec{\mathcal{E}}. \tag{7.74}$$

Equations in the two sets that have the same mathematical form [e.g. (7.61) and (7.67)] are called *dual equations*, and the corresponding quantities in these equations (e.g., $\vec{\mathcal{E}}$ and $c\vec{\mathcal{B}}$) are called *dual quantities.*[11]

Now consider two problems: In the first we have known distributions of electric charge and current (ρ_e, $\vec{\mathcal{J}}_e$); in the second, the *dual problem*, we have identical distributions of magnetic charge and current (ρ_m, $\vec{\mathcal{J}}_m$). The appropriate equation of continuity, (7.65) or (7.70), is satisfied by the charge and current in each problem. From the duality of Maxwell's equations, we automatically know the solution to the second problem if we know the solution to the first problem; we simply make the substitutions (7.71)–(7.74) in the solution to the first problem to obtain the solution to the second problem. Of course, the reverse is also true. Notice the similarity of the relations for the duality principle to those for Babinet's principle, which we discussed in Section 4.4. Both principles result from the fundamental symmetry for electric and magnetic quantities in Maxwell's equations.

We will illustrate the principle of duality by obtaining the electromagnetic field of an *infinitesimal magnetic dipole* from that of an infinitesimal electric dipole. The picture in Figure 7.1a also applies to the magnetic dipole, only the electric point charges q are replaced by magnetic point charges q_m, and the electric current I is replaced by a magnetic current I_m.

The *magnetic dipole moment* is defined to be

$$\vec{m}(t) = m(t)\hat{i} = \left(\frac{1}{\mu_o}\right) q_m(t)\Delta l\,\hat{i}, \tag{7.75}$$

[11] We have chosen the dual quantities so that they have the same electrical units (e.g., $\vec{\mathcal{E}}$ and $c\vec{\mathcal{B}}$, $\zeta_o\vec{\mathcal{J}}_e$ and $\vec{\mathcal{J}}_m$). Other dual quantities are sometimes used (e.g., $\vec{\mathcal{E}}$ and $\vec{\mathcal{H}}$, $\vec{\mathcal{J}}_e$ and $\vec{\mathcal{J}}_m$). Although our discussion is for sources in free space, the duality can be extended to include material regions. One then has dual materials, with special cases such as dual perfect conductors (electric perfect conductors and magnetic perfect conductors) [16].

where, as for the electric dipole, we are interested in the limit as $\Delta l \to 0$.[12] The electromagnetic field of the magnetic dipole is obtained from that of the electric dipole on making the interchanges (7.73), (7.74), and

$$p = q\Delta l \to \frac{1}{\zeta_o}(q_m \Delta l) = \frac{1}{\zeta_o}(\mu_o m) = \frac{1}{c} m \tag{7.76}$$

in the formulas in Table 7.1; thus,

$$\vec{\mathcal{E}}(\vec{r}, t) = \frac{\mu_o}{4\pi}\left(\frac{1}{r^2}\hat{r} \times \dot{\vec{m}} + \frac{1}{cr}\hat{r} \times \ddot{\vec{m}}\right)_{t_r}, \tag{7.77}$$

$$\vec{\mathcal{B}}(\vec{r}, t) = -\frac{\mu_o}{4\pi}\left[(1 - 3\hat{r}\hat{r}\cdot)\left(\frac{1}{r^3}\vec{m} + \frac{1}{cr^2}\dot{\vec{m}}\right) - \frac{1}{c^2 r}\hat{r} \times (\hat{r} \times \ddot{\vec{m}})\right]_{t_r}. \tag{7.78}$$

All of the other formulas for the magnetic dipole follow from the same interchanges; for example, the formula for the Poynting vector is unchanged on making the interchanges (7.73) and (7.74),

$$\vec{\mathcal{S}} = \frac{1}{\mu_o}(\vec{\mathcal{E}} \times \vec{\mathcal{B}}) \to \frac{1}{\mu_o}\left[(c\vec{\mathcal{B}}) \times (-\vec{\mathcal{E}}/c)\right] = \frac{1}{\mu_o}(\vec{\mathcal{E}} \times \vec{\mathcal{B}}),$$

so the total power radiated by the infinitesimal magnetic dipole is simply (7.34) with the interchange (7.76):

$$\mathcal{P}_{\text{rad}}(t_r) = \frac{\mu_o}{6\pi c^3}\left[\ddot{m}(t_r)\right]^2. \tag{7.79}$$

The directional characteristics for the radiation from the magnetic dipole are the same as those for the electric dipole (Figure 7.2).

The magnetic dipole with the structure of Figure 7.1a is a fictitious entity due to the nonexistence of magnetic charge in nature. The field away from the magnetic dipole, however, can be duplicated by a realizable structure – a small loop of electric current. This equivalence is most easily seen by considering the magnetostatic case where the dipole has a fixed moment, that is, the current in the loop is invariant in time. In Figure 7.16 the magnetic field lines for the infinitesimal magnetic dipole are shown on the right, while those for the small loop of current are shown on the left. Away from the loop, the lines are clearly the same as those for the dipole, so in the limit as the size of the loop goes to zero (an infinitesimal current loop) the lines will be equivalent for the two structures, except at the origin.

Although the above pictorial argument based on magnetostatics is convincing, we wish to show the equivalence of the infinitesimal current loop and infinitesimal magnetic dipole for dynamic fields. We will do this by first obtaining the field of a small loop as the superposition of the fields from a group of electric current elements, and then we will show that this field is equivalent to that of the magnetic dipole.

[12] Logically, to completely parallel the electric case, we should define the magnetic dipole moment as $\vec{p}_m = q_m \Delta l\,\hat{i}$. However, we use $\vec{m}$ for the moment and include the factor $1/\mu_o$ in (7.75) so that our later results will agree with long established convention.

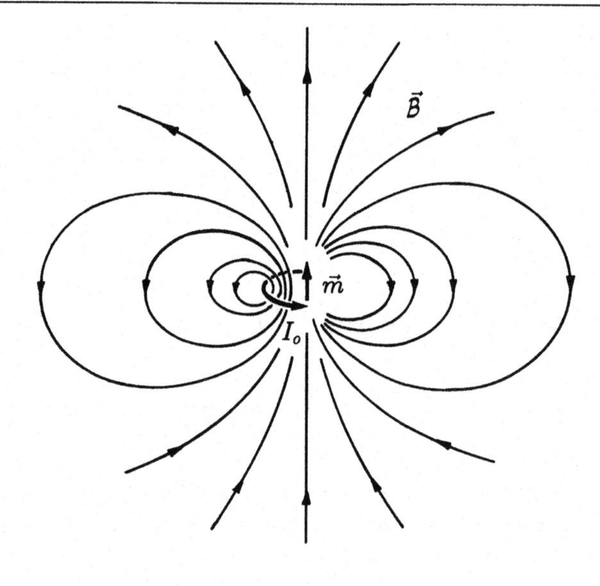

Fig. 7.16. Comparison of magnetostatic fields for infinitesimal magnetic dipole (right) and small loop of electric current (left).

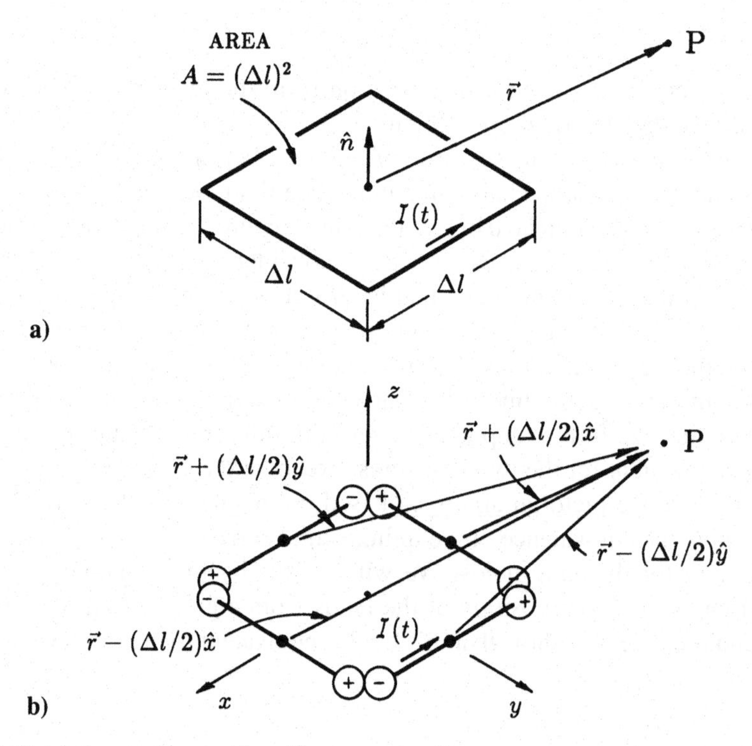

Fig. 7.17. a) Square loop with uniform current $\mathcal{I}(t)$. b) Superposition of four current elements, each with uniform current, to form the square loop.

Consider the square loop shown in Figure 7.17a with sides of length Δl and uniform current $\mathcal{I}(t)$. This loop can be viewed as the superposition of the four current elements shown in Figure 7.17b. Notice that because the charges at the ends of adjacent current elements cancel, there is no charge on the loop. The vector potential for this combination of current elements is the superposition of four terms, one for each element. After using (7.3) and (7.17), the vector potential due to the $\hat{y}$-directed elements at $(\pm\Delta l/2, 0, 0)$ is

$$\mathcal{A}_y(\vec{r}, t) = \frac{\mu_o \Delta l}{4\pi}\left\{\frac{\mathcal{I}[t - |\vec{r} - (\Delta l/2)\hat{x}|/c]}{|\vec{r} - (\Delta l/2)\hat{x}|} - \frac{\mathcal{I}[t - |\vec{r} + (\Delta l/2)\hat{x}|/c]}{|\vec{r} + (\Delta l/2)\hat{x}|}\right\}$$

$$= \frac{\mu_o \Delta l}{4\pi}\frac{\partial}{\partial r}\left[\frac{\mathcal{I}(t - r/c)}{r}\right]\left[-\Delta l\frac{d|\vec{r}\,|}{dx}\right] = -\frac{\mu_o(\Delta l)^2}{4\pi}\frac{\partial}{\partial r}\left[\frac{\mathcal{I}(t - r/c)}{r}\right]\frac{x}{r},$$

where we have used the fact that we are interested in the limit as $\Delta l \to 0$ to write the ratio of differences as a derivative. The vector potential due to the $\hat{x}$-directed elements at $(0, \pm\Delta l/2, 0)$ is obtained in a similar manner:

$$\mathcal{A}_x(\vec{r}, t) = \frac{\mu_o(\Delta l)^2}{4\pi}\frac{\partial}{\partial r}\left[\frac{\mathcal{I}(t - r/c)}{r}\right]\frac{y}{r}.$$

After combining these results, we have

$$\vec{\mathcal{A}}(\vec{r}, t) = \left\{\frac{-\mu_o(\Delta l)^2}{4\pi}\left(\frac{\mathcal{I}}{r^2} + \frac{\dot{\mathcal{I}}}{cr}\right)\left(\frac{y}{r}\hat{x} - \frac{x}{r}\hat{y}\right)\right\}_{t_r}. \tag{7.80}$$

The last factor in this equation is

$$\frac{y}{r}\hat{x} - \frac{x}{r}\hat{y} = \frac{1}{r}\vec{r} \times \hat{z} = \hat{r} \times \hat{z} = \hat{r} \times \hat{n},$$

where we have let the normal to the plane of the loop be $\hat{n}$ ($\hat{n} = \hat{z}$). Thus, the vector potential becomes

$$\vec{\mathcal{A}}(\vec{r}, t) = -\frac{\mu_o}{4\pi}\left\{\frac{1}{r^2}\hat{r} \times \left[\mathcal{I}(\Delta l)^2\hat{n}\right] + \frac{1}{cr}\hat{r} \times \left[\dot{\mathcal{I}}(\Delta l)^2\hat{n}\right]\right\}_{t_r},$$

and the electric field is

$$\vec{\mathcal{E}}(\vec{r}, t) = -\nabla\Phi - \frac{\partial\vec{\mathcal{A}}}{\partial t} = -\frac{\partial\vec{\mathcal{A}}}{\partial t}$$

$$= \frac{\mu_o}{4\pi}\left(\frac{1}{r^2}\hat{r} \times \left[\dot{\mathcal{I}}(\Delta l)^2\hat{n}\right] + \frac{1}{cr}\hat{r} \times \left[\ddot{\mathcal{I}}(\Delta l)^2\hat{n}\right]\right)_{t_r}. \tag{7.81}$$

The absence of charge on the loop made the scalar potential zero in the above calculation.

A comparison of (7.81) with (7.77) shows that the electric field of the current loop is the same as that for the infinitesimal magnetic dipole when we define the magnetic moment of the loop to be, in the limit as $\Delta l \to 0$,

$$\vec{m}(t) = \mathcal{I}(t)(\Delta l)^2\hat{n},$$

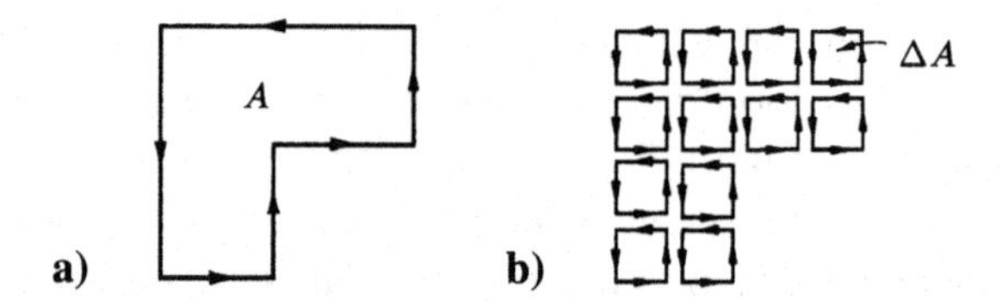

Fig. 7.18. The plane loop of general shape (a) as a superposition of smaller square loops (b).

or

$$\vec{m}(t) = \mathcal{I}(t)A\hat{n}, \tag{7.82}$$

where $A = (\Delta l)^2$ is the area of the loop.

In summary, the infinitesimal current loop is a loop whose area goes to zero ($A \to 0$) while its current becomes infinite ($|\mathcal{I}(t)| \to \infty$) such that their product remains finite. It is equivalent to an infinitesimal magnetic dipole with the moment (7.82), which is the product of the current, area of the loop, and the unit vector normal to the plane of the loop. For this equivalence, the loop need not be square, as in our example; it can be any closed plane curve. To see this, notice that a plane loop of general shape can be viewed as a superposition of small, square loops, as illustrated in Figure 7.18. The currents on the adjacent sides of a pair of square loops cancel, leaving only the current around the periphery; the area, A, of the loop of general shape is the sum of the areas, ΔA, of the small, square loops. Therefore, the magnetic moment of the general loop is the same as the sum of the magnetic moments for the small, square loops.

The results for the electromagnetic field of the infinitesimal magnetic dipole or infinitesimal current loop, for both general and harmonic time dependences, are summarized in Table 7.2.

7.4 Electrically small loop antennas

The multiturn loop antenna shown in Figure 7.19 consists of a tightly wound coil of n turns of metallic wire, fed at its terminals by a voltage source $\mathcal{V}$. The loop is circular with radius b in this figure, but in general it can have any shape (square, rectangular, etc.). The radius of the wire, a, is assumed to be much smaller than the radius of the loop ($a \ll b$).

When the loop is electrically small, the current is approximately uniform (the same at any point along the wire) with the value $\mathcal{I}_o(t)$. Loosely speaking, this occurs whenever the temporal variation in the source $\mathcal{V}(t)$ is negligible during the time for light to travel a distance equal to the total length of the wire ($n2\pi b$). At any cross section of the multiturn loop the total current is $n\mathcal{I}_o(t)$, and so its magnetic dipole moment is the same as that of a single-turn loop carrying the current $n\mathcal{I}_o(t)$:

$$m(t) = n\mathcal{I}_o(t)A = n\pi b^2\mathcal{I}_o(t). \tag{7.83}$$

The electrically small loop antenna can be modeled as an infinitesimal magnetic dipole, and the electromagnetic field can be determined by substituting (7.83) into

Table 7.2. *Electromagnetic field of infinitesimal magnetic dipole or current loop*

General time dependence

$$\vec{\mathcal{E}}(\vec{r},t) = \frac{\mu_o}{4\pi}\left(\frac{1}{r^2}\hat{r}\times\dot{\vec{m}} + \frac{1}{cr}\hat{r}\times\ddot{\vec{m}}\right)_{t_r=t-r/c}$$

$$= \frac{-\mu_o}{4\pi}\left[\left(\frac{1}{r^2}\dot{m} + \frac{1}{cr}\ddot{m}\right)\sin\theta\,\hat{\phi}\right]_{t_r}$$

$$\vec{\mathcal{B}}(\vec{r},t) = \frac{-\mu_o}{4\pi}\left[(1-3\hat{r}\hat{r}\cdot)\left(\frac{1}{r^3}\vec{m} + \frac{1}{cr^2}\dot{\vec{m}}\right) - \frac{1}{c^2 r}\hat{r}\times(\hat{r}\times\ddot{\vec{m}})\right]_{t_r=t-r/c}$$

$$= \frac{\mu_o}{4\pi}\left[\left(\frac{1}{r^3}m + \frac{1}{cr^2}\dot{m}\right)(2\cos\theta\,\hat{r} + \sin\theta\,\hat{\theta}) + \frac{1}{c^2 r}\ddot{m}\sin\theta\,\hat{\theta}\right]_{t_r}$$

$$\vec{m}(t) = \lim_{\Delta l\to 0}\left[\frac{1}{\mu_o}q_m(t)\Delta l\,\hat{i}\right] = \lim_{A\to 0}\left[\mathcal{I}(t)A\,\hat{n}\right]$$

Harmonic time dependence

$$\vec{E}(\vec{r}) = \frac{\zeta_o k_o^3}{4\pi}\left[\frac{j}{(k_o r)^2} - \frac{1}{k_o r}\right]\hat{r}\times\vec{m}\,e^{-jk_o r}$$

$$= \frac{-\zeta_o k_o^3 m}{4\pi}\sin\theta\left[\frac{j}{(k_o r)^2} - \frac{1}{k_o r}\right]\hat{\phi}\,e^{-jk_o r}$$

$$\vec{B}(\vec{r}) = \frac{-k_o^3\mu_o}{4\pi}\left\{\left[\frac{1}{(k_o r)^3} + \frac{j}{(k_o r)^2}\right](1-3\hat{r}\hat{r}\cdot)\vec{m} + \left(\frac{1}{k_o r}\right)\hat{r}\times(\hat{r}\times\vec{m})\right\}e^{-jk_o r}$$

$$= \frac{k_o^3\mu_o m}{4\pi}\left\{2\cos\theta\left[\frac{1}{(k_o r)^3} + \frac{j}{(k_o r)^2}\right]\hat{r} + \sin\theta\left[\frac{1}{(k_o r)^3} + \frac{j}{(k_o r)^2} - \frac{1}{k_o r}\right]\hat{\theta}\right\}e^{-jk_o r}$$

the expressions in Table 7.2. The inequalities that must be satisfied for this model are again (7.52)–(7.54), only with $2h$ replaced by $2b$.

This antenna behaves essentially as an inductor L_a; thus,

$$\mathcal{V}(t) \approx L_a\frac{d\mathcal{I}_o(t)}{dt},$$

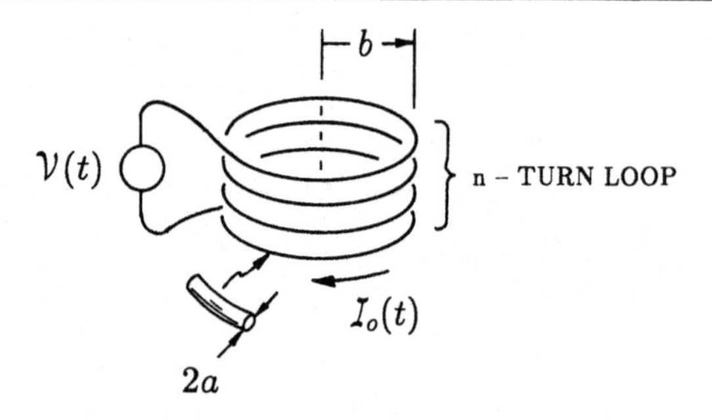

Fig. 7.19. Multiturn, circular loop antenna.

or

$$\mathcal{I}_o(t) \approx \frac{1}{L_a} \int_{-\infty}^{t} \mathcal{V}(t')dt'. \tag{7.84}$$

The radiated field, (7.77) and (7.78), is proportional to $\ddot{m}$, which in terms of the current and voltage is

$$\ddot{m} = n\pi b^2 \ddot{\mathcal{I}}_o \approx \frac{n\pi b^2}{L_a} \dot{\mathcal{V}}. \tag{7.85}$$

So the radiated field is proportional to the second temporal derivative of the current or the first temporal derivative of the voltage. This is to be compared with the radiated field of the electrically short dipole, which is proportional to the first temporal derivative of the current or the second temporal derivative of the voltage (7.55).

For the case of *harmonic time dependence*, the voltage and current are represented by the phasors V and I, respectively. The magnetic dipole moment for the loop is then the phasor

$$m = nI_oA = n\pi b^2 I_o, \tag{7.86}$$

and the time-average power radiated by the loop is

$$\langle \mathcal{P}_{\text{rad}} \rangle = \frac{\zeta_o}{12\pi} n^2 k_o^4 A^2 |I_o|^2 = \frac{\pi \zeta_o}{12} n^2 (k_o b)^4 |I_o|^2$$

$$= \frac{4\pi^3 \zeta_o}{3} \frac{n^2 A^2}{\lambda_o^4} |I_o|^2 = \frac{4\pi^5 \zeta_o}{3} n^2 \left(\frac{b}{\lambda_o}\right)^4 |I_o|^2. \tag{7.87}$$

The simple series circuit of Figure 7.20 applies to the electrically small loop antenna [17]. In addition to the inductance L_a, this circuit contains two small resistors: the resistance due to the ohmic loss in the wire, R_{ohm}, and the radiation resistance, R_{rad}. An expression for the radiation resistance can be obtained in the same manner as for the electrically short linear antenna. The time-average power supplied to the loop is

$$\langle \mathcal{P}_{in} \rangle = \frac{1}{2} |I_o|^2 R_{\text{ohm}} + \frac{1}{2} |I_o|^2 R_{\text{rad}}. \tag{7.88}$$

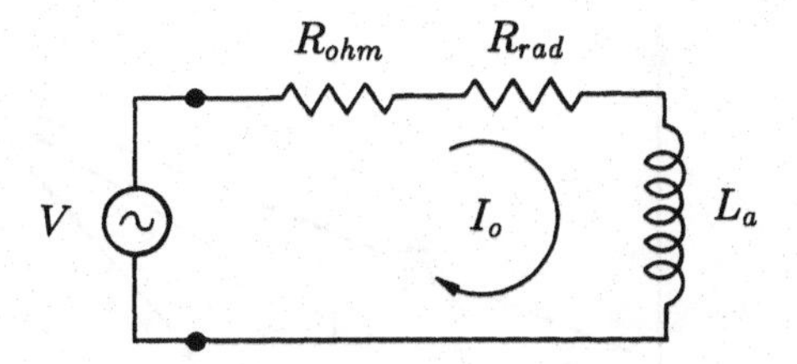

Fig. 7.20. Equivalent circuit for the electrically small loop antenna.

The radiation resistance is determined by equating (7.87) to the second term in (7.88):

$$R_{\text{rad}} = \frac{8\pi^3\zeta_o}{3}\frac{n^2A^2}{\lambda_o^4} = \frac{8\pi^5\zeta_o}{3}n^2\left(\frac{b}{\lambda_o}\right)^4. \tag{7.89}$$

The radiation resistance and the power radiated increase as the square of the number of turns, n, and as the fourth power of the ratio of the radius to the wavelength, b/λ_o (as the square of the area A). Consequently, the performance of the loop (the radiation) can be improved significantly by simply increasing the number of turns or the radius of the loop. Of course, these dependences only apply as long as the loop is electrically small $[k_o(n2\pi b) = n4\pi^2 b/\lambda_o \ll 1]$. Notice that the radiation resistance for the loop antenna is a much stronger function of the electrical size than it is for the linear antenna [proportional to $(b/\lambda_o)^4$ versus $(h/\lambda_o)^2$].

The directional characteristics of the radiation from the electrically small loop antenna are the same as those for the electrically short linear antenna and the infinitesimal electric dipole; thus, the directivity in the horizontal plane is again $D = 3/2 = 1.76$ dB. The electrically small loop antenna, like the electrically short linear antenna, often has an ohmic resistance that is much greater than the radiation resistance, and so the radiating efficiency (7.60) can be very small [18].

7.5 Simple arrays of electrically short linear antennas

The pattern shown in Figure 7.2 for the radiation from the electrically short linear antenna, or dipole, is uniform (omnidirectional) in the horizontal plane ($\theta = \pi/2$). This is a direct consequence of the rotational symmetry of the structure about the z axis. In certain applications, higher directivity for the radiation is desirable. For example, in communications systems, transmission may be between two antennas at fixed locations. In this case, one would like the radiation to be directed along the line joining the antennas. Directive radiation can be achieved by using two or more antennas (elements) in an array. The basic idea is to choose the physical arrangement and the excitation for the elements so as to produce constructive interference of the radiation from the individual elements in a particular direction and destructive interference in other directions.

In the simplest theory for arrays, the current distributions for all of the antennas are assumed to have the same functional form. For example, for an array of electrically short linear antennas, the current distributions in all of the elements are assumed to be either uniform (with top loading) or triangular (without top loading) functions

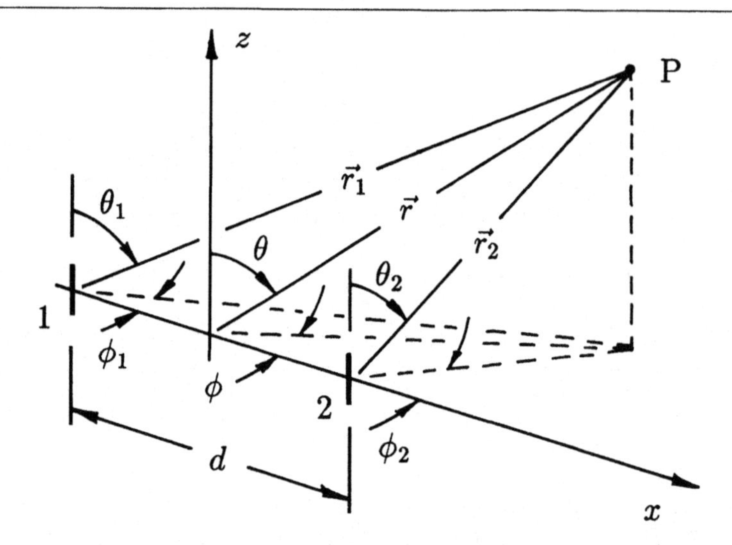

Fig. 7.21. Coordinates for the couplet of electrically short linear antennas.

of the position z, as shown in Figures 7.12 and 7.13. The currents at the terminals of the antennas, $z = 0$, are specified independently. The radiation from the array is then just a superposition of the radiation from the individual antennas, each treated as if it were isolated from the others. In this simple theory, the effect of mutual coupling on the current distributions is ignored. Mutual coupling is the result of every element in the array being in the field of the other elements. In general this field will be different for elements at different locations in the array; hence, the component of the current distribution due to mutual coupling will not be the same for all elements [19].

The science for the design of antenna arrays is highly developed, and there are numerous in-depth treatments addressing the subject [19–25]. Here we will only use the simple theory described above to introduce a few of the basic concepts associated with arrays. We will restrict our discussion to simple *linear arrays* of equally spaced, identical, electrically short linear antennas.[13] First, we will examine the case where the elements are excited by "pulse-like" signals; later, we will consider the special, practically important case of time-harmonic excitation.

7.5.1 General time dependence

For the two element array or couplet shown in Figure 7.21, the antennas are located on the x axis at $x = -d/2$ (antenna 1) and at $x = d/2$ (antenna 2), and their axes are parallel. The currents at the terminals of the antennas are $\mathcal{I}_{o1}(t)$ and $\mathcal{I}_{o2}(t)$, respectively. The radiated electric field from antenna 1, obtained by using (7.51)

[13] A linear array is one in which the elements lie along a straight line. Other geometrical arrangements may be used; for example, in a *circular array* the elements are placed along the circumference of a circle.

with (7.27), is

$$\vec{\mathcal{E}}_1^r(\vec{r}_1, t) = \frac{\mu_o(2\gamma h)}{4\pi r_1} \sin\theta_1 \dot{\mathcal{I}}_{o1}(t - r_1/c)\hat{\theta}_1, \tag{7.90}$$

and similarly for antenna 2

$$\vec{\mathcal{E}}_2^r(\vec{r}_2, t) = \frac{\mu_o(2\gamma h)}{4\pi r_2} \sin\theta_2 \dot{\mathcal{I}}_{o2}(t - r_2/c)\hat{\theta}_2. \tag{7.91}$$

We will apply the approximations for the radiated field that we have used before to this problem. We take $r_1 \approx r_2 \approx r$, $\theta_1 \approx \theta_2 \approx \theta$, etc., except in the retarded times (the arguments of $\mathcal{I}_o$), in which we let

$$r_1 = \left|\vec{r} + (d/2)\hat{x}\right| = \left[r^2 + (d/2)^2 + dr\sin\theta_1\cos\phi_1\right]^{1/2}$$
$$\approx r + (d/2)\sin\theta\cos\phi$$

and

$$r_2 = \left|\vec{r} - (d/2)\hat{x}\right| = \left[r^2 + (d/2)^2 - dr\sin\theta_2\cos\phi_2\right]^{1/2}$$
$$\approx r - (d/2)\sin\theta\cos\phi. \tag{7.92}$$

With these approximations, the radiated electric field of the couplet, the sum of (7.90) and (7.91), becomes

$$\vec{\mathcal{E}}^r(\vec{r}, t) = \frac{\mu_o(2\gamma h)}{4\pi r} \sin\theta\,\hat{\theta}\Big\{\dot{\mathcal{I}}_{o1}\big[t - r/c - (d/2c)\sin\theta\cos\phi\big]$$
$$+ \dot{\mathcal{I}}_{o2}\big[t - r/c + (d/2c)\sin\theta\cos\phi\big]\Big\}. \tag{7.93}$$

We will assume that the currents in the two elements, $\mathcal{I}_{o1}$ and $\mathcal{I}_{o2}$, are the same except for a shift in time,

$$\mathcal{I}_{o2}(t) = \mathcal{I}_{o1}(t - \Delta t); \tag{7.94}$$

that is, the current waveform at the terminals of antenna 2 is the same as that at the terminals of antenna 1 except that it is delayed in time by the amount Δt. The spacing d and time delay Δt will be chosen to increase the directivity of the couplet in the horizontal plane ($\theta = \pi/2$), for which (7.93) becomes

$$\mathcal{E}_\theta^r(r, \theta = \pi/2, \phi; t) = \frac{\mu_o(2\gamma h)}{4\pi r}\Big\{\dot{\mathcal{I}}_{o1}\big[t - r/c - (d/2c)\cos\phi\big]$$
$$+ \dot{\mathcal{I}}_{o1}\big[t - r/c + (d/2c)\cos\phi - \Delta t\big]\Big\}. \tag{7.95}$$

To illustrate this formula, we will let the current $\mathcal{I}_{o1}(t)$ be the Gaussian function:[14]

$$\mathcal{I}_{o1}(t) = I_o e^{-(t/\tau)^2}. \tag{7.96}$$

[14] This current will result when the electrically short antenna is excited by a voltage that is a "step-like" function of time, such as one of the waveforms shown in Figure 5.5b.

The radiated electric field from each antenna is then a differentiated Gaussian function, and the field of the couplet (7.95) is

$$\begin{aligned}\mathcal{E}_\theta^r(r, \theta = \pi/2, \phi; t) &\\ = \frac{-\mu_o(2\gamma h)I_o}{2\pi\tau r}\Big\{&\big[t/\tau - r/c\tau - (d/2c\tau)\cos\phi\big]e^{-[t/\tau - r/c\tau - (d/2c\tau)\cos\phi]^2} \\ &+ \big[t/\tau - r/c\tau + (d/2c\tau)\cos\phi - \Delta t/\tau\big]e^{-[t/\tau - r/c\tau + (d/2c\tau)\cos\phi - \Delta t/\tau]^2}\Big\},\end{aligned} \tag{7.97}$$

which in normalized form becomes[15]

$$\begin{aligned}\mathcal{E}_{\theta n}^r(\rho, \theta = \pi/2, \phi; t) = \sqrt{2}\,e^{1/2}\Big\{&\big[\rho + (d/2c\tau)\cos\phi\big]e^{-[\rho + (d/2c\tau)\cos\phi]^2} \\ &+ \big[\rho - (d/2c\tau)\cos\phi + \Delta t/\tau\big]e^{-[\rho - (d/2c\tau)\cos\phi + \Delta t/\tau]^2}\Big\},\end{aligned} \tag{7.98}$$

where

$$\rho = (r - ct)/c\tau. \tag{7.99}$$

Figure 7.22a is a graph of the radiated field (7.98) as a function of the normalized position ρ (7.99) along the axis of the array.[16] The right half of the figure shows the field for $\phi = 0$ (the forward direction), whereas the left half of the figure shows the field for $\phi = \pi$ (the backward direction). For this graph, the spacing d and the time shift Δt were chosen to separate in space the pulses (differentiated Gaussian functions) radiated by the individual elements. The pulses from each element, 1 or 2, are marked in the figure.

When the couplet is designed to produce maximum radiation in the forward direction ($\phi = 0$) it is called a *unilateral end-fire array*. This requires superposition of the pulses for $\phi = 0$, which from (7.98) means

$$\rho + d/2c\tau = \rho - d/2c\tau + \Delta t/\tau,$$

or

$$\Delta t/\tau = d/c\tau. \tag{7.100}$$

The excitation of element 2 is delayed by the time required for the signal from element 1 to propagate to element 2. We also want the radiation in the backward direction ($\phi = \pi$) to cancel as much as possible. Partial cancellation is obtained by choosing (Problem 7.6)

$$d/c\tau + \Delta t/\tau = 2d/c\tau = 2/\sqrt{2},$$

[15] The radiated field is normalized so that the maximum field for a single element is one.

[16] These graphs can be viewed as plots of the field in space at a fixed time, say $t = t_o$. The point $\rho = 0$ is then at the point $r_o = ct_o$, and $\rho = (r - ct_o)/c\tau = (r - r_o)/c\tau$.

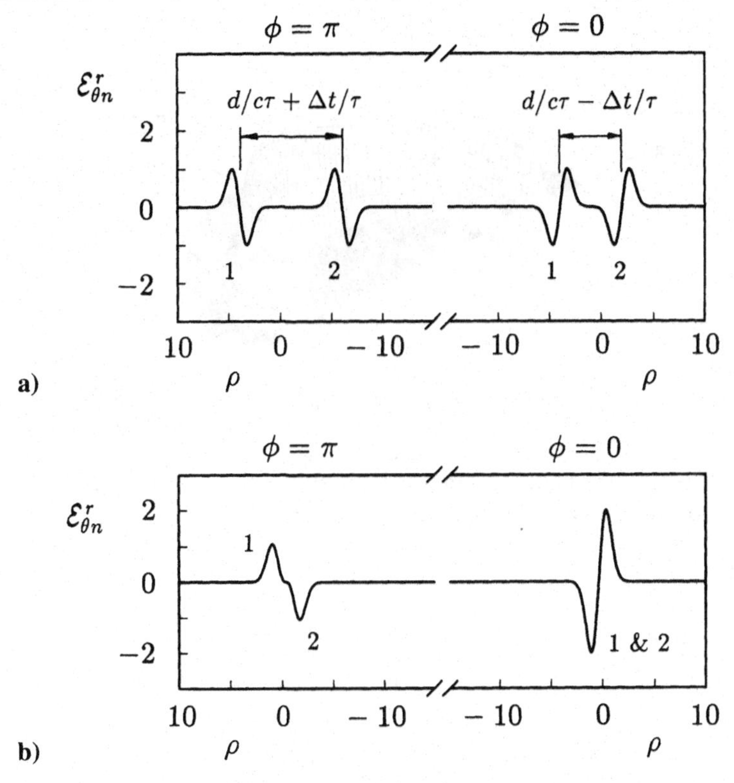

Fig. 7.22. The radiated electric field along the axis of a couplet when the current in the elements is a Gaussian function of time. a) General spacing and time shift. b) Spacing and time shift chosen to make a unilateral end-fire couplet: $d/c\tau = \Delta t/\tau = 1/\sqrt{2}$.

or

$$d/c\tau = 1/\sqrt{2}, \tag{7.101}$$

so that the peak of the positive lobe of the signal from element 2 is superimposed on the peak of the negative lobe of the signal from element 1. This situation is shown in Figure 7.22b. For this arrangement, the maximum field in the forward direction is about twice that in the backward direction. Figure 7.23 shows the radiated electric field (7.98) of this end-fire couplet for all of the angles in the range $0 \leq \phi \leq \pi$. This diagram is essentially a collection of graphs similar to Figure 7.22b, each drawn at the appropriate angle ϕ.

The couplet can also be designed to produce maximum radiation in the directions normal to the line of the array ($\phi = \pi/2,\ 3\pi/2$); it is then called a *bilateral broadside array*. This requires superposition of the pulses for both $\phi = \pi/2$ and $\phi = 3\pi/2$, which from (7.98) means

$$\rho = \rho + \Delta t/\tau,$$

or

$$\Delta t/\tau = 0. \tag{7.102}$$

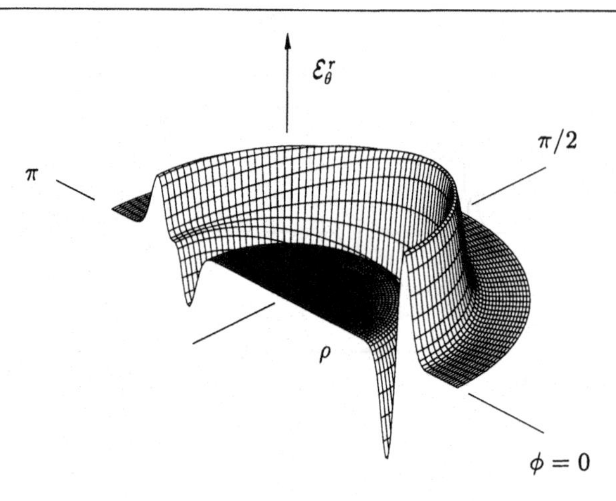

Fig. 7.23. Radiated electric field in horizontal plane, $\theta = \pi/2$, for unilateral end-fire couplet with currents that are Gaussian functions of time: $d/c\tau = \Delta t/\tau = 1/\sqrt{2}$.

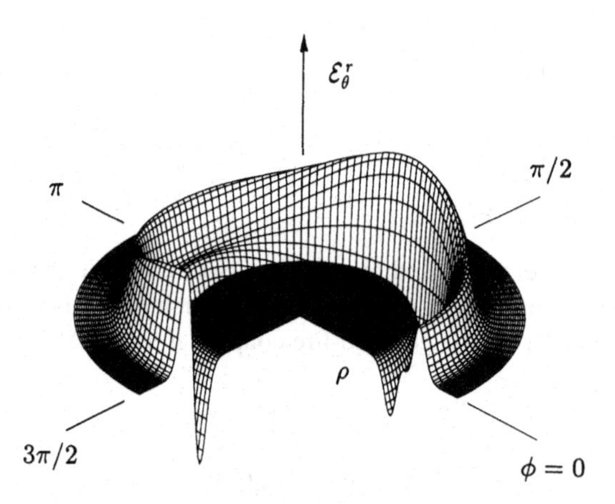

Fig. 7.24. Radiated electric field in horizontal plane, $\theta = \pi/2$, for bilateral broadside couplet with currents that are Gaussian functions of time: $d/c\tau = \sqrt{2}$, $\Delta t/\tau = 0$.

The excitation of element 2 occurs at the same time as that of element 1. For partial cancellation of the radiation in the directions $\phi = 0$ and $\phi = \pi$, we choose

$$\rho + d/2c\tau = \rho - d/2c\tau + 2/\sqrt{2},$$

or

$$d/c\tau = \sqrt{2}. \tag{7.103}$$

Figure 7.24 shows the radiated electric field (7.98) of this broadside couplet for all of the angles in the range $0 \leq \phi \leq 3\pi/2$. The bilateral nature of the pattern (maximum radiation for $\phi = \pi/2$ and $\phi = 3\pi/2$) is clearly evident. The maximum value of $\mathcal{E}_{\theta n}^r$ at $\phi = 3\pi/2$ is about twice that at $\phi = 0$.

An N-element linear array is formed by locating N equally spaced antennas symmetrically along the x axis, as shown in Figure 7.25. When the currents in all

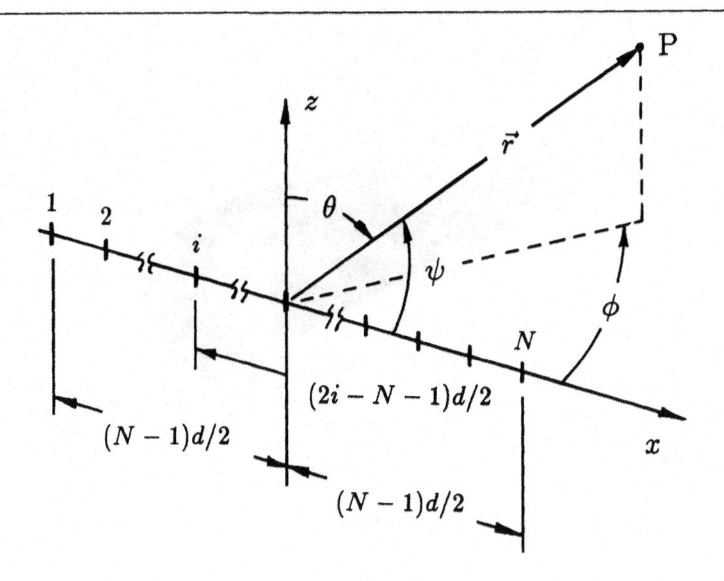

Fig. 7.25. Placement of elements in the N-element array. The element-to-element spacing is d.

elements are the same except for a progressive, element-to-element time shift Δt,

$$\mathcal{I}_{on}(t) = \mathcal{I}_{o1}\big[t - (n-1)\Delta t\big], \qquad n = 1, 2, \ldots N, \tag{7.104}$$

this arrangement is called a *uniform array*. The radiated electric field is then

$$\vec{\mathcal{E}}^r(\vec{r}, t) = \frac{\mu_o(2\gamma h)}{4\pi r} \sin\theta\, \hat{\theta} \sum_{n=1}^{N} \dot{\mathcal{I}}_{o1}$$
$$\big[t - r/c - (d/2c)(N - 2n + 1)\sin\theta\cos\phi - (n-1)\Delta t\big], \tag{7.105}$$

or, in the horizontal plane ($\theta = \pi/2$),

$$\mathcal{E}^r_\theta(r, \theta = \pi/2, \phi; t)$$
$$= \frac{\mu_o(2\gamma h)}{4\pi r} \sum_{n=1}^{N} \dot{\mathcal{I}}_{o1}\big[t - r/c - (d/2c)(N - 2n + 1)\cos\phi - (n-1)\Delta t\big]. \tag{7.106}$$

For currents that are Gaussian functions (7.96), Equation (7.106) becomes, in normalized form,

$$\mathcal{E}^r_{\theta n}(\rho, \theta = \pi/2, \phi; t) = \sqrt{2}\, e^{1/2} \sum_{n=1}^{N} \big[\rho + (d/2c\tau)(N - 2n + 1)\cos\phi$$
$$+ (n-1)(\Delta t/\tau)\big] e^{[\rho + (d/2c\tau)(N-2n+1)\cos\phi + (n-1)(\Delta t/\tau)]^2}. \tag{7.107}$$

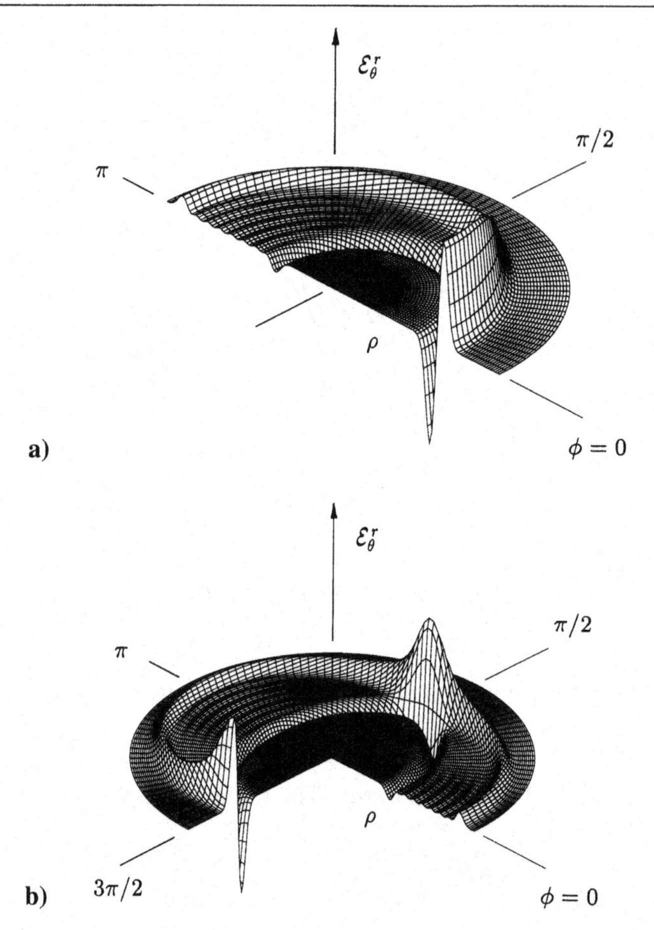

Fig. 7.26. a) Radiated electric field in the horizontal plane, $\theta = \pi/2$, for an 8-element, uniform, unilateral end-fire array with currents that are Gaussian functions of time: $d/c\tau = \Delta t/\tau = 1/\sqrt{2}$. b) Radiated electric field in the horizontal plane, $\theta = \pi/2$, for an 8-element, uniform, bilateral broadside array with currents that are Gaussian functions of time: $d/c\tau = \sqrt{2}$, $\Delta t/\tau = 0$.

Figure 7.26a shows this field for an eight-element ($N = 8$), unilateral end-fire array. The spacing and time delay are again given by (7.100) and (7.101). Figure 7.26b is a similar graph for an eight-element, bilateral broadside array, with the spacing and time delay given by (7.102) and (7.103). A comparison of these figures with those for the couplets ($N = 2$), Figures 7.23 and 7.24, shows the increase in directivity obtained by adding more elements to the array. For the eight-element end-fire array, the maximum field in the forward direction is now about eight times that in the backward direction.

The results presented above show that the ability to design a directive array depends greatly upon the temporal behavior of the radiated signals. The differentiated Gaussian functions radiated by different elements in the array could be superimposed to produce partial cancellation of the field in a particular direction. For signals that do not have the odd symmetry of the differentiated Gaussian, this would not be possible.

7.5.2 Harmonic time dependence

In most practical applications, the excitation for the array is essentially time harmonic: The current at the terminals of all elements is a cosinusoidal function of time. For the N-element, uniform array, we then have (7.104)

$$\begin{aligned}\mathcal{I}_{on}(t) &= I_o \cos\left\{\omega\left[t - (n-1)\Delta t\right]\right\} \\ &= I_o \cos\left[\omega t - (n-1)\delta\right] \\ &= \mathrm{Re}(I_{on} e^{j\omega t}), \qquad n = 1, 2, \ldots N, \end{aligned} \tag{7.108}$$

where we have introduced the element-to-element phase shift $\delta = \omega \Delta t$ and the phasor for the current

$$I_{on} = I_o e^{-j(n-1)\delta}. \tag{7.109}$$

The radiated electric field is obtained by substituting (7.108) into (7.105). For the vector phasor of the electric field, we find

$$\begin{aligned}\vec{E}^r(\vec{r}) = \frac{j\omega\mu_o(2\gamma h)I_o}{4\pi r} e^{-jk_o r} e^{-jk_o(d/2)(N-1)\sin\theta\cos\phi} \\ \sin\theta\,\hat{\theta} \sum_{n=1}^{N} e^{j(n-1)(k_o d \sin\theta\cos\phi - \delta)},\end{aligned} \tag{7.110}$$

which is a geometric sum that is easily evaluated:

$$\begin{aligned}\vec{E}^r(\vec{r}) &= \frac{j\omega\mu_o(2\gamma h)NI_o}{4\pi r} e^{-jk_o r} e^{-j(N-1)\delta/2} \\ &\quad \sin\theta \left\{ \frac{\sin[N(k_o d \sin\theta\cos\phi - \delta)/2]}{N\sin[(k_o d\sin\theta\cos\phi - \delta)/2]} \right\} \hat{\theta} \\ &= \frac{j\omega\mu_o(2\gamma h)NI_o}{4\pi r} e^{-jk_o r} e^{-j(N-1)\delta/2}\, \mathrm{EF}(\theta)\, \mathrm{AF_N}(\theta,\phi)\hat{\theta}. \end{aligned} \tag{7.111}$$

In the last line, we have followed convention and expressed the field in terms of the *element factor* EF and the *array factor* $\mathrm{AF_N}$:

$$\mathrm{EF}(\theta) = \sin\theta, \tag{7.112}$$

$$\begin{aligned}\mathrm{AF_N}(\theta,\phi) &= \frac{\sin[N(k_o d\sin\theta\cos\phi - \delta)/2]}{N\sin[(k_o d\sin\theta\cos\phi - \delta)/2]}. \\ &= \frac{\sin[N(k_o d\cos\psi - \delta)/2]}{N\sin[(k_o d\cos\psi - \delta)/2]}.\end{aligned} \tag{7.113}$$

The element factor describes the directional characteristics (pattern) of the individual isolated element, in our case the electrically short linear antenna, whereas the array factor describes the directional characteristics that result from combining the elements in the array. The fact that EF and $\mathrm{AF_N}$ are simply multiplied to obtain $\vec{E}^r$

is referred to as *pattern multiplication*.[17] Notice that the array factor (7.113) has been written as a function of the two angles θ and ϕ and also as a function of the single angle ψ:

$$\cos\psi = \sin\theta\cos\phi.$$

Here ψ is the angle $\vec{r}$ makes with the x axis in Figure 7.25; hence, the array factor is rotationally symmetric about the x axis.

The time-average power per unit solid angle radiated by the array in the direction θ, ϕ is

$$\frac{d\langle\mathcal{P}_{\text{rad}}\rangle}{d\Omega} = \text{Re}(r^2\hat{r}\cdot\vec{S}_c^r) = \frac{1}{2\zeta_o}|r\vec{E}^r|^2$$

$$= \frac{\zeta_o}{32\pi^2}(k_o 2\gamma h)^2 N^2 |I_o|^2 \left|\text{EF}(\theta)\right|^2 \left|\text{AF}_\text{N}(\theta,\phi)\right|^2, \tag{7.114}$$

and the total time-average power radiated is

$$\langle\mathcal{P}_{\text{rad}}\rangle = \int_{\phi=0}^{2\pi}\int_{\theta=0}^{\pi}\frac{d\langle\mathcal{P}_{\text{rad}}\rangle}{d\Omega}\sin\theta d\theta d\phi. \tag{7.115}$$

After substituting (7.112)–(7.114) and making a change of variables, the integrals in (7.115) can be evaluated in closed form to give (Problem 7.7) [26, 27]:

$$\langle\mathcal{P}_{\text{rad}}\rangle = \frac{\zeta_o}{4\pi}(k_o 2\gamma h)^2 N |I_o|^2 \left(\frac{1}{3} + \left(\frac{1}{Nk_o d}\right)\sum_{m=1}^{N-1}\left(\frac{N}{m} - 1\right)\cos(m\delta)\right.$$
$$\left.\times\left\{\left[1 - \frac{1}{(mk_o d)^2}\right]\sin(mk_o d) + \left(\frac{1}{mk_o d}\right)\cos(mk_o d)\right\}\right). \tag{7.116}$$

The directivity, as defined in (3.115), of the uniform array of electrically short linear antennas is easily determined from (7.114) and (7.116). At the angle ϕ in the horizontal plane ($\theta = \pi/2$), the directivity is

$$D(\theta = \pi/2, \phi) = \frac{4\pi}{\langle\mathcal{P}_{\text{rad}}\rangle}\left[\frac{d\langle\mathcal{P}_{\text{rad}}(\theta = \pi/2, \phi)\rangle}{d\Omega}\right] = \left\{\frac{\sin[N(k_o d\cos\phi - \delta)/2]}{\sin[(k_o d\cos\phi - \delta)/2]}\right\}^2$$
$$\times\left[2N\left(\frac{1}{3} + \left(\frac{1}{Nk_o d}\right)\sum_{m=1}^{N-1}\left(\frac{N}{m} - 1\right)\cos(m\delta)\right.\right.$$
$$\left.\left.\times\left\{\left[1 - \left(\frac{1}{mk_o d}\right)^2\right]\sin(mk_o d) + \left(\frac{1}{mk_o d}\right)\cos(mk_o d)\right\}\right)\right]^{-1}. \tag{7.117}$$

[17] When mutual coupling is important, the current distributions for the individual elements are different. The radiated fields (patterns) for the individual elements are then different, and we no longer have simple pattern multiplication for the array.

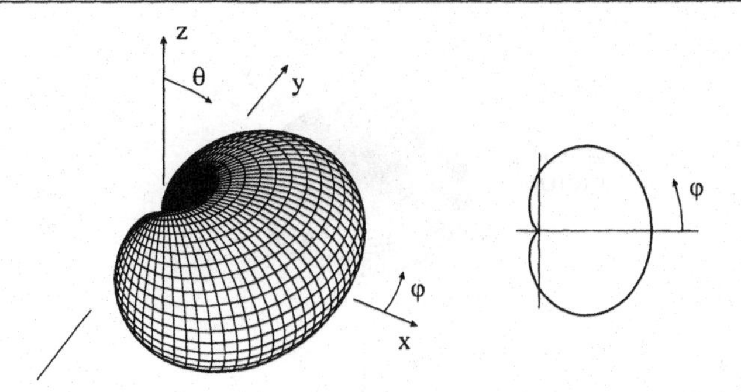

Fig. 7.27. Directional characteristics of the radiation from an ordinary end-fire couplet of electrically short linear antennas, $d\langle\mathcal{P}_{\text{rad}}\rangle/d\Omega$: $\delta = k_o d = \pi/2$.

We will first consider the directive properties of the simple couplet ($N = 2$), for which the electric field in the horizontal plane ($\theta = \pi/2$) is (7.110)

$$\vec{E}^r = \frac{j\omega\mu_o(2\gamma h)I_o}{2\pi r} e^{-jk_o r} e^{-j\delta/2} \cos\left[(k_o d\cos\phi - \delta)/2\right]\hat{\theta}. \qquad (7.118)$$

For a *unilateral end-fire array*, we want the field of the couplet (7.118) to be maximum for $\phi = 0$, so we make the argument of the cosine function zero for this angle by choosing

$$\delta = k_o d. \qquad (7.119)$$

A uniform array, of any number of elements, with this choice for δ is referred to as an *ordinary end-fire array*. If we also require the field to be zero in the backward direction ($\phi = \pi$) we must have

$$(-k_o d - \delta)/2 = -\pi/2,$$

which on substituting (7.119) requires

$$k_o d = \pi/2.$$

With these values of δ and $k_o d$, we are delaying the radiation from element 2 so that in the forward direction ($\phi = 0$) each cycle is exactly superimposed on the radiation from element 1. In the backward direction ($\phi = \pi$) the radiation from element 2 is delayed exactly one half cycle from that of element 1 ($2\delta = 2k_o d = \pi$), and thus there is total cancellation of the field.

The directive characteristics of this array are shown in Figure 7.27.[18] For the picture on the left the radial distance in the direction θ, ϕ, measured from the origin, is proportional to $d\langle\mathcal{P}_{\text{rad}}\rangle/d\Omega$ (7.114). The graph on the right shows the pattern in the horizontal plane ($\theta = \pi/2$); the null in the backward direction is evident.

The directivity (7.117) for this array in the forward direction ($\phi = 0$) is $D = 3.0$ (4.77 dB), which is twice that of a single element ($D = 1.5$). The pattern multiplication for this array is shown on the left-hand side of Figure 7.28. The

[18] These patterns and all similar ones to be presented are normalized so that the maximum value is one.

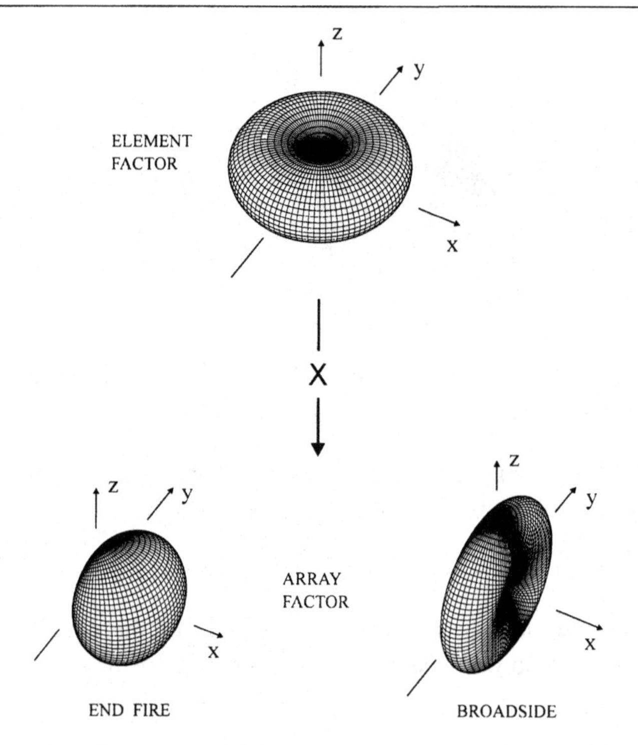

Fig. 7.28. Illustration of pattern multiplication for couplet of electrically short linear antennas. Left side is for the end-fire couplet shown in Fig. 7.27, while the right side is for the broadside couplet shown in Fig. 7.29.

element factor, which is doughnut shaped, is multiplied by the array factor, which is a beam in the direction $\phi = 0$, to obtain the result in Figure 7.27. A pattern that has a single, rotationally symmetric lobe in the direction for maximum radiation, like the one for the end-fire array factor, is referred to as a "pencil beam."

For a *bilateral broadside array*, we want the field of the couplet (7.118) to be maximum for both $\phi = \pi/2$ and $\phi = 3\pi/2$, so we make the argument of the cosine function zero for these angles by choosing

$$\delta = 0. \tag{7.120}$$

If we also require the field to be zero in the directions $\phi = 0$ and $\phi = \pi$, we must have

$$k_o d = \pi.$$

The directive characteristics of this array are shown in Figure 7.29. The directivity (7.117) is $D = 3.57$ (5.53 dB) for either $\phi = \pi/2$ or $\phi = 3\pi/2$; this is about 2.4 times that of a single element. The pattern multiplication for this array is shown on the right-hand side of Figure 7.28. Notice that the broadside array factor is not maximum for a single direction, as is the end-fire array factor. Rather, it obtains its maximum for all values of θ when $\phi = \pi/2$ or $3\pi/2$. A pattern with this characteristic is referred to as a "fan beam."

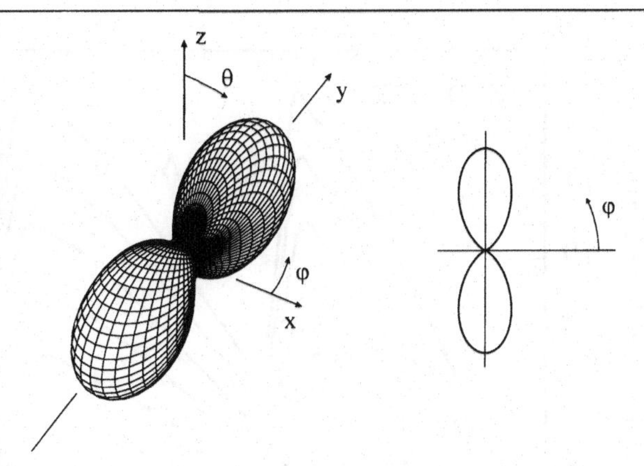

Fig. 7.29. Directional characteristics of the radiation from a broadside couplet of electrically short linear antennas, $d\langle \mathcal{P}_{\text{rad}}\rangle/d\Omega$: $\delta = 0$, $k_o d = \pi$.

The directivity for the broadside couplet ($D = 3.57$) is higher than that for the end-fire couplet ($D = 3.0$), even though the former has two directions for maximum radiation, whereas the latter has only one. The increased directivity of the broadside array is due to its narrower beamwidth, which can be clearly seen by comparing the graphs on the right of Figures 7.27 and 7.29.

Now we will consider uniform arrays that do not have the spacing between the elements ($k_o d$) adjusted to obtain a particular result (e.g., a null in the pattern for a particular direction). The directivities (7.117) for the uniform, ordinary end-fire array ($\delta = k_o d$) and the uniform broadside array ($\delta = 0$) are graphed as functions of $k_o d$ in Figure 7.30 for arrays with two to eight elements. The directivity of the end-fire arrays increases with increasing $k_o d$ until a maximum is reached for $k_o d$ slightly less than π. After this maximum, there is a decrease in the directivity with increasing $k_o d$, until a minimum occurs for $k_o d$ slightly greater than π. This decrease in the directivity is caused by the pattern for the radiation becoming bilateral end fire, with large lobes in the two directions $\phi = 0$ and $\phi = \pi$, rather than in just the one direction $\phi = 0$. With a further increase in $k_o d$, the directivity increases until a second maximum, of nearly equal value to the first, is reached when $k_o d$ is slightly less than 2π. Past this point the directivity decreases. This is caused by the pattern for the radiation again becoming bilateral and, in addition, developing large lobes in the broadside directions: $\phi = \pi/2$ and $\phi = 3\pi/2$.

The directivity for the broadside arrays increases with increasing $k_o d$ until a maximum is reached for $k_o d$ slightly less than 2π. Past this point there is a decrease in the directivity. This is the result of the pattern for the radiation developing additional large lobes in the end-fire directions: $\phi = 0$ and $\phi = \pi$.

The maximum directivities, $D_{\max}$, for these arrays along with the values of $k_o d$ at which they occur are given in Table 7.3. The half-power beamwidths in the horizontal plane ($\theta = \pi/2$) are also given in this table. Recall from Section 3.7 that the half-power beamwidth, $2\phi_{\text{BW}}$, is the angular separation between the points at which the power per unit area has dropped to one half of its maximum value.

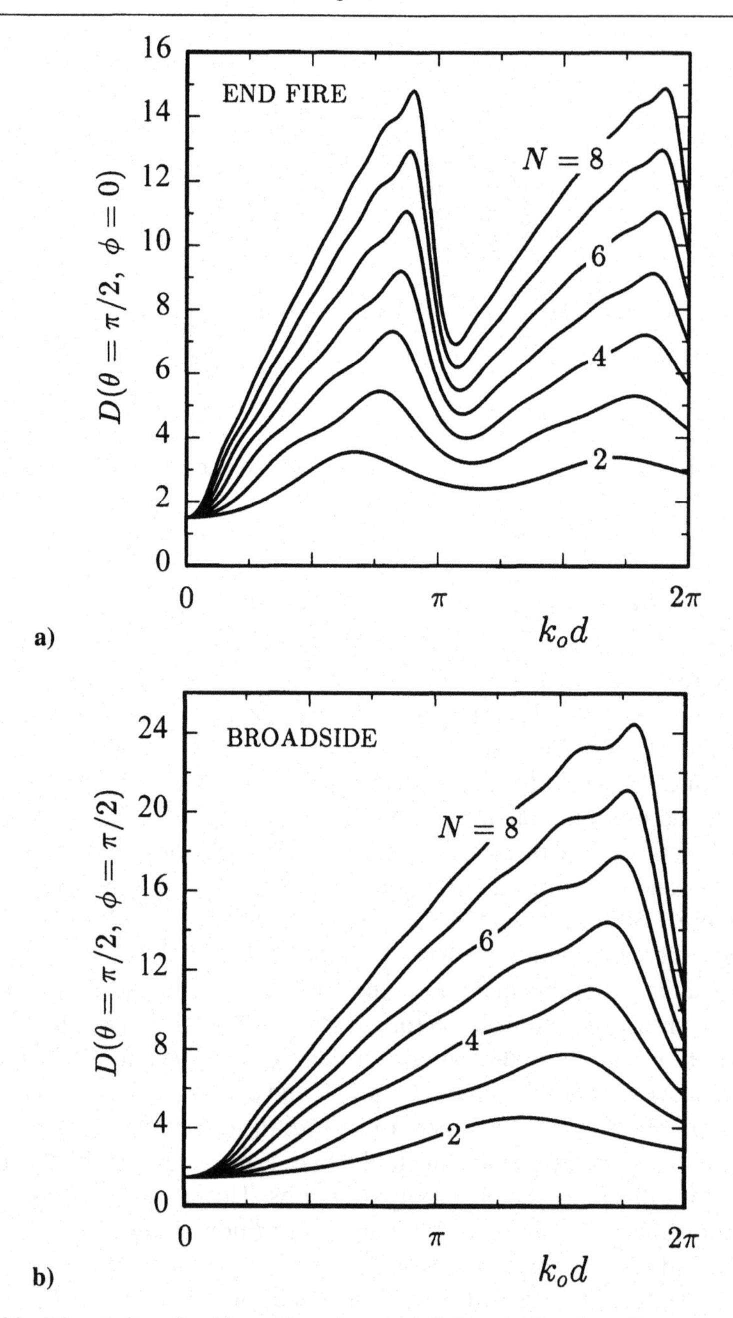

Fig. 7.30. Directivity of uniform linear arrays of electrically short linear antennas versus the spacing between the elements $k_o d$. a) Ordinary end-fire arrays; $\delta = k_o d$. b) Broadside arrays; $\delta = 0$.

Figure 7.31 shows patterns in the horizontal plane ($\theta = \pi/2$) for arrays with 2, 4, and 8 elements. For each case, the spacing $k_o d$ is that given in Table 7.3 for maximum directivity. The decrease in beamwidth, hence the increase in directivity, that accompanies an increase in the number of elements is clearly seen.

Table 7.3. *Uniform arrays of electrically short linear antennas (dipoles)*

		Ordinary end-fire array, $\delta = k_o d$		
N	$k_o d$	d/λ_o	D_{max}	$2\phi_{BW}$
2	2.11	0.336	3.55	150°
3	2.42	0.385	5.43	107°
4	2.58	0.411	7.31	87.4°
5	2.68	0.426	9.19	75.9°
6	2.75	0.437	11.1	67.8°
7	2.80	0.445	12.9	62.1°
8	2.84	0.451	14.8	57.6°
		Broadside array, $\delta = 0$		
N	$k_o d$	d/λ_o	D_{max}	$2\phi_{BW}$
2	4.23	0.674	4.51	43.6°
3	4.80	0.763	7.73	23.5°
4	5.11	0.813	11.0	16.1°
5	5.31	0.846	14.4	12.2°
6	5.46	0.868	17.7	9.88°
7	5.56	0.885	21.1	8.27°
8	5.64	0.898	24.4	7.12°

When the number of elements, N, is large, we can obtain a simple estimate for the half-power beamwidth in the horizontal plane ($\theta = \pi/2$). From (7.111), at the angle ϕ_{BW}

$$\left|\frac{\sin(Nu)}{N \sin u}\right| = \frac{1}{\sqrt{2}}, \tag{7.121}$$

where for the end-fire array

$$u = k_o d(\cos\phi_{BW} - 1)/2 \tag{7.122}$$

and for the broadside array

$$u = k_o d \cos(\pi/2 + \phi_{BW})/2. \tag{7.123}$$

We will now assume that for large N the beamwidth is small; hence $u \ll 1$.[19] The

[19] This assumption can be checked by substituting values from Table 7.3, say for $N = 8$, into (7.122) and (7.123).

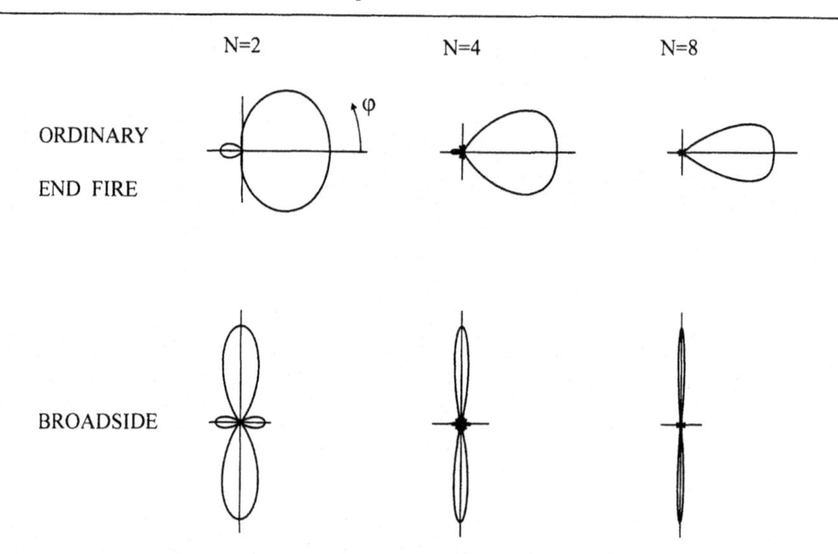

Fig. 7.31. Patterns for the radiation in the horizontal plane ($\theta = \pi/2$) of uniform linear arrays of electrically short linear antennas, $d\langle\mathcal{P}_{\text{rad}}\rangle/d\Omega$. Each pattern is for the element spacing, $k_o d$, that gives maximum directivity, as given in Table 7.3.

left-hand side of (7.121) then can be approximated as

$$\left|\frac{\sin(Nu)}{N\sin u}\right| \approx \frac{Nu - \frac{1}{6}(Nu)^3}{Nu} = 1 - \frac{1}{6}(Nu)^2,$$

therefore

$$|u| \approx \frac{1}{N}\sqrt{6(1 - 1/\sqrt{2})} = 1.33/N.$$

Substituting (7.122) and (7.123), we find the half-power beamwidths: for the end-fire array

$$\left|k_o d(\cos\phi_{\text{BW}} - 1)/2\right| \approx \frac{k_o d}{4}\phi_{\text{BW}}^2 \approx 1.33/N$$

$$2\phi_{\text{BW}} \approx \frac{4.61}{\sqrt{Nk_o d}} \text{ radians} = \frac{264}{\sqrt{Nk_o d}} \text{ degrees}, \tag{7.124}$$

and for the broadside array

$$\left|k_o d\cos(\pi/2 + \phi_{\text{BW}})/2\right| \approx \frac{k_o d}{2}\phi_{\text{BW}} \approx 1.33/N$$

$$2\phi_{\text{BW}} \approx \frac{5.30}{Nk_o d} \text{ radians} = \frac{304}{Nk_o d} \text{ degrees}. \tag{7.125}$$

Notice that beamwidth for the end-fire array is proportional to $(Nk_o d)^{-1/2}$, whereas, that for the broadside array is proportional to $(Nk_o d)^{-1}$. This explains the narrower beamwidth for the broadside array. The approximations (7.124) and (7.125) are fairly good whenever N is large; for example, for $N = 8$ the exact values from

Table 7.3 are 57.6° for end fire and 7.12° for broadside, and the approximations are 55.4° and 6.74°.

We can use the beamwidths (7.124) and (7.125) to estimate the behavior of the directivity for large N. Recall from Section 3.7 that the directivity of an electrically large, uniformly illuminated, circular aperture was (3.129b)

$$D_A(2\theta_{\mathrm{BW}})^2 \approx \pi^2.$$

Now, if we assume that a similar relationship holds for these arrays, we will have[20]

$$D \propto \frac{1}{(2\theta_{\mathrm{BW}})(2\phi_{\mathrm{BW}})}. \tag{7.126}$$

For the end-fire array, the pattern is approximately symmetric about the x axis; hence, $\theta_{\mathrm{BW}} \approx \phi_{\mathrm{BW}}$, and from (7.124) and (7.126)

$$D \propto \frac{1}{(2\phi_{\mathrm{BW}})^2} \propto Nk_o d.$$

For the broadside array, the half-power beamwidth in the vertical plane ($\phi = \pi/2$, $3\pi/2$), i.e. $2\theta_{\mathrm{BW}}$, is determined by the element factor; hence it is independent of N and $k_o d$ (for the electrically short linear antenna, $\sin^2\theta_{\mathrm{BW}} = 1/2$, $2\theta_{\mathrm{BW}} = \pi/2$). From (7.125) and (7.126), the directivity is then

$$D \propto \frac{1}{2\phi_{\mathrm{BW}}} \propto Nk_o d.$$

For both the uniform ordinary end-fire and broadside arrays, the directivity is seen to be proportional to the product of the number of elements N and the electrical spacing between the elements $k_o d = 2\pi(d/\lambda_o)$. This behavior is evident in Figure 7.30. For large N, say $N = 8$, the directivities for both arrays are seen to be approximately linear functions of $k_o d$, provided $k_o d$ is less than the value for D_{max}.

Let us now return to our discussion of the simple couplet ($N = 2$). From Figure 7.30 and Table 7.3, we see that the first maximum in the directivity is obtained for end fire when $k_o d = 2.11$ ($D_{\mathrm{max}} = 3.55$) and for broadside when $k_o d = 4.23$ ($D_{\mathrm{max}} = 4.51$). The patterns for the radiation from these arrays are shown in Figures 7.32 and 7.33, respectively. These couplets are more directive than the ones we discussed earlier (Figures 7.27 and 7.29). However, the nulls that were present in the latter (at $\phi = \pi$ for end fire, and at $\phi = 0, \pi$ for broadside) are not present in the former. We have chosen the spacing of the elements, $k_o d$, to produce maximum directivity rather than to obtain the specified nulls.

For the end fire couplet ($N = 2$), the second maximum in the directivity occurs at $k_o d \approx 5.4$. The pattern for this array is shown in Figure 7.34. Notice that in

[20] Here we have recognized that the beam for the electrically large, circular aperture is nearly symmetrical; thus, $(2\theta_{BW})^2$ is the product of the half-power beamwidths in the two orthogonal planes.

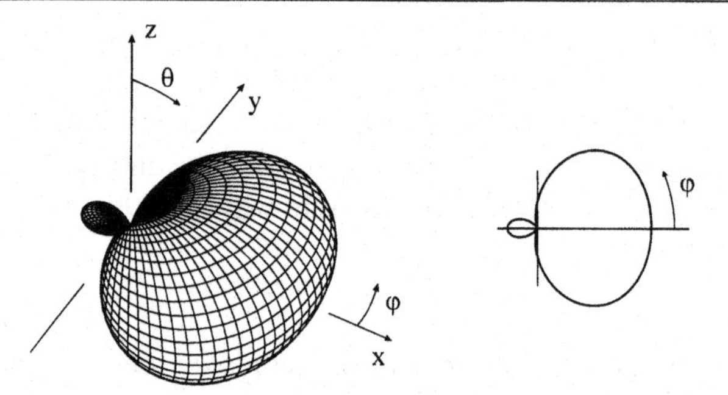

Fig. 7.32. Directional characteristics of the radiation from an ordinary end-fire couplet of electrically short linear antennas, $d\langle\mathcal{P}_{\text{rad}}\rangle/d\Omega$. The spacing has been adjusted for the first maximum in directivity: $\delta = k_o d = 2.11$.

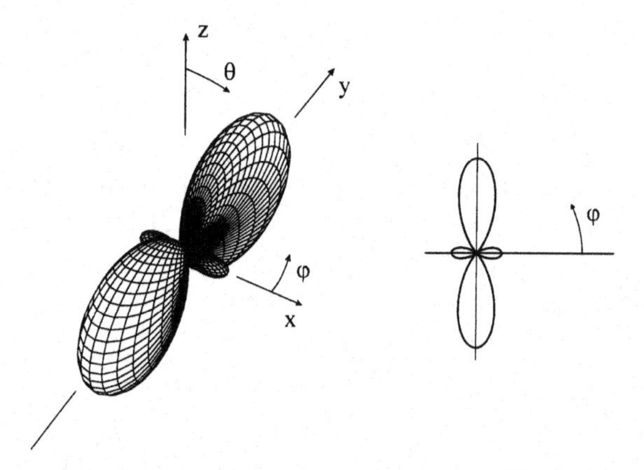

Fig. 7.33. Directional characteristics of the radiation from a broadside couplet of electrically short linear antennas, $d\langle\mathcal{P}_{\text{rad}}\rangle/d\Omega$. The spacing has been adjusted for the first maximum in directivity: $\delta = 0$, $k_o d = 4.23$.

addition to the main beam at $\phi = 0$, there are two additional beams, side lobes, whose magnitudes are nearly equal to that of the main beam. In many applications, the presence of these side lobes is undesirable; hence, the larger spacing between the elements, associated with the second maximum in the directivity, is often avoided.

To complete our discussion of arrays, we will briefly examine a phenomenon known as *superdirectivity*. Again, we will use the simple couplet as an example. So far, we have constrained one parameter, δ, when determining the maximum directivity for the arrays: For ordinary end fire $\delta = k_o d$, and for broadside $\delta = 0$. When we remove this restriction, we can view the directivity as a function of the two parameters, $k_o d$ and δ, as in Figure 7.35. Here contours of constant directivity are shown on the $k_o d$-δ plane for the end-fire direction, $D(\theta = \pi/2,\ \phi = 0)$, and for the broadside direction, $D(\theta = \pi/2, \phi = \pi/2)$.

The dashed line in Figure 7.35a is the locus of points for the ordinary end-fire couplet, $\delta = k_o d$. The directivities for the couplets we have discussed so far are

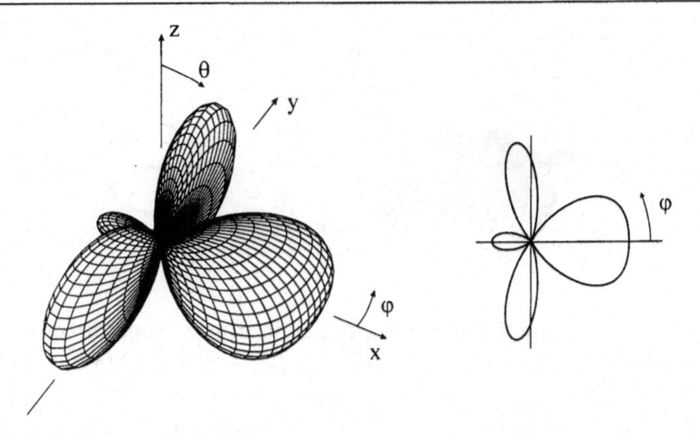

Fig. 7.34. Directional characteristics of the radiation from an ordinary end-fire couplet of electrically short linear antennas, $d\langle \mathcal{P}_{\text{rad}} \rangle / d\Omega$. The spacing has been adjusted for the second maximum in directivity: $\delta = k_o d \approx 5.4$.

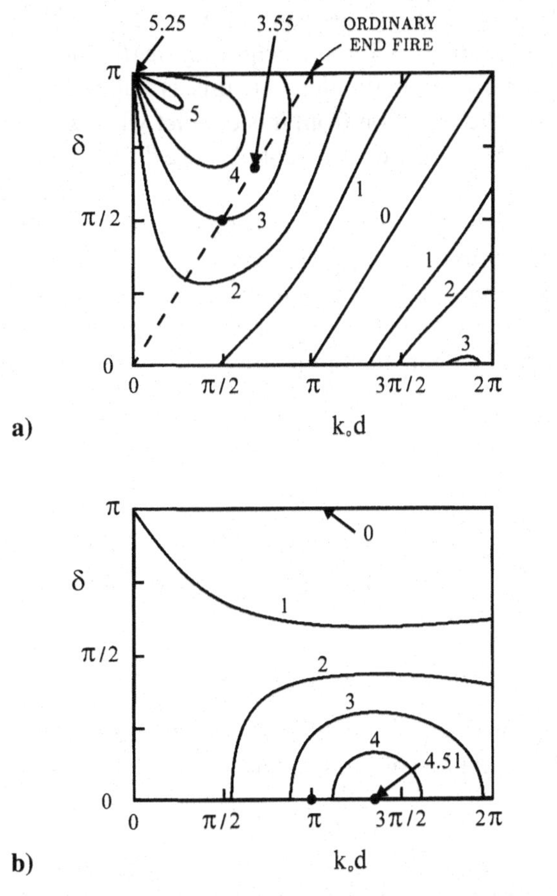

Fig. 7.35. Directivity for couplet of electrically short linear antennas as a function of the element spacing, $k_o d$, and the phase shift between elements, δ. a) End-fire directivity, $D(\theta = \pi/2, \phi = 0)$. b) Broadside directivity, $D(\theta = \pi/2, \phi = \pi/2)$.

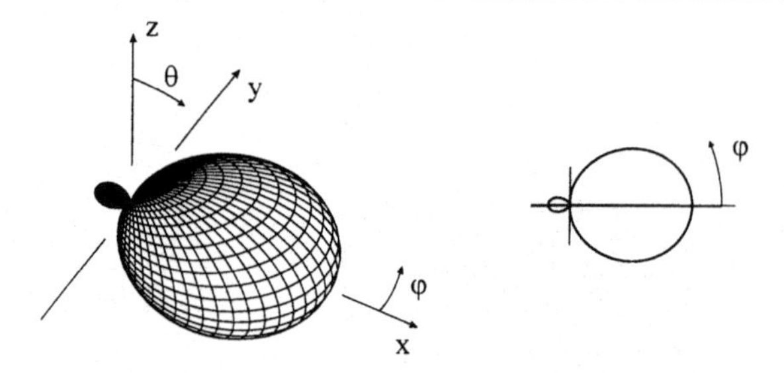

Fig. 7.36. Directional characteristics of the radiation from a "superdirective" end-fire couplet of electrically short linear antennas, $d\langle\mathcal{P}_{\text{rad}}\rangle/d\Omega$: $\delta = \pi - (2/5)k_o d$, $k_o d = 0.1$.

indicated by solid dots ($k_o d = \pi/2$, $D = 3.0$, Figure 7.27; and $k_o d = 2.11$, $D = 3.55$, Figure 7.32). Notice that substantially higher directivity is obtained when $k_o d$ approaches zero while δ approaches π. In fact, if we take $\delta = \pi - (2/5)k_o d$, in the limit as $k_o d \to 0$, the directivity is $D = 5.25$ (7.20 dB) (Problem 7.8). Since the two elements of the couplet essentially merge in this limit, we should compare its directivity with that of a single element, for which $D = 1.5$. We have increased the directivity by a factor of 3.5 while requiring negligible additional space for the couplet over that for a single element; this is superdirectivity. Figure 7.36 shows the pattern for the radiation from this couplet when $k_o d = 0.1$. It is clearly more directive than the ordinary end-fire couplets we discussed earlier (Figures 7.27 and 7.32).

Although superdirectivity is an interesting phenomenon, it is often impractical. To see why this is true, we first note that the directivity is only a measure of the relative distribution of radiation in space. It says nothing about the actual magnitude of the radiation; for this we must consider the gain of the couplet. Recall from Section 3.7 (3.138) that the gain of an antenna or an array is the product of the efficiency and the directivity:

$$G(\theta, \phi) = \eta D(\theta, \phi), \tag{7.127}$$

where

$$\eta = \frac{\text{time-average power radiated by antenna}}{\text{time-average power supplied to antenna}}$$

$$= \frac{\langle\mathcal{P}_{\text{rad}}\rangle}{\langle\mathcal{P}_{in}\rangle} = \frac{\langle\mathcal{P}_{\text{rad}}\rangle}{\langle\mathcal{P}_{\text{rad}}\rangle + \langle\mathcal{P}_{\text{ohm}}\rangle}. \tag{7.128}$$

For the couplet, the power lost due to the finite conductivity of the elements $\langle\mathcal{P}_{\text{ohm}}\rangle$ is just twice that for a single isolated element (7.58); that is, for the couplet[21]

$$\langle\mathcal{P}_{\text{ohm}}\rangle = |I_o|^2 (R_{\text{ohm}})_1, \tag{7.129}$$

where $(R_{\text{ohm}})_1$ is the ohmic resistance of the single, isolated element. On substituting (7.116), (7.117), (7.128), and (7.129) into (7.127), we obtain the gain of the end-fire

[21] This is a consequence of assuming the same distribution of current for the elements when isolated or combined in the couplet and exciting the elements of the couplet so that $|I_{o1}| = |I_{o2}| = |I_o|$.

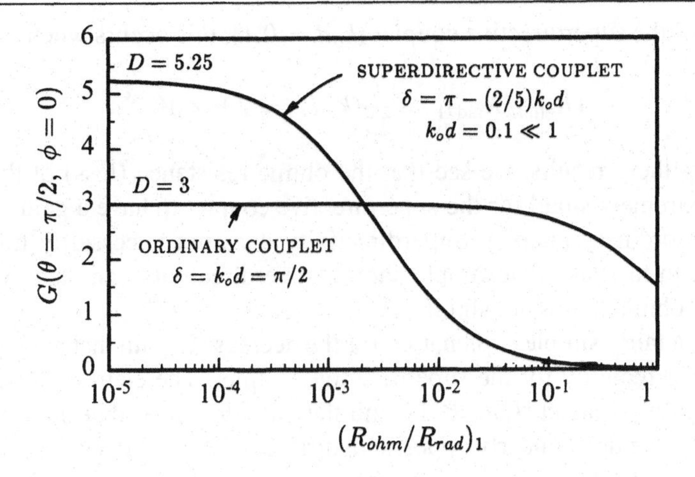

Fig. 7.37. Gain of end-fire couplets versus the parameter $(R_{\mathrm{ohm}}/R_{\mathrm{rad}})_1$.

couplet:

$$G(\theta=\pi/2,\ \phi=0)=\frac{3}{4}\left|\frac{\sin(k_od-\delta)}{\sin[(k_od-\delta)/2]}\right|^2\left(1+\frac{3}{2}\frac{1}{k_od}\cos\delta\left\{\left[1-\frac{1}{(k_od)^2}\right]\right.\right.$$

$$\left.\left.\times\sin(k_od)+\frac{1}{k_od}\cos(k_od)\right\}+(R_{\mathrm{ohm}}/R_{\mathrm{rad}})_1\right)^{-1},\quad (7.130)$$

where the parameter

$$(R_{\mathrm{ohm}}/R_{\mathrm{rad}})_1$$

is the ratio of the ohmic resistance to the radiation resistance (7.59) of a single, isolated element. For the ordinary end-fire couplet with $\delta=k_od=\pi/2$, Equation (7.130) reduces to

$$G(\theta=\pi/2,\ \phi=0)=\frac{3}{1+(R_{\mathrm{ohm}}/R_{\mathrm{rad}})_1},\quad (7.131)$$

whereas for the superdirective end-fire couplet with $\delta=\pi-(2/5)k_od,\ k_od\ll 1$, it reduces to

$$G(\theta=\pi/2,\ \phi=0)=\frac{5.25}{1+\left(\frac{25}{7}\right)\frac{1}{(k_od)^2}(R_{\mathrm{ohm}}/R_{\mathrm{rad}})_1}.\quad (7.132)$$

To illustrate these results, we have graphed the gain as a function of the parameter $(R_{\mathrm{ohm}}/R_{\mathrm{rad}})_1$ in Figure 7.37. For the superdirective couplet, we chose $k_od=0.1$. When the ohmic resistance is extremely small (left side of graph), the gain for both couplets is nearly equal to the directivity (the efficiency η is nearly 100%). For the ordinary couplet, the gain drops to one half the directivity ($\eta=50\%$) when

$$(R_{\mathrm{ohm}}/R_{\mathrm{rad}})_1=1,$$

whereas for the superdirective couplet ($k_o d = 0.1$), this occurs when

$$(R_{\text{ohm}}/R_{\text{rad}})_1 = \frac{7}{25}(k_o d)^2 = 2.8 \times 10^{-3}.$$

Comparing these results, we see that the ohmic resistance (loss) of the elements must be extremely small for the superdirective couplet to have a gain greater than that of the ordinary couplet. Superdirectivity does not necessarily translate into supergain. In fact, as this example shows, superdirectivity can result in very low gain when ohmic losses are significant.

There is a fairly simple explanation for the decrease in gain that accompanies the increase in directivity for the superdirective couplet. The elements of the couplet are very closely spaced ($k_o d \ll 1$), and they are driven so that their currents are equal in magnitude but nearly opposite in direction ($\delta \approx \pi$, as seen in Figure 7.35). Hence, the fields of the elements nearly cancel, and very large currents (large $|I_o|$) are required to produce significant radiation. The large currents produce large ohmic losses in the elements (7.129) and, consequently, low efficiency and gain for the couplet.

Notice from Figure 7.35b that there is no superdirective arrangement for the broadside couplet. Maximum directivity is for the case $\delta = 0$, $k_o d = 4.23$ shown in Figure 7.33.

Additional discussions of superdirectivity and supergain can be found in References [28–33]. We only mention that a linear array of finite electrical length can have unlimited directivity, provided we increase the number of elements in the array indefinitely and precisely adjust the phases of their currents. Of course, as we have seen, the array will not have unlimited gain if there are ohmic losses in the elements.

7.6 Scattering by electrically small objects

Earlier, in Section 4.5, we introduced a few of the basic concepts associated with the scattering of electromagnetic waves, such as the backscattering and total scattering cross sections. At that time we only considered scattering by electrically large objects, ones whose characteristic length d, for example the diameter of a disc or sphere, satisfies the inequality $k_o d = 2\pi d/\lambda_o \gg 1$, where λ_o is the free space wavelength for the time-harmonic field. With this restriction on the size of the object, it was possible to infer certain general characteristics for the scattering.

Now we will consider scattering in the opposite limit, that is, for electrically small objects, ones whose maximum length d satisfies the inequality $k_o d = 2\pi d/\lambda_o \ll 1$. With this restriction on the size of the object, it will again be possible to infer certain general characteristics for the scattering. We will confine our discussion to time-harmonic fields.

Figure 7.38 illustrates the basic ideas we will be using. Here, a perfectly conducting sphere of diameter d is illuminated by the incident plane wave

$$\vec{E}^i = \vec{E}^i_o e^{-jk_o r\hat{k}_i \cdot \hat{r}}, \qquad \vec{B}^i = \vec{B}^i_o e^{-jk_o r\hat{k}_i \cdot \hat{r}}. \tag{7.133}$$

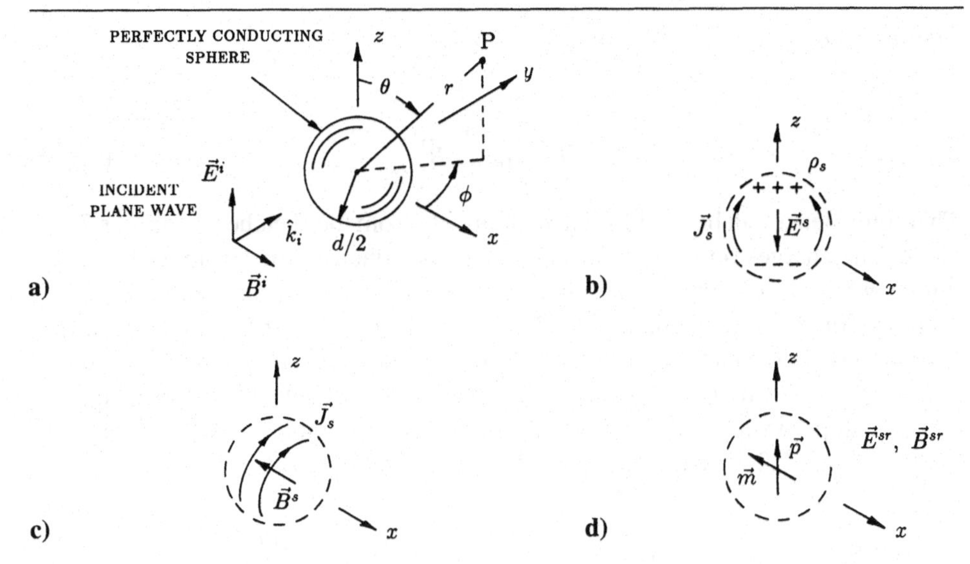

Fig. 7.38. Sketches illustrating the scattering from an electrically small, perfectly conducting sphere.

To simplify the pictorial representation, in Figure 7.38a the plane wave is taken to be linearly polarized with $\hat{k}_i = \hat{y}$ and $\vec{E}^i_o = E^i_o\hat{z}$. Since the sphere is assumed to be electrically small ($k_o d \ll 1$), the incident field is approximately uniform throughout its volume:

$$\vec{E}^i \approx \vec{E}^i_o, \qquad \vec{B}^i \approx \vec{B}^i_o = \frac{1}{c}\hat{k}_i \times \vec{E}^i_o, \qquad r \leq d/2. \tag{7.134}$$

The incident field produces charge and current on the surface of the perfectly conducting sphere. We replace the sphere by the surface densities of charge and current, ρ_s and $\vec{J}_s$, to calculate the scattered electromagnetic field, $\vec{E}^s$, $\vec{B}^s$. These surface densities are sketched in Figures 7.38b and c. The vertical currents shown in Figure 7.38b periodically deposit charge at the top and bottom of the sphere. The sketch is for a time when positive charge is on the top and negative charge is on the bottom. In addition, the closed currents shown in Figure 7.38c circulate around the sphere in the clockwise sense with respect to $\hat{x}$. Within the volume of the sphere, the total field, $\vec{E}^t$, $\vec{B}^t$, must be zero, so the scattered field, which is due to the charge and current, must cancel the incident field:

$$\vec{E}^t = \vec{E}^i + \vec{E}^s = 0, \qquad \vec{B}^t = \vec{B}^i + \vec{B}^s = 0, \qquad r \leq d/2,$$

or

$$\vec{E}^s = -\vec{E}^i, \qquad \vec{B}^s = -\vec{B}^i, \qquad r \leq d/2. \tag{7.135}$$

The directions shown in Figures 7.38b and c for the scattered electric and magnetic fields are seen to be correct for this cancellation.

For the purpose of calculating the scattered electromagnetic field at a large distance from the sphere, $r \gg d/2$, the charge and current on the surface of the sphere are approximately equivalent to the electric and magnetic dipoles shown in

Figure 7.38d:

$$\vec{p} = \alpha_e \varepsilon_o \vec{E}^i, \tag{7.136}$$

$$\vec{m} = \alpha_m \frac{1}{\mu_o} \vec{B}^i. \tag{7.137}$$

This equivalence can be established qualitatively from the sketches in Figures 7.38b and c. The charges on the top and bottom of the sphere form a z-directed electric dipole, while the current circulating around the sphere forms an electrically small loop or x-directed magnetic dipole. The constants of proportionality in the above expressions are the *electric polarizability* α_e and the *magnetic polarizability* α_m.[22]

With the equivalent dipoles (7.136) and (7.137) known, the scattered field is simply a superposition of the fields of the individual dipoles, which are given in Tables 7.1 and 7.2. The radiated part of the scattered field is

$$\vec{E}^{sr}(\vec{r}) = \frac{-k_o^2}{4\pi\varepsilon_o r} e^{-jk_o r} \left[\hat{r} \times (\hat{r} \times \vec{p}) + \frac{1}{c} \hat{r} \times \vec{m} \right], \tag{7.138}$$

$$\vec{B}^{sr}(\vec{r}) = \frac{-k_o^2 \mu_o}{4\pi r} e^{-jk_o r} \left[-c\hat{r} \times \vec{p} + \hat{r} \times (\hat{r} \times \vec{m}) \right], \tag{7.139}$$

with the complex Poynting vector

$$\vec{S}_c^{sr}(\vec{r}) = \frac{1}{2\mu_o} \vec{E}^{sr}(\vec{r}) \times \left[\vec{B}^{sr}(\vec{r}) \right]^*$$

$$= \frac{ck_o^4}{32\pi^2 \varepsilon_o r^2} \left[\left| \hat{r} \times (\hat{r} \times \vec{p}) \right|^2 + \frac{1}{c^2} |\hat{r} \times \vec{m}|^2 + \frac{2}{c} \hat{r} \cdot (\vec{p} \times \vec{m}) \right] \hat{r} \tag{7.140}$$

and the time-average power scattered per unit solid angle given by

$$\frac{d\langle \mathcal{P}^s \rangle}{d\Omega} = \mathrm{Re}[r^2 \hat{r} \cdot \vec{S}_c^{sr}]. \tag{7.141}$$

Other quantities describing the scattering, such as the cross sections, are easily determined from these results.[23]

From the above discussion, the problem of scattering from an electrically small object is seen to reduce to one of determining the polarizabilities α_e and α_m. Because the object is electrically small, the time-harmonic incident field is practically uniform throughout its volume (7.134), and the field varies little during the time for light to cross the object (for the sphere, $d/c \ll T$, where $T = 2\pi/\omega$ is the period of the oscillation). At any instant the response of the object to this dynamic field is then very similar to that for a uniform static field, and the polarizabilities for the dynamic field are essentially the same as those for the static field. Therefore, as a first approximation, we can use the polarizabilities for the static field in the procedure we have outlined above.

[22] The SI units for both α_e and α_m are m^3.

[23] The physical argument we have presented (replacing the sphere by equivalent dipoles, etc.) is somewhat heuristic; mathematical justification for these steps can be found in Reference [34].

For the perfectly conducting sphere, the calculation of the polarizabilities for the static field is particularly simple. This is because everywhere outside the sphere, the electrostatic field of the charge on the sphere is exactly equal to the field of an electric dipole. Similarly, everywhere outside the sphere, the magnetostatic field of the current on the sphere is exactly equal to the field of a magnetic dipole.

First, we will determine the electric polarizability. The sphere of diameter d is placed in the uniform "electrostatic" incident field $\vec{E}^i$. It is to be represented by an equivalent electric dipole, one that produces the same electric field as the sphere for $r \geq d/2$. For this equivalence, the boundary condition at the surface of the perfectly conducting sphere must be satisfied by the field of the dipole; that is, the tangential component of the total electric field must be zero (Table 1.6):

$$\hat{r} \times \vec{E}^t = \hat{r} \times (\vec{E}^i + \vec{E}^s) = 0,$$

or

$$\hat{r} \times \vec{E}^s = -\hat{r} \times \vec{E}^i, \qquad r = d/2. \tag{7.142}$$

In this equation, $\vec{E}^s$ is the "electrostatic" field of the dipole. It is given by the expression in Table 7.1 in the limit as $\omega \to 0$ or $k_o = \omega/c \to 0$ (the term proportional to $1/r^3$). After substituting this field into the boundary condition (7.142), we obtain

$$\frac{-2}{\pi \varepsilon_o d^3} \hat{r} \times \vec{p} = -\hat{r} \times \vec{E}^i. \tag{7.143}$$

Since the directions of the vectors $\vec{p}$ and $\vec{E}^i$ are fixed, and the unit vector $\hat{r}$ can point in any direction, this result implies that

$$\vec{p} = \frac{\pi d^3}{2} \varepsilon_o \vec{E}^i \tag{7.144}$$

or that the electric polarizability is

$$\alpha_e = \frac{\pi d^3}{2}. \tag{7.145}$$

The magnetic polarizability can be determined in a similar manner. The sphere is placed in the uniform "magnetostatic" field $\vec{B}^i$, and the equivalent magnetic dipole is to produce the same magnetic field as the sphere for $r \geq d/2$. For this equivalence, the boundary condition at the surface of the perfectly conducting sphere must be satisfied by the field of the dipole; that is, the normal component of the total magnetic field must be zero (Table 1.6):

$$\hat{r} \cdot \vec{B}^t = \hat{r} \cdot (\vec{B}^i + \vec{B}^s) = 0,$$

or

$$\hat{r} \cdot \vec{B}^s = -\hat{r} \cdot \vec{B}^i, \qquad r = d/2. \tag{7.146}$$

In this equation, $\vec{B}^s$ is the "magnetostatic" field of the dipole. It is given by the expression in Table 7.2 in the limit as $k_o \to 0$ (the term proportional to $1/r^3$). After substituting this field into the boundary condition (7.146), we obtain

$$\frac{4\mu_o}{\pi d^3}\hat{r}\cdot\vec{m} = -\hat{r}\cdot\vec{B}^i, \tag{7.147}$$

or

$$\vec{m} = -\frac{\pi d^3}{4}\frac{1}{\mu_o}\vec{B}^i. \tag{7.148}$$

Thus, the magnetic polarizability is

$$\alpha_m = -\frac{\pi d^3}{4}. \tag{7.149}$$

When these values for the dipoles are substituted into (7.138)–(7.140) and the configuration shown in Figure 7.38a is assumed (linearly polarized incident field and spherical coordinates), the scattered electromagnetic field and Poynting vector become

$$\vec{E}^{sr}(\vec{r}) = \frac{-d(k_o d)^2 E_o^i}{8r}e^{-jk_o r}\left[\left(\sin\theta - \frac{1}{2}\sin\phi\right)\hat{\theta} - \frac{1}{2}\cos\theta\cos\phi\,\hat{\phi}\right], \tag{7.150}$$

$$\vec{B}^{sr}(\vec{r}) = \frac{-d(k_o d)^2 E_o^i}{8cr}e^{-jk_o r}\left[\frac{1}{2}\cos\theta\cos\phi\,\hat{\theta} + \left(\sin\theta - \frac{1}{2}\sin\phi\right)\hat{\phi}\right], \tag{7.151}$$

and

$$\vec{S}_c^{sr}(\vec{r}) = \frac{d^2(k_o d)^4}{128\zeta_o r^2}|E_o^i|^2\left[\sin^2\theta - \sin\theta\sin\phi + \frac{1}{4}(1 - \sin^2\theta\cos^2\phi)\right]\hat{r}. \tag{7.152}$$

The backscattering and total scattering cross sections, (4.49) and (4.50), are

$$\sigma_B = \frac{4\pi r^2\hat{r}\cdot\vec{S}_c^{sr}(\theta = \pi/2,\ \phi = 3\pi/2)}{\hat{k}_i\cdot\vec{S}_c^i}$$

$$= \frac{9\pi d^2}{64}(k_o d)^4 = \frac{9\pi^5 d^2}{4}(d/\lambda_o)^4 \tag{7.153}$$

and

$$\sigma_T = \frac{\displaystyle\int_{\phi'=0}^{2\pi}\int_{\theta'=0}^{\pi}\hat{r}\cdot\vec{S}_c^{sr}(\theta',\ \phi')r^2\sin\theta' d\theta' d\phi'}{\hat{k}_i\cdot\vec{S}_c^i}$$

$$= \frac{5\pi d^2}{96}(k_o d)^4 = \frac{5\pi^5 d^2}{6}(d/\lambda_o)^4. \tag{7.154}$$

The directional characteristics for the scattered radiation are shown in Figure 7.39, where the time-average power per unit solid angle scattered by the sphere (7.141) is graphed as a function of the angles θ and ϕ. Half of the figure has been removed to show the pattern on the plane $\phi = \pi/2, 3\pi/2$. We see that the electrically small, perfectly conducting sphere scatters energy primarily in the backward

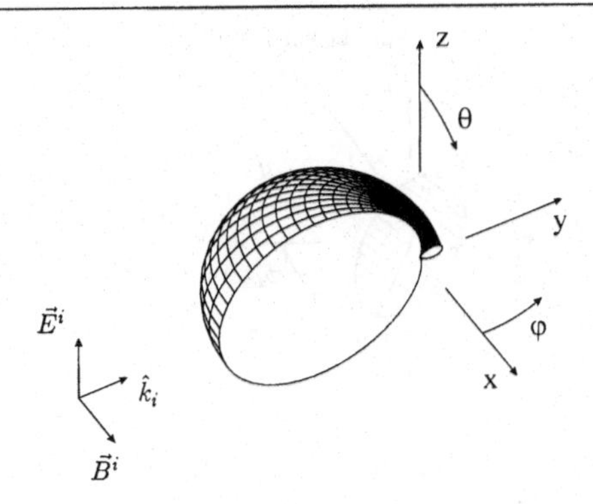

Fig. 7.39. Directional characteristics of the scattered radiation from an electrically small, perfectly conducting sphere, $d\langle \mathcal{P}^s \rangle / d\Omega$.

direction ($\theta = \pi/2, \phi = 3\pi/2$). This is because the fields of the equivalent electric and magnetic dipoles add constructively in the backward direction while adding destructively in the forward direction. These results for the electrically small sphere are to be compared with those for the electrically large sphere, which we briefly discussed in Section 4.5. The electrically large sphere scatters energy primarily in the forward direction (Figure 4.17b).[24]

The scattered electric and magnetic fields are seen to increase as the square of the frequency (ω^2) or as the inverse square of the wavelength ($1/\lambda_o^2$); thus, the scattered power and the cross sections increase as the fourth power of the frequency (ω^4) or as the inverse fourth power of the wavelength ($1/\lambda_o^4$). This behavior is characteristic of the scattering from most electrically small objects.[25] The objects need not be perfect conductors, but they can be dielectric ($\varepsilon \neq \varepsilon_o$) or magnetic ($\mu \neq \mu_o$) materials. The scattering from electrically small objects, in particular the dependence with frequency noted above, is often referred to as *Rayleigh scattering* after Lord Rayleigh (John William Strutt, 1842–1919), who studied this subject in detail [35–38]. We will say more about Lord Rayleigh's contributions in the next section.

As a second example, we will consider the scattering from an electrically small, perfectly conducting, circular disc of diameter d. The incident field will again be the plane wave (7.133). Through the use of Babinet's principle (Section 4.5) the results for the disc will also be used to determine the transmission through an electrically small, circular aperture in a plane, perfectly conducting screen.

To simplify the pictorial representation, in Figure 7.40 the disc is assumed to lie in the x-z plane, and the incident plane wave is taken to be linearly polarized, with the vector wave number in the x-y plane,

$$\hat{k}_i = \cos\phi_i \hat{x} + \sin\phi_i \hat{y}, \tag{7.155}$$

[24] Notice that the orientation of the incident field in the rectangular Cartesian coordinate system (x, y, z) differs from that used in Section 4.5.

[25] An exception occurs when both the equivalent electric and magnetic dipoles for the object are zero, in which case the scattering cross sections depend upon a higher power of the frequency.

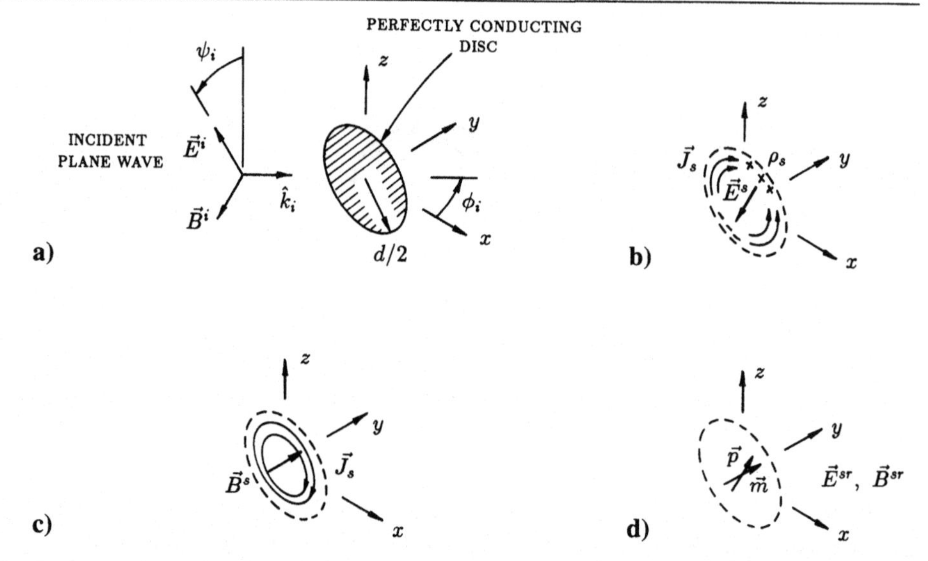

Fig. 7.40. Sketches illustrating the scattering from an electrically small, perfectly conducting, circular disc. The disc lies in the x-z plane, and the vector $\hat{k}_i$ is in the x-y plane.

and the incident electric field at the angle ψ_i to the z axis:

$$\vec{E}^i = E_o^i e^{-jk_o(x\cos\phi_i + y\sin\phi_i)}(\sin\psi_i \sin\phi_i \hat{x} - \sin\psi_i \cos\phi_i \hat{y} + \cos\psi_i \hat{z}), \tag{7.156}$$

$$\vec{B}^i = \frac{1}{c} E_o^i e^{-jk_o(x\cos\phi_i + y\sin\phi_i)}(\cos\psi_i \sin\phi_i \hat{x} - \cos\psi_i \cos\phi_i \hat{y} - \sin\psi_i \hat{z}). \tag{7.157}$$

The surface densities of charge and current for the disc are sketched in Figures 7.40b and c, and the equivalent electric and magnetic dipoles for calculating the scattered field are shown in Figure 7.40d. Analysis, which we will not reproduce here, shows that the equivalent electric dipole is proportional to the component of the incident electric field that is in the plane of the disc, whereas the equivalent magnetic dipole is proportional to the component of the incident magnetic field that is normal to the plane of the disc [39–41]:

$$\vec{p} = \alpha_e \varepsilon_o\big[\vec{E}^i - (\hat{n}\cdot\vec{E}^i)\hat{n}\big] = -\alpha_e \varepsilon_o\big[\hat{n}\times(\hat{n}\times\vec{E}^i)\big], \tag{7.158}$$

$$\vec{m} = \alpha_m \frac{1}{\mu_o}(\hat{n}\cdot\vec{B}^i)\hat{n}, \tag{7.159}$$

with the polarizabilities

$$\alpha_e = \frac{2d^3}{3} \tag{7.160}$$

and

$$\alpha_m = -\frac{d^3}{3}. \tag{7.161}$$

Here $\hat{n}$ is the unit vector normal to the plane of the disc; $\hat{n} = \hat{y}$ in Figure 7.40.

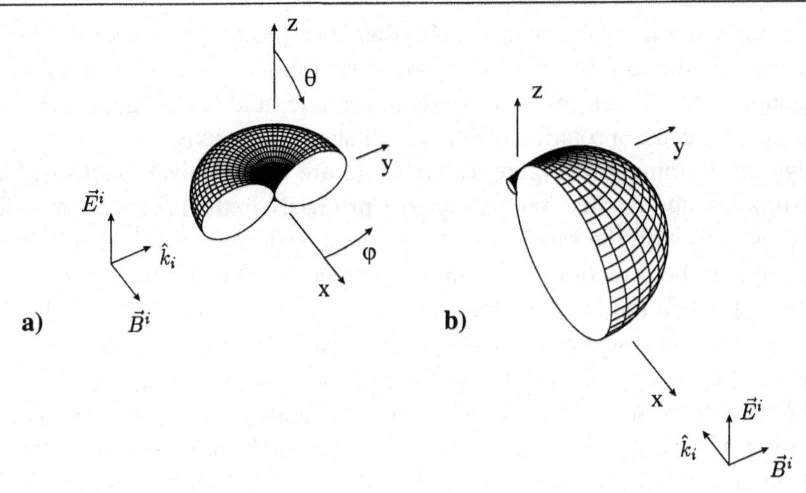

Fig. 7.41. Directional characteristics of the scattered radiation from an electrically small, perfectly conducting, circular disc, $d\langle\mathcal{P}^s\rangle/d\Omega$. a) Broadside illumination: $\phi_i = \pi/2$, $\psi_i = 0$. b) Edge-on illumination: $\phi_i = \pi$, $\psi_i = 0$.

When these values for the dipoles are substituted into the expressions for the complex Poynting vector and the scattering cross sections, (7.140), (4.49), and (4.50), and the configuration in Figure 7.40a is assumed (linearly polarized incident field and spherical coordinates) the following results are obtained:

$$\begin{aligned}\vec{S}_c^{sr}(\vec{r}) = \frac{d^2(k_o d)^4}{72\pi^2\zeta_o r^2}|E_o^i|^2\Big[&\sin^2\psi_i\sin^2\phi_i(1-\sin^2\theta\cos^2\phi)+\cos^2\psi_i\sin^2\theta\\ &-2\sin\psi_i\cos\psi_i\sin\phi_i\sin\theta\cos\theta\cos\phi+\sin\psi_i\cos\psi_i\sin\phi_i\cos\phi_i\cos\theta\\ &-\cos^2\psi_i\cos\phi_i\sin\theta\cos\phi+\frac{1}{4}\cos^2\psi_i\cos^2\phi_i(1-\sin^2\theta\sin^2\phi)\Big]\hat{r}\end{aligned} \tag{7.162}$$

and

$$\sigma_B(\phi_i,\psi_i) = \frac{d^2(k_o d)^4}{9\pi}\left[\sin^2\psi_i\sin^4\phi_i+\cos^2\psi_i\left(1+\frac{1}{2}\cos^2\phi_i\right)^2\right], \tag{7.163}$$

$$\sigma_T(\phi_i,\psi_i) = \frac{2d^2(k_o d)^4}{27\pi}\left(\sin^2\phi_i+\frac{5}{4}\cos^2\psi_i\cos^2\phi_i\right). \tag{7.164}$$

The time-average power scattered per unit solid angle (7.141) is now a function of the orientation of the incident wave, the angles ϕ_i and ψ_i. In Figure 7.41, the directional characteristics of the scattered radiation are shown for two cases: broadside illumination of the disc ($\phi_i = \pi/2$) and edge-on illumination of the disc ($\phi_i = \pi$). For both cases, the incident electric field is in the $\hat{z}$ direction ($\psi_i = 0$). The scales for the two graphs have been made the same so that the scattering for the two cases can be compared.

For broadside illumination (Figure 7.41a) there is no equivalent magnetic dipole, and the pattern is the familiar one for the electric dipole: the fat doughnut or fat tire. A change in the direction of the incident electric field, a change in the angle ψ_i, would merely cause a rotation of this graph about the y axis.

For edge-on illumination (Figure 7.41b) there are both equivalent electric and magnetic dipoles, and the disc scatters energy primarily in the backward direction ($\theta = \pi/2$, $\phi = 0$) with the same pattern as for the sphere (Figure 7.39). For this case, a change in the direction of the incident electric field can greatly affect the pattern. In fact, when the incident electric field is normal to the plane of the disc ($\psi_i = \pi/2$), there is no scattering from the disc. The total scattering cross section is zero, as can be seen by substituting $\phi_i = \pi$, $\psi_i = \pi/2$ into (7.164).

An examination of the expressions for the backscattering and total scattering cross sections (Problem 7.11) shows that they are both maximum for edge-on illumination of the disc with the incident electric field parallel to the plane of the disc: $\phi_i = \pi$, $\psi_i = 0$ (the case shown in Figure 7.41b). For this illumination, σ_B is 9/4 times larger and σ_T is 5/4 times larger than for broadside illumination. These results are most likely contrary to your intuition, which probably says that the disc should block or scatter the most energy at broadside illumination. This contradiction arises because our intuition is usually based on the optical case, that is, the case where the disc is electrically large ($k_o d \gg 1$), not electrically small ($k_o d \ll 1$) as it is here. Indeed, for the electrically large disc, the scattering cross sections are maximum for broadside illumination.

The above results for the disc can be used with Babinet's principle to solve the complementary problem: the transmission of a plane wave through an electrically small, circular aperture in a plane, perfectly conducting screen. The geometry for the complementary problem is shown in Figure 7.42a. In this figure, the quantities for the aperture have the subscript a while those for the disc have the subscript d. Notice that the incident magnetic field is now at the angle ψ_i to the z axis. The directional characteristics for the radiation from the aperture into the right half space, $y > 0$, are the same as those for the scattered field of the disc in Figure 7.41 [recall that the total field of the aperture is simply related to the scattered field of the disc: (4.40a), (4.40b)].

The transmission coefficient for the aperture is simply obtained from the total scattering cross section of the disc (7.164) using Equation (4.53):

$$\tau(\phi_i, \psi_i) = \frac{\sigma_T(\phi_i, \psi_i)}{2A} = \frac{4(k_o d)^4}{27\pi^2}\left(\sin^2\phi_i + \frac{5}{4}\cos^2\psi_i\cos^2\phi_i\right). \quad (7.165)$$

Maximum transmission through the electrically small, circular aperture occurs for grazing incidence on the screen ($\phi_i = \pi$, $\hat{k}_i$ parallel to the plane of the screen) with the incident electric field normal to the screen ($\psi_i = 0$). Again, this result is most likely contrary to your intuition, which is probably based on the optical case and says that the aperture should transmit the most energy for normal incidence.

The radiated field of the electrically small aperture can also be described in terms of the equivalent electric and magnetic dipoles shown in Figure 7.42b. Notice that the dipoles are in free space with no screen present. These dipoles are easily

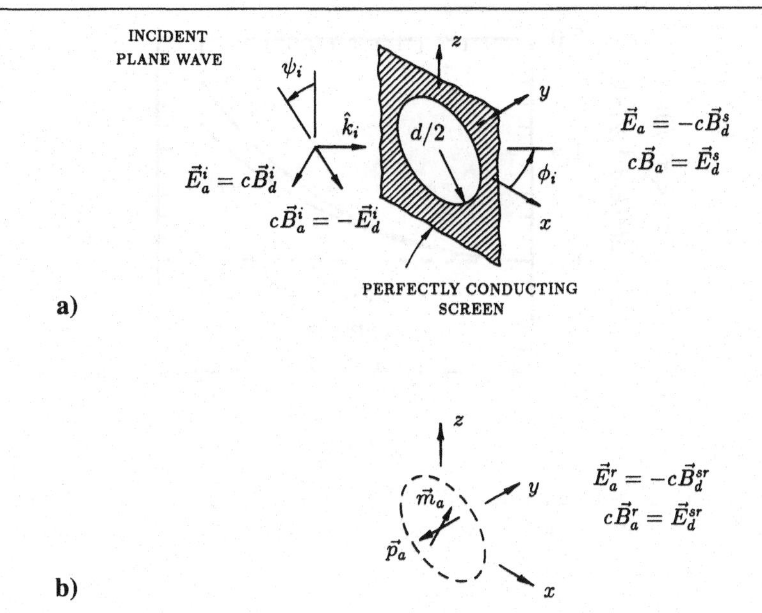

Fig. 7.42. a) The complementary problem – transmission through an electrically small, circular aperture in a plane, perfectly conducting screen. The screen is in the x-z plane, and the vector $\hat{k}_i$ is in the x-y plane. b) Equivalent dipoles for calculating the total field in the right half space, $y > 0$.

determined from the equivalent dipoles for the complementary disc (Problem 7.12):

$$\vec{p}_a = -\frac{1}{c}\vec{m}_d = -\alpha_m \varepsilon_o(\hat{n} \cdot \vec{E}^i_a)\hat{n}, \tag{7.166}$$

$$\vec{m}_a = c\vec{p}_d = -\alpha_e \frac{1}{\mu_o}\left[\vec{B}^i_a - (\hat{n} \cdot \vec{B}^i_a)\hat{n}\right] = \alpha_e \frac{1}{\mu_o}\left[\hat{n} \times (\hat{n} \times \vec{B}^i_a)\right]. \tag{7.167}$$

In microwave engineering, such dipoles are often used to model the coupling through electrically small holes in metallic walls, such as those in a common wall separating two metallic waveguides [42–44].

For the special case of broadside illumination or normal incidence ($\phi_i = \pi/2$), the cross sections and transmission coefficient, (7.163)–(7.165), become

$$\sigma_B(\phi_i = \pi/2) = \frac{d^2}{9\pi}(k_o d)^4 = \frac{16\pi^3 d^2}{9}(d/\lambda_o)^4, \tag{7.168}$$

$$\sigma_T(\phi_i = \pi/2) = \frac{2d^2}{27\pi}(k_o d)^4 = \frac{32\pi^3 d^2}{27}(d/\lambda_o)^4, \tag{7.169}$$

and

$$\tau(\phi_i = \pi/2) = \frac{4}{27\pi^2}(k_o d)^4 = \frac{64\pi^2}{27}(d/\lambda_o)^4. \tag{7.170}$$

These approximate results are compared with exact results in Figure 7.43. There is reasonable agreement, as we would expect, for small $k_o d$: $k_o d \lesssim 0.6$ ($d/\lambda_o \lesssim 0.1$).

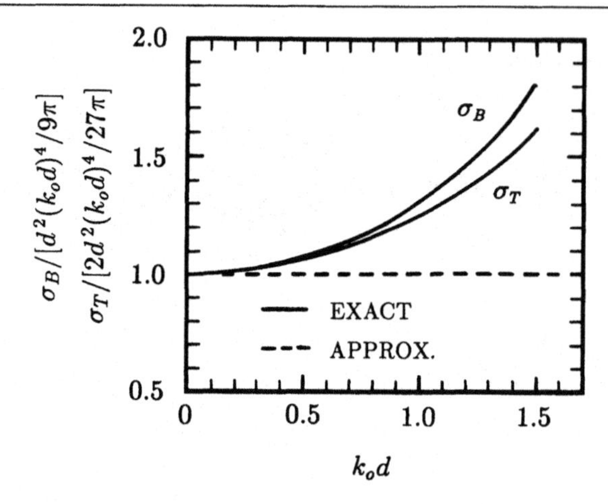

Fig. 7.43. Normalized backscattering and total scattering cross sections for a perfectly conducting disc at normal incidence. Approximations are for an electrically small disc, $k_od \ll 1$.

We have now examined two approaches for determining the scattering from the circular disc (transmission through the circular aperture): The one in Chapter 4, which is based on Kirchhoff's approximation and applies to electrically large discs ($k_od \gg 1$), and the one presented here, which is based on the dipole approximation and applies to electrically small discs ($k_od \ll 1$). In the region of k_od separating these two extremes, the diameter of the disc is comparable to a wavelength, and the scattering cross sections oscillate as k_od is varied. This can be seen in the exact results for normal incidence presented in Figure 4.15. Both σ_B and σ_T have a maximum for $k_od \approx 3$ ($d/\lambda_o \approx 0.5$). This behavior is the result of traveling waves moving across the face of the disc and being reflected at the edges. Strong standing waves develop for certain values of k_od, leading to the oscillations in the scattered energy. The standing wave is evident in Figure 3.26, which shows the field measured in a circular aperture.

Scattering in this intermediate region is very geometry dependent, and there are no simple, general techniques available for analysis, such as the ones we have examined for the extremes $k_od \gg 1$ and $k_od \ll 1$. In this region, the disc has been analyzed using series expansions based on special functions (spheroidal wave functions); in fact, the exact results we have been quoting for the disc are from this analysis [45–47].

As our final example, we will consider the scattering from an electrically short, straight wire. The wire is perfectly conducting, of length $2h$ and radius a, and it is very thin ($a/h \ll 1$). The incident field is again given by Equation (7.133); however, to simplify the pictorial representation, in Figure 7.44 it is shown as being linearly polarized. The vector wave number, $\hat{k}_i$, is in the y-z plane, and the incident electric field is at the angle ψ_i to this plane. For the thin, straight wire, the equivalent electric dipole is directed along the axis of the wire and is proportional to the component of the incident electric field in that direction:

$$\vec{p} = \alpha_e \varepsilon_o (\hat{z} \cdot \vec{E}^i)\hat{z}. \tag{7.171}$$

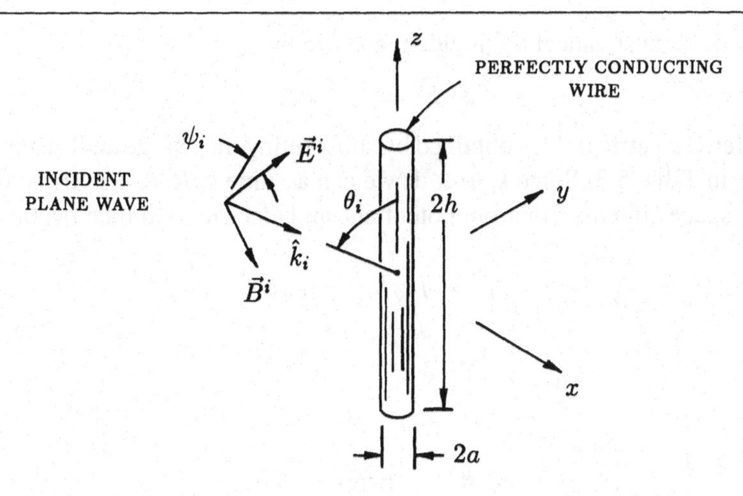

Fig. 7.44. Scattering from a perfectly conducting, thin wire. The vector $\hat{k}_i$ is in the y-z plane.

The equivalent magnetic dipole is insignificant (Problem 7.13).

Now we must determine the electric polarizability α_e. One might think that this can be readily done for such a simple geometry; however, this is not the case. Here we will introduce several simplifying assumptions in order to obtain an analytical result. Our procedure, albeit crude, will illustrate the physical aspects of the problem. At the end, we will compare our result to those from more accurate analyses.

We will ignore the charge and current on the small end caps of the wire, and we will assume that the current density on the cylindrical surface is in the axial direction:

$$\vec{J}_s(z) = \frac{1}{2\pi a} I(z)\hat{z}. \tag{7.172}$$

The incident electromagnetic field is approximately uniform throughout the electrically short wire (7.134); hence, $I(z)$ must be an even function of z. In addition, the current must be approximately zero at the ends of the wire, $z = \pm h$. We will assume that the current distribution is the following quadratic function of the position z satisfying both of these conditions:[26]

$$I(z) = I_o\big[1 - (z/h)^2\big]. \tag{7.173}$$

From the equation of continuity, the accompanying surface charge density is

$$\rho_s(z) = \frac{1}{2\pi a} Q(z), \tag{7.174}$$

with the charge distribution (charge per unit length)

$$Q(z) = \frac{-j2I_o}{\omega h}(z/h). \tag{7.175}$$

The total electric field must be zero within the volume of the wire; that is, the

[26] We cannot assume the current is the "triangular distribution" $I(z) = I_o(1 - |z|/h)$ that we used earlier for the electrically short linear antenna, because the accompanying charge distribution is discontinuous at $z = 0$. This is not a problem for the linear antenna, because the terminals are located at $z = 0$.

scattered field must cancel the incident electric field:

$$\hat{z} \cdot \vec{E}^s = -\hat{z} \cdot \vec{E}^i. \tag{7.176}$$

The scattered electric field is obtained by substituting the charge and current into the formulas in Table 5.3. Since $k_o h \ll 1$, we can assume $k_o R \ll 1$ in these formulas. The free-space Green's function, potentials, and electric field then become

$$G_o(\vec{r}, \vec{r}') = \frac{e^{-jk_oR}}{4\pi R} \approx \frac{1}{4\pi R}\left[1 - jk_oR - \frac{1}{2}(k_oR)^2 + \frac{j}{6}(k_oR)^3 + \cdots\right], \tag{7.177}$$

$$\begin{aligned}\Phi(\vec{r}) &= \frac{1}{\varepsilon_o}\iint_S \rho_s(\vec{r}')G_o(\vec{r}, \vec{r}')dS' \\ &\approx \frac{1}{4\pi\varepsilon_o}\iint_S \frac{\rho_s(\vec{r}')}{R}\left[1 - jk_oR - \frac{1}{2}(k_oR)^2 + \frac{j}{6}(k_oR)^3 + \cdots\right]dS',\end{aligned} \tag{7.178}$$

$$\begin{aligned}\vec{A}(\vec{r}) &= \mu_o \iint_S \vec{J}_s(\vec{r}')G_o(\vec{r}, \vec{r}')dS' \\ &\approx \frac{\mu_o}{4\pi}\iint_S \frac{\vec{J}_s(\vec{r}')}{R}\left[1 - jk_oR - \frac{1}{2}(k_oR)^2 + \frac{j}{6}(k_oR)^3 + \cdots\right]dS',\end{aligned} \tag{7.179}$$

and

$$\begin{aligned}\vec{E}^s(\vec{r}) &= -\nabla\phi(\vec{r}) - j\omega\vec{A}(\vec{r}) \\ &\approx \frac{1}{4\pi\varepsilon_o}\iint_S \rho_s(\vec{r}')\left[\frac{1}{R^2} + \frac{1}{2}k_o^2 - \frac{j}{3}k_o^3R + \cdots\right]\hat{R}dS' \\ &\quad - \frac{j\omega\mu_o}{4\pi}\iint_S \vec{J}_s(\vec{r}')\left[\frac{1}{R} - jk_o - \frac{1}{2}k_o^2R + \frac{j}{6}k_o^3R^2 + \cdots\right]dS'.\end{aligned} \tag{7.180}$$

Now we will make an additional approximation. We will enforce the requirement (7.176) at only one point within the wire, at the center ($r = 0$). The distance R in the above equations is then

$$R = |\vec{r} - \vec{r}'| = \sqrt{a^2 + (z')^2}. \tag{7.181}$$

After substituting (7.172), (7.174), and (7.181) into (7.180) and introducing the result into (7.176), we have

$$\begin{aligned}\frac{jI_o}{\pi\varepsilon_o\omega h^2}\Bigg\{&\int_0^1 \left[\frac{1}{R_n^3} + \frac{1}{2}\frac{1}{R_n}(k_oh)^2 - \frac{j}{3}(k_oh)^3\right]z_n^2dz_n \\ &- \frac{1}{2}(k_oh)^2\int_0^1\left[\frac{1}{R_n} - jk_oh\right](1 - z_n^2)dz_n\Bigg\} = -\hat{z} \cdot \vec{E}^i,\end{aligned} \tag{7.182}$$

with

$$R_n = \sqrt{(a/h)^2 + z_n^2}.$$

Here we have dropped all terms smaller than $(k_o h)^3$. After evaluating the remaining integrals, we obtain

$$\frac{-jI_o}{\pi\varepsilon_o\omega h^2}\left(\left\{\frac{1}{2}\ln\left[\frac{\sqrt{1+(a/h)^2}+1}{\sqrt{1+(a/h)^2}-1}\right]-\frac{1}{\sqrt{1+(a/h)^2}}\right\}\right.$$

$$\left.\times\left\{1-\frac{1}{2}(k_o h)^2\left[1+(a/h)^2\right]\right\}+j\frac{2}{9}(k_o h)^3\right)=\hat{z}\cdot\vec{E}^i. \qquad (7.183)$$

Since the wire is thin $[(a/h)^2 \ll 1]$ Equation (7.183) simplifies to become[27]

$$I_o=\frac{j\pi\varepsilon_o\omega h^2}{\{[\ln(2h/a)-1][1-\frac{1}{2}(k_o h)^2]+j\frac{2}{9}(k_o h)^3\}}(\hat{z}\cdot\vec{E}^i)$$

$$\approx\frac{j\pi\varepsilon_o\omega h^2}{[\ln(2h/a)-1]}(\hat{z}\cdot\vec{E}^i). \qquad (7.184)$$

After substituting (7.184) into (7.175), the charge distribution on the wire becomes

$$Q(z)=\frac{2\pi\varepsilon_o h(\hat{z}\cdot\vec{E}^i)}{[\ln(2h/a)-1]}(z/h), \qquad (7.185)$$

with the electric dipole moment

$$\vec{p}=\hat{z}\int_{-h}^{h} zQ(z)dz=\frac{4\pi h^3}{3[\ln(2h/a)-1]}\varepsilon_o(\hat{z}\cdot\vec{E}^i)\hat{z}. \qquad (7.186)$$

The electric polarizability of the electrically short wire (7.171) is then

$$\alpha_e\approx\frac{4\pi h^3}{3[\ln(2h/a)-1]}. \qquad (7.187)$$

A review of the above calculation of α_e reveals that several simplifying assumptions were made. Naturally, we would like to know the effect of these assumptions on the final result. This can be estimated by comparing our result (7.187) with those obtained by other methods. When the wire is modeled by a thin, perfectly conducting prolate spheroid, an object formed by rotating an ellipse about its major axis, an electrostatic calculation yields the same expression for α_e, only with a as the minor semiaxis and h the major semiaxis of the ellipse ($a/h \ll 1$). A more elaborate electromagnetic calculation obtains an α_e that is the same as (7.187), except the -1 in the denominator is replaced by -0.307 [48]. When the wire is very thin so that $\ln(2h/a) \gg 1$, only a negligible difference in α_e results.

[27] Notice that we have kept terms as small as $(k_o h)^2$ and $(k_o h)^3$ in the denominator of the first line of (7.184), even though they are not required for the argument presented in the remainder of this section. These terms are needed for Problem 7.17.

Table 7.4. *Equivalent dipole moments for electrically small objects*

Perfectly conducting sphere (diameter $= d$)			
$\vec{p} = \alpha_e \varepsilon_o \vec{E}^i$	$\vec{m} = \alpha_m \frac{1}{\mu_o} \vec{B}^i$	$\alpha_e = \frac{\pi d^3}{2}$	$\alpha_m = \frac{-\pi d^3}{4}$
Dielectric sphere (diameter $= d$, $\varepsilon = \varepsilon_r \varepsilon_o$, $\mu = \mu_o$)			
$\vec{p} = \alpha_e \varepsilon_o \vec{E}^i$	$\vec{m} \approx 0$	$\alpha_e = \frac{\pi d^3}{2}\left(\frac{\varepsilon_r - 1}{\varepsilon_r + 2}\right)$	–
Magnetic sphere (diameter $= d$, $\varepsilon = \varepsilon_o$, $\mu = \mu_r \mu_o$)			
$\vec{p} \approx 0$	$\vec{m} = \alpha_m \frac{1}{\mu_o} \vec{B}^i$	–	$\alpha_m = \frac{\pi d^3}{2}\left(\frac{\mu_r - 1}{\mu_r + 2}\right)$
Perfectly conducting disc (diameter $= d$, normal $\hat{n}$)			
$\vec{p} = -\alpha_e \varepsilon_o (\hat{n} \times \hat{n} \times \vec{E}^i)$	$\vec{m} = \alpha_m \frac{1}{\mu_o} (\hat{n} \cdot \vec{B}^i)\hat{n}$	$\alpha_e = \frac{2d^3}{3}$	$\alpha_m = \frac{-d^3}{3}$
Perfectly conducting wire (length $= 2h$, radius $= a$, axial direction $\hat{z}$)			
$\vec{p} = \alpha_e \varepsilon_o (\hat{z} \cdot \vec{E}^i)\hat{z}$	$\vec{m} \approx 0$	$\alpha_e \approx \frac{4\pi h^3}{3[\ln(2h/a) - 1]}$	–

The complex Poynting vector and scattering cross sections for the configuration shown in Figure 7.44 (linearly polarized incident field and spherical coordinates) are

$$\vec{S}_c^{sr}(\vec{r}) = \frac{h^2(k_o h)^4}{18\zeta_o[\ln(2h/a) - 1]^2 r^2} |E_o^i|^2 \cos^2\psi_i \sin^2\theta_i \sin^2\theta \hat{r}, \tag{7.188}$$

$$\sigma_B(\theta_i, \psi_i) = \frac{4\pi h^2(k_o h)^4}{9[\ln(2h/a) - 1]^2} \cos^2\psi_i \sin^4\theta_i, \tag{7.189}$$

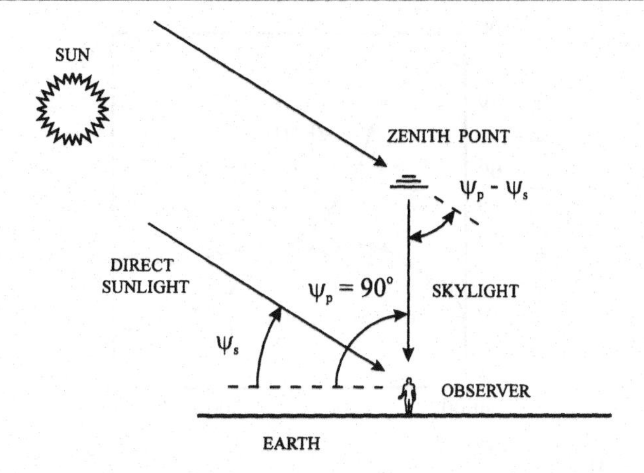

Fig. 7.45. Observer on the Earth viewing direct sunlight and light from the zenith sky.

and

$$\sigma_T(\theta_i, \psi_i) = \frac{8\pi h^2 (k_o h)^4}{27[\ln(2h/a) - 1]^2} \cos^2 \psi_i \sin^2 \theta_i. \quad (7.190)$$

The directional characteristics of the scattered radiation from the electrically short wire are those of the electric dipole, which are shown in Figure 7.2.

For future reference, the equivalent dipole moments and polarizabilities for the perfectly conducting sphere, circular disc, and wire are summarized in Table 7.4. In addition, results are given for electrically small, dielectric and magnetic spheres. For the former $\varepsilon = \varepsilon_r \varepsilon_o$, $\mu = \mu_o$, where $kd = \sqrt{\varepsilon_r} k_o d \ll 1$; for the latter $\varepsilon = \varepsilon_o$, $\mu = \mu_r \mu_o$, where $kd = \sqrt{\mu_r} k_o d \ll 1$. Notice that the dielectric sphere only has an equivalent electric dipole moment, whereas the magnetic sphere only has a magnetic dipole moment, unlike the perfectly conducting sphere, which has both moments.

7.7 The color and polarization of skylight

Undoubtedly, daylight is our most familiar form of electromagnetic radiation. We can roughly divide daylight into two components: direct sunlight and skylight. The latter includes the blue observed for the sky overhead on a clear day.

The theory of light scattering by particles plays an important role in the scientific explanations for skylight. Several of the interesting observed characteristics are explained by applying the material on dipole scattering we developed in the last section. Here we will only be concerned with two of these characteristics: the blue color of the light from a clear sky and the state of polarization for this light. The blue color, of course, has been known since the earliest of times. The polarization of this light, however, was first scientifically observed by Dominique-François Arago (1786–1853) in about 1809 [49].

Consider the arrangement shown in Figure 7.45. An observer on the Earth is viewing the sky when the sun is at the angle ψ_s above the horizon. Using a spectrometer,

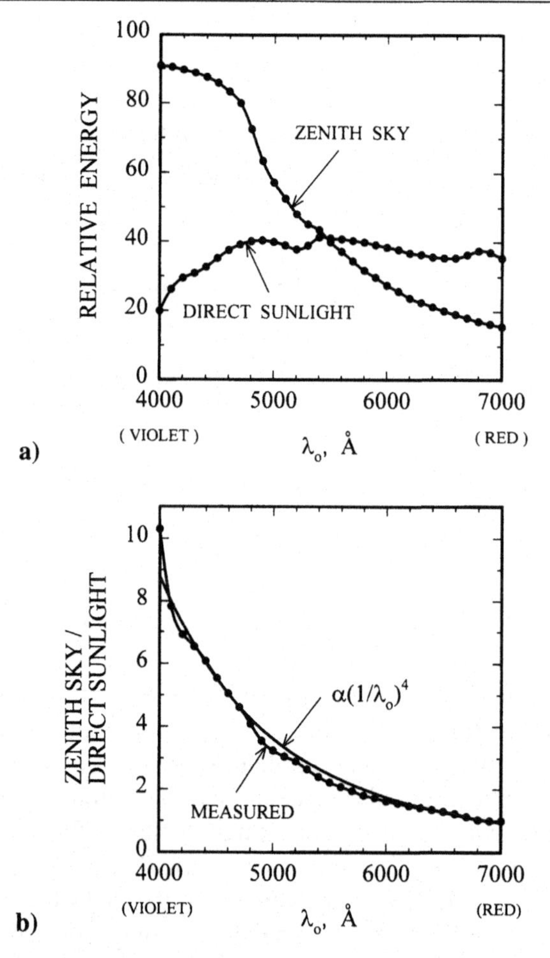

Fig. 7.46. a) Relative energy versus wavelength for direct sunlight and light from the zenith sky. (Measured data from Taylor and Kerr [50].) b) Ratio of energies for light from the zenith sky and direct sunlight versus wavelength.

he measures the relative energy of the light coming from two directions: direct sunlight at the angle ψ_s and skylight from a point directly overhead, $\psi_p = 90°$ (zenith). Results from such measurements are shown in Figure 7.46a; the relative energy of the light is graphed as a function of the wavelength in the visible region: 4,000 Å $\leq \lambda_o \leq$ 7,000 Å [50].[28] When we take the ratio of the two energies (zenith sky/direct sunlight) we get the curve (dots) shown in Figure 7.46b. These quantitative results clearly support the observation we make directly with our eyes: Skylight at zenith on a clear day is much bluer than direct sunlight.

The observer makes an additional measurement. Using a polarimeter, he measures the state of polarization of the light at zenith, $\psi_p = 90°$, as a function of the angle of elevation of the sun, ψ_s. He finds that this light contains a component that is linearly polarized with its electric field normal to the plane passing through the

[28] These results were measured at Cleveland, Ohio on clear days during 1939. Reference [50] contains additional data that show the effects of cloud cover and smoke on these distributions.

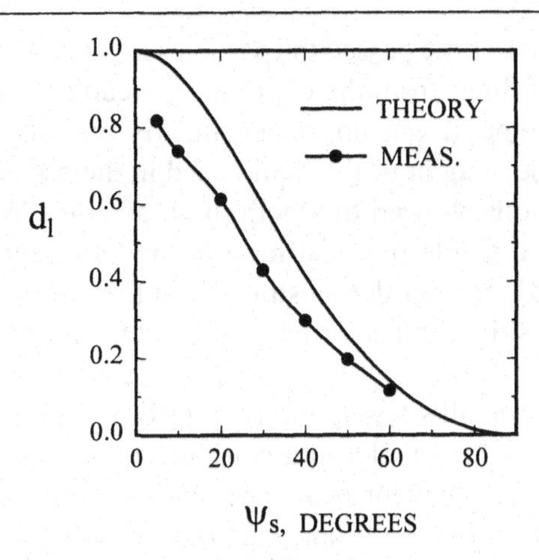

Fig. 7.47. Degree of linear polarization for light from the zenith sky versus the angle of elevation of the sun. (Measured data from Richardson and Hulbert [51].)

sun, the observer, and the zenith point in the sky (this is called the scattering plane, and it is the plane of the page in Figure 7.45). Results from such measurements (dots) are shown in Figure 7.47 [51].[29] Here the degree of linear polarization, d_l (a quantity we will have more to say about later), is graphed as a function of the angle ψ_s. At sunrise ($\psi_s \approx 0°$), the light is about 80% linearly polarized. As the sun continues to rise, the degree of linear polarization decreases, with the light becoming unpolarized as the sun approaches zenith.

When our observer measures the state of polarization for the light from the rest of the sky, he finds a complex pattern that changes with the elevation of the sun. Repeated measurements on clear days demonstrate that this pattern is predictable [52, 53].

During the first half of the nineteenth century, several physical explanations were proposed for the two characteristics of skylight described above [54–57]. All of these explanations failed at one point or another to be consistent with observations. In 1868 John Tyndall (1820–1893) performed a series of careful experiments that further clarified the mechanism responsible for these characteristics [58, 59]. Tyndall placed an evacuated glass tube (about 1 m long and 7 cm in diameter) in a darkened room. A powerful beam of light passed through the tube; the axis of the beam was parallel to the axis of the tube. Tyndall then produced an aerosol in the tube (a suspension of small liquid particles in air). As the aerosol formed, a "blue cloud" appeared. When viewed from the side of the tube (a 90° angle between the axis of the beam and the direction of observation), the light from this cloud was linearly polarized with the electric field perpendicular to the axis of the beam. Notice that this is in complete agreement with the observations for skylight. The light from the zenith sky is nearly linearly polarized at sunrise ($\psi_s \approx 0°$). The angle between the ray of sunlight illuminating the zenith sky and the direction of observation is then 90°.

[29] These results were measured for a cloudless sky at Bocaiuva, Brazil at an altitude of 671 m during 1947.

Tyndall noticed that as time passed the particles in the aerosol coalesced, becoming larger in size. The light from the cloud then became white, and when viewed from the side of the tube, it was no longer linearly polarized. Tyndall found that these effects were independent of the liquid used in the aerosol.

Tyndall's observations showed that both the blue color and the polarization of skylight were due to a single physical mechanism – the interaction of light with suspended particles. He further demonstrated that this interaction was fairly independent of the composition of the particles and that the particles had to be small to produce these effects.

Taking clues from Tyndall's observations, in 1871 Lord Rayleigh used an elastic-solid theory of light, not Maxwell's equations, to explain the observed characteristics of skylight [60]. Both the color and the polarization were due to the scattering of light by objects (particles) in the atmosphere that are *small compared to the wavelength.* He showed that the energy scattered from the particles was proportional to the inverse fourth power of the wavelength ($\propto 1/\lambda_o^4$). Thus, when white light is incident on the particles, the violet scattered light ($\lambda_o = 4{,}000$ Å) is about nine times greater in intensity than the red scattered light ($\lambda_o = 7{,}000$ Å).[30] Rayleigh performed an experiment, much like the one we described earlier, in which he compared skylight with direct sunlight. He found that his experimental results were in reasonable agreement with his theory. We have drawn a curve proportional to $1/\lambda_o^4$ on Figure 7.46b to show this agreement for our results.

In the years that followed, Lord Rayleigh published additional results on the scattering of light in the atmosphere (atmospheric optics). In his first investigation [60], he suspected that the particles responsible for the scattering were common salt; later he proposed that the molecules of the air were the scatterers [61], which is the explanation accepted today.

We will now leave this brief historical account and apply the results we have obtained for dipole scattering to explain these observations. In this process we will introduce additional information about the nature of sunlight (natural light) and the scattering of light by molecules.

7.7.1 Natural light

Before we can examine the scattering of sunlight by the molecules in the Earth's atmosphere, we must describe the electromagnetic properties of this light. The sun's electromagnetic radiation is the sum of independent emissions from myriad atoms. We believe that a measurement of the electric field of this radiation would show

[30] Rayleigh's results explain the blue cloud initially observed by Tyndall. However, they do not explain why the cloud turned white as the size of the particles in the aerosol became larger. When the dimensions of a particle are much larger than the wavelength, the amount of energy scattered by the particle is nearly independent of the wavelength. Since the incident light in Tyndall's experiment was white, the scattered light was also white when the particles were electrically large. We observed this phenomenon in Section 4.5 when we considered the scattering from a disc. The total scattering cross section, shown in Figure 4.15a, is nearly independent of the wavelength when the diameter of the disc is large compared to the wavelength ($k_o d \gg 1$).

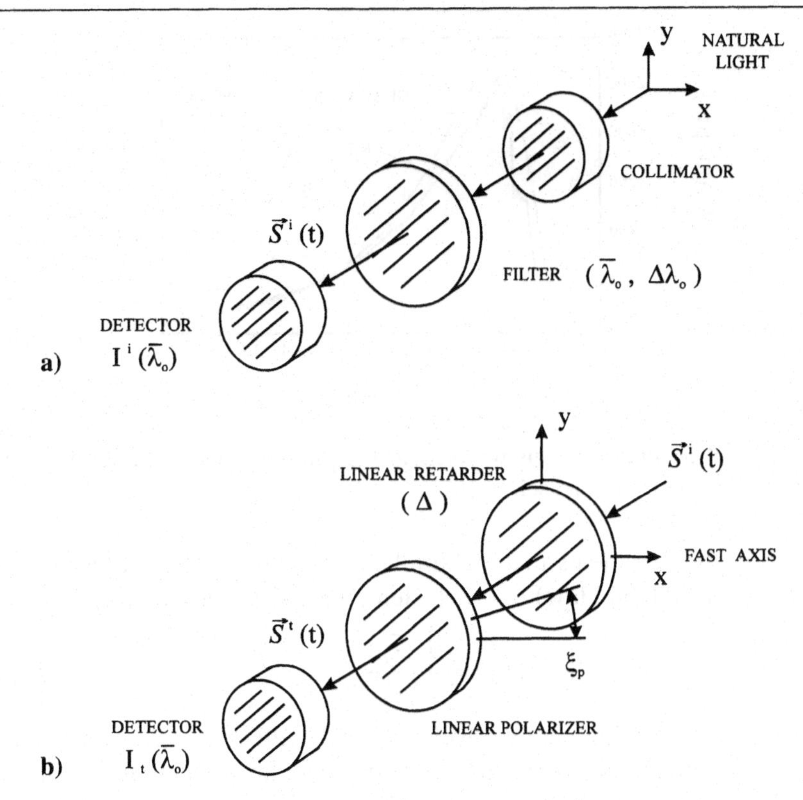

Fig. 7.48. Arrangement of optical elements for measuring the properties of natural light.

a chaotic, noise-like waveform. Unfortunately, we cannot make this measurement; no instrument currently available has a response time short enough to resolve the expected temporal variations. We must settle for measurements that are averages over time of certain properties of the light. We will confine our description of sunlight to a discussion of these measurements – an empirical description. Theoretical descriptions based on the statistical properties of the light are available; however, these are more involved than what we require [62, 63].

The first measurement we will consider uses the arrangement shown in Figure 7.48a. The sunlight is collimated to produce an approximately plane electromagnetic wave. This wave then passes through a filter with center frequency $\overline{\omega}$ and a very narrow bandwidth $\Delta\omega$; $\Delta\omega \ll \overline{\omega}$. In terms of the free-space wavelength, the parameters for the filter are $\bar{\lambda}_o = 2\pi c/\overline{\omega}$ and $\Delta\lambda_o = 2\pi c \Delta\omega/\overline{\omega}^2$. Finally, a detector measures the time-average power per unit area, or the irradiance I^i, of the filtered light:

$$I^i(\bar{\lambda}_o) = \left|\langle \vec{S}^i(t)\rangle\right| = \left|\frac{1}{T_D}\int_0^{T_D} \vec{S}^i(t)dt\right|. \tag{7.191}$$

The time interval T_D in (7.191) must be long enough to make the average independent of T_D.

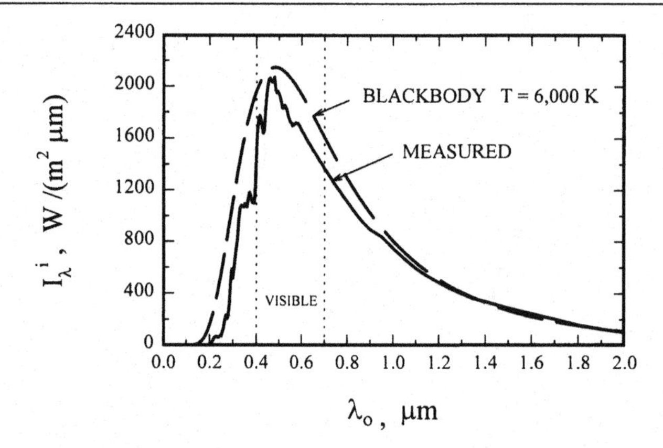

Fig. 7.49. Spectral irradiance versus wavelength for sunlight outside the Earth's atmosphere. (Measured data from Thekaekara [65].)

Results from measurements that are similar to the one we have described are shown in Figure 7.49 (solid line) [64, 65]. Here, the spectral irradiance,

$$I_\lambda^i(\bar{\lambda}_o) = \frac{I^i(\bar{\lambda}_o)}{\Delta\lambda_o}, \tag{7.192}$$

is graphed as a function of the center wavelength of the filter. This curve represents a composite of measurements made with several different instruments aboard aircraft at altitudes high enough for the measurements to be practically free of atmospheric absorption.

If we consider the sun to radiate as a blackbody, the spectral irradiance at the sun's surface is given by Planck's radiation law [66]:

$$I_\lambda^i(\lambda_o)\big|_s = \frac{2\pi c^2 h}{\lambda_o^5}\left(\frac{1}{e^{hc/\lambda_o kT} - 1}\right), \tag{7.193}$$

where $h = 6.63 \times 10^{-34}$ J·s is Planck's constant and $k = 1.38 \times 10^{-23}$ J/K is Boltzmann's constant. Because the sun radiates isotropically, the spectral irradiance at the Earth is simply Equation (7.193) multiplied by the ratio $(r_s/r_{se})^2$, where $r_s \approx 6.96 \times 10^5$ km is the radius of the sun and $r_{se} \approx 1.496 \times 10^8$ km is the distance from the sun to the Earth:

$$I_\lambda^i(\lambda_o)\big|_e = \frac{2\pi c^2 h}{\lambda_o^5}\left(\frac{1}{e^{hc/\lambda_o kT} - 1}\right)\left(\frac{r_s}{r_{se}}\right)^2. \tag{7.194}$$

The dashed curve in Figure 7.49 is Equation (7.194) for the temperature $T = 6{,}000$ K. Within the range of wavelengths shown, the sun is seen to radiate approximately as a blackbody at a temperature of $T = 6{,}000$ K.[31] The extraterrestrial

[31] The sun is an opaque body enveloped by a gas of low emissivity. The opaque body radiates a continuous spectrum (blackbody radiation). The gas absorbs this radiation at certain wavelengths. Thus, if we made a graph like Figure 7.49 with finer resolution, the continuous spectrum would be crossed by dark lines (Fraunhofer lines) [67].

spectral irradiance peaks near the wavelength for blue light ($\lambda_o \approx 4{,}700$ Å), and it varies by about 30% over the range of visible wavelengths.

For our second measurement, we place two additional optical elements between the filter and the detector: an ideal linear retarder and an ideal linear polarizer.[32] Figure 7.48b shows this arrangement. The retarder shifts the phase of the vertical component (y) of the electric field by the amount $-\Delta$ relative to that of the horizontal component (x). The transmission axis of the linear polarizer is at the angle ξ_p to the horizontal axis (x). The properties of both elements are assumed to be constant over the frequencies within the narrow bandwidth of the filter.

Measurements made with this arrangement reveal the following interesting property for the transmitted light: The irradiance, I^t, is the same for any value of the retardance Δ and for any angle ξ_p of the linear polarizer. It always equals one half of the irradiance of the incident light (7.191):

$$\begin{aligned} I^t(\overline{\lambda}_o) &= \left|\langle \vec{S}^t(t)\rangle\right| = \left|\frac{1}{T_D}\int_0^{T_D} \vec{S}^t(t)dt\right| \\ &= \frac{1}{2} I^i(\overline{\lambda}_o). \end{aligned} \tag{7.195}$$

Light with this property is called *natural light* or unpolarized light [68, 69].

For our purposes, we require a simple representation equivalent to natural light that can be used in our scattering calculations. Consider the time-harmonic (frequency $\overline{\omega}$) plane wave with the electric field (phasor)

$$\vec{E}^i_o = E^i_o(\cos\gamma\hat{x} + \sin\gamma\hat{y}). \tag{7.196}$$

This is a linearly polarized electric field at the angle γ to the horizontal axis (x). We now claim that our observations for natural light are reproduced by this wave when we average over all states of linear polarization (average over all angles, $0 \le \gamma < 2\pi$). To see this, we will determine the irradiance of the wave for the two measurements shown in Figure 7.48. The electric field of the wave is unchanged on passing through the filter (center frequency $\overline{\omega}$), so[33]

$$\begin{aligned} \overline{I^i}(\overline{\lambda}_o) &= \frac{1}{2\pi}\int_0^{2\pi} \left|\langle \vec{S}^i(t)\rangle\right| d\gamma = \frac{1}{2\pi}\int_0^{2\pi} \left|\mathrm{Re}(\vec{S}^i_c)\right| d\gamma \\ &= \frac{|E^i_o|^2}{4\pi\zeta_o}\int_0^{2\pi} d\gamma = \frac{|E^i_o|^2}{2\zeta_o}. \end{aligned} \tag{7.197}$$

After passing through the retarder and linear polarizer, the electric field of the wave becomes

$$\vec{E}^t_o = E^i_o(\cos\gamma\cos\xi_p + \sin\gamma\sin\xi_p e^{-j\Delta})(\cos\xi_p\hat{x} + \sin\xi_p\hat{y}), \tag{7.198}$$

[32] The properties of these elements are described in Section 2.5.

[33] The overbar on the irradiance is used to indicate the average over the angle γ.

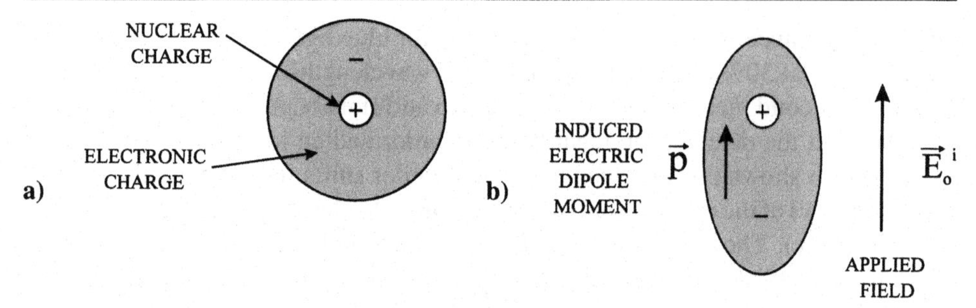

Fig. 7.50. Schematic drawing for the electric dipole moment induced in an atom or molecule by an applied electric field (electronic polarization).

and thus

$$\overline{I^t}(\overline{\lambda}_o) = \frac{1}{2\pi}\int_0^{2\pi} \left|\langle \vec{S}^t(t)\rangle\right| d\gamma = \frac{1}{2\pi}\int_0^{2\pi} \left|\mathrm{Re}(\vec{S}^t_c)\right| d\gamma$$

$$= \frac{|E^i_o|^2}{4\pi\zeta_o}\int_0^{2\pi}(\cos^2\gamma\cos^2\xi_p + \sin^2\gamma\sin^2\xi_p$$

$$+ 2\cos\gamma\sin\gamma\cos\xi_p\sin\xi_p\cos\Delta)d\gamma = \frac{|E^i_o|^2}{4\zeta_o} = \frac{1}{2}\overline{I^i}(\overline{\lambda}_o). \quad (7.199)$$

Notice that this is the same result we obtained for natural light (7.195): The irradiance of the transmitted wave (7.199) is always equal to one half of the irradiance of the incident wave (7.197) and is independent of both the retardance Δ and the angle ξ_p of the linear polarizer.[34]

7.7.2 Molecular scattering

An applied electric field can induce an electric dipole moment in an atom or molecule, as indicated schematically in Figure 7.50 [70]. With no applied electric field (Figure 7.50a) the centers for the nuclear charge and the surrounding spherical "cloud" of electronic charge coincide, and the atom has no dipole moment. When the electric field is applied (Figure 7.50b) the centers for these charges separate, and the atom acquires an electric dipole moment. The electric field and dipole moment are simply related; for a time-harmonic field

$$\vec{p} = \alpha_e\varepsilon_o\vec{E}^i_o, \quad (7.200)$$

where α_e is the electric polarizability of the atom or molecule (for the illustration in Figure 7.50, it would be called the electronic polarizability). When there is a negligible time delay between a change in the field and the corresponding change in the dipole moment, $\vec{E}_i$ and $\vec{p}$ are in phase and α_e is a real number.

The illustration in Figure 7.50b is for an isotropic polarizability; the applied electric field and the induced dipole moment are in the same direction. In some

[34] In Problem 7.14 this equivalence is utilized further; the Stokes parameters are determined for natural light.

cases, the structure of the molecule causes an anisotropic polarizability; the applied electric field and the induced dipole moment are then in different directions.

The Earth's atmosphere is a gas composed mainly of oxygen and nitrogen molecules; in the lower atmosphere, the composition is about 21% O_2 and 78% N_2 by volume. The polarizabilities for these two molecules are comparable, and they are also slightly anisotropic.[35] However, in our discussion we will ignore these factors and assume an average, isotropic polarizability for the molecules. Sunlight induces dipole moments in these molecules. These time-varying dipole moments radiate electromagnetic energy. This "scattered" energy is the skylight we observe on a clear day.

In Figure 7.51a, we show the coordinates associated with one molecule. The incident, time-harmonic electric field is

$$\vec{E}_o^i = E_o^i(\cos\gamma\hat{x} + \sin\gamma\hat{z}), \tag{7.201}$$

with the irradiance

$$\overline{I^i}(\lambda_o) = \frac{|E_o^i|^2}{2\zeta_o}. \tag{7.202}$$

Notice that we have changed coordinates from those in Figure 7.48 so that we can make use of our earlier results for dipole scattering (the incident wave is now propagating in the y direction rather than in the z direction). The induced electric dipole moment of the molecule is

$$\vec{p} = \alpha_e\varepsilon_o E_o^i(\cos\gamma\hat{x} + \sin\gamma\hat{z}), \tag{7.203}$$

and, from (7.138) and (7.140), the radiated electric field and complex Poynting vector of this moment are

$$\begin{aligned}\vec{E}^{sr} &= \frac{-k_o^2}{4\pi\varepsilon_o r}e^{-jk_o r}(\hat{r}\times\hat{r}\times\vec{p})\\ &= \frac{-k_o^2\alpha_e E_o^i}{4\pi r}e^{-jk_o r}\left[(\sin\gamma\sin\theta - \cos\gamma\cos\theta\cos\phi)\hat{\theta} + \cos\gamma\sin\phi\hat{\phi}\right]\end{aligned} \tag{7.204}$$

and

$$\begin{aligned}\vec{S}_c^{sr}(\vec{r}) &= \frac{ck_o^4}{32\pi^2\varepsilon_o r^2}|\hat{r}\times\hat{r}\times\vec{p}|^2\hat{r}\\ &= \frac{k_o^4|\alpha_e|^2|E_o^i|^2}{32\pi^2\zeta_o r^2}(\sin^2\gamma\sin^2\theta + \cos^2\gamma\cos^2\theta\cos^2\phi\\ &\quad - 2\sin\gamma\cos\gamma\sin\theta\cos\gamma\cos\phi + \cos^2\gamma\sin^2\phi)\hat{r}.\end{aligned} \tag{7.205}$$

[35] The average polarizabilities for these molecules are $\alpha_e \approx 2.00\times10^{-29}$ m^3 for O_2 and $\alpha_e \approx 2.21\times10^{-29}$ m^3 for N_2 (SI Units) [71]. The anisotropy of the polarizabilities has a small effect on our calculations. It causes about a 4% change in the degree of linear polarization at $\psi_s = 0°$ in Figure 7.47 [56, 72].

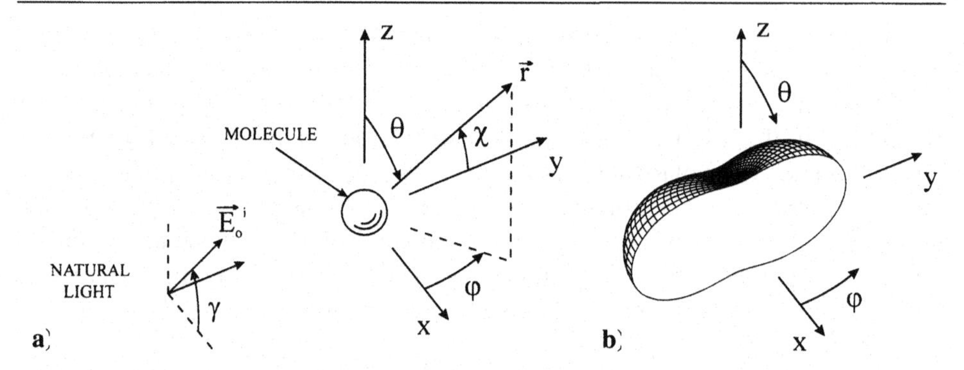

Fig. 7.51. a) Coordinates used to determine scattering of natural light by a molecule. b) Directional characteristics of the scattered radiation, $\overline{I^{sr}}$.

To obtain the irradiance for the scattered sunlight, we must average the magnitude of (7.205) over all states of linear polarization (over all values of the angle γ):

$$\overline{I^{sr}}(\lambda_o) = \frac{1}{2\pi}\int_0^{2\pi} \left|\mathrm{Re}(\vec{S}_c^{sr})\right| d\gamma = \frac{k_o^4|\alpha_e|^2|E_o^i|^2}{64\pi^2\zeta_o r^2}(1+\sin^2\theta\sin^2\phi)$$

$$= \frac{\pi^2|\alpha_e|^2}{2r^2}\left(\frac{1}{\lambda_o}\right)^4 \overline{I^i}(\lambda_o)(1+\cos^2\chi). \qquad (7.206)$$

In the last line, we have simplified the result by introducing the angle χ between the position vector $\vec{r}$ and the y axis ($\cos\chi = \sin\theta\sin\phi$). The directional characteristics of the scattered radiation from the molecule are shown in Figure 7.51b; half of the figure has been removed to show the pattern on the plane $\phi = \pi/2, 3\pi/2$. Notice that the natural light is scattered in all directions; there is no null in the pattern, as there is when the incident light is linearly polarized (Figure 7.2).

Two factors are seen to determine the dependence of the scattered irradiance (7.206) upon the wavelength: the spectrum of the incident sunlight, $\overline{I^i}(\lambda_o)$ shown in Figure 7.49, and the term $(1/\lambda_o^4)$ from the dipole scattering, shown in Figure 7.46b.[36] The latter is seen to be the most significant factor at visible wavelengths; it is the principal cause for the blue color of the light from a clear sky.

We will now determine the polarization of skylight. We will confine our discussion to the situation used for Figure 7.45, an observation of the light at zenith as the sun rises. The geometry is shown in Figure 7.52. The plane y–z (x'–z') is the scattering plane; it contains the sun, the observer, and the molecule producing the scattered light. The observer examines the scattered light with a linear polarizer. The scattered electric field in the direction of the observer is (7.204) with $\theta = \pi - \psi_s$

[36] Of course, when we observe the color of the sky with our eyes there is a third factor – the response of the human eye to the various wavelengths in the visible spectrum [73].

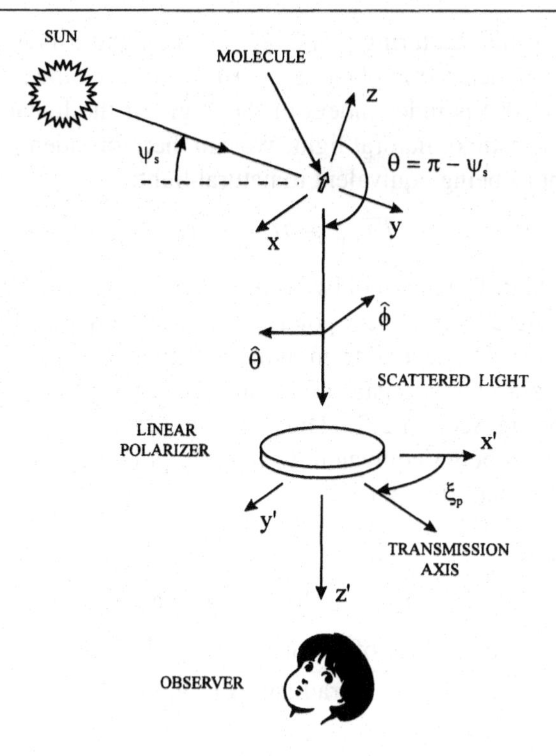

Fig. 7.52. Observation of the polarization of light from the zenith sky.

and $\phi = \pi/2$:

$$\begin{aligned}\vec{E}^{sr} &= \frac{-k_o^2 \alpha_e E_o^i}{4\pi r} e^{-jk_o r}(\sin\gamma \sin\psi_s \hat{\theta} + \cos\gamma \hat{\phi})\\ &= \frac{k_o^2 \alpha_e E_o^i}{4\pi r} e^{-jk_o r}(\sin\gamma \sin\psi_s \hat{x}' + \cos\gamma \hat{y}'). \end{aligned} \tag{7.207}$$

In the last line, we have expressed the field in terms of the local coordinates $(\hat{x}', \hat{y}')$ at the polarizer. After passing through the linear polarizer, this field becomes

$$\vec{E}^t = \frac{k_o^2 \alpha_e E_o^i}{4\pi r} e^{-jk_o r}(\cos\xi_p \sin\gamma \sin\psi_s + \sin\xi_p \cos\gamma)(\cos\xi_p \hat{x}' + \sin\xi_p \hat{y}'), \tag{7.208}$$

where ξ_p is the angle the transmission axis of the polarizer makes with the x' axis. The irradiance for the transmitted sunlight is then

$$\overline{I^t} = \frac{1}{2\pi}\int_0^{2\pi} \left|\mathrm{Re}(\vec{S}_c^t)\right| d\gamma = \frac{\pi^2 |\alpha_e|^2}{2r^2}\left(\frac{1}{\lambda_o}\right)^4 \overline{I^i}(\sin^2\psi_s + \sin^2\xi_p \cos^2\psi_s). \tag{7.209}$$

The irradiance (7.209) is seen to change as the linear polarizer is rotated, that is, as the angle ξ_p of the transmission axis is varied. It is maximum when the axis is perpendicular to the scattering plane ($\xi_p = \pi/2$), and it is minimum when the axis is parallel to the scattering plane ($\xi_p = 0$). Notice that the first term within the parentheses of (7.209) is independent of the angle of the linear polarizer. Recall that this a characteristic of natural light. We can therefore identify a portion of the scattered sunlight as being equivalent to natural light:

$$\overline{I^{sr}}_{\text{nat}} = 2C \sin^2 \psi_s,$$

where C is a constant. The factor of two is required in the above formula because the irradiance transmitted by the linear polarizer is always one half that of the incident natural light (7.199). The second term in the parentheses of (7.209) is proportional to $\sin^2 \xi_p$. This is a characteristic of linearly polarized light (the law of Malus that we mentioned in Section 2.5). Thus, we can identify a second portion of the scattered sunlight as being equivalent to linearly polarized light with the electric field normal to the scattering plane:

$$\overline{I^{sr}}_{\text{lin}} = C \cos^2 \psi_s.$$

The degree of linear polarization for the scattered sunlight is defined to be [66]

$$\begin{aligned} d_l &= \frac{\text{irradiance of linearly polarized component}}{\text{irradiance of total}} \\ &= \frac{\overline{I^{sr}}_{\text{lin}}}{\overline{I^{sr}}_{\text{nat}} + \overline{I^{sr}}_{\text{lin}}} = \frac{\overline{I^t}(\xi_p = \pi/2) - \overline{I^t}(\xi_p = 0)}{\overline{I^t}(\xi_p = \pi/2) + \overline{I^t}(\xi_p = 0)} \\ &= \frac{\cos^2 \psi_s}{1 + \sin^2 \psi_s}. \end{aligned} \tag{7.210}$$

This result is compared with the measurements in Figure 7.47 (solid line). The agreement is only fair but can be improved by including several factors that we have omitted in our simple analysis. One of these factors we have already mentioned: the anisotropic polarizability of the molecules. Other factors include the sunlight reflected from the ground, multiple scattering between molecules, and the scattering from haze and dust in the air [52, 74].

The theory predicts that the skylight will be completely polarized ($d_l = 1$) when $\psi_s = 0$. This is easily understood when we consider the electric field of the incident light (7.201) to be composed of a component parallel to the scattering plane (z) and a component normal to the scattering plane (x). Patterns for the scattered irradiances of these two components are shown in Figure 7.53. For $\psi_s = 0$, we are interested in the scattered electric field in the direction of the negative z axis. For the z component of the incident field (Figure 7.53a), there is a null in this direction, whereas, for the x component of the incident field (Figure 7.53b), the scattered electric field is linearly polarized in the x direction (normal to the scattering plane). Since this is true no matter what values the relative amplitudes of the two components of the incident field have (for any angle γ), it is true for natural light.

Our discussion of polarization has been confined to the observation of skylight from directly overhead (zenith). However, it is easily extended to include

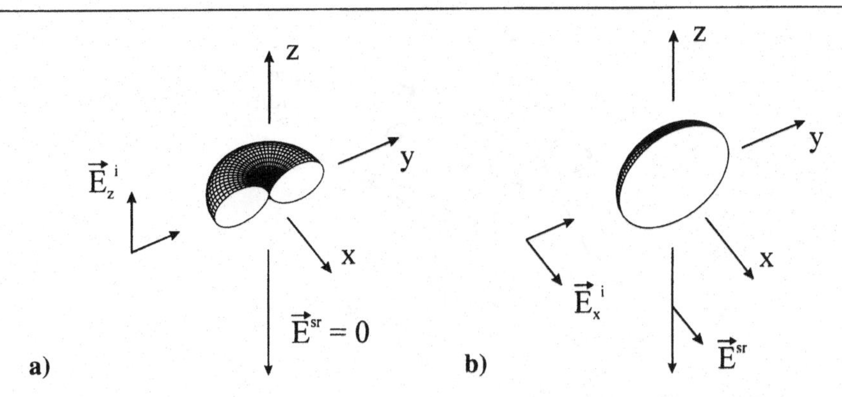

Fig. 7.53. Directional characteristics of scattered radiation from a molecule. a) Incident electric field linearly polarized parallel to scattering plane: $\vec{E}_z^i$. b) Incident electric field linearly polarized normal to scattering plane: $\vec{E}_x^i$.

observations at other points of the sky. From the patterns in Figure 7.53, we see this simple theory predicts that skylight is always completely polarized ($d_l = 1$) at 90° to the sun, that is, when the ray of sunlight to the point in the sky and the ray of scattered light from that point to the observer form a right angle. The electric field is then normal to the scattering plane.

Until now, we have only considered the scattering from a single molecule, and we have seen that this explains, reasonably well, the characteristics of skylight: the blue color and polarization. Of course, it is the scattering from a large number of molecules in the atmosphere that produces these effects, and so to complete our discussion, we must determine the relationship between the scattering from a single molecule and the scattering from a collection of molecules. We will use the approximation known as *single scattering* [75–77]. That is, we will assume that each molecule is polarized by only the electric field of the incident sunlight, and that the scattered light we observe is simply the sum of the scattered electric fields of the individual molecules, each treated as if it were isolated from the others. We are ignoring the additional dipole moment induced in a molecule by the electric fields of the other polarized molecules; hence, we are ignoring the scattered field of this additional moment (ignoring multiple scattering). This is a reasonable assumption when the molecules are far apart, as in a gas.

We will consider a collection of N molecules. The schematic drawing in Figure 7.54 shows the scattering by two of these molecules, 1 and 2. The scattered electric fields are the same for the two molecules except for a phase difference, which is due to the difference in the paths (s_{21}) light travels in going to and from these molecules:

$$\vec{E}_2^{sr} = \vec{E}_1^{sr} e^{j\varphi_2},$$

where the phase difference φ_2 is

$$\varphi_2 = -k_o s_{21}.$$

For the collection of molecules, the scattered electric field, complex Poynting vector,

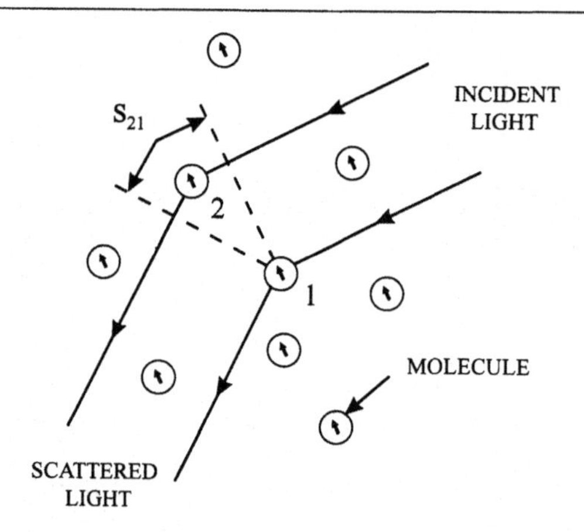

Fig. 7.54. Schematic drawing showing scattering of incident light by two molecules (1 and 2).

and irradiance are

$$\vec{E}_N^{sr} = \sum_{n=1}^{N} \vec{E}_n^{sr} = \vec{E}_1^{sr} \sum_{n=1}^{N} e^{j\varphi_n},$$

$$\vec{S}_{cN}^{sr} = \vec{S}_{c1}^{sr} \left| \sum_{n=1}^{N} e^{j\varphi_n} \right|^2,$$

and

$$\overline{I_N^{sr}} = \overline{I_1^{sr}} \left| \sum_{n=1}^{N} e^{j\varphi_n} \right|^2, \tag{7.211}$$

with

$$\varphi_n = -k_o s_{n1} = -2\pi \left(\frac{s_{n1}}{\lambda_o} \right) \tag{7.212}$$

and $s_{11} = 0$.

Now we must evaluate the sum in (7.211). A volume of atmospheric gas several wavelengths on a side contains a large number of randomly spaced molecules.[37] Consequently, the phases of the scattered fields from these molecules (7.212) are random and uniformly distributed over the range $-\pi < \varphi_n \leq \pi$.[38] The sum in (7.211) is then a collection of N phasors, each of unit amplitude and random phase; this is shown diagrammatically in Figure 7.55.[39] The square of the magnitude of

[37] At the surface of the Earth, there are about 2.5×10^{25} molecules in one cubic meter of air. Thus, a cube one wavelength on a side ($\lambda_o = 5{,}300$ Å for green light) contains about 3.7×10^6 molecules.

[38] Here we are excluding the special case where the ray for the incident light and the ray for the scattered light in Figure 7.54 are parallel. The scattered fields from all of the molecules are then in phase.

[39] This is the "random-walk problem" [78]. A person starting at the origin takes consecutive steps of equal length but in random directions. Each step would be one of the arrows in Figure 7.55. The problem is to determine statistics for the location of the person, relative to the origin, after a large number of steps. The square of the magnitude of the sum (7.213) corresponds to the square of the distance from the origin.

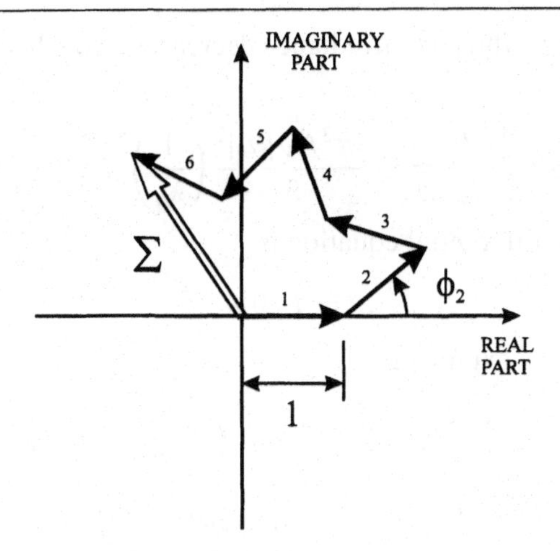

Fig. 7.55. Illustration for sum of phasors, each of unit amplitude and random phase.

the sum can be expressed as

$$\left|\sum_{n=1}^{N} e^{j\varphi_n}\right|^2 = \sum_{n=1}^{N} e^{j\varphi_n} \sum_{m=1}^{N} e^{-j\varphi_m} = N + \sum_{\substack{n=1\\ n\neq m}}^{N} \sum_{m=1}^{N} e^{j(\varphi_n-\varphi_m)}$$

$$= N + \sum_{\substack{n=1\\ n\neq m}}^{N} \sum_{m=1}^{N} \cos(\varphi_n - \varphi_m) = N. \tag{7.213}$$

In the last line, we have recognized that, due to the random nature of φ_n and φ_m, $\cos(\varphi_n - \varphi_m)$ is as likely to be a particular positive number, say u ($0 < u \leq 1$), as the negative of that number, $-u$. Thus, when N is very large, the contribution of the double sum in (7.213) is negligible, and the "most probable" value for (7.213) is N.

The irradiance of the scattered light from the N randomly spaced molecules (7.211) is now seen to be just N times the irradiance of the scattered light from a single molecule:

$$\overline{I_N^{sr}} = N\overline{I_1^{sr}}. \tag{7.214}$$

This is why we could explain the observed characteristics of skylight using the scattering of sunlight by a single molecule.

As a beam of sunlight passes through the Earth's atmosphere, the molecules scatter energy from the beam and it is attenuated. The time-average power scattered per unit volume of the gas is the product of the number of molecules per unit volume, N_V, and the total power scattered by a single molecule (the integral of (7.206) over a spherical surface surrounding the molecule):

$$2\pi r^2 N_V \int_{\chi=0}^{\pi} \overline{I^{sr}} \sin\chi \, d\chi = \pi^3 N_V |\alpha_e|^2 \left(\frac{1}{\lambda_o}\right)^4 \overline{I^i} \int_{\chi=0}^{\pi} (1+\cos^2\chi)\sin\chi \, d\chi$$

$$= \frac{8\pi^3 N_V |\alpha_e|^2}{3} \left(\frac{1}{\lambda_o}\right)^4 \overline{I^i}.$$

This is the amount by which the irradiance decreases as the beam propagates a unit distance (dz):

$$\frac{d\overline{I^i}}{dz} = -\frac{8\pi^3 N_V |\alpha_e|^2}{3}\left(\frac{1}{\lambda_o}\right)^4 \overline{I^i}. \tag{7.215}$$

The solution to this differential equation is

$$\overline{I^i}(z) = \overline{I^i}(0)e^{-\gamma z}, \tag{7.216}$$

where the factor in the exponent,

$$\gamma = \frac{8\pi^3 N_V |\alpha_e|^2}{3}\left(\frac{1}{\lambda_o}\right)^4, \tag{7.217}$$

is called the *extinction coefficient*. When the relationship between the polarizability of the molecules, α_e, and the index of refraction, n, is introduced, the extinction coefficient becomes[40]

$$\gamma = \frac{32\pi^3 |n-1|^2}{3N_V}\left(\frac{1}{\lambda_o}\right)^4. \tag{7.218}$$

This famous result was obtained by Lord Rayleigh in 1899 and used by him to determine the "degree of transparency of air" [61].

When the sun is high in the sky, there is slight attenuation for light passing through the atmosphere.[41] At sunrise and sunset the situation is quite different. The path for light through the atmosphere is much longer, and the attenuation is much larger and more easily observed (Problem 7.16). Because the extinction coefficient (7.218) is proportional to $(1/\lambda_o^4)$, light at the violet end of the visible spectrum is attenuated much more than light at the red end. This is the reason for the red sky at sunrise and sunset.

It goes without saying that the blue of the sky and the red of the sunset have inspired humankind, but what of the polarization of skylight? This phenomenon, which under normal conditions is not observable by human beings, is sensed by many insects. As an example, we will briefly describe the honeybee's use of polarization for navigation [80, 81].

[40] The polarization (dipole moment per unit volume) is

$$\vec{P} = N_V \vec{p} = \varepsilon_o(\varepsilon_r - 1)\vec{E}.$$

For the single scattering approximation

$$\vec{p} \approx \alpha_e \varepsilon_o \vec{E},$$

so

$$\alpha_e = (\varepsilon_r - 1)/N_V.$$

The square of the index of refraction is $n^2 = \varepsilon_r$; thus,

$$\alpha_e = (n^2 - 1)/N_v = (n+1)(n-1)/N_V \approx 2(n-1)/N_V.$$

In the last step we have used the fact that $n \approx 1$ for air.

[41] The ratio of transmitted irradiance to incident irradiance at zenith is about 0.96 for red light and 0.70 for violet light [79].

A forager bee returning to her hive communicates the location of a distant food source to the other bees. She does this through a dance, the "tail-wagging dance." When the bee is on a horizontal surface and can see the sun, the direction of the "tail-wagging run" indicates the bearing with respect to the sun of the food source.[42] New recruits will maintain this bearing when flying to the food source. The biologist Karl Ritter von Frisch (1886–1982) noticed that bees did not have to see the sun but only a small patch of blue sky to perform the dance correctly. Through a series of elegant experiments, he showed that the bees also use the polarization pattern of skylight and its relationship to the sun to determine their bearing. In one experiment, a hive was enclosed except for a window that made a small patch of blue sky visible to the bees. A linear polarizer (sheet) was placed over the window. When the polarizer was oriented so it did not alter the state of polarization of the skylight, the bees' dances pointed toward the food source. When the polarizer was rotated from this position, however, the bees changed the direction of their dances.

Experiments performed by von Frisch and others showed that it is primarily the direction of the linearly polarized component of skylight at ultraviolet wavelengths that the bees use for navigation. The degree of polarization is much less important; a value as small as $d_l \approx 10\,\%$ is sufficient. This makes sense from the biological point of view, because the direction of polarization varies less with atmospheric changes than the degree of polarization, particularly at ultraviolet wavelengths, and the relationship between the direction of polarization and the sun's position is more precise at ultraviolet wavelengths than at longer wavelengths.

A theory that explains how bees detect the polarization of skylight is given in References [82] and [83].

References

[1] H. Hertz, "Die Kräfte electrischer Schwingungen, behandelt nach der Maxwell'schen Theorie," *Ann. Physik*, Vol. 36, pp. 1–22, 1889.

[2] —, *Electric Waves*, Macmillan, London, 1893. Republication, Dover Publications, New York, 1962.

[3] K. Pearson and A. Lee, "On the Vibrations in the Field round a Theoretical Hertzian Oscillator," *Philos. Trans. Roy. Soc. London*, Series A, Vol. 193, pp. 159–88, 1900.

[4] S. A. Schelkunoff and H. T. Friis, *Antennas, Theory and Practice*, Wiley, New York, 1952.

[5] G. Marconi, "Transatlantic Wireless Telegraphy," Friday Evening Discourse at the Royal Institution, March 13, 1908, in *Wireless Telegraphy*, E. Eastwood, editor, Wiley, New York, 1974.

[6] —, "Wireless Telegraphic Communication," Nobel Lecture, December 11, 1909, in *Nobel Lectures, Physics 1901–1921*, Elsevier, New York, 1967.

[7] W. J. Baker, *A History of the Marconi Company*, St. Martin's Press, New York, 1971.

[8] T. L. Eckersley, "An Investigation of Transmitting Aerial Resistances," *J. IEE*, Vol. 60, pp. 581–604, 1922.

[9] J. Zenneck, *Wireless Telegraphy*, McGraw-Hill, New York, 1915.

[42] When the bee is on a vertical surface in the dark, the angle between the run and the direction of gravity is used to indicate the bearing with respect to the sun of the food source.

[10] A. Fürst, *Im Bannkreis Von Nauen*, Deutsche Verlags-Anstalt, Berlin, 1922.
[11] A. E. Kennelly, "On the Elevation of the Electrically-Conducting Strata of the Earth's Atmosphere," *Electrical World and Engineer*, Vol. 39, p. 473, March 15, 1902.
[12] O. Heaviside, "Theory of Electric Telegraphy," *Encyclopedia Britannica*, 10th Edition, 1902. Also in *Electromagnetic Theory*, Vol. III, p. 335, "The Electrician" Printing and Publishing Co., London, 1912. Republication, Chelsea, New York, 1971.
[13] E. V. Appleton and M. A. F. Barnett, "Local Reflection of Wireless Waves from the Upper Atmosphere," *Nature*, Vol. 115, pp. 333–4, March 7, 1925.
[14] —, "On Some Direct Evidence for Downward Atmospheric Reflection of Electric Rays," *Proc. Roy. Soc. London*, A, Vol. 109, pp. 621–41, 1925.
[15] E. V. Appleton, "Wireless Studies of the Ionosphere," *J. IEE*, Vol. 71, pp. 642–50, 1932.
[16] R. F. Harrington, *Time-Harmonic Electromagnetic Fields*, McGraw-Hill, New York, 1961.
[17] G. S. Smith, "Loop Antennas," in *Antenna Engineering Handbook*, 3rd Edition, R. C. Johnson and H. Jasik, editors, McGraw-Hill, New York, 1993.
[18] —, "Radiation Efficiency of Electrically Small Loop Antennas," *IEEE Trans. Antennas Propagat.*, Vol. AP-20, pp. 656–7, September 1972.
[19] R. W. P. King, R. B. Mack, and S. S. Sandler, *Arrays of Cylindrical Dipoles*, Cambridge University Press, Cambridge, 1968.
[20] S. A. Schelkunoff, "A Mathematical Theory of Linear Arrays," *Bell Syst. Tech. J.*, Vol. 22, pp. 80–107, January 1943.
[21] H. Bach and J. E. Hansen, "Uniformly Spaced Arrays," in R. E. Collin and F. J. Zucker, editors, *Antenna Theory*, Part I, Ch. 5, McGraw-Hill, New York, 1969.
[22] M. T. Ma, *Theory and Application of Antenna Arrays*, Wiley, New York, 1974.
[23] W. L. Stutzman and G. A. Thiele, *Antenna Theory and Design*, Ch. 3, Wiley, New York, 1981.
[24] C. A. Balanis, *Antenna Theory: Analysis and Design*, Ch. 6, Harper and Row, New York, 1982.
[25] J. D. Kraus, *Antennas*, 2nd Edition, McGraw-Hill, New York, 1988.
[26] H. Bach, "Directivity of Basic Linear Arrays," *IEEE Trans. Antennas Propagat.*, Vol. AP-18, pp. 107–10, January 1970.
[27] C. T. Tai, "The Nominal Directivity of Uniformly Spaced Arrays of Dipoles," *Microwave Journal*, Vol. 9, pp. 51–5, September 1964.
[28] S. A. Schelkunoff and H. T. Friis, *Antennas Theory and Practice*, Ch. 6, Wiley, New York, 1952.
[29] C. J. Bouwkamp and N. G. de Bruijn, "The Problem of Optimum Antenna Current Distribution," *Philips Res. Rep.*, Vol. 1, pp. 135–58, 1945/46.
[30] R. M. Wilmotte, "Note on Practical Limitations in Directivity of Antennas," *Proc. IRE*, Vol. 36, p. 878, July 1948.
[31] N. Yaru, "A Note on Super-Gain Antenna Arrays," *Proc. IRE*, Vol. 39, pp. 1,081–5, September 1951.
[32] A. Bloch, R. G. Medhurst, and S. D. Pool, "A New Approach to the Design of Super-Directive Aerial Arrays," *Proc. IEE*, Vol. 100, Pt. III, pp. 303–14, September 1953.
[33] H. B. G. Casimir, "On Supergain Antennae," in *Old and New Problems in Elementary Particles*, G. Puppi, editor, Academic Press, New York, 1968.
[34] R. E. Kleinman, "The Rayleigh Region," *Proc. IEEE*, Vol. 53, pp. 848–56, August, 1965.

[35] Lord Rayleigh, "On the Electromagnetic Theory of Light," *Philos. Mag.*, Vol. 12, pp. 81–101, August 1881.
[36] —, "On the Passage of Waves through Apertures in Plane Screens, and Allied Problems," *Philos. Mag.*, Vol. 43, pp. 259–72, April 1897.
[37] —, "On the Incidence of Aerial and Electric Waves upon Small Obstacles in the form of Ellipsoids or Elliptic Cylinders, and on the Passage of Electric Waves through a Circular Aperture in a Conducting Screen," *Philos. Mag.*, Vol. 44, pp. 28–52, July 1897.
[38] V. Twersky, "Rayleigh Scattering," *Appl. Opt.*, Vol. 3, pp. 1,150–62, October 1964.
[39] C. J. Bouwkamp, "On the Diffraction of Electromagnetic Waves by Small Circular Disks and Holes," *Philips Res. Rep.*, Vol. 5, pp. 401–22, December 1950.
[40] J. S. Hey, G. S. Stewart, J. T. Pinson, and P. E. V. Prince, "The Scattering of Electromagnetic Waves by Conducting Spheres and Discs," *Proc. Phys. Soc.* (London), Vol. 69B, pp. 1,038–49, October 1956.
[41] W. H. Eggimann, "Higher-Order Evaluation of Electromagnetic Diffraction by Circular Disks," *IRE Trans. Microwave Theory and Tech.*, Vol. MTT-9, pp. 408–18, September 1961.
[42] H. A. Bethe, "Theory of Diffraction by Small Holes," *Phys. Rev.*, Vol. 66, pp. 163–82, October 1944.
[43] C. J. Bouwkamp, "On Bethe's Theory of Diffraction by Small Holes," *Philips Res. Rep.*, Vol. 5, pp. 321–32, October 1950.
[44] R. E. Collin, *Field Theory of Guided Waves*, 2nd Edition, Section 7.3, IEEE Press, Piscataway, NJ, 1991.
[45] J. Meixner and W. Andrejewski, "Strenge Theorie der Beugung ebener elektromagnetischer Wellen an der vollkommen leitenden Kreisscheibe und an der kreisförmigen Öffnung im vollkommen leitenden ebenen Schirm," *Ann. Physik*, Vol. 7, pp. 157–68, 1950.
[46] W. Andrejewski, "Die Beugung elektromagnetischer Wellen an der leitenden Kreisscheibe und an der kreisförmigen Öffnung im leitenden ebenen Schirm," *Z. Angew. Phys.* Vol. 5, pp. 178–86, May 1953.
[47] R. DeVore, D. B. Hodge, and R. G. Kouyoumjian, "Backscattering Cross Sections of Circular Disks for Arbitrary Incidence," *J. Appl. Phys.*, Vol. 42, pp. 3,075–83, July 1971.
[48] J. H. Van Vleck, F. Bloch, and M. Hamermesh, "Theory of Radar Reflection from Wires or Thin Metallic Strips," *J. Appl. Phys.*, Vol. 18, pp. 274–94, March 1947.
[49] D. -F. Arago, *Oeuvres Complétes de François Arago*, Vol. 7, pp. 394–5, Gide, Paris, 1858.
[50] A. H. Taylor and G. P. Kerr, "The Distribution of Energy in the Visible Spectrum of Daylight," *J. Opt. Soc. Am.*, Vol. 31, pp. 3–8, January 1941.
[51] R. A. Richardson and E. O. Hulbert, "Sky-Brightness Measurements Near Bocaiuva, Brazil," *J. Geophys. Res.*, Vol. 54, pp. 215–27, September 1949.
[52] S. Chandrasekhar and D. D. Elbert, "The Illumination and Polarization of the Sunlit Sky on Rayleigh Scattering," *Trans. Am. Phil. Soc.*, Vol. 44, pp. 643–54, 1954.
[53] Z. Sekera, "Recent Developments in the Study of the Polarization of Sky Light," in *Advances in Geophysics*, Vol. 3, H. E. Landsberg, editor, pp. 43–77, Academic Press, New York, 1956.
[54] G. V. Rozenberg, "Light Scattering in the Earth's Atmosphere," *Usp. Fiz. Nauk*, Vol. 71, pp. 173–213, June 1960.
[55] S. T. Henderson, *Daylight and its Spectrum*, 2nd Edition, Wiley, New York, 1977.
[56] A. T. Young, "Rayleigh Scattering," *Physics Today*, Vol. 35, pp. 2–8, January 1982.

[57] C. F. Bohren, editor, *Selected Papers on Scattering in the Atmosphere*, SPIE, Bellingham, WA, 1989. This reference contains papers [52], [54], [56], [60], [61], [72], and [79].
[58] J. Tyndall, "On the Blue Colour of the Sky, the Polarization of Skylight, and on the Polarization of Light by Cloudy Matter Generally," *Proc. Roy. Soc. London*, Vol. 17, pp. 223–33, 1869.
[59] —, "On the Action of Rays of High Refrangibility upon Gaseous Matter," *Phil. Trans. Roy. Soc. London*, Vol. 160, pp. 333–65, 1870.
[60] J. W. Strutt, "On the Light from the Sky, its Polarization and Colour," *Philos. Mag.*, Vol. 41, pp. 107–20, February 1871, and pp. 274–9, April 1871.
[61] Lord Rayleigh, "On the Transmission of Light Through an Atmosphere Containing Small Particles in Suspension, and on the Origin of the Blue Sky," *Philos. Mag.*, Vol. 47, pp. 375–84, April 1899.
[62] H. Hurwitz, Jr., "The Statistical Properties of Unpolarized Light," *J. Opt. Soc. Am.*, Vol. 35, pp. 525–31, August 1945.
[63] J. W. Goodman, *Statistical Optics*, Wiley, New York, 1985.
[64] M. P. Thekaekara, R. Kruger, and C. H. Duncan, "Solar Irradiance Measurements from a Research Aircraft," *Appl. Opt.*, Vol. 8, pp. 1,713–32, August 1969.
[65] M. P. Thekaekara, "Extraterrestrial Solar Spectrum, 3000–6100 Å at 1-Å Intervals," *Appl. Opt.*, Vol. 13, pp. 518–22, March 1974.
[66] J. D. Kraus, *Radio Astronomy*, McGraw-Hill, New York, 1966.
[67] H. Zirin, *Astrophysics of the Sun*, Cambridge University Press, Cambridge, 1988.
[68] S. Chandrasekhar, *Radiative Transfer*, Oxford University Press, 1950. Republication, Dover Publications, New York, 1960.
[69] E. Wolf, "Coherence Properties of Partially Polarized Electromagnetic Radiation," *Nuovo Cimento*, Vol. 13, pp. 1,165–81, September 1959.
[70] A. R. von Hippel, *Dielectrics and Waves*, M. I. T. Press, Cambridge, MA, 1954.
[71] P. M. Banks and G. Kockarts, *Aeronomy*, Part A, Academic Press, New York, 1973.
[72] R. Tousey and E. O. Hulbert, "Brightness and Polarization of the Daylight Sky at Various Altitudes above Sea Level," *J. Opt. Soc. Am.*, Vol. 37, pp. 78–92, February 1947.
[73] R. L. Gregory, *Eye and Brain*, 2nd Edition, McGraw-Hill, New York, 1973.
[74] T. Gehrels, "Wavelength Dependence of the Polarization of the Sunlit Sky," *J. Opt. Soc. Am.*, Vol. 52, pp. 1,164–73, October 1962.
[75] H. C. van de Hulst, *Light Scattering by Small Particles*, Wiley, New York, 1957. Republication, Dover Publications, New York, 1981.
[76] M. Kerker, *The Scattering of Light and Other Electromagnetic Radiation*, Academic Press, New York, 1969.
[77] C. F. Bohren and D. R. Huffman, *Absorption and Scattering of Light by Small Particles*, Wiley, New York, 1983.
[78] G. Gamow, *One Two Three . . . Infinity: Facts and Speculations of Science*, Viking Press, New York, 1948.
[79] R. Penndorf, "Tables of the Refractive Index for Standard Air and the Rayleigh Scattering Coefficient for the Spectral Region between 0. 2 and 20. 0 μ and Their Application to Atmospheric Optics," *J. Opt. Soc. Am.*, Vol. 47, pp. 176–82, February 1957.
[80] K. von Frisch, "Gelöste und ungelöste Rätsel der Bienensprache," *Naturwissenschaften*, Vol. 35, No. 2, pp. 38–43, 1948.

[81] —, *The Dance Language and Orientation of Bees*, Harvard University Press, Cambridge, MA, 1967.

[82] R. Wehner and S. Rossel, "The Bee's Celestial Compass – A Case Study in Behavioural Neurobiology," in *Experimental Behavioral Ecology and Sociobiology*, B. Hölldobler and M. Lindauer, editors, Sinauer Associates, Sunderland, MA, 1985.

[83] S. Rossel and R. Wehner, "Polarization Vision in Bees," *Nature*, Vol. 323, pp. 121–31, September 1986.

Problems

7.1 Write expressions for the volume densities of charge and current, ρ and $\vec{\mathcal{J}}$, of the electric dipole shown in Figure 7.1a. Make use of the Dirac delta function and the Heaviside unit-step function. Show that (7.1) and (7.2) follow from application of the equation of continuity for electric charge to these expressions.

7.2 a) Starting with the Taylor series (7.10), fill in all of the steps required to obtain the vector potential for the electric dipole (7.13).
b) Obtain the magnetic field (7.24) for the electric dipole from the expression for the vector potential (7.17).

7.3 Obtain expressions (7.29) and (7.30) for the Poynting vector and the power instantaneously passing outward through a spherical surface surrounding the electric dipole.

7.4 For the special case of harmonic time dependence, evaluate the inequalities (7.53) and (7.54) in the root-mean-square sense and show that they imply that $2k_o h = 4\pi h/\lambda_o \ll 1$.

7.5 Show that Maxwell's equations with only electric sources, (7.61)–(7.65), become Maxwell's equations with only magnetic sources, (7.66)–(7.70), on the substitution of (7.71)–(7.74).

7.6 Consider the couplet of electrically short linear antennas, each excited with a current that is a Gaussian function of time. For the unilateral end fire couplet, show that the choice (7.101) for the spacing between elements will cause partial cancellation of the field in the backward direction ($\phi = \pi$). This choice causes the peak of the positive lobe of the signal from element 2 to be superimposed on the peak of the negative lobe of the signal from element 1.

7.7 Obtain the expression (7.116) for the total time-average power radiated by the uniform array of electrically short linear antennas. Hint: In the integral (7.115), change the coordinates to those shown in Figure P7.1 (ψ, χ), and use the relationship [26]

$$\left[\frac{\sin(Nx/2)}{N\sin(x/2)}\right]^2 = \frac{1}{N} + \frac{2}{N^2}\sum_{m=1}^{N-1}(N-m)\cos mx$$

to rewrite the integrand.

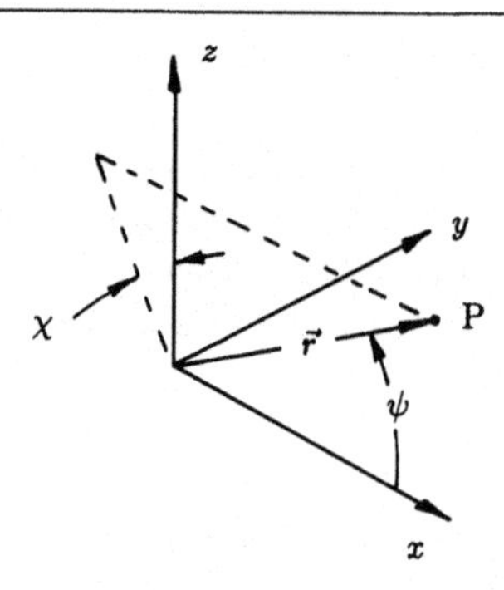

Fig. P7.1. New coordinates.

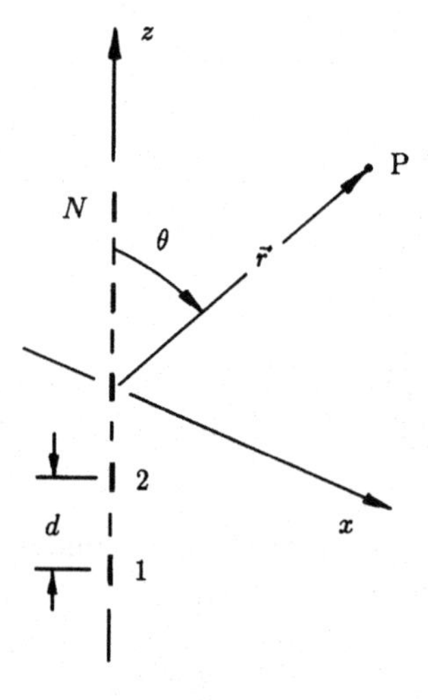

Fig. P7.2. Collinear array of linear antennas.

7.8 Consider the expression for the directivity, $D(\theta = \pi/2, \phi = 0)$, of the end fire couplet in the limit as $k_o d \to 0$ and $\delta \to \pi$. Show that maximum directivity ($D = 5.25$) is obtained when this limit is taken such that $\delta = \pi - (2/5)k_o d$, $k_o d \to 0$.

7.9 A uniform *collinear array* is formed by placing N equally spaced, electrically short linear antennas along the z axis. All of the antennas are parallel to the z axis, as shown in Figure P7.2, and they are excited with the time-harmonic currents given in (7.109).

a) Determine general expressions for the field radiated by the array and the directivity of the array.
b) For the case $\delta = 0$, make a graph, like the ones in Figure 7.30, for the directivity in the horizontal plane, $D(\theta = \pi/2)$, versus $k_o d$.
c) What is the maximum directivity for the collinear couplet ($N = 2$)? Make a graph (pattern) showing the directive characteristics of the radiation for this case.

7.10 The electrostatic polarizability of a dielectric sphere of relative permittivity ε_r and diameter d is

$$\alpha_e = \frac{\pi d^3}{2}\left(\frac{\varepsilon_r - 1}{\varepsilon_r + 2}\right).$$

When the sphere is electrically small ($kd = \sqrt{\varepsilon_r} k_o d \ll 1$), the polarizability for the time-harmonic field is assumed to be the same as that for the static field.

a) Derive the above result for α_e. The sphere is placed in the uniform incident electrostatic field $\vec{E}_i = E_o^i \hat{z}$. Assume that the total electric field inside the sphere is also uniform and points in the z direction. Let the total electric field outside the sphere be the incident field plus the field of an electric dipole $\vec{p} = p\hat{z}$. Determine p, thus α_e, by satisfying the boundary conditions at the surface of the sphere. (Note that the magnetic polarizability α_m for the electrically small, dielectric sphere is insignificant).
b) What are the scattering cross sections, σ_B and σ_T, for the electrically small, dielectric sphere?

7.11 a) Determine the pair of angles ϕ_i and ψ_i for the incident plane wave that gives the maximum backscattering cross section for the electrically small, perfectly conducting, circular disc. What is the backscattering cross section for this incident wave?
b) Determine ϕ_i and ψ_i as in part a but for the case where the total scattering cross section is maximized. What is the total scattering cross section for this incident wave?

7.12 Let the equivalent dipoles for the electrically small, perfectly conducting, circular disc be $\vec{p}_d$ and $\vec{m}_d$. These dipoles are used to calculate the radiated part of the scattered field for the disc. The complementary problem of an electrically small, circular aperture in a plane, perfectly conducting screen can be analyzed using Babinet's principle with the results for the disc. The total field for the aperture in the right half space, $y > 0$ in Figure 7.42a, is simply obtained from the scattered field of the disc.

Obtain the expressions in (7.166) and (7.167) for the equivalent electric and magnetic dipoles, $\vec{p}_a$ and $\vec{m}_a$, that produce the same field in the right half space as the aperture. First express $\vec{p}_a$ and $\vec{m}_a$ in terms of $\vec{p}_d$ and $\vec{m}_d$ using Babinet's principle; then determine the relationship of these dipoles to the field of the plane wave incident on the aperture.

7.13 In this problem you are to make an estimate of the equivalent magnetic dipole moment for the electrically short, perfectly conducting, thin wire and show that it is insignificant. Use the configuration in Figure 7.44 and let $\theta_i = \pi/2,\ \psi_i = 0$.

Assume that an additional surface current $\vec{J}_s'$ is on the cylinder. It can be modeled as a current I_o' that runs up the front of the cylinder ($x = 0,\ y = -a$), over the top, down the back ($x = 0,\ y = a$), and across the bottom. This current forms a closed loop in the y-z plane. The magnetic field of this current is to cancel the incident magnetic field at the center of the wire ($x = 0,\ y = 0,\ z = 0$).

Determine the equivalent magnetic dipole moment $\vec{m}$ of this current (loop). Compare the radiated electric field, $\vec{E}^{sr}$, of the equivalent magnetic dipole moment with that for the equivalent electric dipole moment, and show that the former is insignificant when the wire is thin ($a/h \ll 1$).

7.14 *Stokes parameters for natural light*

a) The definitions for the Stokes parameters of a time-harmonic field are given in Equations (2.72a–d). Use the equivalent, time-harmonic representation for natural light [(7.197) with the average over γ] to show that the Stokes parameters for natural light are $\overline{S_0^i} = \overline{I^i},\ \overline{S_1^i} = 0,\ \overline{S_2^i} = 0$, and $\overline{S_3^i} = 0$.

b) Consider a wave composed of a totally polarized component (p) plus natural light (n). The electric fields (phasors) for the two components are

$$\vec{E}^p = E_x^p \hat{x} + E_y^p \hat{y}$$

and

$$\vec{E}^n = E^n(\cos\gamma\hat{x} + \sin\gamma\hat{y}).$$

Show that each of the Stokes parameters for this wave is the sum of the corresponding Stokes parameters for the individual components (e.g., $\overline{S_0} = S_0^p + \overline{S_0^n}$).

7.15 For an ideal gas

$$pV = n_m \mathcal{R} T,$$

where p, V, n_m, and T are the absolute pressure, volume, number of moles, and the absolute temperature, respectively. Assume that the *gas constant*, $\mathcal{R} = 8.315\,\mathrm{J/mol{\cdot}K}$, is known.

a) A gas is contained in a vessel, and the pressure and temperature are measured. Measurements are also made of the extinction coefficient γ and the index of refraction n at the wavelength λ_o. Show that these measured quantities (p, T, γ, n, and λ_o) can be used to determine *Avogadro's number* $N_A = 6.022 \times 10^{23}\,\mathrm{mol}^{-1}$, the number of elementary entities in a mole.

This is one of the early methods used to determine Avogadro's number; a similar calculation was made by Lord Rayleigh in 1899 [61].

b) Assume that the measurements are made for air and the following values are obtained: $p = 1.010 \times 10^5\,\mathrm{N/m^2}$, $T = 273.2\,\mathrm{K}$, $\gamma = 1.226 \times 10^{-5}\,\mathrm{m^{-1}}$, $n - 1 = 2.931 \times 10^{-4}$, and $\lambda_o = 5{,}500\,\text{Å}$.

What do these measurements predict for Avogadro's number?

7.16 In the atmosphere, the concentration of molecules $N_V(h)$ varies with the height h above the surface of the Earth. Assume the total height of the atmosphere is h_{top} and the irradiance of the light incident at the top of the atmosphere is $\overline{I^i}(h_{\mathrm{top}})$.

a) Show that when the sun is at zenith, the irradiance at the surface of the Earth is

$$\overline{I^i}(0) = \overline{I^i}(h_{\text{top}})e^{-\gamma_o H},$$

where γ_o is the extinction coefficient at the surface and

$$H = \int_{h=0}^{h_{\text{top}}} \frac{N_V(h)}{N_V(0)} dh$$

is called the "reduced height" of the atmosphere. This corresponds to the height of a column of air with constant concentration equal to $N_V(0)$ that gives the same attenuation as the actual atmosphere.

b) Assume that the atmosphere can be modeled as a homogeneous layer of concentration $N_V(0)$ and height H surrounding the Earth. A typical value is $H \approx 8\,\text{km}$, whereas the radius of the Earth is $r_e \approx 6.37 \times 10^3\,\text{km}$; thus, $H \ll r_e$. Using this inequality, obtain an expression for the irradiance at the surface of the Earth at sunset. Compare the lengths of the paths light travels through the atmosphere when the sun is at zenith and at sunset.

c) Estimate the ratio of transmitted irradiance to incident irradiance for red and violet light when the sun is at zenith and at sunset. Use the following parameters:

$$n - 1 = 2.94 \times 10^{-4}$$

$$N_V(0) = 2.5 \times 10^{25}\,\text{m}^{-3}.$$

7.17 Starting with the current given in (7.184), determine the scattered electric field $\vec{E}^{sr}$ of the electrically short wire. Use this electric field with the forward scattering theorem, Equation (4. 67), to determine the total scattering cross section σ_T of the wire. Compare your result with that given for σ_T in (7.190).

Notice that (7.190) was obtained after neglecting the small term $(k_o h)^3$ in the denominator of (7.184), whereas this term was required for the use of the forward scattering theorem.

8

Radiation from thin-wire antennas

Many practical antennas are formed by simply bending a piece of thin wire into an appropriate shape. We have seen examples in the previous chapter where we discussed linear (straight wire) and loop (circle of wire) antennas. There we were concerned with the radiation from such antennas when they are electrically small, that is, when the temporal variation of all signals is negligible during the time for light to travel across the antenna. Now we will remove that restriction and consider such antennas for a general excitation.

The arrangement is shown schematically in Figure 8.1. Here the perfectly conducting wire is bent so that its axis lies along the curve C. The wire need not form a closed curve as in the figure, but it can be open as is the case for the linear dipole. The antenna is characterized by the dimension d, which may be the length of the wire, the diameter of the loop, etc.

The wire is assumed to be both physically and electrically thin. By this we mean that the radius of the wire a is much smaller than the dimension of the

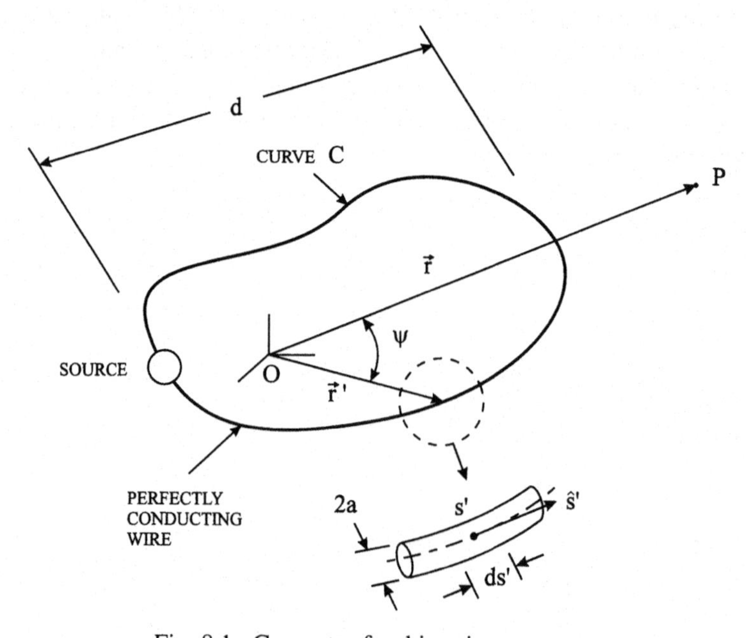

Fig. 8.1. Geometry for thin-wire antenna.

antenna, $a \ll d$, and that all signals vary negligibly during the time a/c for light to travel a distance equal to the radius of the wire (for harmonic time dependence, $a/\lambda_o \ll 1$).

We will be concerned with the calculation of the electromagnetic field at points away from the wire, such as at the point marked P in Figure 8.1. In this case, the thin-wire assumption greatly simplifies the analysis. The current, which is actually on the surface of the perfectly conducting wire, is replaced by a filament of current on the axis of the wire. The filamentary current $\mathcal{I}(s, t)$ is a function of the arc length s along the curve C, and it is in the direction of the unit tangent vector to the curve, $\hat{s}$. The charge per unit length $\mathcal{Q}(s, t)$ is related to the current through the equation of continuity (1.54):

$$\frac{\partial \mathcal{I}(s,t)}{\partial s} = -\frac{\partial \mathcal{Q}(s,t)}{\partial t}. \tag{8.1}$$

The radiated or far-zone electromagnetic field of the antenna is obtained by inserting the current into expressions (5.147) and (5.148). Since the filamentary current exists only on the axis of the wire, the volume integrals in these expressions reduce to line integrals:

$$\vec{\mathcal{E}}^r(\vec{r}, t) = \frac{-\mu_o}{4\pi r} \int_C \left[\frac{\partial \mathcal{I}(s', t')}{\partial t'} \right]_{t'=t_r} \left[\hat{s}' - \hat{r}(\hat{r} \cdot \hat{s}') \right] ds', \tag{8.2}$$

$$\vec{\mathcal{B}}^r(\vec{r}, t) = \frac{1}{c} \hat{r} \times \vec{\mathcal{E}}^r(\vec{r}, t) = \frac{-\mu_o}{4\pi c r} \int_C \left[\frac{\partial \mathcal{I}(s', t')}{\partial t'} \right]_{t'=t_r} (\hat{r} \times \hat{s}') ds'. \tag{8.3}$$

In these expressions, the retarded time is

$$t_r = t - (r - \hat{r} \cdot \vec{r}')/c = t - r/c + (r' \cos \psi)/c. \tag{8.4}$$

The radiated field is seen to be readily calculated once the current distribution in the wire is known.

8.1 Charge and current: Physical arguments

In general, accurate values for the charge and current on a thin-wire antenna cannot be obtained by simple analysis. One must resort to numerical methods [1–4]. We have already seen an example of this in Section 1.6. There we showed how the charge and current on a monopole antenna can be obtained by numerically solving a discretized form of Maxwell's equations.

Recall that the monopole is a vertical metallic rod placed over a metallic image plane. It is fed through the image plane by a coaxial transmission line, as in Figures 1.44 and 1.45. From the method of images presented in Section 4.7, we know that the monopole is equivalent to the dipole structure shown in Figure 8.2. Here we have not been concerned with the details of the imaging near the feed point (coaxial aperture). This simplification will not affect our discussion of the physical

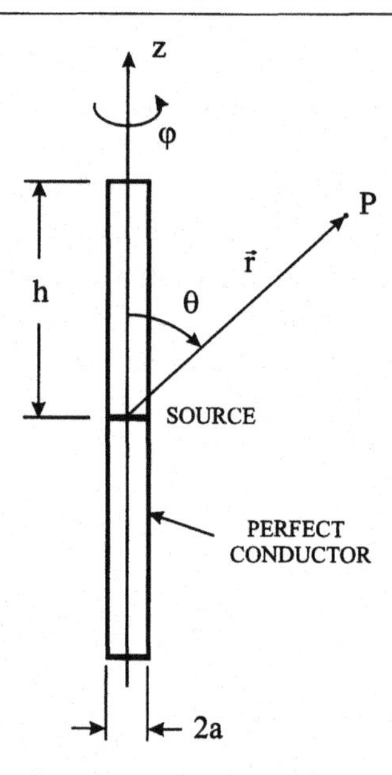

Fig. 8.2. Cylindrical dipole antenna.

principles. Since the radius of the dipole is assumed to be much smaller than its half length ($a \ll h$), this is a thin-wire antenna.

The results presented in Section 1.6 for the monopole also apply to the dipole. For example, the plot of the surface charge density on the monopole (Figure 1.47) now applies to the top half of the dipole ($0+ \leq z \leq h$). The charge on the bottom half of the dipole ($-h \leq z \leq 0-$) is the negative of that on the top half. The electric field surrounding the monopole can be imaged to obtain the field surrounding the dipole. This is illustrated in Figure 8.3, where the result shown in Figure 1.48c has been imaged. Notice that the wavefront W_{2R}, which for the monopole was interpreted as the reflection of the wavefront W_2 at the image plane, is now the wavefront that originates at the bottom end of the dipole.

For our purposes, we would like to have simple, approximate expressions for the charge and current on wire antennas that can be used with Equations (8.2) and (8.3) to gain physical insight into the process of radiation. We can obtain these approximations by examining the accurate results mentioned above for the monopole/dipole. From Figures 1.47–1.49, we see that the electric field moves out along the dipole roughly at the speed of light, producing charge on the conductors as it goes. The field is "guided" by the conductors. When this field reaches the ends of the dipole there is a reflection, and this moves back down the dipole, roughly at the speed of light, again producing charge on the conductors as it goes. Radiation occurs each time the field encounters a discontinuity in the conductor, such as at the source and ends (the spherical wavefronts marked W_1, W_2, W_{2R}, etc. in Figures 1.48, 1.49, and 8.3).

Now we will use these observations to construct approximations for the charge and current on the dipole. The source at the center of the dipole places charge on the two conductors of the dipole. We will let the charge per unit length deposited

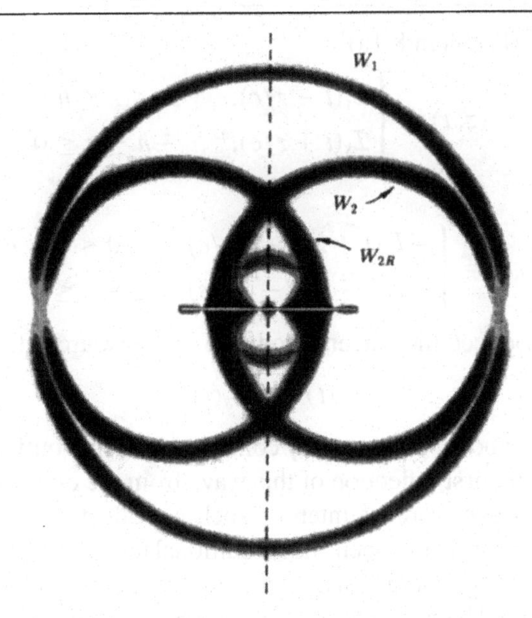

Fig. 8.3. Radiation of a Gaussian pulse from a cylindrical dipole antenna. Gray scale shows magnitude of electric field: $h/a = 65.8$, $\tau/\tau_a = 0.114$. (After Maloney et al. [4], © 1990 IEEE.)

by the source be $\mathcal{Q}_s(t)$ at $z = 0+$ (on the top conductor) and $-\mathcal{Q}_s(t)$ at $z = 0-$ (on the bottom conductor). For the example discussed in Section 1.6, this charge would be a Gaussian pulse in time. The charge per unit length on the dipole associated with the outgoing field is then approximately

$$\mathcal{Q}^+(z,t) = \begin{cases} \mathcal{Q}_s(t - z/c), & 0+ \le z \le h \\ -\mathcal{Q}_s(t + z/c), & -h \le z \le 0-. \end{cases} \tag{8.5}$$

Similarly, the charge per unit length associated with the first reflection of this field at the ends of the dipole is approximately

$$\mathcal{Q}^-(z,t) = \begin{cases} \mathcal{Q}_s(t + z/c - 2h/c), & 0+ \le z \le h \\ -\mathcal{Q}_s(t - z/c - 2h/c), & -h \le z \le 0-. \end{cases} \tag{8.6}$$

Here the + superscript indicates a wave traveling away from the source, while the − superscript indicates a wave traveling toward the source. An examination of Figures 1.47 and 1.48 clearly shows the approximate nature of (8.5) and (8.6). The charge density actually decreases in amplitude as the field travels along the antenna, and there is some change in shape of the pulse (distortion), particularly on reflection from the ends. Neither of these effects is included in (8.5) and (8.6). These effects generally become smaller as the wire is made thinner (as $a/h \to 0$) [5].

The current in the dipole in the direction $\hat{z}$ can be obtained by inserting the charge per unit length into the equation of continuity (8.1). For the two terms given in (8.5) and (8.6) we have (Problem 8.1a)[1]

$$\mathcal{I}^+(z,t) = \begin{cases} \mathcal{I}_s(t - z/c), & 0 \le z \le h \\ \mathcal{I}_s(t + z/c), & -h \le z \le 0 \end{cases} \tag{8.7}$$

and

$$\mathcal{I}^-(z,t) = \begin{cases} -\mathcal{I}_s(t + z/c - 2h/c), & 0 \le z \le h \\ -\mathcal{I}_s(t - z/c - 2h/c), & -h \le z \le 0. \end{cases} \tag{8.8}$$

Here we have introduced the current produced by the source at $z = 0$:

$$\mathcal{I}_s(t) = c\mathcal{Q}_s(t). \tag{8.9}$$

In the discussion above, we have only considered the first outward traveling wave on the dipole and the first reflection of this wave from the open ends. If subsequent reflections of the wave were of interest, such as reflections from the source or additional reflections from the open ends, additional terms like (8.5) and (8.6) would be needed to describe the charge. In this case, we would call the charge (8.5) $\mathcal{Q}_1^+$ and the charge (8.6) $\mathcal{Q}_1^-$. The next reflection would be from the source, and we would call it $\mathcal{Q}_2^+$. This would be followed by the second reflection from the open ends, $\mathcal{Q}_2^-$, and so on. The *total charge* on the dipole would be the sum of all of these terms:

$$\mathcal{Q}(z,t) = \mathcal{Q}_1^+ + \mathcal{Q}_1^- + \mathcal{Q}_2^+ + \mathcal{Q}_2^- + \cdots.$$

The motivation for our choice of approximations for the charge and current on the dipole antenna (8.5–8.9) came from our earlier investigation of the monopole antenna. However, these expressions are actually the charge and current on a section (length h) of air-filled transmission line terminated with an open circuit. Thus, if we argue that the dipole is similar to an open-circuited section of transmission line, transmission line theory also provides motivation for our choice of approximations.[2]

Now we wish to offer physical explanations for two of the interesting phenomena observed above [5–11]. First, we noticed that the source connected to the terminals of the dipole introduced pulses of charge (current) onto the conductors and that these pulses then appeared to travel along the conductors at the speed of light, being reflected at discontinuities, such as at the ends of the conductors. Of course, the

[1] It is important to recognize that there are two directions associated with each of these currents: the direction of travel for the *wave*, which is determined by the argument of $\mathcal{I}_s$, and the direction of the *current*, which is determined by the sign of $\mathcal{I}_s$. For example, consider the current on the bottom half of the dipole:

$$\mathcal{I}^+(z,t) = \mathcal{I}_s(t + z/c), \qquad -h \le z \le 0.$$

This is a wave traveling down the dipole, that is, a wave traveling in the $-\hat{z}$ direction. When $\mathcal{I}_s$ is positive, it is a current in the $+\hat{z}$ direction. The charge associated with this current is

$$\mathcal{Q}^+(z,t) = -\mathcal{Q}_s(t + z/c), \qquad -h \le z \le 0-.$$

When $\mathcal{Q}_s$ is positive [which occurs when $\mathcal{I}_s$ is positive; see (8.9)], a wave of negative charge travels in the $-\hat{z}$ direction; hence, the current is in the $+\hat{z}$ direction.

[2] The relationship between simple wire antennas and transmission lines is examined in detail in Reference [5].

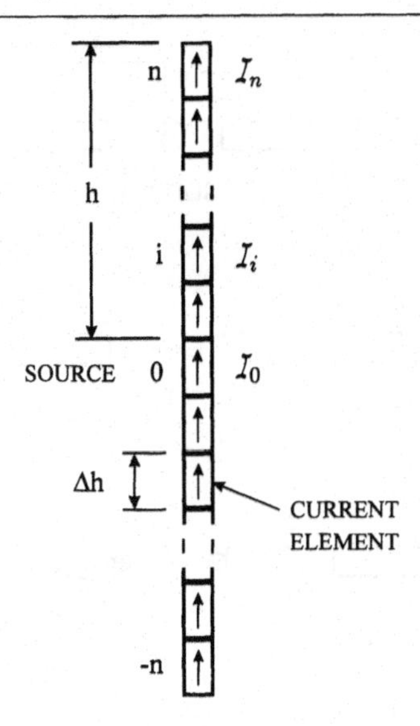

Fig. 8.4. Dipole antenna divided into $2n + 1$ small segments.

individual charges within the conductors – the electrons – do not travel the length of the conductor at the speed of light. What then is the mechanism that causes the pulses of charge to appear to move along the conductors at the speed of light? Second, we noticed that radiation originated at the points where a pulse of charge encountered a discontinuity in the conductor, such as at the ends. Why does the radiation appear to occur only at these points?

We will use the very simple model shown in Figure 8.4 for our argument. Here we have divided the top half of the dipole into a large number, n, of small segments ($i = 1, 2, \ldots, n$), each of length $\Delta h = h/n$. The bottom half of the dipole is divided similarly ($i = -1, -2, \ldots, -n$), and there is one additional segment ($i = 0$) for the source. Thus, there is a total of $2n + 1$ segments.

Our qualitative explanation for the apparent motion of the charge in the conductors will be based on Figures 8.5 and 8.6. Consider Figure 8.5a, which applies for times $t < 0$. Our source is simply a pair of stationary positive and negative charges of equal magnitude. They are superimposed so as to produce no electric field, and the charge and current in the conductors are zero.

At time $t = 0$ (Figure 8.5b) the two charges of the source are separated, producing the current $\mathcal{I}_0$. One can think of a nonelectrical source producing the charge separation, such as a mechanical force. The electromagnetic field produced by this current travels at the speed of light; it first reaches the centers of the segments 1 and -1 at the time $t = \Delta t = \Delta h/c$. In response to this field, now indicated as $\mathcal{E}_1$ in Figure 8.5c, a current $\mathcal{I}_1$ is produced in segment 1.[3] Because the electric field within the conductor must be zero, this current must be in a direction so as to produce a separation of charge whose electric field will cancel the field $\mathcal{E}_1$.

[3] The drawings for charge and current in these figures are only symbolic. The charge and current are actually on the surface of each segment of the perfect conductor.

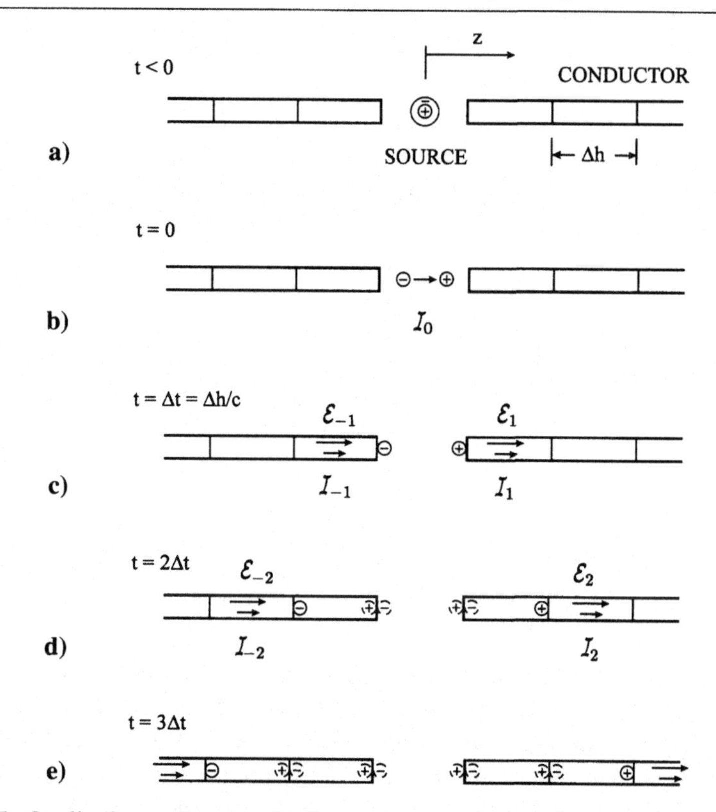

Fig. 8.5. Qualitative explanation for the apparent motion of charge along the conductors of the dipole antenna for the region near the source.

At the time $\Delta t = 2\Delta t$ (Figure 8.5d) an electric field, $\mathcal{E}_2$, has reached the center of segment 2, and a current is being produced in this segment. The electric field $\mathcal{E}_2$ is the result of the current $\mathcal{I}_1$ as well as the current $\mathcal{I}_0$. The aforementioned separation of charge has started to occur in segments 1 and -1. The negative charge at the left-hand end of segment 1 roughly cancels the positive charge of the source (these charges are drawn with dashed lines in the figure). There is a net positive charge at the right-hand end of segment 1.

The process described above is repeated as the field moves out along the conductors. At time step j, there is a net positive charge at the end of segment $j - 1$. For example, at $t = 3\Delta t$ (Figure 8.5e) there is a net positive charge at the right-hand end of segment 2. In this way charge appears to move at the speed of light out along the conductors.

A similar process occurs at the ends of the dipole, as illustrated for the right end ($z = h$) in Figure 8.6. At time $t = (n + 1)\Delta t$ (Figure 8.6b) a net positive charge is at the right-hand end of segment n. This charge now produces an electric field in the segment. The charge separation that results (Figure 8.6c) causes a net positive charge to occur at the left-hand end of segment n at time $t = (n + 2)\Delta t$. This process is repeated, and charge appears to move at the speed of light back down the conductor toward the source.

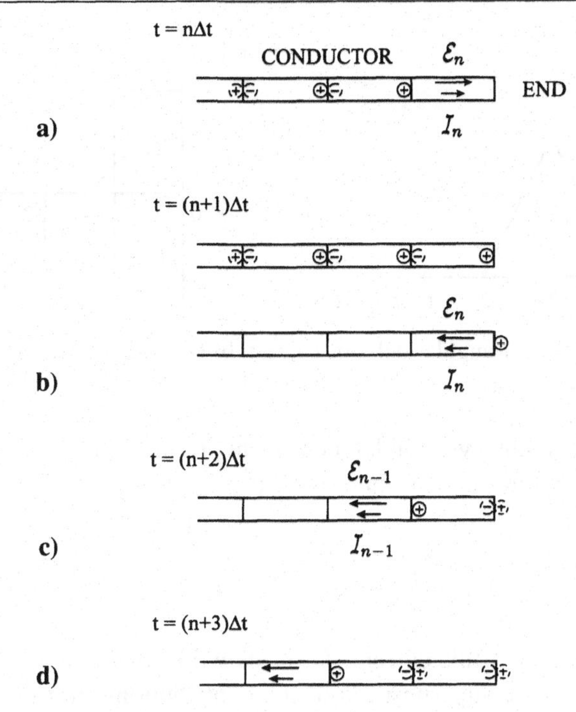

Fig. 8.6. Qualitative explanation for the apparent motion of charge along the conductors of the dipole antenna for the region near the end.

In the qualitative argument presented above, the source placed a small amount of positive charge on the end of the right-hand conductor at time $t \approx 0$. Of course, an actual source would continually place charge on the conductor, and this would be a definite function of time, $\mathcal{Q}_s(t)$, such as a Gaussian pulse. A pulse of charge, $\mathcal{Q}_s(t - z/c)$, would then appear to move out along the conductor at the speed of light.

The radiation from the dipole antenna is the superposition of the radiation from the $2n + 1$ individual segments shown in Figure 8.4. When the number of segments is large, each segment can be treated as an infinitesimal current element and the radiated field can be obtained from the results in Table 7.1. Here we will only consider the radiation due to the outgoing current (8.7). The current moment of the i-th segment is then

$$j_i(t) = \mathcal{I}_i(t)\Delta h = \mathcal{I}_s\big(t - |i|\Delta t\big)\Delta h, \tag{8.10}$$

and the radiated electric field in the broadside direction ($\theta = 90°$) of the entire antenna is

$$\begin{aligned}\vec{\mathcal{E}}^r(\vec{r}, t) &= \frac{\mu_o \Delta h}{4\pi r} \sum_{i=-n}^{n} \dot{\mathcal{I}}_s\big(t - r/c - |i|\Delta t\big)\hat{\theta} \\ &= \frac{\mu_o \Delta h}{4\pi r}\left[\dot{\mathcal{I}}_s(t - r/c) + 2\sum_{i=1}^{n} \dot{\mathcal{I}}_s(t - r/c - i\Delta t)\right]\hat{\theta}.\end{aligned} \tag{8.11}$$

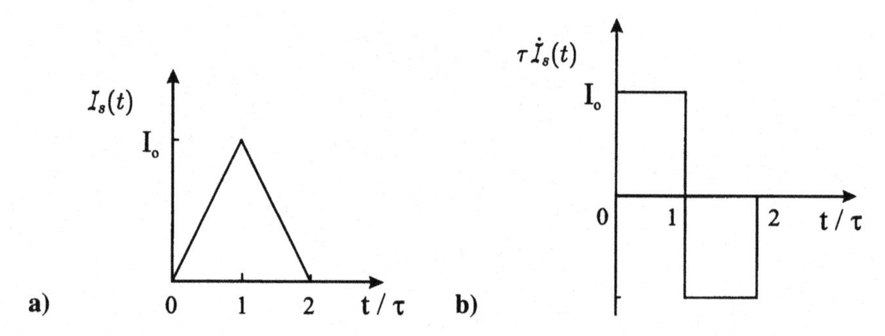

Fig. 8.7. a) Triangular pulse of current. b) Derivative of pulse.

To illustrate these results, we will let the current produced by the source at $z = 0$ be the triangular function shown in Figure 8.7a:

$$\mathcal{I}_s(t) = \begin{cases} 0, & |t/\tau - 1| > 1 \\ I_o\big(1 - |t/\tau - 1|\big), & |t/\tau - 1| < 1. \end{cases} \tag{8.12}$$

The temporal derivative of this current, $\dot{\mathcal{I}}_s$, is shown in Figure 8.7b. The radiated electric field (8.11) is the superposition of the contributions from all of the individual segments. Each contribution is proportional to $\dot{\mathcal{I}}_s$ and is shifted by the small increment of time Δt from the previous one. This is shown schematically for the first few segments in Figure 8.8a and for the entire dipole in Figure 8.8b. Here the time is normalized to $\tau_a = h/c$, the time for light to travel the half length of the dipole, and the duration of the current (8.12) is $2\tau = \tau_a/4$. Notice that a radiated field exists only near the times $t - r/c = 0$ and $t - r/c = \tau_a$; that is, radiation originates only from segments near the source and the ends of the dipole.[4] The radiated fields from the segments away from the source and ends of the dipole cancel. This is illustrated in the inset of Figure 8.8b, where portions of the radiated fields from segments i and $i + \tau/\Delta t$ are shown to cancel.[5] It may seem that this cancellation is a fortuitous result of our choice of current, the triangular function in Figure 8.7a. This effect, however, will occur for any current that is a smooth function of time, for the piecewise linear approximation to the current can be represented by a sum of overlapping triangles, as in Figure 8.9, and the cancellation will occur for each of the triangles.

Each pulse of the radiated field in Figure 8.8b is seen to resemble the current of the source, the triangular function shown in Figure 8.7a. In fact, this observation can be verified analytically by taking the limit as $n \to \infty$ ($\Delta t \to 0$). For the outgoing current we have been discussing, the radiated field in the direction $\theta = 90°$ (8.11)

[4] The numerical results in Figure 8.8 are for $2n + 1 = 65$ segments. The small differences in the waveforms at $t - r/c = 0$ and $t - r/c = \tau_a$ are caused by the two factors in (8.11) (the single term and the sum) adding differently at the two times.

[5] Here we have assumed that $i + \tau/\Delta t = i + n(\tau/\tau_a)$ is an integer. For the results shown in Figure 8.8, $n = 32$, $\tau = \tau_a/8$, so $i + \tau/\Delta t = i + 4$.

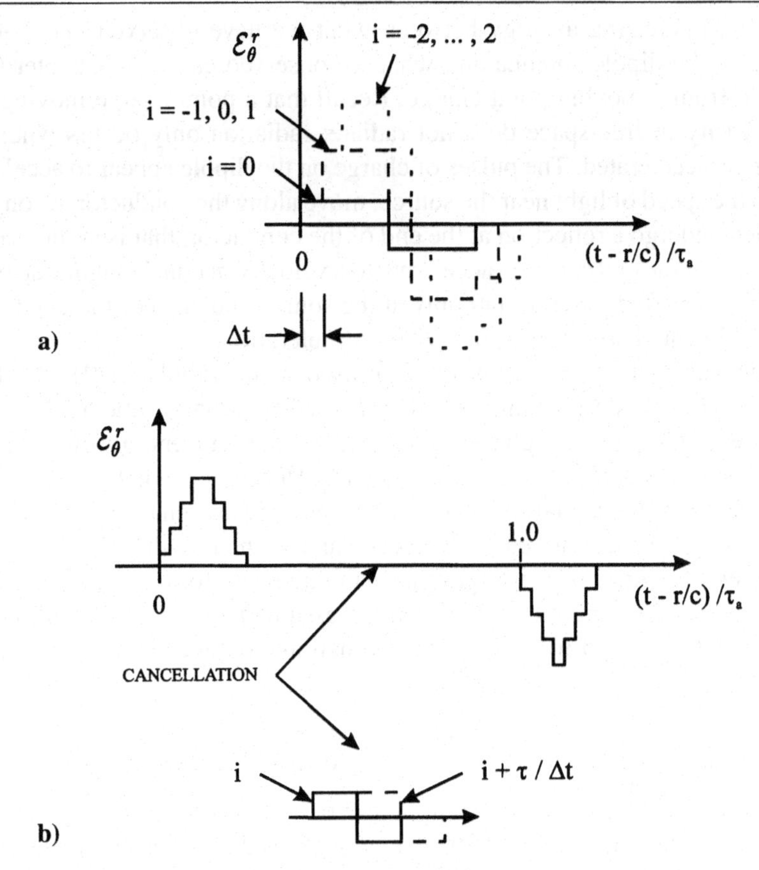

Fig. 8.8. Radiated electric field broadside to the dipole antenna ($\theta = 90°$). a) Contributions from individual segments. b) Total field for outgoing wave.

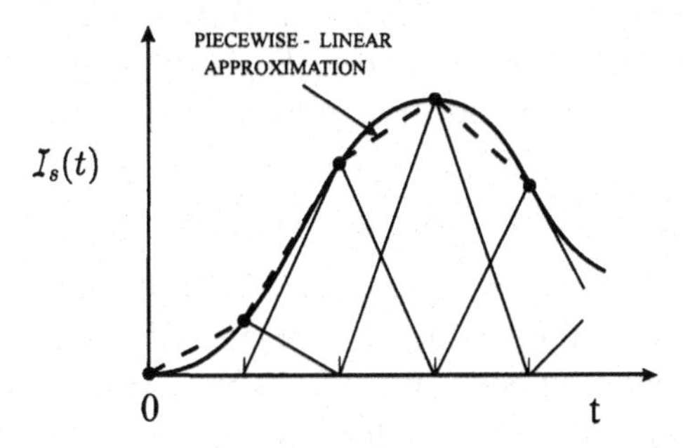

Fig. 8.9. Piecewise linear approximation of a general current.

becomes (Problem 8.2)

$$\vec{\mathcal{E}}^r(\vec{r}, t) = \frac{\mu_o c}{4\pi r} \lim_{n\to\infty} \left\{ \Delta t \left[\dot{\mathcal{I}}_s(t - r/c) + 2 \sum_{i=1}^{n} \dot{\mathcal{I}}_s(t - r/c - i\,\Delta t) \right] \right\} \hat{\theta}$$

$$= \frac{\mu_o c}{2\pi r} \left[\mathcal{I}_s(t - r/c) - \mathcal{I}_s(t - r/c - h/c) \right] \hat{\theta}. \tag{8.13}$$

There is an interesting analogy between what we have observed here for the radiation from the dipole antenna and what we observed earlier in Chapter 6 for the radiation from a moving point charge. Recall that a point charge moving at a constant velocity in free space does not radiate; radiation only occurs when the point charge is accelerated. The pulses of charge on the dipole appear to accelerate from rest to the speed of light near the source, move along the conductor at constant velocity, then undergo a reflection at the end of the conductor, that is, to decelerate to zero velocity at the end and then accelerate to a velocity of equal magnitude in the opposite direction. Radiation occurs only at the source and ends of the conductor, where the pulses of charge appear to undergo acceleration.

When currents such as (8.7) and (8.8) are used with formulas (8.2) and (8.3), we obtain an accurate representation for the radiated electromagnetic field of these currents. However, we must emphasize that these assumed currents are only approximations to the actual currents on the wire antenna. Hence, the radiated field is only an approximation to the actual radiated field of the wire antenna. This situation is similar to the one we encountered in Chapters 3 and 4 when we discussed radiation from apertures. Huygens' principle gave an accurate result for the radiated field of an assumed aperture distribution. However, the assumed aperture distribution and the resulting radiated field were only approximations to the results for the actual aperture.

8.2 Basic traveling-wave element; dipole antennas

In the rest of this chapter, we will examine the radiation from different thin-wire antennas. We will assume a current on an antenna, based on what we have learned from our discussion in the last section. The radiated electric field could be determined by simply inserting this current into Equation (8.2) and performing the indicated integration. However, it is more instructive, from the point of view of understanding the mechanism for radiation, to first study the radiation from a simple structure, which we will call the *basic traveling-wave element*, and then to view a more general antenna as a superposition of these basic elements.

8.2.1 Description of element

The basic traveling-wave element is shown schematically in Figure 8.10. A source is connected to one end of a straight, thin wire of length h; a reflectionless termination is connected to the opposite end.[6] We will assume that the source produces the following traveling wave of current in the wire:

$$\mathcal{I}(z,t) = \mathcal{I}_s(t - z/c)\big[U(z) - U(z-h)\big]. \tag{8.14}$$

Here we have used the Heaviside unit-step function to show that the current is confined to the length of wire $0 \le z \le h$. This current is essentially the outgoing

[6] The basic traveling-wave element is a model for the simplest traveling-wave antennas [12]. It can only be realized approximately in practice, because of our inability to construct a purely reflectionless termination.

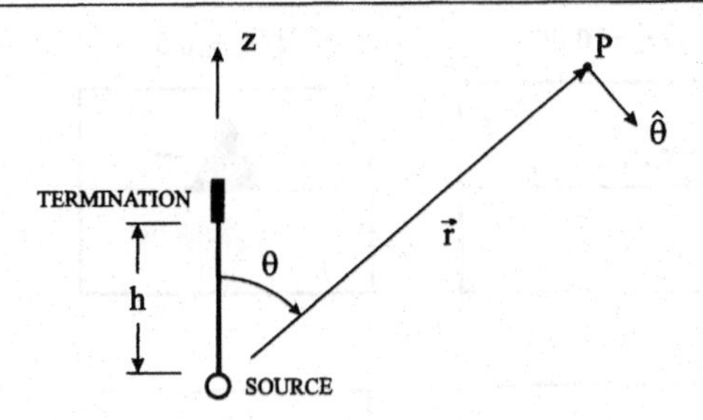

Fig. 8.10. Basic traveling-wave element.

wave on the top conductor of the dipole we discussed in the last section (8.7).[7] The charge per unit length that accompanies the current is

$$\begin{aligned}\mathcal{Q}(z,t) &= \mathcal{Q}_s(t-z/c)\big[U(z)-U(z-h)\big]\\ &\quad -\delta(z)\int_{t'=-\infty}^{t}\mathcal{I}_s(t')dt' + \delta(z-h)\int_{t'=-\infty}^{t}\mathcal{I}_s(t'-h/c)dt', \qquad (8.15)\end{aligned}$$

where $\mathcal{Q}_s$ and $\mathcal{I}_s$ satisfy (8.9). The terms in (8.15) that contain Dirac delta functions are required for conservation of charge at the ends of the wire (Problem 8.1b). As positive charge leaves the source ($z = 0$) and moves onto the wire, an equal amount of negative charge is left behind. Similarly, as positive charge leaves the end of the wire and enters the termination, positive charge accumulates at the termination ($z = h$). We did not require these terms in our discussion of the dipole. For the dipole, the amount of negative charge on the bottom conductor is equal to the amount of positive charge on the top conductor, and there is total reflection (no accumulation) of the charge at the open ends.

The radiated electric field for the basic traveling-wave element is obtained by inserting the current (8.14) into Equation (8.2). The various terms in this equation are: $s' = r' = z'$, $\hat{s}' = \hat{z}$, $\psi = \theta$, and

$$\hat{s}' - \hat{r}(\hat{r}\cdot\hat{s}') = \hat{z} - \hat{r}(\hat{r}\cdot\hat{z}) = -\sin\theta\,\hat{\theta};$$

so the radiated electric field becomes

$$\vec{\mathcal{E}}^r(\vec{r},t) = \frac{\mu_o}{4\pi r}\sin\theta\,\hat{\theta}\int_{z'=0}^{h}\left[\frac{\partial\mathcal{I}_s(t'-z'/c)}{\partial t'}\right]_{t'=t_r} dz', \qquad (8.16)$$

with

$$t_r = t - r/c + (z'/c)\cos\theta. \qquad (8.17)$$

Now we notice that the integrand in the above equation can be expressed as

$$\left[\frac{\partial\mathcal{I}_s(t'-z'/c)}{\partial t'}\right]_{t'=t_r} = \left(\frac{-c}{1-\cos\theta}\right)\frac{d\mathcal{I}_s[t-r/c-(z'/c)(1-\cos\theta)]}{dz'} \qquad (8.18)$$

[7] Notice that we didn't put the superscript + on this current. Since we only have one traveling wave on the antenna, it comprises the total current: $\mathcal{I}(z,t) = \mathcal{I}^+(z,t)$.

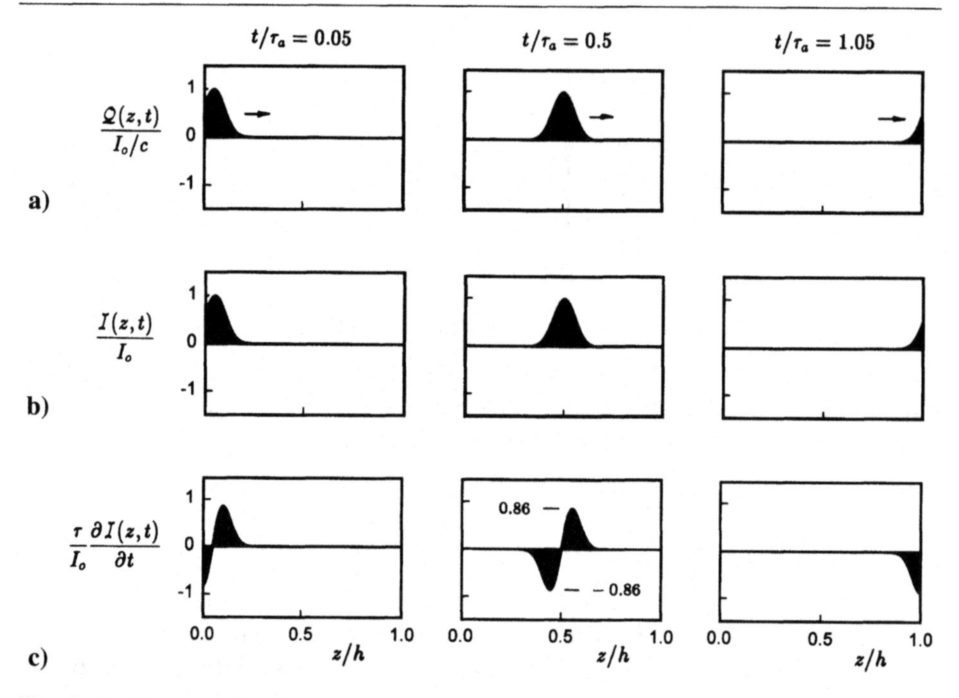

Fig. 8.11. Charge per unit length (a), current (b), and derivative of current (c) along basic traveling-wave element. Excitation is a Gaussian pulse of current; $\tau/\tau_a = 0.076$.

(see Problem 8.3). On substitution of (8.18) into (8.16), the radiated electric field becomes[8]

$$\vec{\mathcal{E}}^r(\vec{r}, t) = \frac{\mu_o c \sin\theta}{4\pi r(1 - \cos\theta)} \Big\{ \mathcal{I}_s(t - r/c) - \mathcal{I}_s\big[t - r/c - (h/c)(1 - \cos\theta)\big]\Big\}\hat{\theta}. \tag{8.19}$$

To illustrate this result, we will let the current of the source be a Gaussian function with characteristic time τ:

$$\mathcal{I}_s(t) = I_o e^{-(t/\tau)^2}. \tag{8.20}$$

A Gaussian pulse of charge then moves at the speed of light along the wire from the source to the termination. This charge, the current, and the temporal derivative of the current are shown in Figure 8.11 at three times: when the pulse is near the source, $t/\tau_a = 0.05$, when it is at the center of the wire, $t/\tau_a = 0.5$, and when it is near the termination, $t/\tau_a = 1.05$. For this illustration, $\tau/\tau_a = 0.076$, so roughly three pulses fit along the length of the wire.

The radiated electric field of this current is shown in Figure 8.12. We will frequently use diagrams like this to describe the radiated fields of antennas, so it is important that we understand their construction. At each of the nine angles shown

[8] At broadside, $\theta = 90°$, this is one half of the result we obtained earlier for the radiated electric field of the outward-going current on the dipole (8.13). As we shall show, the dipole with only outward-going currents consists of essentially two basic traveling-wave elements. Hence, at broadside it radiates twice the field of a single element.

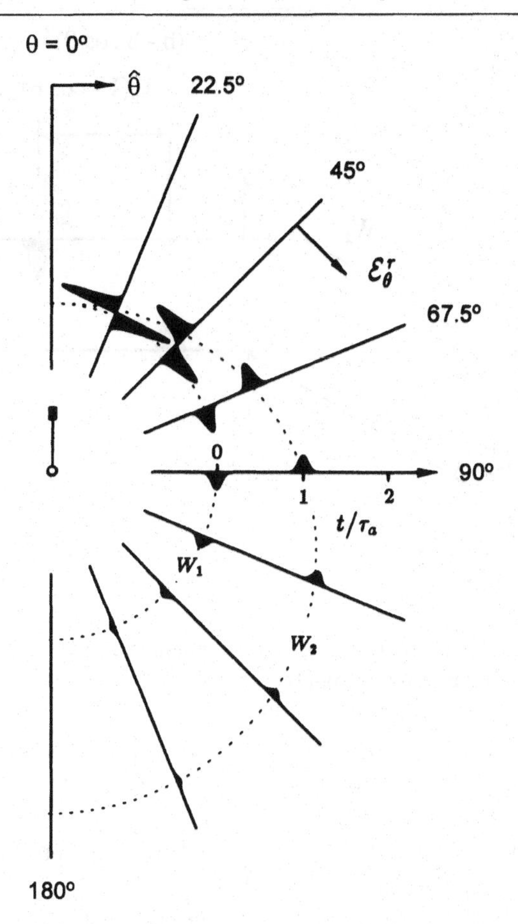

Fig. 8.12. Radiated electric field of basic traveling-wave element. Excitation is a Gaussian pulse of current; $\tau/\tau_a = 0.076$. Each graph shows the field as a function of the normalized time, t/τ_a, for a particular angle of observation θ.

$(0 \leq \theta \leq 180°)$ a graph is drawn for the radiated electric field $\mathcal{E}_\theta^r$ as a function of the normalized time t/τ_a. Each graph is the field that would be measured in the particular direction θ by an observer on the surface of a large sphere centered on the source. For each of the graphs, the electric field is positive on the side of the time axis for which $\hat{\theta}$ points away from the axis (the clockwise direction measured from the axis). For the time axis at the angle $\theta = 45°$, this is indicated by an arrow. The antenna is rotationally symmetric about the z axis; thus, Figure 8.12 applies for any angle ϕ.

Notice that the type of graph shown earlier in Figure 8.3 is distinctly different from the type of graph shown in Figure 8.12. Figure 8.3 shows the magnitude of the electric field throughout space for a single instant of time. In this representation, the spherical wavefronts centered on the source and the ends of the antenna (W_1, W_2, etc.) are circles centered on these same points. Figure 8.12 shows the electric field as a function of time for nine different points fixed in space. In this representation, the dashed lines connect all of the fields (times of arrival) associated with a particular

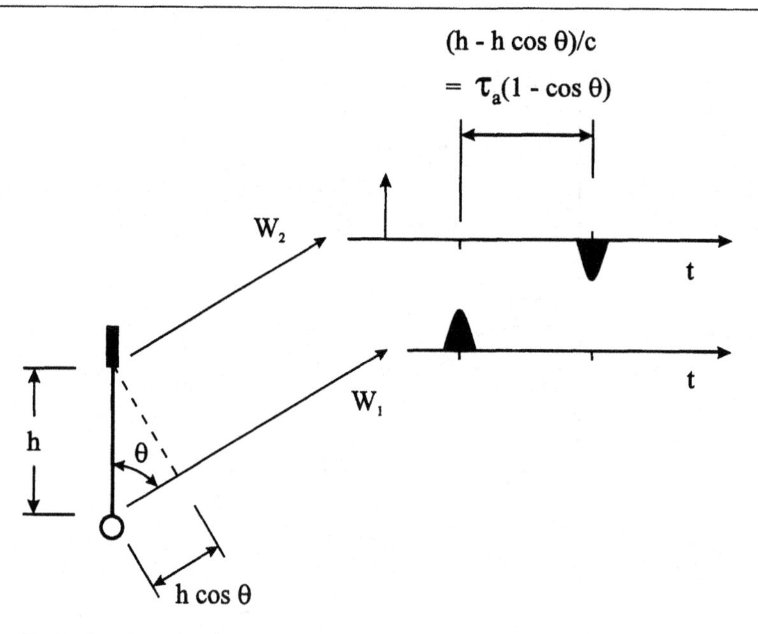

Fig. 8.13. Radiation from basic traveling-wave element. Spherical wavefronts are centered on the source (W_1) and the termination (W_2).

wavefront. For example, the dashed line marked W_1 connects all fields associated with the spherical wavefront centered on the source, while the dashed line marked W_2 connects all fields associated with the spherical wavefront centered on the termination. The dashed line W_1 is a circle, because the sphere for observation is centered on the source. The dashed line W_2 is not a circle.[9]

The field of each of the wavefronts, W_1 and W_2 in Figure 8.12, is seen to be a Gaussian pulse. The pulse for wavefront W_2 is delayed in time from that for wavefront W_1 by the amount $\tau_a(1-\cos\theta)$. As illustrated in Figure 8.13, this delay is due to the longer path for a signal traveling from the source to W_2 as compared to W_1. The signal for W_2 must travel from the source to the end of the element before traveling to the observation point, whereas the signal for W_1 travels from the source directly to the observation point. Notice that the sign for the pulse on W_2 (negative) is opposite to the sign for the pulse on W_1 (positive). This is analogous to what we would have for the radiated field of a point charge accelerated at the source and decelerated at the termination.

An examination of the graph for the temporal derivative of the current, $\partial\mathcal{I}/\partial t$ in Figure 8.11c, shows why radiation only originates near the ends of the element. Consider the case of radiation in the broadside direction ($\theta = 90°$). The electric field (8.16) then depends upon the spatial integral of

$$\left[\frac{\partial\mathcal{I}_s(t'-z'/c)}{\partial t'}\right]_{t'=t_r} = \frac{\partial\mathcal{I}_s(t-r/c-z'/c)}{\partial t}. \tag{8.21}$$

[9] The small picture of the antenna at the center of the drawing is included to show the orientation of the antenna. The length of the antenna in this picture is not related to the scales on the drawing.

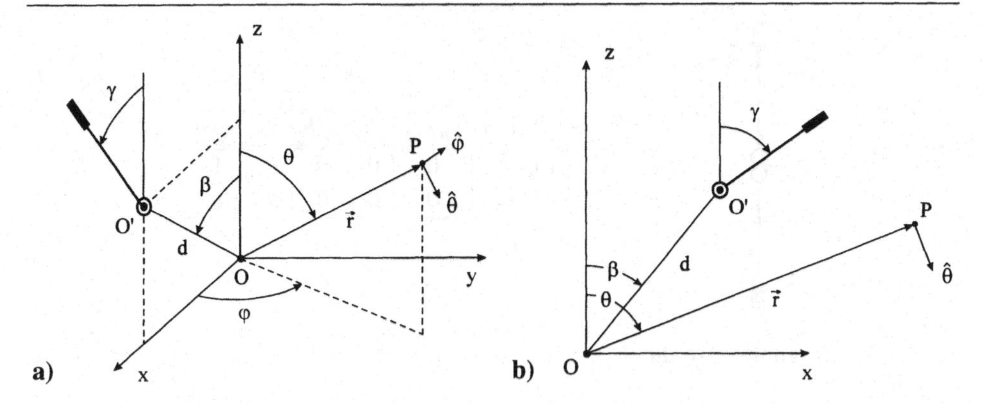

Fig. 8.14. Displaced basic traveling-wave element. a) General observation point P. b) Observation point P in x-z plane.

For a time when the entire pulse of current is out on the element, such as at $t/\tau_a = 0.5$ in the figure, the integral is zero; the areas under the positive and negative portions of the curve are then equal. However, for times when the pulse is near the ends, such as $t/\tau_a = 0.05$ and 1.05 in the figure, this integral is clearly nonzero.

For future use, we will need to know the radiated field of the basic traveling-wave element when the element is displaced from the z axis. In Figure 8.14a the element has been moved so that it lies in the x-z plane with the source end at O′. This translation is described by the distance d and the angle β ($0 \leq \beta < 2\pi$). In addition, the element has been rotated through the angle γ ($0 \leq \gamma < 2\pi$) with respect to the z axis. We will describe the radiated field of the displaced element in terms of the coordinates r, θ, and ϕ (Problem 8.4):

$$\vec{\mathcal{E}}^r(\vec{r}, t) = \frac{\mu_o c}{4\pi r(1 - \sin\gamma \sin\theta \cos\phi - \cos\gamma \cos\theta)}$$

$$\Big[(\cos\gamma \sin\theta - \sin\gamma \cos\theta \cos\phi)\hat{\theta} + (\sin\gamma \sin\phi)\hat{\phi}\Big]$$

$$\Big(\mathcal{I}_s\Big\{t - t_o - \big[r - d(\sin\beta \sin\theta \cos\phi + \cos\beta \cos\theta)\big]/c\Big\}$$

$$- \mathcal{I}_s\Big\{t - t_o - \big[r - d(\sin\beta \sin\theta \cos\phi + \cos\beta \cos\theta)\big]/c$$

$$- (h/c)(1 - \sin\gamma \sin\theta \cos\phi - \cos\gamma \cos\theta)\Big\}\Big). \tag{8.22}$$

Here t_o is a relative offset time associated with the source. When several elements are used, they may be excited at different times.

For the special case where the point of observation, P, lies in the plane of the element (x-z plane, $\phi = 0$), the simplified diagram in Figure 8.14b applies and

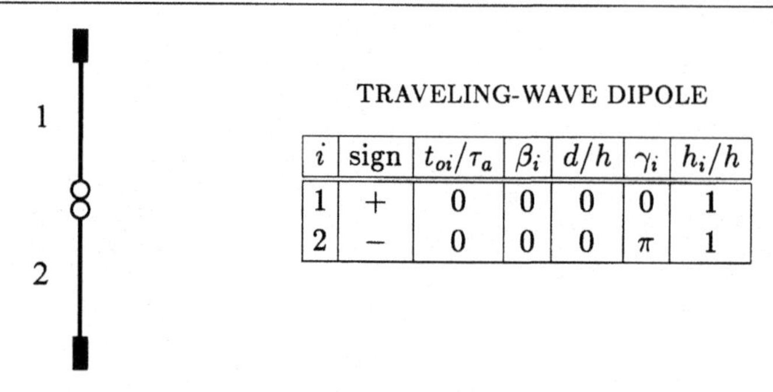

TRAVELING-WAVE DIPOLE

i	sign	t_{oi}/τ_a	β_i	d/h	γ_i	h_i/h
1	+	0	0	0	0	1
2	−	0	0	0	π	1

Fig. 8.15. Traveling-wave dipole antenna as a combination of two basic traveling-wave elements.

(8.22) reduces to

$$\vec{\mathcal{E}}^r(\vec{r},t) = \frac{\mu_o c \sin(\theta-\gamma)}{4\pi r[1-\cos(\theta-\gamma)]}\Big(\mathcal{I}_s\Big\{t - t_o - \big[r - d\cos(\theta-\beta)\big]/c\Big\}$$
$$- \mathcal{I}_s\Big\{t - t_o - \big[r - d\cos(\theta-\beta)\big]/c - (h/c)\big[1-\cos(\theta-\gamma)\big]\Big\}\Big)\hat{\theta}. \tag{8.23}$$

Our displacement has destroyed the symmetry of the field about the z axis, and so we have to be careful about our definition for the angle θ in (8.23). This angle will now be in the range $0 \le \theta < 2\pi$, and the unit vector $\hat{\theta}$ will always point in the clockwise direction with respect to a rotation about O.

8.2.2 Dipole antennas

We have already discussed some of the properties of the simple dipole antenna. Now we will show that the radiated field of this antenna can be obtained by simply superimposing the fields of basic traveling-wave elements (8.23). First we will consider the *traveling-wave dipole*. This is a dipole with reflectionless terminations at both ends. As shown in Figure 8.15, it can be viewed as the combination of two basic traveling-wave elements. The parameters that must be used in (8.23) for each of the two elements ($i = 1, 2$) are given in the accompanying table. Here the lengths and times for each element are specified in terms of the length h and time $\tau_a = h/c$. The parameter "sign" is positive when the source places a positive pulse of charge on the element, and it is negative when the source places a negative pulse of charge on the element.

The radiated electric field for the traveling-wave dipole is

$$\vec{\mathcal{E}}^r(\vec{r},t) = \sum_{i=1}^{2} \vec{\mathcal{E}}_i^r(\vec{r},t)$$
$$= \frac{\mu_o c}{4\pi r}\Big(\frac{\sin\theta}{1-\cos\theta}\Big\{\mathcal{I}_s(t-r/c) - \mathcal{I}_s\big[t - r/c - (h/c)(1-\cos\theta)\big]\Big\}$$

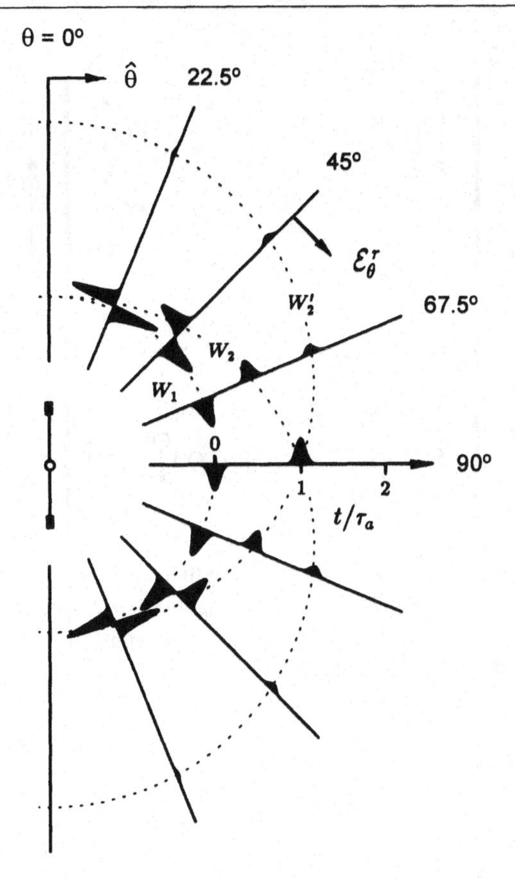

Fig. 8.16. Radiated electric field of traveling-wave dipole antenna. Excitation is a Gaussian pulse of current; $\tau/\tau_a = 0.076$.

$$+\frac{\sin\theta}{1+\cos\theta}\left\{\mathcal{I}_s(t-r/c)-\mathcal{I}_s\big[t-r/c-(h/c)(1+\cos\theta)\big]\right\}\Bigg)\hat{\theta}$$

$$=\frac{\mu_o c}{4\pi r}\sin\theta\left\{\frac{2}{\sin^2\theta}\mathcal{I}_s(t-r/c)-\frac{1}{1-\cos\theta}\mathcal{I}_s\big[t-r/c-(h/c)(1-\cos\theta)\big]\right.$$

$$\left.-\frac{1}{1+\cos\theta}\mathcal{I}_s\big[t-r/c-(h/c)(1+\cos\theta)\big]\right\}\hat{\theta}. \tag{8.24}$$

This field is shown in Figure 8.16 for the Gaussian pulse of current (8.20) with $\tau/\tau_a = 0.076$. It is composed of three spherical wavefronts that are centered on the source (W_1), on the top termination (W_2), and on the bottom termination (W_2').

Next we will consider a dipole that has total reflection of the charge at the ends: a *standing-wave dipole*.[10] As we saw earlier, this is approximately the situation for the simple dipole with open ends (Figure 8.2). When reflection occurs at the ends, waves of charge travel on the wire toward the source. To complete our description,

[10] This terminology is associated with a dipole that is excited by a time-harmonic source, and its meaning will be made clear later when we discuss that method of excitation.

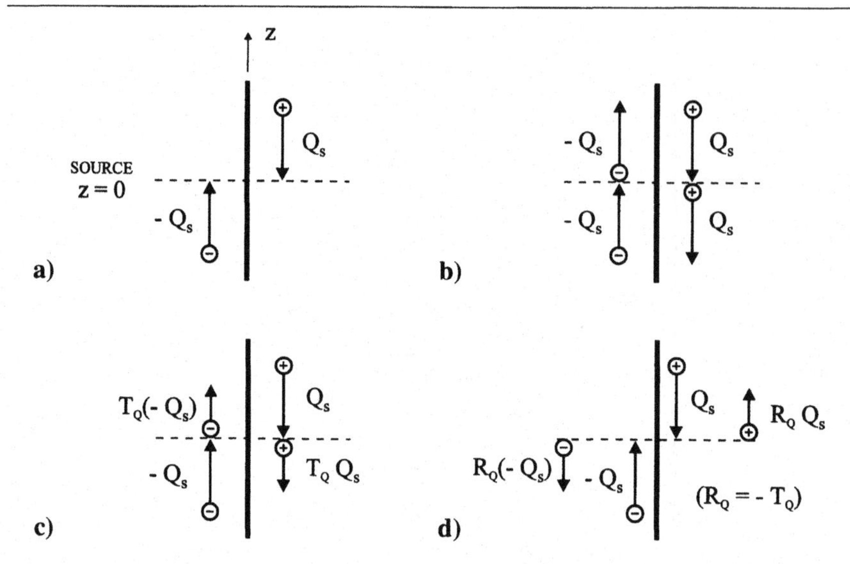

Fig. 8.17. Traveling waves of charge approaching the source after reflection from the open ends of the dipole. Arrows show direction of propagation. a) Waves stop at source. b) Waves pass source unperturbed. c) Combination of (a) and (b): partial transmission. d) Alternate interpretation of (c): partial reflection.

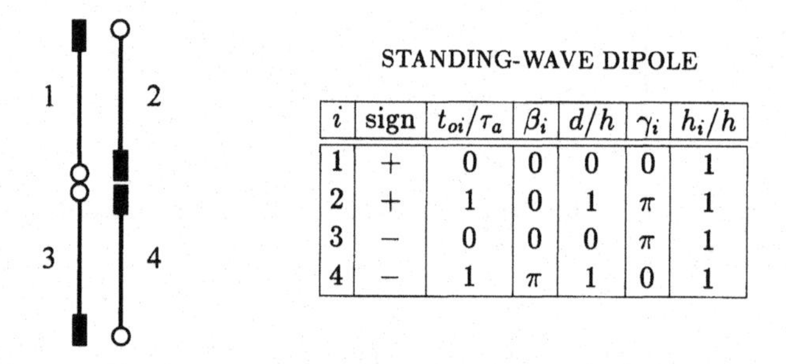

STANDING-WAVE DIPOLE

i	sign	t_{oi}/τ_a	β_i	d/h	γ_i	h_i/h
1	$+$	0	0	0	0	1
2	$+$	1	0	1	π	1
3	$-$	0	0	0	π	1
4	$-$	1	π	1	0	1

Fig. 8.18. Standing-wave dipole antenna as a combination of four basic traveling-wave elements.

we must decide what these waves will do when they reach the source. First, we will assume that these waves of charge simply stop at the source, as shown schematically in Figure 8.17a. For this case, positive charge will accumulate at $z = 0+$ due to the wave coming down the top conductor, and an equal amount of negative charge will accumulate at $z = 0-$ due to the wave coming up the bottom conductor. Thus, there will be no net accumulation of charge at the source. There also will be no net accumulation of charge at the open ends of the conductors, since the currents entering and leaving these points are the same.

This standing-wave dipole can be viewed as the combination of the four basic traveling-wave elements shown in Figure 8.18. The parameters that must be used in (8.23) for each of the elements are given in the accompanying table. Notice that the excitation of element 2 (4) is delayed by $\tau_a = h/c$, because the wave of charge must travel the length of element 1 (3) before it reaches the source end of element

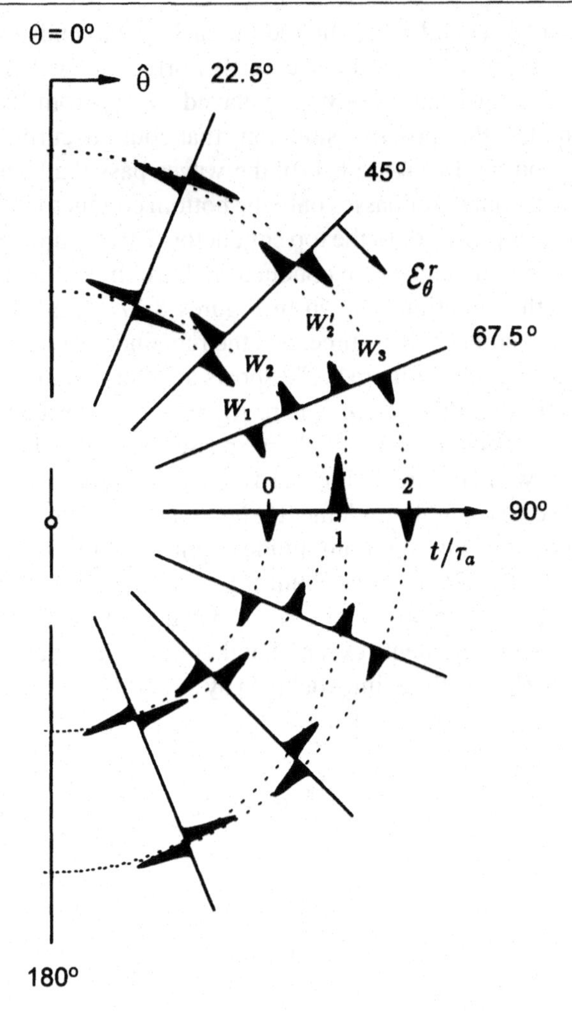

Fig. 8.19. Radiated electric field of standing-wave dipole antenna. After reflection from open ends, waves of charge stop at source. Excitation is a Gaussian pulse of current; $\tau/\tau_a = 0.076$.

2 (4). The "sign" for element 2 (4) is the same as the "sign" for element 1 (3), because, on reflection, the positive (negative) charge entering the termination of element 1 (3) leaves the source of element 2 (4).

The radiated electric field of this standing-wave dipole is

$$\vec{\mathcal{E}}^r(\vec{r},t) = \sum_{i=1}^{4} \vec{\mathcal{E}}_i^r(\vec{r},t) = \frac{\mu_o c}{2\pi r \sin\theta}\Big\{\mathcal{I}_s(t - r/c) + \mathcal{I}_s(t - r/c - 2h/c)$$
$$- \mathcal{I}_s\big[t - r/c - (h/c)(1-\cos\theta)\big] - \mathcal{I}_s\big[t - r/c - (h/c)(1+\cos\theta)\big]\Big\}\hat{\theta} \tag{8.25}$$

(see Problem 8.13). This field, shown in Figure 8.19 for the Gaussian pulse of current (8.20), is composed of four spherical wavefronts: two that are centered on the source (W_1 and W_3) and one centered on each of the ends (W_2 and W_2'). W_1 originates when the pulse leaves the source; W_3 originates when the pulse enters the source; hence, for all angles (θ) W_3 is delayed by $2\tau_a$ from W_1.

Figure 8.17 shows other possible situations that could occur when the waves of charge reach the source. In Figure 8.17b, the waves pass the source unperturbed: The wave on the top conductor passes onto the bottom conductor, while the wave on the bottom conductor passes onto the top conductor. This would occur if the source appeared as a continuous, unbroken conductor (short circuit). Figure 8.17c shows a combination of the situations shown in Figures 8.17a and 8.17b. A portion of each wave of charge stops at the source, and the remainder (the fraction T_Q) passes onto the adjacent conductor. Figure 8.17d shows an alternate interpretation for this situation. A portion (the fraction R_Q, where $R_Q = -T_Q$) of each wave of charge is reflected at the source, and no wave of charge passes onto the adjacent conductor.[11]

The situation shown in Figures 8.17c and d is easily modeled using basic traveling-wave elements. We have the four elements shown in Figure 8.18. To these we add four identical elements, with their amplitudes multiplied by $R_Q = -T_Q$ and their times delayed by $2\tau_a = 2h/c$. To account for additional reflections of the charge at the source, we repeat this procedure. At each stage we add four more elements, identical to the previous elements except that their amplitudes are multiplied by an additional factor of R_Q and their times delayed by an additional amount $2\tau_a = 2h/c$. The resulting radiated electric field is

$$\begin{aligned}\vec{\mathcal{E}}^r(\vec{r}, t) = \frac{\mu_o c}{2\pi r \sin\theta}\,\hat{\theta} \sum_{n=0}^{\infty} (R_Q)^n \Big\{ &\mathcal{I}_s(t - r/c - 2nh/c) \\ &+ \mathcal{I}_s(t - r/c - 2h/c - 2nh/c) \\ &- \mathcal{I}_s\big[t - r/c - (h/c)(1 - \cos\theta) - 2nh/c\big] \\ &- \mathcal{I}_s\big[t - r/c - (h/c)(1 + \cos\theta) - 2nh/c\big]\Big\}. \end{aligned} \tag{8.26}$$

This field is shown in Figure 8.20 for the Gaussian pulse of current (8.20). The parameters used for the graph are $R_Q = -T_Q = -0.5$ and $\tau/\tau_a = 0.114$. The results are plotted so that they can be directly compared with the very accurate results for the monopole antenna we discussed earlier (Figure 1.49). The predictions of our very simple model are seen to be in good qualitative agreement with the more accurate results. The location, sense, and relative amplitude of the pulses are roughly correct, although, clearly, there is some change in shape (distortion) of the pulses not accounted for by the simple model.

[11] Here we have assumed that the shape of the pulse of charge does not change (distort) on transmission and reflection. This is the same assumption we made for the reflections at the open ends.
T_Q is the transmission coefficient for a wave of charge passing through the source, and R_Q is the reflection coefficient for a wave of charge at the source. For the accompanying waves of current, the transmission coefficient is $T_I = T_Q$; however, the reflection coefficient is $R_I = -R_Q$.

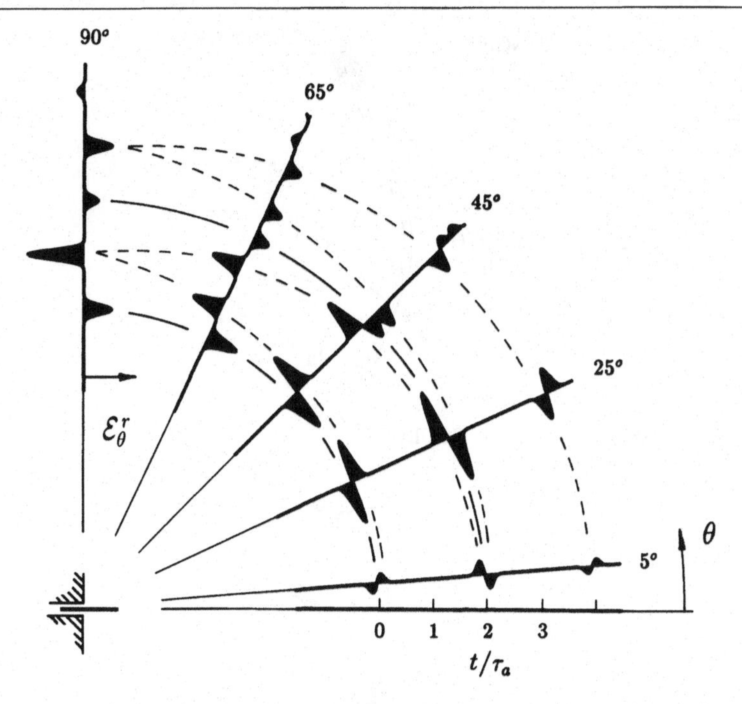

Fig. 8.20. Radiated electric field for standing-wave monopole antenna. After reflection from open ends, waves of charge are reflected at the source with reflection coefficient $R_Q = -0.5$. Excitation is a Gaussian pulse of current; $\tau/\tau_a = 0.114$.

8.3 Traveling-wave bends and loops

So far we have only examined antennas that are straight wires. We have found that radiation appears to originate from the ends of the wire, where a source or termination may be connected. Now we will examine antennas that have simple bends in the wire. Again, we will be able to approximately determine the radiation from the antennas by using a combination of basic traveling-wave elements. We will find that radiation now appears to originate at additional points on the antennas – at the bends.

8.3.1 Simple bend

Figure 8.21a shows a thin wire of length $2h$ and radius a with a simple bend of angle α (obtuse) at its center. A source is connected to one end of the wire, and a reflectionless termination is connected to the other end. This is a bent-wire, traveling-wave antenna. We will model this antenna with the two basic traveling-wave elements shown in Figure 8.21b. The parameters for the elements are given in the accompanying table. Notice that the origin for the coordinates is at the bend and that $t_{o1} = -\tau_a$ and $t_{o2} = 0$, so the pulse is at the bend when $t - r/c = 0$. In this simple model we assume that there is no reflection of the pulse at the bend and that there is no accumulation of charge at the bend.

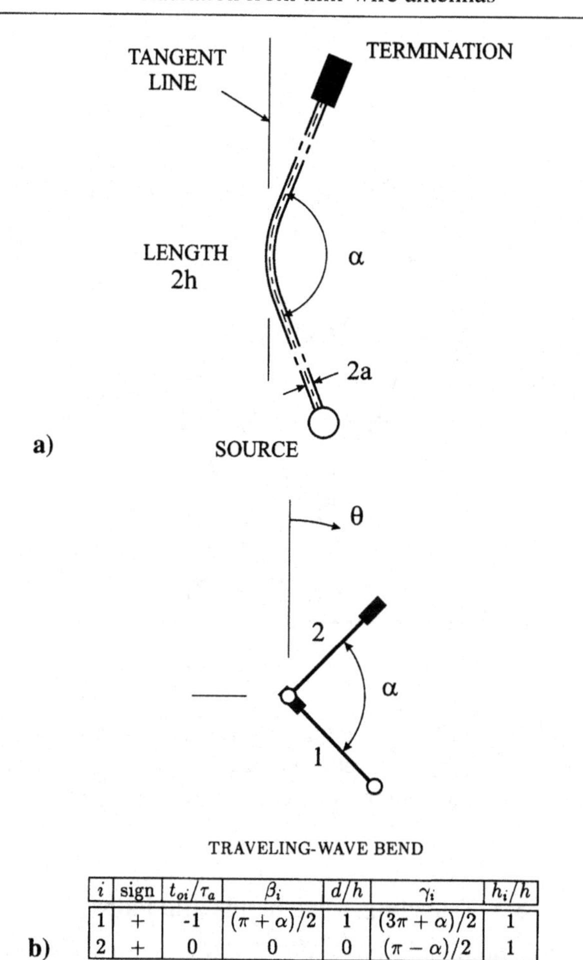

i	sign	t_{oi}/τ_a	β_i	d/h	γ_i	h_i/h
1	+	-1	$(\pi+\alpha)/2$	1	$(3\pi+\alpha)/2$	1
2	+	0	0	0	$(\pi-\alpha)/2$	1

Fig. 8.21. a) Bent-wire, traveling-wave antenna. b) Antenna as a combination of two basic traveling-wave elements.

The radiated electric field in the plane of the bent wire is obtained by combining the fields of the two basic traveling-wave elements (Problem 8.17):

$$\vec{\mathcal{E}}^r(\vec{r},t) = \sum_{i=1}^{2} \vec{\mathcal{E}}_i^r(\vec{r},t) = \frac{\mu_o c}{4\pi r}\Bigg(\frac{-2\cos(\alpha/2)}{\cos\theta - \sin(\alpha/2)}\mathcal{I}_s(t - r/c)$$
$$+ \frac{\cos(\theta - \alpha/2)}{1+\sin(\theta-\alpha/2)}\mathcal{I}_s\Big\{t - r/c + (h/c)\big[1+\sin(\theta-\alpha/2)\big]\Big\}$$
$$+ \frac{\cos(\theta + \alpha/2)}{1-\sin(\theta+\alpha/2)}\mathcal{I}_s\Big\{t - r/c - (h/c)\big[1-\sin(\theta+\alpha/2)\big]\Big\}\Bigg)\hat{\theta}. \tag{8.27}$$

This field is graphed in Figure 8.22 for a bend of angle $\alpha = 135°$. The excitation is the Gaussian pulse of current (8.20) with $\tau/\tau_a = 0.076$. There are three spherical

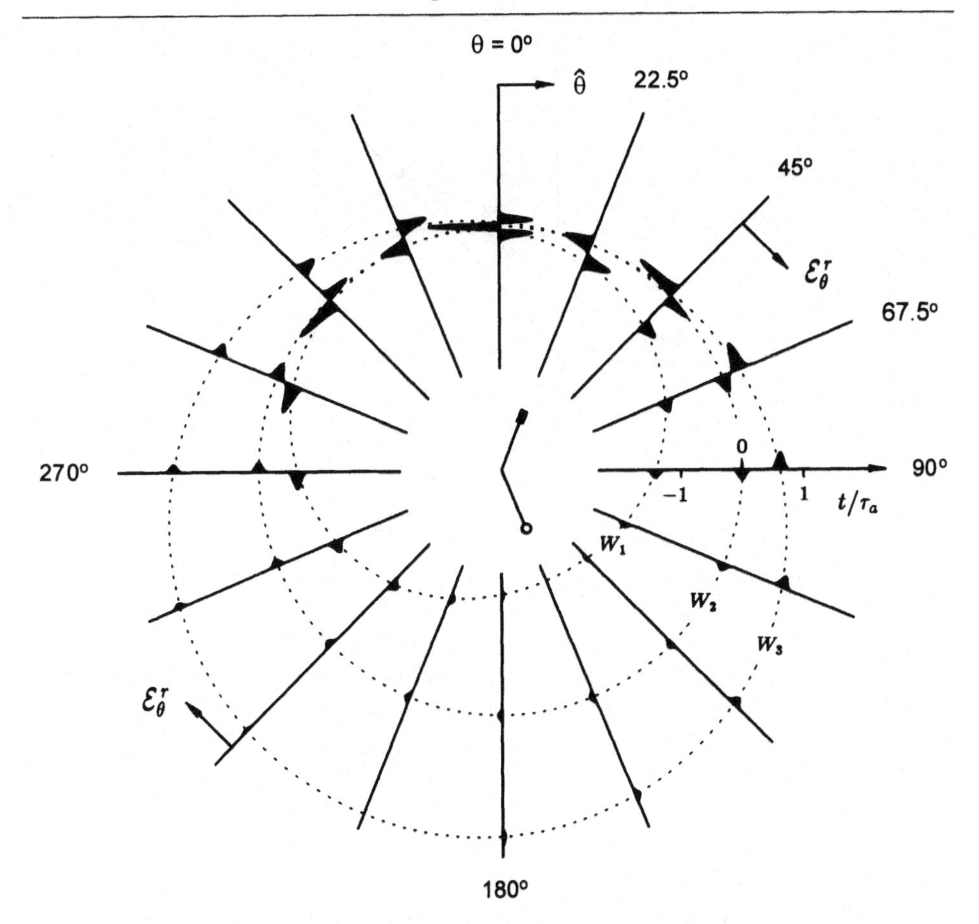

Fig. 8.22. Radiated electric field of bent-wire, traveling-wave antenna with $\alpha = 135°$. Excitation is a Gaussian pulse of current; $\tau/\tau_a = 0.076$.

wavefronts for this field: W_1 centered on the source, W_2 centered on the bend, and W_3 centered on the termination. Because of our choice for the origin of the coordinates and the time of the excitation, the dashed line for W_2 is a circle centered on the bend. The field of each wavefront is a Gaussian pulse, and the relative positions in time of these pulses change with the angle θ. Notice that there is a radiated field in the direction $\theta = 0°$. The radiation in this direction would be zero for a straight wire, as in Figure 8.12; consequently, this radiation is attributed to the bend.

For this example, the maximum and minimum amplitudes for the radiation occur along the tangent line to the center of the bend, which is marked in Figure 8.21a. The maximum is in the direction of travel for the pulse of charge ($\theta = 0°$, upward in Figure 8.22), whereas the minimum is in the opposite direction ($\theta = 180°$, downward in Figure 8.22). This situation is analogous to what we observed for a moving point charge in Chapter 6. Recall that a charge moving at a constant speed on a circular path radiates energy (synchrotron radiation). This is due to the acceleration of the charge, which in this case is simply a change in the direction of the velocity. When the speed of the charge is close to the speed of light, the

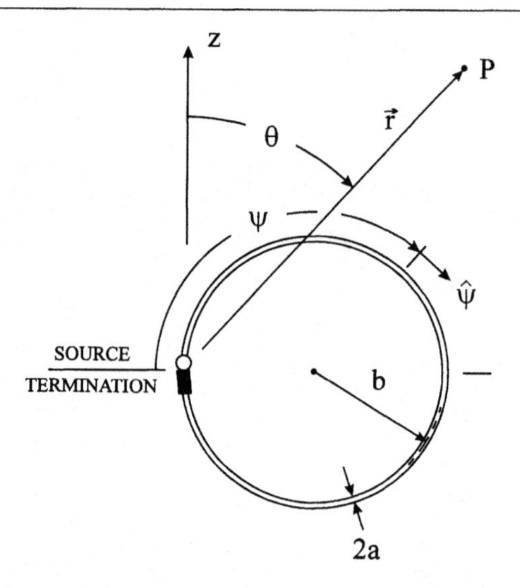

Fig. 8.23. Geometry for traveling-wave, circular loop antenna.

radiation is concentrated in the forward direction along the line tangent to the curve (circle) (Figure 6.27). Now the pulse of charge on the bent wire appears to travel with the speed of light. The direction of its apparent velocity (the direction of the current) changes at the bend, and radiation appears to originate at this point.

The simple model we have used for the bend assumes that the pulse of current travels along the axis of the curved wire at the speed of light. This is a reasonable assumption when the bend is not too sharp, such as in our example where α is a large obtuse angle [13, 14]. For a sharp bend, however, the current on one side of the bend can directly induce a current on the other side of the bend, and this can occur in a time that is less than the time for light to travel along the axis of the curved wire. Our simple model does not account for this phenomenon.

8.3.2 Circular loop

Figure 8.23 shows a traveling-wave loop antenna. The thin wire of radius a is bent into a circle of radius b, with a source at one end and a reflectionless termination at the other end. The angle ψ $(0 \leq \psi \leq 2\pi)$ determines the position along the loop, and the direction for positive current is $\hat{\psi}$. We will again assume that the source produces a traveling wave of current in the wire. For a pulse traveling in the $\hat{\psi}$ direction (clockwise), the current is

$$\mathcal{I}_{cw}(\psi, t) = \mathcal{I}_s(t - b\psi/c)\big[U(b\psi) - U(b\psi - 2\pi b)\big], \tag{8.28}$$

with the charge per unit length

$$\mathcal{Q}_{cw}(\psi, t) = \mathcal{Q}_s(t - b\psi/c)\big[U(b\psi) - U(b\psi - 2\pi b)\big] - \delta(b\psi)\int_{t'=-\infty}^{t} \mathcal{I}_s(t')dt'$$
$$+\,\delta(b\psi - 2\pi b)\int_{t'=-\infty}^{t} \mathcal{I}_s(t' - 2\pi b/c)dt', \tag{8.29}$$

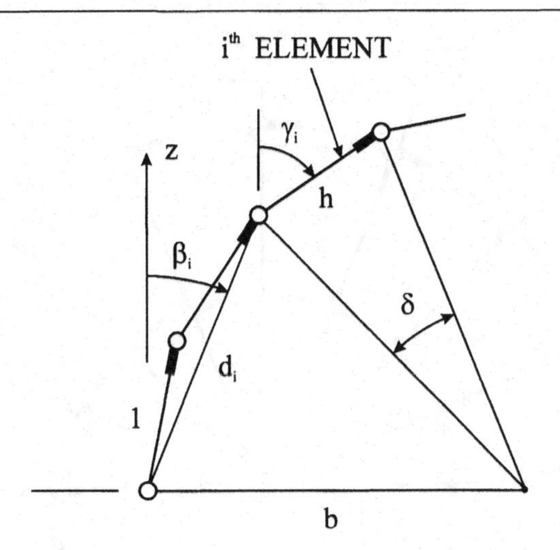

Fig. 8.24. Traveling-wave, circular loop antenna as a combination of basic traveling-wave elements.

where $\mathcal{Q}_s$ and $\mathcal{I}_s$ satisfy (8.9). The radiated electric field of the loop can be calculated by inserting this current directly into Equation (8.2). The resulting integral then has to be evaluated numerically [15, 16]. We will follow a simpler approach: We will consider the loop to be composed of a large number of basic traveling-wave elements and obtain the radiated field of the loop by summing the fields of the individual elements. We will confine our discussion to the field in the plane of the loop and describe this field in terms of the coordinates r, θ with origin at the source.

The loop is modeled by a large number, n, of basic traveling-wave elements placed end to end as in Figure 8.24. The length of each element is

$$h = 2b\sin(\delta/2), \tag{8.30}$$

with

$$\delta = 2\pi/n. \tag{8.31}$$

The other parameters needed to determine the field (8.23) of an element are

$$d_i = 2b\sin\left[(i-1)\delta/2\right], \qquad i = 1, \ldots, n, \tag{8.32}$$

$$\beta_i = (i-1)\delta/2, \tag{8.33}$$

$$\gamma_i = (i-1/2)\delta, \tag{8.34}$$

and

$$t_{oi} = \tau_a(i-1)\sin(\delta/2). \tag{8.35}$$

Here the characteristic time for the antenna is taken to be $\tau_a = 2b/c$. As with the model for the simple bend discussed earlier, we assume no reflection of the pulse along the loop and no accumulation of charge along the loop.

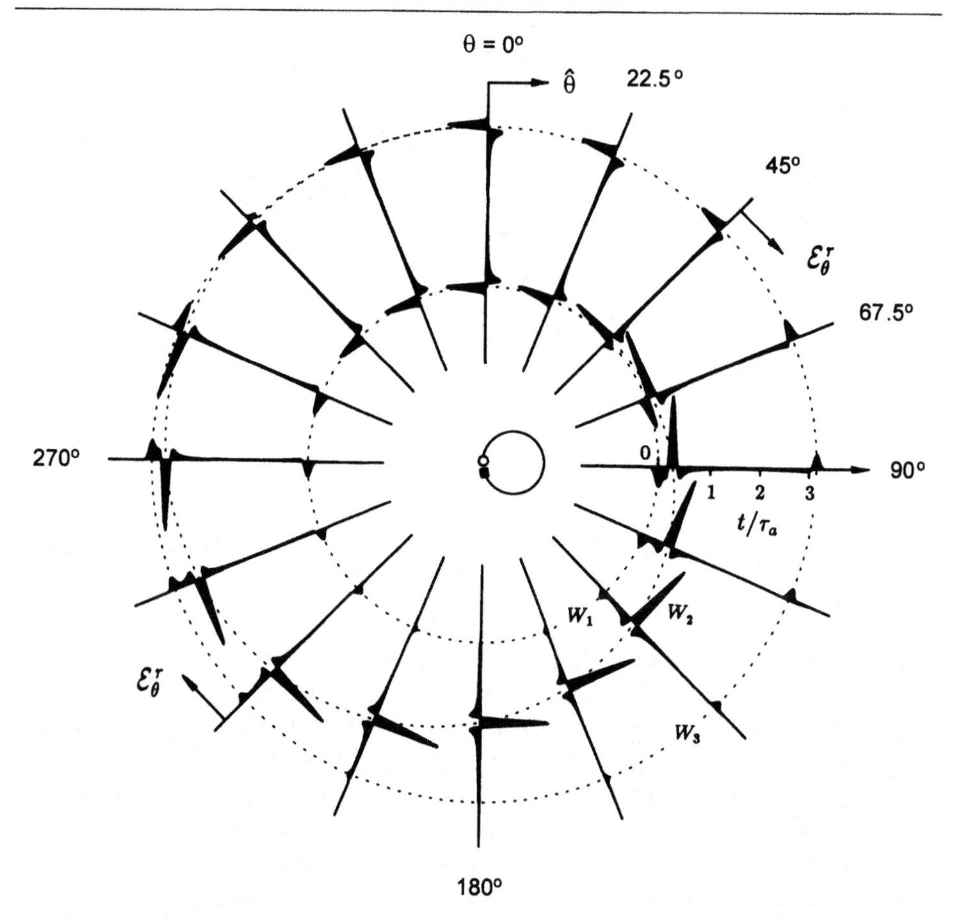

Fig. 8.25. Radiated electric field of traveling-wave, circular loop antenna. Excitation is a Gaussian pulse of current: $\tau/\tau_a = 0.076$, $\tau_a = 2b/c$.

The radiated electric field in the plane of the traveling-wave loop is the sum of the fields of the individual elements:

$$\vec{\mathcal{E}}^r_{cw}(\vec{r}, t) = \frac{\mu_o c}{4\pi r} \sum_{i=1}^{n} \frac{\sin(\theta - \gamma_i)}{1 - \cos(\theta - \gamma_i)} \Bigg(\mathcal{I}_s \Big\{ t - t_{oi} - \big[r - d_i \cos(\theta - \beta_i) \big]/c \Big\}$$

$$- \mathcal{I}_s \Big\{ t - t_{oi} - \big[r - d_i \cos(\theta - \beta_i) \big]/c - (h/c)\big[1 - \cos(\theta - \gamma_i) \big] \Big\} \Bigg) \hat{\theta}. \tag{8.36}$$

This field is graphed in Figure 8.25 for a loop excited by a Gaussian pulse of current (8.20) with $\tau/\tau_a = 0.076$. For this numerical example, the loop is composed of $n = 65$ elements; however, the results change negligibly when more elements are used. There are three wavefronts for the radiation. The spherical wavefront W_1 is centered on the source, and the spherical wavefront W_3 is centered on the termination, which is at the same location as the source. Hence, the dashed lines for these two wavefronts lie on concentric circles, and they are always separated

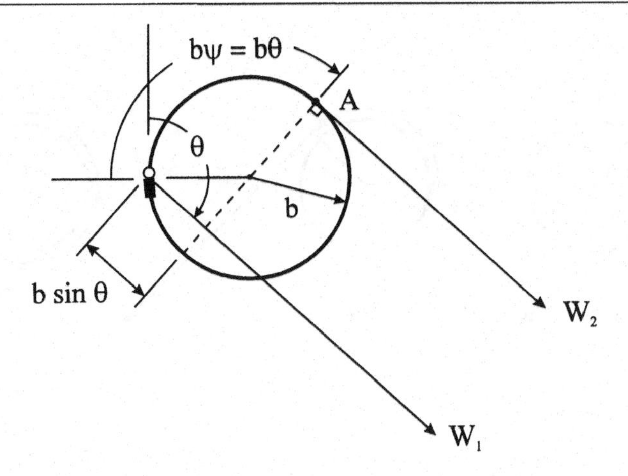

Fig. 8.26. Radiation from traveling-wave, circular loop antenna.

by the time it takes for light to travel the circumference of the loop, $\pi\tau_a = 2\pi b/c$. The remaining wavefront W_2 is a result of the curvature of the wire. Its time of arrival relative to W_1 changes with the angle of observation, θ.

Wavefront W_2 appears to originate at the point on the loop where the tangent line to the loop points in the direction of observation, as indicated in Figure 8.26. There are actually two diametrically opposite points that fit this description. It is the point at which the current pulse is traveling toward the observer that concerns us, the point marked A. From this figure, we see that W_2 is delayed in time from W_1 by the amount $(\theta - \sin\theta)b/c = \tau_a(\theta - \sin\theta)/2$.

To further illustrate this case, the results for the angle of observation $\theta = 135°$ are shown in detail in Figure 8.27. The peaks in the electric field associated with the three wavefronts are well separated at this angle of observation. Notice that the field is spread out in time. This is because the radiation occurs continuously along the loop (there are no uncurved portions of wire). Again, we call attention to the similarity of this radiation to synchrotron radiation from a point charge moving on a circular path.

Just as for the dipole, the loop can be configured to support a standing wave. The antenna would then be a continuous wire with the source inserted at one point: the structure shown in Figure 8.23 without the termination and with the end of the wire at $\psi = 2\pi$ connected to the source. The source would introduce positive charge onto the wire at $\psi = 0+$ and negative charge onto the wire at $\psi = 2\pi-$, and waves of charge would travel around the loop in both the clockwise and the counterclockwise directions. If we assume that the waves of charge stop at the source after traveling once around the loop (a situation analogous to that shown in Figure 8.17a for the dipole), then the current for the clockwise traveling wave would be given by (8.28). The current for the counterclockwise traveling wave (pulse traveling in the $-\hat{\psi}$ direction) would be

$$\mathcal{I}_{ccw}(\psi, t) = \mathcal{I}_s\big[t - (2\pi - \psi)b/c\big]\big[U(b\psi) - U(b\psi - 2\pi b)\big], \qquad (8.37)$$

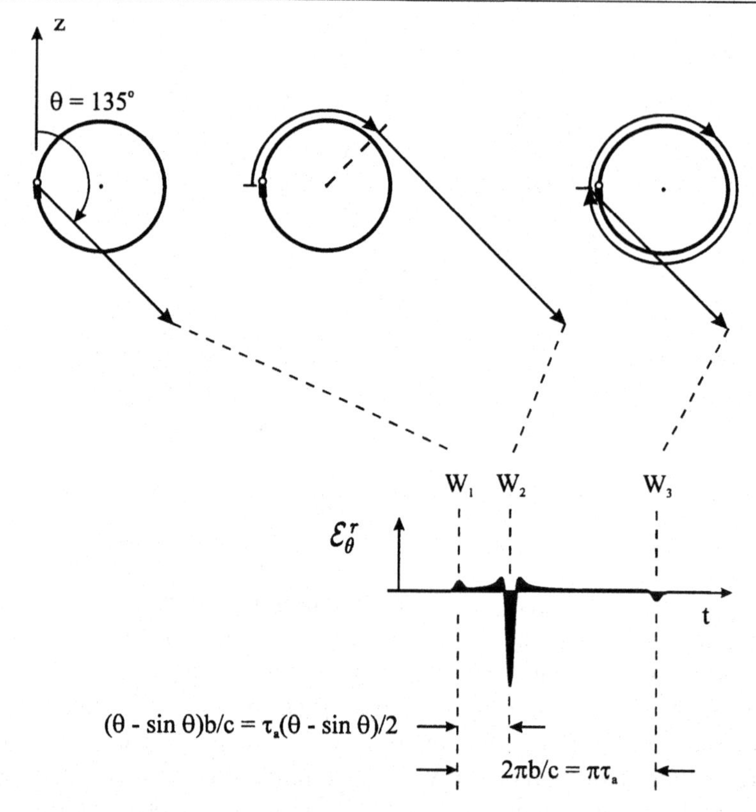

Fig. 8.27. Traveling-wave, circular loop antenna. Explanation for radiation at angle $\theta = 135°$.

with the charge per unit length[12]

$$\mathcal{Q}_{ccw}(\psi, t) = -\mathcal{Q}_s\big[t - (2\pi - \psi)b/c\big]\big[U(b\psi) - U(b\psi - 2\pi b)\big]$$
$$- \delta(b\psi)\int_{t'=-\infty}^{t} \mathcal{I}_s(t' - 2\pi b/c)dt' + \delta(b\psi - 2\pi b)\int_{t'=-\infty}^{t} \mathcal{I}_s(t')dt'. \tag{8.38}$$

The radiated electric field for this current is (see Problem 8.9)

$$\vec{\mathcal{E}}^r_{ccw}(\vec{r}, t) = \frac{\mu_o c}{4\pi r}\sum_{i=1}^{n}\frac{\sin(\theta + \gamma_i)}{1 + \cos(\theta + \gamma_i)}\bigg(\mathcal{I}_s\Big\{t - t_{oi} - \big[r + d_i\cos(\theta + \beta_i)\big]/c\Big\}$$
$$- \mathcal{I}_s\Big\{t - t_{oi} - \big[r + d_i\cos(\theta + \beta_i)\big]/c - (h/c)\big[1 + \cos(\theta + \gamma_i)\big]\Big\}\bigg)\hat{\theta}, \tag{8.39}$$

where the parameters $h, \delta, d_i, \ldots, t_{oi}$ are the same as in (8.30)–(8.35). The total field for this standing-wave loop is the sum of (8.36) and (8.39). When the waves

[12] When both the clockwise and the counterclockwise currents are present, no net charge accumulates at the source ($\psi = 0, 2\pi$). The terms in (8.29) and (8.38) that contain Dirac delta functions cancel.

of charge are partially reflected at the source, they can be treated as for the dipole. Additional terms similar to those in (8.36) and (8.39) then would be added to the field, with their amplitudes multiplied by the appropriate reflection coefficients and their times delayed by the appropriate amounts. A result analogous to (8.26) for the field of the standing-wave dipole antenna would be obtained.

A glance at Figure 8.25 shows that the graphs for the field are quite complex even for a loop with a single (clockwise) traveling wave. When the additional traveling waves are added to represent the standing-wave loop, the graphs are even more complex, and it is difficult to obtain any additional physical information from them. Therefore, we will refrain from presenting graphs for the field of the standing-wave loop.

8.4 Other traveling-wave antennas

The basic structure of the traveling-wave antenna can be altered in many different ways to change the radiated field [12]. In this way the performance of the antenna can be tailored to various applications. These alterations generally consist of changes in the geometry and material properties along the length of the antenna. This affects the current distribution and hence the radiated field. In this section we will examine two modifications of the basic traveling-wave antenna: the dipole with continuous resistive loading and the insulated linear antenna. Both of these antennas are intended for very special applications and are not in common use. They are chosen for examination because of the interesting physical principles they illustrate. In addition, their examination further strengthens the analogy we have observed between the radiation from a moving point charge and the radiation from simple wire antennas.

8.4.1 Dipole antenna with continuous resistive loading

We have seen that radiation occurs from a dipole antenna when the pulse of charge encounters the source and ends of the wire. The radiated electric field then consists of a series of pulses whose relative location in time changes with the angle of observation, as in Figures 8.19 and 8.20. In certain applications, one would like the radiated field to be a single pulse; generally, this is to be the first pulse from the dipole, the one associated with the charge leaving the source. To eliminate the other pulses in the radiated field, the antenna must be modified to prevent abrupt changes in the waves of charge at the ends of the wire. Various methods have been proposed for this purpose. The one we will discuss places continuous resistive loading along the length of the conductors [10, 17–20]. The resistance per unit length has the value R_o at the source and increases to infinite value at the open ends according to the formula

$$R(z) = \frac{R_o}{1 - |z|/h}, \qquad -h \le z \le h. \tag{8.40}$$

For the proper choice of R_o, the current on the top conductor of the dipole is approximately [17]

$$\mathcal{I}(z,t) = \mathcal{I}_s(t - z/c)(1 - z/h)\big[U(z) - U(z-h)\big], \tag{8.41}$$

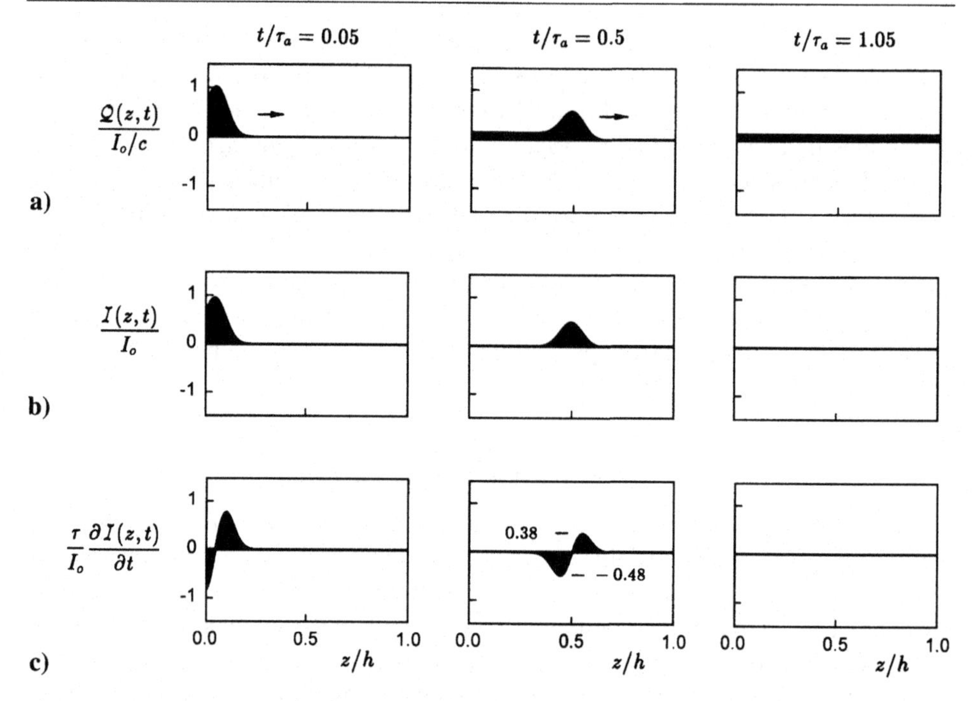

Fig. 8.28. Charge per unit length (a), current (b), and derivative of current (c) along dipole antenna with continuous resistive loading. Excitation is a Gaussian pulse of current; $\tau/\tau_a = 0.076$.

with the accompanying charge per unit length

$$\mathcal{Q}(z,t) = \frac{1}{c}\mathcal{I}_s(t - z/c)(1 - z/h)\big[U(z) - U(z-h)\big] + \frac{1}{h}\int_{t'=-\infty}^{t} \mathcal{I}_s(t' - z/c)dt'\,\big[U(z) - U(z-h)\big] - \delta(z)\int_{t'=-\infty}^{t} \mathcal{I}_s(t')dt'. \tag{8.42}$$

These are actually the current and charge for a basic traveling-wave element with continuous resistive loading. Again, the dipole can be constructed from two basic elements, as in Figure 8.15. The radiated electric field of the dipole is easily calculated as the sum of the fields from the two basic elements (Problem 8.10):

$$\vec{\mathcal{E}}^r(\vec{r},t) = \frac{\mu_o c}{4\pi r}\sin\theta\Bigg\{\frac{2}{\sin^2\theta}\mathcal{I}_s(t - r/c) - \frac{1}{h(1-\cos\theta)}\int_{z'=0}^{h}\mathcal{I}_s\big[t - r/c - (z'/c)(1-\cos\theta)\big]dz' - \frac{1}{h(1+\cos\theta)}\int_{z'=0}^{h}\mathcal{I}_s\big[t - r/c - (z'/c)(1+\cos\theta)\big]dz'\Bigg\}\hat{\theta}. \tag{8.43}$$

The charge per unit length, current, and the temporal derivative of the current on the top conductor of the dipole are shown in Figure 8.28 when the excitation is the

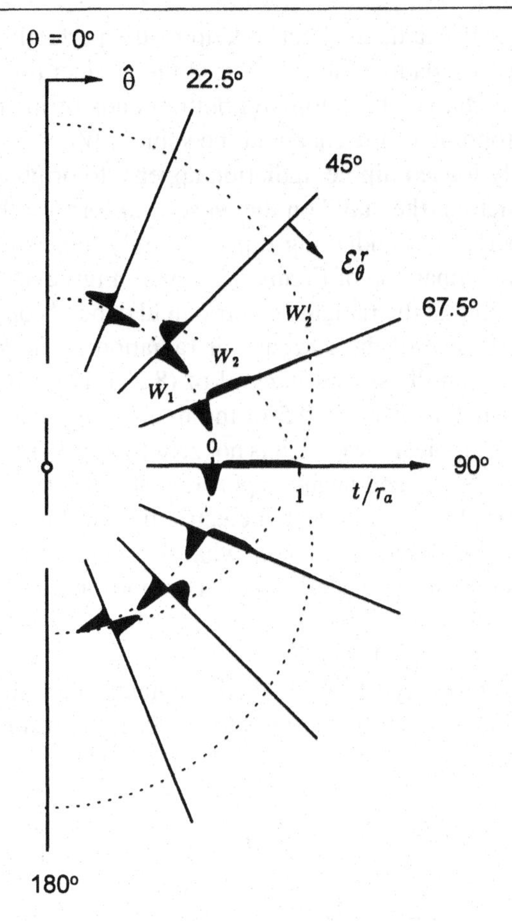

Fig. 8.29. Radiated electric field of dipole antenna with continuous resistive loading. Excitation is a Gaussian pulse of current; $\tau/\tau_a = 0.076$.

Gaussian pulse (8.20) with $\tau/\tau_a = 0.076$. The three times shown are for the pulse at the source, $t/\tau_a = 0.05$, at the center of the wire, $t/\tau_a = 0.5$, and at the open end, $t/\tau_a = 1.05$. Notice that the current is a traveling-wave pulse that decreases linearly in amplitude as it moves along the conductor. This outward-going current is zero at the open end of the conductor ($z = h$); thus, there is no reflection at this point. The charge also contains a traveling-wave pulse that decreases linearly in amplitude as it moves along the conductor. However, after the pulse of charge passes, charge remains on the conductor. Charge appears to leave the pulse to be deposited on the conductor. When the pulse reaches the open end ($z = h$), the total charge that has been deposited is equal to the charge that was initially in the pulse ($z = 0$).

The radiated electric field for this case is shown in Figure 8.29. The dipole radiates a spherical wavefront, W_1, centered on the source. The electric field of this wavefront, particularly at broadside ($\theta = 90°$), resembles the Gaussian pulse of the excitation with a "tail" appended. This is what we would expect from Equation (8.43). The first term in this equation is proportional to the excitation; the second and third terms are proportional to the integral of the excitation. A comparison with Figure 8.19 for the radiated electric field of the perfectly conducting dipole shows that the resistive loading has eliminated the distinct pulses associated with the reflection of the charge at the top and bottom ends of the conductor, W_2 and W_2', and with the stopping of the charge at the source, W_3.

For the resistively loaded dipole, radiation appears to originate along the entire length of the conductors (the "tail" on the waveform for $\theta = 90°$ in Figure 8.29), whereas, for the perfectly conducting dipole, it only appears to originate at the source and ends. A comparison of Figure 8.28c with Figure 8.11c helps to explain this difference. Recall that the radiated electric field depends on the spatial integral of the temporal derivative of the current; for radiation in the broadside direction ($\theta = 90°$) it depends on the spatial integral of (8.21). For a time when the pulse is out on the wire, such as $t/\tau_a = 0.5$ in the figures, this integral is zero for the perfectly conducting dipole. However, it is not zero for the resistively loaded dipole; the areas under the positive and negative portions of the curve are not equal.

Notice that the graphs for the radiated electric field in Figure 8.29 always have positive and negative portions, no matter what the angle of observation. A close examination shows that the areas under these two portions are equal. Another way to state this is to say that the radiated electric field has no zero-frequency ($\omega = 0$) component to its Fourier transform.[13]

This is a general property of any realistic radiated field. To see this, we will evaluate the temporal integral of the radiated electric field for a wire antenna of general shape (8.2):

$$\begin{aligned}\int_{t=-\infty}^{\infty} \vec{\mathcal{E}}^r(\vec{r}, t)dt &= \frac{-\mu_o}{4\pi r}(1 - \hat{r}\hat{r}\cdot) \int_C \int_{t=-\infty}^{\infty} \left[\frac{\partial \mathcal{I}(s', t')}{\partial t'}\right]_{t'=t_r} dt\, \hat{s}'ds' \\ &= \frac{-\mu_o}{4\pi r}(1 - \hat{r}\hat{r}\cdot) \int_C \left[\mathcal{I}(s', t = \infty) - \mathcal{I}(s', t = -\infty)\right]\hat{s}'ds'.\end{aligned} \tag{8.44}$$

From this result, we see that there will be no zero-frequency component to the radiated field provided the total current on the antenna is initially zero (at $t = -\infty$) and finally zero (at $t = +\infty$). For all of the examples we have considered, this has been the case. However, examples can be set up where this is not the case. Consider a basic traveling-wave element with the step-function excitation $\mathcal{I}_s(t) = I_o U(t)$ (Problem 8.5). Now $\mathcal{I}(s', t = -\infty) = 0$, whereas $\mathcal{I}(s', t = \infty) = I_o$; thus, there is a zero-frequency component to the radiated field. This is an unrealistic example. After the time $t = 0$, the source continually places charge at the ends of the element. An infinite amount of charge will be deposited at the ends by the time $t = \infty$.

The radiation from the resistively loaded dipole is also analogous to the radiation from a moving point charge. Consider a point charge (or group of point charges) initially at rest (at $z = 0$). It is quickly accelerated; then it moves with constant

[13] From (1.201), the zero frequency component of the Fourier transform is

$$\tilde{\vec{E}}^r(\omega = 0) = \int_{-\infty}^{\infty} \vec{\mathcal{E}}^r(t)dt.$$

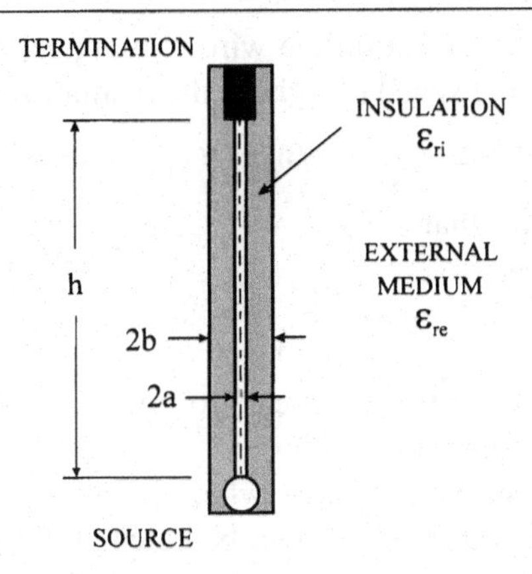

Fig. 8.30. Insulated, traveling-wave linear antenna.

velocity. At each point along the path of motion (a straight line), a small amount of the charge is brought to rest. By the end of the path (at $z = h$) all of the charge is stationary. As the charge is brought to rest along the path, it is decelerated, and it radiates. Thus, in addition to the radiation that occurs from the initial acceleration, radiation occurs continuously along the path.

8.4.2 Insulated linear antenna

The wire antennas we have been discussing have been surrounded by free space. When a pulse of charge is introduced on the antenna, it appears to move along the wire at a velocity equal to the speed of light in free space. In some applications, the antenna may be surrounded by some other medium; for example, it could be buried in earth or immersed in lake water or sea water. For our purpose, we will assume the loss in this medium to be negligible so that we can model it as a dielectric material with the electrical parameters $\varepsilon_e = \varepsilon_{re}\varepsilon_o$, $\mu_e = \mu_o$, and $\sigma_e \approx 0$.[14] Here the subscript *e* indicates the "external" medium. The speed of light in this medium is

$$v_e = c/\sqrt{\varepsilon_{re}}. \tag{8.45}$$

When we place our basic traveling-wave element in this medium, we expect the pulse of charge to appear to move along the wire at the velocity v_e.

Now we will modify the traveling-wave element by placing a concentric dielectric sheath around it, as in Figure 8.30. This configuration is called an *insulated linear antenna*. We will assume that the sheath has radius b and the following electrical parameters: $\varepsilon_i = \varepsilon_{ri}\varepsilon_o$, $\mu_i = \mu_o$, and $\sigma_i \approx 0$. Here the subscript i indicates the "insulation." The speed of light in the insulation is

$$v_i = c/\sqrt{\varepsilon_{ri}}. \tag{8.46}$$

A thorough analysis of this antenna shows that when

$$\varepsilon_{re} \gg \varepsilon_{ri}, \tag{8.47}$$

[14] For natural materials like soil and water, this may be a good assumption at some frequencies and a bad assumption at others [21].

the pulses of charge and current on the wire will appear to travel with a velocity approximately equal to the speed of light in the insulation, v_i [21]:

$$\mathcal{I}(z,t) = \mathcal{I}_s(t - z/v_i)\big[U(z) - U(z-h)\big]. \qquad (8.48)$$

Notice that (8.47) implies that

$$v_i/v_e = \sqrt{\frac{\varepsilon_{re}}{\varepsilon_{ri}}} > 1, \qquad (8.49)$$

indicating that the pulse of charge appears to travel at a velocity greater than the speed of light in the surrounding medium.

A thorough analysis also shows that when the insulation is not too thick, the electric field in the external medium can be calculated directly from the current (8.48); that is, the presence of the insulation can be ignored in the calculation [21]. The radiated electric field is then (see Problem 8.11)

$$\vec{\mathcal{E}}^r(\vec{r},t) = \frac{\mu_o v_e \sin\theta}{4\pi r(v_e/v_i - \cos\theta)}\Big\{\mathcal{I}_s(t - r/v_e) - \mathcal{I}_s\big[t - r/v_e - (h/v_e)(v_e/v_i - \cos\theta)\big]\Big\}\hat{\theta}. \qquad (8.50)$$

Figure 8.31 shows this field for the case where $\varepsilon_{re}/\varepsilon_{ri} = 6.83$, or $v_i/v_e = 2.61$. The excitation is the Gaussian pulse (8.20) with $\tau/\tau_{ai} = \tau/(h/v_i) = 0.076$. Notice that the field is fairly small except near the angle $\theta = 67.5°$, and at that angle it looks like the derivative of the excitation (the derivative of a Gaussian pulse). We will call this angle χ_c for reasons that will become apparent later. This observation is easily explained once we recognize that $\cos\chi_c = 0.383 = v_e/v_i$, which makes the denominator of (8.50) equal to zero at this angle.[15]

A careful examination of (8.50) shows that the electric field at $\theta = \chi_c$ is (see Problem 8.12)

$$\mathcal{E}^r_\theta(\theta = \chi_c) = \frac{\mu_o v_i}{4\pi r}\tau_{ai}\sqrt{1 - (v_e/v_i)^2}\,\frac{d\mathcal{I}_s(t - r/v_e)}{dt}, \qquad (8.51)$$

whereas that for the first pulse of the electric field at $\theta = 90°$ is

$$\mathcal{E}^r_\theta(\theta = 90°) = \frac{\mu_o v_i}{4\pi r}\mathcal{I}_s(t - r/v_e). \qquad (8.52)$$

For the Gaussian pulse (8.20), the ratio of the maximum values for these two fields is

$$\frac{\max|\mathcal{E}^r_\theta(\theta = \chi_c)|}{\max|\mathcal{E}^r_\theta(\theta = 90°)|} = \sqrt{1 - (v_e/v_i)^2}\left(\frac{\tau_{ai}}{\tau}\right)\frac{\max\left|\tau\dfrac{d\mathcal{I}_s(t)}{dt}\right|}{\max|\mathcal{I}_s(t)|}$$

$$= \sqrt{1 - (v_e/v_i)^2}\left(\frac{\tau_{ai}}{\tau}\right)\sqrt{2}\,e^{-1/2}. \qquad (8.53)$$

The longer the antenna, the greater the value of $\tau_{ai}/\tau = h/v_i\tau$ and the more

[15] The ratio v_i/v_e was purposefully chosen to be 2.61, so that the maximum for the radiation would occur at one of the angles used for the graphs of the radiated field ($\theta = 67.5°$).

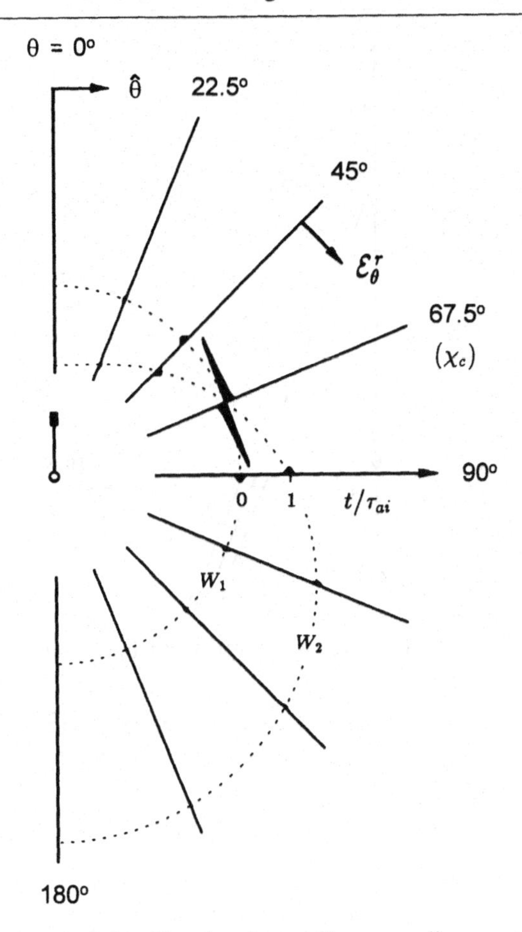

Fig. 8.31. Radiated electric field of insulated, traveling-wave linear antenna with $\varepsilon_{re}/\varepsilon_{ri} =$ 6.83. Excitation is a Gaussian pulse of current; $\tau/\tau_{ai} = 0.076$.

concentrated the radiation is about the angle χ_c. For our example $\tau_{ai}/\tau = 13.2$, and the ratio (8.53) equals 10.4.

Notice that the spherical wavefronts W_1 and W_2 in Figure 8.31 cross at the angle for maximum radiation, $\theta = \chi_c$, despite the fact that the signal for wavefront W_2 travels a longer path than the signal for wavefront W_1 [longer by the distance $h(1 - \cos\chi_c)$]. This crossing of wavefronts is a consequence of the velocity for the charge on the antenna being greater than the speed of light in the surrounding medium. For the traveling-wave element in free space (Figure 8.12) these wavefronts cross at the angle $\theta = 0°$, where the radiated field is zero.

The geometry associated with this example is shown in Figure 8.32. The direction for maximum radiation is the angle $\theta = \chi_c = \cos^{-1}(v_e/v_i)$. The wavefront associated with this radiation makes the obtuse angle $\theta = \theta_c = \pi/2 + \chi_c$ or the acute angle $\psi_c = \pi/2 - \chi_c$ with the direction of the current (z axis). Recall that we have seen this geometry before, when we discussed Cherenkov radiation in Section 6.4. A point charge moving in a straight line with a velocity v greater than the speed of light c_n in the surrounding medium produced radiation at the angle

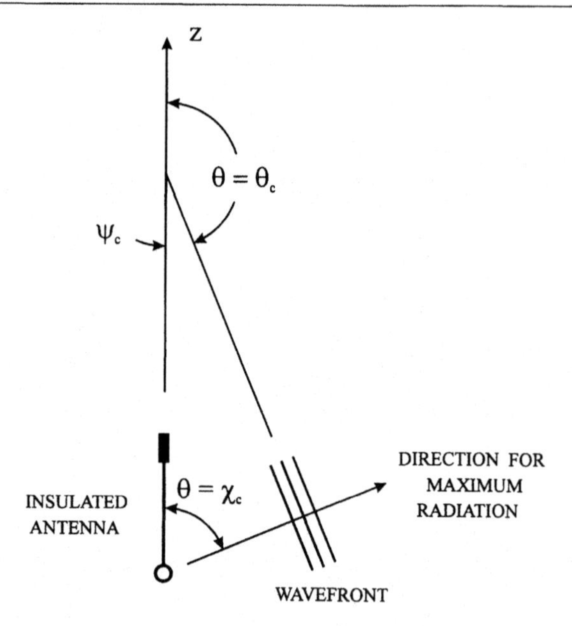

Fig. 8.32. Geometry associated with radiation from insulated traveling-wave linear antenna.

$\chi_c = \cos^{-1}(c_n/v)$ to the path of the particle (see Figure 6.38a). The radiation from the pulse of charge on the insulated antenna is thus analogous to the Cherenkov radiation from the point charge.

This analogy is developed further by considering the energy radiated by the antenna. With

$$\mathcal{I}_s(t) \leftrightarrow I_s(\omega), \tag{8.54}$$

the Fourier transform of the radiated electric field (8.50) is

$$\vec{E}^r(\vec{r}, \omega) = \frac{j\mu_o h}{4\pi r} \sin\theta \, e^{-jk_e[r+h/2(v_e/v_i - \cos\theta)]} \text{sinc}\big[k_e(h/2)(v_e/v_i - \cos\theta)\big]\omega I_s(\omega)\hat{\theta}, \tag{8.55}$$

where

$$k_e = \omega/v_e. \tag{8.56}$$

Here the sinc function is as defined in (3.62). The total energy radiated per unit frequency by the antenna is[16]

$$\begin{aligned} \frac{dW}{d\omega} &= \frac{2}{\pi} \int_{\phi=0}^{2\pi} \int_{\theta=0}^{\pi} \text{Re}\left[\vec{S}_c^r(\vec{r}, \omega)\right] \cdot \hat{r} r^2 \sin\theta d\theta d\phi \\ &= \frac{\mu_o^2 h^2}{8\pi^2 \zeta_e} \omega^2 \left|I_s(\omega)\right|^2 \int_{\theta=0}^{\pi} \left\{\text{sinc}\big[k_e(h/2)(v_e/v_i - \cos\theta)\big]\right\}^2 \sin^3\theta d\theta. \end{aligned} \tag{8.57}$$

[16] The factor of $2/\pi$ appears because of the relationship between the energy per unit frequency and the Poynting vector expressed in terms of the Fourier transform of the field (1.216).

With the change of variable

$$u = k_e(h/2)(v_e/v_i - \cos\theta),$$

this becomes

$$\frac{dW}{d\omega} = \frac{\mu_o h}{4\pi^2}\omega\left|I_s(\omega)\right|^2 \int_{-k_e(h/2)(1-v_e/v_i)}^{k_e(h/2)(1+v_e/v_i)} \left|\mathrm{sinc}(u)\right|^2 \left\{1 - \left[v_e/v_i - u/(k_e h/2)\right]^2\right\} du. \tag{8.58}$$

Now we will assume that the antenna is very long, so that $k_e h \gg 1$ at all of the significant frequencies for (8.58). Then we can approximate the integrand and the limits to obtain

$$\frac{dW}{d\omega} \approx \frac{\mu_o h}{4\pi^2}\left[1 - (v_e/v_i)^2\right]\omega\left|I_s(\omega)\right|^2 \int_{-\infty}^{\infty} \left|\mathrm{sinc}(u)\right|^2 du. \tag{8.59}$$

After evaluating the definite integral (its value is π [22]) and dividing by the length h, we obtain our final result, the energy radiated per unit length per unit frequency by the antenna:

$$\frac{dW}{h d\omega} \approx \frac{\mu_o}{4\pi}\left[1 - (v_e/v_i)^2\right]\omega\left|I_s(\omega)\right|^2. \tag{8.60}$$

Notice that we have the factor ω in this expression, which is characteristic of Cherenkov radiation; compare (8.60) with (6.158).

For a Gaussian pulse of current, we have

$$\mathcal{I}_s(t) = I_o e^{-(t/\tau)^2} \leftrightarrow I_s(\omega) = q e^{-(\omega\tau/2)^2}, \tag{8.61}$$

where q is the total charge in the pulse:

$$q = \int_{-\infty}^{\infty} \mathcal{I}_s(t)dt = \sqrt{\pi}\tau I_o. \tag{8.62}$$

After substituting for $I_s(\omega)$, (8.60) becomes

$$\frac{dW}{h d\omega} \approx \frac{\mu_o q^2}{4\pi}\left[1 - (v_e/v_i)^2\right]\omega e^{-(\omega\tau)^2/2}. \tag{8.63}$$

On letting $v_i/v_e \to \beta_n$ and $\tau \to T_o$, this is seen to be the same as our earlier result for the energy per unit path length per unit frequency of Cherenkov radiation from a bunch of charges (6.166).[17]

8.5 Harmonic time dependence; examples

The results we have obtained for thin-wire antennas are easily specialized for the case of harmonic time dependence. When the current of the source varies harmonically in time, all of the other electromagnetic quantities associated with the antenna

[17] Our calculation for the radiation from the insulated antenna is essentially the same as that used by others to estimate the Cherenkov radiation from a charge moving with a constant velocity on a path of finite length [23–25].

also vary harmonically in time, and they can be expressed in terms of phasors. The total current at any point s on the thin-wire antenna of Figure 8.1 is given by the phasor

$$I(s) = \left|I(s)\right| e^{j\psi(s)}; \tag{8.64}$$

that is,

$$\mathcal{I}(s,t) = \mathrm{Re}\left[I(s)e^{j\omega t}\right] = \left|I(s)\right| \cos\left[\omega t + \psi(s)\right]. \tag{8.65}$$

The vector phasors for the radiated field are obtained by inserting (8.65) into (8.2) and (8.3):

$$\vec{E}^r(\vec{r}) = \frac{-j\omega\mu_o}{4\pi r} e^{-jk_o r} \int_C I(s') e^{jk_o \hat{r}\cdot\vec{r}'} \left[\hat{s}' - \hat{r}(\hat{r}\cdot\hat{s}')\right] ds', \tag{8.66}$$

$$\vec{B}^r(\vec{r}) = \frac{1}{c}\hat{r} \times \vec{E}^r(\vec{r}) = \frac{-j\omega\mu_o}{4\pi c r} e^{-jk_o r} \int_C I(s') e^{jk_o \hat{r}\cdot\vec{r}'} (\hat{r} \times \hat{s}') ds'. \tag{8.67}$$

These expressions can be used to determine the radiated field of a wire antenna of general shape once the current distribution is known.

8.5.1 Basic traveling-wave element

For the basic traveling-wave element, the current associated with source (8.9) is now[18]

$$\mathcal{I}_s(t) = \mathrm{Re}\left[I_s e^{j\omega t}\right] = I_p \cos(\omega t + \psi_s), \tag{8.68}$$

where the phasor is

$$I_s = I_p e^{j\psi_s}. \tag{8.69}$$

The total current on the element is just the outward-going wave associated with $\mathcal{I}_s(t)$, (8.14):

$$\mathcal{I}(z,t) = \mathrm{Re}\left[I(z)e^{j\omega t}\right] = I_p \cos(\omega t - k_o z + \psi_s)\left[U(z) - U(z-h)\right], \tag{8.70}$$

where the phasor is

$$I(z) = I_s e^{-jk_o z}\left[U(z) - U(z-h)\right]. \tag{8.71}$$

The vector phasor for the radiated electric field can be obtained by inserting this current into (8.66) and performing the indicated integration, or, more simply, by using (8.68) in (8.19):

$$\begin{aligned} \vec{E}^r(\vec{r}) &= \frac{\mu_o c \sin\theta}{4\pi r(1-\cos\theta)} e^{-jk_o r} I_s\left[1 - e^{-jk_o h(1-\cos\theta)}\right]\hat{\theta} \\ &= \frac{j\mu_o c k_o h I_s}{4\pi r} e^{-jk_o r} e^{-jk_o h(1-\cos\theta)/2} \sin\theta \,\mathrm{sinc}\left[k_o h(1-\cos\theta)/2\right]\hat{\theta}. \end{aligned} \tag{8.72}$$

[18] Here the subscript p is used to indicate the peak value for the time-varying current.

The time-average power per unit solid angle radiated by the element is then

$$\frac{d\langle \mathcal{P}_{\text{rad}}\rangle}{d\Omega} = \text{Re}(r^2\hat{r}\cdot\vec{S}_c^r) = \frac{1}{2\zeta_o}|r\vec{E}^r|^2$$

$$= \frac{\zeta_o(k_oh)^2 I_p^2}{32\pi^2}\sin^2\theta\,\text{sinc}^2\big[k_oh(1-\cos\theta)/2\big], \tag{8.73}$$

and the total time-average power radiated is (see Problem 8.14a)

$$\langle \mathcal{P}_{\text{rad}}\rangle = \int_{\phi=0}^{2\pi}\int_{\theta=0}^{\pi}\frac{d\langle \mathcal{P}_{\text{rad}}\rangle}{d\Omega}\sin\theta d\theta d\phi$$

$$= \frac{\zeta_o I_p^2}{4\pi}\big[\gamma - 1 + \ln(2k_oh) + \text{sinc}(2k_oh) - \text{Ci}(2k_oh)\big], \tag{8.74}$$

where $\gamma = 0.57721\ldots$ is Euler's constant and Ci is the cosine integral [26]:

$$\text{Ci}(z) = \gamma + \ln(z) + \int_0^z \frac{\cos u - 1}{u}du. \tag{8.75}$$

The directivity of the element at the angle θ is readily obtained from these results and (3.115):

$$D(\theta) = \frac{4\pi}{\langle \mathcal{P}_{\text{rad}}\rangle}\left[\frac{d\langle \mathcal{P}_{\text{rad}}(\theta)\rangle}{d\Omega}\right]$$

$$= \frac{(k_oh)^2\sin^2\theta\,\text{sinc}^2[k_oh(1-\cos\theta)/2]}{2[\gamma - 1 + \ln(2k_oh) + \text{sinc}(2k_oh) - \text{Ci}(2k_oh)]}. \tag{8.76}$$

An examination of (8.73) shows that an electrically short element ($k_oh \ll 1$) has the pattern of the infinitesimal electric dipole ($d\langle \mathcal{P}_{\text{rad}}\rangle/d\Omega \propto \sin^2\theta$). As the electrical length of the element is increased, the major lobe of the pattern in the vertical plane moves from $\theta \approx \pi/2$ to smaller angles. The radiation is said to become more "end fire" as the electrical length of the element is increased. This trend is illustrated in Figure 8.33, where patterns are shown for elements with lengths in the range $\pi \le k_oh \le 8\pi$ ($0.5 \le h/\lambda_o \le 4.0$). The directional characteristics of the radiation for the element with $k_oh = 8\pi$ ($h/\lambda_o = 4.0$) are shown in more detail in Figure 8.34. Here the radial distance in the direction θ, ϕ, measured from the origin, is proportional to $d\langle \mathcal{P}_{\text{rad}}\rangle/d\Omega$. The radiation from the electrically long element is seen to be a hollow, conical beam centered on the axis of the element.

Figure 8.35 shows the maximum directivity (D_{max}), the angle at which this directivity occurs (θ_{max}), and the half-power beamwidth for the lobe in this direction (BW), all as functions of the electrical length of the element k_oh (h/λ_o). The trend mentioned above is evident in these results: The radiation monotonically becomes more "end fire" with a higher directivity and a narrower beamwidth as the electrical length of the element is increased. Notice that, as expected, in the limit as $k_oh \to 0$, the directivity becomes that of the infinitesimal electric dipole ($D_{\text{max}} \approx 1.5$ at $\theta_{\text{max}} = 90°$).

The directive characteristics for the element with time-harmonic excitation can be explained using the ideas we introduced earlier when we discussed thin-wire

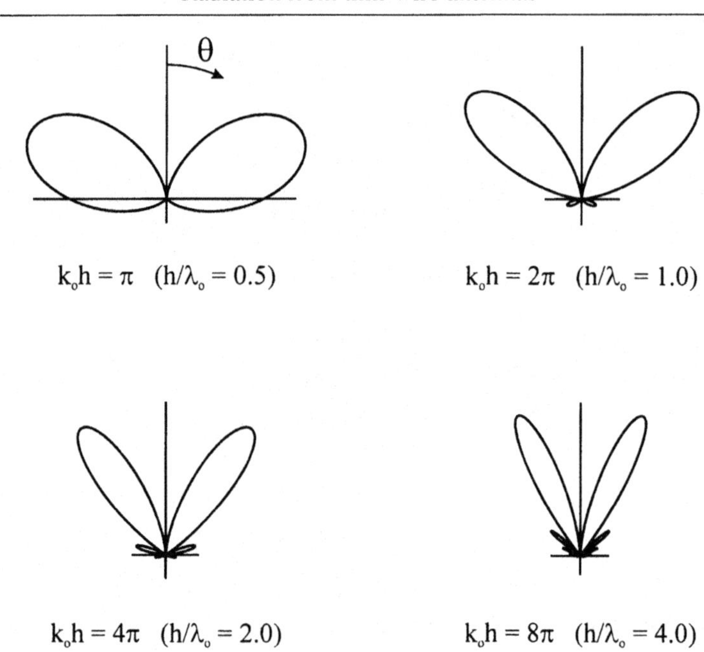

Fig. 8.33. Patterns for the radiation in the vertical plane for basic traveling-wave elements, $d\langle\mathcal{P}_{\text{rad}}\rangle/d\Omega$.

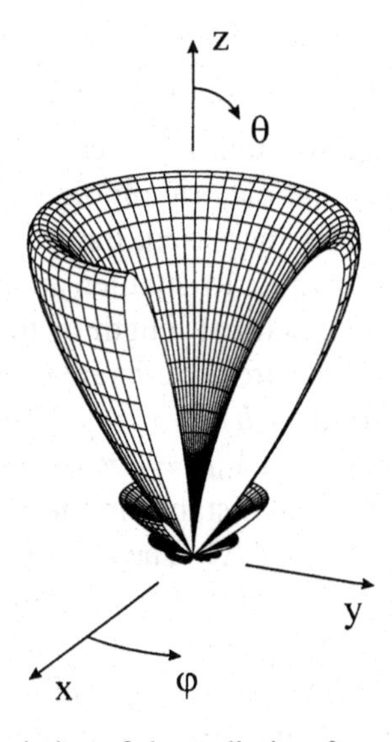

Fig. 8.34. Directional characteristics of the radiation from basic traveling-wave element, $d\langle\mathcal{P}_{\text{rad}}\rangle/d\Omega$: $k_o h = 8\pi$ $(h/\lambda_o = 4.0)$.

antennas with pulse excitation. The multilobed structure of the pattern is a result of radiation from the source end of the element interfering, constructively and destructively, with radiation from the termination end of the element. This is illustrated schematically in Figure 8.36 for destructive interference. The drawing shows the

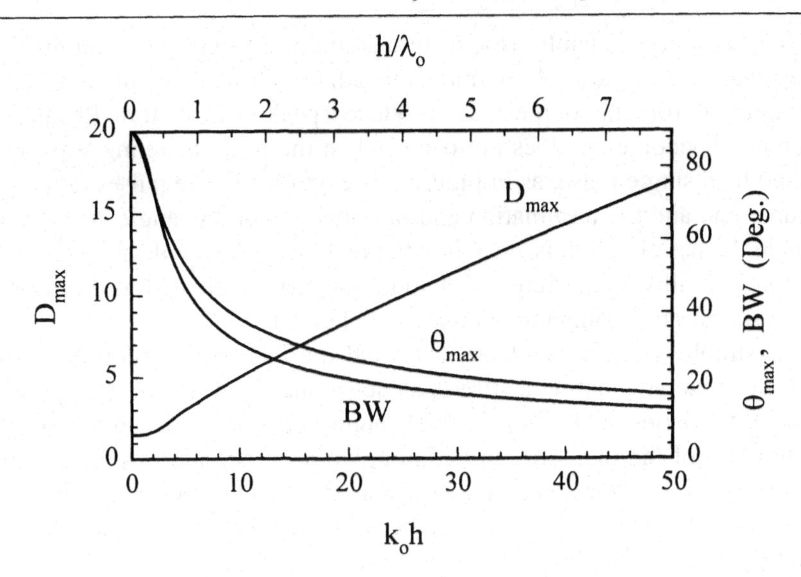

Fig. 8.35. Maximum directivity, D_{max}, of basic traveling-wave element versus electrical length, $k_o h$ (h/λ_o). θ_{max}: angle for maximum directivity. BW: half-power beamwidth.

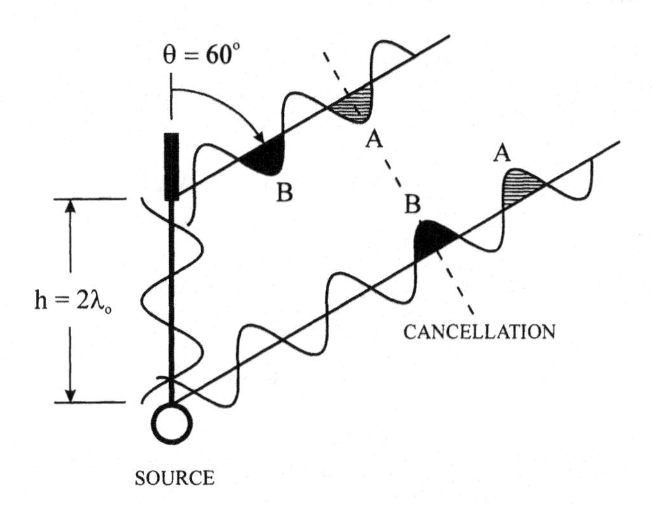

Fig. 8.36. Schematic drawing showing radiation from source end and termination end of basic traveling-wave element. Drawing is for the field in space at a fixed time.

variation of the field in space at a fixed time. For this example, the length of the element is $h/\lambda_o = 2.0$ and the angle of observation is $\theta = 60°$. The half cycles of the cosinusoid marked A in the figure left the source at the same time, as did the half cycles marked B at a later time. The radiation from the termination end is delayed in time from that from the source end, because the former has to travel the extra length $h(1 - \cos\theta)$. This is why the two half cycles marked A in the figure do not line up. Notice that there is a sign reversal for the radiation from the termination end.

For the case shown in Figure 8.36, the radiation of the half cycle A from the termination end cancels the radiation of the half cycle B from the source end, and so on. This produces the null in the pattern at this angle of observation. In a similar manner, at another angle of observation, the radiation from the termination end adds to the radiation from the source end to produce a peak in the pattern. Recall that this interference phenomenon does not occur when the basic traveling-wave element is excited by a single pulse, as in Figures 8.12 and 8.13. The pulses radiated from the source end and the termination end only overlap at one angle, $\theta = 0°$, and at that angle the net radiation is zero. It is the repetitive nature of the time-harmonic excitation that allows interference of radiation that originated at different times from the source and termination ends of the element.

In our simple analysis, we have neglected the fact that the current wave will experience some attenuation as it propagates along the element. This attenuation will be most pronounced for very long elements and will cause some change in the characteristics of the radiation. In addition, we have not addressed the important practical problems of how a source will be constructed to launch a traveling wave on the element and how a termination will be constructed to absorb the wave without reflection [12, 27].

When the basic traveling-wave element is displaced from the z axis, as in Figure 8.14, the radiated electric field is given by (8.22) and (8.23). For harmonic time dependence, these equations become:

$$\vec{E}^r(\vec{r}) = \frac{j\omega\mu_o h I_s}{4\pi r}\left[(\cos\gamma\sin\theta - \sin\gamma\cos\theta\cos\phi)\hat{\theta} + (\sin\gamma\sin\phi)\hat{\phi}\right]e^{-j(k_o r+\omega t_o)}$$
$$e^{jk_o d(\sin\beta\sin\theta\cos\phi+\cos\beta\cos\theta)}e^{-jk_o h(1-\sin\gamma\sin\theta\cos\phi-\cos\gamma\cos\theta)/2}$$
$$\mathrm{sinc}\left[k_o h(1-\sin\gamma\sin\theta\cos\phi-\cos\gamma\cos\theta)/2\right] \tag{8.77}$$

and

$$\vec{E}^r(\vec{r}) = \frac{j\omega\mu_o h I_s}{4\pi r}e^{-j(k_o r+\omega t_o)}e^{j\{k_o d\cos(\theta-\beta)-k_o h[1-\cos(\theta-\gamma)]/2\}}$$
$$\sin(\theta-\gamma)\mathrm{sinc}\left\{k_o h\left[1-\cos(\theta-\gamma)\right]/2\right\}\hat{\theta}. \tag{8.78}$$

8.5.2 Long-wire antennas

The field pattern of the electrically long, basic traveling-wave element has the desirable feature of being almost unilateral with moderate directivity. However, the hollow, conical shape of the pattern is usually not desired; one would rather have a solid beam. In the early days of high-frequency radio communications, a number of antennas based on the traveling-wave element were proposed and tested [28]. These antennas usually combined a few traveling-wave elements to produce a pattern with the desired shape. As a class, they are often referred to as "long-wire antennas" [29].

The basic idea of combining traveling-wave elements to synthesize a directive pattern is illustrated in Figure 8.37, where results are shown for a traveling-wave V dipole [30, 31]. This is essentially a dipole with the arms inclined to form the interior angle α, and it is easily analyzed using two basic traveling-wave elements (Problem 8.6). When $\alpha = 2\theta_{\max}$, the conical beams of the two arms (basic traveling-

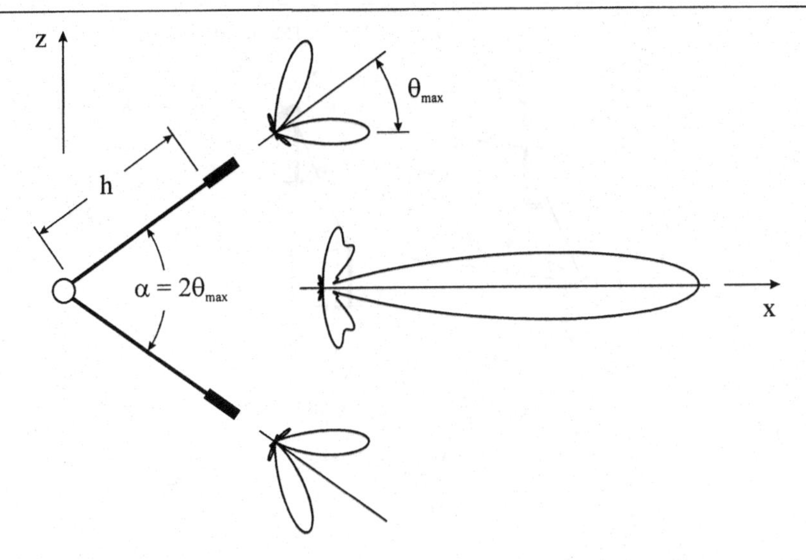

Fig. 8.37. Traveling-wave V dipole antenna. Fields from two basic traveling-wave elements combine to produce the directive pattern in the vertical plane. Patterns are for $d\langle\mathcal{P}_{\text{rad}}\rangle/d\Omega$: $h/\lambda_o = 2.0$, $\alpha = 70°$.

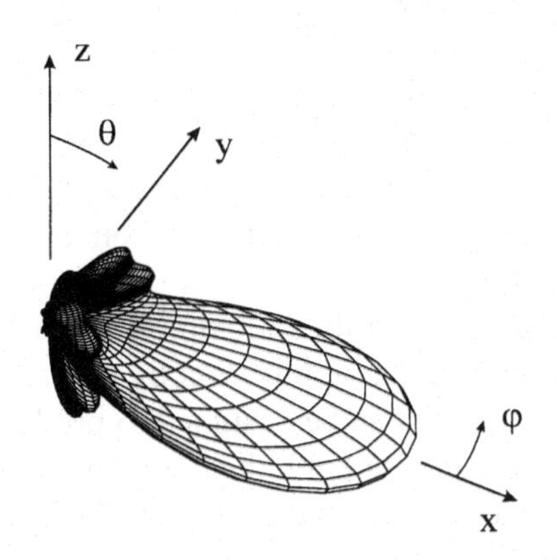

Fig. 8.38. Directional characteristics of the radiation from the traveling-wave V dipole antenna, $d\langle\mathcal{P}_{\text{rad}}\rangle/d\Omega$: $h/\lambda_o = 2.0$, $\alpha = 70°$. The V dipole is in the x-z plane.

wave elements) are superimposed in the vertical plane (plane of the elements) to increase the directivity. The results in Figure 8.37 are for a V with $h/\lambda_o = 2$ and $\alpha \approx 2\theta_{\text{max}} = 70°$. The full pattern for this antenna is shown in Figure 8.38. Notice that the beamwidth in the horizontal plane (x-y plane) is much larger than that in vertical plane (x-z plane, the plane of the V).

One of the most successful and historically most important long-wire antennas is the horizontal rhombic shown in the sketch of Figure 8.39a [5, 32–38]. It is formed by placing the wire on the perimeter of a rhombus of side length h and acute, interior angle α (obtuse, interior angle $\pi - \alpha$). The source is located at one corner and a termination, usually a resistance, is located at the opposite corner. The plane of the rhombus is at the height s above the surface of the Earth.

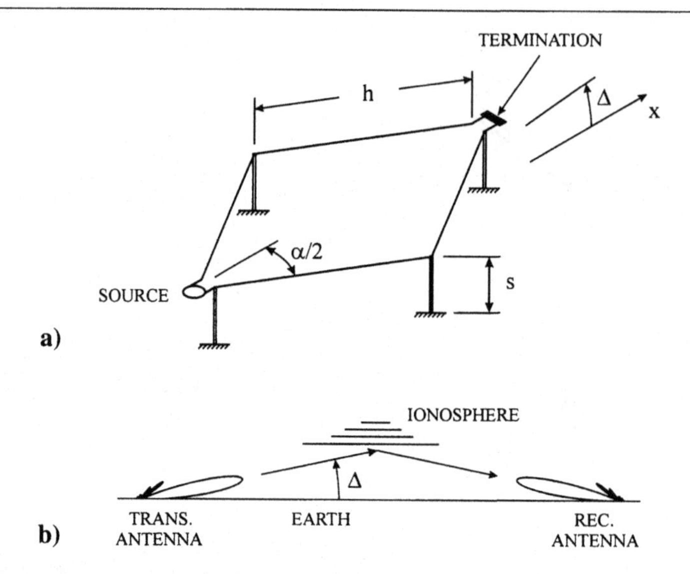

Fig. 8.39. a) Geometry for horizontal rhombic antenna. b) Schematic drawing showing communication by ionospheric reflection.

The isolated rhombic antenna is analyzed as four basic traveling-wave elements (Problem 8.15). The Earth is modeled as a perfect electric conductor, and its effect is included using image theory. The rhombic and its image (a similar antenna with all currents in the opposite direction to those in the rhombic) are treated as an array. The field of the rhombic over the Earth is obtained by multiplying the field of the isolated rhombic by an array factor similar to the ones used for the dipole arrays discussed in Section 7.5.

The directional characteristics of the radiation from the horizontal rhombic antenna can be tailored to a particular application by adjusting the three parameters h/λ_o, s/λ_o, and α. The antenna can be designed to produce a directive beam at a low angle of elevation Δ (in Figure 8.39, the angle in the vertical plane measured from the x axis). This characteristic makes this relatively simple antenna well suited for high-frequency (3 MHz to 30 MHz) communication by ionospheric reflection (Figure 8.39b).[19] It was for this application that the rhombic antenna was popular in the early days of radiotelephone communications [35].

In Figure 8.40 patterns for the horizontal component of the electric field calculated using the simple theory based on basic-traveling wave elements are compared with measurements. The measured data are for a model antenna studied during the early development of the horizontal rhombic [32].[20] The parameters for the antenna are $h/\lambda_o = 3.25$, $s/\lambda_o = 0.5$, and $\alpha = 64°$, and the pattern in Figure 8.40b is for the

[19] The vertical plane pattern shown in Figure 8.39b is for the magnitude of the horizontal component of the electric field of an antenna with $h/\lambda_o = 6$, $s/\lambda_o = 1.1$, and $\alpha = 40°$.

[20] These are patterns for the magnitude of the horizontal component of the electric field, not for the time-average power per unit solid angle. In the original graphs [32], the theoretical curve for the vertical plane pattern was corrected for the finite distance between the rhombic antenna and the dipole antenna used to measure the pattern. Here the experimental points have been corrected for this effect, and the theoretical curve is for the ideal situation.

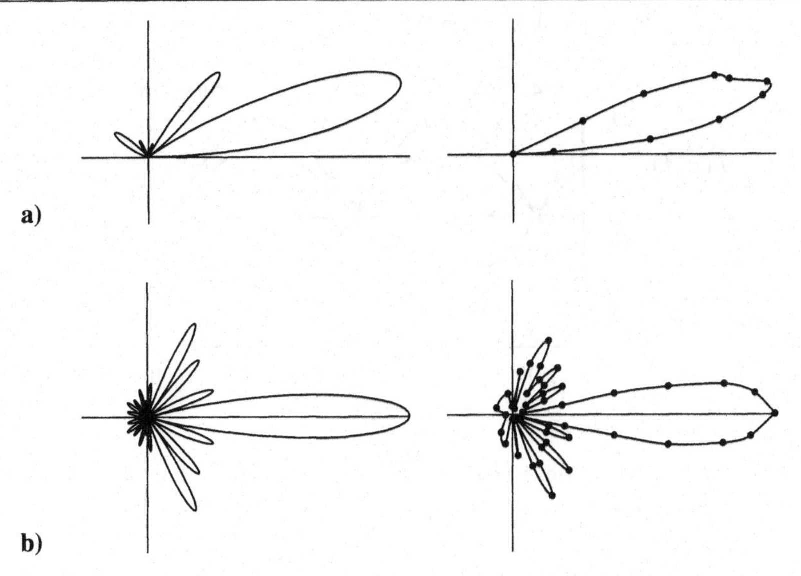

Fig. 8.40. Comparison of theoretical (left) and measured (right) electric field patterns for horizontal rhombic antenna: $h/\lambda_o = 3.25$, $s/\lambda_o = 0.5$, $\alpha = 64°$. a) Vertical plane. b) Horizontal plane, with angle of elevation $\Delta = 2.7°$. (Measured data from Bruce et al. [32].)

angle of elevation $\Delta = 2.7°$. The reasonable agreement demonstrates the adequacy of using the simple theory for describing the radiation from this antenna. In fact, this theory was used by early investigators to produce extensive design data for the horizontal rhombic antenna [35].

8.5.3 Standing-wave dipole

Recall that the standing-wave dipole is one with open ends, at which outgoing waves of charge and current are assumed to be totally reflected. When the first reflected waves from the open ends are assumed to stop at the source, as in Figure 8.17a, the total current on the top conductor of the dipole is simply the sum of one outgoing wave and one reflected wave:

$$\mathcal{I}(z,t) = \left[\mathcal{I}_s(t - z/c) - \mathcal{I}_s(t + z/c - 2h/c)\right]\left[U(z) - U(z-h)\right]. \quad (8.79)$$

For a time-harmonic source (8.68), this becomes

$$\mathcal{I}(z,t) = \mathrm{Re}\left[I(z)e^{j\omega t}\right] = -2I_p \sin\left[k_o(h-z)\right]\sin(\omega t - k_o h + \psi_s) \times \left[U(z) - U(z-h)\right], \quad (8.80)$$

where the phasor is

$$I(z) = 2jI_s e^{-jk_o h}\sin\left[k_o(h-z)\right]\left[U(z) - U(z-h)\right]. \quad (8.81)$$

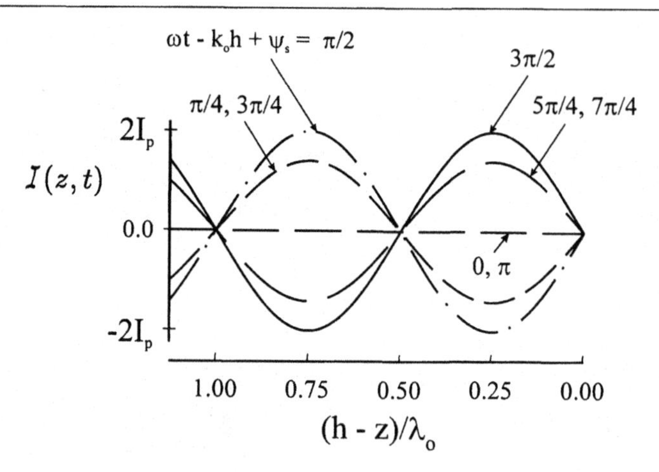

Fig. 8.41. Current distribution on top conductor of standing-wave dipole antenna; $h/\lambda_o = 1.125$.

It is convenient to express the current distribution (8.81) in terms of the phasor I_o for the total current at the source:[21]

$$I(z) = I_o \frac{\sin[k_o(h - z)]}{\sin(k_o h)} \big[U(z) - U(z - h)\big], \tag{8.82}$$

where

$$I_o = I(0) = 2j I_s e^{-jk_o h} \sin(k_o h). \tag{8.83}$$

If we had assumed that a portion of each wave of charge approaching the source was reflected, as in Figure 8.17d, only the term I_o would be modified in the distribution for the current (8.82) (Problem 8.16). This term would then become

$$I_o = 2j I_s \sin(k_o h) \left[\frac{e^{-jk_o h}}{1 - R_Q e^{-j2k_o h}} \right], \tag{8.84}$$

where R_Q is again the reflection coefficient for a wave of charge at the source.

Figure 8.41 is a sketch of the current (8.80) on the top conductor of a dipole antenna of length $h/\lambda_o = 1.125$. The distribution of current along the conductor is shown for several times during one cycle of a temporal oscillation. The relative spatial distribution of the current is seen to be fixed, with the peak values (the maxima of $|\mathcal{I}|$, which equal $2I_p$) and the nulls (the zeroes of $|\mathcal{I}|$) always occurring at the same locations. Adjacent peaks are separated by the distance $\lambda_o/2$, as are adjacent nulls. This behavior is caused by the factor $\sin[k_o(h - z)]$ in (8.80). The distribution shown in Figure 8.41 is known as a standing wave, which explains the name standing-wave dipole.

[21] It is important to distinguish the various currents we have introduced. I_s is the phasor for the current associated with the outward-going wave from the source (8.69), and $I_p = |I_s|$ is the modulus of this current. I_o is the phasor for the total current at the source, that is, the sum of the currents associated with all of the outward- and inward-going waves at the source.

The vector phasor for the radiated electric field is most easily obtained by using (8.68) with (8.83) in (8.25). After combining terms, we find

$$\vec{E}^r(\vec{r}) = \frac{j\mu_o c I_o}{2\pi r} e^{-jk_o r} \left[\frac{\cos(k_o h \cos\theta) - \cos(k_o h)}{\sin(k_o h)\sin\theta} \right] \hat{\theta}. \tag{8.85}$$

The time-average power per unit solid angle radiated by the dipole is then

$$\frac{d\langle \mathcal{P}_{\text{rad}} \rangle}{d\Omega} = \frac{\zeta_o |I_o|^2}{8\pi^2} \left[\frac{\cos(k_o h \cos\theta) - \cos(k_o h)}{\sin(k_o h)\sin\theta} \right]^2, \tag{8.86}$$

and the total time-average power radiated is (see Problem 8.14b)

$$\begin{aligned} \langle \mathcal{P}_{\text{rad}} \rangle = \frac{\zeta_o |I_o|^2}{4\pi \sin^2(k_o h)} \Big\{ & \gamma + \ln(2k_o h) - \text{Ci}(2k_o h) \\ & + \frac{1}{2}\cos(2k_o h)\big[\gamma + \ln(k_o h) + \text{Ci}(4k_o h) - 2\text{Ci}(2k_o h)\big] \\ & + \frac{1}{2}\sin(2k_o h)\big[\text{Si}(4k_o h) - 2\text{Si}(2k_o h)\big] \Big\}, \end{aligned} \tag{8.87}$$

where Ci is the previously defined cosine integral (8.75) and Si is the sine integral [26]:

$$\text{Si}(z) = \int_0^z \frac{\sin u}{u} du. \tag{8.88}$$

The directivity of the standing-wave dipole at the angle θ is readily obtained from these results and (3.115):

$$\begin{aligned} D(\theta) &= \frac{4\pi}{\langle \mathcal{P}_{\text{rad}} \rangle} \left[\frac{d\langle \mathcal{P}_{\text{rad}}(\theta) \rangle}{d\Omega} \right] \\ &= 2\left[\frac{\cos(k_o h \cos\theta) - \cos(k_o h)}{\sin\theta} \right]^2 \Big\{ \gamma + \ln(2k_o h) - \text{Ci}(2k_o h) \\ &\quad + \frac{1}{2}\cos(2k_o h)\big[\gamma + \ln(k_o h) + \text{Ci}(4k_o h) - 2\text{Ci}(2k_o h)\big] \\ &\quad + \frac{1}{2}\sin(2k_o h)\big[\text{Si}(4k_o h) - 2\text{Si}(2k_o h)\big] \Big\}^{-1}. \end{aligned} \tag{8.89}$$

The directive properties of the radiation from the standing-wave dipole are examined in Figures 8.42 and 8.43. The former shows vertical plane patterns for dipoles of various length; the latter shows the maximum directivity (D_{max}), the angle at which this directivity occurs (θ_{max}), and the half-power beamwidth for the lobe in this direction (BW), all as functions of the electrical length of the element $k_o h$ (h/λ_o). The patterns are seen to be maximum in the horizontal plane ($\theta_{\text{max}} = 90°$) for lengths $0 < k_o h \lesssim 4.4$ ($0 < h/\lambda_o \lesssim 0.7$). In this range of $k_o h$, the directivity steadily increases from the value $D_{\text{max}} = 1.5$ (1.76 dB) for the electrically short

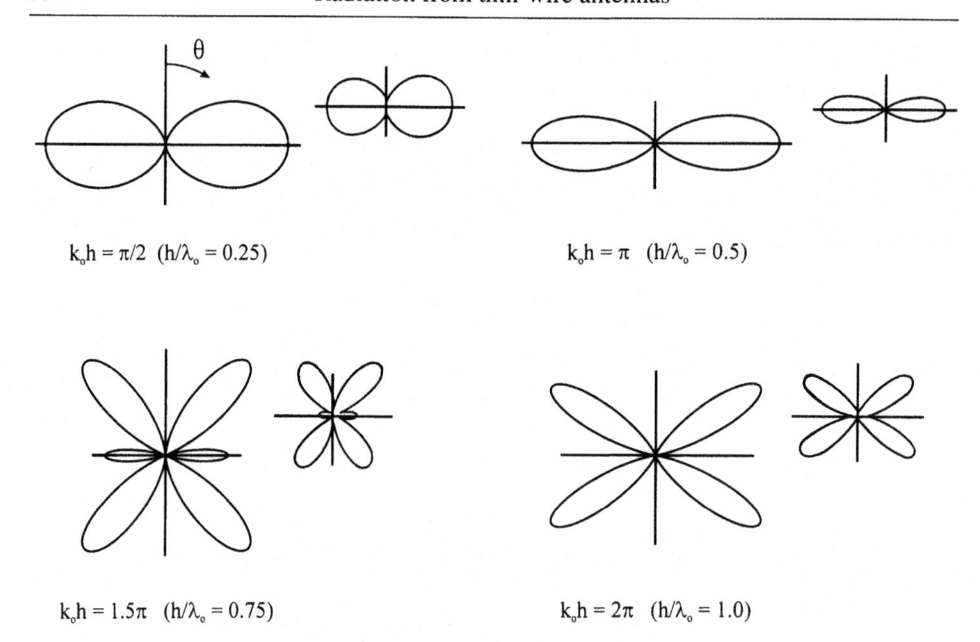

Fig. 8.42. Patterns for the radiation in the vertical plane for standing-wave dipole antennas, $d\langle \mathcal{P}_{\text{rad}}\rangle/d\Omega$. Insets show measured patterns. (Measured data from Reich [39].)

dipole, and the beamwidth steadily decreases. For the *half-wave dipole* ($k_o h = \pi/2$, $h/\lambda_o = 0.25$), the directivity is $D_{\max} = 1.64$ (2.15 dB), whereas for the *full-wave dipole* ($k_o h = \pi$, $h/\lambda_o = 0.5$), the directivity is $D_{\max} = 2.41$ (3.82 dB).[22]

At a length near $h/\lambda_o = 0.7$, the curves for $\theta_{\max}$ and BW in Figure 8.43 are discontinuous. This is caused by the maximum of the pattern abruptly shifting from $\theta_{\max} = 90°$ to a smaller angle; compare the patterns for $h/\lambda = 0.5$ and $h/\lambda = 0.75$ in Figure 8.42. The pattern is seen to change from one with a single lobe at $\theta = 90°$ to one with large lobes at $\theta \approx 42°$ and $138°$ and a small lobe at $\theta = 90°$. For the length $h/\lambda = 0.75$, the radiated energy is spread over several lobes; hence, the directivity $D_{\max}$ is less than that for the length $h/\lambda_o = 0.5$, where the radiated energy is concentrated in one lobe. This behavior is repeated periodically as the length of the dipole is increased, with the angle $\theta_{\max}$ abruptly changing from a larger value to a smaller value.

When we ignore the variations that occur over regions roughly $\lambda_o/2$ in extent, we see that the overall trend for the directive properties of the standing-wave dipole is similar to that for the basic traveling-wave element (compare Figures 8.35 and 8.43). The radiation becomes more end fire with a higher directivity and a narrower beamwidth as the length of the dipole is increased. However, the patterns are symmetric about $\theta = 90°$ for the standing-wave dipole, whereas they are not for the basic traveling-wave element (compare Figures 8.33 and 8.42).

The inset in the upper right of each drawing in Figure 8.42 shows a measured pattern for the dipole [39]. For these measurements, the wire forming the dipole was

[22] The terminology "half-wave dipcle" and "full-wave dipole" refers to the total length of the dipole, $2h$, being one half wavelength long and one wavelength long, respectively.

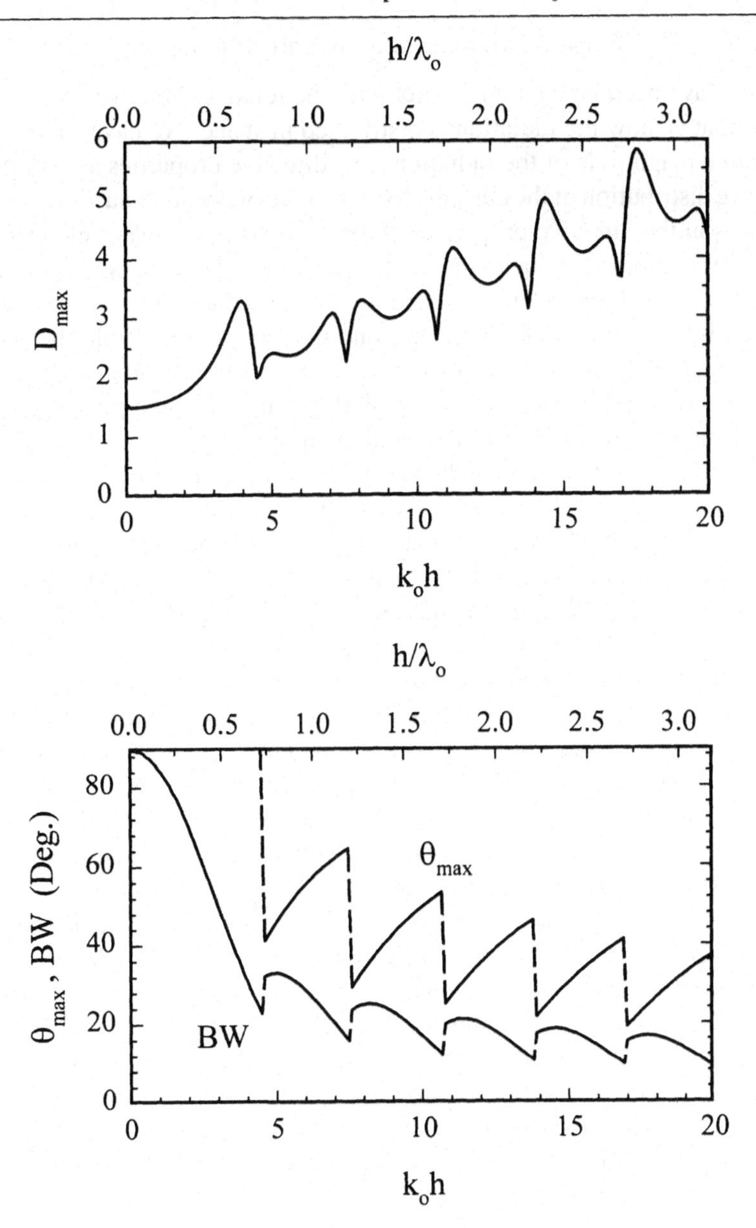

Fig. 8.43. Maximum directivity, D_{max}, of standing-wave dipole antenna versus electrical length, $k_o h$ (h/λ_o). θ_{max}: angle for maximum directivity. BW: half-power beamwidth.

very thin $(a/h = 1.81 \times 10^{-3})$. The reasonable agreement between the theoretical and measured patterns shows that the simple theory is adequate for describing the radiation from these dipole antennas.

In practical applications, one usually wants the maximum of the pattern for the radiation to be in the horizontal plane $(\theta_{\text{max}} = 90°)$. Hence, standing-wave dipoles with length $h/\lambda_o \leq 0.5$ are generally used; the half-wave dipole, with $h/\lambda_o = 0.25$, is the most popular.

8.5.4 Resonance and the half-wave dipole

So far we have been concerned mainly with the relative directive properties of an antenna, that is, how the radiation is distributed in space. We have said little about the absolute magnitude of the radiation. The directive properties are controlled by the relative distribution of the current along the wire of the antenna. For example, the differences in the directive properties of the basic traveling-wave element and the standing-wave dipole are due to the differences in the traveling-wave and standing-wave current distributions on the straight wire. The magnitude of the radiation is controlled by the magnitude of the current. Clearly, if we double the magnitude of the current while the relative distribution of the current remains constant, the radiated electric field (8.2) will double, and the time-average power radiated per unit solid angle [e.g. (8.73) or (8.86)] will quadruple.

To obtain a physical understanding of how the magnitude of the current on the standing-wave dipole can be increased, we will first consider the case of pulse excitation. In Figure 8.44 we show the charge distribution (charge per unit length) on the conductors of the dipole at several different times. When the dipole is excited by a single Gaussian pulse of peak amplitude Q_o, the distribution is the one shown in black in the figure. The source places charge on the conductors: positive charge on the right conductor ($z > 0$) and negative charge on the left conductor ($z < 0$). The pulses travel out along the conductors, as in Figures 8.44a and b, until they reach the open ends where they are reflected; they then travel in toward the source, as in Figures 8.44c and d. At the source, the pulses are reflected; recall the discussion associated with Figure 8.17. After reflection, pulses of peak amplitude $R_Q Q_o$ (black in the figure) travel out along the conductors, as in Figures 8.44e and f. In the figure, we have assumed that the reflection coefficient for the pulse is $R_Q = -0.5$.

Now we will consider the case where the excitation is the train of Gaussian pulses shown in Figure 8.45a. The period of the waveform is T, and each period consists of a positive and a negative pulse separated by the time $T/2$. We can maximize the waves of charge and current on the dipole by choosing the period of this waveform correctly. We should excite the dipole with the negative pulse just as the reflection of the positive pulse (which is negative, because R_Q is negative) is leaving the source. This situation is shown by the white pulses in Figures 8.44e and f. Because the time for a pulse to travel from the source to the open end of the dipole and back is $2\tau_a = 2h/c$, this means that we should choose $T/2 = 2h/c$, or $T = 4h/c$. Every pulse we apply to the dipole will then be timed so as to reinforce the wave of charge traveling along the conductors.

The waveform for time-harmonic excitation is the cosinusoid shown in Figure 8.45b. For reinforcement of the wave of charge on the dipole, we must choose the period of this waveform as we did for the train of Gaussian pulses: $T = 4h/c$. Since $T = 1/f = \lambda_o/c$ for the cosinusoid, the length h of one arm of the dipole must be $h = \lambda_o/4$. This is the half-wave dipole; the total length of the dipole is one half of a wavelength: $2h = \lambda_o/2$. The condition we have described in which each cycle of the excitation reinforces the waves of charge and current on the dipole is called *resonance*.

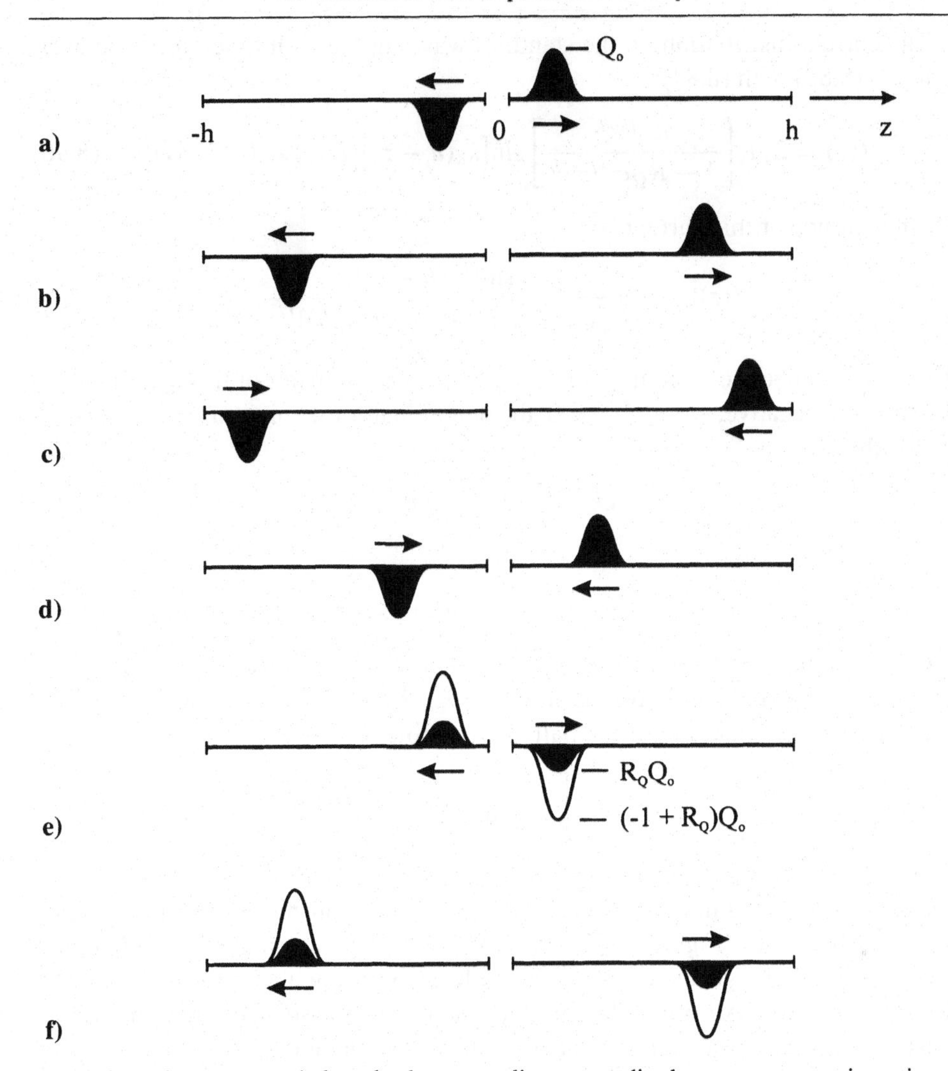

Fig. 8.44. Charge per unit length along standing-wave dipole antenna at various times. Excitation is a Gaussian pulse (black) or train of Gaussian pulses (black for (a)–(d), white for (e) and (f)); $R_Q = -0.5$.

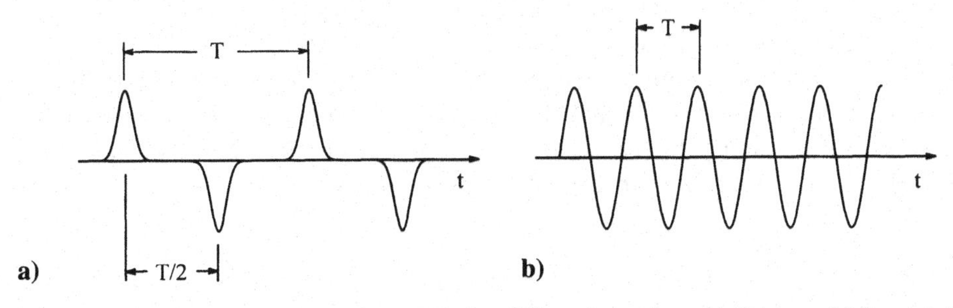

Fig. 8.45. Waveforms for excitation. a) Train of Gaussian pulses. b) Cosinusoid (harmonic time dependence).

The current distribution on the standing-wave dipole with time-harmonic excitation is (8.82) with (8.84):

$$I(z) = 2jI_s\left[\frac{e^{-jk_oh}}{1 - R_Q e^{-j2k_oh}}\right]\sin\big[k_o(h-z)\big]\big[U(z) - U(z-h)\big]. \qquad (8.90)$$

The magnitude of this current is[23]

$$|I(z)| = 2I_p\frac{|\sin[k_o(h-z)]|}{[1 + R_Q^2 - 2R_Q\cos(2k_oh)]^{1/2}}. \qquad (8.91)$$

We can maximize the current by minimizing the denominator of this expression.[24] When R_Q is negative, $-1 \leq R_Q < 0$, the denominator is a minimum for $\cos(2k_oh) = -1$; that is,

$$2k_oh = (2n+1)\pi, \qquad (8.92)$$

or

$$h = (2n+1)\lambda_o/4, \qquad n = 0, 1, 2, \ldots. \qquad (8.93)$$

For $n = 0$, this is just the condition for resonance we obtained from the physical argument above (the half-wave dipole): $h = \lambda_o/4$.

The current distribution for the half-wave dipole is

$$I(z) = 2I_s\cos(k_oz)\left[\frac{1}{1+R_Q}\right]\big[U(z) - U(z-h)\big]. \qquad (8.94)$$

Notice that there is a problem with this expression when $R_Q = -1$ ($T_Q = 1$): The current becomes infinite. This is the case of total transmission for the wave of charge at the source (Figure 8.17b). It corresponds to an antenna with no damping of the waves of charge and current. This failure points out a major shortcoming of this simple model. All of the damping in the model is associated with the partial transmission (partial absorption) at the source; the damping due to radiation of energy from the antenna is not included.

For harmonic time dependence, the input impedance, Z, of an antenna proves to be a quantity of practical interest. It is a measure of how easily an antenna can be excited by a voltage source, since $I(0) = V/Z$, where V is the total voltage. In Figure 8.46 we graph the impedance for a standing-wave monopole antenna, $Z_{\text{mono}} = R_{\text{mono}} + jX_{\text{mono}}$, as a function of the electrical length, k_oh (h/λ_o). The configuration is that in Figure 1.45 with $b/a = 2.30$ and $h/a = 65.8$. Recall from the discussion in Section 7.2 that the impedance of the dipole is approximately twice that of the monopole ($Z_{\text{dip}} \approx 2Z_{\text{mono}}$); therefore, these results can also be used to estimate the impedance of the dipole. The solid lines in the figure are

[23] Here, as in our earlier discussion (Section 8.2), we are assuming that R_Q is a real number, $-1 \leq R_Q \leq 1$.

[24] Here we have assumed that the range for z includes a value for which $|\sin[k_o(h-z)]| = 1$.

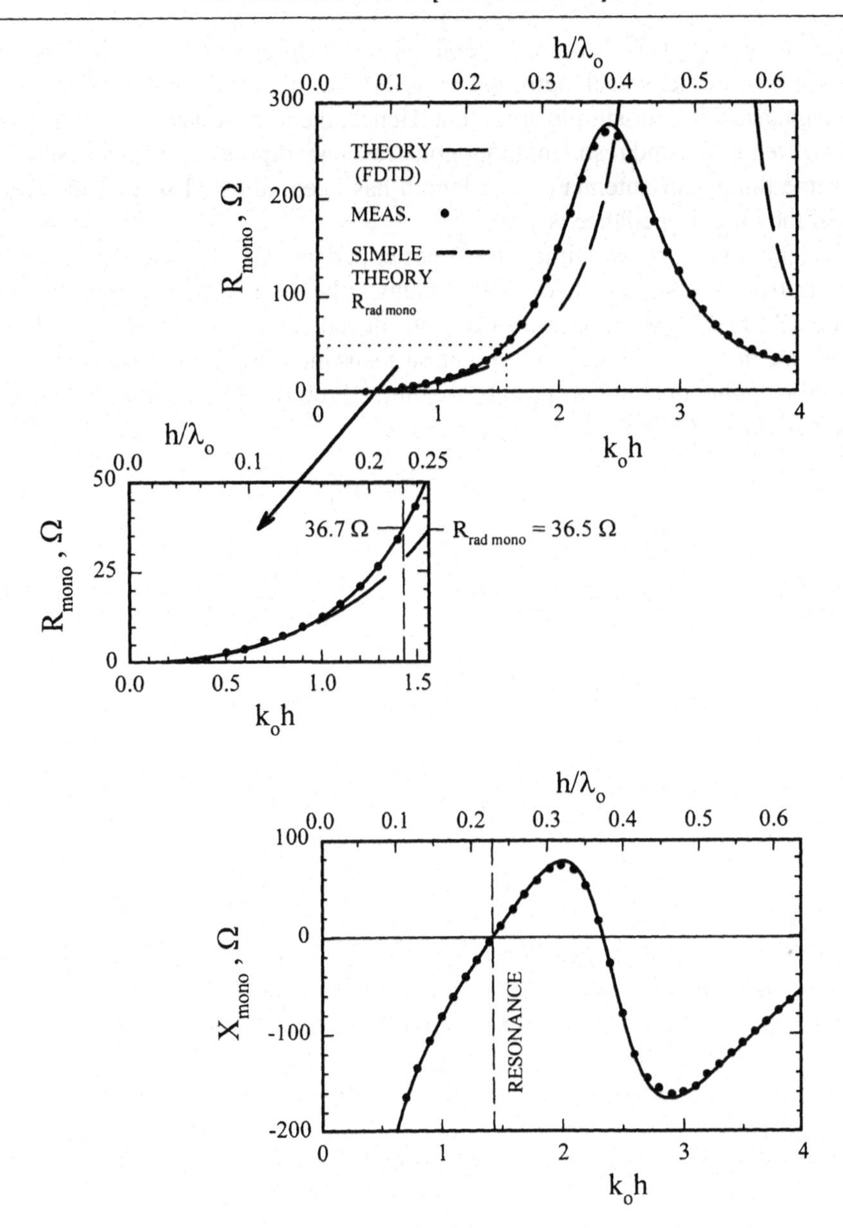

Fig. 8.46. Input impedance of cylindrical monopole antenna versus electrical length $k_o h$ (h/λ_o): $b/a = 2.30$, $h/a = 32.8$. (Theoretical results from Maloney [40]. Measured data from Scott [41].)

accurate calculations made with the finite-difference time-domain (FDTD) method discussed in Section 1.6, and the solid dots are measured data [40, 41].

Notice that the first zero for the reactance, X_{mono}, occurs at $k_o h \approx 1.44$ $(h/\lambda_o \approx 0.229)$. This is a convenient point at which to operate the antenna, because the input impedance is then a pure resistance of reasonable value. This length is called the "resonant length" and the corresponding frequency the "resonant frequency" of the monopole [42].[25] As the antenna is made thinner ($a/h \to 0$) this length approaches a quarter wavelength, $k_o h = \pi/2$ ($h/\lambda_o = 0.25$), the length we obtained for resonance with our simple argument. Hence, the terms quarter-wave monopole (half-wave dipole) and resonant monopole (resonant dipole) are often used to mean the same thing – an antenna whose length has been adjusted so that the reactive part of the input impedance is zero.

It is tempting to try to estimate the input impedance of the standing-wave dipole antenna from the simple theory that assumes the sinusoidal distribution for the current (8.81). Using the approach we introduced earlier for the electrically short linear antenna (Section 7.2), the radiation resistance (the input resistance for the perfectly conducting antenna) can be obtained directly from the total time-average radiated power (8.87):

$$\frac{1}{2}|I_o|^2 R_{\text{rad dip}} = \langle \mathcal{P}_{\text{rad}} \rangle, \tag{8.95}$$

or

$$\begin{aligned} R_{\text{rad dip}} &= \frac{2\langle \mathcal{P}_{\text{rad}} \rangle}{|I_o|^2} \\ &= \frac{\zeta_o}{2\pi \sin^2(k_o h)} \bigg\{ \gamma + \ln(2k_o h) - \text{Ci}(2k_o h) \\ &\quad + \frac{1}{2}\cos(2k_o h)\big[\gamma + \ln(k_o h) + \text{Ci}(4k_o h) - 2\text{Ci}(2k_o h)\big] \\ &\quad + \frac{1}{2}\sin(2k_o h)\big[\text{Si}(4k_o h) - 2\text{Si}(2k_o h)\big] \bigg\}. \end{aligned} \tag{8.96}$$

This result is graphed as a dashed line in Figure 8.46 (actually one half of this result, since the figure is for a monopole). It is seen to disagree with the measured results and the accurate theoretical results except for electrically short monopoles, that is, for $k_o h \lesssim 1.0$ ($h/\lambda_o \lesssim 0.16$). An obvious problem is that the value of Equation (8.96) becomes infinite whenever $\sin(k_o h) = 0$; in the figure this occurs for the half-wave monopole [$k_o h = \pi$ ($h/\lambda_o = 0.5$)].

For the quarter-wave monopole, this formula gives $R_{\text{rad mono}} \approx 36.5\ \Omega$ (for the half-wave dipole $R_{\text{rad dip}} \approx 73\ \Omega$). From Figure 8.46, we see that this resistance is close to the actual resistance at resonance ($R_{\text{mono}} \approx 36.7\ \Omega$), and so it is often used as an estimate for the actual resistance at resonance.[26]

[25] Resonance for the antenna is defined as in circuit theory to be the frequency (value of $k_o h = \omega h/c$) at which the input impedance is purely real. Near resonance the curves for R_{mono} and X_{mono} in Figure 8.46 are similar to those for the impedance of a series resonant circuit (a resistor, inductor, and capacitor in series).

[26] A detailed analysis for the thin-wire, linear antenna shows that the input resistance is generally sensitive to the radius of the wire; an exception is the resistance near resonance [42]. As we have already mentioned, the characteristics of an actual antenna approach those of our simple model in the limit as the wire becomes vanishingly thin ($a/h \to 0$). Hence, the resistance at resonance given by the simple theory is a fair estimate for the resistance at resonance of actual, thin-wire, linear antennas.

We should not be surprised by the failure of the simple theory to accurately predict the input resistance of the dipole antenna. After all, the theory is based on an "educated guess" for the current distribution on the antenna. Because the input impedance is directly related to the current distribution, $Z = V/I(0)$, errors in the current distribution directly affect the input impedance (resistance). The radiated field, however, is the integral of the current distribution, (8.66) and (8.67). Errors in the current distribution are "smoothed out" by the integration; consequently, the radiated field is fairly insensitive to the precise details of the current distribution. This explains the fairly good agreement of the radiated fields calculated from the simple theory with measurements (Figures 8.40 and 8.42).

8.6 Perspective

In this chapter and Chapter 1 we have seen the extremes of antenna analysis. The finite-difference time-domain (FDTD) method, a purely numerical method based on a direct discretization of Maxwell's equations, was applied to the monopole antenna, and the results were shown to be in excellent agreement with measurements for both the radiated field and the input impedance. In addition, a simple analysis, based on an assumed distribution of current, was applied to a number of thin-wire antennas including the monopole. It was shown to produce reasonable results for the radiated field but generally unacceptable results for quantities like the input impedance. A legitimate question is: Why do we need both methods of analysis? After all, the numerical method produced results that were essentially identical to the measurements.

The answer to this question has as much to do with how we think about electromagnetic fields as to do with the accuracy of computations. Numerical methods, like the FDTD method, simply provide us with a table of numbers for the electromagnetic response of a particular antenna, such as values of the electric field at various times for a fixed point in space. Once the numbers are displayed on a graph, it is possible to identify special points or trends for the particular antenna under investigation. However, it is generally not possible to directly use the results to predict, in a simple manner, what the response of a different antenna will be. In contrast, the simple analysis for thin-wire antennas provides us with models, such as the basic traveling-wave element. The physical concepts and equations associated with these models can be used, with reasonable accuracy, to predict the radiation from many different antennas. We can think about how radiation occurs in terms of these models. They also provide us with a tie-in to other fundamental electromagnetic problems, such as the radiation from a moving point charge.

Therefore, the answer to the above question is that we really need both types of analyses. We need the simple methods to provide us with a framework for thinking about how radiation occurs from antennas, and we need the numerical methods for obtaining accurate results for use in optimizing the design of a particular antenna.

In our discussion of thin-wire antennas, we have made no attempt to provide a comprehensive review of the many antennas that can be placed in this category. Rather, we have tried to provide a basic physical understanding of the process of radiation for such antennas. There are a number of books that provide fairly complete discussions of the various thin-wire antennas, and these should be consulted when selecting an antenna for a particular application [42–50].

References

[1] E. P. Sayre and R. Harrington, "Time Domain Radiation and Scattering by Thin Wires," *Applied Scientific Research*, Vol. 26, pp. 413–44, September 1972.

[2] E. K. Miller, A. J. Poggio, and G. J. Burke, "An Integro-Differential Equation Technique for the Time-Domain Analysis of Thin Wire Structures. I. The Numerical Method," *J. Computational. Phys.*, Vol. 12, pp. 24–48, May 1973.

[3] A. J. Poggio, E. K. Miller, and G. J. Burke, "An Integro-Differential Equation Technique for the Time-Domain Analysis of Thin Wire Structures. II. Numerical Results," *J. Computational. Phys.*, Vol. 12, pp. 210–33, June 1973.

[4] J. G. Maloney, G. S. Smith, and W. R. Scott, Jr., "Accurate Computation of the Radiation from Simple Antennas Using the Finite-Difference Time-Domain Method," *IEEE Trans. Antennas Propagat.*, Vol. 38, pp. 1,059–68, July 1990.

[5] S. A. Schelkunoff and H. T. Friis, *Antennas, Theory and Practice*, Wiley, New York, 1952.

[6] C. Manneback, "Radiation from Transmission Lines," *J. AIEE*, Vol. 42, pp. 95–105, February 1923.

[7] H. J. Schmidt, "Transients in Cylindrical Antennae," *Proc. IEE*, Pt. C, Monograph No. 377 E, pp. 292–8, April 1960.

[8] —, "The Transient Response of Linear Antennas and Loops," *IRE Trans. Antennas Propagat.*, Vol. AP-10, pp. 222–8, May 1962.

[9] G. Franceschetti and C. H. Papas, "Pulsed Antennas," *IEEE Trans. Antennas Propagat.*, Vol. AP-22, pp. 651–61, September 1974.

[10] D. L. Sengupta and C. -T. Tai, "Radiation and Reception of Transients by Linear Antennas," Ch. 4 in *Transient Electromagnetic Fields*, L. B. Felsen, editor, Springer-Verlag, Berlin, 1976.

[11] E. K. Miller and J. A. Landt, "Direct Time-Domain Techniques for Transient Radiation and Scattering from Wires," *Proc. IEEE*, Vol. 68, pp. 1,396–423, November 1980.

[12] C. H. Walters, *Traveling Wave Antennas*, McGraw-Hill, New York, 1965. Republication, Dover Publications, New York, 1970.

[13] G. A. Burrell, "Propagation of Current on Bent Wire Antennas – An Experimental Study," *IEEE Trans. Antennas Propagat.*, Vol. AP-26, pp. 427–34, May 1978.

[14] G. Mönich and N. Scheffer, "Impulse Radiation by Wire Bends," *Sixth International Conference on Antennas and Propagation*, Pt. 1: Antennas, pp. 307–11, 1989.

[15] A. M. Abo-Zena and R. E. Beam, "Transient Radiation Field of a Circular Loop Antenna," *IEEE Trans. Antennas Propagat.*, Vol. AP-20, pp. 380–83, May 1972.

[16] K. J. Langenberg, "Pulsed Loop Antennas," *Applied Physics*, Vol. 10, pp. 309–16, 1976.

[17] T. T. Wu and R. W. P. King, "The Cylindrical Antenna with Nonreflecting Resistive Loading," *IEEE Trans. Antennas Propagat.*, Vol. AP-13, pp. 369–73, May 1965.

[18] L. C. Shen and R. W. P. King, "The Cylindrical Antenna with Nonreflecting Resistive Loading," *IEEE Trans. Antennas Propagat.*, Vol. AP-13, p. 998, November 1965.

[19] L. C. Shen and T. T. Wu, "Cylindrical Antenna with Tapered Resistive Loading," *Radio Science*, Vol. 2, pp. 191–201, February 1967.

[20] J. G. Maloney and G. S. Smith, "A Study of Transient Radiation from the Wu-King Resistive Monopole – FDTD Analysis and Experimental Measurements," *IEEE Trans. Antennas Propagat.*, Vol. 41, pp. 668–76, May 1993. Correction, Vol. 43, p. 226, February 1995.

[21] R. W. P. King and G. S. Smith, *Antennas in Matter: Fundamentals, Theory, and Applications*, MIT Press, Cambridge, MA, 1981.

[22] I. S. Gradshteyn and I. W. Ryzhik, *Tables of Integrals, Series, and Products*, Academic Press, New York, 1980.

[23] J. D. Lawson, "On the Relation Between Čerenkov Radiation and Bremsstrahlung," *Philos. Mag.*, Vol. 45, pp. 748–50, July 1954.

[24] —, "Cherenkov Radiation, 'Physical' and 'Unphysical,' and its Relation to Radiation from an Accelerated Electron," *Am. J. Phys.*, Vol. 33, pp. 1,002–5, December 1965.

[25] P. C. Clemmow and J. P. Dougherty, *Electrodynamics of Particles and Plasmas*, Addison-Wesley, Reading, MA, 1969.

[26] M. Abramowitz and I. A. Stegun, *Handbook of Mathematical Functions*, U.S. Government Printing Office, Washington, DC, 1964.

[27] E. E. Altshuler, "The Traveling-Wave Linear Antenna," *IRE Trans. Antennas and Propagat.*, Vol. 9, pp. 324–9, July 1961.

[28] E. Bruce, "Developments in Short-Wave Directive Antennas," *Proc. IRE*, Vol. 19, pp. 1,406–33, August 1931.

[29] E. A. Laport, "Long-Wire Antennas," in *Antenna Engineering Handbook*, 3rd Edition, R. C. Johnson and H. Jasik, editors, Ch. 11, McGraw-Hill, New York, 1993.

[30] K. Iizuka, "The Traveling-Wave V-Antenna and Related Antennas," *IEEE Trans. Antennas and Propagat.*, Vol. AP-15, pp. 236–43, March 1967.

[31] G. A. Thiele and E. P. Ekelman, "Design Formulas for Vee Dipoles," *IEEE Trans. Antennas and Propagat.*, Vol. AP-28, pp. 588–90, July 1980.

[32] E. Bruce, A. C. Beck, and L. R. Lowry, "Horizontal Rhombic Antennas," *Proc. IRE*, Vol. 23, pp. 24–46, January 1935.

[33] D. Foster, "Radiation from Rhombic Antennas," *Proc. IRE*, Vol. 25, pp. 1,327–53, October 1937.

[34] L. Lewin, "Discussion on 'Radiation from Rhombic Antennas'," *Proc. IRE*, Vol. 29, p. 523, September 1941.

[35] A. E. Harper, *Rhombic Antenna Design*, D. Van Nostrand, Inc., New York, 1941.

[36] E.-G. Hoffmann, "Der Einfluss der exponentiellen Stromverteilung auf die Strahlungseigenschaften der Rhombusantenne," *Hochfrequenztechnik und Elektroakustik*, Vol. 62, pp. 15–20, July 1943.

[37] W. N. Christiansen, "Directional Patterns of Rhombic Antennae," *A. W. A. Tech. Rev.*, Vol. 7, No. 1, pp. 33–51, 1946.

[38] E. A. Laport, "Design Data for Horizontal Rhombic Antennas," *RCA Review*, Vol. 13, pp. 71–94, March 1952.

[39] H. J. Reich, editor, *Very High-Frequency Techniques*, Vol. 1, Ch. 4, McGraw-Hill, New York, 1947.

[40] J. G. Maloney, "Analysis and Synthesis of Transient Antennas using the Finite-Difference Time-Domain (FDTD) Method," Ph. D. Dissertation, School of Elec. Engr., Georgia Inst. Technol., Atlanta, GA, 1992.

[41] W. R. Scott, Jr., "Dielectric Spectroscopy Using Open-Circuited Coaxial Lines and Monopole Antennas of General Length," Ph. D. Dissertation, School of Elec. Engr., Georgia Inst. Technol., Atlanta, GA, 1985.
[42] R. W. P. King, *The Theory of Linear Antennas*, Harvard University Press, Cambridge, MA, 1956.
[43] S. A. Schelkunoff, *Advanced Antenna Theory*, Wiley, New York, 1952.
[44] R. E. Collin and F. J. Zucker, editors, *Antenna Theory*, Parts I and II, McGraw-Hill, New York, 1969.
[45] R. F. Harrington, *Field Computation by Moment Methods*, Macmillan, New York, 1968.
[46] R. C. Hansen, editor, *Moment Methods in Antennas and Scattering*, Artech House, Norwood, MA, 1990.
[47] W. L. Weeks, *Antenna Engineering*, McGraw-Hill, New York, 1968.
[48] W. L. Stutzman and G. A. Thiele, *Antenna Theory and Design*, Wiley, New York, 1981.
[49] C. A. Balanis, *Antenna Theory: Analysis and Design*, Harper and Row, New York, 1982.
[50] J. D. Kraus, *Antennas*, 2nd Edition, McGraw-Hill, New York, 1988.

Problems

8.1 a) Use the equation of continuity for electric charge to obtain the current on the dipole antenna, (8.7) and (8.8), from the charge per unit length, (8.5) and (8.6).

b) Starting from the current on the basic traveling-wave element (8.14), use the equation of continuity for electric charge to obtain the charge per unit length on the element (8.15).

8.2 Show that, in the limit as the number of segments on the dipole antenna becomes infinite ($n \to \infty$), the radiated electric field at broadside becomes Equation (8.13).

8.3 Derive the expression given in (8.18) that relates the derivatives with respect to time and space of the current on the basic traveling-wave element.

8.4 Obtain the expression (8.22) for the radiated electric field of the displaced basic traveling-wave element shown in Figure 8. 14a.

8.5 The excitation for a basic traveling-wave element is the current

$$\mathcal{I}_s(t) = I_o U(t),$$

where $U(t)$ is the Heaviside unit-step function.

a) What is the charge per unit length on the element?

b) Obtain an expression for the radiated electric field of the element and make a graph like Figure 8.12 for the field.

c) For any direction, θ, is the area under the graph for the electric field, $\mathcal{E}_\theta^r$, zero?

8.6 A traveling-wave V antenna has the angle α between the two arms (Figure 8.37). The antenna is to be used for pulse radiation.

a) Consider the V antenna to be composed of two basic traveling-wave elements. Determine the parameters that describe these elements: t_{oi}/τ_a, γ_i, β_i, d_i, etc.
b) Make a graph like Figure 8.22 for the radiated electric field in the plane of the V antenna by directly summing the fields of the individual elements. Let $\alpha = 60°$ and the excitation be a Gaussian pulse with $\tau/\tau_a = 0.076$.
c) Obtain an analytical expression for the radiated electric field $\vec{\mathcal{E}}^r$ in the plane of the V.
d) Specialize your result from part c to the direction $\theta = 90°$.
e) Assume that the angle of the V is small ($\alpha \ll 1$), and show from part d that the radiated electric field in the direction $\theta = 90°$ is proportional to the derivative of the current $\dot{\mathcal{I}}(t - r/c)$. Is this result consistent with your graph from part b?

8.7 In this problem we will examine the radiation from a standing-wave dipole antenna (8.25) in the limit as $h \to 0$. This will be compared with the radiation from an electrically short linear antenna with a triangular current distribution, which was discussed in Section 7.2. Notice that the results for the dipole antenna are described in terms of the current $\mathcal{I}_s(t)$ placed on the conductors by the source, whereas the results for the electrically short linear antenna are described in terms of the total current at the source $\mathcal{I}_o(t) = \mathcal{I}(z = 0, t)$.

a) Obtain an expression for the total current, $\mathcal{I}(z, t)$, on the dipole antenna. Use this result to express the current $\mathcal{I}_s(t)$ in terms of the total current at the source $\mathcal{I}_o(t)$.
b) Let the length of the dipole, h, become infinitesimally small ($\lim h \to 0$) in the expressions for $\vec{\mathcal{E}}^r$ and $\mathcal{I}_o$. Express $\vec{\mathcal{E}}^r$ in terms of $\mathcal{I}_o$ using your result from part a.
c) Compare your result for $\vec{\mathcal{E}}^r$ from part b with the radiated electric field of the electrically short linear antenna with a triangular current distribution: (7.27) with (7.51), and $\gamma = 1/2$.

8.8 Draw pictures like those in Figure 8.27 for the radiated electric field of a traveling-wave, circular loop antenna excited by a Gaussian pulse of current with $\tau/\tau_a = 0.076$. Make these pictures for the angles of observation $\theta = 0°$ and $\theta = 180°$. Compare your results for the two angles.

8.9 Derive the expression (8.39) for the radiated electric field of the traveling-wave, circular loop antenna with a counterclockwise traveling wave of current.

8.10 For the dipole antenna with continuous resistive loading:

a) Obtain the expression for the charge per unit length (8.42) on the top conductor of the dipole (basic traveling-wave element) from the current (8.41).

b) Determine the radiated electric field of the basic traveling-wave element; then use this expression to obtain the radiated electric field of the dipole (8.43).

8.11 Starting with the current on the insulated linear antenna (8.48), obtain the expression for the radiated electric field (8.50).

8.12 Show that the radiated electric field of the insulated linear antenna is given by Equation (8.51) at the angle $\theta = \chi_c$, and that it is given by Equation (8.52) for the first pulse at the angle $\theta = 90°$.

8.13 Obtain the radiated electric field of the standing-wave dipole (8.25) by combining the radiated electric fields of the four basic traveling-wave elements shown in Figure 8.18.

8.14 Obtain the expression for the total time-average power radiated, $\langle \mathcal{P}_{\text{rad}} \rangle$, for

a) the basic traveling-wave element (8.74),
b) the standing-wave dipole antenna (8.87).

8.15 In this problem we will examine the radiation in the vertical plane from a horizontal rhombic antenna.

a) An isolated rhombic antenna (no Earth) can be modeled by four basic traveling-wave elements. Determine the parameters ωt_{oi}, γ_i, β_i, d_i, etc. for these elements. Let the plane of the rhombic be the x-z plane, and place the source and termination on the x axis.
b) Assume that the horizontal rhombic antenna is placed over a perfectly conducting Earth. Show how the Earth can be included in the calculation of the radiated electric field using image theory and an array factor.
c) Write a computer program to determine the magnitude of the radiated electric field in the vertical plane ($\theta = \pi/2$) for the horizontal rhombic antenna over a perfectly conducting Earth. Make a pattern for the electric field of an antenna with the parameters given in Figure 8. 40. Compare your results with those in the figure.

8.16 Consider the standing-wave dipole antenna that has reflection of the traveling waves of charge and current at the source (Figure 8.17d).

a) Obtain an expression for the time-varying current distribution on the top conductor of the dipole. Assume that the initial outward-going wave of current from the source is $\mathcal{I}_s(t - z/c)$.
b) Specialize your result from part a to the case of harmonic time dependence. Obtain the phasor for the current distribution, $I(z)$.
c) Show that Equation (8.84) follows from your result in part b.

8.17 The bent-wire, traveling-wave antenna is to modeled as the two basic traveling-wave elements shown in Figure 8.21b.

a) Combine the radiated electric fields of the two basic elements to obtain Equation (8.27).

b) Show that this field reduces to be that of a single, basic traveling-wave element (straight, traveling-wave antenna) when $\alpha = 180^\circ$.
c) Specialize your result from part a to the direction $\theta = 0^\circ$.
d) Let the wire be nearly straight, so that $\alpha = \pi - \varepsilon$, $\varepsilon \ll 1$. Use the result from part c to show that the radiated electric field in the direction $\theta = 0^\circ$ is proportional to the second derivative of the current, $\ddot{\mathcal{I}}(t - r/c)$. Comment on the waveform for the radiated electric field in the direction $\theta = 0^\circ$ in Figure 8.22.

Appendix A

Units and dimensions

The equations of electromagnetism are a mathematical representation of observed physical phenomena. For practical use, the quantities in these equations are measured in terms of units that are embodied in physical standards. To be of value, the concepts associated with the quantities, the units, the physical standards, etc., must have international acceptance. Several different systems of units are in use with these equations, and none is superior to all others for all applications.

The different philosophical approaches to this subject can be quite involved [1–4]. It is not our purpose here to delve into such questions. Instead, we wish to present a brief argument that clarifies our particular choice for the form of the equations of electromagnetism (Table 1.3) and the system of units we have adopted, the *International System of Units* (le Système International d'Unités, abbreviated SI).

We will begin our discussion by considering familiar results from classical mechanics. Physical phenomena are described by relations such as Newton's second law, which says that the force on a particle of constant mass m is the product of the particle's mass and acceleration,

$$\vec{\mathcal{F}} = m\vec{a}, \tag{A.1}$$

or the law of universal gravitation, which states that the force of attraction of a particle of mass m_1 on a particle of mass m_2 is

$$\vec{\mathcal{F}} = -\frac{Gm_1m_2}{r^2}\hat{r}, \tag{A.2}$$

where $\hat{r}$ is a vector in the direction from m_1 to m_2.

These *quantity equations* have the general form

$$w = kuv, \tag{A.3}$$

where w, u, and v are mathematical quantities, and k is a constant of proportionality. In (A.1) $k = 1$, while in (A.2) $k = G$, with G conventionally determined by experimentation.

Each term in (A.3) has the representation

$$w = \{w\}\langle w\rangle, \tag{A.4}$$

where $\langle w\rangle$ is the *unit* of w, and $\{w\}$ is a numerical value that is the *measure* of w

in terms of the unit.[1] Thus, (A.3) is equivalent to

$$\{w\}\langle w\rangle = \{k\}\langle k\rangle\{u\}\langle u\rangle\{v\}\langle v\rangle. \tag{A.5}$$

This equation is now separated into a *measure equation*, which contains only numerical values,

$$\{w\} = \{k\}\{u\}\{v\}, \tag{A.6}$$

and a *unit equation*

$$\langle w\rangle = \langle k\rangle\langle u\rangle\langle v\rangle. \tag{A.7}$$

Here the system of units is assumed to be *coherent*, that is, the relationship between the units contains no numerical factors.

The measure and unit equations corresponding to Newton's second law (A.1) and the law of universal gravitation (A.2) are

$$\{\mathcal{F}\} = \{m\}\{a\} \tag{A.8}$$

$$\langle\mathcal{F}\rangle = \langle m\rangle\langle a\rangle \tag{A.9}$$

and

$$\{\mathcal{F}\} = -\{G\}\{m_1\}\{m_2\}\{r\}^{-2} \tag{A.10}$$

$$\langle\mathcal{F}\rangle = \langle G\rangle\langle m\rangle^2\langle r\rangle^{-2}. \tag{A.11}$$

In classical mechanics three units are customarily selected as *base units*: the unit for length $\langle\ell\rangle$, the unit for mass $\langle m\rangle$, and the unit for time $\langle t\rangle$. The units for the other quantities are *derived units* that are expressible in terms of the three base units. For example, the unit for acceleration is

$$\langle a\rangle = \langle\ell\rangle\langle t\rangle^{-2}, \tag{A.12}$$

and the unit for force, obtained from (A.9), is

$$\langle\mathcal{F}\rangle = \langle\ell\rangle\langle m\rangle\langle t\rangle^{-2}. \tag{A.13}$$

In the International System of Units, the three base mechanical units are the unit for length, the meter (m); the unit for mass, the kilogram (kg); and the unit for time, the second (s). Table A.1 is a complete list of the SI base units. Each of the SI base units is embodied in a physical standard, and the relationship between the unit and the standard is specified. The official definitions for the meter, the kilogram, and the second are [5]:

The *meter* is the length of the path traveled by light in a vacuum during a time interval of 1/299792458 of a second.

The *kilogram* is equal to the mass of the international prototype of the kilogram – a platinum-iridium cylinder.

[1] The brackets ⟨⟩ are not to be confused with the same notation used for the average value of a quantity.

Table A.1. *SI base units*

Quantity	Name of unit	Symbol
Length	meter	m
Mass	kilogram	kg
Time	second	s
Electric current	ampere	A
Temperature	kelvin	K
Luminous intensity	candela	cd
Amount of substance	mole	mol

> The *second* is the duration of 9192631770 periods of the radiation corresponding to the transition between the two hyperfine levels of the ground state of the cesium-133 atom.

The SI derived unit for the force is the newton; it is expressed in terms of the base units by (A.13)

$$\text{newton} = \text{m} \cdot \text{kg} \cdot \text{s}^{-2}.$$

Table A.2 lists some of the more common SI derived units.

The *dimension* is an additional label that can be attached to a mathematical quantity, such as those appearing in Equations (A.1) and (A.2). "The dimension of a given quantity can be thought of as a sort of generalized symbolic unit which retains some of the information specified by the latter but which is not limited to any particular choice of the sizes of the basic units" [4]. As with the units, it is customary in classical mechanics to select three base dimensions: the length L, the mass M, and the time T. The dimensions for other quantities are then expressed in terms of these base dimensions; for example, the dimensions for acceleration are

$$[a] = [LT^{-2}]$$

and the dimensions for force are

$$[\mathcal{F}] = [LMT^{-2}].$$

Here the brackets are read "dimensions of."

The equations of classical electrodynamics contain quantities in addition to those used in classical mechanics, for example, the electric current and the electric charge. In the SI the unit for one of these quantities, the electric current, is chosen to be a base unit. This base unit is the ampere (A) with the official definition [5]:

> The *ampere* is the constant current which, if maintained in two straight parallel conductors of infinite length, of negligible circular cross section, and placed

Table A.2. *SI derived units*

Quantity	Name of unit	Symbol for unit	In terms of other units	In terms of base units
Frequency	hertz	Hz		s^{-1}
Force	newton	N		$m \cdot kg \cdot s^{-2}$
Energy	joule	J	$N \cdot m$	$m^2 \cdot kg \cdot s^{-2}$
Power	watt	W	J/s	$m^2 \cdot kg \cdot s^{-3}$
Charge	coulomb	C	$A \cdot s$	$s \cdot A$
Electric potential	volt	V	J/C	$m^2 \cdot kg \cdot s^{-3} \cdot A^{-1}$
Magnetic flux	weber	Wb	$V \cdot s$	$m^2 \cdot kg \cdot s^{-2} \cdot A^{-1}$
Resistance	ohm	Ω	V/A	$m^2 \cdot kg \cdot s^{-3} \cdot A^{-2}$
Conductance	siemens	S	A/V	$m^{-2} \cdot kg^{-1} \cdot s^3 \cdot A^2$
Capacitance	farad	F	C/V	$m^{-2} \cdot kg^{-1} \cdot s^4 \cdot A^2$
Inductance	henry	H	Wb/A	$m^2 \cdot kg \cdot s^{-2} \cdot A^{-2}$
Electric field ($\vec{\mathcal{E}}$)			V/m	$m \cdot kg \cdot s^{-3} \cdot A^{-1}$
Magnetic field ($\vec{\mathcal{B}}$)	tesla	T	Wb/m^2	$kg \cdot s^{-2} \cdot A^{-1}$
Electric excitation ($\vec{\mathcal{D}}$)			C/m^2	$m^{-2} \cdot s \cdot A$
Magnetic excitation ($\vec{\mathcal{H}}$)			A/m	$m^{-1} \cdot A$

one meter apart in vacuum, would produce between these conductors a force equal to 2×10^{-7} newton per meter of length.

The base dimension of current is indicated by $[I]$.

Now we will examine the experimental basis for electromagnetism, which we have discussed in Chapters 1 and 2, and show how the particular form of Maxwell's equations with SI units (Tables 1.2 and 1.3) arises [4, 6–10]. We will limit our discussion to electromagnetism in a vacuum or free space.

Coulomb's experimental observations show that the force between two balls carrying the charges q_1 and q_2 is (1.17)

$$\vec{\mathcal{F}} = k_1 \frac{q_1 q_2}{4\pi r_{12}^2} \hat{r}_{12}, \tag{A.14}$$

where k_1 is a constant of proportionality.[2] The electric field is defined to be proportional to the force per unit charge; thus, from (A.14) the electric field of the ball

[2] In Equations (A.14) and (A.17) the factors $1/4\pi$ and $1/2\pi$ are written separately, rather than being included in the constants k_1 and k_3, respectively. This process, known as *rationalization*, is used to keep factors of 4π from appearing in Maxwell's equations.

with charge q_1 is

$$\vec{\mathcal{E}}_1 = k_2 \frac{\vec{\mathcal{F}}}{q_2} = k_2 k_1 \frac{q_1}{4\pi r_{12}^2} \hat{r}_{12}. \tag{A.15}$$

Here a second constant of proportionality k_2 has been introduced. This result suggests the following Maxwell's equation:

$$\nabla \cdot \vec{\mathcal{E}} = k_1 k_2 \rho. \tag{A.16}$$

Ampère's experimental observation for the force between current-carrying wires is similar in form to Coulomb's result. The force on a length ℓ of wire 2 carrying current I_2 due to a parallel wire 1 carrying current I_1 is (1.22)

$$\vec{\mathcal{F}} = -k_3 \frac{I_1 I_2 \ell}{2\pi d} \hat{x}. \tag{A.17}$$

The magnetic field is defined to be proportional to the force per unit length, per unit current; thus, from (A.17) the magnetic field of the current-carrying wire 1 is

$$|\vec{\mathcal{B}}_1| = k_4 \frac{|\vec{\mathcal{F}}|}{\ell I_2} = k_4 k_3 \frac{|I_1|}{2\pi d}. \tag{A.18}$$

This result suggests the following Maxwell's equation:

$$\nabla \times \vec{\mathcal{B}} = k_3 k_4 \vec{\mathcal{J}}.$$

Recall that Maxwell added an additional term to the right-hand side of this equation, for which we will introduce the constant of proportionality k_5:

$$\nabla \times \vec{\mathcal{B}} = k_3 k_4 \vec{\mathcal{J}} + k_5 \frac{\partial \vec{\mathcal{E}}}{\partial t}. \tag{A.19}$$

Faraday's experimentation with time-varying currents showed that (1.23)

$$\oint_C \vec{\mathcal{E}} \cdot d\vec{\ell} = -k_6 \iint_S \frac{\partial \vec{\mathcal{B}}}{\partial t} \cdot d\vec{S},$$

which is equivalent to the differential form

$$\nabla \times \vec{\mathcal{E}} = -k_6 \frac{\partial \vec{\mathcal{B}}}{\partial t}. \tag{A.20}$$

We will enforce conservation of electric charge with the equation of continuity (1.54)

$$\nabla \cdot \vec{\mathcal{J}} = -\frac{\partial \rho}{\partial t}. \tag{A.21}$$

After inserting (A.16) and (A.19) into this equation, we find that

$$\frac{k_5}{k_3 k_4} = \frac{1}{k_1 k_2}, \tag{A.22}$$

which makes (A.19)

$$\nabla \times \vec{\mathcal{B}} = k_3 k_4 \left(\vec{\mathcal{J}} + \frac{1}{k_1 k_2} \frac{\partial \vec{\mathcal{E}}}{\partial t} \right). \tag{A.23}$$

Let us summarize the observations we have made so far. The experimental results suggest the following Maxwell's equations in differential form:

$$\nabla \times \vec{\mathcal{E}} = -k_6 \frac{\partial \vec{\mathcal{B}}}{\partial t} \tag{A.24}$$

$$\frac{1}{k_3}\nabla \times \vec{\mathcal{B}} = k_4 \left(\vec{\mathcal{J}} + \frac{1}{k_1 k_2}\frac{\partial \vec{\mathcal{E}}}{\partial t} \right) \tag{A.25}$$

$$\nabla \cdot \vec{\mathcal{E}} = k_1 k_2 \rho \tag{A.26}$$

$$\nabla \cdot \vec{\mathcal{B}} = 0 \tag{A.27}$$

and the Lorentz force density

$$\vec{f} = \frac{1}{k_2}\rho\vec{\mathcal{E}} + \frac{1}{k_4}\vec{\mathcal{J}} \times \vec{\mathcal{B}}, \tag{A.28}$$

which follows from (A.15) and (A.18). The five remaining constants of proportionality (k_1, k_2, k_3, k_4, and k_6) in these equations are to be determined from additional experimental observations or by convention.

In Chapter 2 we showed that Equations (A.24), (A.25), and (A.26) could be combined to obtain a wave equation for the electric field in a region of space containing no sources ($\rho = 0$, $\vec{\mathcal{J}} = 0$):

$$\nabla^2 \vec{\mathcal{E}} - \frac{k_3 k_4 k_6}{k_1 k_2}\frac{\partial^2 \vec{\mathcal{E}}}{\partial t^2} = 0. \tag{A.29}$$

The combination of constants in this equation clearly has the dimensions of velocity squared,

$$\left[\frac{k_1 k_2}{k_3 k_4 k_6}\right] = [L^2 T^{-2}].$$

Indeed, the experiments of Hertz, discussed in Section 2.7, show that this combination is equal to the square of the speed of light in a vacuum or free space:

$$\frac{k_1 k_2}{k_3 k_4 k_6} = c^2. \tag{A.30}$$

Equation (A.14) is the electrostatic force $\vec{\mathcal{F}}_e$ between charges, whereas Equation (A.17) is the magnetostatic force $\vec{\mathcal{F}}_m$ between currents. When the charges and the currents in these equations are related through the equation of continuity (A.21), a measurement of these forces, or of related quantities, can be used to obtain an additional relationship between the constants k_i. Several clever experiments have been designed for this purpose [11]. Here we will only describe a simple thought experiment to illustrate the relationship; it is not a practical experiment.

Two identical metallic spheres are separated by a distance d that is much greater than their radius. The spheres are connected to a constant current source by two thin, parallel wires of length ℓ (Figure A.1). When the switch is closed, equal currents in opposite directions are in the two wires, and the two spheres obtain charges of equal magnitude but of opposite sign. The charge on the spheres increases linearly

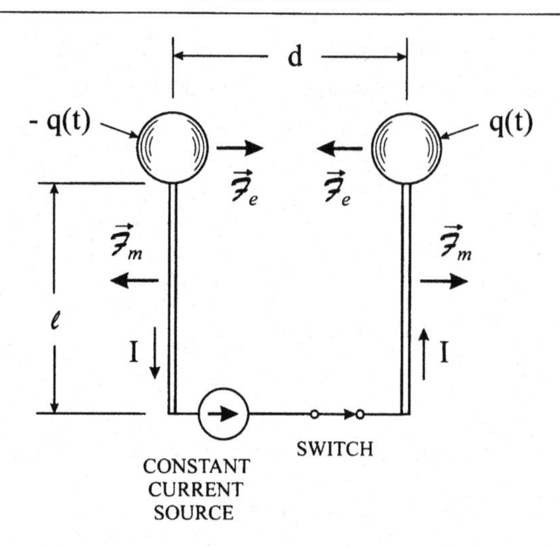

Fig. A.1. Schematic drawing for experimental arrangement.

with time:

$$q(t) = It.$$

We will assume that the charge on the thin wires is negligible and that the charging is carried out slowly so that the electrostatic and magnetostatic formulas, (A.14) and (A.17), can be used. The attractive force between the spheres (charges) is then

$$\mathcal{F}_e(t) = k_1 \frac{q^2(t)}{4\pi d^2} = k_1 \frac{I^2 t^2}{4\pi d^2},$$

and the repulsive force between the wires (currents) is

$$\mathcal{F}_m = k_3 \frac{I^2 \ell}{2\pi d}.$$

When the length ℓ is chosen to be $d/2$, the ratio of these two forces is

$$\frac{\mathcal{F}_e(t)}{\mathcal{F}_m} = \left(\frac{k_1}{k_3}\right)\left(\frac{t}{d}\right)^2.$$

Now we will let the spheres charge for the length of time t_1 that makes the ratio of the forces equal to one:

$$\frac{\mathcal{F}_e(t_1)}{\mathcal{F}_m} = 1 = \left(\frac{k_1}{k_3}\right)\left(\frac{t_1}{d}\right)^2,$$

which gives

$$\left(\frac{k_1}{k_3}\right) = \left(\frac{d}{t_1}\right)^2.$$

This ratio of constants has the dimensions of velocity squared,

$$\left[\frac{k_1}{k_3}\right] = [L^2 T^{-2}],$$

and a number of experiments show that the ratio is equal to the square of the speed of light in a vacuum or free space [11]:[3]

$$\frac{k_1}{k_3} = c^2. \tag{A.31}$$

With (A.30) and (A.31), Maxwell's equations (A.24)–(A.28) become

$$\nabla \times \vec{\mathcal{E}} = -\frac{k_2}{k_4}\frac{\partial \vec{\mathcal{B}}}{\partial t} \tag{A.32}$$

$$\frac{1}{k_3}\nabla \times \vec{\mathcal{B}} = k_4\left(\vec{\mathcal{J}} + \frac{1}{k_2 k_3 c^2}\frac{\partial \vec{\mathcal{E}}}{\partial t}\right) \tag{A.33}$$

$$\nabla \cdot \vec{\mathcal{E}} = k_2 k_3 c^2 \rho \tag{A.34}$$

$$\nabla \cdot \vec{\mathcal{B}} = 0 \tag{A.35}$$

$$\vec{f} = \frac{1}{k_2}\rho\vec{\mathcal{E}} + \frac{1}{k_4}\vec{\mathcal{J}} \times \vec{\mathcal{B}}. \tag{A.36}$$

These equations now contain the three constants k_2, k_3, and k_4.

It is customary to choose the constant k_2 to be unity and dimensionless, which is indicated by $\{k_2\} = 1$ and $[k_2] = [1]$. This makes the electric field (A.15) equal to the force per unit charge. Similarly, the constant k_4 is chosen such that $\{k_4\} = 1$ and $[k_4] = [1]$, which makes the magnetic field (A.18) the force per unit length, per unit current.

The definition for the ampere in the SI determines the remaining constant k_3. From (A.17)

$$k_3 = \frac{2\pi d}{I_1 I_2}\left(\frac{|\vec{\mathcal{F}}|}{\ell}\right),$$

and the definition of the ampere [that is, $I_1 = I_2 = 1$ A, $d = 1$ m, and $\mathcal{F}/\ell = 2 \times 10^{-7}$ kg $\cdot$ s^{-2} (N/m)] gives

$$\{k_3\} = 4\pi \times 10^{-7}$$

$$\langle k_3 \rangle = \mathrm{m} \cdot \mathrm{kg} \cdot \mathrm{s}^{-2} \cdot \mathrm{A}^{-2} = \mathrm{H/m}.$$

The convention is to call this constant the *permeability of free space* μ_o:

$$\mu_o = k_3 = 4\pi \times 10^{-7}\ \mathrm{H/m}. \tag{A.37}$$

The measure and the units for the combination of constants $1/k_2 k_3 c^2$ that appears in the two Maxwell's equations (A.33) and (A.34) are now fixed. By convention this combination of constants is called the *permittivity of free space* ε_o:

$$\varepsilon_o = \frac{1}{k_2 k_3 c^2} = \frac{1}{\mu_o c^2} = 8.8541\ldots \times 10^{-12}\ \mathrm{F/m}. \tag{A.38}$$

[3] In 1889 the American physicist Henry Augustus Rowland (1848–1901) reported the value 2.9815×10^8 m/s for this ratio [12]. The closeness of this value to the currently accepted value for the speed of light $c = 2.9979 \cdots \times 10^8$ m/s shows the accuracy of Rowland's measurements (within 0.5%).

Maxwell's equations (A.32)–(A.36) now assume their familiar form for free space and SI units:[4]

$$\nabla \times \vec{\mathcal{E}} = -\frac{\partial \vec{\mathcal{B}}}{\partial t}$$

$$\frac{1}{\mu_o}\nabla \times \vec{\mathcal{B}} = \vec{\mathcal{J}} + \varepsilon_o \frac{\partial \vec{\mathcal{E}}}{\partial t}$$

$$\nabla \cdot \vec{\mathcal{E}} = \frac{1}{\varepsilon_o}\rho$$

$$\nabla \cdot \vec{\mathcal{B}} = 0$$

$$\vec{f} = \rho\vec{\mathcal{E}} + \vec{\mathcal{J}} \times \vec{\mathcal{B}}.$$

From these equations and (A.37) and (A.38), the SI units for all electromagnetic quantities can be derived in terms of the four base units (m, kg, s, A) or in terms of other derived units. Examples are in Table A.2.

The reader is referred to the references for a discussion of other systems of units and for the conversions between these systems and the SI [4,10].

References

[1] P. W. Bridgman, *Dimensional Analysis*, Yale University Press, New Haven, 1931.

[2] A. G. McNish, "Dimensions Units and Standards," *Physics Today*, Vol. 10, pp. 19–25, April 1957.

[3] C. H. Page, "Physical Entities and Mathematical Representation," *J. Res. Natl. Bur. Stand.*, Vol. 65B (Mathematics and Mathematical Physics), pp. 227–35, October–December 1961.

[4] F. B. Silsbee, "Systems of Electrical Units," *J. Res. Natl. Bur. Stand.*, Vol. 66C (Engineering and Instrumentation), pp. 137–78, April–June 1962.

[5] R. A. Nelson, *SI: The International System of Units*, Revised 2nd Edition, American Association of Physics Teachers, Stony Brook, NY, 1985.

[6] R. T. Birge, "On Electric and Magnetic Units and Dimensions," *Am. Phys. Teacher*, Vol. 2, pp. 41–8, May 1934.

[7] —, "On the Establishment of Fundamental and Derived Units, with Special Reference to Electric Units, Parts I and II," *Am. Phys. Teacher*, Vol. 3, pp. 102–9, September 1935; Vol. 3, pp. 171–9, December 1935.

[8] W. F. Brown, Jr., "Field Vectors and Unit Systems in the Theory of Electricity," *Am. J. Phys.*, Vol. 8, pp. 338–45, December 1940.

[9] G. L. Trigg, "Electromagnetic Equations in Generalized Units," *Am. J. Phys.*, Vol. 27, pp. 515–6, October 1959.

[10] J. D. Jackson, *Classical Electrodynamics*, 2nd Edition, Wiley, New York, 1975.

[11] A. Gray, *Absolute Measurements in Electricity and Magnetism*, 2nd Edition, Macmillan, London, 1921. Republication, Dover Publications, New York, 1967.

[12] H. A. Rowland, "On the Ratio of the Electrostatic to the Electromagnetic Units of Electricity," *Philos. Mag.*, Vol. 28, pp. 304–15, October 1889.

[4] These equations with the four base units (m, kg, s, A) are sometimes referred to as "the rationalized MKSA system" or "the Giorgi System."

Appendix B

Review of vector analysis

The development of vector analysis during the last half of the nineteenth century paralleled that of classical electrodynamics (we will have more to say about the common history of these two subjects at the end of this appendix). By the beginning of the twentieth century, the elements of vector analysis needed for electrodynamics were essentially in the form we use today. Since then, numerous books on vector analysis and sets of tables summarizing useful formulas from vector analysis have been published. A sampling of these is given in the list of references [1–8].

Our purpose here is to provide a short review suitable for those who have prior knowledge of the subject and to summarize in tables results that are used throughout this book.

Vector algebra

Vectors are used to represent quantities that have a magnitude and a direction, such as the electric and magnetic fields. The geometrical representation for the vector $\vec{A}$ is the arrow or directed line segment shown in Figure B.1. The length of the arrow indicates the magnitude of the vector, which is $A = |\vec{A}|$, while the direction of the arrow indicates the direction of the vector. The direction is the same as that for the unit vector $\hat{A} = \vec{A}/|\vec{A}|$ (a vector of unit magnitude in the direction of $\vec{A}$).

The sum of the two vectors $\vec{A}$ and $\vec{B}$ in Figure B.1 is the vector $\vec{C}$, the diagonal of the parallelogram with the sides $\vec{A}$ and $\vec{B}$ or the resultant formed by placing $\vec{A}$ and $\vec{B}$ head-to-tail. The laws that apply to the addition of vectors are presented in Table B.1.

There are two different products defined for the vectors $\vec{A}$ and $\vec{B}$. The geometrical interpretations for these products are given in Figure B.2. The *scalar* or *dot product* is

$$\vec{A} \cdot \vec{B} = |\vec{A}||\vec{B}| \cos \psi, \qquad 0 \le \psi \le \pi, \tag{B.1}$$

where ψ is the angle between $\vec{A}$ and $\vec{B}$. The result of this product is a scalar (a real number), not a vector. The scalar product can be viewed as the projection of the vector $\vec{A}$ onto the vector $\vec{B}$ times the magnitude of the vector $\vec{B}$, or vice versa.

The *vector* or *cross product* is

$$\vec{A} \times \vec{B} = |\vec{A}||\vec{B}| \sin \psi \, \hat{n}, \qquad 0 \le \psi \le \pi, \tag{B.2}$$

where $\hat{n}$ is a unit vector normal to the plane of $\vec{A}$ and $\vec{B}$. The direction of $\hat{n}$ is

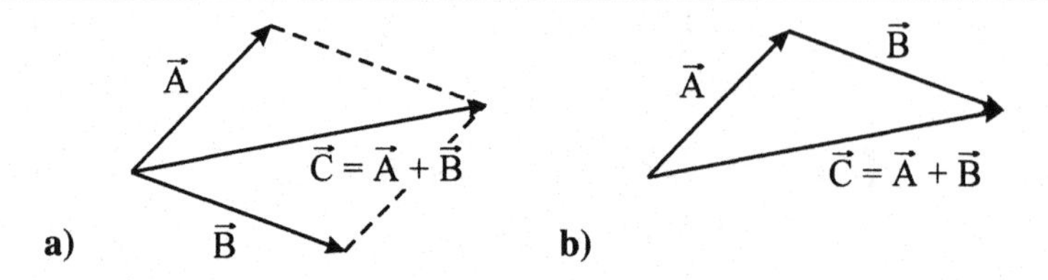

Fig. B.1. Addition of the two vectors $\vec{A}$ and $\vec{B}$. a) $\vec{C}$ is the diagonal of the parallelogram with sides $\vec{A}$ and $\vec{B}$. b) $\vec{C}$ is the resultant formed by placing $\vec{A}$ and $\vec{B}$ head-to-tail.

Table B.1. *Vector relations*

Addition
$\vec{A}, \vec{B}, \vec{C}$ vectors, α real number
$\vec{A} + \vec{B} = \vec{B} + \vec{A}$, commutative law
$\vec{A} + (\vec{B} + \vec{C}) = (\vec{A} + \vec{B}) + \vec{C}$, associative law
$\alpha(\vec{A} + \vec{B}) = \alpha\vec{A} + \alpha\vec{B}$, distributive law
Scalar or dot product
$\vec{A} \cdot \vec{B} = \vec{B} \cdot \vec{A}$, commutative law
$\vec{A} \cdot (\vec{B} + \vec{C}) = \vec{A} \cdot \vec{B} + \vec{A} \cdot \vec{C}$, distributive law
$\alpha(\vec{A} \cdot \vec{B}) = (\alpha\vec{A}) \cdot \vec{B} = \vec{A} \cdot (\alpha\vec{B})$, associative law
$\vec{A} \cdot \vec{B} = 0$, $\lvert\vec{A}\rvert \neq 0$, $\lvert\vec{B}\rvert \neq 0$, $\vec{A}$ and $\vec{B}$ are perpendicular vectors
Vector or cross product
$\vec{A} \times \vec{B} = -\vec{B} \times \vec{A}$, anticommutative law
$\vec{A} \times (\vec{B} + \vec{C}) = \vec{A} \times \vec{B} + \vec{A} \times \vec{C}$, distributive law
$\alpha(\vec{A} \times \vec{B}) = (\alpha\vec{A}) \times \vec{B} = \vec{A} \times (\alpha\vec{B})$, associative law
$\vec{A} \times \vec{B} = 0$, $\lvert\vec{A}\rvert \neq 0$, $\lvert\vec{B}\rvert \neq 0$, $\vec{A}$ and $\vec{B}$ are parallel vectors
Triple products
$\vec{A} \cdot (\vec{A} \times \vec{B}) = 0$
$\vec{A} \cdot (\vec{B} \times \vec{C}) = \vec{B} \cdot (\vec{C} \times \vec{A}) = \vec{C} \cdot (\vec{A} \times \vec{B})$
$\vec{A} \times (\vec{B} \times \vec{C}) = \vec{B} \times (\vec{A} \times \vec{C}) + \vec{C} \times (\vec{B} \times \vec{A}) = (\vec{A} \cdot \vec{C})\vec{B} - (\vec{A} \cdot \vec{B})\vec{C}$

given by the right-hand rule: with the curled fingers of the right hand pointing from $\vec{A}$ to $\vec{B}$, the thumb points in the direction of $\hat{n}$. The vectors $\vec{A}$, $\vec{B}$, and $\hat{n}$ form a right-handed system. The result of the vector product is a vector whose magnitude is equal to the area of the parallelogram with sides $\vec{A}$ and $\vec{B}$.

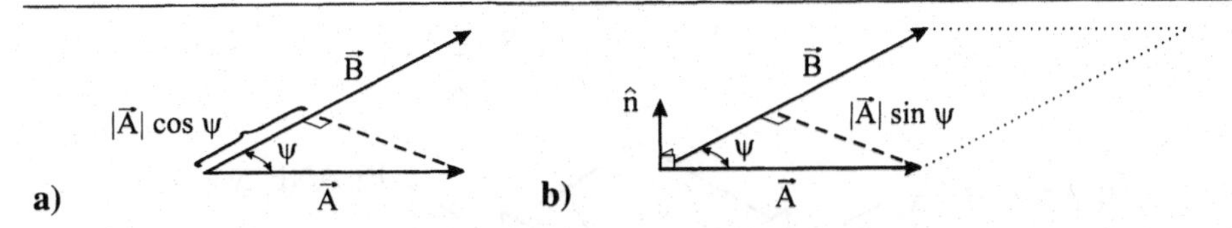

Fig. B.2. Details for products of two vectors. a) Scalar or dot product $\vec{A} \cdot \vec{B} = |\vec{A}||\vec{B}| \cos \psi$. b) Vector or cross product $\vec{A} \times \vec{B} = |\vec{A}||\vec{B}| \sin \psi \, \hat{n}$.

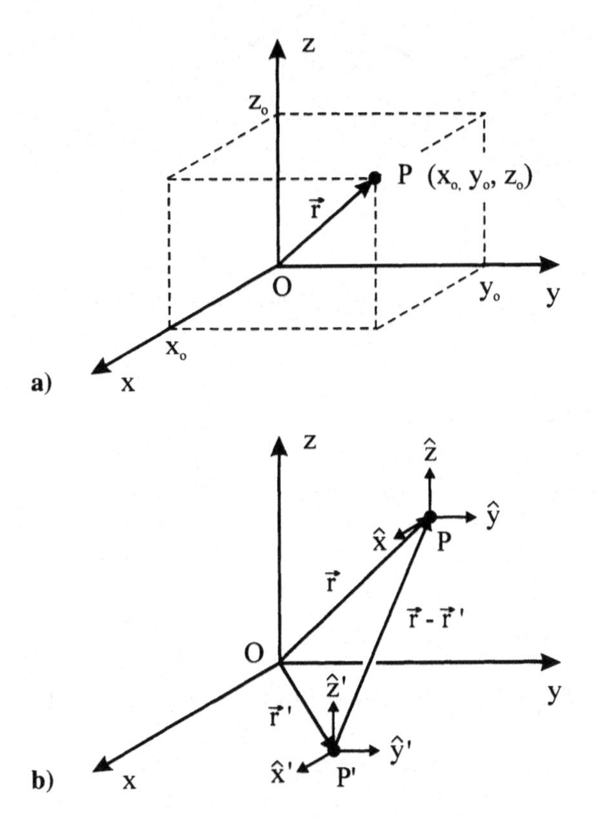

Fig. B.3. Rectangular Cartesian coordinate system.

Laws that apply to the scalar and vector products are presented in Table B.1. Results are also given for various products of three vectors – triple products, formed using the scalar and vector products.

In practical problems involving vectors, we often introduce a coordinate system to facilitate computation. The most familiar coordinate system is the rectangular Cartesian system.

Rectangular Cartesian coordinates

The diagrams in Figure B.3 describe the rectangular Cartesian coordinate system. The three mutually perpendicular coordinate axes (x, y, and z) meet at the point O, the origin. The unit of measure is common for all axes. Each point in three-dimensional space corresponds to an ordered triple of real numbers (x, y, z),

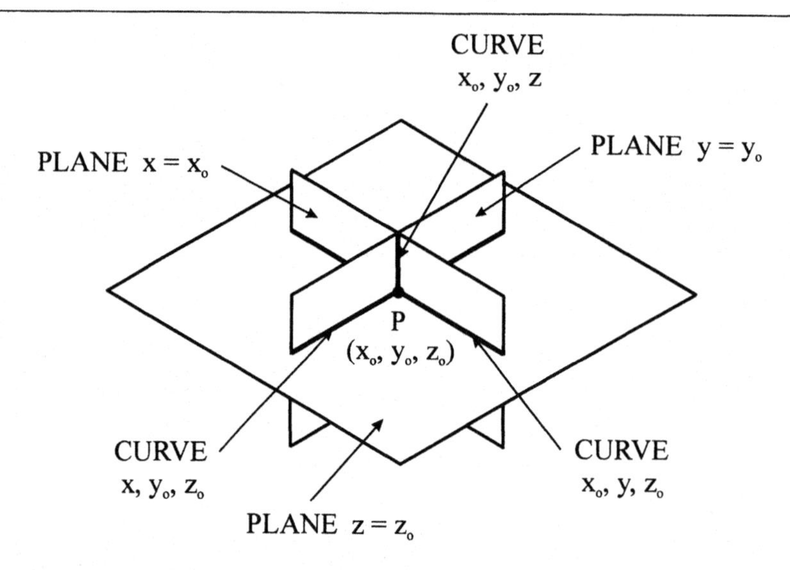

Fig. B.4. Rectangular Cartesian coordinates showing coordinate surfaces and coordinate curves through point P.

the coordinates of the point. For example, for the origin we have (0, 0, 0), and for the point P, located by the position vector $\vec{r}$ from the origin, we have (x_o, y_o, z_o). The orthogonal unit vectors $\hat{x}$, $\hat{y}$, and $\hat{z}$ are in the directions of the three coordinate axes, and they form a right-handed system:

$$\hat{x} \cdot \hat{x} = \hat{y} \cdot \hat{y} = \hat{z} \cdot \hat{z} = 1, \qquad \hat{x} \cdot \hat{y} = \hat{y} \cdot \hat{z} = \hat{z} \cdot \hat{x} = 0,$$
$$\hat{x} \times \hat{y} = \hat{z}, \qquad \hat{y} \times \hat{z} = \hat{x}, \qquad \hat{z} \times \hat{x} = \hat{y}. \tag{B.3}$$

Notice that the unit vectors are in the same directions at all points in space; compare the unit vectors at the points P and P′ in Figure B.3b.

Figure B.4 shows another way of viewing this coordinate system. The coordinate system is composed of three families of orthogonal planes, the *coordinate surfaces*. A point such as P at (x_o, y_o, z_o) is determined by the intersection of three coordinate surfaces: the three planes along which $x = x_o$, $y = y_o$, and $z = z_o$. The *coordinate curves* are the lines along which two coordinate surfaces intersect. For example, the coordinate curve for x passing through the point P is the straight line with $y = y_o$ and $z = z_o$. The unit vectors are tangent to the coordinate curves. For example, at the point P, $\hat{x}$ is tangent to the curve for x.

A general vector $\vec{A}$, when expressed in terms of these coordinates, becomes

$$\vec{A} = A_x\hat{x} + A_y\hat{y} + A_z\hat{z}, \tag{B.4}$$

with the magnitude determined from the Pythagorean theorem:

$$A = |\vec{A}| = \sqrt{\vec{A} \cdot \vec{A}} = \sqrt{A_x^2 + A_y^2 + A_z^2}. \tag{B.5}$$

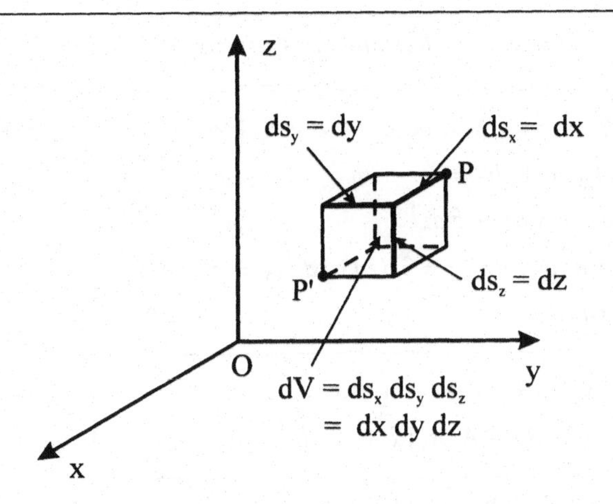

Fig. B.5. Rectangular Cartesian coordinates showing the differential distances along the coordinate curves and the differential volume.

The scalar and vector products of the two vectors $\vec{A}$ and $\vec{B}$ follow from (B.1) through (B.4):

$$\vec{A} \cdot \vec{B} = A_x B_x + A_y B_y + A_z B_z \tag{B.6}$$

and

$$\vec{A} \times \vec{B} = (A_y B_z - A_z B_y)\hat{x} + (A_z B_x - A_x B_z)\hat{y} + (A_x B_y - A_y B_x)\hat{z}. \tag{B.7}$$

The position vector $\vec{r}$ from the origin O to the point P is

$$\vec{r} = x\hat{x} + y\hat{y} + z\hat{z}, \tag{B.8}$$

with

$$r = |\vec{r}| = \sqrt{\vec{r} \cdot \vec{r}} = \sqrt{x^2 + y^2 + z^2}, \tag{B.9}$$

and the distance between the two points P and P′ in Figure B.3b is

$$|\vec{r} - \vec{r}'| = \sqrt{(\vec{r} - \vec{r}') \cdot (\vec{r} - \vec{r}')} = \sqrt{(x - x')^2 + (y - y')^2 + (z - z')^2}. \tag{B.10}$$

When we let the separation between the two points P and P′ become infinitesimal, as in Figure B.5, (B.8) becomes the *differential distance vector*,

$$d\vec{r} = \hat{x}\, ds_x + \hat{y}\, ds_y + \hat{z}\, ds_z = \hat{x}\, dx + \hat{y}\, dy + \hat{z}\, dz, \tag{B.11}$$

where

$$ds_x = dx, \qquad ds_y = dy, \qquad ds_z = dz \tag{B.12}$$

are the *differential distances* along the coordinate curves. The *differential distance* or *differential arc length* between the points is

$$ds = \sqrt{d\vec{r} \cdot d\vec{r}} = \sqrt{(ds_x)^2 + (ds_y)^2 + (ds_z)^2} = \sqrt{(dx)^2 + (dy)^2 + (dz)^2}. \tag{B.13}$$

Table B.2. *Rectangular Cartesian coordinates*

(x, y, z): $-\infty < x < \infty, \quad -\infty < y < \infty, \quad -\infty < z < \infty$

$h_x = 1,\ h_y = 1,\ h_z = 1$

$d\vec{r} = \hat{x}\,dx + \hat{y}\,dy + \hat{z}\,dz$

$d\vec{S}_{yz} = \hat{x}\,dydz,\ d\vec{S}_{zx} = \hat{y}\,dzdx,\ d\vec{S}_{xy} = \hat{z}\,dxdy$

$dV = dxdydz$

$\vec{A}\cdot\vec{B} = A_xB_x + A_yB_y + A_zB_z$

$\vec{A}\times\vec{B} = (A_yB_z - A_zB_y)\hat{x} + (A_zB_x - A_xB_z)\hat{y} + (A_xB_y - A_yB_x)\hat{z}$

$$\nabla\Phi = \frac{\partial\Phi}{\partial x}\hat{x} + \frac{\partial\Phi}{\partial y}\hat{y} + \frac{\partial\Phi}{\partial z}\hat{z}$$

$$\nabla\cdot\vec{A} = \frac{\partial A_x}{\partial x} + \frac{\partial A_y}{\partial y} + \frac{\partial A_z}{\partial z}$$

$$\nabla\times\vec{A} = \left(\frac{\partial A_z}{\partial y} - \frac{\partial A_y}{\partial z}\right)\hat{x} + \left(\frac{\partial A_x}{\partial z} - \frac{\partial A_z}{\partial x}\right)\hat{y} + \left(\frac{\partial A_y}{\partial x} - \frac{\partial A_x}{\partial y}\right)\hat{z}$$

$$\nabla^2\Phi = \frac{\partial^2\Phi}{\partial x^2} + \frac{\partial^2\Phi}{\partial y^2} + \frac{\partial^2\Phi}{\partial z^2}$$

$$\nabla^2\vec{A} = \nabla^2 A_x\,\hat{x} + \nabla^2 A_y\,\hat{y} + \nabla^2 A_z\,\hat{z}$$

The *differential volume*

$$dV = ds_x ds_y ds_z = dxdydz \tag{B.14}$$

is shown in Figure B.5. There is a *differential surface area* for each face of this volume. For example, on the front of the volume (y, z surface) the differential surface area is

$$dS_{yz} = ds_y ds_z = dydz. \tag{B.15}$$

The *differential surface area vector* associated with this surface is

$$d\vec{S}_{yz} = \hat{x}\,dS_{yz} = \hat{x}\,dydz. \tag{B.16}$$

Results for the rectangular Cartesian coordinate system are summarized in Table B.2.

Differential relations

A scalar field $\Phi(\vec{r})$ assigns a number to each point throughout space; similarly, a vector field $\vec{A}(\vec{r})$ assigns a vector (three numbers) to each point throughout space. For example, in the rectangular Cartesian coordinate system, we would have $\Phi(x, y, z)$ for the scalar field and $A_x(x, y, z)$ for the x component of the vector field.

When certain operations on scalar and vector fields repeatedly occur in calculations for physical systems, it is convenient to assign to them a notation and a formal definition [1–10]. The three operations we will discuss, the gradient, the divergence, and the curl, occur and have a physical interpretation in several areas of physics, e.g., electrodynamics and fluid mechanics.

Gradient The scalar field Φ is continuous in the neighborhood of the point P located by the position vector $\vec{r}$ shown in Figure 1.16a. The element of volume ΔV containing P has the piecewise smooth surface S with outward-pointing, unit normal vector $\hat{n}$. The gradient of the scalar field at P is the limiting value of the following integral over the surface S divided by the volume ΔV, as the volume approaches zero about P:

$$\text{grad}\,\Phi(\vec{r}) \equiv \lim_{\Delta V \to 0} \left(\frac{1}{\Delta V} \oiint_S \Phi \hat{n} dS \right). \tag{B.17}$$

Divergence The components of the vector field $\vec{A}$ are continuous in the neighborhood of the point P located by the position vector $\vec{r}$ shown in Figure 1.16a. The element of volume ΔV containing P has the piecewise smooth surface S with outward-pointing, unit normal vector $\hat{n}$. The divergence of the vector field at P is the limiting value of the net outward flux of $\vec{A}$ through the surface S divided by the volume ΔV, as the volume approaches zero about P:

$$\text{div}\,\vec{A}(\vec{r}) \equiv \lim_{\Delta V \to 0} \left(\frac{1}{\Delta V} \oiint_S \hat{n} \cdot \vec{A} dS \right). \tag{B.18}$$

Curl The components of the vector field $\vec{A}$ are continuous in the neighborhood of the point P located by the position vector $\vec{r}$ shown in Figure 1.15a. The element of surface area ΔS on which P lies is bounded by the piecewise smooth, simple, closed curve C (a simple curve is one that does not cross itself). At P the unit vector normal to the surface is $\hat{n}$. The circulation of the vector field $\vec{A}$ around the curve C is defined to be

$$\Gamma \equiv \oint_C \vec{A} \cdot d\vec{\ell}, \tag{B.19}$$

where $d\vec{\ell}$ is the vector differential length along C (an infinitesimal vector locally tangent to C). The sense in which the curve C is traversed relative to $\hat{n}$ is determined by the right-hand rule. The component in the direction $\hat{n}$ of the curl of the vector field at P is the limiting value of the circulation of $\vec{A}$ divided by the surface area ΔS, as the area approaches zero about P:

$$\hat{n} \cdot \text{curl}\,\vec{A}(\vec{r}) \equiv \lim_{\Delta S \to 0} \left(\frac{\Gamma}{\Delta S} \right) = \lim_{\Delta S \to 0} \left(\frac{1}{\Delta S} \oint_C \vec{A} \cdot d\vec{\ell} \right). \tag{B.20}$$

These definitions are independent of the coordinate system. To obtain results for the rectangular Cartesian coordinate system, we let the element of surface area and the element of volume be those shown in Figures 1.15b and 1.16b, respectively. After assuming that the partial derivatives for the scalar field and for the components

Table B.3. *Differential relations*

Φ and Ψ are scalar fields; $\vec{A}$ and $\vec{B}$ are vector fields.
$\nabla(\Phi + \Psi) = \nabla\Phi + \nabla\Psi$
$\nabla(\Phi\Psi) = \Phi(\nabla\Psi) + \Psi(\nabla\Phi)$
$\nabla^2\Phi \equiv \nabla \cdot (\nabla\Phi)$, scalar Laplacian
$\nabla^2(\Phi\Psi) = \Psi(\nabla^2\Phi) + 2(\nabla\Psi)\cdot(\nabla\Phi) + \Phi(\nabla^2\Psi)$
$\nabla \times (\nabla\Phi) = 0$
$\nabla \times [\Psi(\nabla\Phi)] = (\nabla\Psi) \times (\nabla\Psi)$
$\nabla \cdot (\vec{A} + \vec{B}) = \nabla \cdot \vec{A} + \nabla \cdot \vec{B}$
$\nabla \times (\vec{A} + \vec{B}) = \nabla \times \vec{A} + \nabla \times \vec{B}$
$\nabla \cdot (\Phi\vec{A}) = (\nabla\Phi)\cdot\vec{A} + \Phi(\nabla\cdot\vec{A})$
$\nabla \times (\Phi\vec{A}) = (\nabla\Phi)\times\vec{A} + \Phi(\nabla\times\vec{A})$
$\nabla(\vec{A}\cdot\vec{B}) = (\vec{A}\cdot\nabla)\vec{B} + (\vec{B}\cdot\nabla)\vec{A} + \vec{A}\times(\nabla\times\vec{B}) + \vec{B}\times(\nabla\times\vec{A})$
$\nabla\cdot(\vec{A}\times\vec{B}) = \vec{B}\cdot(\nabla\times\vec{A}) - \vec{A}\cdot(\nabla\times\vec{B})$
$\nabla\times(\vec{A}\times\vec{B}) = \vec{A}(\nabla\cdot\vec{B}) - \vec{B}(\nabla\cdot\vec{A}) + (\vec{B}\cdot\nabla)\vec{A} - (\vec{A}\cdot\nabla)\vec{B}$
$\nabla\cdot(\nabla\times\vec{A}) = 0$
$\nabla^2\vec{A} \equiv \nabla(\nabla\cdot\vec{A}) - \nabla\times(\nabla\times\vec{A})$, vector Laplacian
$\nabla^2(\Phi\vec{A}) = \Phi\nabla^2\vec{A} + (\nabla^2\Phi)\vec{A} + 2\,[(\nabla\Phi)\cdot\nabla]\,\vec{A}$

of the vector field are continuous, the expressions above can be evaluated, yielding

$$\operatorname{grad}\Phi = \nabla\Phi = \frac{\partial\Phi}{\partial x}\hat{x} + \frac{\partial\Phi}{\partial y}\hat{y} + \frac{\partial\Phi}{\partial z}\hat{z}, \tag{B.21}$$

$$\operatorname{div}\vec{A} = \nabla\cdot\vec{A} = \frac{\partial A_x}{\partial x} + \frac{\partial A_y}{\partial y} + \frac{\partial A_z}{\partial z}, \tag{B.22}$$

$$\operatorname{curl}\vec{A} = \nabla\times\vec{A} = \left(\frac{\partial A_z}{\partial y} - \frac{\partial A_y}{\partial z}\right)\hat{x} + \left(\frac{\partial A_x}{\partial z} - \frac{\partial A_z}{\partial x}\right)\hat{y} + \left(\frac{\partial A_y}{\partial x} - \frac{\partial A_x}{\partial y}\right)\hat{z}. \tag{B.23}$$

Here we have introduced the notation ∇, $\nabla\cdot$, and $\nabla\times$ to indicate the differential operations shown on the right-hand sides of (B.21), (B.22), and (B.23), respectively. The derivations for the divergence and the curl are carried out in detail in Section 1.2; that for the gradient is left as an exercise (Problem B.2).

Table B.3 lists several frequently used relations that involve ∇, $\nabla\cdot$, and $\nabla\times$. In this table, operations such as $(\vec{A}\cdot\nabla)\vec{B}$ are to be evaluated in the following manner:

$$(\vec{A}\cdot\nabla)\vec{B} = A_x\left(\frac{\partial B_x}{\partial x}\hat{x} + \frac{\partial B_y}{\partial x}\hat{y} + \frac{\partial B_z}{\partial x}\hat{z}\right) + A_y\left(\frac{\partial B_x}{\partial y}\hat{x} + \frac{\partial B_y}{\partial y}\hat{y} + \frac{\partial B_z}{\partial y}\hat{z}\right)$$

$$+ A_z\left(\frac{\partial B_x}{\partial z}\hat{x} + \frac{\partial B_y}{\partial z}\hat{y} + \frac{\partial B_z}{\partial z}\hat{z}\right). \quad \text{(B.24)}$$

Integral relations

Several relations that apply to integrals of scalar and vector fields can also be developed [1–10]. The most important of these are the divergence theorem (or Gauss' theorem) and Stokes' theorem:

Divergence Theorem (Gauss' Theorem) The volume V shown in Figure 1.1b has the piecewise smooth surface S with outward-pointing, unit normal vector $\hat{n}$. The components of the vector field $\vec{A}$ are continuous and have continuous first partial derivatives within and on the surface of V. Then the divergence theorem states that

$$\iiint_V \nabla\cdot\vec{A}\,dV = \oiint_S \hat{n}\cdot\vec{A}\,dS. \quad \text{(B.25)}$$

Stokes' Theorem The piecewise smooth, open surface S shown in Figure 1.1a has the boundary C, which is a piecewise smooth, simple, closed curve. The unit vector normal to the surface is $\hat{n}$. The components of the vector field $\vec{A}$ are continuous and have continuous first partial derivatives within and on the boundary of S. Then Stokes' theorem states that

$$\iint_S (\nabla\times\vec{A})\cdot\hat{n}\,dS = \oint_C \vec{A}\cdot d\vec{\ell}, \quad \text{(B.26)}$$

where $d\vec{\ell}$ is the vector differential length along C. The sense in which the curve C is traversed relative to $\hat{n}$ is determined by the right-hand rule.

These results and other related integral formulas are summarized in Table B.4. For Green's theorems, the scalar fields (Φ and Ψ) and the components of the vector fields ($\vec{A}$ and $\vec{B}$) must be continuous and have continuous first and second partial derivatives.

More general requirements for the use of both the differential relations and the integral relations are given in references [9] and [10].

Orthogonal curvilinear coordinates

In the rectangular Cartesian coordinate system (x, y, z), points in space, such as P in Figure B.4, are determined by the intersection of three families of orthogonal coordinate surfaces, which are planes (the planes $x = x_o$, $y = y_o$, and $z = z_o$ for P), and the coordinate curves are straight lines. In a general orthogonal curvilinear coordinate system (u, v, w), the orthogonal coordinate surfaces need not be planes; they may be families of circular cylinders, elliptic cylinders, spheres, cones,

Table B.4. *Integral relations*

Φ and Ψ are scalar fields; $\vec{A}$ and $\vec{B}$ are vector fields.

Divergence theorem (Gauss' theorem)

$$\iiint_V \nabla \cdot \vec{A}\, dV = \oint\!\!\oint_S \hat{n} \cdot \vec{A}\, dS$$

Related theorems

$$\iiint_V \nabla \times \vec{A}\, dV = \oint\!\!\oint_S \hat{n} \times \vec{A}\, dS$$

$$\iiint_V \nabla \Phi\, dV = \oint\!\!\oint_S \Phi \hat{n}\, dS$$

Stokes' theorem

$$\iint_S (\nabla \times \vec{A}) \cdot \hat{n}\, dS = \oint_C \vec{A} \cdot d\vec{\ell}$$

Related theorems

$$\iint_S \hat{n} \times (\nabla \Phi)\, dS = \oint_C \Phi\, d\vec{\ell}$$

$$\iint_S [(\hat{n} \times \nabla) \times \vec{A}]\, dS = -\oint_C \vec{A} \times d\vec{\ell}$$

Green's theorems

First scalar theorem

$$\iiint_V \left[\Psi(\nabla^2 \Phi) + (\nabla \Psi) \cdot (\nabla \Phi)\right] dV = \oint\!\!\oint_S \Psi\left[\hat{n} \cdot (\nabla \Phi)\right] dS$$

First vector theorem

$$\iiint_V \left\{(\nabla \times \vec{A}) \cdot (\nabla \times \vec{B}) - \vec{A} \cdot \left[\nabla \times (\nabla \times \vec{B})\right]\right\} dV = \oint\!\!\oint_S \hat{n} \cdot \left[\vec{A} \times (\nabla \times \vec{B})\right] dS$$

paraboloids, etc., and the coordinate curves need not be straight lines; they may be circles, ellipses, parabolas, etc. [7]. The most familiar curvilinear coordinate systems are the circular cylindrical coordinates shown in Figure B.6 and the spherical coordinates shown in Figure B.7. Results for these two systems are summarized in Tables B.5 and B.6, respectively.[1]

[1] The small tables within Tables B.5 and B.6 give the dot products of the unit vectors in the coordinate system (circular cylindrical or spherical) with the unit vectors $\hat{x}$, $\hat{y}$, and $\hat{z}$.

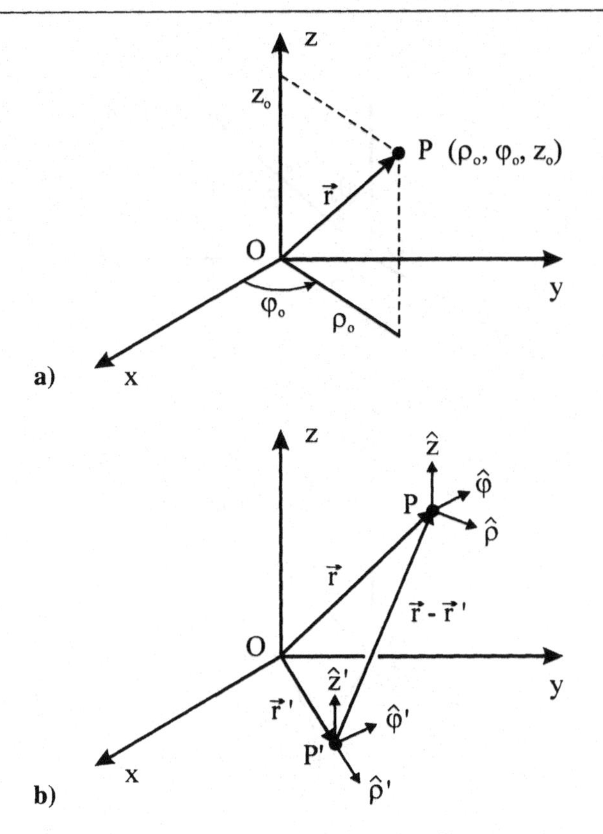

Fig. B.6. Circular cylindrical coordinate system.

For the circular cylindrical coordinate system, $u = \rho, v = \phi$, and $w = z$; the coordinate surfaces are families of circular cylinders, half planes, and planes. The point P in Figure B.8 is determined by the intersection of the circular cylinder $\rho = \rho_o$ with the half plane $\phi = \phi_o$ and the plane $z = z_o$. The coordinate curves are straight lines (ρ and z curves) and circles (ϕ curves).

For the spherical coordinate system, $u = r, v = \theta$, and $w = \phi$; the coordinate surfaces are families of spheres, cones, and half planes. The point P in Figure B.9 is determined by the intersection of the sphere $r = r_o$ with the cone $\theta = \theta_o$ and the half plane $\phi = \phi_o$. The coordinate curves are straight lines (r curves) and circles (θ and ϕ curves).

The differential distance vector (B.11) is now

$$d\vec{r} = \frac{\partial \vec{r}}{\partial u} du + \frac{\partial \vec{r}}{\partial v} dv + \frac{\partial \vec{r}}{\partial w} dw = ds_u \hat{u} + ds_v \hat{v} + ds_w \hat{w}, \tag{B.27}$$

where the differential distances along the coordinate curves and the unit vectors have been introduced. For the u curve, these quantities are

$$ds_u = h_u du \tag{B.28}$$

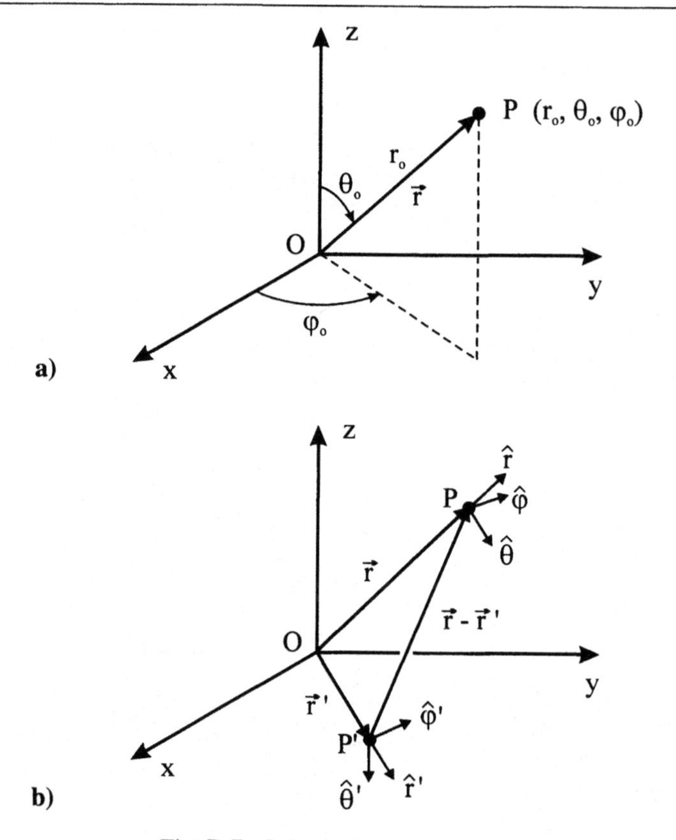

Fig. B.7. Spherical coordinate system.

and

$$\hat{u} = \frac{1}{h_u}\frac{\partial \vec{r}}{\partial u} = \frac{1}{h_u}\left[\frac{\partial x}{\partial u}\hat{x} + \frac{\partial y}{\partial u}\hat{y} + \frac{\partial z}{\partial u}\hat{z}\right], \tag{B.29}$$

where

$$h_u = \left|\frac{\partial \vec{r}}{\partial u}\right| = \left[\left(\frac{\partial x}{\partial u}\right)^2 + \left(\frac{\partial y}{\partial u}\right)^2 + \left(\frac{\partial z}{\partial u}\right)^2\right]^{1/2} \tag{B.30}$$

is called the *scale factor* for the coordinate u. Notice that ds_u must have the units of length. These quantities for the v and w curves are obtained by replacing u by v and u by w, respectively, in (B.28)–(B.30).

The unit vectors are orthogonal and form a right-handed system, so they satisfy the same relations we had earlier for the unit vectors in the rectangular Cartesian system (B.3):

$$\hat{u}\cdot\hat{u} = \hat{v}\cdot\hat{v} = \hat{w}\cdot\hat{w} = 1, \qquad \hat{u}\cdot\hat{v} = \hat{v}\cdot\hat{w} = \hat{w}\cdot\hat{u} = 0,$$
$$\hat{u}\times\hat{v} = \hat{w}, \qquad \hat{v}\times\hat{w} = \hat{u}, \qquad \hat{w}\times\hat{u} = \hat{v}. \tag{B.31}$$

The unit vectors for the circular cylindrical and spherical coordinate systems are shown in Figures B.6b and B.7b, respectively. Compare the unit vectors at the points P and P′ in these figures; notice that some of the unit vectors change with the

Table B.5. *Circular cylindrical coordinates*

$(\rho,\ \phi,\ z)$: $0 \le \rho < \infty,\ 0 \le \phi < 2\pi,\ -\infty < z < \infty$

$h_\rho = 1,\ h_\phi = \rho,\ h_z = 1$

$x = \rho\cos\phi,\ y = \rho\sin\phi,\ z;\ \rho = \sqrt{x^2+y^2},\ \phi = \tan^{-1}(y/x),\ z$

$\cdot$	$\hat{x}$	$\hat{y}$	$\hat{z}$
$\hat{\rho}$	$\cos\phi$	$\sin\phi$	0
$\hat{\phi}$	$-\sin\phi$	$\cos\phi$	0
$\hat{z}$	0	0	1

$d\vec{r} = \hat{\rho}\,(d\rho) + \hat{\phi}\,(\rho d\phi) + \hat{z}\,(dz)$

$d\vec{S}_{\phi z} = \hat{\rho}\,(\rho d\phi)(dz),\ d\vec{S}_{z\rho} = \hat{\phi}\,(dz)(d\rho),\ d\vec{S}_{\rho\phi} = \hat{z}\,(d\rho)(\rho d\phi)$

$dV = (d\rho)(\rho d\phi)(dz) = \rho d\rho d\phi dz$

$\vec{A}\cdot\vec{B} = A_\rho B_\rho + A_\phi B_\phi + A_z B_z$

$\vec{A}\times\vec{B} = (A_\phi B_z - A_z B_\phi)\,\hat{\rho} + (A_z B_\rho - A_\rho B_z)\,\hat{\phi} + (A_\rho B_\phi - A_\phi B_\rho)\,\hat{z}$

$$\nabla\Phi = \frac{\partial\Phi}{\partial\rho}\hat{\rho} + \frac{1}{\rho}\frac{\partial\Phi}{\partial\phi}\hat{\phi} + \frac{\partial\Phi}{\partial z}\hat{z}$$

$$\nabla\cdot\vec{A} = \frac{1}{\rho}\frac{\partial(\rho A_\rho)}{\partial\rho} + \frac{1}{\rho}\frac{\partial A_\phi}{\partial\phi} + \frac{\partial A_z}{\partial z}$$

$$\nabla\times\vec{A} = \left(\frac{1}{\rho}\frac{\partial A_z}{\partial\phi} - \frac{\partial A_\phi}{\partial z}\right)\hat{\rho} + \left(\frac{\partial A_\rho}{\partial z} - \frac{\partial A_z}{\partial\rho}\right)\hat{\phi} + \left[\frac{1}{\rho}\frac{\partial(\rho A_\phi)}{\partial\rho} - \frac{1}{\rho}\frac{\partial A_\rho}{\partial\phi}\right]\hat{z}$$

$$\nabla^2\Phi = \frac{1}{\rho}\frac{\partial}{\partial\rho}\left(\rho\frac{\partial\Phi}{\partial\rho}\right) + \frac{1}{\rho^2}\frac{\partial^2\Phi}{\partial\phi^2} + \frac{\partial^2\Phi}{\partial z^2}$$

$$\nabla^2\vec{A} = \left(\nabla^2 A_\rho - \frac{A_\rho}{\rho^2} - \frac{2}{\rho^2}\frac{\partial A_\phi}{\partial\phi}\right)\hat{\rho} + \left(\nabla^2 A_\phi - \frac{A_\phi}{\rho^2} + \frac{2}{\rho^2}\frac{\partial A_\rho}{\partial\phi}\right)\hat{\phi} + (\nabla^2 A_z)\hat{z}$$

position in space ($\hat{\rho}$ and $\hat{\phi}$ change in the circular cylindrical system, while $\hat{r}$, $\hat{\theta}$, and $\hat{\phi}$ change in the spherical system). Recall that the unit vectors in the rectangular Cartesian system were independent of the position in space.

The differential arc length (B.13) is

$$ds = \sqrt{d\vec{r}\cdot d\vec{r}} = \sqrt{(ds_u)^2 + (ds_v)^2 + (ds_w)^2}$$
$$= \sqrt{g_{uu}(du)^2 + g_{vv}(dv)^2 + g_{ww}(dw)^2}, \qquad \text{(B.32)}$$

where the factors $g_{uu} = h_u^2$, $g_{vv} = h_v^2$, and $g_{ww} = h_w^2$ are called the *metric coefficients*.[2]

[2] Some authors call the factors h_u, etc. the metric coefficients.

Table B.6. *Spherical coordinates*

$(r,\ \theta,\ \phi)$: $0 \le r < \infty,\ 0 \le \theta \le \pi,\ 0 \le \phi < 2\pi$

$h_r = 1,\ h_\theta = r,\ h_\phi = r\sin\theta$

$x = r\sin\theta\cos\phi,\ y = r\sin\theta\sin\phi,\ z = r\cos\theta$

$r = \sqrt{x^2 + y^2 + z^2},\ \theta = \tan^{-1}\left(\sqrt{x^2 + y^2}/z\right),\ \phi = \tan^{-1}(y/x)$

$\cdot$	$\hat{x}$	$\hat{y}$	$\hat{z}$
$\hat{r}$	$\sin\theta\cos\phi$	$\sin\theta\sin\phi$	$\cos\theta$
$\hat{\theta}$	$\cos\theta\cos\phi$	$\cos\theta\sin\phi$	$-\sin\theta$
$\hat{\phi}$	$-\sin\phi$	$\cos\phi$	0

$d\vec{r} = \hat{r}\,(dr) + \hat{\theta}\,(rd\theta) + \hat{\phi}\,(r\sin\theta d\phi)$

$d\vec{S}_{\theta\phi} = \hat{r}\,(rd\theta)(r\sin\theta d\phi) = \hat{r}\,r^2 d\Omega,\ d\vec{S}_{\phi r} = \hat{\theta}\,(r\sin\theta d\phi)(dr),\ d\vec{S}_{r\theta} = \hat{\phi}\,(dr)(rd\theta)$

$dV = (dr)(rd\theta)(r\sin\theta d\phi) = r^2\sin\theta dr d\theta d\phi$

$\vec{A}\cdot\vec{B} = A_r B_r + A_\theta B_\theta + A_\phi B_\phi$

$\vec{A}\times\vec{B} = (A_\theta B_\phi - A_\phi B_\theta)\hat{r} + (A_\phi B_r - A_r B_\phi)\hat{\theta} + (A_r B_\theta - A_\theta B_r)\hat{\phi}$

$$\nabla\Phi = \frac{\partial\Phi}{\partial r}\hat{r} + \frac{1}{r}\frac{\partial\Phi}{\partial\theta}\hat{\theta} + \frac{1}{r\sin\theta}\frac{\partial\Phi}{\partial\phi}\hat{\phi}$$

$$\nabla\cdot\vec{A} = \frac{1}{r^2}\frac{\partial(r^2 A_r)}{\partial r} + \frac{1}{r\sin\theta}\frac{\partial(\sin\theta A_\theta)}{\partial\theta} + \frac{1}{r\sin\theta}\frac{\partial A_\phi}{\partial\phi}$$

$$\nabla\times\vec{A} = \frac{1}{r\sin\theta}\left[\frac{\partial(\sin\theta A_\phi)}{\partial\theta} - \frac{\partial A_\theta}{\partial\phi}\right]\hat{r} + \frac{1}{r}\left[\frac{1}{\sin\theta}\frac{\partial A_r}{\partial\phi} - \frac{\partial(rA_\phi)}{\partial r}\right]\hat{\theta}$$
$$+\frac{1}{r}\left[\frac{\partial(rA_\theta)}{\partial r} - \frac{\partial A_r}{\partial\theta}\right]\hat{\phi}$$

$$\nabla^2\Phi = \frac{1}{r^2}\frac{\partial}{\partial r}\left(r^2\frac{\partial\Phi}{\partial r}\right) + \frac{1}{r^2\sin\theta}\frac{\partial}{\partial\theta}\left(\sin\theta\frac{\partial\Phi}{\partial\theta}\right) + \frac{1}{r^2\sin^2\theta}\frac{\partial^2\Phi}{\partial\phi^2}$$

$$\nabla^2\vec{A} = \left(\nabla^2 A_r - \frac{2A_r}{r^2} - \frac{2\cot\theta}{r^2}A_\theta - \frac{2}{r^2}\frac{\partial A_\theta}{\partial\theta} - \frac{2}{r^2\sin\theta}\frac{\partial A_\phi}{\partial\phi}\right)\hat{r}$$
$$+\left(\nabla^2 A_\theta + \frac{2}{r^2}\frac{\partial A_r}{\partial\theta} - \frac{A_\theta}{r^2\sin^2\theta} - \frac{2\cos\theta}{r^2\sin^2\theta}\frac{\partial A_\phi}{\partial\phi}\right)\hat{\theta}$$
$$+\left(\nabla^2 A_\phi + \frac{2}{r^2\sin^2\theta}\frac{\partial A_r}{\partial\phi} - \frac{1}{r^2\sin^2\theta}A_\phi + \frac{2\cos\theta}{r^2\sin^2\theta}\frac{\partial A_\theta}{\partial\phi}\right)\hat{\phi}$$

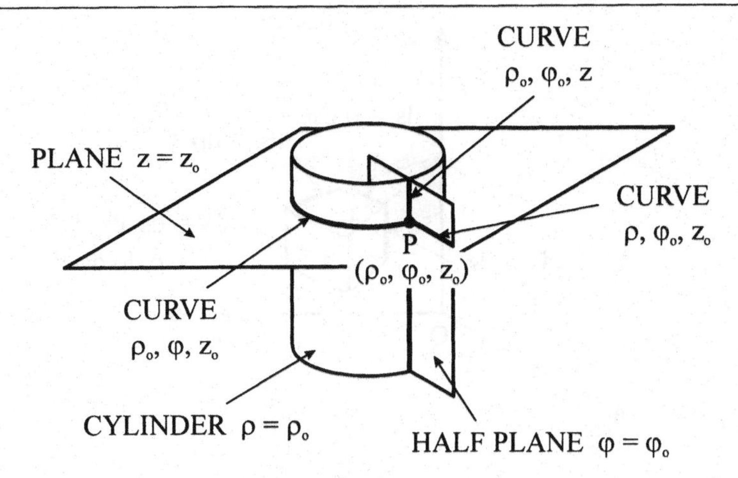

Fig. B.8. Circular cylindrical coordinates showing coordinate surfaces and coordinate curves through point P.

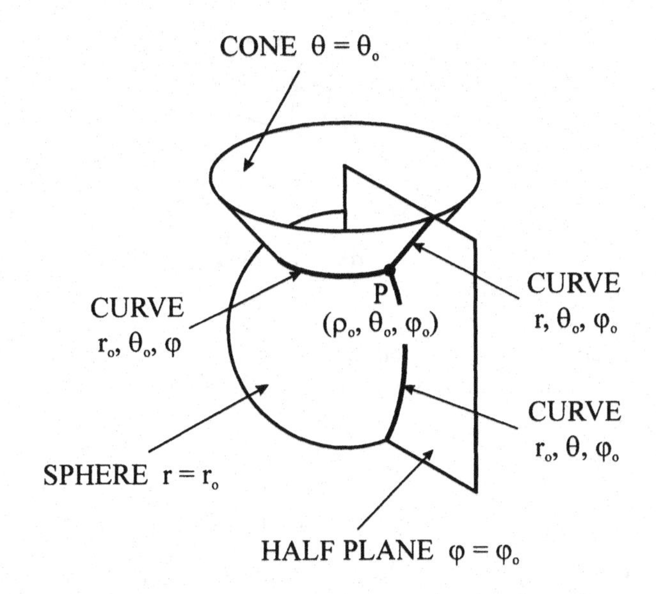

Fig. B.9. Spherical coordinates showing coordinate surfaces and coordinate curves through point P.

The differential volume is

$$dV = ds_u ds_v ds_w = h_u h_v h_w du dv dw, \tag{B.33}$$

and the differential surface area and differential surface area vector associated with the v, w surface are

$$dS_{vw} = ds_v ds_w = h_v h_w dv dw \tag{B.34}$$

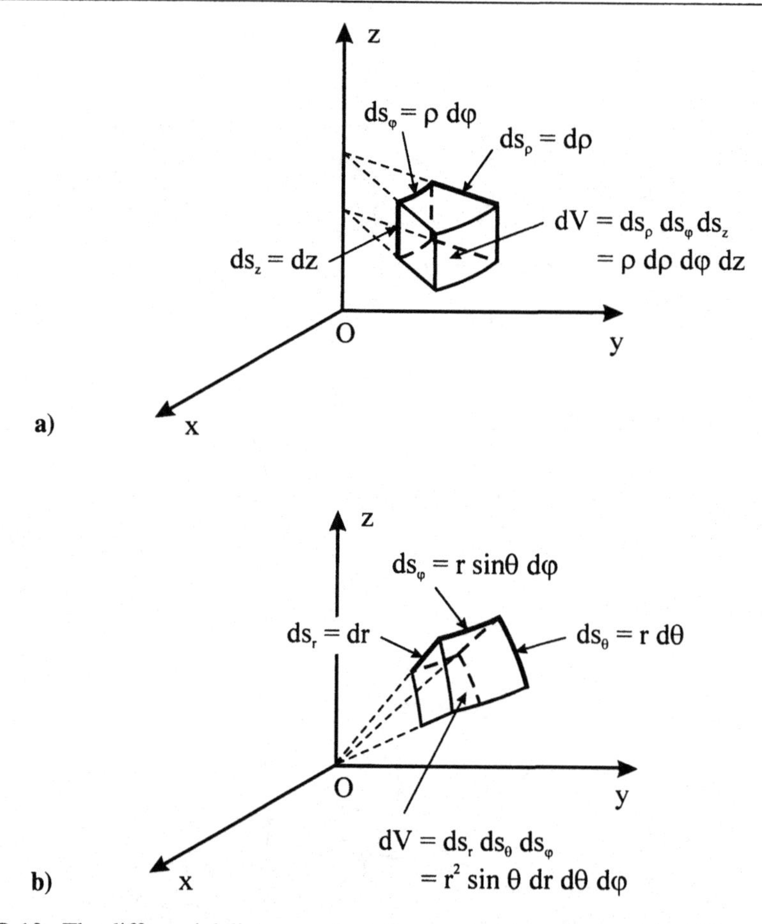

Fig. B.10. The differential distances along the coordinate curves and the differential volume for a) circular cylindrical coordinates and b) spherical coordinates.

and

$$d\vec{S}_{vw} = \hat{u} dS_{vw} = \hat{u} h_v h_w dv dw. \tag{B.35}$$

In Figure B.10, the differential volume, dV, and the differential lengths along the coordinate curves, ds_u, etc., are shown for the circular cylindrical and spherical coordinate systems. In the spherical coordinate system (Figure B.10b) the differential surface area on the spherical surface of radius r is

$$dS_{\theta\phi} = r^2 \sin\theta d\theta d\phi. \tag{B.36}$$

The *differential solid angle* subtended by this area is

$$d\Omega = \frac{dS_{\theta\phi}}{r^2} = \sin\theta d\theta d\phi, \tag{B.37}$$

so (B.36) is sometimes written as

$$dS_{\theta\phi} = r^2 d\Omega. \tag{B.38}$$

Table B.7. *Relations for distance between two points*

$\vec{R} = \vec{r} - \vec{r}\,' = R\hat{R},\ R = \lvert\vec{r} - \vec{r}\,'\rvert$
$\nabla(R) = \hat{R}$
$\nabla(R^n) = nR^{n-1}\hat{R}$
$\nabla[f(R)] = \dfrac{df(R)}{dR}\hat{R}$
$\nabla \cdot \hat{R} = \dfrac{2}{R}$
$\nabla \cdot [f(R)\hat{R}] = \dfrac{2f(R)}{R} + \dfrac{df(R)}{dR}$
$\nabla \times \hat{R} = 0$
$\nabla \times [f(R)\hat{R}] = 0$

The differential relations ∇, $\nabla\cdot$, and $\nabla\times$ can be obtained for an orthogonal curvilinear coordinate system by applying expressions (B.17)–(B.20) to the appropriate elements of volume and surface area [3]:

$$\text{grad}\,\Phi = \nabla\Phi = \frac{1}{h_u}\frac{\partial\Phi}{\partial u}\hat{u} + \frac{1}{h_v}\frac{\partial\Phi}{\partial v}\hat{v} + \frac{1}{h_w}\frac{\partial\Phi}{\partial w}\hat{w}, \tag{B.39}$$

$$\text{div}\,\vec{A} = \nabla\cdot\vec{A} = \frac{1}{h_u h_v h_w}\left[\frac{\partial(h_v h_w A_u)}{\partial u} + \frac{\partial(h_w h_u A_v)}{\partial v} + \frac{\partial(h_u h_v A_w)}{\partial w}\right], \tag{B.40}$$

and

$$\text{curl}\,\vec{A} = \nabla\times\vec{A} = \frac{1}{h_v h_w}\left[\frac{\partial(h_w A_w)}{\partial v} - \frac{\partial(h_v A_v)}{\partial w}\right]\hat{u} + \frac{1}{h_w h_u}\left[\frac{\partial(h_u A_u)}{\partial w} - \frac{\partial(h_w A_w)}{\partial u}\right]\hat{v} + \frac{1}{h_u h_v}\left[\frac{\partial(h_v A_v)}{\partial u} - \frac{\partial(h_u A_u)}{\partial v}\right]\hat{w}. \tag{B.41}$$

Operations such as $\nabla^2\Phi \equiv \nabla\cdot\nabla\Phi$ can be obtained by successively applying the above results (Problem B.7). Tables B.5 and B.6 give specific results for the differential relations in the circular cylindrical and spherical coordinate systems.

The distance between two points $R = \lvert\vec{r} - \vec{r}\,'\rvert$ often occurs in calculations for physical systems, so it helpful to know, beforehand, the results for differential operations applied to the distance and related quantities. These are listed in Table B.7.

Historical note

Today, we generally first encounter vector analysis when dealing with mechanics: the vectors for velocity, force, etc. Interestingly, the motivation for the development of modern vector analysis was electrodynamics (Maxwell's equations), not mechanics.

In 1843, the famous Irish mathematician William Rowan Hamilton (1805–1865) invented *quaternions* [11–14]. At the time, Hamilton was attempting to extend the algebra of complex numbers to three dimensions ("triplets"). Instead, he discovered the quaternion of four numbers. Hamilton continued to work on quaternions until his death in 1865. His massive two volume treatise (1,096 pages) on the subject was published posthumously [13].

The quaternion is a hypercomplex number, which is written as

$$Q = w + \alpha\,\mathbf{i} + \beta\,\mathbf{j} + \gamma\,\mathbf{k}, \tag{B.42}$$

where w, α, β, and γ are real numbers. The **i**, **j**, and **k** in this equation obey the relations

$$\mathbf{ii} = \mathbf{jj} = \mathbf{kk} = -1; \qquad \mathbf{ij} = -\mathbf{ji} = \mathbf{k}, \qquad \mathbf{jk} = -\mathbf{kj} = \mathbf{i}, \qquad \mathbf{ki} = -\mathbf{ik} = \mathbf{j}. \tag{B.43}$$

Notice the similarity of **i**, **j**, and **k** to the familiar i ($ii = -1$) from the theory of complex numbers.

Hamilton introduced the now familiar terminology scalar and vector. He wrote the quaternion as

$$Q = SQ + VQ,$$

and he called $SQ = w$ the "scalar part" and $VQ = \alpha\,\mathbf{i} + \beta\,\mathbf{j} + \gamma\,\mathbf{k}$ the "vector part." The vector part corresponds to the three rectangular Cartesian components of a modern vector. The **i**, **j**, and **k** are akin to (but clearly not equal to) the modern unit vectors $\hat{x}$, $\hat{y}$, and $\hat{z}$. In fact, some authors use **i**, **j**, and **k** for $\hat{x}$, $\hat{y}$, and $\hat{z}$.

The product of the two quaternions

$$A = \alpha\,\mathbf{i} + \beta\,\mathbf{j} + \gamma\,\mathbf{k}$$

and

$$B = \delta\,\mathbf{i} + \epsilon\,\mathbf{j} + \zeta\,\mathbf{k},$$

each with zero scalar part (each has only a vector part), is

$$AB = -(\alpha\delta + \beta\epsilon + \gamma\zeta) + (\beta\zeta - \gamma\epsilon)\,\mathbf{i} + (\gamma\delta - \alpha\zeta)\,\mathbf{j} + (\alpha\epsilon - \beta\delta)\,\mathbf{k}, \tag{B.44}$$

a quaternion with both a scalar part and a vector part. When we write A and B in modern vector notation,

$$\vec{A} = \alpha\hat{x} + \beta\hat{y} + \gamma\hat{z}$$
$$\vec{B} = \delta\hat{x} + \epsilon\hat{y} + \zeta\hat{z},$$

we have for the two modern products of these vectors, (B.6) and (B.7),

$$\vec{A} \cdot \vec{B} = \alpha\delta + \beta\epsilon + \gamma\zeta \tag{B.45}$$

and

$$\vec{A} \times \vec{B} = (\beta\zeta - \gamma\epsilon)\hat{x} + (\gamma\delta - \alpha\zeta)\hat{y} + (\alpha\epsilon - \beta\delta)\hat{z}. \tag{B.46}$$

Notice that the scalar part of the quaternion (B.44) is just the negative of the modern scalar or dot product (B.45) and that the vector part of the quaternion (B.44) is just the modern vector or cross product (B.46). Thus, the quaternion product produces the modern scalar and vector products in one operation.

Hamilton and his followers introduced other notation that we still use today, for example, the operator

$$\lhd = \mathbf{i}\frac{\partial}{\partial x} + \mathbf{j}\frac{\partial}{\partial y} + \mathbf{k}\frac{\partial}{\partial z},$$

which we now write as ∇ in rectangular Cartesian coordinates.

In his famous paper on electrodynamics, published in 1865, Maxwell expressed all field quantities in terms of their components in the rectangular Cartesian coordinate system [15]. Later, in his treatise, he expressed these quantities in both component and quaternion notation [16]. Apparently, Maxwell felt that quaternions were of value in representing physical phenomena, but he was not as enthusiastic about their use as a method of calculation [11, 17]. Maxwell called the operation that is equivalent to the modern $-\nabla\cdot$ the "convergence." Of course, today we call $\nabla\cdot$ the divergence; the difference is that (B.18) with the negative sign indicates an inward flux rather than an outward flux. He called the operation that is equivalent to the modern $\nabla\times$ the "rotation," "version," or "curl;" the last name is the one we still use [16, 18].

Maxwell's use of quaternions in his treatise served as motivation for others interested in electromagnetic theory to investigate quaternions. Among those who took up the subject were the American physicist Josiah Willard Gibbs (1839–1903) at Yale University and Oliver Heaviside in England.[3] They both recognized that it was not necessary to combine the scalar and vector parts into a single entity, the quaternion, but that the two parts could be treated more simply if separated. Gibbs and Heaviside went on to develop the modern system of vector analysis that we use today [19, 20]. Their departure from the methodology of quaternions did not go uncontested. A heated debate between the followers of Hamilton and those of Gibbs and Heaviside continued into the early twentieth century [11].

References

[1] W. Kaplan, *Advanced Calculus*, Chapters 1–5, Addison-Wesley, Reading, MA, 1952.
[2] M. R. Spiegel, *Vector Analysis*, McGraw-Hill, New York, 1959.
[3] G. Arfken, *Mathematical Methods for Physicists*, 3rd Edition, Chapters 1 and 2, Academic Press, New York, 1985.
[4] A. L. Borisenko and I. E. Tarapov, *Vector and Tensor Analysis with Applications*, Prentice-Hall, Englewood Cliffs, NJ, 1968.
[5] R. C. Wrede, *Introduction to Vector and Tensor Analysis*, Wiley, New York, 1963. Republication, Dover Publications, New York, 1972.
[6] J. Van Bladel, *Electromagnetic Fields*, Appendices 1 and 2, McGraw-Hill, New York, 1964.
[7] P. Moon and D. E. Spencer, *Field Theory Handbook*, Springer-Verlag, Berlin, 1971.

[3] In 1863 Gibbs received the first doctorate in engineering issued in the United States.

[8] C. T. Tai, *Generalized Vector and Dyadic Analysis*, IEEE Press, Piscataway, NJ, 1992.
[9] O. D. Kellogg, *Foundations of Potential Theory*, Springer-Verlag, Berlin, 1929.
[10] C. Müller, *Foundations of the Mathematical Theory of Electromagnetic Waves*, Springer-Verlag, Berlin, 1969.
[11] M. J. Crowe, *A History of Vector Analysis*, University of Notre Dame Press, South Bend, IN, 1967. Republication, Dover Publications, New York, 1985.
[12] T. L. Hankins, *Sir William Rowan Hamilton*, Johns Hopkins Press, Baltimore, MD, 1980.
[13] W. R. Hamilton, *Elements of Quaternions*, Volumes I and II, Edited by C. J. Joly, 3rd Edition, Chelsea, New York, 1969. Originally published in 1866.
[14] P. G. Tait, *An Elementary Treatise on Quaternions*, 3rd edition, Cambridge University Press, Cambridge, 1890.
[15] J. C. Maxwell, "A Dynamical Theory of the Electromagnetic Field," *Phil. Trans. Roy. Soc. London*, Vol. 155, pp. 459–512, 1865. Paper read December 8, 1864. Also reprinted in book form, *A Dynamical Theory of the Electromagnetic Field*, Scottish Academic Press, Edinburgh, 1982.
[16] —, *A Treatise on Electricity and Magnetism*, 3rd Edition, Vol. 1, pp. 9–31, Vol. 2, pp. 257–9, Clarendon Press, Oxford, 1891. Republication, Dover Publications, New York, 1954.
[17] J. Hendry, *James Clerk Maxwell and the Theory of the Electromagnetic Field*, Adam Hilger, Bristol, 1986.
[18] J. C. Maxwell, "On the Mathematical Classification of Physical Quantities," *Proc. London Math. Soc.*, Vol. 3, pp. 224–32, 1871.
[19] E. B. Wilson, *Vector Analysis, A Textbook for use by Students of Mathematics and Physics, Founded upon the Lectures of J. W. Gibbs*, Yale University Press, New Haven, CT, 1901.
[20] O. Heaviside, *Electromagnetic Theory*, Vol. I, Chapter III, "The Elements of Vectorial Algebra and Analysis," pp. 132–305, "The Electrician" Printing and Publishing Co., London, 1893. Republication, Chelsea, New York, 1971.

Problems

B.1 Use the triple products in Table B.1 to show the following:

a) $\vec{A} \times (\vec{B} \times \vec{C}) \neq (\vec{A} \times \vec{B}) \times \vec{C}$,

b) $(\vec{A} \times \vec{B}) \cdot (\vec{C} \times \vec{D}) = (\vec{A} \cdot \vec{C})(\vec{B} \cdot \vec{D}) - (\vec{A} \cdot \vec{D})(\vec{B} \cdot \vec{C})$,

c) $(\vec{A} \times \vec{B}) \times (\vec{C} \times \vec{D}) = \left[(\vec{A} \times \vec{B}) \cdot \vec{D}\right]\vec{C} - \left[(\vec{A} \times \vec{B}) \cdot \vec{C}\right]\vec{D}$.

B.2 Apply (B.17) to the element of volume in the rectangular Cartesian coordinate system (Figure 1.16b) to obtain (B.21).

B.3 Using the results for ∇, $\nabla\cdot$, and $\nabla\times$ in the rectangular Cartesian coordinate system, obtain the expression for $\nabla^2\vec{A}$ in Table B.2.

B.4 Starting with (B.27)–(B.30), obtain the expressions for the differential distance vector ($d\vec{r}$), the unit vectors ($\hat{r}$, $\hat{\theta}$, $\hat{\phi}$), and the scale factors (h_r, h_θ, h_ϕ) for the spherical coordinate system. Check your answers by comparing them with the results in Table B.6.

B.5 Make a table, like the one in Table B.5, that shows the relations between the unit vectors in the circular cylindrical and spherical coordinate systems ($\hat{\rho} \cdot \hat{r}$, etc.).

B.6 a) Show that

$$\int_{\phi=0}^{2\pi} \int_{\theta=0}^{\pi} d\Omega = 4\pi \quad \text{(steradians)}.$$

b) A circle of radius a is drawn on the surface of a sphere of radius $r > a$. What is the area A on the surface of the sphere within the circle? What is the solid angle subtended by this area, $\Omega = A/r^2$ (steradians)?

B.7 Using the results for ∇ and $\nabla\cdot$ in Equations (B.39) and (B.40), obtain a general expression for $\nabla^2\Phi$ in an orthogonal curvilinear coordinate system. Use your expression to determine $\nabla^2\Phi$ in the spherical coordinate system, and compare your result with that in Table B.6.

B.8 Obtain the expression for $\nabla\Phi$ in circular cylindrical coordinates (Table B.5) by converting the expression in rectangular Cartesian coordinates (Table B.2) to circular cylindrical coordinates. Use the chain rule for computing derivatives, for example,

$$\frac{\partial \Phi}{\partial x} = \frac{\partial \Phi}{\partial \rho}\frac{\partial \rho}{\partial x} + \frac{\partial \Phi}{\partial \phi}\frac{\partial \phi}{\partial x} + \frac{\partial \Phi}{\partial z}\frac{\partial z}{\partial x}.$$

B.9 Obtain the results for $\nabla(R^n)$ and $\nabla \times \hat{R}$ given in Table B.7. What is $\nabla \cdot \vec{R}$?

Supplemental references

History – general

H. G. J. Aitken, *Syntony and Spark*, Princeton University Press, Princeton, NJ, 1985.

J. Z. Buchwald, *From Maxwell to Microphysics*, University of Chicago Press, Chicago, IL, 1985.

J. Z. Buchwald, *The Rise of the Wave Theory of Light*, University of Chicago Press, Chicago, IL, 1989.

B. J. Hunt, *The Maxwellians*, Cornell University Press, Ithaca, NY, 1991.

P. F. Mottelay, *Bibliographical History of Electricity and Magnetism*, Charles Griffin, London, 1922.

R. A. R. Tricker, *Early Electrodynamics, The First Law of Circulation,* Pergamon Press, Oxford, 1965.

R. A. R. Tricker, *The Contributions of Faraday and Maxwell to Electrical Science,* Pergamon Press, Oxford, 1966.

E. Whittaker, *A History of the Theories of Aether and Electricity*, Volumes I and II, Tomash Publ. /American Institute of Physics, New York, 1987. Originally published, Nelson, London, 1951, 1953.

History – biographical

J. Z. Buchwald, *The Creation of Scientific Effects: Heinrich Hertz and Electric Waves*, University of Chicago Press, Chicago, IL, 1994.

B. Dibner, *Oersted and the Discovery of Electromagnetism,* Burndy Library, Norwalk, CT, 1961.

C. S. Gillmor, *Coulomb and the Evolution of Physics and Engineering in Eighteenth-Century France*, Princeton University Press, Princeton, NJ, 1971.

J. Hendry, *James Clerk Maxwell and the Theory of the Electromagnetic Field*, Adam Higler, Bristol, 1986.

J. R. Hofmann, *André–Marie Ampère, Enlightenment and Electrodynamics*, Cambridge University Press, Cambridge, 1996.

P. J. Nahin, *Oliver Heaviside: Sage in Solitude*, IEEE Press, New York, 1988.

J. G. O'Hara and W. Pricha, *Hertz and the Maxwellians*, Peter Peregrinus, London, 1987.

S. P. Thompson, *Michael Faraday, His Life and Works,* Cassell, Ltd., London, 1901.

I. Tolstoy, *James Clerk Maxwell: A Biography*, University of Chicago Press, Chicago, IL, 1981.

L. P. Williams, *Michael Faraday,* Basic Books, New York, 1965.

Electromagnetic theory – classic texts

R. Becker and F. Sauter, *Electromagnetic Fields and Interactions, Volume I, Electromagnetic Theory and Relativity*, Blaisdell Publishing, New York, 1964. The history of this volume can be traced to the work A. Föppl, *Introduction to Maxwell's Theory*, 1894.

M. Born and E. Wolf, *Principles of Optics: Electromagnetic Theory of Propagation, Interference and Diffraction of Light*, 3rd Edition, Pergamon Press, New York, 1965. First published in 1959, it incorporates portions of Born's earlier (1933) German text *Optik*.

J. R. Oppenheimer, *Lectures on Electrodynamics*, Gordon and Breach, New York, 1970. Based on lecture notes prepared at the University of California, Berleley, CA in 1939.

L. Page and N. I. Adams, Jr., *Electrodynamics*, D. Van Nostrand, New York, 1940. Republication, Dover Publications, New York, 1965.

W. Pauli, *Pauli Lectures on Physics: Volume 1. Electrodynamics*, MIT Press, Cambridge, MA, 1973. Based on lecture notes prepared at ETH, Zürich, Switzerland in 1949.

W. Pauli, *Pauli Lectures on Physics: Volume 2. Optics and the Theory of Electrons*, MIT Press, Cambridge, MA, 1973. Based on lecture notes prepared at ETH, Zürich, Switzerland in 1948.

A. Sommerfeld, *Electrodynamics*, Academic Press, 1952. German edition 1948. Based in part on lectures from 1934 at the University of Munich, Munich, Germany.

A. Sommerfeld, *Optics*, Academic Press, 1954. German edition 1949. Based in part on lectures from 1933/34 at the University of Munich, Munich, Germany.

W. R. Smythe, *Static and Dynamic Electricity*, 3rd Edition, McGraw-Hill, New York, 1968. First published in 1939.

J. A. Stratton, *Electromagnetic Theory*, McGraw-Hill, New York, 1941.

I. E. Tamm, *Fundamentals of the Theory of Electricity*, 9th Edition, Mir Publishers, Moscow, 1979. First Russian edition 1929.

Electromagnetic theory – introductory level

(References in this category provide suitable preparation for this book.)

D. K. Cheng, *Field and Wave Electromagnetics*, 2nd Edition, Addison-Wesley, Reading, MA, 1989.

P. C. Clemmow, *An Introduction to Electromagnetic Theory*, Cambridge University Press, Cambridge, 1973.

R. S. Elliot, *Electromagnetics*, McGraw-Hill, New York, 1966.

R. P. Feynman, R. P. Leighton, and M. Sands, *The Feynman Lectures on Physics*, Volumes I and II, Addison-Wesley, Reading, MA, 1964.

D. J. Griffiths, *Introduction to Electrodynamics*, 2nd Edition, Prentice-Hall, Englewood Cliffs, NJ, 1989.

O. D. Jefimenko, *Electricity and Magnetism*, 2nd Edition, Electret Scientific, Star City, WV, 1989.

J. D. Kraus, *Electromagnetics*, 4th Edition, McGraw-Hill, New York, 1992.

D. T. Paris and F. K. Hurd, *Basic Electromagnetic Theory*, McGraw-Hill, New York, 1969.

E. M. Purcell, *Electricity and Magnetism*, Berkeley Physics Course, Vol. 2, 2nd Edition, McGraw-Hill, New York, 1985.

L. C. Shen and J. A. Kong, *Applied Electromagnetism*, 3rd Edition, PWS Publishing, Boston, MA, 1995.

Electromagnetic theory – intermediate level

(References in this category are, very roughly, at about the same level as this book.)

C. A. Balanis, *Advanced Engineering Electromagnetics*, Wiley, New York, 1989.
L. Eyges, *The Classical Electromagnetic Field*, Addison-Wesley, Reading, MA, 1972. Republication, Dover Publications, New York, 1980.
R. F. Harrington, *Time-Harmonic Electromagnetic Fields*, McGraw-Hill, New York, 1961.
H. A. Haus, *Waves and Fields in Optoelectronics*, Prentice-Hall, Englewood Cliffs, NJ, 1984.
E. C. Jordan and K. G. Balmain, *Electromagnetic Waves and Radiating Systems*, 2nd Edition, Prentice-Hall, Englewood Cliffs, NJ, 1968.
R. W. P. King and S. Prasad, *Fundamental Electromagnetic Theory and Applications*, Prentice-Hall, Englewood Cliffs, NJ, 1986.
C. H. Papas, *Theory of Electromagnetic Wave Propagation*, McGraw-Hill, New York, 1965. Republication, Dover Publications, New York, 1988.
S. Ramo, J. R. Whinnery, and T. Van Duzer, *Fields and Waves in Communication Electronics*, 3rd Edition, Wiley, New York, 1994.
J. R. Wait, *Electromagnetic Wave Theory*, Harper and Row, New York, 1985.

Electromagnetic theory – advanced level

(References in this category go beyond the presentation in this book.)

W. C. Chew, *Waves and Fields in Inhomogeneous Media*, Van Nostrand Reinhold, New York, 1990.
S. K. Cho, *Electromagnetic Scattering*, Springer–Verlag, New York, 1990.
P. C. Clemmow, *The Plane Wave Spectrum Representation of Electromagnetic Fields*, Pergamon Press, New York, 1966.
P. C. Clemmow and J. P. Dougherty, *Electrodynamics of Particles and Plasmas*, Addison-Wesley, Reading, MA, 1969.
R. E. Collin, *Field Theory of Guided Waves*, 2nd Edition, IEEE Press, Piscataway, NJ, 1991.
D. G. Dudley, *Mathematical Foundations for Electromagnetic Theory*, IEEE Press, Piscataway, NJ, 1994.
L. B. Felsen and N. Marcuvitz, *Radiation and Scattering of Waves*, Prentice-Hall, Englewood Cliffs, NJ, 1973.
J. D. Jackson, *Classical Electrodynamics*, 2nd Edition, Wiley, New York, 1975.
D. S. Jones, *The Theory of Electromagnetism*, Pergamon Press, New York, 1964.
D. S. Jones, *Acoustic and Electromagnetic Waves*, Oxford University Press, Oxford, 1986.
D. S. Jones, *Methods in Electromagnetic Wave Propagation*, 2nd Edition, Oxford University Press, Oxford, 1994.
J. A. Kong, *Electromagnetic Wave Theory*, 2nd Edition, Wiley, New York, 1990.
L. D. Landau and E. M. Lifshitz, *The Classical Theory of Fields*, 4th Edition, Pergamon Press, New York, 1975.

L. D. Landau, E. M. Lifshitz, and L. P. Pitaevskiĭ, *Electrodynamics of Continuous Media*, 2nd Edition, Pergamon Press, New York, 1984.

I. V. Lindell, *Methods for Electromagnetic Field Analysis*, Oxford University Press, Oxford, 1992.

C. Müller, *Foundations of the Mathematical Theory of Electromagnetic Waves*, Springer-Verlag, Berlin, 1969.

W. K. H. Panofsky and M. Phillips, *Classical Electricity and Magnetism*, 2nd Edition, Addison-Wesley, Reading, MA, 1962.

F. N. H. Robinson, *Macroscopic Electromagnetism*, Pergamon Press, New York, 1973.

C. -T. Tai, *Dyadic Green Function in Electromagnetic Theory*, 2nd Edition, IEEE Press, Piscataway, NJ, 1994.

J. Van Bladel, *Electromagnetic Fields*, McGraw-Hill, New York, 1964.

Index

Page numbers for illustrations and tables are in italic.

acceleration field of moving point charge, 368, *376*
advanced potentials, 343, 344
Aharonov-Bohm effect, 331
Airy pattern for circular aperture, 231–2, *231*, *232*
Airy, G.B., 231
Ampère, A.M., *frontispiece*, 1, 15–8
ampère, SI unit, physical standard for, 610–1, *610*
Ampère-Maxwell law
  for general time dependence
    differential form of, 31, *33*
    integral form of, 4, *33*
  for harmonic time dependence
    differential form of, *94*
    integral form of, *94*
Ampère's experiments, 15–9, *15*, *18*, 107–10, *108*, *109*, 612
antenna arrays, *see* arrays of electrically short linear antennas
antennas
  basic traveling-wave element, 556–62, 584–8
  bent-wire, traveling wave, 567–70, *568*, 606–7
  complementary, 294–9
  dipole
    Hertz's, 162, *162*, 469
    resistively loaded, 575–9, 605–6
    slot, 294–9, *294*, 316–7
    standing-wave, 563–7, 591–601
    strip, 294–9, *294*, 316–7
    top-loaded, 470, *470*
    traveling-wave, 562–3
    traveling-wave V, 588–9, 605
  linear
    electrically short, 469–77
    insulated, 579–83, *579*, 606
  long-wire, 588–91
  loop
    electrically small, 482–5
    Hertz's, 163–4, *165*
    standing-wave, 573–5
    traveling-wave, 570–3
  monopole, 81–8, *81*, 475–6, 547–8, 566–7, 598–600
  reflector
    Cassagrain, 252–4, *253*
    NASA Goldstone, 232, *233*, 234
    paraboloidal, 233–5, *235*
  rhombic, horizontal, 589–91, *590*, 606
  self-complementary, 315
  thin-wire, 546–7
  umbrella, 476
aperture, circular
  Airy pattern for, 231–2, *231*, *232*
  analyzed with Huygens' principle, 272, 308–10
  analyzed with plane-wave spectrum, 218–22
  directivity of, 224, *226*, 227–8
  electrically large, 225–8
  electrically small, 514–6
    equivalent dipoles for transmission, 515, 543–4
  excited by Gaussian pulse in time, 308–10
  experimental results for, 228–31, *229*, 230, 285, 290
  Fresnel zones for, 273–7
  half-power beamwidth of, 224, *226*, 227, 228
  radiated field of, 222, 272
  relation to circular disc, 285–6, 288–9, 514, 543
  side lobes, 225, *226*, 227
  transmission coefficient of, 289, *290*, 514–5
  with tapered illumination, 234–7
Arago, D.F.J., 521
arrays of electrically short linear antennas
  bilateral broadside, 489, 496
  circular, 486n
  collinear, 542
  couplet, 486
    directivity, *503*
  excited by "pulse-like" signals, 486–92
  for harmonic time dependence, 493–506
  linear, 486
  neglect of mutual coupling for, 486, 494n
  ordinary end-fire, 495
  superdirectivity for, 502–6
  supergain for, 506
  uniform, 491, 493–502
    array factor for, 493
    directivity of, 494
    element factor for, 493
    half-power beamwidth of, 497, *499*
    maximum directivity of, 497, *498*, *499*
    pattern multiplication for, 494, *496*

radiated electric field of, 493
with large number of elements, 499–501
unilateral end-fire, 488, 495
astatic combination, Ampère's, 107–10, *108*
asymptotic approximation, 199
asymptotic field, 199; *see also* radiated field
Avogadro's number, value of, 544

Babinet, J., 282
Babinet's principle
electromagnetic formulation, 283–5, *283*
experimental verification, 285–6, *285*
for circular aperture and obstacle, 288–9, 514–5
optical formulation, 281–3, *282*
basic traveling-wave element
charge per unit length on, 557
current in, 556, 584
directivity of, 585, *587*
displaced from coordinate axis, 561–2, 588
for harmonic time dependence, 584–8
maximum directivity and beamwidth for, 585, *587*
radiated electric field of, 558, 584
with source that is Gaussian pulse, 558–61
bent-wire, traveling-wave antenna, 567–70, *568*, 606–7
Bessel function
modified, 430
of first kind, 221, 258
of third kind, Hankel function, 316
Biot, J.B., 14
Biot-Savart formula, 111, 369
Boltzmann's constant, value of, 526
Booker's relation for impedances, 298
boundary conditions, electromagnetic, 35, 41–8, 115
at perfect conductor, *47*, 48, *95*, 116
for double layers of charge and current, 120–2
for harmonic time dependence, 94, *95*, 117
for normal components of field, 47
for tangential components of field, 44
summary of, *47*, *95*
boundary value problem, electromagnetic, 65
bremsstrahlung (breaking radiation), 391

Cavendish, H., 9
charge, electric
surface density of, 38, *6*
volume density of, 4, *6*
charge, magnetic
and Maxwell's equations, 112–4, 477–8
lack of experimental observation of, 11–2, 112
volume density of, 12, 112, 477
Cherenkov detectors, 434, *436*, 449
Cherenkov radiation
and flashes in human eye, 434–5
and insulated linear antenna, 426n, 582–3
angular dependence of, 413–4, *414*, *421*, 426, *433*
characteristics of, 413–5
electromagnetic field of, 422, 425, 430, *433*
energy per unit path length per unit frequency for, 428, 432, *432*
experimental results for, *414*, *416*
for charge of finite size or bunch of charges, 429–34
for point charge, 415–29
from nuclear reactor, 412n
history of, 412
in real materials versus ideal materials, 429
on path of finite length, 426n, 583n
polarization of, 414, 426
potential functions for, 420–1
Poynting vector for, 424
retarded times for, *418*, 419
similarity to other phenomena, 433–4, *435*
spectrum of energy for, 415, *416*, 427–8, 432, *432*
Cherenkov, P.A., 412
circular aperture, *see* aperture, circular
classical electron, 8n, 439
complementary antennas, 294–9
Booker's relation for impedances of, 298
self-complementary, 315
conduction current, 6, 53, 96
conductivity, electrical, 6
constitutive relations, electromagnetic
for simple materials, 5–6, 429
Coolidge X-ray tube, 391, *392*
coordinate curves, 620, 626
coordinate surfaces, 620, 625
coordinates
circular cylindrical, 627, *627*, *629*, *631*, *632*
orthogonal curvilinear, 625–33
metric coefficients for, 629
scale factors for, 628
rectangular Cartesian, 619–22, *619*, *620*, *621*, *622*
spherical, 628, *628*, *630*, *631*, *632*
cosine integral, 585
Coulomb gauge, 330–1, 356
Coulomb, C.A., 9
Coulomb's experiments, 9–12, *10*, 611–2
Coulomb's law, 9
Courant-Friedrichs-Lewy condition, 77
Crab Nebula, 408–11
curl of vector
definition of, 25–6, 623
in circular cylindrical coordinates, *629*
in orthogonal curvilinear coordinates, 633
in rectangular Cartesian coordinates, 27–28, *622*, 624
in spherical coordinates, *630*
current, displacement, 25, 110
current, electric
surface density of, 41, *6*
volume density of, 4, *6*
current, magnetic
volume density of, 112, 477

degree of linear polarization, 409, 532
differential distance vector, 621, 627

differential solid angle, 632
differential surface area, 4, 622, 631
differential surface area vector, 4, 622, 631–2
differential volume, 4, 622, 631
diffracted field, 197n
dipole antenna
  analogy to radiation from moving charge, 556
  approximate charge per unit length on, 549
  approximate current in, 550
  as monopole imaged in perfect conductor, 547–8
  description of wavefronts radiated by, 548
  FDTD results for, *549*
  Hertz's, 162, *162*, 469
  polarization for
    on reception, 156
    on transmission, 155
  radiated electric field of, 553–5
  slot, 294–9, *294*, 316–7
  strip, 294–9, *294*, 316–7
  see also linear antenna
dipole antenna, standing-wave
  as combination of basic traveling-wave elements, *564*, 564
  current distribution on, 591–2, *592*
  description of wavefronts radiated by, 565–6
  directivity of, 593, *595*
  for harmonic time dependence, 591–601
  full-wave, 594
  half-wave, 594, 595, 596, 598, 600
  input impedance for, 598–601
  maximum directivity and beamwidth for, 593–5, *595*
  measured radiation patterns for, *594*
  radiated electric field of, 565, 566, 593
  radiation resistance of, 600
  reflection of wave at drive point of, 563–4, *564*, 566, 592, 596, *597*
  reflection of wave at open ends of, 552–3
  relation to monopole antenna, 547–8
  resonance for, 596, 599–600
  with source that is Gaussian pulse, 565–7
dipole antenna, traveling-wave, 562–3
dipole antenna, with resistive loading, 575–9, 605–6
dipole moment, electric, 451, *458*, 465
dipole moment, magnetic, 478, 481–2, *483*
dipole, infinitesimal electric
  as dual of magnetic dipole, 477–8
  complex Poynting vector for, 467
  construction of field lines for, 460–1, *460*
  current moment for, 451, *458*
  dipole moment for, 451, *458*, 465
  directional characteristics of radiation from, *460*
  directivity of, 467
  electric field of, 457, *458*, 465
  field lines for, *463*, *464*, *468*, *470*
  for harmonic time dependence, 465–9
  frequency dependence of radiation from, 467
  magnetic field of, 457, *458*, 465
  Poynting vector for, 459
  radiated field of, 458
  summary of results for, *458*
  with moment that is Gaussian function, 461–5
dipole, infinitesimal magnetic
  as dual of electric dipole, 477–8
  as infinitesimal current loop, 479–82
  dipole moment for, 478, 481–2, *483*
  electric field of, 479, 481, *483*
  for harmonic time dependence, *483*
  magnetic field of, 479, *483*
  Poynting vector for, 479
  summary of results for, *483*
Dirac delta function
  one-dimensional, 41, 190, 332–5
    as sequence of ordinary functions, 332–3
    dimensions for, 332n
    Fourier transform of, 119
    summary of relationships for, *335*
  three-dimensional, 335–7
Dirac, P.A.M., 41
directivity of antenna (aperture), definition of, 223, 286
disc, perfectly conducting, circular
  electrically large
    general discussion of scattering from, 288–93
    scattering cross sections for, 289, *290*, *292*
  electrically small, 511–4
    directional characteristics of scattering from, *513*
    equivalent dipoles for scattering from, 512, *520*
    scattering cross sections for, 513, 515, *516*
  experimental results for, *285*, *290*
  relation to circular aperture, 288–9, 514, 543
displacement current, 25, 110
divergence of vector
  definition of, 28, 623
  in circular cylindrical coordinates, *629*
  in orthogonal curvilinear coordinates, 633
  in rectangular Cartesian coordinates, 29–31, *622*, 624
  in spherical coordinates, *630*
  on surface, 45–6
divergence theorem (Gauss' theorem), 51, 625, *626*
domain of dependence
  for difference equations, 77, *78*
  for differential equations, 77, *78*
domain of dependence condition, 77
Doppler effect for moving point charge, 388–9, 404
Doppler, C.J., 388
double layer
  of charge, 120–2
  of current, 120–2
duality
  and Maxwell's equations, 477–8
  applied to infinitesimal electric dipole, 478–9
Duane-Hunt limit for X-ray radiation, 392

E-plane, principal, 222

Earth's atmosphere
composition of, 529, 534n
reduced height of, 544–5
effective length of dipole antenna, 155, 472n
efficiency factor for antenna, 238; *see also* radiating efficiency of antenna
electric charge, *see* charge, electric
electric current, *see* current, electric
electric dipole, *see* dipole, infinitesimal electric
electric displacement, 4, *6*, 24, 110
electric excitation, 4, *6*
electric field strength, 4, *6*
electromagnetic induction, 19
electromagnetic mass of charged particle, 439
electromagnetic potentials, 320–31
electronic charge, value of, 392
electrostatic field, 321–2
energy of electromagnetic field
density of, 56
stored, 55
time-average of stored, 97–8
equation of continuity for electric charge
for general time dependence
differential form of, 32, *33*
integral form of, 4, *33*
on surface, 46, *46*, *47*
for harmonic time dependence
differential form of, *94*
integral form of, *94*
on surface, *95*
error function, 334
Euler's constant, value of, 585
evanescent plane wave, 187; *see also* plane wave, electromagnetic, inhomogeneous
extinction coefficient, 536

fan beam for antenna, 496
far-zone field, 199; *see also* radiated field
Faraday rotation in ionospheric plasma, 173
Faraday, M., *frontispiece*, 1, 13, 19–23
Faraday's experiments, 13, *14*, 19–23, *21*, 612
Faraday's law
for general time dependence
differential form of, 31, *33*
integral form of, 4, *33*
for harmonic time dependence
differential form of, *94*
integral form of, *94*
Feynman, R.P., 369
finite-difference time-domain method (FDTD), 71
applied to cylindrical monopole antenna, 81–8
applied to plane wave, 71–8
domain of dependence condition for, 77–8, 80–1
finite-difference equations for
one spatial dimension, 74–5
three spatial dimensions, 78–80
two spatial dimensions, 83–4
Yee lattice for, 78, *79*
force, electromagnetic, 7
forward scattering theorem (optical theorem), 293
applied to electrically short wire, 545
Fourier optics, 250
Fourier transformation
definition of, 101
of Dirac delta function, 119
of Heaviside unit step function, 257
of Maxwell's equations, 102
Parseval's formula for, 102
time convolution theorem for, 431
Fourier, J.B.J., 101
Frank, I.M., 412
Fraunhofer diffraction pattern
for circular aperture, *230*, 231
Fraunhofer field, 199; *see also* radiated field
Fraunhofer lines, 526n
Fraunhofer, von, J., 199
free space (vacuum), 5
Fresnel half-period zones, 274
Fresnel integrals, 256, 258
Fresnel zones
as basis for zone plate, 313, 317
for circular aperture, 273–7
Fresnel, A.J., 273
Frisch, von, K.R., 537
frustrated total internal reflection, 258–60

gain of antenna, definition of, 237–8, 504–6
gamma function, 430
gas constant, value of, 544
gauge transformation, 324, 326, 330
restricted, 325, 330
Gauss' electric law
for general time dependence
differential form of, 32, *33*
integral form of, 4, *33*
for harmonic time dependence
differential form of, *94*
integral form of, *94*
Gauss' magnetic law
for general time dependence
differential form of, 32, *33*
integral form of, 4, *33*
for harmonic time dependence
differential form of, *94*
integral form of, *94*
Gauss' theorem (divergence theorem), 51, 625, *626*
Gaussian beam
analyzed with plane wave spectrum, 240–7
in laser cavity, 247–50
paraxial approximation for, 242
Gaussian function and its derivatives, 461–2, *462*
Gaussian pulse, 72, *72*
geometrical optics, 197n, 218
geometrical rays, 215–8
geometrical theory of diffraction, 309
geometrical wavefronts, 215–8

Gibbs, J.W., 635
gradient of scalar
  definition of, 623
  in circular cylindrical coordinates, *629*
  in orthogonal curvilinear coordinates, 633
  in rectangular Cartesian coordinates, *622*, 624
  in spherical coordinates, *630*
Gray, S., 35
Green, G., 265
Green's function, for free-space
  as causal solution to scalar wave equation, 347n
  for general time dependence, 347, *348*
  for harmonic time dependence, 265–6, *268*, 353, *354*
    two-dimensional, 316
  Fourier transform of, 354
Green's theorems, 625, *626*

H-plane, principal, 222
half-power beamwidth of antenna, definition of, 224
half-wave plate, 147, *150*
half plane, conducting, diffraction by, 255–8
Hamilton, W.R., 634–5
Hankel function, 316
harmonic time dependence
  general discussion of, 88–91
  with Maxwell's equations, 92–4, *94*
Heaviside-Feynman field of moving charge, 369–70
Heaviside unit-step function, 256, 334
  Fourier transform of, 257
Heaviside, O., 1, 59, 61n, 369, 477
  quoted, 331n
Helmholtz equation, vector, 177
Helmholtz, von, H.L.F., 177
Hermitian magnitude, of complex vector, 97
Hertz, H.R., *frontispiece*, 1, 24, 161–8, 468–9, *470*
  quoted, 167–8
Hertz's experiments, 24, 161–8, 613
history
  of Cherenkov radiation, 412
  of early radio communication, 474–7
  of electromagnetic theory, 1–25, 2, *3*, 161–8
  of explanation for properties of skylight, 521–4
  of Huygens' principle, 263, *264*
  of synchrotron radiation, 407–8
  of vector analysis, 633–5
  of X-ray radiation, 391–2
homogeneous plane wave, 134, 178
honeybee
  navigation using polarization of skylight, 536–7
Huygens' principle
  and plane wave spectrum, 262–3
  applied to circular aperture, 272, 308–10
  for general time dependence, 306–7, *307*
  for general volume, 272–3
  for harmonic time dependence, 267, *268*
  for planar surface, 264–8, 305–8
  for radiated field, 268–72, *271*
  history of, 263, *264*
  summary of, *268*, *271*, *307*
Huygens, C., 263

ideal gas, 544
images
  in electromagnetics, 300–5
  in optics, 299–300
  method of, 302, *303*, 475, 547
impressed current, 54, 96
index of refraction, 411
  relation to polarizability of molecule, 536n
inhomogeneous plane wave, 179
International System of Units (SI), 4, *6*, 608, 609–10, 610–1, *610*, *611*
ionosphere of Earth
  and Kennelly-Heaviside layer, 477
  Faraday rotation in, 173
  reflection from, 477, 590
iron filings, to show magnetic lines of force, 13, *14*

Jones matrix, 145
  eigenvalues for, 149–51, *150*
  eigenvectors for, 149–51, *150*
  for half-wave plate, 147, *150*
  for ideal circular polarizer, 148, *150*
  for ideal linear polarizer, 146, *150*
  for ideal linear retarder, 146, *150*
  for quarter-wave plate, 147, *150*
  normalized, 146
Jones vector, 145
  for various states of polarization, *143*
  normalized, 145
Jones, R.C., 145
Joule heat, 53
Joule, J.P., 53

Kelvin, Lord (W. Thomson), 200, 305
Kennelly, A.E., 477
Kennelly-Heaviside layer, 477
kilogram, SI unit, physical standard for, 609, *610*
Kirchhoff, G.R., 193
Kirchhoff's approximation, 193
  and physical optics current, 291–3
  for circular aperture, 218, 219
  for slit, 193, 194

Land, H.L., 147
Laplace, P.S., 59
Laplace's partial differential equation, 59
Larmor frequency, 397
Larmor, J., 378
Larmor's formula, 378
laser
  gas, 218, *220*
    Gaussian beam in cavity of, 248–50
  helium-neon, 254
Lawrence, E.O., 396
Liénard, A., 360

Liénard-Wiechert potentials, 359–64
light
natural (unpolarized), 524–8
speed of in dielectric, 411
speed of in free space (vacuum), 5, *6*
linear antenna
electrically short, 469–77
arrays of, 485–506
radiating efficiency of, 476
radiation resistance of, 474
top-loaded, 470, *470*
insulated, 579–83, *579*, 606
see also dipole antenna
long-wire antennas
rhombic, horizontal, 589–91, *590*
traveling-wave V dipole, 588–9
loop antenna
electrically small, 482–5
radiation resistance of, 485
Hertz's, 163–4, *165*
standing-wave, circular, 573–5
traveling-wave, circular, 570–3
Lorentz condition, 327, 330
Lorentz force density, 7
modified to include magnetic charge, 113
Lorentz force expression, 7, *33*, 117
Lorentz force for charged particle, 8
Lorentz gauge, 330
Lorentz, H.A., 1, 8n
Lyot filter, 174–5

Mach cone for projectile in air, 433, *435*
magnetic charge, *see* charge, magnetic
magnetic current, *see* current, magnetic
magnetic dipole, *see* dipole, infinitesimal magnetic
magnetic excitation, 4, *6*
magnetic field strength, 4, *6*
magnetic pole strength, 11
magnetostatic field, 322–5
main lobe of antenna pattern, 223
Malus, E.L., 146
law of, 146, 532
Marconi, G., 474
antenna at Glace Bay, Nova Scotia, 475–6, *476*
Maxwell, J.C., *frontispiece*, 1, 161, 305, 635
and electric displacement, 24, 612
quoted, xi–xii, 17, 24
Maxwell's equations
for general time dependence
differential form of, 31–2, *33*
integral form of, 4, *33*
for harmonic time dependence
differential form of, 93, *94*
integral form of, *94*
Fourier transform of, 102
independent versus dependent, 32–4
modified to include magnetic charge, 112–4, 477–8
summary of, *33*, *94*
McMillan, E.M., 399n
mean value theorem, 27
mean value theorem for integrals, 27
meter, SI unit, physical standard for, 609, *610*
method of images, 302, *303*
applied to monopole antenna, 475, *476*, 547
experimental application of, 303–5, *304*
for point charge and sphere, 318
modified Bessel functions, 430
monopole antenna, cylindrical
description of wavefronts radiated by, 86–7
electromagnetic field of, *85*, *87*, *89*
experimental results for, *89*
FDTD analysis of, 81–8
input impedance versus electrical length of, *599*
quarter-wave, 600
radiation resistance of, 475, 600
relation to dipole antenna, 475, *476*, 547–8
resonance for, 599–600
surface charge density on, *84*
moving point charge, *see* point charge, moving in free space
mutual coupling for antenna arrays, 486, 494n

Napier's rules for spherical triangles, 140
natural light, 524–8; *see also* sunlight
Newton, I., 48
Newton's second law of motion, 48
Newton's third law of motion
and self force of charged particle, 438
nonuniform plane wave, 179; *see also* plane wave, electromagnetic, inhomogeneous
notation used for field quantities, 4, *6*, *103*, 104

observed mass of charged particle, 439
Oersted, H.C., *frontispiece*, 1, 12–4
Oersted's experiment, 12–4, *13*
Oppenheimer, J.R., quoted, xvi

paraxial approximation
applied to Gaussian beam, 242
Parseval's formula, 102
particle
kinetic energy of, 49, 397
linear momentum, 48
rest energy of, 397
total energy of, 397
work done on, 49, 440
particle-induced visual sensation, 434
pencil beam for antenna, 232, 496
perfect conductor, 48
boundary conditions for, *47*, 48, *95*
permeability
of free space, 5, *6*, 615
relative, 6
permittivity
and index of refraction, 411, 536n
of free space, 5, *6*, 615

relative, 6
phosphenes, 434
physical optics current, 293
and Kirchhoff's approximation, 291–3
Planck, M.K.E.L., 393
Planck's constant, value of, 393, 526
Planck's radiation law, 526
plane-wave spectrum
for three-dimensional fields, 207–15
and Huygens' principle, 262–3
applied to circular aperture, 218–32
applied to Gaussian beam, 240–50
applied to perfectly conducting half plane, 255–8
applied to reflector antennas, 232–40
evanescent waves for, 208, 211
for radiated field, 210–5, *215*
propagating waves for, 208, 211
spectral functions for, 209, *211*
summary for, *211*, *215*
for two-dimensional fields, 184–92
applied to slit, 192–9, 205–6
evanescent waves for, 187, *189*, 191
for radiated field, 199–207, *203*
propagating waves for, 187, *189*, 191
spectral function for, 185, 186, *188*
summary for, *188*, *203*
transverse electric (TE) problem, 185–6
transverse magnetic (TM) problem, 186–7
plane wave, electromagnetic
complex Poynting vector for, 133–4
degree of linear polarization for, 409, 532
for Gaussian pulse, 76, 126–7, *127*
homogeneous (uniform, propagating), 134, 178
impedance for, in free space, 128
inhomogeneous (nonuniform, evanescent), 179
Poynting vector for, 128, 133
propagating in general direction, 127–8, *128*, 132–4
relation between electric and magnetic fields of, 127, 132
state of polarization for, 131–3, *133*
time-harmonic (monochromatic), 128–40
transverse electric (TE), 183
transverse electromagnetic (TEM), 126
transverse magnetic (TM), 184
Poincaré sphere, 140–2, *140*, *141*
for optical systems, 148
for transmitting and receiving antennas, 159–61
relation to Stokes parameters, 141
Poincaré, J.H., 140
point charge, electrostatic, 337–41
point charge, moving in free space
acceleration field of, 368, *376*
and bremsstrahlung, 391
and X-ray radiation, 391–4
at low velocity, 377–8, *376*
Cherenkov radiation from, *see* Cherenkov radiation
construction of field lines for, 385–7
directional characteristics of radiation from, *378*, *390*, *395*
Doppler effect for field of, 388–9
electromagnetic field of, 366, *376*
Heaviside-Feynman form for field of, 370
Larmor's formula for total power radiated by, 378
Liénard-Wiechert potentials for, 360
power radiated per unit solid angle for, 373, *376*
Poynting vector for radiation from, 374
relation between electric and magnetic fields for, 366, 368
retarded potentials for, 358
retarded time for, 360
summary of results for, *376*
synchrotron radiation from, *see* synchrotron radiation
total power radiated by, 374, *376*
velocity field for, 368, *376*
with constant velocity, 379–83
with uniform acceleration, 383–9
with velocity parallel to acceleration, 378–94
with velocity perpendicular to acceleration, 394–6
point charge, time-varying
advanced and retarded scalar potentials for, 343
point current, time-varying
advanced and retarded vector potentials for, 344
point of stationary phase, 200
Poisson, S.D., 321
Poisson's partial differential equation, 321
polarimeter, 151, *152*
polarizability, electric
for atom or molecule, 528–9
for dielectric sphere, *520*, 521, 543
for perfectly conducting sphere, 509, *520*
for perfectly conducting, circular disc, 512, *520*
for perfectly conducting, straight wire, 519, *520*
polarizability, magnetic
for magnetic sphere, *520*, 521
for perfectly conducting circular disc, 510, *520*
for perfectly conducting sphere, 512, *520*
for perfectly conducting straight wire, 543–4
polarization ellipse, 129, *130*, *132*, *133*
polarization state, for antenna, 153–161, 173, 175
and use of Poincaré sphere, 159–61
mismatch factor for, 158
on reception, 156, 171–2
on transmission, 155
polarization state, for plane wave, 129–40
and Poincaré sphere, 140–1
and Stokes parameters, 141–2
angles defining, 131, *132*, 134–5, 139
chart for, *138*, 139–40, 170
circular, 132, *133*
elliptical, 131, *133*
IEEE convention for, 131
linear, 131, *133*
polarizer
circular, 148

construction of, 148, *149*, 170–1
Jones matrix for, 148, *150*
linear, 145
construction of, 147
dichroic, 147
Jones matrix for, 146, *150*
Polaroid, 147
wire grid, 166–7, *168*
potential
electrostatic scalar, 59, 321
magnetostatic vector, 322
gauge transformation for, 324
restricted gauge transformation for, 325
potentials, electrodynamic
advanced, 343, 344
Coulomb gauge for, 330–1
field calculated from, 325–6, *348*, 353, *354*
gauge transformation for, 326, 330
Lorentz condition for, 327, 330
Lorentz gauge for, 330
restricted gauge transformation for, 330
retarded, 343, 344
scalar, 326
vector, 325
Poynting vector, *6*, *33*, 51
alternate, 58
complex, for harmonic time dependence, *94*, 96
example for: current-carrying wire, 59–64
time-average of, 96
Poynting, J.H., 51, 59, 61n
Poynting's theorem, *33*, 51
complex, for harmonic time dependence, *94*, 96
physical interpretation of, 53–8, *57*
Preistly, J., 9
propagating plane wave, 187; *see also* plane wave, electromagnetic, homogeneous
Pythagorean theorem, 620

quarter-wave plate, 147, *150*
quaternions
relation to modern vector analysis, 634–5

radar, 278, *278*
radiated field (asymptotic field, Fraunhofer field, far-zone field), 153
criteria for
general time dependence, 349–50
harmonic time dependence, 207, 215, 270, 271–2, 355
for distribution of charge and current
general time dependence, *348*, 352
harmonic time dependence, *354*, 355
for electric dipole, 458
for Huygens' principle, 271, *271*
for moving point charge (acceleration field), 368, 376
for plane wave spectrum
three-dimensional, 210–5, *215*
two-dimensional, 199–207, *203*
for thin-wire antenna, 547, 584
radiating efficiency of antenna, 238–41, 476, 485, 504
radiation-reaction force for charged particle, 439
radiation resistance
of dipole antenna, 600
of linear antenna, electrically small, 474, 475
of loop antenna, electrically small, 485
random-walk problem
and scattering from molecules, 533–5
Rayleigh scattering, 511
frequency dependence for, 511, 524
Rayleigh, Lord (J.W. Strutt), 511
and color and polarization of skylight, 524
and degree of transparency of air, 536, 544
and scattering from electrically small objects, 511
rectangular function, 332
reflector antenna
Cassagrain, 252–4, *253*
directivity of, 237
gain of, 237–8
NASA Goldstone, 232, *233*, *234*
paraboloidal, 233–5, *235*
radiating efficiency of, 238–40
with prime focus feed, 233, *235*
resonance for dipole (monopole) antenna, 596, 599–600
retarded potentials, scalar and vector
as causal solution, 343
electromagnetic field calculated from, 347, *348*
for distributed sources, 345, 347, *348*
for point sources, 343, 344
retarded time, *307*, 308, 343, *348*, 351, 360, *376*, 415, 547
for Cherenkov radiation, *418*, 419
retarder, linear, 146
construction of, 147
half-wave plate, 147, *150*
Jones matrix for, 146, *150*
quarter-wave plate, 147, *150*
rhombic antenna, horizontal, 589–91, *590*, 606
right-hand rule, 4
Röntgen, W.C., 391
rotation matrix
for left-handed rotation, 136, 145
for right-handed rotation, 145
Rowland, H.A., 615n
Ruhmkorff coil, 162
runaway solution for motion of charge, 441, 450

Savart, F., 14
scalar potential, 59, 321, 326
scale model, electromagnetic, 303–5, 318
scattered field
definition of, 277
for radar, 278
scattering by
electrically large object

perfectly conducting sphere, 290–1, *292*, 314
perfectly conducting, circular disc, 288–91, *290*, *292*
electrically small object, 506–21
complex Poynting vector for, 508
dielectric sphere, *520*, 521, 543
electric polarizability for, 508, *520*
equivalent electric dipole for, 508, *520*
equivalent magnetic dipole for, 508, *520*
magnetic polarizability for, 508, *520*
magnetic sphere, *520*, 521
perfectly conducting sphere, *507*, 506–11, *520*
perfectly conducting, circular disc, 511–4, *512*, *516*, *520*
perfectly conducting, straight wire, 516–21, *520*, 543–4
qualitative discussion of, 506–8
scattered electromagnetic field for, 508
summary of results for, *520*
scattering cross section of obstacle
bistatic or differential, 287
monostatic or backscattering, 287
radar, 287n
relation to transmission coefficient of aperture, 288
total, 287–8
scattering, molecular
and single scattering approximation, 533–5
induced electric dipole moment for, 528, *528*
of sunlight (natural light), 529–32
screen
opaque in optics, 281
perfectly conducting with aperture, 278–81
second, SI unit, physical standard for, 610, *610*
self field of charged particle, 7, 436
self force of charged particle, 7, 436–41, 449, 450
Shklovsky, I.S., 408
side lobes of antenna pattern, 223, 502
similarity transformation for matrix, 136
simple materials, 6, 429
sinc function, 195, *195*
sine integral, 593
single scattering approximation, 533–5
skin depth for good conductor, 38–9
skin effect approximation, 39–41
skylight
and molecular scattering, 529
and single scattering approximation, 533
blue color of, 521–4
degree of linear polarization for, 521–4
dependence on wavelength, 524, 530
experimental results for energy of, *522*
experimental results for polarization of, *523*
use for navigation by honeybees, 536–7
slit in perfectly conducting screen
analyzed with plane wave spectrum, 192–9
radiated field of, 205, *206*
with linear phase for illumination, 254
with uniform illumination, 192–9
slot dipole antenna, *294*, 316–7
relation to strip dipole antenna, 294–9
solid angle, 632, 637
Sommerfeld, A.J.W., 4, 255, 391, 412
source region, 54
spark-gap transmitter, 162–4
special theory of relativity, 361, 397, 412
spectral density for energy, 104
speed of light
and index of refraction, 411
in dielectric materials, 411
in free space (vacuum), 5, *6*
sphere, dielectric, scattering from electrically small, *520*, 521, 543
sphere, magnetic, scattering from electrically small, *520*, 521
sphere, perfectly conducting
electrically large
scattering cross sections for, *292*, 314
similarity of scattering to that from disc, 290–1
electrically small
and Rayleigh scattering, 506–11, 507
directional characteristics of scattering from, 510–1, *511*
equivalent dipoles for scattering from, 509, 510, *520*
frequency dependence of scattering from, 511
scattering cross sections for, 510
standing-wave antennas
dipole, 563–7, 591–601
loop, 573–5
standing wave, electromagnetic, 165, *166*
stationary phase, method of, 200, 258
for calculation of radiated field, 200–3, 213–4
Stirling's formula, 199
Stokes parameters, 141–2
for natural light, 528n, 544
normalized, 142
phenomenological definitions for, 153
relation to Poincaré sphere, 141
Stokes' theorem, 625, *626*
Stokes vector, 142
for various states of polarization, *143*
Stokes, G.G., 141, 200
strip dipole antenna, *294*, 316–7
relation to slot dipole antenna, 294–9
Strutt, J.W., *see* Rayleigh, Lord
sunlight (natural light), 524–8
as blackbody radiation, 526
at sunrise and sunset, 536, 544–5
attenuation on passing through atmosphere, 535–6
empirical description of, 525–7
experimental results for energy of, *522*
Fraunhofer lines for, 526n
measured extraterrestrial spectral irradiance of, *526*
Stokes parameters for, 528n, 544
superdirectivity for antenna array, 502–6
supernova, 408

surface density of charge, *6*, 35–8
surface density of current, *6*, 38–41
surface divergence of vector, 45–6
surface equation of continuity for electric charge
  for general time dependence, 46, *46*, *47*
  for harmonic time dependence, *95*
synchrotron radiation
  characteristics of, 407, *408*
  critical wavelength of, 406, *408*
  electric field of, 394, *395*, 401–5
  from Crab Nebula, 408–11
  from ultra-relativistic charged particles, 407, *408*
  history of, 407–8
  polarization of, 394, *395*, 407
synchrotron, particle accelerator, 396–401
  for electrons, 398n, 407
  for protons, 398n, 407, 447–8
  fraction of energy lost to radiation, 407
  maximum energy of particle in, 401
  phase stability for, 399–401, *400*
  radiation from particle within, 401–7
  resonance for, 398
system
  linear, 88–92
    eigenfunction for, 90
  real, 91
  time-invariant, 90

tachyons, 361n
Tamm, I.E., 412
thin-wire antennas
  charge per unit length on, 547
  current in, 547, 584
  for harmonic time dependence, 583–4
  inequalities defining, 546–7
  radiated field for, 547, 578, 584
Thomson, J.J., 389
Thomson, W., *see* Kelvin, Lord
total field, definition of, 277
transmission coefficient of aperture, 286
  relation to scattering cross sections of obstacle, 288
traveling-wave antennas
  basic element, 556–62, 584–8
  bent-wire, 567–70
  dipole, 562–3
  insulated, 579–83, *579*, 606
  loop, 570–3
  resistively loaded, 575–579, 605–6
  rhombic, horizontal, 589–91, *590*, 606
  V, 588–9, 605
triangular function, 554
two-dimensional fields
  transverse electric (TE), 181–4
  transverse magnetic (TM), 181–4
Tyndall, J., 523

umbrella antenna, 476
uniform plane wave, 134, 178
  see also plane wave, electromagnetic, homogeneous
uniqueness theorem for Maxwell's equations
  and causality, 68n
  applied to radiator (antenna), 70–1
  for general time dependence, 65–71
  for harmonic time dependence, 98–100
units, 608
  and Maxwell's equations, 611–6
  base, 609, *610*
  derived, 609, *611*
  dimension, 610
  for electromagnetic quantities, *6*
  International System (SI), 4, *6*, 608, *610*, *611*

V dipole, traveling-wave, 588–9, 605
Vavilov, S.I., 412n
vector analysis
  differential relations from, *624*
  history of, 633–5
  integral relations from, *626*
  relations for addition, *618*
  relations for products, *618*
vector differential length, 4
vector operations
  addition, 617, *618*
  circulation, 25–6, 623
  curl, 25–6, 623
  divergence, 28, 623
  gradient, 623
  scalar (dot) product, 617, *618*
  triple products, *618*, 619
  vector (cross) product, 617–8, *618*
vector phasor, complex, 92
vector potential, 322, 325
Veksler, V.I., 399n
velocity field of moving point charge, 368, *376*
volume density of charge, *see* charge, volume density of
volume density of current, *see* current, volume density of

wave equation
  scalar, 124, 327, 342
  vector, 124
wave impedance
  for free space, 128
wave number
  complex vector, 178
  for free space, 129
  vector, 133
wavelength
  for free space, 129
Weyl, H., 266
Wiechert, E., 360
wire, electrically short, straight, 516–21

and forward scattering theorem, 545
directional characteristics of scattering from, 521
equivalent electric dipole for scattering from, 516, 519, *520*
equivalent magnetic dipole for scattering from, 543–4
scattering cross sections for, 520–1

X-ray radiation, 391–4
  characteristic spectrum of, 391
  continuous spectrum of, 391

Young, T., 263, 309

zone plate, 277, 313, *313*, 317

Printed in the United States
By Bookmasters